W9-ABN-185

MATHEMATICAL SURVEYS AND MONOGRAPHS SERIES LIST

Volume

AMENABILITY

MATHEMATICAL SURVEYS
AND MONOGRAPHS

NUMBER 29

AMENABILITY

ALAN L. T. PATERSON

American Mathematical Society
Providence, Rhode Island

1980 *Mathematics Subject Classification* (1985 *Revision*). Primary 43–02, 43A07, 47H10, 22D25, 22D40, 22E15, 22E25, 03E10: Secondary 03E25, 16A61, 20F05, 20F16, 20F18, 20F19, 20F24, 22M10, 22D10, 22D15, 22E27, 28C15, 28D05, 43A60, 43A65, 43A80, 46B30, 46H05, 46H25, 46J10, 46L05, 46L10, 46P05, 47A35, 53C42, 54D35, 60B05, 60J15, 62F03.

Library of Congress Cataloging-in-Publication Data

Paterson, Alan L. T., 1944-
Amenability.
(Mathematical surveys and monographs; no. 28)
Includes bibliographies and index.
1. Harmonic analysis. 2. Locally compact groups. I. Title. II. Series.
QA403.P37 1988 515′.2433 88-14485
ISBN 0-8218-1529-6 (als. paper)

The paper used in this book is acid-free and falls within the guidelines
established to ensure permanence and durability. ∞

Dedicated
to

MAHLON M. DAY

Nicht die Neugierde, nicht die Eitelkeit, nicht die Betrachtung der Nützlichkeit, nicht die Pflicht und Gewissenhaftigkeit, sondern ein unauslöschlicher, unglücklicher Durst, der sich auf keinen Vergleich enlässt, führt uns zur Wahrheit.

G. W. F. Hegel

Contents

Preface

The subject of amenability essentially begins with Lebesgue (1904). One of the properties of his integral is a version of the Monotone Convergence Theorem, and Lebesgue asked if this property was really fundamental, that is, if the property follows from the more familiar integral axioms. Now the Monotone Convergence Theorem is equivalent to countable additivity, and so the question is concerned with the existence of a positive, finitely (but not countably) additive, translation invariant measure μ on $\mathbb{R}$ with $\mu([0,1]) = 1$.

The *classical period* (1904–1938) is therefore concerned with the study of finitely additive, invariant measure theory. The study of *isometry*-invariant measures led to the Banach-Tarski Theorem (1924) and the theory of paradoxical decompositions. The class of amenable groups was introduced and studied by von Neumann (1929) and used to explain why the Banach-Tarski Paradox occurs only for dimension greater than or equal to three.

The *modern period* begins in the 1940s and continues with increasing energy to the present. The main shift is from finitely additive measures to *means*: integrating a positive, finitely additive measure μ on a set X with $\mu(X) = 1$ gives rise to mean m on X, that is, a continuous linear functional on $l_\infty(X)$ such that $m(1) = 1 = \|m\|$. The connection between μ and m is given by $\mu(E) = m(\chi_E)$, and the correspondence $\mu \to m$ is bijective.

This shift is of fundamental importance, for it makes available the substantial resources of functional analysis (and eventually of abstract harmonic analysis) to the study of amenability. Von Neumann's definition translates into the language of means and gives the familiar definition of an amenable group: *a group G is amenable if and only if there exists a left invariant mean m on G.* Here, a mean m is left invariant if $m(\phi x) = m(\phi)$, where $\phi x(y) = \phi(xy)$ $(y \in G)$ for $\phi \in l_\infty(G)$, $x \in G$. The above definition also applies to semigroups, a semigroup admitting a left invariant mean being called *left amenable*.

In the 1940s and 1950s, the subject of amenable groups and amenable semigroups was studied by M. M. Day, and his 1957 paper on amenable semigroups is a major landmark.

Since that time the subject has developed at a fast pace, as the bibliography will testify. The plan at the beginning of this book will give some idea of the range of mathematics in which the amenability phenomenon has been observed and proved relevant. The amenability phenomenon is protean and is not readily "pigeon-holed". However, the roots of the subject lie in functional analysis and abstract harmonic analysis, and it draws its coherence from these.

The ubiquity of amenability ideas and the depth of the mathematics with which the subject is involved seems evidence to the author that here we have a topic of fundamental importance in modern mathematics, one that deserves to be more widely known than it is at present.

Until 1984, the only book available on the subject was the influential short account by F. P. Greenleaf [**2**] (1969). We should also mention Chapter 8 of Reiter's book [**R**] (1968) and Day's indispensable survey papers [**4**], [**9**] of 1957 and 1968 respectively. In addition, §17 of the work on abstract harmonic analysis [**HR1**] by Hewitt and Ross discusses amenable semigroups and groups.

More recently (1984), a book by Jean-Paul Pier on amenable locally compact groups has appeared. As one might expect, there is overlap between Professor Pier's book and the present one. However, the points of view adopted in the two books are very different, as also are much of the contents.

Another work that has recently appeared (1985), relevant to amenability, is the elegant book by Stan Wagon dealing with the Banach-Tarski Theorem.

The main objectives of the present work are to provide an introduction to the subject as a whole and to go into many of its topics in some depth. While the main area of amenability lies in analysis on locally compact groups, the subject applies to a much wider range of mathematics than that. We have tried to bear this in mind. In particular, applications in the areas of statistics, differential geometry, and operator algebras will be found in Chapters 4, 7, and 1, 2 respectively. Chapter 4 also contains discussions of the two most outstanding theorems in amenability established since 1980—these are in the areas of von Neumann's Conjecture and the Banach-Ruziewicz Problem.

We have attempted to describe the main lines of development of the subject, showing what progress or lack of progress has been made in solving the main problems, and raising a number of open problems. It is hoped that the present work will stimulate further research in the subject.

The introduction gives an informal, nontechnical discussion of amenability as a whole. It is recommended that the reader read this chapter carefully before going on with the rest of the book.

Chapters 1 and 2 establish the basic theory of amenability. Chapter 3 investigates amenability for Lie groups. (Lie theory is very important in Chapters 3 and 6. The Lie theory we need is briefly sketched in the Appendix B.) The last three chapters deal with amenability and ergodic theorems, polynomial growth, and invariant mean cardinalities respectively.

The main prerequisites for reading the present work are a knowledge of abstract harmonic analysis (such as is contained in [**HR1**]) and of functional analysis. Of course, we also presuppose a sound understanding of undergraduate mathematics. At the same time, we hope that the book is accessible to workers in other areas of mathematics.

Chapters 1, 2 and 4 could be used for a graduate course in amenability, assuming a background in basic functional and harmonic analysis, but not in Lie theory. Some of the results in Chapters 5 and 7 could also be included in such a course. A course on groups of polynomial growth could be given based on Chapters 6 and 3, and presupposing acquaintance with the Lie theory sketched in Appendix B.

Throughout the book we use results from certain standard textbooks. These textbooks are listed, together with their abbreviated forms of reference, at the beginning of the Bibliography.

Each chapter is divided into three parts. The first and main part is intended to give a coherent account of the area dealt with by the chapter. The second part ("Further Results") deals with important developments of the ideas in the first part. The third part is a list of problems; these vary from the routine to the quite difficult, and substantially increase the scope of the book. The author hopes that readers will try to solve some of the problems. References and sketched solutions (where appropriate) appear in the "Sketched Solutions to Problems" at the end of the book. Problem b of Chapter a is referred to in the text as: Problem a–b.

In addition to the main bibliography, there is a supplementary bibliography [**S**] which contains many of the most recent papers in the subject.

In view of the tangled and wide-ranging nature of the subject, it seems likely that inaccuracies will have occurred. Of course, absence of a specific reference for a result does not imply any claim to originality on the part of the author. The author will be grateful to readers who take the trouble to point out to him errors in the text.

I am very grateful to many mathematicians for mathematical discussions and advice. These include Rob Archbold, James Bondar, Ching Chou, Mahlon Day, G. Dales, John Duncan, David Edwards, William Emerson, Edmond Granirer, John Hubbuck, Barry Johnson, Anthony Lau, Michael Leinert, Mick MacCrudden, Paul Milnes, Isaac Namioka, A. Yu. Ol′shanskii, John Pym, Michael Rains, Guyan Robertson, A. van Rooij, Joseph Rosenblatt, Stan Wagon, David Wallace, Alan White, John Williamson, and James Wong. Thanks are also due to Louise Thompson, who coped heroically with masses of handwritten pages and typed most of the book. I am especially indebted to my friends John Duncan, Ed Granirer, Anthony Lau, Paul Milnes, and John Pym, whose kindness and interest in this work encouraged me to keep writing when the going was difficult. My thanks are also due to the editorial staff of the American Mathematical Society for their helpfulness and skill. Finally, I am grateful to my wife Christine, whose encouragement and support have meant more than I can say.

University of Mississippi *Alan L. T. Paterson*
August 1988

List of Further Results

PLAN

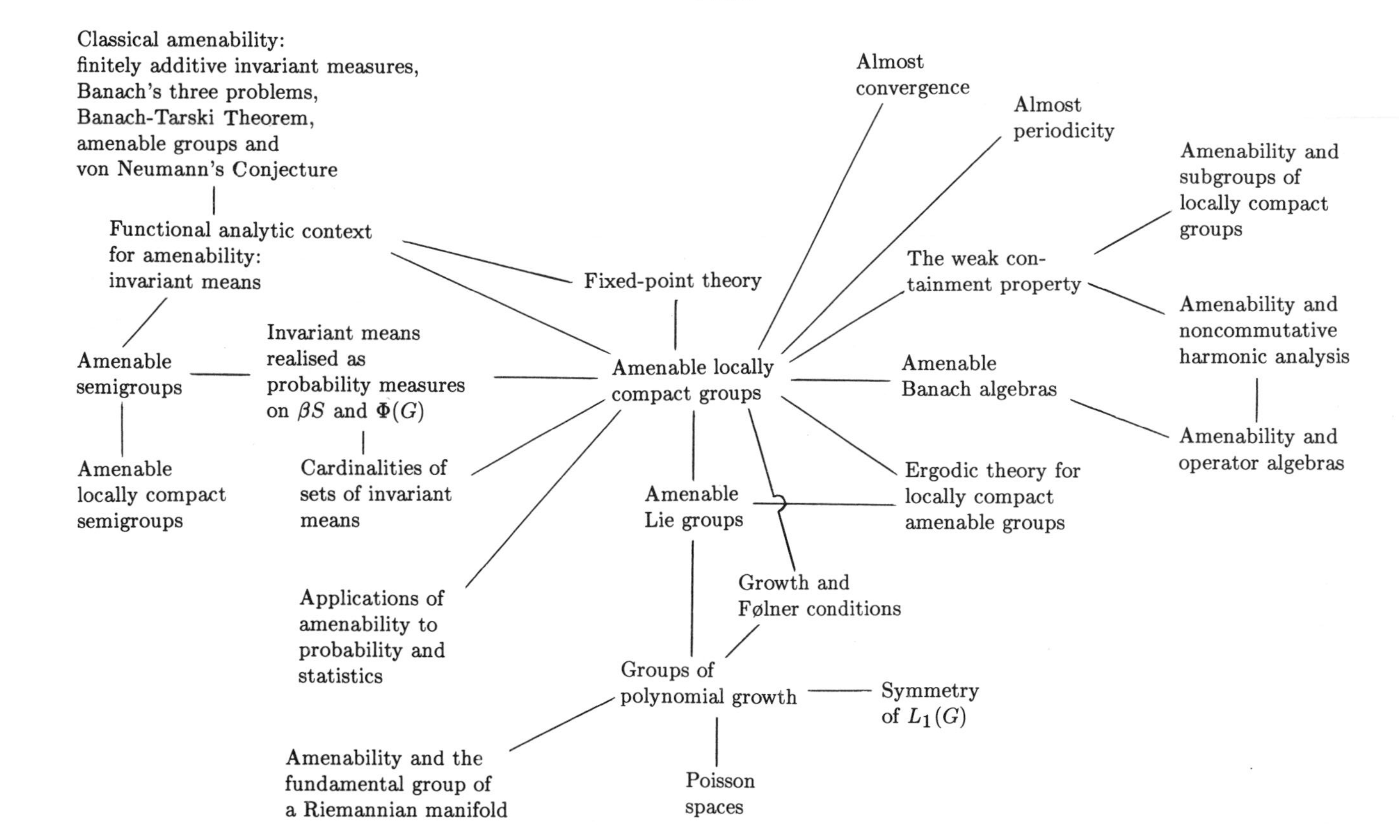

This plan is an attempt to give a reasonable indication of the range of contexts to which amenability applies. It is not comprehensive. A strong relationship between two topics is indicated by a line joining them. The topics vary greatly in scope.

AMENABILITY

CHAPTER 0

Introduction: Basic Concepts and Problems of Amenability Theory

Amenability essentially began in 1904 when Lebesgue gave a list of properties that uniquely specified his integral on $\mathbb{R}$. All of these, except one, are elementary properties of the Riemann integral. The exceptional property is a version of the Monotone Convergence Theorem (MCT). Lebesgue naturally asked if the integral was still uniquely specified if MCT is dropped. A moment's thought will convince the reader that MCT is actually equivalent to countable additivity, and Lebesgue's question can be rephrased: *Is the Lebesgue integral still unique if the MCT condition is replaced by mere finite additivity?*

Banach later answered this question negatively by producing a finitely additive, invariant integral on $\mathbb{R}$ different from the Lebesgue integral. (As we shall see in (4.29), for certain compact groups the answer to the corresponding question is positive.) In his proof, Banach used the existence of a finitely additive, invariant, positive integral of total mass one on the circle group $\mathbb{T}$.

It turned out that there exists such an invariant measure μ on the real line $\mathbb{R}$ as well. Two remarkable facts about μ are, first, that it is defined on *all* subsets of $\mathbb{R}$ (not just on the Lebesgue measurable ones) and, second, that $\mathbb{R}$ has finite μ-measure ($\mu(\mathbb{R}) = 1$), contrasting with the fact that the Lebesgue measure of $\mathbb{R}$ is ∞! In modern language, $\mathbb{R}$ is "amenable as a discrete group," and the measure μ is an *invariant mean.*

In the 1920s and 1930s, the question of the existence of an invariant mean for a group G acting on a set X was investigated by Banach and Tarski. Tarski (1938) showed that such a mean exists if and only if X does not admit a "G-paradoxical decomposition." This deep theorem will be discussed in (3.15).

In his study of the Banach-Tarski Theorem, von Neumann introduced and studied the class of amenable groups (amenable = mittelbar (German) = moyennable (French) = аменабельная (Russian)). The term *amenable* was introduced by M. M. Day (as a pun) in 1950. In more recent times, invariant means have proved relevant for answering an amazing variety of questions (especially in analysis on locally compact groups).

We therefore start by investigating some of the more basic properties of invariant means. These means relate to most of the results discussed in the rest of this work, and this enables the present chapter to serve as an introduction to the book as a whole. Thus we will take the opportunity to introduce at a basic level ideas and results which will later be developed in greater depth. It is hoped that the present chapter will also serve as a (somewhat informal) motivation for the rest of the work.

The classical period of amenability (1904–1939) defined (left) invariant means as we have done above, that is, as finitely additive measures on a group G. This is still useful, sometimes, in the modern theory, but in general, suffers from serious drawbacks. For a start, the powerful theorems of countably additive measure theory are no longer available, and the finitely additive versions of these theorems are pale reflections of the originals. (We note (Problem 1-10) that there are *no* countably additive invariant means on noncompact groups.) M. M. Day revolutionised the subject by showing that an invariant mean can be regarded as a continuous linear functional on $L_\infty(G)$.

Before discussing how this is done, let us recall some basic facts from abstract harmonic analysis. (Our standard reference for this area of mathematics is [**HR1**].) Let G be a locally compact group. There exists a left Haar measure λ on G: so λ is a nonzero, positive, regular Borel measure on G, and $\lambda(xE) = \lambda(E)$ for all $x \in G$ and all $E \in \mathscr{B}(G)$, the σ-algebra of Borel subsets of G. The space $L_1(G) = L_1(G, \lambda)$ of complex-valued, integrable functions on G is a Banach space with norm given by

$$\|f\|_1 = \int_G |f|\, d\lambda.$$

A very important fact for our purposes is that the dual of $L_1(G)$ can be identified with the space $L_\infty(G)$ of bounded, λ-measurable, complex-valued functions ϕ on G with the "ess sup"-norm:

$$\|\phi\|_\infty = \operatorname*{ess\,sup}_{x\in G} |\phi(x)|.^1$$

(Here, if $\psi\colon G \to \mathbb{R}$ is bounded, then

$$\operatorname*{ess\,sup}_{x\in G} \psi(x) = \inf\left\{ \sup_{x\in G\sim N} \psi(x) : N \text{ is locally null}\right\},$$

a subset N of G being *locally null* if it is λ-measurable and $\lambda(N \cap C) = 0$ for all $C \in \mathscr{C}(G)$, the family of compact subsets of G. $\operatorname*{Ess\,inf}_{x\in G} \psi(x)$ is defined similarly.) The above correspondence between $L_\infty(G)$ and the dual $L_1(G)'$ of $L_1(G)$ is given by associating $\phi \in L_\infty(G)$ with the functional $f \to \int \phi f\, d\lambda$ $(f \in L_1(G))$. The family of λ-measurable subsets of G is denoted by $\mathscr{M}(G)$.

Let μ be a positive, finitely additive measure on $\mathscr{M}(G)$ vanishing on locally null sets with $\mu(G) = 1$. We can regard μ as a member m of $L_\infty(G)'$ as follows. Obviously we must define $m(\chi_E) = \mu(E)$ for $E \in \mathscr{M}(G)$. More generally (Problem 0-2) we can define m on the span A of $\{\chi_E : E \in \mathscr{M}(G)\}$ by defining

$$m\left(\sum_{i=1}^n \alpha_i \chi_{E_i}\right) = \sum_{i=1}^n \alpha_i \mu(E_i) \qquad (\alpha_i \in \mathbb{C},\ E_i \in \mathscr{M}(G)).$$

[1] Two functions in $L_\infty(G)$ are identified if they differ only on a locally null set.

Now m is continuous on A and since A is norm dense in $L_\infty(G)$, we can extend m to all of $L_\infty(G)$. The functional m is a *mean* (or *state*) on $L_\infty(G)$, that is, an element $p \in L_\infty(G)'$ such that $p(1) = 1 = \|p\|$. Obviously m is determined by μ, and, conversely, μ is determined by m: $\mu(E) = m(\chi_E)$.

The advantages of looking at means from the point of view of m rather than μ are obvious. The machinery of dual Banach spaces (e.g., weak*-compactness) becomes available, and means fit profitably into functional and harmonic analysis. Like most fundamental ideas, the change in perspective from measures to functionals, due to Day, is natural in retrospect, but without it, amenability could not have developed as it has done.

When we do wish to regard m from the measure point of view—this will normally be when G is a discrete group (or semigroup)—we write $m(E)$ in place of $m(\chi_E)$ $(E \in \mathscr{M}(G))$.

The set of means on $L_\infty(G)$ will be denoted by $\mathfrak{M}(G)$. A serious problem is *how can we get hold of elements of* $\mathfrak{M}(G)$? The reason why this problem is serious is because means are elements of $L_\infty(G)'$, and the space $L_\infty(G)'$ is a "very large" space that is badly understood. To get a grip on $\mathfrak{M}(G)$ we need to pick out a subset that is accessible and yet, in a suitable sense, is big enough for us to go from it to any mean. This subset is $P(G)^\wedge$, which we now introduce.

Recall that $L_\infty(G) = L_1(G)'$. Of course, $L_1(G)$ is isometrically embedded in its second dual $L_1(G)'' = L_\infty(G)'$ by the map $f \to \hat{f}$, where $\hat{f}(\phi) = \phi(f)$ $(\phi \in L_\infty(G))$. We are therefore led to ask *which functions* $f \in L_1(G)$ *are such that* $\hat{f} \in \mathfrak{M}(G)$? The set of such functions is denoted by $P(G)$. Here are some fundamental facts about $\mathfrak{M}(G)$ and $P(G)$.

(0.1) PROPOSITION. (i) *An element* $n \in L_\infty(G)'$ *is a mean if and only if* $n(1) = 1$ *and* $n(\phi) \geq 0$ *whenever* $\phi \geq 0$ *in* $L_\infty(G)$; *further, if* $n \in \mathfrak{M}(G)$ *and* $\phi \in L_\infty(G)$ *is real-valued, then*

$$\operatorname*{ess\,sup}_{x \in G} \phi(x) \leq n(\phi) \leq \operatorname*{ess\,sup}_{x \in G} \phi(x)$$

(ii) $\mathfrak{M}(G)$ *is a weak* compact, convex spanning subset of* $L_\infty(G)'$;

(iii) $P(G) = \{f \in L_1(G)\colon f \geq 0,\ \int f\,d\lambda = 1\}$ *and* $P(G)^\wedge$ *is weak* dense in* $\mathfrak{M}(G)$.

PROOF. The space $L_\infty(G)$ is a commutative C^*-algebra under the pointwise product, and in C^*-terminology, mean = state. The results (ii) and the first part of (i) are well known to be true for states on any (unital) C^*-algebra [**D2**, 2.1.9, 2.6.4]. The last assertion of (i) follows from the first since both

$$\left(\phi - \left(\operatorname*{ess\,sup}_{x \in G} \phi(x)\right) 1\right), \left(\left(\operatorname*{ess\,sup}_{x \in G} \phi(x)\right) 1 - \phi\right) \geq 0.$$

We now prove (iii). Using (i),

$$P(G) = \left\{f \in L_1(G)\colon \int f\,d\lambda = 1 \text{ and } \int f\phi\,d\lambda \geq 0 \text{ if } \phi \geq 0 \text{ in } L_\infty(G)\right\}.$$

Now if $\int f\phi\, d\lambda \geq 0$ for all $\phi \geq 0$ in $L_\infty(G)$, it follows that $f \geq 0$ a.e., that is, that $f \geq 0$ in $L_1(G)$. The converse is obviously true, and the first assertion of (iii) follows.

For the second assertion, suppose that $\mathfrak{M}(G)$ does not equal the weak* closure K of $P(G)^\wedge$. Let $m_0 \in \mathfrak{M}(G) \sim K$. Separating m_0 from K [**DS**, V.2.10, V.3.9], we can find $\phi_0 \in L_\infty(G)$ such that

$$\text{(1)} \qquad \operatorname{Re} \hat{f}(\phi_0) \leq \operatorname{Re} m_0(\phi_0) - \varepsilon$$

for all $f \in P(G)$. We can suppose (by (i)) that ϕ_0 is real-valued. Let $k_0 = \operatorname{ess\,sup}_{x\in G} \phi_0$. Then we can find a compact, nonnull subset C of G such that $\phi_0(x) \geq k_0 - \frac{1}{2}\varepsilon$ for all $x \in C$. Taking $f = \chi_C/\lambda(C)$ in (1) we obtain $m_0(\phi_0) \geq k_0 + \frac{1}{2}\varepsilon$. This is impossible by (i). □

An important special case of (0.1(iii)) occurs when G is discrete. In this case $L_1(G) = l_1(G)$, $L_\infty(G) = l_\infty(G)$, and λ is counting measure. Each $x \in G$ can be regarded as an element of $l_1(G)$ via the correspondence $x \to \chi_{\{x\}} = \delta_x$. It is easy to see that $P(G)$ is the norm closure of the convex hull $\operatorname{co} G$ in $l_1(G)$ and we have $\operatorname{co} \hat{G}$ *is weak* dense in* $\mathfrak{M}(G)$. (The same conclusion holds when G is replaced by *any* set X.)

We have shown how a mean as a measure μ can be switched to a mean m as a linear functional on $L_\infty(G)$. How does the invariance of μ translate in the m-context? To answer this, if $E \in \mathscr{M}(G)$ and $\mu(x^{-1}E) = \mu(E)$ $(x \in G)$, then $m(\chi_E x) = m(\chi_E)$, where $\chi_E x(y) = \chi_E(xy)$ for all $y \in G$ (since $\chi_E x = \chi_{x^{-1}E}$). This leads to considering the natural actions of G on $L_\infty(G)$ and its dual.

First, let us establish some terminology. A set A is called a *right G-set* if for each $x \in G$, we are given a transformation $a \to ax$ of A such that $(ax)y = a(xy)$ for all $a \in A$, $x, y \in G$. (So G has a *right action* on A.) Left and two-sided G-sets are defined in the obvious way, and the same definitions apply for semigroup actions.

For $\phi \in L_\infty(G)$, $x \in G$, define $\phi x \in L_\infty(G)$ by $\phi x(y) = \phi(xy)$. (A commonly used notation for our ϕx is ${}_x\phi$, but we prefer to use ϕx since it makes explicit the right action involved and is consistent with the terminology of Banach G-modules.) Then $L_\infty(G)$ is a right G-set under the transformations $\phi \to \phi x$. The action fits in well with the Banach space structure of $L_\infty(G)$: indeed each map $\phi \to \phi x$ is an isometry in $\mathbf{B}(L_\infty(G))$. (If Y is a Banach space, $\mathbf{B}(Y)$ is the Banach algebra of bounded linear operators on Y.) (For future use, we note that there is a natural left action $\phi \to x\phi$ of G on $L_\infty(G)$, where $x\phi(y) = \phi(yx)$.) The relevance of this G-action on $L_\infty(G)$ for amenability is as follows. If μ, m are as earlier, then μ is left invariant (that is, $\mu(xE) = \mu(E)$ for all $x \in G$, $E \in \mathscr{M}(G)$) if and only if

$$m(\phi x) = m(\phi) \qquad (\phi \in L_\infty(G),\ x \in G).$$

This equality can be reformulated as follows. The right action of G on $L_\infty(G)$ gives rise to a dual left action on $L_\infty(G)'$:

$$xn(\phi) = n(\phi x) \qquad (x \in G,\ n \in L_\infty(G)',\ \phi \in L_\infty(G)).$$

So the above equality can be expressed *n is left invariant for G*. This leads us to the most important definition in the book.

(0.2) DEFINITION. The locally compact group G is called (left) *amenable* if there exists a left invariant mean for G.

The set of left invariant means on G is denoted by $\mathfrak{L}(G)$. (A more commonly used notation is $Ml(G)$, but we have chosen to use $\mathfrak{L}(G)$ in the interest of succinctness.) It is almost trivial to check that $\mathfrak{L}(G)$ is a weak* compact, convex subset of $\mathfrak{M}(G)$. Every compact group is obviously amenable: in that case, the map $\phi \to \int \phi \, d\lambda$ is in $\mathfrak{L}(G)$.

Before launching into the theory of amenable groups, it is clearly desirable to look at some special, well-understood groups to see if they are amenable or not and how invariant means can be produced. This will also help to motivate a number of the important questions discussed in the later chapters. The five groups we will look at now are (a) $\mathbb{Z}$, (b) $\mathbb{R}$, (c) the "$ax + b$" group, (d) F_2, the free group on two generators, (e) the group $\mathrm{SL}(2, \mathbb{R})$ of real 2×2 matrices of determinant 1. The results below follow from more general theorems proved later.

(0.3) EXAMPLE: $G = \mathbb{Z}$. Is there an invariant mean on $\mathbb{Z}$? Since $l_\infty(\mathbb{Z})'$ $(= L_\infty(\mathbb{Z})')$ is inaccessible, we concentrate on exploiting $P(G)$ with (0.1(iii)) in mind. A typical element of $P(G)$ is an infinite sum of the form $\sum_{-\infty}^{\infty} a_r \delta_r$, where $a_r \geq 0$ for all r and $\sum_{-\infty}^{\infty} a_r = 1$. What we want to do is to construct a sequence $\{f_n\}$ in $P(G)$ with at least one of its weak* cluster points an invariant mean. Clearly, we need the f_n's to become "more and more" invariant under $\mathbb{Z}$-translates as n gets large. A reasonable choice for f_n would be the "Cesàro sum":

$$f_n = \frac{1}{2n+1} \sum_{r=-n}^{n} \delta_r.$$

Then if $\phi \in l_\infty(\mathbb{Z})$ and $s \geq 0$ in $\mathbb{Z}$, we have

$$\begin{aligned}
|\hat{f}_n(\phi s) - \hat{f}_n(\phi)| &= \left| \frac{1}{2n+1} \left(\sum_{r=-n}^{n} (\phi(r+s) - \phi(r)) \right) \right| \\
&= \frac{1}{2n+1} \left| \left(- \sum_{-n}^{-n+s-1} \phi(r) + \sum_{n+1}^{n+s} \phi(r) \right) \right| \\
&\leq \frac{2s\|\phi\|}{2n+1} \to 0 \quad \text{as } n \to \infty.
\end{aligned}$$

A similar result holds if $s < 0$, and we see that *every weak* cluster point of $\{\hat{f}_n\}$ in $\mathfrak{M}(\mathbb{Z})$ is a left invariant mean*! In particular, *$\mathbb{Z}$ is amenable.*

(0.4) EXAMPLE: $G = \mathbb{R}$. Modifying (0.3), we take $f_n = \chi_{[-n,n]}/2n$. Then for $x \geq 0$ and $\phi \in L_\infty(\mathbb{R})$, we have, as in (0.3),

$$
\begin{aligned}
|\hat{f}_n(\phi x) - \hat{f}_n(\phi)| &= \frac{1}{2n}\left|\int_{-n}^{n} (\phi(x+t) - \phi(t))\, dt\right| \\
&\leq \frac{2x\|\phi\|}{2n} \to 0 \quad \text{as } n \to \infty.
\end{aligned} \tag{2}
$$

Again we have that *every weak* cluster point of* $\{\hat{f}_n\}$ *is a (left) invariant mean and so* $\mathbb{R}$ *is amenable.*

(0.5) EXAMPLE: G is the "$ax + b$" group. This is the affine group S_2 of $\mathbb{R}$ (Appendix A). So $S_2 = \mathbb{R} \times_\rho (0, \infty)$ with multiplication given by

$$(b, a)(b', a') = (b + ab', aa').$$

It is not so clear how to construct the f_n's in this case since the multiplication is getting "twisted" in the first component. The following construction was shown to the author by Paul Milnes.

Let A_n be the shaded subset of S_2 shown below:

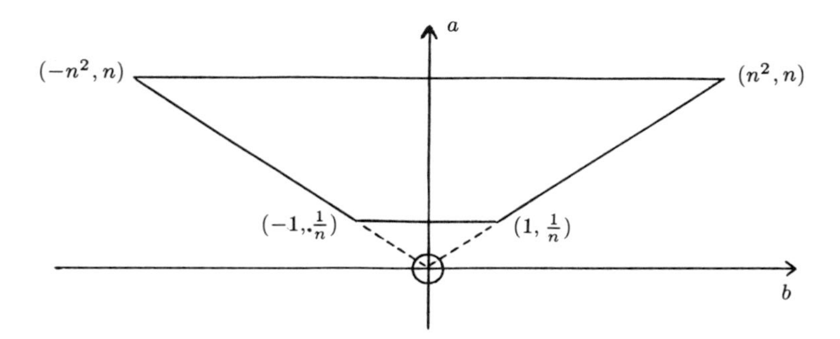

A Haar measure λ on S_2 is given by $\frac{dx\,dy}{y^2}$ and direct integration shows that if we take $f_n = \chi_{A_n}/\lambda(A_n) \in P(G)$, then we again have that *every weak* cluster point of* $\{\hat{f}_n\}$ *is left invariant, so that* S_2 *is amenable.* (See Problem 0-8.)

(0.6) EXAMPLE: $G = F_2$. Let F_2 be free on the generators u, v (with, of course, the discrete topology). We will show that F_2 is *not* amenable.

To this end, as usual, we identify F_2 with the set of reduced words in $\{u, v, u^{-1}, v^{-1}\}$. For $x \in \{u, v, u^{-1}, v^{-1}\}$, let E_x be the set of elements in F_2 beginning with x. If $m \in \mathfrak{L}(F_2)$, then, regarding m as a measure on F_2, $m(\{e\}) + m(E_u) + m(E_{u^{-1}}) + m(E_v) + m(E_{v^{-1}}) = m(G)$, and $m(G) = m(E_u) + m(uE_{u^{-1}}), m(G) = m(E_v) + m(vE_{v^{-1}})$. Now $m(G) = 1$ and a contradiction follows using the invariance of m.

(0.7) EXAMPLE: $G = \mathrm{SL}(2, \mathbb{R})$. This locally compact group is not amenable either. To show this, we can use the result (an easy consequence of (3.2)) that $G = \mathrm{SL}(2, \mathbb{R})$ contains a discrete subgroup H isomorphic to F_2. (Discreteness here means that there exists a neighbourhood U of the identity e in G with

$U \cap H = \{e\}$. This implies that H is closed in G.) Using Appendix C, Corollary (with $G = X$, $G \backslash H = Y$), we can find a Borel cross-section B for the space $G \setminus H$ of right H-cosets in G. So $B \subset G$ is Borel, $G = HB$, and $hB \cap h'B = \varnothing$ if $h, h' \in H$, $h \neq h'$. Since H is countable, EB is Borel for every subset E of H. If $m \in \mathfrak{L}(G)$, then we can define $n \in \mathfrak{L}(H)$ by setting $n(E) = m(EB)$. This is impossible by (0.6).

The above examples "point beyond themselves" and suggest a number of research directions that will be developed later. Most of the rest of this chapter will be devoted to discussing briefly some of these directions; these include Følner conditions, topologically left invariant means, the sizes of sets of invariant means, polynomial growth, fixed-point theorems, the construction of new amenable groups from old, and the relationship between F_2 and nonamenability. We concentrate first on Examples (0.3)–(0.5).

If we look closely at how $\mathbb{Z}$, $\mathbb{R}$, and S_2 were shown to be amenable, it becomes clear that the arguments relied on constructing a sequence $\{K_n\}$ of compact subsets of G with the property

$$\lambda(xK_n \bigtriangleup K_n)/\lambda(K_n) \to 0 \quad \text{as } n \to \infty \tag{3}$$

for each $x \in G$. Such a sequence is called a *summing sequence* for G. For each n, the sets K_n of (0.3), (0.4), and (0.5) are $[-n; n]$ $(= \{r \in \mathbb{Z}\colon -n \leq r \leq n\})$, $[-n, n]$, and A_n respectively. We will show in Chapter 4 that a σ-compact amenable group always admits such a sequence $\{K_n\}$ and that amenability is equivalent to the "Følner condition": *given $\varepsilon > 0$ and $C \in \mathscr{C}(G)$, the family of compact subsets of G, there exists a nonnull set $K \in \mathscr{C}(G)$ such that*

$$\lambda(xK \bigtriangleup K)/\lambda(K) < \varepsilon \qquad (x \in C). \tag{4}$$

What this remarkable condition asserts is that, roughly, there is a nonnull compact subset of G which is "almost invariant" under left translates by elements of C. The condition illustrates two fundamental options in amenability theory. We can either have *precise* invariance and pay for it by having to work in the difficult space $L_\infty(G)'$, or we can work in a "down to earth" setting with compact subsets of G and pay for it by having only "almost" invariance.

The Følner conditions are among the most useful in amenability. For example, as in (0.3)–(0.5), they can be used to produce invariant means with little effort. Deriving such conditions is, understandably, not easy. However, the first stage in such a derivation is a version of (4) for L_1-functions known as *Reiter's condition.* To formulate this condition, we require a left action of G on $L_1(G)$ whose right dual action is just that given by the maps $\phi \to \phi x$ discussed earlier. For $x \in G$, $f \in L_1(G)$, we define $x * f \in L_1(G)$ by

$$x * f(y) = f(x^{-1}y) \qquad (y \in G).$$

This is clearly an isometric left action, and since, for $\phi \in L_\infty(G)$, we have

$$(\phi, x * f) = \int f(x^{-1}y)\phi(y)\,d\lambda(y) = \int f(y)\phi(xy)\,d\lambda(y) = (\phi x, f)$$

the above claim is justified. Of course, $x\hat{f} = (x * f)^\wedge$ (in $L_\infty(G)'$). There is a corresponding right action of G on $L_1(G)$ given by

$$f * x(y) = f(yx^{-1})\Delta(x^{-1}) \qquad (y \in G),$$

where Δ is the modular function on G.

Now if $f = \chi_K/\lambda(K)$, then

$$\|x * f - f\|_1 = (\lambda(K))^{-1} \int |\chi_K(x^{-1}y) - \chi_K(y)|\, d\lambda(y) = \lambda(xK \,\Delta\, K)/\lambda(K).$$

This suggests a weaker version of (4) in which $\chi_K/\lambda(K)$ is replaced by a function in $P(G)$, and such a version ("Reiter's condition") is proved in (4.4). A related, easier condition will now be proved (Day [**4**], Namioka [**1**]). The elegant argument depends on switching from $L_\infty(G)'$ with the weak* topology to $L_1(G)$ with the weak topology using the left invariance of an element of $\mathfrak{L}(G)$.

(0.8) PROPOSITION. *The group G is amenable if and only if there exists a net $\{f_\delta\}$ in $P(G)$ such that*

$$\|x * f_\delta - f_\delta\|_1 \to 0 \tag{5}$$

for all $x \in G$.

PROOF. Let $\{f_\delta\}$ be a net in $P(G)$ such that $\|x * f_\delta - f_\delta\|_1 \to 0$ for all $x \in G$. Then $(x\hat{f}_\delta - \hat{f}_\delta) \to 0$ weak* for each x, and every weak* cluster point of $\{\hat{f}_\delta\}$ is in $\mathfrak{L}(G)$, so that G is left amenable.

Conversely, suppose that G is left amenable. Let $m \in \mathfrak{L}(G)$ and find a net $\{g_\sigma\}$ $(\sigma \in I)$ in $P(G)$ such that $\hat{g}_\sigma \to m$ weak*. Let $x_1, \ldots, x_n \in P(G)$. Since $((x_1 * g_\sigma - g_\sigma), \ldots, (x_n * g_\sigma - g_\sigma))$ converges weakly to 0 in $L_1(G)^n$, we can find [**DS**, V.3.13] a net $\{h_\gamma\}$ with each $h_\gamma \in \operatorname{co}\{g_\sigma : \sigma \in I\} \subset P(G)$ and such that $\|x_i * h_\gamma - h_\gamma\|_1 \to 0$ $(1 \le i \le n)$. The existence of the required net $\{f_\delta\}$ now follows. □

We now consider those elements m that arise as weak* cluster points of a net $\{\hat{f}_\delta\}$, where $\{f_\delta\}$ is as in (0.8). It turns out that, for nondiscrete G, such an element m is rather special—it is *topologically left invariant.* Generally speaking, topologically left invariant means are much more manageable than means that are merely left invariant.

Before defining topologically left invariant means, we recall that $L_1(G)$ is a Banach algebra under convolution, where for $f, g \in L_1(G)$, the convolution product $f * g$ is given by

$$f * g(x) = \int_G f(t)g(t^{-1}x)\, d\lambda(t).$$

The left action by multiplication of $L_1(G)$ on itself dualizes to give a right action of $L_1(G)$ on $L_\infty(G) = L_1(G)'$:

$$\phi * f(g) = \phi(f * g) \qquad (\phi \in L_\infty(G),\ f, g \in L_1(G)).$$

(Our definition of $\phi * f$ differs from that in [**HR1**, §20]. However, the two versions of $\phi * f$ are closely related, and ours is natural from a duality viewpoint and fits in well with our requirements.)

Note that $\|\phi * f\|_\infty \leq \|\phi\|_\infty \|f\|_1$. As a function, we have

$$\phi * f(x) = \int \phi(yx) f(y)\, d\lambda(y) = (x\phi)(f).$$

This right action dualizes to give a left action of $L_1(G)$ on $L_\infty(G)'$ written $n \to f * n$ (or fn) ($f \in L_1(G)$, $n \in L_\infty(G)'$). One readily checks that $x(f * n) = (x * f) * n$ for all $x \in G$.

(0.9) DEFINITION. A mean m on $L_\infty(G)$ is called *topologically left invariant* (Hulanicki [**4**]) if $fm = m$ for all $f \in P(G)$, that is, if

$$m(\phi * f) = m(\phi) \qquad (\phi \in L_\infty(G),\ f \in P(G)). \tag{6}$$

The set of topologically left invariant means is denoted by $\mathfrak{L}_t(G)$, and is clearly a weak* compact, convex subset of $\mathfrak{M}(G)$. Further, if $m \in \mathfrak{L}_t(G)$ and $x \in G$, then $xm = x * (f * m) = (x * f) * m = m$ since $x * f \in P(G)$. So $\mathfrak{L}_t(G) \subset \mathfrak{L}(G)$.

It is readily checked that if G is discrete, then every function in $P(G)$ is a sum of point masses so that $\mathfrak{L}_t(G) = \mathfrak{L}(G)$. However as we shall see in Chapter 7, the two sets are very different in general for nondiscrete locally compact groups.

Why are means satisfying (6) called *topologically* left invariant? The reason for this is that the means in (6) are $P(G)$-invariant, and $P(G)$ depends on λ which in turn is determined by the topology of G. The fact that $P(G) \subset L_1(G)$ whereas $G \not\subset L_1(G)$ in the nondiscrete case will be usefully exploited later.

To illustrate the relevance of $\mathfrak{L}_t(G)$, we can ask the following question: *what can one say about those means arising as weak* cluster points of the net* $\{\hat{f}_\delta\}$ *in* (0.8)? *Can we get all of* $\mathfrak{L}(G)$ *in this way?*

Given that we know that $\mathfrak{L}(G) \neq \mathfrak{L}_t(G)$ in many cases ((7.21)) and that we assume (as we may) that the convergence in (5) is uniform on compacta, the answer to the second question is *no*; all such weak* cluster points m are in $\mathfrak{L}_t(G)$! Further, as we shall see (cf. (4.17)), in a certain sense, we obtain all of $\mathfrak{L}_t(G)$ from the weak* cluster points of $\{\hat{f}_\delta\}$.

We now illustrate these comments in the case $G = \mathbb{R}$: the same proof applies under much more general circumstances.

(0.10) PROPOSITION. *Let* $f_n = \chi_{[-n,n]}/2n$, *and let* m *be a weak* cluster point of* $\{\hat{f}_n\}$ *in* $L_\infty(\mathbb{R})'$. *Then* $m \in \mathfrak{L}_t(G)$.

PROOF. Let $\phi \in L_\infty(\mathbb{R})$, $f \in P(\mathbb{R})$. Using the norm continuity of the map $g \to \phi * g$ ($g \in L_1(G)$), we can suppose that f vanishes outside a compact set C. Let N be such that $C \subset [-N, N]$. Then for $n > N$, we have, using Fubini's

Theorem and (2),

$$\begin{aligned}|\hat{f}_n(\phi * f - \phi)| &= \left|\int f_n(x)\,dx \int [\phi(y+x)f(y) - \phi(x)f(y)]\,dy\right| \\ &= \left|\int_C f(y)\,dy \int_{\mathbb{R}} f_n(x)[\phi(y+x) - \phi(x)]\,dx\right| \\ &\le \int_C |f(y)|\,dx \cdot \frac{N\|\phi\|}{n} \le \frac{N\|\phi\|\,\|f\|_1}{n} \to 0\end{aligned}$$

as $n \to \infty$. It follows that $m(\phi * f) = m(\phi)$, and $m \in \mathfrak{L}_t(G)$ as required. □

We shall see in Chapter 1 that there always exist topologically left invariant means on an amenable locally compact group.

Let us return again to the notion of a summing sequence $\{K_n\}$ with Example (0.3) in mind. In this example $K_n = [-n; n]$ and integrating $\phi \in l_\infty(\mathbb{Z})$ over $\chi_{K_n}/\lambda(K_n)$ gives a Cesàro average. Since such averages appear in summability theory, one might wonder if summability ideas can be developed for general amenable groups using the K_n's. This is in fact the case; in particular, the theory of almost convergence has been developed in the amenability setting (see, for example, Problems 0-13, 4-14, 4-15, 5-8). Another context in which Cesàro averages occur is in the mean and pointwise ergodic theorems. We will show in Chapter 5 that a version of the mean ergodic theorem, involving the K_n, does hold for amenable locally compact groups. The question of a pointwise ergodic theorem for amenable groups is, understandably, rather more subtle, but using Lie group techniques, we will establish such a theorem in Chapter 5.

Another interesting issue arising naturally from Examples (0.3)–(0.5) is the question of the cardinality $|\mathfrak{L}(G)|$ of $\mathfrak{L}(G)$ for an amenable group G. The fact that in all three examples, *any* weak* cluster point of $\{\hat{f}_n\}$ is in $\mathfrak{L}(G)$ suggests that $\mathfrak{L}(G)$ is probably rather large. For the present, let us give support to this suggestion by showing that $\mathfrak{L}(\mathbb{Z})$ is infinite (Mitchell [**2**]). (Our estimate will be greatly improved in Chapter 7.)

(0.11) PROPOSITION. *There are infinitely many left invariant means on* $\mathbb{Z}$.

PROOF. One might think of trying to show this by proving that there exist infinitely many weak* cluster points of $\{\hat{f}_n\}$ (where, of course, $f_n = (1/(2n+1)) \cdot (\sum_{r=-n}^{n} \delta_r)$. This can be done, but requires some functional analytic machinery (cf. (7.28)). Instead we use the following elementary argument (which prefigures the theory of "left thick" sets developed in Chapters 1 and 7). The main idea here is one already operative in (0.3)–(0.5), viz., if I_n is a sequence of intervals of increasing length in $\mathbb{Z}$, then, for fixed r, the r-translate of $\chi_{I_n}/|I_n|$ does not differ much from $\chi_{I_n}/|I_n|$ for large n. This gives that every weak* cluster point of $\{(\chi_{I_n}/|I_n|)^\wedge\}$ is in $\mathfrak{L}(\mathbb{Z})$, and to show that $\mathfrak{L}(\mathbb{Z})$ is infinite, we only have to produce an infinite number of disjoint families $\{I_n\}$. We now proceed with the proof.

Divide $\mathbb{P}$ into "intervals" $I_1^1, I_2^1, I_2^2, I_3^1, I_3^2, I_3^3, I_4^1, \dots$ as follows: $I_1^1 = \{1\}$, $I_2^1 = \{2,3\}$, $I_2^2 = \{4,5\}$, $I_3^1 = \{6,7,8\}$, $I_3^2 = \{9,10,11\}$, $I_3^3 = \{12,13,14\}$, $I_4^1 =$

$\{15,16,17,18\}$, etc. For each r, let m_r be a weak* cluster point of $\{(\chi_{I_n^r}/n)^\wedge\}$ in $\mathfrak{L}(\mathbb{Z})$. Let $A_r = \bigcup_{n=r}^{\infty} I_n^r$. If $s \neq r$, then $A_s \cap A_r = \emptyset$ and since $m_r(A_r) = 1 = m_r(\mathbb{Z})$, we have $m_r \neq m_s$ if $r \neq s$. So $|\mathfrak{L}(\mathbb{Z})|$ is infinite. □

In fact $|\mathfrak{L}(\mathbb{Z})|$ is considerably bigger than $\aleph_0$. As we shall see in Chapter 7, $|\mathfrak{L}(G)| = 2^{2^{|G|}}$ if G is infinite, discrete, and amenable, while, more generally, if G is amenable and noncompact, then $|\mathfrak{L}_t(G)| = 2^{2^{\mathfrak{m}}}$, where $\mathfrak{m}$ is the smallest possible cardinality of a cover of G by compact sets. There are three essential ingredients in the proofs of such results. The first involves delicate, transfinite induction arguments, which are used to construct certain special subsets of the group that, in a sense, generate "many" left invariant means.

The second involves interpreting means as (genuine!) probability measures on a compact space. To motivate this, consider the argument of (0.11). Each mean m_r can be regarded as "supported" on the set $A_r = \bigcup_{n=r}^{\infty} I_n^r$ in the sense that $m_r(A_r) = 1$ or, alternatively, that $m_r(\phi) = m_r(\phi\chi_{A_r})$ $(\phi \in l_\infty(\mathbb{Z}))$. Since $A_r \cap A_s = \emptyset$ if $r \neq s$, the m_r's are "disjointly supported" and are therefore different. The trouble is that any partition of $\mathbb{Z}$ has to be countable so that there is no hope of getting an uncountable set of invariant means by this method. Another unsatisfactory feature is that there is no unique "support" of a mean m regarded as a finitely additive measure on $\mathbb{Z}$.

To overcome these difficulties, we can proceed as follows. For simplicity, let us suppose that G is discrete and let βG be the Stone-Cěch compactification of G. The points of βG are the complex-valued, continuous homomorphisms of $l_\infty(G)$ and βG is given the relative weak* topology which it inherits as a subset $l_\infty(G)'$. In the obvious way, $G \subset \beta G$. In fact G is open in βG, and the closure of any subset of G in βG is both open and closed. The map $\phi \to \hat{\phi}$, where $\hat{\phi}(\alpha) = \alpha(\phi)$ $(\alpha \in \beta G)$, is an isometric isomorphism from $l_\infty(G)$ onto $C(\beta G)$, the algebra of continuous, complex-valued functions on βG. (Of course, $\phi \to \hat{\phi}$ is just the Gelfand transform of the commutative C^*-algebra $l_\infty(G)$, and βG is its carrier space.)

Why is it worth looking at βG to obtain information about invariant means? The crucial point here is that $l_\infty(G)$ is identified with $C(\beta G)$ and *we know what* $C(\beta G)'$ *is*: by the Riesz representation theorem, $C(\beta G)'$ is just $M(\beta G)$, the space of regular, complex-valued, Borel measures on βG. Since $l_\infty(G)'$ "=" $C(\beta G)'$ and $\mathfrak{L}(G) \subset l_\infty(G)'$, it readily follows that every mean m can be regarded as a "genuine" (that is, countably additive) probability measure $\hat{m}$ on βG. As such, $\hat{m}$ has a (unique) support—the smallest closed subset C of βG such that $\hat{m}|_{\beta G \sim C} = 0$. Further, G has a natural left action on βG, and $m \in \mathfrak{L}(G) \Leftrightarrow \hat{m}$ is a probability measure that is invariant under the action of G. The whole philosophy is simple: we shift from studying a bad (finitely additive) measure on a good set (G) to studying a good (that is, countably additive) measure on a complicated space βG. (In the latter connection, βG is much bigger than G: if G is infinite, then $|\beta G| = 2^{2^{|G|}}$.)

Reverting to the m_r's and A_r's above, we readily see that the fact that $m_r(A_r) = 1$ implies that the support of $\hat{m}_r$ is contained in the closure A_r^- of A_r in βG. The trouble is that A_r^- is too "big"—it is actually an open set. What we must do is to try to construct invariant means which have *much smaller* supports in βG so that more means can be "squeezed in." We will see how this can be done in Chapter 7. The idea is to obtain "small" closed, invariant sets in βG by forming large intersections of closed sets of the form A^- $(A \subset G)$ and then finding means in $\mathfrak{L}(G)$ supported inside these small closed sets.

The third essential ingredient is an amenability fixed-point theorem that can be used to show that, for every closed invariant subset C of βG, there exists $m \in \mathfrak{L}(G)$ with $\hat{m}$ supported inside C. Then if we have a large, disjoint collection of closed invariant subsets of βG, we automatically also have a large set of left invariant means.

The fixed-point theorems of amenability are of great importance and of wide application. (For this reason, we have included an early discussion of these theorems in Chapters 1 and 2.) A completely trivial (but significant) observation is that amenability is equivalent to the existence in $\mathfrak{M}(G)$ of a G-fixed-point—such a fixed-point is, of course, a left invariant mean! Note that $\mathfrak{M}(G)$ is a compact convex subset of a locally convex space ($L_\infty(G)'$ with the weak* topology) and that the elements of G act affinely on $\mathfrak{M}(G)$ (that is, preserve convex combinations). Not surprisingly, amenability fixed-point theorems work in the context of affine actions on compact convex sets. The theorems are proved basically by lifting a G-action to an action of a set of means that possesses a natural semigroup structure. Multiplying by a left invariant mean then gives G-fixed-points. It turns out that the fixed-point property for a reasonable class of group actions is equivalent to the existence of a left invariant mean on a certain space of functions on G. The algebraic structure behind this and other parts of amenability theory is related to the "Arens product construction" familiar in the theory of Banach algebras, and this is discussed in Chapter 2.

Reverting back to Examples (0.3)–(0.5), closer inspection shows that there exists a distinction between, on the one hand, (0.3) and (0.4), and, on the other hand, (0.5). In the former cases, the K_n's satisfying (3) are of the form C^n for some compact neighbourhood of the identity. In the case of $\mathbb{Z}$, $C = [-1; 1]$, while in the case of $\mathbb{R}$, $C = [-1, 1]$. This is not the case in (0.5). We also notice that in (0.3) and (0.4), respectively, we have $\lambda(C^n) = 2n + 1$ and $\lambda(C^n) = 2n$. In other words, $\lambda(C^n)$ grows in a polynomial way, and this is true for *every* compact subset C of $\mathbb{Z}$ or $\mathbb{R}$. The "$ax + b$" group does not share this property. This can be checked directly; a different approach is given in Chapter 6. It turns out that this group is a "fundamental" solvable, simply connected Lie group that does *not* have polynomial growth. We now formally define polynomial growth for a group. Let $\mathscr{C}_e(G)$ be the family of compact neighbourhoods of e in a locally compact group G.

(0.12) DEFINITION. A locally compact group G is said to have *polynomial growth* if, for every $C \in \mathscr{C}_e(G)$, there exists a polynomial p such that $\lambda(C^n) \leq p(n)$ for all $n \geq 1$.

We will see ((6.17)) that every nilpotent G has polynomial growth. The theory of groups of polynomial growth is impressive, and Chapter 6 is devoted to this topic. These groups can be regarded as "super-amenable" and have rather special amenability properties. We will be content, for the present, to prove that polynomial growth implies amenability.

(0.13) PROPOSITION. *Every locally compact group of polynomial growth is amenable.*

PROOF. Let G have polynomial growth. Let $x_1, \ldots, x_r \in G$, and let $C \in \mathscr{C}_e(G)$ be such that each $x_i, x_i^{-1} \in C$. Let $a_n = \lambda(C^n)$. There exists a polynomial p such that $0 < a_n \leq p(n)$. We can suppose that $p(n) \leq Kn^r$ for some $K > 0$ and some fixed $r \in \mathbb{P}$, the set of positive integers. Note that $a_n \leq a_{n+1}$ since C^{n+1} contains a left translate of C^n and λ is left invariant. Now $0 < \lambda(C)^{1/n} \leq a_n^{1/n} \leq K^{1/n}(n^{1/n})^r$, so that $a_n^{1/n} \to 1$. An elementary result on convergent sequences gives $\liminf_{n\to\infty}(a_{n+1}/a_n) = 1$. Let $\varepsilon > 0$. Then there exists N such that $\lambda(C^{N+1})/\lambda(C^N) < 1 + \varepsilon/2$. So $\lambda(C^{N+1} \,\Delta\, C^N)/\lambda(C^n) = (\lambda(C^{N+1}) - \lambda(C^N))/\lambda(C^N) < \varepsilon/2$. For any i, $x_iC^N \sim C^N \subset C^{N+1} \sim C^N$, and so $\lambda(x_iC^N \sim C^N)/\lambda(C^N) < \varepsilon/2$. Similarly $\lambda(C^N \sim x_iC^N) = \lambda(x_i^{-1}C^N \sim C^N) < \frac{1}{2}\varepsilon\lambda(C^N)$, we have $\lambda(x_iC^N \,\Delta\, C^N)/\lambda(C^N) < \varepsilon$, that is, $\|x_i * f - f\|_1 < \varepsilon$ $(1 \leq i \leq n)$, where $f = \chi_{C^N}/\lambda(C^N) \in P(G)$. By (0.8), G is amenable. □

We note that when G has polynomial growth, we can answer a natural question concerning the set K of the Følner condition (4). The question is *given C, can we construct a set K satisfying this condition*? The way in which K is derived is highly nonconstructive, so that the method of proof is not helpful for this question. When G has polynomial growth, we can give a reasonable answer to this question: *the set K can be taken to be C^N for some N.*

Returning once again to (0.3)–(0.5), we note that two of the (amenable) groups $(\mathbb{Z}, \mathbb{R})$ are abelian, while the other group S_2 is solvable and so is "built up" out of abelian groups. (In fact, S_2 is a semidirect product $\mathbb{R} \times_\rho \mathbb{R}$.) This raises the question *is every abelian G amenable*? The answer is "yes" and is a simple consequence of the Markov-Kakutani Theorem whose proof we now give ([**DS**]).

(0.14) PROPOSITION. *Let K be a compact convex subset of a locally convex space E. Let S be an abelian semigroup of continuous transformations $T\colon K \to K$ such that each T is affine (in the sense that $T(\alpha k_1 + (1-\alpha)k_2) = \alpha Tk_1 + (1-\alpha)Tk_2$ for all $\alpha \in [0,1]$, $k_1, k_2 \in K$). Then the elements of S have a common fixed point.*

PROOF. Let $T \in S$ and F_T be the set of fixed points for T in K. We show first that $F_T \neq \varnothing$. Let $k \in K$ and $T_n = (n+1)^{-1}\sum_{r=0}^{n} T^r$. Then $T_n\colon K \to K$. Let $k_n = T_nk$ and k_0 be a cluster point of $\{k_n\}$ in K. We prove that $Tk_0 = k_0$, that is, that $k_0 \in F_T$. To this end, we have, using the affineness of T, that

$(T-I)k_n = (n+1)^{-1}(T^{n+1}-I)k_0 \in (n+1)^{-1}(K-K)$. Since $K-K$ is compact in E and $(T-I)$ is continuous, we have $(T-I)k_0 = 0$ and $k_0 \in F_T$. Note that F_T is a compact convex set.

Let $Z = \bigcap\{F_T : T \in S\}$. If $T, T' \in S$, then $T'(F_T) \subset F_T$, and applying the preceding result to $T'|_{F_T}$, we obtain that $F_T \cap F_{T'} \neq \varnothing$. Thus $\{F_T : T \in S\}$ has the finite intersection property, and $Z \neq \varnothing$. Any element of Z is a common fixed point for S. □

(0.15) PROPOSITION. *Every abelian locally compact group is amenable.*

PROOF. Let G be an abelian locally compact group. For the purposes of applying (0.14), take $E = L_\infty(G)'$ with the weak* topology, $K = \mathfrak{M}(G)$, and let S be the group of transformations $m \to xm$ $(x \in G)$ of $\mathfrak{M}(G)$. The common fixed-point set of S is just $\mathfrak{L}(G)$. □

The group S_2 is an extension of an amenable group by an amenable group, and (0.5) tells us that S_2 is amenable. Further, the argument of (0.7) proves that *a discrete subgroup of a (separable) amenable locally compact group is itself amenable.* These two observations suggest the question: *under which algebraic processes is amenability preserved?* The topological version of this question requires some work to answer—see Chapter 1. However, von Neumann [**1**] showed that the question can be readily answered for discrete groups, and we give the proof here. For more examples of amenable groups see the Problems at the end of this chapter.

(0.16) PROPOSITION. *Amenability for discrete groups is preserved by the following five processes:*

(1) *taking subgroups;*
(2) *forming quotient groups;*
(3) *forming group extensions;*
(4) *forming upward directed unions of amenable groups;*
(5) *forming a direct limit of amenable groups.*

PROOF. (1) This follows as in (0.7)—merely delete any "Borel" requirements.

(2) Let G be amenable, H a normal subgroup, and $\Phi : l_\infty(G/H) \to l_\infty(G)$ given by $\Phi(\phi)(x) = \phi(xH)$. If $m \in \mathfrak{L}(G)$, then $m \circ \Phi \in \mathfrak{L}(G/H)$, so that G/H is amenable.

(3) Let H be a normal subgroup of a group G such that both H and G/H are amenable. We have to show that G is amenable. Let $\psi \in l_\infty(G)$ and $p \in \mathfrak{L}(H)$, $q \in \mathfrak{L}(G/H)$. Let $\psi_p : G/H \to \mathbf{C}$ be given by $\psi_p(tH) = p((\psi t)|_H)$ $(t \in G)$. Since p is left invariant, the function ψ_p is well-defined and belongs to $l_\infty(G/H)$. Now check that $(\Psi t_0)_p = \Psi p t_0 H$ $(t_0 \in G)$ and the functional $\psi \to q(\psi_p)$ belongs to $\mathfrak{L}(G)$.

(4) Suppose that $\{H_\delta : \delta \in \Delta\}$ is an upwards directed family of amenable subgroups of G such that $G = \bigcup_{\delta \in \Delta} H_\delta$. (So Δ is a directed set and $H_\gamma \subset H_\delta$ if $\gamma \leq \delta$.) For each δ, choose $m_\delta \in \mathfrak{L}(H_\delta)$, and define $n_\delta \in \mathfrak{M}(G)$ by setting $n_\delta(\phi) = m_\delta(\phi|_{H_\delta})$. We can suppose that $n_\delta \to n$ weak* in $\mathfrak{M}(G)$. If $x \in G$, then

for some $\gamma \in \Delta$, $x \in H_\delta$ whenever $\delta \geq \gamma$, so that $n_\delta(\phi x) = m_\delta(\phi|_{H_\delta} x) = n_\delta(\phi)$. It follows that $n \in \mathfrak{L}(G)$ and G is amenable. (The proof of the corresponding result for semigroups is similar.)

(5) Let Δ be a directed set, let $\{G_\delta : \delta \in \Delta\}$ be a family of left amenable groups, and for each pair $(\sigma, \delta) \in \Delta \times \Delta$ with $\sigma \leq \delta$, let $\rho_{\delta,\sigma} : G_\sigma \to G_\delta$ be a homomorphism such that whenever $\eta \leq \varsigma \leq \theta$, we have $\rho_{\theta,\eta} = \rho_{\theta,\varsigma} \circ \rho_{\varsigma,\eta}$. The groups G_δ and homomorphisms $\rho_{\delta,\sigma}$ form a direct system of groups, and if G is the direct limit of this system then, for each δ, there is a canonical homomorphism $\rho_\delta : G_\delta \to G$ such that $\rho_\eta = \rho_\varsigma \circ \rho_{\varsigma,\eta}$ whenever $\eta \leq \varsigma$, and $G = \bigcup_{\delta \in \Delta} \rho_\delta(G_\delta)$. The amenability of G follows from (2) and (4). □

The above results together with Example (0.6) suggest the following important questions. We introduce three classes of (discrete) groups. The first of these is EG ("elementary groups"): this is the smallest class of groups that contains all finite and all abelian groups and that is closed under the four processes of taking subgroups and of forming factor groups, group extensions, and upwards directed unions. The class AG ("amenable groups") is the class of amenable groups. The class NF ("no free") is the class of those groups not containing F_2 as a subgroup. Since AG contains all finite and all abelian groups and is closed under the above four processes, we have EG $\subset$ AG. Further, we have AG $\subset$ NF ((0.6), (0.16(1))). We therefore have

$$\mathrm{EG} \subset \mathrm{AG} \subset \mathrm{NF}.$$

The questions referred to above are: *does* EG = AG *and does* AG = NF?

The conjecture that AG = NF (or, equivalently, that a discrete group is *not* amenable if and only if it contains F_2 as a subgroup) is known as *von Neumann's Conjecture*. (It is sometimes referred to as *Day's Conjecture*.) It has been shown (Ol′shanskii [**4**]) that, in fact, AG $\neq$ NF. As might be expected, the proof is difficult. It is discussed in (4.30)–(4.33). R. I. Grigorchuk [**S1**] has recently shown that EG $\neq$ AG. We will show in (3.11) that EG $\neq$ NF.

Turning to (0.7), the nonamenability of $\mathrm{SL}(2, \mathbb{R})$ depends on its containing F_2 as a discrete subgroup. This raises the speculation that a connected Lie group is nonamenable if and only if it contains F_2 as a discrete subgroup. We will show (Chapter 3) that this result is true for almost connected groups (that is, for those groups that are extensions of a compact group by a connected group). The proof involves detailed Lie theory.

Having discussed a number of the directions that (0.3)–(0.7) lead to, we now mention some extensions of the amenability notions of this chapter that often arise in practice.

(0.17) Other types of invariant mean. We start by commenting on the "left-wing" prejudice shown in our definition of amenability in (0.2). We can, after all, define a right invariant mean on G in a natural way. In fact, we define a right action $n \to nx$ ($x \in G$) of G on $L_\infty(G)'$ by requiring $nx(\phi) = n(x\phi)$, where $x\phi(y) = \phi(yx)$ ($x, y \in G$, $\phi \in L_\infty(G)$). A mean m on G is called *right invariant* if $mx = m$ for all $x \in G$, and the set of right invariant means on G is denoted by $\mathfrak{R}(G)$. If $m \in \mathfrak{L}(G)$, then we can obtain $n \in \mathfrak{R}(G)$ by setting $n(\phi) = m(\phi^*)$, where $\phi^*(x) = \phi(x^{-1})$. So left and right amenability for G are equivalent.

The set of means on G which are *invariant* (that is, both left and right invariant) is denoted by $\mathfrak{I}(G)$. We will see (Problem 1-26) that amenability for G is equivalent to the existence of an *invariant* mean on G. This is not immediately clear and in fact depends on the "Arens product" structure of $L_\infty(G)'$.

We can also define the set $\mathfrak{R}_t(G)$ of topologically right invariant means on G in the obvious way. Similarly, we can define $\mathfrak{I}_t(G)$.

Of course, results for left invariant means have right invariant versions.

It is often important to study invariant means on subspaces of $L_\infty(G)$. Much of the theory we have discussed extends trivially when $L_\infty(G)$ is replaced by a subspace A of $L_\infty(G)$ that contains 1. (In some cases, for example, in (0.1(i)), we also require A to be closed under complex conjugation.) We say that A is *right invariant* if $\phi x \in A$ whenever $\phi \in A$, $x \in G$. A mean on A is left invariant if it is invariant under the dual left action of G (just as in the $L_\infty(G)$ case). The set of left invariant means on A is denoted by $\mathfrak{L}(A)$. So $\mathfrak{L}(L_\infty(G)) = \mathfrak{L}(G)$. The sets $\mathfrak{R}(A)$, $\mathfrak{I}(A)$ are defined in the obvious way for A with the appropriate kind of G-invariance. Similar results apply if A is invariant under the action of $P(G)$. For example, if A is right invariant for $P(G)$ (that is, if $\phi * f \in A$ if $\phi \in A$, $f \in P(G)$), then $\mathfrak{L}_t(A)$ is defined in the obvious way. Similar considerations apply to $\mathfrak{R}_t(A)$ and $\mathfrak{I}_t(A)$.

The advantage of studying invariant means on subspaces of $L_\infty(G)$ is that these subspaces can have "nice" properties (e.g., continuity) not possessed by $L_\infty(G)$.

(0.18) Semigroups and amenability. Let S be a discrete semigroup. Recall that $l_1(S)$ is the space of functions $f\colon S \to \mathbb{C}$ such the $\sum_{s\in S} |f(s)| < \infty$. It is often convenient to write $f = \sum_{s\in S} f(s)s$. The norm $\|f\|_1$ of f is, of course, just $\sum_{s\in S} |f(s)|$. We make $l_1(S)$ into a convolution Banach algebra in the obvious way:

$$\left(\sum_{s\in S} a_s s\right)\left(\sum_{s\in S} b_s s\right) = \sum_{u\in S}\left(\sum_{st=u} a_s b_t\right)u \quad \left(= \sum_{s\in S}\sum_{t\in S} a_s b_t st\right)$$

($\sum_{s\in S} a_s s, \sum_{s\in S} b_s s \in l_1(S)$). Of course $S \subset l_1(S)$, and S has a left action by convolution on $l_1(S)$. The dual right action of S on $l_\infty(S) = l_1(S)'$ is given as for groups: $\phi s(t) = \phi(st)$ $(s, t \in S)$. As in the group case, S is *left amenable* if there exists a mean on $l_\infty(S)$ invariant under the dual left action of S on $l_\infty(S)'$. The set of left invariant means on S is denoted by $\mathfrak{L}(S)$. *Right amenability* and (two-sided) *amenability* are defined in the obvious ways, and the corresponding sets of means are denoted by $\mathfrak{R}(S)$ and $\mathfrak{I}(S)$, respectively. As in (0.15), it follows from (0.14) that *every abelian semigroup is amenable.*

The study of left amenable semigroups goes back to the 1920s. In fact, Banach's well-known *limits* [**DS**, p. 73] are nothing other than invariant means on $\mathbb{P}$, the additive semigroup of positive integers.

There are a number of results in the theory of amenable semigroups that are straightforward generalisations of corresponding group results. However there are also a number of results for groups that do not readily extend to semigroups, the reason for this being that the algebraic structure of a semigroup is much more complicated, in general, than that of a group. At the same time, the semigroup situation allows for more variety than the group one; for example, no nontrivial group admits a multiplicative left invariant mean, yet many semigroups do ((2.29), Problem 2-36). The main interest in amenable semigroups, therefore, lies in the *algebraic* side.

An example of an amenable group result that does not extend to the semigroup case is the Følner condition of (3): it is not the case that a semigroup S is left amenable if and only if, for any $\varepsilon > 0$ and finite subset C of S, there exists a finite, nonempty subset K of S such that $|xK \bigtriangleup K|/|K| < \varepsilon$. (Here, $|A|$ is the cardinality of a set A.) The reason for this is the lack of cancellation in S in general. In this connection, see (4.9), (4.22).

Another example is that of the cardinality question. The question of the cardinality of $\mathfrak{L}(S)$ is substantially more difficult than the corresponding group question. (See (7.22)–(7.27).)

The theory of left amenable semigroups is often very elegant and has a combinatorial flavor to it. In our view, it does have a legitimate place in amenability theory.

There is a not inconsiderable literature on amenability for topological semigroups (cf. Problem 1-30, Solution). Invariant means on such semigroups do arise naturally in the context of weakly almost periodic compactifications. However, it has to be admitted that the study of amenable topological semigroups is very much in its infancy. One serious problem is the lack of good concrete examples of such semigroups to serve as a basis for the theory. (There is a multitude of readily available examples of *discrete* amenable semigroups). For these reasons, we do not discuss amenable topological semigroups in detail.

(0.19) The scope of amenability. One of the interesting features of amenability is the wide range of mathematics with which it is involved. Another related feature is the way in which its ideas can be fruitfully modified to enable it to jump categories. These two features suggest that amenability is more than just a phenomenon in which analysts are interested—it does have a wider significance for mathematics as a whole. We now mention briefly some of the contexts involving amenability ideas that are discussed in this work and that we have not so far mentioned in this chapter.

Amenability ideas arise in the following classical problem: *when does a group G acting on a set X* not *admit a Banach-Tarski paradox*? The answer is: *when X admits a G-invariant mean.* In particular, *G is amenable if and only if it does not admit a "paradoxical decomposition."* This beautiful theorem by Tarski characterises amenability purely in terms of translates of subsets of G—no mention of means, measures, or functional analysis! The result is discussed in (3.12)–(3.15). Problem 3-15 gives a topological version of the above characterisation of amenability.

Two major classical problems in amenability have been settled in the 1980s: these questions are the *Banach-Ruziewicz Problem* and *von Neumann's Conjecture*. Detailed discussions of these solutions are given in (4.27)–(4.29) and (4.30)–(4.33), respectively. The version of the Banach-Ruziewicz result that we discuss reads as follows: *Let μ_n be the natural normalised $O(n+1)$-invariant measure on $\mathbb{S}^n$ obtained from Lebesgue measure on $\mathbb{R}^{n+1}$. Then μ_n is the unique $O(n+1)$-invariant, finitely additive, positive measure on $\mathbb{S}^n$ of total mass 1 if and only if $n \geq 2$.* The proof of this theorem uses the Kazhdan Property T as well as the theory of algebraic groups. A consequence of the theorem is that the only invariant mean on the orthogonal group O_n ($n \geq 5$) is Haar measure.

We have already met von Neumann's conjecture in (0.16). The resolution of the conjecture falls into two parts. The first part is the construction of a group that is a candidate for a nonamenable group not containing F_2 as a subgroup. The construction of the group, due to A. Yu. Ol′shanskii, involves complicated and lengthy combinatorial group theory, and we do not attempt it. The second part is of great interest from an amenability viewpoint and characterises amenable, finitely generated groups in terms of combinatorial group theory. This characterisation (due to Grigorchuk and Cohen) is used by Ol′shanskii to show that his group is not amenable, and it runs roughly as follows. Let G be a group on k generators $u_1, \ldots, u_k$. For each n, let N_n be the number of words of length n in u_i, u_i^{-1} whose product in G is e. Then (roughly) *G is amenable if and only if $|N_n|^{1/n} \to (2k-1)$.*

Amenability applies to the statistical theory of testing hypotheses. The Hunt-Stein Theorem ensures the existence of invariant minimax tests when the group of an invariant test problem is amenable. This and other statistical applications of amenability are discussed in (4.24)–(4.26), Problem 4-19. Amenability also proves relevant to the study of transience/recurrence of random walks on locally compact groups and to Poisson spaces ((6.46), (6.47)).

Amenability's impact on the representation theory of locally compact groups is largely through the remarkable "weak containment property". Let G be a locally compact group. The representation theory of G is equivalent to that of its enveloping C^*-algebra $C^*(G)$, the completion of $L_1(G)$ under its largest C^*-norm. Unfortunately, $C^*(G)$, though categorically "nice", is hard to get hold of in general. On the other hand, there is available a natural, accessible C^*-algebra $C_l^*(G)$ associated with G: $C_l^*(G)$ is the C^*-subalgebra of $\mathbf{B}(L_2(G))$ generated by $\pi_2(L_1(G))$, where π_2 is the left regular representation of G. One can do calculations with $C_l^*(G)$! Unfortunately, the representation theory of $C_l^*(G)$ picks up only *some* of the irreducible, unitary representations of G, viz., those which are "weakly contained" in π_2. In the best of both worlds, we would have $C^*(G) = C_l^*(G)$. When is this true? The answer ((4.21)) is that $C_l^*(G) = C^*(G)$

if and only if G is amenable. Polynomial growth has, as might be expected, an even stronger impact on the representation theory of solvable Lie groups ((6.45)).

Indeed, polynomial growth has a particularly rich range of applications. For example, if a complete Riemannian manifold M has nonnegative mean curvature, then its fundamental group $\pi_1(M)$ has polynomial growth ((6.41)). Symmetry of the group algebra $L_1(G)$ is related to amenability: in fact, *every almost connected, locally compact group with symmetric group algebra is amenable.* The converse is false. However, a beautiful theorem due to Ludwig ((6.50)) states *if G is connected and has polynomial growth, then G is symmetric.* A good example of how amenability ideas can cross mathematical categories is afforded by the theory of *amenable Banach algebras.* This theory, which is largely the creation of B. E. Johnson, is discussed in (1.30). A Banach algebra A is called *amenable* if the first cohomology group $H^1(A, X') = \{0\}$ for every dual Banach A-module X'. This means that every continuous derivation $D\colon A \to X'$ is inner. At first sight, there does not seem to be any connection between invariant means and derivations. However, Johnson proved that *if G is a locally compact group, then G is amenable if and only if the Banach algebra $L_1(G)$ is amenable.* This justifies the terminology *amenable Banach algebra.*

Banach algebra amenability has proved very fruitful when specialized to operator algebras. Amenability for a C^*-algebra A is equivalent to a strong "tensor product" property of A known as *nuclearity.* A brief discussion of amenable C^*-algebras is given in (1.31). Amenability for a von Neumann algebra, obtained by modifying Banach algebra amenability in a sensible way, is equivalent to a number of very important properties: *injectivity, semidiscreteness, Property P,* and *approximate finite-dimensionality.* A discussion of amenable von Neumann algebras is given in (2.35). It was not realistic to give a thorough account of amenable von Neumann algebras, and instead, we have tried to show how the various formulations of amenability in that context relate to one another and to group amenability.

In the study of the Fourier algebra $A(G)$ of a locally compact group G, amenability of G translates into a familiar Banach algebra property. When G is abelian, $A(G)$ can be identified with $L_1(\hat{G})$; more generally, the commutative Banach algebra $A(G)$ is the predual of the von Neumann algebra $\mathrm{VN}(G)$ generated by the left regular representation of G with a natural "pointwise" multiplication. It turns out ((4.34)) that *G is amenable if and only if $A(G)$ contains a bounded approximate identity.*

A recent development of amenability is that of K-amenability. In Kasparov's KK-theory, a (countable discrete) group G is called *K-amenable* if $K^0(C^*(G)) = K^0(C_l^*(G))$, where K^0 denotes K-homology. So K-amenability is a homological version of the weak containment property. Remarkably, F_2—the bête noire of amenability—is K-amenable! (References: Kasparov [**S1**], Cuntz [**S2**], Julg and Valette [**S1**].)

We started this chapter with Lebesgue's question on the uniqueness of his integral. The reader will probably agree that in the last few pages we have "come a long way." We hope to have demonstrated the range of amenability ideas.

References

Banach [**1**], [**2**], Banach and Tarski [**1**], Chou [**9**], Day [**4**], [**9**], Følner [**2**], Greenleaf [**2**], Guivarc'h [**1**], Hulanicki [**2**], Lebesgue [**1**], Milnes [**S1**], Namioka [**1**], von Neumann [**1**], Reiter [**R**], and Tarski [**2**].

Problems 0

1. Let X be a set and for each $m \in \mathfrak{M}(X)$, define a set function μ_m by $\mu_m(E) = m(\chi_E)$ $(E \subset X)$. Prove that the map $m \to \mu_m$ is a bijection from $\mathfrak{M}(X)$ onto the set of positive, finitely additive measures on X of total mass 1.

2. This is the topological version of Problem 1 above. Let X be a locally compact space and μ be a positive, regular Borel measure on X. Let $\mathscr{M}(X)$ be the σ-algebra of μ-measurable subsets of X. Let $m \to \mu_m$ be the map defined as in Problem 1 from $\mathfrak{M}(X)$, the set of means on $L_\infty(X, \mu)$, into the set of measures on $\mathscr{M}(X)$. Prove that this map is a bijection onto the set of positive, finitely additive measures of total mass 1 on $\mathscr{M}(X)$ that vanish on μ-locally-null sets.

Deduce that the set of left invariant means on a locally compact group can be identified with the set of left invariant, positive, finitely additive measures on $\mathscr{M}(G)$ of total mass 1 that vanish on λ-locally-null sets.

3. Let A be a subspace of $L_\infty(G)$ containing 1 and closed under complex conjugation. Establish the analogue of (0.1) with A in place of $L_\infty(G)$.

4. An element $F \in L_\infty(G)'$ is called *left invariant* if $xF = F$ for all $x \in G$. The set of left invariant functionals F is denoted by $\mathfrak{J}_l(G)$.

Show that $\mathfrak{J}_l(G)$ is an invariant subspace of $L_\infty(G)'$ containing $\mathfrak{L}(G)$. What is $\mathfrak{J}_l(G)^\perp \cap L_\infty(G)^\wedge$?

5. Let G be a group and X be a left G-set with $ex = x$ for all $x \in X$. Let $A \subset X$ and suppose that there exists a G-invariant finitely additive, $[0, \infty]$-valued measure μ on X such that $\mu(A) = 1$. Prove that if $n \geq 1$, $s_1, \ldots, s_n \in G$, $\alpha_1, \ldots, \alpha_n \in \mathbb{R}$, and $\sum_{i=1}^n \alpha_i \chi_{s_i A} \geq 0$, then $\sum_{i=1}^n \alpha_i \geq 0$. [This and its converse is proved in (2.32).]

6. Let $\{A_n\}$ be a sequence of nonnull measurable sets in a locally compact group G. Prove that for $x \in G$, $\lambda(xA_n \,\triangle\, A_n)/\lambda(A_n) \to 0$ if and only if $\lambda(xA_n \cap A_n)/\lambda(A_n) \to 1$.

7. Let G be a locally compact group. A mean m on G is called *symmetrically invariant* if $\frac{1}{2}(xm + x^{-1}m) = m$ for all $x \in G$. Prove that every symmetrically invariant mean on G is actually a left invariant mean.

8. Let A_n be as in (0.5). Show that $\lambda(xA_n \triangle A_n)/\lambda(A_n) \to 0$ for all $x \in S_2$.

9. Let G be an amenable locally compact group. Prove that there exists a net $\{g_\delta\}$ in $P(G)$ such that $\{g_\delta * \phi\}$ converges pointwise to a constant function on G for each $\phi \in L_\infty(G)$.

10. Prove that if G is noncompact and $m \in \mathfrak{L}(G)$, then $m(C_0(G)) = \{0\}$, where $C_0(G)$ is the space of continuous complex-valued functions on G vanishing at infinity. Deduce that $m(C) = 0$ for all $C \in \mathscr{C}(G)$. Prove that $\mathfrak{L}(G) \cap P(G)^\wedge \neq \varnothing$ if and only if G is compact (c.f. (1.20)).

11. Let G be abelian and $\{f_\delta\}$ satisfy the condition of (0.8). Let Γ be the dual group of G. Prove that $\{\hat{f}_\delta\}$ converges pointwise on Γ to the function that is 1 at the identity and 0 everywhere else.

12. Prove that if G is compact then $\mathfrak{L}(C(G))$ is the singleton set $\{\lambda\}$.

13. A sequence $\phi\colon \mathbb{P} \to \mathbb{C}$ is called *almost convergent* (with *limit* l) if $\lim_{p\to\infty} p^{-1} \sum_{r=n+1}^{n+p} \phi(r) = l$ uniformly in n. The set of such sequences is denoted by AC. Show that AC is a closed subspace of l_∞ containing 1, and that if $\phi \in \mathrm{AC}$ has limit l, then $\{l\} = \{m(\phi)\colon m \in \mathfrak{L}(\mathbb{P})\}$. (See Problem 4-15 for the converse.) Give examples of sequences in AC, and show that AC is not norm separable.

14. Prove that $\mathfrak{L}_t(\mathbb{R})$, $\mathfrak{L}(\mathbb{P})$, and $\mathfrak{L}_t(S_2)$ are infinite.

15. Let G be an infinite discrete group, $m \in \mathfrak{L}(G)$, and $E \subset G$ be such that $|E| < |G|$. Prove that $m(E) = 0$.

16. Let G be a connected locally compact group and $C \in \mathscr{C}_e(G)$. Prove that G has polynomial growth if and only if there exists a polynomial p such that $\lambda(C^n) \leq p(n)$ for all $n \geq 1$.

17. Let G be a (discrete) finitely generated group. Let F be a finite set of generators for G. Prove that G has polynomial growth if and only if there exists a polynomial p such that $|F^n| \leq p(n)$ for all $n \geq 1$.

18. Prove that every (discrete) abelian group has polynomial growth. Let G be the free abelian group on two generators x, y and $E = \{x, y, x^{-1}, y^{-1}\}$. Calculate $|E^s|$.

19. Prove that S_2 does not have polynomial growth.

20. A locally compact group G is said to be *exponentially bounded* if $\lambda(C^n)^{1/n} \to 1$ for all $C \in \mathscr{C}_e(G)$. Prove that if G has polynomial growth, then G is exponentially bounded. Show that exponential boundedness implies amenability.

21. A discrete group G is said to be *locally finite* if every finite subset of G generates a finite subgroup. Prove that every locally finite group is amenable.

22. Prove that the group of those permutations of a set which leave all but a finite number of elements fixed is amenable.

23. Let G be a discrete group for which there is a normal series

$$G = G_1 \rhd G_2 \rhd \cdots \rhd G_n = \{e\}$$

such that G_i/G_{i+1} is amenable for $1 \leq i \leq n-1$. Show that G is amenable. Deduce that a semidirect product of amenable groups is amenable and that every solvable group is amenable.

24. Let $\{G_i \colon i \in I\}$ be a family of finite groups such that $\sup_{i\in I} |G_i| = K < \infty$. Prove that the Cartesian product $\prod_{i\in I} G_i$ is amenable.

25. Prove that the isometry group G_n of $\mathbb{R}^n$, regarded as discrete, is amenable if $n = 1, 2$. (G_n is not amenable as discrete if $n > 2$—cf. Problem 3-5.)

26. Show that every amenable subgroup of a group G is contained in a maximal amenable subgroup of G and that there is a normal amenable subgroup N of G which contains every normal amenable subgroup of G and is contained in every maximal amenable subgroup of G.

27. Prove that if every nontrivial, homomorphic image of a group G contains a nontrivial, amenable, normal subgroup, then G is amenable. (Supersolvable groups (Baer [**1**], M. Hall [**1**]) satisfy this property.)

28. Let FS_2 be the free semigroup on two generators x, y. (So a typical element of FS_2 is uniquely of the form $u_1 u_2 \cdots u_n$, where $u_i \in \{x, y\}$.) Prove that FS_2 is not left amenable.

29. Is every subsemigroup of an amenable group necessarily left amenable?

30. Prove semigroup versions of (2), (4), and (5) of (0.16).

31. Give an example of a semigroup that is left amenable but not right amenable.

32. Let S be a semigroup. Prove that the map $m \to \mu_m$ (cf. Problem 1 above) is a bijection from $\mathfrak{L}(S)$ onto the set of finitely additive, positive measures μ on S of mass 1 such that

$$\mu(s^{-1}E) = \mu(E)$$

for all $E \subset S$, $s \in S$. [Note: $s^{-1}E = \{t \in S \colon st \in E\}$].

33. Show that if S is a semigroup and $m \in \mathfrak{L}(S)$, then $m(I) = 1$ for every right ideal I in S.

34. Let G be a locally compact group and π a strongly continuous representation of G on a Hilbert space $\mathfrak{H}$ such that $\sup_{x\in G} \|\pi(x)\| < \infty$ and each $\pi(x)$ is an invertible element of $B(\mathfrak{H})$. Show that there exists a unitary representation π' of G on $\mathfrak{H}$ such that π and π' are *equivalent*, that is, there exists an invertible element $A \in B(\mathfrak{H})$ such that $\pi(x) = A\pi'(x)A^{-1}$ for all $x \in G$ (cf. Problem 1-40).

35. Suppose that the locally compact group G is the union of an upwards directed family of open, amenable subgroups. Show that G is amenable.

CHAPTER 1

Amenable Locally Compact Groups and Amenable Semigroups

(1.0) Introduction. In (0.16) we showed that amenability is preserved for discrete groups under the processes of taking subgroups and forming quotient groups, group extensions, and direct limits. We saw in Problem 0-35 that a version of the direct limit result holds in the nondiscrete case. The other three processes, however, pose problems, and one of the main purposes of this chapter is to show that topological versions of these three processes preserve amenability in the nondiscrete case also. The proofs depend on two crucial facts. The first is that the amenability of G is equivalent to the existence of a left invariant mean on any one of a wide range of subspaces of $L_\infty(G)$. The second is that the amenability of G is equivalent to the existence of a topologically left invariant mean ((0.9)) on $L_\infty(G)$.

We finally investigate some of the properties of left amenable semigroups.

Let us start by looking at the difficulties that arise when we try to extend (1) and (2) of (0.16) to the locally compact group case. Let G be an amenable locally compact group and H be a closed subgroup of G. If G is discrete, then we obtain $n \in \mathfrak{L}(H)$ by writing $n(\psi) = m(\psi')$, where $m \in \mathfrak{L}(G)$, $\psi \in l_\infty(H)$, and $\psi'(hb) = \psi(h)$ $(b \in B)$, where B is a transversal for the set of right H-cosets. In the nondiscrete case we run into serious measurability problems for ψ': it is not clear that B can be taken to be even Borel in G. The problems arising with (2) of (0.16) are not so serious, but nevertheless non-trivial results are needed to push through the discrete argument. In both cases, life would be made easier by replacing $L_\infty(G)$ by $C(G)$, the space of bounded, continuous, complex-valued functions on G. Of course, $C(G) \subset L_\infty(G)$ (in the natural way), and every $m \in \mathfrak{L}(G)$ restricts to an element of $\mathfrak{L}(C(G))$. The converse is not immediately clear.

It turns out that to solve the above problems, we have to consider not only $C(G)$ but also the space $\mathrm{U}_r(G)$ of right uniformly continuous, complex-valued functions on G. In addition, we will show that G is amenable if and only if there exists a *topologically* left invariant mean on G.

It is convenient, before dealing with these matters, to reformulate the actions of $L_1(G)$ on $L_\infty(G)$ (see (0.8)) in terms of measures. This not only enables us to extend the action to one of $M(G)$ on $L_\infty(G)$ but also clarifies and smooths out many of the arguments involving such an action. Of course, the function viewpoint is also useful.

(1.1) The action of $M(G)$ on $L_\infty(G)$ (cf. [**HR1**, §20]). Let X be a locally compact Hausdorff space and $M(X)$ be the space of complex-valued, regular Borel measures on X. Then $M(X)$ is a Banach space under the total variation norm. By the Riesz Representation Theorem, $M(X)$ is canonically the Banach space dual of $C_0(X)$, the space of functions $\phi \in C(X)$ which vanish at infinity, each $\mu \in M(X)$ corresponding to the functional $\phi \to \int \phi \, d\mu$ on $C_0(X)$. When $X = G$, the space $M(G)$ is a convolution Banach*-algebra, the product being given by

$$\int \phi \, d(\mu * \nu) = \mu * \nu(\phi) = \iint_{G \times G} \phi(xy) \, d\mu(x) \, d\nu(y) \qquad (\phi \in C_0(G)).$$

The involution $\mu \to \mu^\sim$ on $M(G)$ is given by $\mu^\sim(E) = \overline{\mu(E^{-1})}$ $(E \in \mathscr{B}(G))$ and is isometric. Closely related to $\mu^\sim$ is μ^*, where $\mu^*(E) = \mu(E^{-1})$. By the Radon-Nikodým Theorem, $L_1(G)$ is identifiable with the space of measures in $M(G)$ that are absolutely continuous with respect to λ. Thus a function $f \in L_1(G)$ corresponds to the measure $\mu_f \in M(G)$ determined by

$$\mu_f(\phi) = \int_G \phi(x) f(x) \, d\lambda(x) \qquad (\phi \in C_0(G)).$$

We will normally identify $f \in L_1(G)$ with μ_f, the context making clear if we are dealing with a function or a measure. Note that $x * f$, $f * x$ (defined in (0.7)), as measures, are just $\delta_x * \mu_f$, $\mu_f * \delta_x$, where δ_x is the point mass at x. We will often write $x * \mu$, $\mu * x$ or even $x\mu$, μx in place of $\delta_x * \mu$, $\mu * \delta_x$.

An important fact is that $L_1(G)$ is an ideal in $M(G)$, so that $M(G)$ acts on $L_1(G)$ by convolution. The product $f * g$ of functions, $f, g \in L_1(G)$ is given by: $f * g(x) = \int_G f(t) g(t^{-1}x) \, d\lambda(t)$ $(x \in G)$. The dual of this $M(G)$-action leads to an action of $M(G)$ on $L_\infty(G)$, and the formulae for this action are as follows: if $\mu \in M(G)$ and $\phi \in L_\infty(G)$, then $\phi\mu$, $\mu\phi$ as members of $L_1(G)'$ are given by

$$\text{(1)} \qquad \begin{aligned} \phi\mu(\nu) &= \phi(\mu * \nu) = \int \phi \, d(\mu * \nu) \\ \mu\phi(\nu) &= \phi(\nu * \mu) = \int \phi \, d(\nu * \mu) \end{aligned} \qquad (\nu \in L_1(G)),$$

while $\phi\mu$, $\mu\phi$ as functions in $L_\infty(G)$ are given by

$$\text{(2)} \qquad \begin{aligned} \phi\mu(x) &= x\phi(\mu) = \int \phi(yx) \, d\mu(y), \\ \mu\phi(x) &= \mu(\phi x) = \int \phi(xy) \, d\mu(y). \end{aligned}$$

We note that this action of $M(G)$ dualizes to give an action on $L_\infty(G)'$.

The algebra $L_1(G)$ is a $*$-ideal of $M(G)$, the involution on $L_1(G)$ in function terms being given by the map $f \to f^\sim$, where $f^\sim(x) = \overline{f(x^{-1})}\Delta(x^{-1})$ and Δ

is the modular function on G. The "adjoint" of this map on $L_\infty(G)$ is the map $\phi \to \overline{\phi}^*$, where $\phi^*(x) = \phi(x^{-1})$. So

$$\int \overline{\phi}^*(x) f(x)\, d\lambda(x) = \int \phi(x) f^\sim(x)\, d\lambda(x). \tag{3}$$

Here are two illustrations of why it is often convenient to regard elements of $L_1(G)$ as measures. In the first place, it makes clear the essential difference between the actions of G on $L_1(G)$ and $L_\infty(G)$: the one is the dual of the other, and the fact that $x * f$ is defined similarly to ϕx^{-1} is accidental. In the second place, it often obviates the need for calculations involving the modular function. For example, if $f \in L_1(G)$, $\mu \in M(G)$, then, as a function,

$$f * \mu(y) = \int f(yz^{-1})\Delta(z^{-1})\, d\mu(z),$$

whereas the measure version is just the simple convolution $\mu_f * \mu$.

Our next result gives an important property that characterizes $L_1(G)$ as a subset of $M(G)$.

(1.2) PROPOSITION [**HR1**, (19.27)]. *Let $\mu \in M(G)$. Then $\mu \in L_1(G)$ if and only if the map $x \to x * \mu$ is norm continuous from G into $M(G)$.*

We now introduce the spaces $\mathrm{U_r}(G)$, $\mathrm{U}_l(G)$, and $\mathrm{U}(G)$ of uniformly continuous functions. These spaces link up well with topologically invariant means. For example, as we shall see, every left invariant mean on $\mathrm{U_r}(G)$ is actually topologically left invariant!

(1.3) DEFINITIONS. Recall that $C(G)$ is the space of bounded, continuous, complex-valued functions on G. A function $\phi \in C(G)$ is said to be *left* [*right*] *uniformly continuous* if the map $x \to x\phi$ [$x \to \phi x$] is norm continuous from G into $C(G)$. The set of left [right] uniformly continuous functions in $C(G)$ is denoted by $\mathrm{U}_l(G)$ [$\mathrm{U_r}(G)$], and the elements of $\mathrm{U}(G) = \mathrm{U}_l(G) \cap \mathrm{U_r}(G)$ are said to be *uniformly continuous*. Our discussion will proceed in terms of $U_r(G)$; of course, analogous results hold for $U_l(G)$.

These spaces are often discussed in terms of the uniformities on G (cf. [**HR1**, §4]). However, the two approaches are equivalent, and the approach that we adopt is usually easier to work with and applies directly to the semigroup case.

Let $\phi \in \mathrm{U_r}(G)$ and $x_0 \in G$. If $x_\delta \to x$ in G, then $\|(x_0\phi)x_\delta - (x_0\phi)x\| \le \|\phi x_\delta - \phi x\| \to 0$, and $\|(\phi x_0)x_\delta - (\phi x_0)x\| = \|\phi x_0 x_\delta - \phi x_0 x\| \to 0$ so that $x_0\phi, \phi x_0 \in \mathrm{U_r}(G)$. So $\mathrm{U_r}(G)$ is an invariant subspace of $C(G)$. (Of course, $C(G)$ is an invariant subspace of $L_\infty(G)$.)

(1.4) PROPOSITION. *If $\phi \in L_\infty(G)$, $\psi \in \mathrm{U_r}(G)$, and $\mu, \nu \in L_1(G)$, then $\phi\nu \in \mathrm{U_r}(G)$, $\mu\psi \in \mathrm{U_r}(G)$, and $\mu\phi\nu \in \mathrm{U}(G)$.*

PROOF. If $x, y \in G$, then for $z \in G$,

$$|(\phi\nu)x(z) - (\phi\nu)y(z)| \le |(z\phi)(\nu x - \nu y)| \le \|\phi\|\,\|\nu x - \nu y\| \tag{1}$$

and

(2) $$|(\mu\psi)x(z) - (\mu\psi)y(z)| \leq |(z\mu)^\wedge(\psi x - \psi y)| \leq ||\mu||\,||\psi x - \psi y||.$$

The first and second assertions of the proposition follow using (1), (2), and (1.2)[1] Replacing ψ in the second assertion by $\phi\nu$, it follows that $\mu\phi\nu \in \mathrm{U_r}(G)$. Similarly, $\mu\phi\nu \in \mathrm{U}_l(G)$ and so belongs to $\mathrm{U}(G)$. □

(1.5) COROLLARY. *Each of the spaces $C(G)$, $\mathrm{U_r}(G)$, and $\mathrm{U}(G)$ is invariant for $P(G)$.*

In the next result, $\{\mu_\delta\}$ is a bounded, approximate identity for $L_1(G)$ in $P(G)$ such that the support C_δ of each μ_δ is compact, and every neighbourhood of the identity e contains C_δ eventually [**HR1**, (20.15)].

(1.6) LEMMA. *If $\psi \in \mathrm{U_r}(G)$, then $||\psi\mu_\delta - \psi|| \to 0$.*

PROOF. For each δ, we have, using (1.1.(2)),

$$||\psi\mu_\delta - \psi|| = \sup_{x\in G}\left|\int_{C_\delta} (\psi(tx) - \psi(x))\, d\mu_\delta(t)\right| \leq \sup_{t\in C_\delta} ||\psi t - \psi||,$$

and the last term tends to 0 since $\{C_\delta\}$ "contracts" to e and $\psi \in \mathrm{U_r}(G)$. □

The next result is of great importance. Recall ((0.17)) that $\mathfrak{L}(A)$ and $\mathfrak{L}_t(A)$ are the sets of left invariant and topologically left invariant means on a suitable subspace A of $L_\infty(G)$. The next result is the key for relating invariant means on subspaces of $L_\infty(G)$. References are Reiter [**14**], Greenleaf [**2**], Hulanicki [**4**], and Namioka [**2**].

(1.7) PROPOSITION. *Let $\mu_0, \nu_0 \in P(G)$. If $m \in \mathfrak{L}(\mathrm{U}(G))$, then $m' \in \mathfrak{L}_t(G)$, where $m'(\phi) = m(\mu_0\phi\nu_0)$, and m' is independent of ν_0. If $n \in \mathfrak{L}(\mathrm{U_r}(G))$, then $n' \in \mathfrak{L}_t(G)$, where $n'(\phi) = n(\phi\nu_0)$, and n' is independent of ν_0.*

PROOF. Let $\mu_0, \nu_0 \in P(G)$ and $m \in \mathfrak{L}(\mathrm{U}(G))$. Let $\phi_0 \geq 0$ be in $L_\infty(G)$. Define $F \in L_1(G)'$ by

$$F(\nu) = m(\mu_0\phi_0\nu^*),$$

where ν^* is defined in (1.1). Note that F is well defined by (1.4). Now for all $\nu \in L_1(G)$, $x \in G$, we have $(x\nu)^* = \nu^* x^{-1}$. Since $m \in \mathfrak{L}(\mathrm{U}(G))$, it follows that $F(x\nu) = F(\nu)$. Regarding the elements of $L_1(G)$ as functions, we obtain $F(x * f) = F(f)$ $(f \in L_1(G))$. Now $x * f = fx^{-1}$ for $f \in C_c(G)$. So F, restricted to $C_c(G)$, is a positive, left invariant, linear functional on $C_c(G)$ and so is a multiple of the restriction of the left Haar measure λ to $C_c(G)$. Since $C_c(G)$ is dense in $L_1(G)$ and F is continuous, it follows that F is a multiple of λ regarded

[1] We also use the fact that if $\phi' \in l_\infty(G)$ is such that the map $x \to \phi' x$ is norm continuous, then $\phi' \in U_r(G)$. Indeed, the continuity of ϕ' follows from

$$|\phi'(x) - \phi'(y)| = |\phi' x(e) - \phi' y(e)| \leq ||\phi' x - \phi' y||.$$

as a linear functional on $L_1(G)$ and so is constant on $P(G)$. Since $L_\infty(G)$ is the span of such functions ϕ_0 and $P(G) = \{\nu^* : \nu \in P(G)\}$, we obtain

$$m(\mu_0\phi\nu) = m(\mu_0\phi\nu') \qquad (\phi \in L_\infty(G),\ \nu, \nu' \in P(G)).$$

So m' does not change if ν_0 is replaced by another element of $P(G)$, and hence $m'(\phi\nu) = m(\mu_0\phi\nu\nu_0) = m(\mu_0\phi\nu_0) = m'(\phi)$. Since m' trivially belongs to $\mathfrak{M}(G)$, it is clear that $m' \in \mathfrak{L}_t(G)$.

The corresponding assertions for n' are proved similarly. □

The next result is generalized in Problem 2-15.

(1.8) COROLLARY. *If B is $\mathrm{U_r}(G)$ or $\mathrm{U}(G)$, then $\mathfrak{L}(B) = \mathfrak{L}_t(B)$.*

PROOF. Let $B = \mathrm{U_r}(G)$, $m \in \mathfrak{L}(B)$, and $\{\mu_\delta\}$ be as in (1.5). Then we obtain from (1.6) that $\mu_\delta m \to m$ weak* in B'. Now $\mu_\delta m \in \mathfrak{L}_t(B)$ by (1.7). Since $\mathfrak{L}_t(B)$ is weak* closed, $m \in \mathfrak{L}_t(B)$ and $\mathfrak{L}(B) \subset \mathfrak{L}_t(B)$. The reverse inclusion is given by (0.9).

The proof for the case $B = \mathrm{U}(G)$ is similar, with $\mu_\delta m \mu_\delta$ in place of $\mu_\delta m$. □

(1.9) COROLLARY. *Let B be a subspace of $L_\infty(G)$ with $1 \in B$. If $B \supset \mathrm{U_r}(G)$ (so that B is right invariant for $P(G)$ ((1.4))), then the mapping $m \to m|_B$ is a bijection from $\mathfrak{L}_t(G)$ onto $\mathfrak{L}_t(B)$. If $B \supset \mathrm{U_r}(G) \cup \mathrm{U}_l(G)$ (so that B is invariant for $P(G)$), then the mapping $p \to p|_B$ is a bijection from $\mathfrak{I}_t(G)$ onto $\mathfrak{I}_t(B)$.*

PROOF. Suppose that $B \supset \mathrm{U_r}(G)$. Using (1.4), elements $m, n \in \mathfrak{L}_t(G)$ coincide if and only if they coincide on $\mathrm{U_r}(G)$. It follows that the map $m \to m|_B$ is an injection from $\mathfrak{L}_t(G)$ into $\mathfrak{L}_t(B)$. By (1.7), if $p \in \mathfrak{L}_t(B)$ and n is the restriction of p to $\mathrm{U_r}(G)$, then $n' \in \mathfrak{L}_t(G)$, and if $\phi \in B$, then

$$p(\phi) = p(\phi\nu_0) = n(\phi\nu_0) = n'(\phi)$$

so that the map $m \to m|_B$ is a bijection as required. The other assertion of the corollary is proved similarly, noting that if $m \in \mathfrak{I}(\mathrm{U}(G))$, then the mean m' of (1.7) belongs to $\mathfrak{I}_t(G)$. □

As we shall see in (7.17), the map $m \to m|_{C(G)}$ is *not*, in general, a bijection from $\mathfrak{L}(G)$ onto $\mathfrak{L}(C(G))$.

(1.10) COROLLARY. *Let B be a right invariant subspace of $L_\infty(G)$. If $B \supset \mathrm{U}(G)$ $[B \supset \mathrm{U_r}(G)]$, then G is amenable if and only if $\mathfrak{L}(B) \neq \emptyset$ $[\mathfrak{L}_t(B) \neq \emptyset]$.*

In particular, $\mathfrak{L}(G) \neq \emptyset$ if and only if $\mathfrak{L}_t(G) \neq \emptyset$.

Our next objective is to show that every closed subgroup H of an amenable locally compact group G is itself amenable. To prove this, we require some results relating Haar measures λ_G and λ_H on G and H respectively. A good discussion of these results is given in [**R**, Chapter 8]. An important concomitant of these results is *Weil's formula*, which we will need later and is given in (1) below.

(1.11) Weil's formula and Bruhat functions [**R**]. Let λ_G, λ_H be fixed left Haar measures on G and H. If H is normal in G, then it is reasonable to expect, crudely, that there exists a left Haar measure $\lambda_{G/H}$ on G/H such that "$\lambda_G = \lambda_H \times \lambda_{G/H}$". For example, with $G = \mathbb{R}^2$, $H = \mathbb{R} \times \{0\}$, G/H"="$\{0\} \times \mathbb{R}$, we do indeed obtain $dx\,dy = dx \times dy$! Making this precise, it turns out that when H is normal in G (or more generally, when $\Delta_H(h) = \Delta_G(h)$ for all $h \in H$, where Δ_H, Δ_G represent the modular functions on H and G respectively), we have *Weil's formula*:

$$(1) \quad \int_G \phi(x)\, d\lambda_G(x) = \int_{G/H} d\lambda_{G/H}(xH) \int_H \phi(xh)\, d\lambda_H(h) \qquad (\phi \in C_c(G)).$$

Here, $\lambda_{G/H}$ is a positive, regular Borel measure on G/H that is left invariant under the action of G. There are a number of situations that arise in practice in which $\Delta_H \neq \Delta_G|_H$ so that (1) does not apply. However, in general, there exists a continuous function $q\colon G \to (0, \infty)$ and a quasi-invariant measure $\lambda_{G/H}$ such that

$$(2) \quad \int_G \phi(x)q(x)\, d\lambda_G(x) = \int_{G/H} d\lambda_{G/H}(xH) \int_H \phi(xh)\, d\lambda_H(h) \qquad (\phi \in C_c(G)).$$

(Quasi-invariance for $\lambda_{G/H}$ means that $\lambda_{G/H}(xE) = 0$ whenever $x \in G$ and E is $\lambda_{G/H}$-null.) If H is unimodular, then we can take $q(x) = \Delta_G(x)^{-1}$. In general, the function q can be produced using *Bruhat functions* for H, and these are the functions which will concern us in our next result. Roughly, a Bruhat function for H is a function on G that behaves like a continuous function with compact support on left H-cosets and integrates "well" with respect to λ_H. We now give the precise definition.

A Bruhat function for H is a function $\beta\colon G \to \mathbb{R}$ such that:

(i) if $C \in \mathscr{C}(G)$, then there exists $\psi \geq 0$ in $C_c(G)$ such that

$$\beta|_{CH} = \psi|_{CH};$$

(ii) for all $x \in G$, we have

$$\int_H \beta(xh)\, d\lambda_H(h) = 1.$$

Such a function can be shown to exist. (Indeed, the construction uses an argument closely related to that of (4.11).)

Given a Bruhat function β, we can define the function $q\colon G \to \mathbb{R}$ of (2) by

$$q(x) = \int_H \beta(xh)\Delta_G(h)\Delta_H(h^{-1})\, d\lambda_H(h).$$

For future use, we note that if H is also normal, then there is a natural $*$-homomorphism $T_H\colon M(G) \to M(G/H)$ given by $T_H\mu(E) = \mu(Q^{-1}(E))$ $(E \in \mathscr{B}(G/H))$, where $Q\colon G \to G/H$ is the quotient map. It turns out that $Q_H = T_H|_{(L_1(G))}$ is a $*$-homomorphism from $L_1(G)$ onto $L_1(G/H) = L_1(G/H, \lambda_{G/H})$. (In function terms, we have $Q_H f(xH) = \int_H f(xh)\, d\lambda_H(h)$.) So the dual map Q_H^* is an isometric $*$-homomorphism from $L_\infty(G/H)$ into $L_\infty(G)$.

References for the next result are Rickert [**2**] and Greenleaf [**2**]. We follow the proof in [**R**, Chapter 8] using properties (i) and (ii) of Bruhat functions.

(1.12) PROPOSITION. *Every closed subgroup of an amenable locally compact group G is amenable.*

PROOF. Let H be a closed subgroup of G, and let β be a Bruhat function for H. Define $\Phi\colon C(H) \to l_\infty(G)$ by setting

$$\Phi(\phi)(x) = \int_H \beta(x^{-1}h)\phi(h)\, d\lambda_H(h).$$

Using (ii), Φ is a norm-decreasing, unit-preserving, linear map. We claim that the map $x \to (\beta x)|_H$ is norm-continuous from G into $C(H)$.

For suppose otherwise. Then we can find $\varepsilon > 0$, nets $\{x_\delta\}$ in G, $\{h_\delta\}$ in H, and $x \in G$ such that $x_\delta \to x$ and

$$(1) \qquad |\beta(x_\delta h_\delta) - \beta(x h_\delta)| \geq \varepsilon.$$

We can suppose that there exists $C \in \mathscr{C}(G)$ such that $x_\delta \in C$ for all δ. Let L be the compact set $\{z \in CH\colon \beta(z) > 0\}^-$ ((i)). If a subnet of $\{h_\delta\}$ is eventually inside a compact subset of G, then a contradiction of (1) follows easily. Otherwise, we eventually have $Ch_\delta \cap L = \emptyset$, and again (1) is contradicted.

This establishes the norm continuity of the map $x \to (\beta x)|_H$, and using (i) again, it follows that $\Phi(C(H)) \subset C(G)$. From the left invariance of λ_H, we have $\Phi(\phi h_0) = \Phi(\phi)h_0$ for all $\phi \in C(H)$, $h_0 \in H$. Thus $m \circ \Phi \in \mathfrak{L}(C(H))$ whenever $m \in \mathfrak{L}(C(G))$, so that H is amenable ((1.10)). □

(1.13) PROPOSITION. *If H is a closed normal subgroup of G, then G is amenable if and only if both H and G/H are amenable*

PROOF. If G is amenable, then so is H by (1.12). The amenability of G/H follows as in (0.16): $\mathfrak{L}(C(G)) \neq \emptyset$ implies that $\mathfrak{L}(C(G/H)) \neq \emptyset$. Conversely, suppose that both H and G/H are amenable. The argument follows that of its discrete version (0.16.(3)). Note that this works only because we can use continuous and uniformly continuous functions.

Let $p \in \mathfrak{L}(C(H))$, and for $\psi \in \mathrm{U_r}(G)$ define $\psi_p\colon G/H \to \mathbb{C}$ by setting $\psi_p(tH) = p((\psi t)|_H)$ $(t \in G)$. Since the quotient map from G onto G/H is open and the map $t \to (\psi t)|_H \in C(H)$ is norm continuous, it follows that ψ_p belongs to $C(G/H)$. Then for $q \in \mathfrak{L}(C(G/H))$, the map $\psi \to q(\psi_p)$ belongs to $\mathfrak{L}(\mathrm{U_r}(G))$, and G is amenable ((1.10)). □

We now turn to the elegant theory of left amenable semigroups. This area of amenability is largely the creation of M. M. Day, E. E. Granirer, I. Namioka, J. Sorenson, A. H. Frey, T. Mitchell, C. Chou, A. T. Lau, J. C. S. Wong, and M. Klawe.

To make progress with this subject, it is essential to have at our disposal the amenability fixed-point theorem known as *Day's Fixed-Point Theorem* (Day [**5**]). Amenability fixed-point theorems will be studied in detail in the next chapter. However, we feel justified in proving Day's theorem here, partly because we need it now and also because it exhibits the basic ideas of these theorems very clearly.

Let E, X be locally convex spaces and K a compact convex subset of E. A continuous map $T\colon K \to X$ is called *affine* if $T(\alpha k_1 + (1-\alpha)k_2) = \alpha T k_1 + (1-\alpha)Tk_2$ for all $k_1, k_2 \in K$ and $\alpha \in [0,1]$. The set of affine maps from K to $\mathbb{C}$ is denoted by $A_f(K)$. Note that $\{F|_K + \mathbb{C}1\colon F \in E'\} \subset A_f(K)$. The set K is called an *affine left S-set*, where S is a semigroup, if K is a left S-set with each map $k \to sk$ affine. An obvious (and important) affine left S-set is the set of means $\mathfrak{M}(S)$ under the natural action. (In that case, $E = l_\infty(S)'$ with the weak* topology.)

(1.14) THEOREM (DAY'S FIXED-POINT THEOREM). *Let S be a left amenable semigroup and K an affine left S-set. Then there exists an S-fixed-point in K.*

PROOF. Let $m \in \mathfrak{L}(S)$ and let $\{f_\delta\}$ be a net in $P(S)$ such that $\hat{f}_\delta \to m$ weak* in $l_\infty(S)'$. Let $k_0 \in K$ and $k_\delta = \sum_{s\in S} f_\delta(s)(sk_0)$. Since K is convex and S-invariant, it follows that $k_\delta \in K$. We will show that $\{k_\delta\}$ converges to a fixed-point for S. To this end, let $\mathfrak{I}$ be the "weak topology" induced on K by $A_f(K)$: so $a_\delta \to a$ in $(K, \mathfrak{I})$ if and only if $F(a_\delta) \to F(a)$ for all $F \in A_f(K)$. Let $\mathfrak{I}_1$, $\mathfrak{I}_2$ be respectively the relative topologies on K induced by the given and weak topologies of E. Then $\mathfrak{I}_1 \geq \mathfrak{I} \geq \mathfrak{I}_2$ and since $(K, \mathfrak{I}_1)$ is compact and $\mathfrak{I}_2$ is Hausdorff, we have $\mathfrak{I}_1 = \mathfrak{I}$. For $F \in A_f(K)$, let $\phi_F \in l_\infty(S)$ be given by $\phi_F(s) = F(sk_0)$. Then, by the affineness of F,

$$F(k_\delta) = \sum_{s\in S} f_\delta(s)\phi_F(s) = \hat{f}_\delta(\phi_F) \to m(\phi_F),$$

and since $\mathfrak{I}_1 = \mathfrak{I}$, we see that $\{k_\delta\}$ converges to some $k_1 \in K$, and $F(k_1) = m(\phi_F)$. If $s_0 \in S$, then Fs_0, where $Fs_0(k) = F(s_0 k)$ $(k \in K)$, is in $A_f(K)$, and so $F(s_0 k_1) = m(\phi_{Fs_0}) = m(\phi_F s_0) = m(\phi_F) = F(k_1)$. So $s_0 k_1 = k_1$, and k_1 is an S-fixed-point. □

(1.15) Fundamental problems for amenable semigroups. We start our discussion of the basic theory of amenable semigroups by listing a number of natural questions which a satisfactory theory should be able to answer.

(1) *When is a finite semigroup left amenable?* (The answer for groups is *always*.)

(2) *When does a semigroup S admit a left invariant mean in $P(S)^\wedge$?* (An infinite group *never* admits such a mean.)

(3) *When does a subset of a left amenable semigroup "support" a left invariant mean?*

(4) *How "close" is a left amenable semigroup to being a subsemigroup of an amenable group?*

(5) *Which subsemigroups of an amenable group are left amenable?*

(6) *When does a semigroup admit a multiplicative left invariant mean?* (A nontrivial group never admits such a mean.)

We will give reasonable answers to questions (1)–(5) in (1.19), (1.20), (1.21), (1.27), and (1.28). The sixth question will be considered in the next chapter ((2.29)).

We start with our first question. It turns out that with little extra effort, we can answer the topological version of this question, where finite is replaced by compact. Before proving this version, we recall some facts about the structure of compact semigroups.

(1.16) Compact semigroups. Let T be a jointly continuous, compact, Hausdorff semigroup. (By "jointly continuous" we mean that the map $(s,t) \to st$ from $T \times T$ into T is continuous.) It is obvious that T contains a unique minimal, closed ideal K, called the *kernel* of T. A good account of the structure theory of the kernel is given in M. Rosenblatt [**1**, Chapter 5]. The kernel K is the disjoint union of the family of left [right] minimal ideals of T. Every minimal right ideal of T is closed and of the form eT for some idempotent $e \in K$ and conversely. For such an e, $eTe = Te \cap eT$ is a compact group.

The space $C(T)$ is an invariant subspace of $l_\infty(T)$, and so it makes sense to talk of left invariant means on $C(T)$. The following result is due to Rosen [**1**, **2**].

(1.17) PROPOSITION. *There exists a left invariant mean on $C(T)$ if and only if T has exactly one minimal right ideal.*

PROOF. Let $m \in \mathfrak{L}(C(T))$ and suppose that R_1 and R_2 are (closed) minimal right ideals with $R_1 \neq R_2$. Then $R_1 \cap R_2 = \varnothing$, and since T is compact Hausdorff, we can find, using Urysohn's Lemma, functions $\phi_i \in C(T)$ $(i = 1, 2)$ with $0 \leq \phi_i \leq 1$, $\phi_i(R_i) = \{1\}$, $\phi_i(R_j) = \{0\}$ for $i \neq j$, and $0 \leq \phi_1 + \phi_2 \leq 1$. Since $\phi_i x_i = 1$ $(x_i \in R_i)$, we have $m(\phi_1) + m(\phi_2) = 2$. A contradiction results.

Conversely, suppose that T contains precisely one minimal right ideal R. Then R is the kernel of T, and we can find an idempotent e in R with $eT = eR = R$. If $s \in T$, then $se \in R$, so that $se = ese$. Now eRe is a compact group and so has a Haar measure μ. If $\phi \in C(T)$, $s \in T$, and $u \in eRe$, then

$$(\phi s)(u) = \phi(seu) = \phi|_{eRe}(ese)(u).$$

It follows that the functional $\phi \to \mu(\phi|_{eRe})$ belongs to $\mathfrak{L}(C(T))$. □

(1.18) COROLLARY. *There exists an invariant mean on $C(T)$ if and only if the kernel of T is a group.*

(1.19) COROLLARY. *A finite semigroup is left amenable if and only if it contains precisely one minimal right ideal.*

If S is finite, then $l_1(S)$ is finite-dimensional so that $\mathfrak{M}(S) = P(S)^\wedge$. In particular, if S is also left amenable, then $\mathfrak{L}(S) \cap P(S)^\wedge = \mathfrak{L}(S)$. This leads us to our second question which we now answer. (The proof uses the later result (1.22).) References for the following result are Sorenson [**2**], Lau [**2**], and Granirer [**2**].

(1.20) PROPOSITION. *The following are equivalent for a left amenable semigroup S:*

(i) $\mathfrak{L}(S) \cap P(S)^\wedge \neq \varnothing$;

(ii) S *contains a finite left ideal;*

(iii) S *contains a group which is a finite left ideal.*

PROOF. Suppose that $\alpha = \sum_{s \in S} \alpha_s s \in P(S)$ and that $\hat{\alpha} \in \mathfrak{L}(S)$. Note that $\alpha_s \geq 0$ for all s and that $\sum_{s \in S} \alpha_s = 1$. Let $k = \max_{s \in S} \alpha_s$ and $L = \{s \in S : \alpha_s = k\}$. Since $\hat{\alpha} \in \mathfrak{L}(S)$, we have $\hat{\alpha}(s^{-1}E) = \hat{\alpha}(E)$ for all $s \in S$, $E \subset S$. Let $s_0 \in L$. Then

$$k \geq \alpha_{ss_0} = \hat{\alpha}(\{ss_0\}) = \hat{\alpha}(s^{-1}\{ss_0\}) \geq \hat{\alpha}(\{s_0\}) = k.$$

It follows that L is a left ideal in S. Further, since $|L|k \leq \sum_{s \in S} \alpha_s = 1$, we have that L is finite. So (i) implies (ii).

Now suppose that (ii) holds. Let L be a finite left ideal in S. Applying the later result (1.22), we see that L is also left amenable. Hence, by (1.19), it contains exactly one minimal *right* ideal R, and R is the kernel of L. Let e be an idempotent in R. Then $eL = R$ and $Le = Le \cap eL$ is a finite left ideal group in L, so that (ii) implies (iii).

Finally, suppose that (iii) holds. Let G be a finite left ideal group in S and $\beta = |G|^{-1} \sum_{x \in G} x \in P(S)$. Now check that $\hat{\beta} \in \mathfrak{L}(S)$ to give that (iii) implies (i). □

We now characterize those subsets of a left amenable semigroup that support left invariant means. A subset E of a semigroup S is said to be *left thick* if whenever $F \in \mathscr{F}(S)$, the family of finite subsets of S, then there exists $s \in S$ such that $Fs \subset E$. Similarly, E is said to be *right thick* if, for each $F \in \mathscr{F}(S)$, there exists $s \in S$ such that $sF \subset E$. Left thick sets were introduced and studied by Mitchell [**2**], and the following proposition is due to him.

Left thick subsets of an amenable group G are normally big. For example, if G is infinite and E is left thick in G, then $|E| = |G|$ (Problem 1-18). The family of right thick subsets of a semigroup rarely forms a filter (Problem 7-19).

(1.21) PROPOSITION. *A subset E of a left amenable semigroup S is left thick if and only if there exists $m \in \mathfrak{L}(S)$ with $m(E) = 1$.*

PROOF. Suppose that E is left thick in S. If $F \in \mathscr{F}(S)$ and $s_F \in S$ is such that $Fs_F \subset E$, then $\hat{s}_F(\chi_E s) = 1$ for all $s \in F$. It follows that if m_0 is a weak* cluster point of $\{\hat{s}_F\}$ in $\mathfrak{M}(S)$, we have $m_0(\chi_E s) = 1$ for all $s \in S$, or equivalently, $sm_0(\chi_E) = 1$ $(s \in S)$. Let $K = \{n \in \mathfrak{M}(S) : sn(\chi_E) = 1$ for all $s \in S\}$. Clearly, K is a nonvoid weak* compact, convex subset of $l_\infty(S)'$ that is left invariant for S. By Day's Fixed-Point Theorem, there exists $m \in K \cap \mathfrak{L}(S)$. So $m(\chi_E) = m(\chi_E s) = 1$.

Conversely, suppose that $m \in \mathfrak{L}(S)$ is such that $m(E) = 1$ and that E is *not* left thick in S. Then we can find $s_1, \ldots, s_n$ in S such that $\{s_1, \ldots, s_n\}s \not\subset E$ for all $s \in S$. So $\sum_{i=1}^n \chi_E s_i(s) \leq (n-1)$ for all s, and hence $nm(E) =$

$m(\sum_{i=1}^n \chi_E s_i) \leq \| \sum_{i=1}^n \chi_E s_i \| \leq (n-1)$, contradicting the equality $m(E) = 1$. □

(1.22) COROLLARY. *If L is a left ideal in S, then L is left amenable if S is.*

PROOF. The set L is left thick in S, so that there exists $m \in \mathfrak{L}(S)$ with $m(L) = 1$. Now show that $m|_L$ is in $\mathfrak{L}(L)$. □

We note that the converse to (1.22) is also true, as is the (easy) result that left amenability is inherited by right ideals (Problems 0-33, 1-19).

We now consider our fourth problem: How close is a left amenable semigroup S to being a subsemigroup of an amenable group? To tackle this, it is reasonable to try to quotient out by some congruence on S in order to bring some "cancellativeness" to the situation. In the theory of semigroups, the condition that we need to achieve such a quotient is that of left reversibility.

The semigroup S is called *left reversible* [**CP2**, p. 194] if every pair of right ideals in S has nonempty intersection (or equivalently, if the family of right ideals of S has the finite intersection property). Similarly, S is *right reversible* if every pair of left ideals in S has nonempty intersection.

(1.23) PROPOSITION. *Every left amenable semigroup is left reversible.*

PROOF. If $m \in \mathfrak{L}(S)$ and R_1, R_2 are right ideals in S, then, by Problem 0-33, $m(R_1) = 1 = m(R_2)$. So $m(R_1 \cap R_2) = 1$ and $R_1 \cap R_2 \neq \emptyset$. □

Recall that a *congruence* on a semigroup S is an equivalence relation $\sim$ on S such that if $r \sim t$, then $sr \sim st$, $rs \sim ts$ for all $s \in S$. Note that $S/\sim$ is a semigroup in the natural way and that the quotient map is a surjective homomorphism. The following result on left reversible semigroups is given in [**CP1**, p. 35]. The reader may wish to regard it as an exercise.

(1.24) PROPOSITION. *Let S be a left reversible semigroup, and let $\approx$ be the relation on S defined by setting $x \approx y$ if there exists $s \in S$ for which $xs = ys$. Then $\approx$ is a congruence on S, and the semigroup $S/\approx$ is right cancellative.*

The next result, in a sense, reduces the study of left amenable semigroups to the study of right cancellative, left amenable semigroups. (We shall see ((1.29)) that a reduction to cancellative semigroups is not possible.)

(1.25) PROPOSITION. *A left reversible semigroup S is left amenable if and only if $S/\approx$ is left amenable.*

PROOF. The semigroup $S/\approx$ is a homomorphic image of S, so that $S/\approx$ is left amenable if S is left amenable.

Conversely, suppose that $S/\approx$ is left amenable. Let $K = \mathfrak{M}(S)$. If $s_1, \ldots, s_n \in S$, then, since S is left reversible, we can find $t \in \bigcap_{i=1}^n (s_i S)$, so that $\bigcap_{i=1}^n (s_i K) \supset \bigcap_{i=1}^n (s_i SK) \supset tK$. So the family $\{sK : s \in S\}$ has the finite intersection property, and hence $K_0 = \bigcap \{sK : s \in S\}$ is nonempty. Further, if $s_1, s_2 \in S$ and $r_1, r_2 \in S$ are such that $s_1 r_1 = s_2 r_2$, then $s_1 K_0 \subset s_1 r_1 K = s_2 r_2 K \subset s_2 K$.

So $s_1 K_0 \subset K_0$, and K_0 is an affine left S-set. Now if $s_1 u = s_2 u$ in S, then since $K_0 \subset uK$, it follows that $s_1 m = s_2 m$ for all $m \in K_0$. Hence, in a natural way, K_0 is an affine left $(S/{\approx})$-set, and any fixed-point in K_0 for $S/{\approx}$ ((1.14)) is in $\mathfrak{L}(S)$. □

The following well-known theorem (Ore's Theorem) from semigroup theory will be needed in (1.27). References for Ore's Theorem are [**CP1**, p. 35] and Ljapin [**1**, p. 392].

Again the reader may wish to regard the result as an exercise.

(1.26) PROPOSITION. *Every cancellative, left reversible semigroup S can be embedded as a subsemigroup of a group G such that $G = \{st^{-1} : s, t \in S\}$.*

The next result is due to Wilde and Witz [**1**].

(1.27) PROPOSITION. *Every left amenable, cancellative semigroup S is a subsemigroup of an amenable group G such that S is a generating left thick subset of G.*

PROOF. A left amenable, cancellative semigroup is left reversible and so by (1.26) can be embedded as a subsemigroup of a group G, with G generated by S. We now show that G is amenable. In fact, we will show that there exists $m \in \mathfrak{L}(G)$ such that $m(S) = 1$, and the desired result will follow by (1.21).

Let $m_1 \in \mathfrak{L}(S)$, and define $m_0 \in \mathfrak{M}(G)$ by setting $m_0(\phi) = m_1(\phi|_S)$. Then $sm_0 = m_0 = s^{-1}m_0$ for all $s \in S$, and since S generates G, we see that $m_0 \in \mathfrak{L}(G)$. So G is amenable, and since $m_0(S) = 1$, S is left thick in G. □

Problem 2-10 deals with a topological version of the preceding result (and (1.28)).

With the above results in mind, we see that the fourth problem can be settled most satisfactorily if we can bridge the gap between right cancellative + left amenable and full cancellativity. Indeed, from (1.25), S is left amenable if and only if the right cancellative semigroup $S/{\approx}$ is left amenable. If we know that $S/{\approx}$ is actually cancellative, then (1.27) identifies it with an amenable subsemigroup of an amenable group. The conjecture that every right cancellative, left amenable semigroup is cancellative is called *Sorenson's Conjecture*. We will see in (1.29) that the conjecture is not true in general.

Finally, we turn to our fifth problem: *which subsemigroups of an amenable group are left amenable*? The answer is elegant and algebraic—the subsemigroup has to be left reversible! There are familiar amenable groups containing subsemigroups that are not left amenable (Problem 0-29).

The following result is due to Frey [**1**].

(1.28) PROPOSITION. *Let G be amenable and let S be a subsemigroup of G. Then S is left amenable if and only if S is left reversible.*

PROOF. By (0.16), we can suppose that S generates G. Suppose that S is left amenable. By (1.23), S is left reversible.

Conversely, suppose that S is left reversible. By the Hahn-Banach Theorem, the linear functional $\alpha\chi_S \to \alpha$ ($\alpha \in \mathbb{C}$) on the subspace $\mathbb{C}\chi_S$ of $l_\infty(G)$ extends to give a mean $p \in \mathfrak{M}(G)$ with $p(S) = 1$. Let

$$K = \{m \in \mathfrak{M}(G) \colon m(S) = 1\}.$$

Then K is a nonvoid, compact convex subset of $\mathfrak{M}(G)$. One readily checks that $SK \subset K$. Let $K_0 = \bigcap\{sK \colon s \in S\}$. Then K_0 is a nonvoid compact, convex subset of $\mathfrak{M}(G)$. As in the proof of (1.25), the left reversibility of S yields that $K_0 \neq \varnothing$ and that $SK_0 \subset K_0$. We now show that $GK_0 = K_0$. As G is a group, each of the maps $k \to sk$ ($k \in K$) is one-to-one. Hence

$$K \supset sK_0 = s\left(\bigcap_{t \in S} tK\right) = \bigcap_{t \in S}(stK) \supset \bigcap_{t \in S}(tK) = K_0.$$

So $sK_0 = K_0 = s^{-1}K_0$ for all $s \in S$. Since S generates G, we have $GK_0 = K_0$. Now apply Day's Fixed-Point Theorem to obtain $m_0 \in K_0 \cap \mathfrak{L}(G)$. Noting that $m_0 \in K_0 \subset K$, we have $m_0(S) = 1$, and $m|_{l_\infty(S)} \in \mathfrak{L}(S)$, so that S is left amenable. □

References

Chou [**4**], Day [**1**], [**2**], [**4**], [**5**], [**7**], [**9**], Dixmier [**1**], Følner [**2**], Frey [**1**], Granirer [**9**], [**10**], Greenleaf [**2**], Hewitt [**2**], Hulanicki [**4**], Jenkins [**3**], Klawe [**1**], [**2**], Lau [**3**], [**6**], Mitchell [**2**], Namioka [**2**], Reiter [**14**], [**R**], Rickert [**2**], Rosen [**1**], [**2**], Specht [**1**].

Further Results

(1.29) The Sorenson Conjecture. Recall ((1.27)) that Sorenson's Conjecture asserts *every left amenable, right cancellative semigroup is cancellative.* The results (i) and (ii) below give some support to the conjecture. However, (iv) below demonstrates that the conjecture is false. The conjecture still seems to be open for finitely generated semigroups.

We note that the conjecture is equivalent to a conjecture involving the "strong Følner condition" ((4.22)).

(i) *A finite, left amenable, right cancellative semigroup is a group.*

[Since S is right cancellative and finite, $Su = S$ for all $u \in S$. So S is the kernel of S. Using (1.17), S is a left and right minimal ideal in itself and so is a group.]

(ii) (Sorenson [**2**], Klawe [**2**]). *Let S be left amenable, right cancellative, and such that there exists $m \in \mathfrak{M}(S)$ for which $m(sA) = m(A)$ for all $s \in S$, $A \subset S$. Then S is cancellative.*

[Suppose that S is not left cancellative, and find $x, y, s_1 \in S$ such that $x \neq y$ and $s_1x = s_1y$. Using Zorn's Lemma and the fact that S is right cancellative, we can find a maximal, nonvoid subset A of S such that $xA \cap yA = \emptyset$. If $z \in S \sim A$, then, since S is right cancellative and A is maximal, either $xz \in yA$ or $yz \in xA$. So $S = A \cup A_1 \cup A_2$, where $A_1 = x^{-1}(yA)$ and $A_2 = y^{-1}(xA)$. Since $xA_1 \subset yA$ and $yA_2 \subset xA$, we have $m(A_i) \leq m(A)$ $(i = 1, 2)$. So $1 = m(S) \leq 3m(A)$, and $m(A) \geq \frac{1}{3}$. But since $xA \cap yA = \emptyset$ and $s_1 x = s_1 y$, we have

$$2m(A) = m(xA \cup yA) = m(s_1 xA \cup s_1 yA) = m(A)$$

so that $m(A) = 0$ and a contradiction follows.]

A semigroup S is called *left measurable* if there exists $m \in \mathfrak{M}(S)$ such that $m(sA) = m(A)$ for all $s \in S$, $A \subset S$. The properties of left measurable semigroups are explored by Sorenson and Klawe.

(iii) Let S and T be semigroups and let ρ be a homomorphism from T into the semigroup of endomorphisms of S. The semidirect product semigroup $S \times_\rho T$ is defined in the obvious way: as a set, $S \times_\rho T$ is just $S \times T$, and multiplication is given by $(s,t)(s',t') = (s\rho(t)s', tt')$. A semidirect product of locally compact groups is always amenable. A semigroup version of this result is the following, which is due to Klawe [**2**].

Let S and T be left amenable and let $\rho(t)$ be surjective for all $t \in T$. Then $S \times_\rho T$ is left amenable.

[For $\phi \in l_\infty(S)$, $t \in T$, and $m \in \mathfrak{M}(S)$, define $\phi\rho(t) \in l_\infty(S)$ and $\rho(t)m \in \mathfrak{M}(S)$ by setting $\phi\rho(t)(s) = \phi(\rho(t)(s))$ $(s \in S)$ and $(\rho(t)m)(\phi) = m(\phi\rho(t))$. Then $\mathfrak{M}(S)$ becomes an affine left T-set with respect to the map $(t,m) \to \rho(t)m$. Let $s \in S$ and $t \in T$, and find $s_0 \in S$ such that $\rho(t)(s_0) = s$. As $\rho(t)$ is a homomorphism, we have, for $x \in S$, $(\phi s)\rho(t)(x) = \phi(\rho(t)(s_0 x)) = \phi\rho(t)s_0(x)$, and so $(s\rho(t)m)(\phi) = m(\phi\rho(t)s_0) = \rho(t)(s_0 m)(\phi)$. It follows that $\mathfrak{L}(S)$ is also an affine left T-set, and by Day's Fixed-Point Theorem, there exists $m_0 \in \mathfrak{L}(S)$ for which $\rho(t)m_0 = m_0$ for all $t \in T$. For $\psi \in l_\infty(S \times_\rho T)$ and $s \in S$, define $\psi_s \in l_\infty(T)$ by $\psi_s(t) = \psi(s,t)$. Let $m_1 \in \mathfrak{L}(T)$, and define $m_1\psi \in l_\infty(S)$ by $m_1\psi(s) = m_1(\psi_s)$. If $x = (s_1, t_1) \in S \times_\rho T$, then $(\psi x)_s(t) = (\psi_{s_1\rho(t_1)(s)})t_1(t)$. It follows that if $\psi' = \psi x \in l_\infty(S \times_\rho T)$, then $m_1\psi' = ((m_1\psi)s_1)\rho(t_1)$ and the functional $\psi \to m_0(m_1\psi)$ belongs to $\mathfrak{L}(S \times_\rho T)$.]

Maria Klawe also shows that the above result is false in general if the requirement that each $\rho(t)$ be surjective is dropped (Problem 1-29).

(iv) (Klawe [**2**]). *There exists a right cancellative, left amenable semigroup that is not cancellative.*

[Let F be the free commutative semigroup on an infinite, countable set which is enumerated $\{u_n : n \in \mathbb{P}\}$. Define a surjective homomorphism α of F by setting $\alpha(u_1^{n_1} u_2^{n_2} \cdots u_r^{n_r}) = u_1^{n_1+n_2} u_2^{n_3} \cdots u_{r-1}^{n_r}$, and let $\rho(n) = \alpha^n$ $(n \in \mathbb{P})$. Let $S = F \times_\rho \mathbb{P}$. The semigroup S is left amenable by (iii). We now show that S is right cancellative but not cancellative. Suppose that $(w_1, n_1)(w, n) = (w_2, n_2)(w, n)$ in S. Then $(w_1\rho(n_1)(w), n_1 + n) = (w_2\rho(n_2)(w), n_2 + n)$ so that $n_1 = n_2$ and $w_1\rho(n_1)(w) = w_2\rho(n_2)(w)$. Since F is cancellative, $w_1 = w_2$ and S is right cancellative. However $(u_1, 1)(u_1, 1) = (u_1^2, 2) = (u_1, 1)(u_2, 1)$ so that S is not cancellative.]

(v) Klawe [**3**] gives an example of a cancellative right amenable semigroup that is not left amenable, thus answering a question of Granirer. The example is also a semidirect product of semigroups.

(1.30) Amenable Banach algebras. Amenable Banach algebras were introduced and studied by B. E. Johnson in his definitive monograph [**2**]. This class of Banach algebras arises naturally out of the cohomology theory for Banach algebras, the algebraic version of which was developed by Hochschild [**1**, **2**, **3**]. Other relevant papers are Guichardet [**S1**, **S2**], Johnson [**3**, **5**, **S2**, **S3**, **S4**], Kamowitz [**1**], Khelemskii [**S1**, **S2**], Khelemskii and Sheinberg [**1**], Lau [**11**], and Racher [**1**]. The author is grateful to G. Dales for a helpful communication.

Amenability has proved particularly fruitful in the category of operator algebras—amenable operator algebras are briefly discussed in (1.31) and (2.35).

Let A be a Banach algebra and X a Banach space that is an A-module. Then X is called a *Banach A-module* if there exists $K > 0$ such that $||a\xi|| \leq K||a||\,||\xi||$, $||\xi a|| \leq K||\xi||\,||a||$ for all $a \in A$, $\xi \in X$. (A good example of a Banach A-module is afforded by $X = L_1(G)$ and $A = M(G)$, with convolution operation as in (1.1).) The dual space X' of a Banach A-module X is itself a Banach A-module in the natural way: $af(\xi) = f(\xi a)$, $fa(\xi) = f(a\xi)$ ($f \in X'$, $\xi \in X$, $a \in A$). We say that X' is a *dual Banach A-module.*

A derivation from A to a Banach A-module X is a *norm continuous* linear mapping $D\colon A \to X$ such that $D(ab) = aDb + (Da)b$ for all $a, b \in A$.[2] If $\xi \in X$, then it is easily checked that the map D_ξ, where $D_\xi(a) = a\xi - \xi a$, is a derivation on A. Such derivations D_ξ are called *inner.* The set of [inner] derivations $D\colon A \to X$ is a linear subspace of the Banach space $\mathbf{B}(A, X)$ of bounded linear operators from A to X.

We are interested in those Banach algebras A for which "many" of the derivations are inner. It is unrealistic to expect there to be many interesting algebras A with *every* derivation $D\colon A \to X$ inner for *every* Banach A-module X. However, *dual* Banach A-modules have pleasant weak*-compactness properties, and modifying the above suggestion leads to the definition of amenable Banach algebras. The Banach algebra A is called *amenable* if every derivation from A into any dual Banach A-module is inner. The use of the term *amenable* in this context may seem strange since, a priori, there seems to be no obvious connection between invariant means on a locally compact group G and derivations on a Banach algebra. However, Johnson proved the remarkable result ((iv) below) that *G is amenable if and only if $L_1(G)$ is an amenable Banach algebra.*

In Johnson's Banach-algebra version of Hochschild's cohomology theory for associative algebras, the algebra A is amenable if and only if every first cohomology group $H^1(A, Y) = \{0\}$ for every dual Banach A-module Y. The classes of Banach algebras A for which $H^n(A, Y) = \{0\}$ for every dual Banach A-module Y ($n > 1$) does not seem to have been studied in any detail.

---

[2]See Willis [**1**] for automatic continuity results on derivations on amenable group algebras.

We now discuss some of the properties of amenable Banach algebras. The discussion is based on Johnson [**2**].

(i) *Every amenable Banach algebra contains a bounded (two-sided) approximate identity.*

[Let A be an amenable Banach algebra. Let X be the Banach A-module A' with the usual right action and zero-left action. Then $Y = X' = A''$ is a dual Banach A-module with zero-right action, and by amenability, the derivation $a \to \hat{a}$ must be inner. So $\hat{a} = aE$ for some $E \in A''$. Approximating E weak* by a bounded net in $\hat{A}$ and then following (with slight modification) the proof of (0.8), we produce a bounded right approximate identity $\{f_\beta\}$ for A. By taking X with the usual left action and zero-right action, we produce a bounded left approximate identity $\{e_\alpha\}$ for A. Then (as observed by P. G. Dixon) the net $\{e_{\alpha,\beta}\}$, where $e_{\alpha,\beta} = e_\alpha + f_\beta - f_\beta e_\alpha$, is a bounded (two-sided) approximate identity for A.]

The above result (i) means that for developing the theory of amenable Banach algebras, we can restrict attention to those algebras A with a bounded approximate identity. The extension of Cohen's Factorisation Theorem [**HR2**, (32.22)] is thus available, and a consequence of this is that for proving the amenability of such an algebra A, we need only consider dual modules X' where X is *neo-unital* (or *essential*). The Banach A-module X is called *neo-unital* if for all $\xi \in X$, there exist $a, a' \in A$, $\eta, \eta' \in X$ such that $a\eta = \xi = \eta' a'$. The advantage of dealing with a neo-unital module X is that X can be made into a $\Delta(A)$-module, where $\Delta(A)$ is the multiplier (or centraliser) algebra of A (discussed below). For example, this gives that a neo-unital $L_1(G)$-module is also a G-module, and the $L_1(G)$-action is obtained by "integrating up" the G-action (cf. the relationship between the nondegenerate *-representations of $L_1(G)$ and the continuous, unitary representations of G).

(ii) *Let A be a Banach algebra with a bounded approximate identity. Then A is amenable if and only if every derivation $D: A \to X'$ is inner whenever X is a neo-unital Banach A-module.*

[Let X be a Banach A-module. Since A has a bounded approximate identity $\{e_\delta\}$, the extension of Cohen's Theorem applies to give that the set $\{a\xi\colon a \in A,\ \xi \in X\}$ is a closed subspace Z of X. Clearly, Z is also a Banach A-module, and applying the "right-action" version of the above extension gives that $Y = \{a\xi b\colon a, b \in A,\ \xi \in X\}$ is a closed submodule of X. By Cohen's original theorem [HR2, (32.26)] every $z \in A$ is a product ab in A, and it follows that Y is neo-unital. Let D: $A \to X'$ be a derivation and Q_1: $X' \to Z'$ be the restriction map. Since Q_1 preserves the A-module action, it follows that $D_1 = Q_1 \circ D$ is a derivation into Z'. We claim that D is inner if D_1 is inner. For suppose that $\alpha_1 \in Z'$ is such that $D_1(a) = a\alpha_1 - \alpha_1 a (a \in A)$ and let $\alpha \in X'$ be such that $Q_1(\alpha) = \alpha_1$. Let D_α be the inner derivation associated with α. Then $D' = (D - D_\alpha)$ is a derivation into the weak $*$ closed submodule $Z^\perp = \{f \in X'\colon f(Z) = \{0\}\}$ of X'. Since $\beta(a\xi) = 0$ for $\beta \in Z^\perp$, $a \in A$, $\xi \in X$, it follows that $Z^\perp A = \{0\}$. So for all $a \in A$, $D'(ae_\delta) = aD'e_\delta$, and if f_0 is a weak

$*$ cluster point of $\{D'e_\delta\}$, then $f_0 \in Z^\perp$ and

$$D'(a) = af_0 = af_0 - f_0a$$

and D' is inner. Hence $D = D_\alpha + D'$ is inner, and so D is inner if D_1 is inner.

Now follow the right-hand version of the above argument with D, X, Z replaced by D_1, Z, Y to obtain that D_1 is inner if $Q_2 \circ D_1 \colon A \to Y'$ is inner, where $Q_2 \colon Z' \to Y'$ is the restriction map. Since Y is neo-unital, the non-trivial part of the proof is complete.]

The theory of *multipliers* (or centralisers) of a Banach algebra is developed by Johnson [**S1**]; see also Larsen [**S**]. We sketch briefly the relevant facts that we require. Let A be a Banach algebra with a bounded approximate identity $\{e_\delta\}$ with $\sup_\delta ||e_\delta|| = M$ $(< \infty)$. A pair $(T_1, T_2) \in \mathbf{B}(A) \times \mathbf{B}(A)$ is called a *multiplier* if for all $x, y \in A$, we have $(T_1x)y = T_1(xy)$, $xT_2y = T_2(xy)$, and $xT_1y = (T_2x)y$. The set $\Delta(A)$ of multipliers on A is a unital Banach algebra with multiplication and norm given by $(T_1, T_2)(S_1, S_2) = (T_1S_1, S_2T_2)$, $||(T_1, T_2)|| = \max\{||T_1||, ||T_2||\}$. Another natural and useful topology on $\Delta(A)$ is the *multiplier topology*: a net $T_\delta \to T$ in the multiplier topology if $T_\delta a \to Ta$ in norm for all $a \in A$. Here $a \in A$ is identified with the pair (L_a, R_a), which we now introduce. There is a canonical algebra homomorphism $a \to (L_a, R_a)$ from A into $\Delta(A)$, where $L_ax = ax$, $R_ax = xa$ $(x \in A)$. Trivially, this map is onto if A has an identity element. In general, the map is faithful and bicontinuous: indeed $||a|| = \lim ||ae_\delta|| = \lim ||L_ae_\delta|| \leq M||L_a|| \leq M||a||$, and similarly, $||a|| \leq M||R_a|| \leq M||a||$. So A can be identified with a closed ideal of $\Delta(A)$, and the $\Delta(A)$-norm on A coincides with the original A-norm if $M = 1$. The latter happy situation occurs fairly often, in particular, when A is a C^*-algebra, or a group algebra $L_1(G)$, or the Fourier algebra $A(G)$ when G is amenable ((4.34)). The algebra $\Delta(A)$ can be concretely realised in the natural way in a number of cases: for example if $A = C_0(Y)$ (Y locally compact Hausdorff), then $\Delta(A) = C(Y)$, while if A is the C^*-algebra of compact operators on a Hilbert space $\mathfrak{H}$, then $\Delta(A) = \mathbf{B}(\mathfrak{H})$. Further, $\Delta(L_1(G)) = M(G)$ (Wendel's Theorem) and, as we shall see in (4.34), $\Delta(A(G)) = B(G)$ when G is amenable.

After the above digression, we now discuss how a neo-unital Banach A-module X is naturally a Banach $\Delta(A)$-module. Indeed, if $T \in \Delta(A)$, then $Te_\delta, e_\delta T \in A$ and $\{(Te_\delta)\xi\}$, $\{\xi(e_\delta T)\}$ converge in norm for each $\xi \in X$ by the neo-unital property. We simply define $T\xi = \lim(Te_\delta)\xi$, $\xi T = \lim \xi(e_\delta T)$, and X becomes a Banach $\Delta(A)$-module which is unital in the sense that $I\xi = \xi = \xi I$ $(\xi \in X)$, where I is the identity of $\Delta(A)$.

The actions of $\Delta(A)$ on X are point-norm continuous for the multiplier topology, that is, if $T_\delta a \to Ta$ for all $a \in A$, then $T_\delta \xi \to T\xi$ for all $\xi \in X$. A number of familiar properties of amenable locally compact groups have analogues for amenable Banach algebras. The following is an analogue of (1.13).

(iii) *Let A be a Banach algebra and J a closed, two-sided ideal of A with a bounded approximate identity. Then A is amenable if and only if both J and A/J are amenable.*

[We shall prove only one implication, the other being left to Problem 1-32. Suppose that both J and A/J are amenable. Let X be a Banach A-module and $D: A \to X'$ a derivation. Since J is amenable, $D|_J$ is inner, and by subtracting an inner derivation from D, we can suppose that $D(J) = \{0\}$. If $a \in A$, $b \in J$, then $0 = D(ab) = (Da)b$, $0 = D(ba) = bDa$, and it follows that $D(A) \subset X_J^\perp$, where X_J is the closed linear subspace of X spanned by $JX \cup XJ$. Clearly, X/X_J is a Banach A/J-module in the natural way, and so $X_J^\perp \cong (X/X_J)'$ is a dual Banach A/J-module. Now $D_J: A/J \to X_J^\perp$ is a derivation, where $D_J(a + J) = Da$ for all $a \in A$. Since A/J is amenable, D_J is inner, and it immediately follows that D is inner and A is amenable.]

Now let G be a locally compact group and $A = L_1(G)$. Then (neo-unital) Banach $L_1(G)$-modules and $L_1(G)$-derivations can be usefully regarded as G-modules and G-derivations as follows. Let X be a neo-unital Banach $L_1(G)$-module. Since $\Delta(L_1(G)) = M(G)$ [**HR2**, (35.5)], X is a unital Banach $M(G)$-module. In the obvious way, $G \subset M(G)$, and it follows that X is a G-set, with each map $\xi \to x\xi$, $\xi \to \xi x$ a bounded, invertible element of $\mathbf{B}(X)$, and $e\xi = \xi = \xi e$ for all $\xi \in X$. Further, there exists $K > 0$ such that $||x\xi y|| \leq K||\xi||$ for all $\xi \in X$. Also, using (1.2) and the continuity of the $M(G)$-actions for the multiplier topology, we see that the maps $(x, y) \to x\xi y$ are norm-continuous for each $\xi \in X$. A Banach space which is a G-set with the above properties is called a *Banach G-module.* So a neo-unital Banach $L_1(G)$-module is, in the natural way, a Banach G-module. Conversely, a Banach G-module becomes a neo-unital Banach $L_1(G)$-module by integrating the G-action. If the G-action arises from a Banach $L_1(G)$-module X as above, then integration brings us back to where we started. (The proof is formally the same as that establishing the equivalence of nondegenerate *-representations of $L_1(G)$ with the continuous, unitary representations of G [**D2**, 13.3].)

Similarly, a derivation $D: L_1(G) \to X'$, where X is a neo-unital Banach $L_1(G)$-module, determines, and is determined by, a *G-derivation*, that is, a weak* continuous, norm bounded map $\delta: G \to X'$ satisfying $\delta(xy) = \delta(x)y + x\delta(y)$ for all $x, y \in G$. Indeed, if $\{\mu_\delta\}$ is a bounded approximate identity for $L_1(G)$, then, for each $x \in G$, $\{D(x * \mu_\delta)\}$ converges weak*, and we define $\delta(x)$ to be the weak* limit. Very explicitly, we have $\delta(x)(a\xi b) = [bD(x * a) - bxD(a)](\xi)$ ($\xi \in X$, $a, b \in L_1(G)$). Noting that for $\mu \in L_1(G)$, $\int D(x * a)\, d\mu(x) = D(\mu a)$, we see that integrating δ brings us back to D again. More generally, if A is a Banach algebra with a bounded approximate identity, then a derivation $D: A \to X'$, where X is a neo-unital Banach A-module, extends in a natural way to a derivation $D': \Delta(A) \to X'$.

We now come to the remarkable result of Johnson that justifies the use of the term "amenable" as applied to Banach algebras. Khelemskii and Sheinberg [**1**] interpret this result in terms of "flat" modules.

(iv) *Let G be a locally compact group. Then G is amenable if and only if $L_1(G)$ is an amenable Banach algebra.*

[Suppose that G is amenable and let X be a Banach G-module and $\delta\colon G \to X'$ a G-derivation. By (ii) and the preceding comments, the amenability of $L_1(G)$ will follow once we have shown that δ is inner in the sense that, for some $\alpha_0 \in X'$, $\delta(x) = x\alpha_0 - \alpha_0 x$ $(x \in G)$. Define a new Banach G-module structure on X with operations: $\xi . x = x^{-1}\xi x$, $x.\xi = \xi$ $(x \in G,\ \xi \in X)$. The dual actions on X' are given by $x.\alpha = x\alpha x^{-1}$, $\alpha.x = \alpha$ $(\alpha \in X')$. Further, $\delta'\colon G \to X'$, where $\delta'(x) = \delta(x)x^{-1}$, is a G-derivation for this new module structure, and δ is inner if and only if δ' is inner since $\delta(x) = x\alpha - \alpha x \Leftrightarrow \delta'(x) = x.\alpha - \alpha.x$. Define $\beta\colon X \to C(G)$ by $\beta(\xi)(x) = \delta'(x)(\xi)$ $(x \in G)$. Let $m \in \mathfrak{L}(G)$ and $\alpha \in X'$ be given by $\alpha(\xi) = m(\beta(\xi))$. Then for $x_0 \in G$, $(x_0.\alpha - \alpha.x_0)(\xi) = \alpha(\xi.x_0 - x_0.\xi) = m(\beta(\xi.x_0 - \xi))$. Since

$$\begin{aligned}\beta(\xi.x_0 - \xi)(x) &= \delta'(x)(\xi.x_0 - \xi) = (x_0.\delta'(x) - \delta'(x))(\xi)\\ &= (\delta'(x_0x) - \delta'(x_0).x - \delta'(x))(\xi)\\ &= \beta(\xi)x_0(x) - \delta'(x_0)(\xi) - \beta(\xi)(x),\end{aligned}$$

we have

$$\begin{aligned}(x_0.\alpha - \alpha.x_0)(\xi) &= m(\beta(\xi)x_0) - \delta'(x_0)(\xi) - m(\beta(\xi))\\ &= -\delta'(x_0)(\xi),\end{aligned}$$

so that for all $x_0 \in G$, $\delta'(x_0) = x_0.\alpha_0 - \alpha_0.x_0$ where $\alpha_0 = -\alpha$. So δ' is inner. Hence $L_1(G)$ is amenable.

Conversely, suppose that $L_1(G)$ is amenable. The following argument is due to J. R. Ringrose. The space $X = U(G)$ is a Banach G-module with trivial left action and the usual right action: $(\phi, x) \to \phi x$. Since $\mathbb{C}1$ is G-invariant, $Y = X/\mathbb{C}1$ is also a Banach G-module, and $Y' \cong \{\alpha \in X'\colon \alpha(1) = 0\}$. Let $\nu = \delta_e \in X'$ and let $\delta\colon G \to Y'$ be the derivation given by $\delta(x) = x\nu - \nu$. Since $L_1(G)$ is amenable, δ is inner, and so there exists $\mu \in Y'$ such that $x\nu - \nu = \delta(x) = x\mu - \mu$. Then $\xi = (\nu - \mu) \neq 0$ and is left invariant in X'. So there exists a left invariant mean on $U(G)$ (see (2.2)) and G is amenable by (1.10).]

We note that Johnson [**5**] shows that amenability of a Banach algebra A is "stable" under small perturbations of the multiplication.

Khelemskii [**S**] has given characterisations of amenable Banach algebras. The author understands that related work in this area has been done by Curtis and Loy [**S1**] and by Dales and Esterle (in a Monograph under preparation).

Khelemskii's paper also contains the following result due to Steinberg: *a uniform algebra A is amenable if and only if A is a $C(X)$ (X compact Hausdorff).*

The important notion of *weak amenability* for a Banach algebra A was introduced in the commutative case by Bade, Curtis and Dales [**S**]. The notion readily extends to the non-commutative case (Johnson [**S5**]) and is defined as follows: a Banach algebra A is called *weakly amenable* if every derivation $D\colon A \to A'$ is inner. Of course, A' is a dual Banach A-module in the obvious way, and so every amenable Banach algebra is weakly amenable. The converse is false: as Johnson [**S5**] points out, if $1 \leq p < \infty$, then l^p with pointwise multiplication is weakly amenable (using the fact that the span of its idempotents is dense) yet l^p is not amenable since it does not contain a bounded approximate identity.

Bunce and Haagerup have shown that every C^*-algebra is weakly amenable (Haagerup [**3**]). An interesting open question is: *is* $L_1(G)$ *weakly amenable for every locally compact group* G? Johnson [**S5**] has proved that $L_1(G)$ is weakly amenable if G is either discrete or $[SIN]$ or one of the groups $GL(n, \mathbf{C})$. Weak amenability for the Fourier algebra $A(G)$ of G is considered by Forrest [**S**], [**S3**].

Characterisations of weak amenability and applications to semigroup algebras are given in Groenbaek [**S**]. In the latter context, Duncan and Namioka [**1**] have shown that $l^1(S)$ is never amenable if S is an infinite semilattice.

(1.31) Amenable C^*-algebras. The class of amenable C^*-algebras has a particularly rich theory. We shall be content to discuss briefly some of the results of the theory. (Amenable von Neumann algebras are discussed in (2.35).)

Let G be a locally compact group and $C^*(G)$ be the enveloping C^*-algebra of G. There is a canonical, continuous $*$-homomorphism from $L_1(G)$ onto a dense subalgebra of $C^*(G)$. If G is amenable, then ((1.30(iv))) $L_1(G)$ is amenable, and it easily follows (Problem 1-38) that $C^*(G)$ is an amenable C^*-algebra. More generally, if π is a continuous, unitary representation of G, then the norm closure of $\pi(L_1(G))$ (which is also $\pi(C^*(G))$) is amenable. In particular, $C_l^*(G) = \pi_2(C^*(G))$ is amenable, where $\pi_2 \colon G \to \mathbf{U}(L_2(G))$ is the left regular representation: $\pi_2(x)f(y) = f(x^{-1}y)$, and $\mathbf{U}(L_2(G))$ is the unitary group of $\mathbf{B}(L_2(G))$. The above comments and Problem 1-38 then give the following (Johnson [**2**], Bunce [**3**]). *If G is amenable, then $C^*(G)$ is amenable. If G is discrete, then G is amenable if and only if $C_l^*(G)$ is amenable.*

It follows that if G is discrete and nonamenable (for example, if $G = F_2$), then $C_l^*(G)$ is not amenable.

Johnson shows in [**2**] that *every Type* 1 *C^*-algebra is amenable.* Other classes of amenable C^*-algebras are the *UHF-algebras* (Sakai [**1**, p. 73]) and the *Cuntz algebras* $\mathscr{O}_n$ (Cuntz [**S1**]). Further, the class of amenable C^*-algebras is closed under *inductive limits* and *spatial tensor products.* Rosenberg [**2**] has also shown that *amenability is preserved under crossed products by amenable discrete groups.* (A more general result, following from a theorem of Green, is briefly discussed below.)

Johnson [**2**] introduced the class of *strongly amenable* C^*-algebras. For motivation, we note that if A is an amenable C^*-algebra, then every derivation $D \colon A \to X'$, where X is a Banach A-module, is inner. However, a priori, we do not know if there is any control over $K_D = \inf\{\|\alpha_0\| \colon D(a) = a\alpha_0 - \alpha_0 a$ for all $a \in A\}$. Now if G is an amenable locally compact group and $A = C^*(G)$ with X neo-unital then D can be regarded as a G-derivation δ, and it easily follows from the proof of the first part of (1.30(iv)) and (0.1) that there exists such an α_0 with α_0 belonging to the weak* closure of $\mathrm{co}\{-\delta(x)x^{-1} \colon x \in G\}$. This suggests the following definition: a unital C^*-algebra A is called *strongly amenable* if, whenever X is a Banach A-module with $1\xi = \xi = \xi 1$ for all $\xi \in X$ and $D \colon A \to X'$ is a derivation, then there exists α_0 in the weak* closure of $\mathrm{co}\{-D(u)u^*: u \in \mathbf{U}(A)\}$, such that $D(a) = a\alpha_0 - \alpha_0 a$ $(a \in A)$. (Here, $\mathbf{U}(A) = \{u \in A\colon uu^* = 1 = u^*u\}$ is the unitary group of A.) There is a corresponding definition of strong amenability in the nonunital case. Much of the theory of strongly amenable C^*-algebras parallels the amenable theory. However Rosenberg [**2**] showed that the Cuntz algebras $\mathscr{O}_n$ $(n \geq 2)$ are amenable but not strongly amenable. Haagerup [**3**] shows that amenable C^*-algebras are stably isomorphic to strongly amenable C^*-algebras—in fact, if A is amenable, then the C^*-tensor product $A \otimes K$, where K is the algebra of compact operators on an infinite-dimensional Hilbert space, is strongly amenable.

Bunce [**2**] gives the following elegant "Fixed-Point Theorem" characterisation of strong amenability (cf. Problem 1-39): *A unital C^*-algebra A is strongly amenable if and only if, whenever X is a Banach A-module and C is a nonempty, weak* compact, convex subset of X' such that $uCu^* \subset C$ for all $u \in \mathbf{U}(A)$, then there exists $c \in C$ such that $ucu^* = c$ for all $u \in \mathbf{U}(A)$.*

It also follows from the work of Haagerup [**3**] that if A is an amenable C^*-algebra, then K_D of the preceding paragraph is $\leq ||D||$; indeed, we can find an appropriate α_0 in the weak* closure of $\{a^*D(a)\colon a \in A,\ ||a|| \leq 1\}$. The same paper of Haagerup, together with a result of Connes [**4**], establishes the remarkable result that *the classes of amenable and nuclear C^*-algebras coincide.* (An earlier partial result was obtained by Bunce and Paschke [**1**].) The proof uses the deep result (Connes [**1**], Choi and Effros [**3**], Elliott [**2**]) that a nuclear C^*-algebra has approximately finite-dimensional second dual, as well as Pisier's generalisation of Grothendieck's inequality. It seems appropriate, therefore, to discuss briefly the topic of nuclear C^*-algebras. A good reference for this topic is Lance [**1**], to which the reader is referred for more details and references.

Let A, B be C^*-algebras, and let $A \otimes B$ be the algebraic tensor product of A and B. Clearly $A \otimes B$ is a $*$-algebra in the natural way. A *C^*-norm* on $A \otimes B$ is a norm γ on $A \otimes B$ such that $\gamma(x^*x) = \gamma(x)^2$ for all $x \in A \otimes B$. Any such γ is a cross-norm, that is, $\gamma(a \otimes b) = ||a||\,||b||$ for all $a \in A$, $b \in B$. It is easy to see that the projective tensor product $A \hat{\otimes} B$ is a Banach $*$-algebra and has a bounded approximate identity. It therefore has an enveloping C^*-algebra C [**D2**, (2.7)] in which $A \hat{\otimes} B$ is faithfully embedded, and the restriction $||\cdot||_{\max}$ of the norm of C to $A \otimes B$ is a C^*-norm. This is the maximum C^*-norm on $A \otimes B$. The completion of $A \otimes B$ with respect to $||\cdot||_{\max}$ is denoted by $A \otimes_{\max} B$. There is also a minimum C^*-norm $||\cdot||_{\min}$ on $A \otimes B$, and this can be obtained as follows. Realise A, B as C^*-algebras of operators on Hilbert spaces $\mathfrak{H}$, $\mathfrak{K}$. Then $A \otimes B$ is realised as a $*$-algebra of operators on $\mathfrak{H} \otimes \mathfrak{K}$. The operator norm of $A \otimes B \subset B(\mathfrak{H} \otimes \mathfrak{K})$ is $||\cdot||_{\min}$ (which is independent of the particular spaces $\mathfrak{H}$, $\mathfrak{K}$ on which A, B are realised). The completion of $A \otimes B$ with respect to $||\cdot||_{\min}$ is denoted by $A \otimes_{\min} B$. The C^*-algebra A is called *nuclear* if $||\cdot||_{\max} = ||\cdot||_{\min}$ for every C^*-algebra B, or equivalently, if there is exactly *one* C^*-norm on $A \otimes B$ for every C^*-algebra B.

As commented above, a C^*-algebra is amenable if and only if it is nuclear.

Phillip Green [**S**, Proposition 14] has shown, using Rieffel's "tensor product" theory of induced representations, that if H is a closed subgroup of a locally compact group G with G/H amenable (in the sense of Eymard [**2**] that there exists a G-invariant mean on $C(G/H)$) and $(G, A, \mathfrak{I})$ is a "twisted covariant system", then $C^*(G, A, \mathfrak{I})$ is nuclear if $C^*(H, A, \mathfrak{I})$ is nuclear. Since nuclear = amenable, two immediate corollaries are (a) every cross-product $A \times_\alpha G$ is amenable if A is an amenable C^*-algebra and G is amenable, and (b) $C^*(G)$ is amenable if there is a closed subgroup H of G with $C^*(H)$ amenable and G/H amenable. Lau and Paterson [**S3**] give a more general result than Green's above using direct amenability arguments (cf. Rosenberg [**2**]). In particular, these techniques give a very straightforward proof of (b) above (Problem 1-41). The result (b) above suggests a question which is of interest from a locally compact group point of view.

Let $\mathscr{A}$ be the class of locally compact groups G for which $C^*(G)$ is amenable. The class $\mathscr{A}$ is large. Indeed, there seem to be three main subclasses of $\mathscr{A}$. Let G be a locally compact group. Then

(i) $G \in \mathscr{A}$ *if* G *is almost connected.*

[The definition of "almost connected" is given in (3.7). The result when G is separable and connected follows from a deep result of Connes [**1**], and also uses Choi and Effros [**S**] and the fact that nuclear = amenable. The extension to the almost connected case is given in Lau and Paterson [**S3**], and uses results of Lipsman [**S**], Miličić [**S**] and Batty [**S**].]

(ii) $G \in \mathscr{A}$ *if* G *is Type* 1.

(iii) $G \in \mathscr{A}$ *if* G *is amenable.*

[This, and (ii), were discussed above.] In view of (1.13), a natural question is *is it true that if* H *is a closed, normal subgroup of* G, *then* $G \in \mathscr{A}$ *if and only if both* $H, G/H \in \mathscr{A}$?

This seems to be an open question. If it has a positive answer, then we can construct many more groups $G \in \mathscr{A}$ by using group extensions involving (i)–(iii). (Note that (Effros and Lance [**1**]) the Type 1 group $\mathrm{SL}(2, \mathbb{R})$ contains the nonamenable group F_2 as a discrete subgroup ((3.2)) so that closed subgroups do not always inherit the property of being in $\mathscr{A}$.)

Here are three partial results proved in Lau and Paterson [**S3**]. Note that it is easy to show that $G/H \in \mathscr{A}$ if $G \in \mathscr{A}$. As mentioned earlier, Green proved the nuclear version of (a) below. Let H be a closed normal subgroup of G.

(a) *If* $H \in \mathscr{A}$ *and* G/H *is amenable, then* $G \in \mathscr{A}$.

(b) *If* $G \in \mathscr{A}$, *then* $H \in \mathscr{A}$.

(c) *If* G *is separable,* H *is of Type* 1, *and* G/H *is almost connected, then* $G \in \mathscr{A}$.

The results (a) and (the separable case of) (b) are set as Problems 1-41, 1-42. All three results (a), (b), (c) are special cases of more general results.

Problems 1

1. Establish (1) and (2) of (1.1).

2. Show that $C_0(G) \subset \mathrm{U_r}(G)$.

3. Show that $C(G) = \mathrm{U_r}(G)$ if and only if G is either compact or discrete. Prove also that $\mathrm{U_r}(G)$ is finite-dimensional if and only if G is finite.

4. Show that $\mathfrak{L}_t(G) = \{\lambda\}$ if G is compact.

5. Prove that a semidirect product $G \times_\rho H$ of locally compact groups is amenable if and only if both G and H are amenable.

6. Prove that every solvable locally compact group is amenable.

7. Prove that the following statements are equivalent for a locally compact group G:
(i) G is amenable;
(ii) the closure of every compactly generated subgroup of G is amenable;
(iii) the closure of every finitely generated subgroup of G is amenable.

8. Let G be amenable, and let H be an open subgroup of G. Let T be a transversal for the right H-cosets in G and $m_0 \in \mathfrak{L}(C(G))$. For $\phi \in C(H)$, $\psi \in C(G)$ and for $p \in \mathfrak{L}(C(H))$ let $\psi_p \in C(G)$ be given by $\psi_p(x) = p((\psi x)|_H)$, and $\phi_T \in C(G)$ be given by $\phi_T(ht) = \phi(h)$ $(h \in H,\ t \in T)$. Show that we can define maps $\alpha\colon \mathfrak{L}(C(G)) \to \mathfrak{L}(C(H))$ and $\beta\colon \mathfrak{L}(C(H)) \to \mathfrak{L}(C(G))$ by $\alpha(m)(\phi) = m(\phi_T)$, $\beta(p)(\psi) = m_0(\psi_p)$. Show that β is one-to-one and that if G has equivalent left and right uniform structures (that is, $\mathrm{U}_l(G) = \mathrm{U_r}(G)$) then there exists a one-to-one map $\gamma\colon \mathfrak{L}(\mathrm{U}(H)) \to \mathfrak{L}(\mathrm{U}(G))$. Obtain relationships between the cardinals $|\mathfrak{L}(C(G))|$, $|\mathfrak{L}(C(H))|$ and the cardinals $|\mathfrak{L}(\mathrm{U}(H))|$ and $|\mathfrak{L}(\mathrm{U}(G))|$.

9. Let G be a locally compact amenable group, H a closed subgroup of G, and $\lambda_{G/H}$ a quasi-invariant measure on G/H as in (1.11). Show that G/H is amenable (in the sense that on $L_\infty(G/H,\ \lambda_{G/H})$ there exists a G-invariant mean).

10. Let G be a locally compact group and μ a positive, finitely additive, left invariant measure on the family $\mathscr{B}(G)$ of Borel sets with $\mu(G) = 1$. Prove that μ is countably additive if and only if G is compact. (This result shows that invariant means are never countably additive in the noncompact case. Hint: deal with the σ-compact case first.)

11. Let S be a left amenable semigroup with a continuous left action on a compact Hausdorff space X. Show that there exists $\nu \in \mathrm{PM}(X)$ such that $\nu(\phi s) = \nu(\phi)$ for all $s \in S$, $\phi \in C(X)$. [Here, $\mathrm{PM}(X)$ is the (weak* compact, convex) set of probability measures on X.]

12. Let G be a locally compact group that is amenable as discrete. Prove in two ways that G is amenable: (i) using Day's Fixed-Point Theorem, and (ii) using (1.10).

13. Let G be a locally compact group that is amenable as discrete. Let B be a right invariant subspace of $L_\infty(G)$ containing 1. Prove that the map $m \to m|_B$ is a surjection from $\mathfrak{L}(G)$ onto $\mathfrak{L}(B)$.

14. Let S be an amenable semigroup and K a compact convex subset of a locally convex space that is a (two-sided) affine S-set. Show that there exists $k_0 \in K$ such that $sk_0 = k_0 = k_0 s$ for all $s \in S$.

15. Let X be a Banach space and S a left amenable semigroup with a right action on X such that for each $s \in S$, the map $x \to xs$ is linear ($x \in X$), and, for some $M > 0$ we have $||xs|| \leq M||x||$ ($x \in X, s \in S$) (that is, X is a *right Banach S-module*). Let Y be a right invariant subspace of X, and let $f \in Y'$ be left invariant for S. Prove that there exists left invariant $F \in X'$ such that $F|_Y = f$ and $||F|| \leq M||f||$. Establish a converse.

16. Give examples of jointly continuous, compact Hausdorff semigroups T with $\mathfrak{L}(C(T)) \neq \varnothing$ and of left amenable, finite semigroups.

17. Determine the left thick subsets of $\mathbb{N}$.

18. Let G be a discrete group. Prove that every left thick subset of G has cardinality $|G|$.

19. Prove that a semigroup S is left amenable if it contains a left amenable left ideal L.

20. Show that if a semidirect product $S = U \times_\rho T$ of semigroups is left amenable, then so also are U and T.

21. Suppose that T is a subsemigroup of a semigroup S and that $m(T) > 0$ for some $m \in \mathfrak{L}(S)$. Prove that T is left amenable and that $m|_T/m(T) \in \mathfrak{L}(T)$.

22. (Semigroup version of (1.13)). Let T be a subsemigroup of S. Let us say that $s \sim_l s'$ in T if there exists $s_1, \ldots, s_n \in S$ and $t_1, \ldots, t_{n-1}, t'_1, \ldots, t'_{n-1}$ in T such that $s = s_1$, $s' = s_n$, and $s_1 t_1 = s_2 t'_1$, $s_2 t_2 = s_3 t'_2, \ldots, s_{n-1} t_{n-1} = s_n t'_{n-1}$. Show that $\sim_l$ is a left congruence on S. Let $\sim_r$ be the right version of $\sim_l$. Prove that if $\sim_l = \sim_r$ and both T and $S/\sim_l$ are left amenable, then S is left amenable.

23. Let S be a left amenable, cancellative semigroup. Prove that every subsemigroup of S is left amenable if and only if FS_2 (Problem 0-28) is *not* a subsemigroup of S.

24. Let S be a subsemigroup of an amenable discrete group G such that $s^{-1}(Ss) \subset S$ for all $s \in S$. Show that S is left amenable.

25. A semigroup S is called an *inverse semigroup* if for each $s \in S$, there exists a *unique* element $s^* \in S$ such that $ss^*s = s$, $s^*ss^* = s^*$.

(a) (i) Show that the set of partial 1-1 maps on a set is an inverse semigroup in the natural way.

(ii) Prove that a $*$-semigroup of partial isometries on a Hilbert space is an inverse semigroup. (This is why inverse semigroups are of interest to functional analysts.)

(b) Let S be an inverse semigroup and E be the set of idempotents in S.

(i) Show that $ss^* \in E$ for all $s \in S$.

(ii) Show that E is a semilattice (commutative, idempotent semigroup) in S.

(iii) Show that if I, J are right ideals in S, then $I \cap J \neq \varnothing$.

(iv) Show that $x \approx y$ (in the sense of (1.24)) if and only if $ex = ey$ for some $e \in E$.

(v) Deduce that $S/{\approx}$ is a group, the *maximal group homomorphic image of* S.

(c) Prove that an inverse semigroup is left amenable if and only if its maximal group homomorphic image is amenable.

26. By using an "Arens product" type argument (c.f. Problem 2-6), show that an amenable locally compact group always admits a (two-sided) invariant mean.

27. Let G be an amenable locally compact group and E a Borel subset of G. Prove that $m(E) = 1$ for some $m \in \mathfrak{L}_t(G)$ if and only if, given $C \in \mathscr{C}(G)$ and $\varepsilon > 0$, there exists $t \in G$ such that $\lambda(C \sim Et^{-1}) < \varepsilon$. (Assume the Fixed-Point Theorem (2.24).)

28. Let E be an open subset of $\mathbb{R}$. Write $E = \bigcup\{(a_n, b_n) \colon n \in \mathbb{Z},\ a_n < b_n \leq a_{n+1}\}$. Prove that $m(E) = 1$ for some $m \in \mathfrak{L}_t(\mathbb{R})$ if and only if given M, $\varepsilon > 0$, there exist $n_1 \leq n_2$ in $\mathbb{Z}$ such that

$$\sum_{n=n_1}^{n_2} (b_n - a_n) \geq M, \qquad \sum_{n=n_1}^{n_2-1} (a_{n+1} - b_n) < \varepsilon.$$

29. Give an example of a semidirect product $S \times_\rho T$ of left amenable semigroups which is not left amenable. (Of course, by (1.29(iii)), $\rho(t)$ cannot be surjective for all t.)

30. Let G be a locally compact group, and note that $M(G)' = C_0(G)''$ is a unital C^*-algebra. Prove that there exists a left G-invariant state on $M(G)'$ if and only if G is amenable. (Here $M(G)'$ has its natural dual G-action.)

31. Let A, B be Banach algebras and $\Phi \colon A \to B$ a continuous homomorphism with $\Phi(A)$ dense in B. Prove that B is amenable if A is amenable.

32. Complete the proof of (1.30(iii)).

33. Prove that the projective tensor product $A\hat{\otimes}B$ of two amenable Banach algebras A, B is also amenable.

34. Let A be a Banach algebra and $\tilde{A}$ the Banach algebra obtained by adjunction of an identity 1 to A. Prove that A is amenable if and only if $\tilde{A}$ is amenable.

35. Let A be an amenable Banach algebra, X a Banach A-module, and Y a closed submodule of X. Suppose that $h \in Y'$ is such that $ah = ha$ for all $a \in A$. Prove that there exists an extension $k \in X'$ of h such that $ak = ka$ for all $a \in A$.

36. Let A be a Banach algebra and $\pi\colon A\hat{\otimes}A \to A$ the bounded, linear map specified by $\pi(a \otimes b) = ab$. Let $A\hat{\otimes}A$ be the Banach A-module with module action given by $a(b \otimes c) = ab \otimes c$, $(b \otimes c)a = b \otimes ca$, and note that π is an A-module map. A *virtual diagonal* for A is an element $M \in (A\hat{\otimes}A)''$ such that $aM = Ma$, $\pi^{**}(M)a = \hat{a} = a\pi^{**}(M)$ for all $a \in A$.

(i) Show that A has a bounded approximate identity if A has a virtual diagonal.

(ii) Show that if A has a virtual diagonal, then there exists a virtual diagonal M for A such that $M(a \otimes b \to D(ab)(\xi)) = 0$ whenever X is a neo-unital Banach A-module, $\xi \in X$ and $D\colon A \to X'$ is a derivation.

(iii) Prove that A is amenable if and only if A has a virtual diagonal.

In the following problems, H is a closed subgroup of a locally compact group G and $\mathscr{A}$ is the class defined in (1.31).

37. Prove that G/H is amenable (in the sense that there exists a G-invariant mean on $C(G/H)$) if and only if, whenever X is a Banach G-module, then a G-derivation $D\colon G \to X'$ is inner if $D|_H$ is inner.

38. Show that $C^*(G)$ is amenable if G is amenable, and that if G is discrete, then G is amenable if and only if $C_l^*(G)$ is amenable. (Hint: use Problem 35 above.)

39. Show that if A is a unital, strongly amenable C^*-algebra, then there exists a tracial state on A.

40. Let A be a unital, strongly amenable C^*-algebra, $\mathfrak{K}$ a Hilbert space, and $\pi\colon A \to \mathbf{B}(\mathfrak{K})$ a continuous homomorphism with $\pi(1) = I$. Prove that π is similar to a *-representation of A on $\mathfrak{K}$ (cf. Problem 0-34).

41. Prove that if $H \in \mathscr{A}$ and G/H is amenable, then $G \in \mathscr{A}$. (Hint: use Problem 37 above.) Deduce that an extension of an amenable locally compact group by a group in $\mathscr{A}$ is also in $\mathscr{A}$.

42. Let G be separable and let H be normal in G. Prove that $H \in \mathscr{A}$ if $G \in \mathscr{A}$. (Hint: use the theory of induced representations and the fact that nuclear = amenable.)

CHAPTER 2

The Algebra of Invariant Means

(2.0) Introduction. We have already seen how useful Day's Fixed-Point Theorem is. When we try to obtain topological versions of this theorem, we find ourselves involved in consideration of Arens-type products involving means. Such products arise in other amenability matters and are the central theme of this chapter.

The idea of applying such products to amenability is due to Day.

We start by looking at the fixed-point space $\mathfrak{I}_l(G)$ for a locally compact group G acting on $L_\infty(G)'$. The space $\mathfrak{I}_l(G)$ is actually an abstract L-space ((2.31)) and is spanned by $\mathfrak{L}(G)$ ((2.2)). We then introduce the algebraic manipulations of the Arens type in the context of left invariant means. This leads to the theory of left introverted spaces, examples of which are $\mathrm{U_r}(G)$ ((1.3)) and the space $\mathrm{AP}(G)$ of almost periodic functions on G. Left introverted spaces are particularly well suited for Arens-type arguments. After pausing briefly to prove the basic facts concerning amenability and the space of (weakly) almost periodic functions, we prove topological amenability fixed-point theorems. We then discuss how means can be regarded as (countably additive) probability measures on a compact Hausdorff space X. When S is a discrete semigroup, X is just the Stone-Čech compactification βS of S. We conclude by solving the sixth problem of (1.15): when does S admit a *multiplicative* left invariant mean? The elegant solution is given in (2.29): the semigroups in question are those for which every finite subset has a right zero. The class can also be characterised in terms of right zeros in βS.

We first discuss the relationship between $\mathfrak{L}(G)$ and the space $\mathfrak{I}_l(G)$ of left invariant elements of $L_\infty(G)'$. Of course, $\mathfrak{I}_l(G)$ is a vector space containing $\mathfrak{L}(G)$ as convex subset. In some contexts $\mathfrak{I}_l(G)$ is more appropriate than $\mathfrak{L}(G)$; for example, in cardinality questions it is sometimes more useful to look at the dimension of $\mathfrak{I}_l(G)$ than at the cardinal $|\mathfrak{L}(G)|$. As we shall see, however, $\mathfrak{L}(G)$ actually spans $\mathfrak{I}_l(G)$, so that the study of $\mathfrak{I}_l(G)$ usually "boils down" to that of $\mathfrak{L}(G)$.

Our first result is a simple application of the Hahn-Banach Theorem. References are Dixmier [**1**] and Følner [**2**].

(2.1) PROPOSITION. *The space* $\mathfrak{I}_l(G) \neq \{0\}$ *if and only if*

$$D(G) = \operatorname{Span}\{\phi - \phi s \colon \phi \in L_\infty(G), s \in G\}$$

is not *norm dense in* $L_\infty(G)$.

PROOF. Let $F \in L_\infty(G)'$, $\phi \in L_\infty(G)$, and $s \in G$. Since $sF(\phi) = F(\phi)$ if and only if $F(\phi - \phi s) = 0$, it follows that

$$\mathfrak{I}_l(G) = \{F \in L_\infty(G)' \colon F(\psi) = 0 \text{ for all } \psi \in D(G)\}.$$

It follows from the Hahn-Banach Theorem that

$$D(G)^- = \{\psi \in L_\infty(G) \colon F(\psi) = 0 \text{ for all } F \in \mathfrak{I}_l(G)\}. \tag{1}$$

The result now follows. □

Easy modification of the above proof gives that $\mathfrak{L}(G) \neq \varnothing$ *if and only if* $\|1 - \psi\|_\infty \geq 1$ *for all* $\psi \in D(G)$.

The next result is due to Day [**2**] and Namioka [**3**]. The same proof gives a corresponding result for other C^*-subalgebras of $L_\infty(G)$ (such as $U(G)$).

(2.2) PROPOSITION. *The space* $\mathfrak{I}_l(G)$ *is spanned by* $\mathfrak{L}(G)$.

PROOF. Since $L_\infty(G)$ is a commutative C^*-algebra, it can be identified with $C(\Phi(G))$, where $\Phi(G)$ is the carrier space of $L_\infty(G)$. We also identify $L_\infty(G)'$ with $M(\Phi(G))$ by the Riesz Representation Theorem. Let ν be a nonzero element of $\mathfrak{I}_l(G)$ ($\subset M(\Phi(G))$). Write

$$\nu = (\nu_1^+ - \nu_1^-) + i(\nu_2^+ - \nu_2^-),$$

where ν_1 and ν_2 are respectively the real and imaginary parts of ν, and for $i = 1, 2$, ν_i^+ and ν_i^- are respectively the positive and negative variations of ν_i. It is sufficient to show that each nonzero ν_i^+, ν_i^- is a multiple of an element in $\mathfrak{L}(G)$. (Note that since $\nu \neq 0$, at least one of the ν_i^+, ν_i^- is nonzero.) By taking suitable scalar multiples of ν, we need only deal with the case $\nu_1^+ \neq 0$. Suppose, then, that $\nu_1^+ \neq 0$.

It is sufficient to show that $\nu_1^+ \in \mathfrak{I}_l(G)$ since then $\nu_1^+/\|\nu_1^+\|$ will belong to $\mathfrak{L}(G)$. Now if $\phi \in L_\infty(G)$, ϕ real-valued, and $s \in G$, then

$$\nu_1(\phi - \phi s) = \operatorname{Re} \nu(\phi - \phi s) = 0,$$

so that $\nu_1 \in \mathfrak{I}_l(G)$. Further, $\nu_1 = s\nu_1 = s\nu_1^+ - s\nu_1^-$, and $s\nu_1^+$, $s\nu_1^- \geq 0$. By the minimum property of the Jordan decomposition, it follows that $s\nu_1^+ \geq \nu_1^+$, $s\nu_1^- \geq \nu_1^-$. But then

$$\|s\nu_1^+ - \nu_1^+\| = (s\nu_1^+ - \nu_1^+)(1) = 0$$

since $\|F\| = F(1)$ for a positive functional F on $L_\infty(G)$. Thus $s\nu_1^+ = \nu_1^+$, and $\nu_1^+ \in \mathfrak{I}_l(G)$ as required. □

The next result continues with the "Hahn-Banach" theme. It is essentially due to Dixmier [**1**]. We formulate the proof so that it applies in the semigroup case.

(2.3) PROPOSITION. *Let G be a locally compact group and $L_\infty(G,\mathbb{R})$ the space of real-valued functions in $L_\infty(G)$. Then G is amenable if and only if the following condition* $(*)$ *holds:*

$(*)$ *whenever $n \geq 1$, $\phi_1,\ldots,\phi_n \in L_\infty(G,\mathbb{R})$ and $s_1,\ldots,s_n,t_1,\ldots,t_n \in G$, then*

$$\operatorname*{ess\,sup}_{x\in G}\left(\sum_{i=1}^{n}(\phi_i s_i - \phi_i t_i)(x)\right) \geq 0. \tag{1}$$

PROOF. Suppose that $\mathfrak{L}(G) \neq \varnothing$ and let $\phi_1,\ldots,\phi_n \in L_\infty(G,\mathbb{R})$ and $s_1,\ldots,s_n,t_1,\ldots,t_n \in G$. Then for $m \in \mathfrak{L}(G)$, $m(\sum_{i=1}^{n}(\phi_i s_i - \phi_i t_i)) = 0$ and (1) follows from (0.1(i)).

Conversely, suppose that $(*)$ holds. Let A be the span in $L_\infty(G,\mathbb{R})$ of the set of functions $\{(\phi s - \phi t)\colon \phi \in L_\infty(G,\mathbb{R}), s,t \in G\}$. If $\psi \in A$, then $-\psi \in A$, and so is of the form $\sum_{i=1}^{n}(\phi_i s_i - \phi_i t_i)$. From (1), $\|1-\psi\| \geq 1$. By the Hahn-Banach Theorem, we can find $m \in \mathfrak{M}(L_\infty(G,\mathbb{R}))$ with $m(A) = \{0\}$. Let $s_0 \in G$. Then $s_0 m \in \mathfrak{M}(L_\infty(G,\mathbb{R}))$ and for $\phi \in L_\infty(G,\mathbb{R})$, $s \in G$,

$$s_0 m(\phi s) = m(\phi s s_0) = m(\phi s_0) = s_0 m(\phi)$$

so that $s_0 m \in \mathfrak{L}(L_\infty(G,\mathbb{R}))$. Now extend $s_0 m$ to $L_\infty(G)$ in the obvious way to obtain an element of $\mathfrak{L}(G)$. □

We now introduce the Arens-type products mentioned earlier. The natural context for discussing this is that of a semigroup action on a Banach space.

(2.4) Notations. Let S be a semigroup and X be a *right Banach S-space.*[1] This means that X is a right S-set, that each map $\xi \to \xi s$ $(s \in S)$ is in $\mathbf{B}(X)$ and there exists $M > 0$ such that

$$\|\xi s\| \leq M\|\xi\| \tag{1}$$

for all $s \in S$, $\xi \in X$. For each $\xi \in X$ and $F \in X'$, define $F\xi \in l_\infty(S)$ by

$$F\xi(s) = F(\xi s) \qquad (s \in S).$$

Of course, using (1),

$$\|F\xi\| \leq M\|F\|\|\xi\|. \tag{2}$$

Let $X(S)$ be the linear subspace of $l_\infty(S)$ spanned by $\{F\xi\colon F \in X', \xi \in X\}$.

Now let $P \in X(S)'$ and $F \in X'$. Then we define $PF\colon X \to \mathbb{C}$ by

$$PF(\xi) = P(F\xi) \qquad (\xi \in X).$$

It follows from (2) and the linearity of P and F that $PF \in X'$ and $\|PF\| \leq M\|P\|\|F\|$. Let $\mathfrak{J}_l(X) = \{F \in X'\colon sF = F \text{ for all } s \in S\}$.

[1] The space X could also be called a right *Banach S-module* (cf. (1.30)).

(2.5) Some simple facts. We now list some elementary facts about $F\xi$ and PF above, leaving the proofs to the reader.

(i) *The map* $(F, \xi) \to F\xi$ *is bilinear from* $X' \times X$ *into* $X(S)$.

(ii) *The map* $(P, F) \to PF$ *is bilinear from* $X(S)' \times X'$ *into* X'.

(iii) *If* $s \in S$, *then*

$$s(F\xi) = (sF)\xi \quad \text{and} \quad (F\xi)s = F\xi s, \tag{1}$$

so that $X(S)$ *is an invariant subspace of* $l_\infty(S)$.

(iv) *For* $s \in S$,

$$s(PF) = (sP)F \quad \text{and} \quad (PF)s = P(Fs).$$

(v) *Suppose that* $F \in \mathfrak{J}_l(X)$. *Then*

$$F\xi = F(\xi)1,$$

and if $1 \in X(S)$, *then*

$$PF = P(1)F.$$

(vi) *If* $P \in \mathfrak{J}_l(X(S))$ *then* $PF \in \mathfrak{J}_l(X)$.

(vii) *If* $\xi_0 \in X$, *then the map* $F \to F\xi_0$ *is continuous from* X' *into* $X(S)$ *when* X' *is given the weak* topology and* $X(S)$ *is given the topology of pointwise convergence on* S.

(viii) *If* $F_0 \in X'$, *then the map* $P \to PF_0$ *is continuous from* $X(S)'$ *into* X', *both spaces being given the weak* topology.*

(2.6) Discussion. Now suppose that X above is a right invariant subspace of $l_\infty(S)$. Then both X and $X(S)$ are subspaces of $l_\infty(S)$. In favorable circumstances, we might hope that forming $X(S)$ does not take us outside X. This leads to the definition of introversion below. Four examples of left introverted spaces are given in (2.11). If G is a locally compact group, then $C(G)$ is not normally left introverted ((2.33)). The space $L_\infty(G)$, in general, fails spectacularly to be left introverted ((7.29)). However, it is trivial that $l_\infty(S)$ is always left introverted.

As far as amenability theory is concerned, left invariant means are particularly well-behaved in the presence of left introversion. (For example, Problem 2-15 shows that for left introverted spaces, left invariance for S is the same as left invariance for the semigroup PM(S) of probability measures on S, S being a locally compact, separately continuous, Hausdorff semigroup.)

It is convenient to allow the P in the formula PF of (2.4) to be a continuous linear functional on any subspace Y of $l_\infty(S)$ containing $X(S)$; indeed, if P' is the restriction of P to $X(S)$, then we just take PF to be $P'F$.

The following definitions are formulated for a locally compact group, but apply in the obvious way to the case of a discrete semigroup S.

Let G be, as usual, a locally compact group, and let Y be a right invariant subspace of $L_\infty(G)$. We define Y to be *left introverted* if $Y(G) \subset Y$. Note that if $Y \subset C(G)$ is left introverted, then it is also left invariant, since for $\phi \in Y$, $x\phi = \hat{x}\phi \in Y$ ($\hat{x}(\phi) = \phi(x)$ for $\phi \in Y$).

The notion of a *right introverted* subspace Z of $L_\infty(G)$ is defined similarly. Here, Z is left invariant, and in place of $Y(G)$, we put the subspace of $l_\infty(G)$ spanned by functions of the form ξF, where

$$\xi F(s) = F(s\xi).$$

An invariant subspace of $L_\infty(G)$ is said to be *introverted* if it is both left and right introverted. Normally, we will concentrate on the left introverted case.

If Y is a left introverted subspace of $L_\infty(G)$, then, since $Y(G) \subset Y$, we can define, as in the third paragraph of the present (2.6), the "product" $PQ \in Y'$ whenever P, $Q \in Y'$. The next corollary shows that Y' is actually a Banach algebra under this product. In the next result, $PQ \in X(S)'$ is the product of P, $Q \in X(S)'$—this makes sense since $X(S)$ is left introverted.

(2.7) PROPOSITION. *Let S be a semigroup and X be a right Banach S-space. Then $X(S) \subset l_\infty(S)$ is left introverted, and*

$$(PQ)F = P(QF) \tag{1}$$

for all P, $Q \in X(S)'$, $F \in X'$.

PROOF. It follows from (2.5(iii)) that $X(S)$ is right invariant in $l_\infty(S)$. Now if $P \in X(S)'$, $F \in X'$, $\xi \in X$, and $\phi = F\xi \in X(S)$, then

$$P\phi(s) = P((F\xi)s) = P(F(\xi s)) = (PF)\xi(s) \tag{2}$$

so that $P\phi = (PF)\xi \in X(S)$. Thus $X(S)$ is left introverted. Further, using (2), with PQ in place of P,

$$((PQ)F)(\xi) = PQ(F\xi) = P(Q\phi) = P((QF)\xi) = P(QF)(\xi)$$

and (1) follows. □

(2.8) COROLLARY. *If Y is a left introverted subspace of $L_\infty(G)$ or $l_\infty(S)$, then Y' is a Banach algebra with product $(P,Q) \to PQ$.*

The proof of the next proposition is routine.

(2.9) PROPOSITION. *Let Y be a left introverted subspace of $L_\infty(G)$ with $1 \in Y$. Then*

(i) *$1 \in Y(G)$;*

(ii) *if $s \in G$, P, $Q \in Y'$, then*

$$s(PQ) = (sP)Q, \quad P(sQ) = (Ps)Q, \quad \textit{and} \quad P(Qs) = (PQ)s;$$

(iii) *$\mathfrak{M}(Y)$ is a subsemigroup of Y';*

(iv) *if $Q \in \mathfrak{J}_l(Y)$ and $P \in Y'$, then $PQ = P(1)Q$.*

Before going any further, we need examples of left introverted subspaces.

(2.10) Definitions. A *separately* [*jointly*] *continuous* semigroup is a semigroup S which is also a topological space for which the multiplication map $(x, y) \to xy$ is separately [jointly] continuous. Let S be a separately continuous semigroup. As in (1.3), $\mathrm{U_r}(S)$ is the space of functions $\phi \in C(S)$ such that the map $s \to \phi s$ is norm continuous. (Note that $C(S)$ is a right invariant subspace of $l_\infty(S)$.)

The set of continuous functions ϕ for which the map $s \to \phi s$ from S into $C(S)$ is weakly continuous is denoted by $\mathrm{WU_r}(S)$. Thus if $\phi \in C(S)$, then $\phi \in \mathrm{WU_r}(S)$ if and only if $F\phi \in C(S)$ for all $F \in C(S)'$. Once we know that $\mathrm{WU_r}(S)$ is left introverted, then it is evident that $\mathrm{WU_r}(S)$ is the maximal, left introverted subspace of $C(S)$. (When S is a locally compact group, then $\mathrm{WU_r}(S) = \mathrm{U_r}(S)$ ((2.33)).)

A function $\phi \in C(S)$ for which $S\phi = \{s\phi \colon s \in S\}$ is relatively compact in the norm [weak] topology of $C(S)$ is said to be *almost periodic* [*weakly almost periodic*]. The set of almost periodic [weakly almost periodic] functions on S is denoted by $\mathrm{AP}(S)$ [$\mathrm{WP}(S)$].

We now show that the four spaces $\mathrm{U_r}(S)$, $\mathrm{WU_r}(S)$, $\mathrm{AP}(S)$ and $\mathrm{WP}(S)$ are left introverted subspaces of $C(S)$ ($\subset l_\infty(S)$). Note that by the Hahn-Banach Theorem, a right invariant subspace A of $C(S)$ is left introverted if and only if $F\phi \in A$ whenever $F \in C(S)'$, $\phi \in A$. References for the following result are Eberlein [**1**], Glicksberg and de Leeuw [**S**], Burckel [**S**], and Berglund, Junghenn, and Milnes [**1**].

(2.11) PROPOSITION. *Each of the sets* $\mathrm{U_r}(S)$, $\mathrm{WU_r}(S)$, $\mathrm{AP}(S)$, *and* $\mathrm{WP(S)}$ *is an invariant, left introverted, closed subspace of* $C(S)$ *containing* 1.

PROOF. We omit the proof for $\mathrm{U_r}(S)$ as this is an easier version of the $\mathrm{WU_r}(S)$ proof.

It is obvious that $\mathrm{WU_r}(S)$ is a subspace of $C(S)$ containing 1. Let $\{\phi_n\}$ be a sequence in $\mathrm{WU_r}(S)$ with $\phi_n \to \phi$ in norm in $C(S)$. Let $F \in C(S)'$ and $s_\delta \to s$ in S. Then, for each n,

$$\begin{aligned}|F(\phi s_\delta - \phi s)| &\leq |F(\phi s_\delta - \phi_n s_\delta)| + |F(\phi_n s_\delta - \phi_n s)| + |F(\phi_n s - \phi s)| \\ &\leq |F(\phi_n s_\delta - \phi_n s)| + 2\|\phi_n - \phi\|\,\|F\|,\end{aligned}$$

and it readily follows that $F(\phi s_\delta - \phi s) \to 0$. So $\phi \in \mathrm{WU_r}(S)$, which is therefore closed in $C(S)$. Also, if $\psi \in \mathrm{WU_r}(S)$ and $s_0 \in S$, then $F((\psi s_0)s_\delta) = F(\psi s_0 s_\delta) \to F((\psi s_0)s)$, and $F((s_0\psi)s_\delta) = (Fs_0)(\psi s_\delta) \to F((s_0\psi)s)$ so that $\mathrm{WU_r}(S)$ is invariant. Further, if $P \in C(S)'$, then the functional Q, where $Q(\psi) = F(P\psi)$, belongs to $\mathrm{WU_r}(S)'$. Thus

$$F((P\psi)s_\delta) = Q(\psi s_\delta) \to Q(\psi s) = F((P\psi)s)$$

and $P\psi \in \mathrm{WU_r}(S)$. So $\mathrm{WU_r}(S)$ is left introverted.

We now turn to $\mathrm{WP}(S)$. If α_1, $\alpha_2 \in \mathbb{C}$ and ϕ_1, $\phi_2 \in \mathrm{WP}(S)$, then $S(\alpha_1\phi_1 + \alpha_2\phi_2)$ is contained in the sum of two weakly compact subsets of $C(S)$, and it follows that $\mathrm{WP}(S)$ is a subspace of $C(S)$. Clearly, $1 \in \mathrm{WP}(S)$. Also, if $\phi \in \mathrm{WP}(S)$ and $s \in S$, then $S(s\phi) \subset S\phi$, so that $s\phi \in \mathrm{WP}(S)$. Since the map $\psi \to \psi s$ is weakly continuous on $C(S)$, it follows that $\phi s \in \mathrm{WP}(S)$. So $\mathrm{WP}(S)$ is invariant. Now let $\phi_n \to \phi$ in norm in $C(S)$ with $\phi_n \in \mathrm{WP}(S)$ for all n. To show that $\phi \in \mathrm{WP}(S)$, it is sufficient, by the Eberlein-Šmulian Theorem [**DS**, V.6.1] to show that if $\{s_k\}$ is a sequence in S, then $\{s_k\phi\}$ has a weakly convergent subsequence in $C(S)$. The same theorem, coupled with the "Cantor Diagonal Process", gives a subsequence $\{s_{\alpha(l)}\}$ of $\{s_k\}$ and, for each n, a function $\psi_n \in C(S)$ with $s_{\alpha(l)}\phi_n \to \psi_n$ weakly as $l \to \infty$. If $F \in C(S)'$, then $|F(\psi_n - \psi_m)| = \lim_{l\to\infty} |Fs_{\alpha(l)}(\phi_n - \phi_m)| \leq \|F\|\,\|\phi_n - \phi_m\|$, so that $\{\psi_n\}$ is a (norm) Cauchy sequence in $C(S)$ and so converges in norm to some $\psi \in C(S)$. A routine triangular inequality argument gives that $s_{\alpha(l)}\phi \to \psi$ weakly. Thus $\phi \in \mathrm{WP}(S)$ and $\mathrm{WP}(S)$ is closed in $C(S)$.

We now prove that $\mathrm{WP}(S)$ is left introverted. Let $\psi \in \mathrm{WP}(S)$. By the Krein-Šmulian Theorem [**DS**, V.6.4], the weak closure C_ψ of $\mathrm{co}(S\psi)$ in $C(S)$ is weakly compact. Since the norm and weak closures of $\mathrm{co}(S\psi)$ coincide and as $\mathrm{WP}(S)$ is norm closed in $C(S)$, we have $C_\psi \subset \mathrm{WP}(S)$. Let $m \in \mathfrak{M}(S)$, and let $\{\alpha_\delta\}$ be a net in $\mathrm{co}\, S \subset l_1(S)$ such that $\hat{\alpha}_\delta \to m$ weak* ((0.1(i)). Since each $\hat{\alpha}_\delta\psi \in C_\psi$, we can suppose that for some $\psi' \in \mathrm{WP}(S)$, $\{\hat{\alpha}_\delta\psi\}$ converges weakly, and hence pointwise, to ψ'. Since $\hat{\alpha}_\delta\psi \to m\psi$ pointwise, we have $m\psi \in \mathrm{WP}(S)$. It follows ((0.1(ii))) that $\mathrm{WP}(S)$ is left introverted. □

We will discuss the amenability properties of $\mathrm{WP}(S)$ and $\mathrm{AP}(S)$ in (2.36). For the moment, we turn to a result on left introverted spaces essentially due to Mitchell [**2**] and Granirer and Lau [**1**]. The result in question is (2.13).

(2.12) Discussion. Suppose that S is a (discrete) semigroup, and let B be a left introverted subspace of $l_\infty(S)$ containing 1. If $m \in \mathfrak{L}(B)$ and $\phi \in B$, then $m\phi = m(\phi)1$. Approximating m weak* by elements in $\mathrm{co}\,\hat{S}$, we can approximate the *constant* function $m(\phi)1$ by convex combinations of functions of the form $s\phi$ ($s \in S$). This is made precise in the semigroup version of the theorem below, which establishes the (more difficult) converse characterising the "amenability" of B in terms of the existence of a constant function in the closure of $\mathrm{co}\, S\phi$ in the topology of pointwise convergence on S.

In the case where B is a subspace of $L_\infty(G)$, with G a locally compact group, there is an obvious problem with this result since we cannot define the pointwise topology on $L_\infty(G)$ (as L_∞-functions are equivalence classes of "genuine" functions on G). This is overcome by replacing $\mathrm{co}\{\delta_s : s \in G\}$ by $P(G)$.

(2.13) THEOREM. *Let B be a left introverted subspace of $L_\infty(G)$ with $1 \in B$. Then*

(i) *$\mathfrak{L}(B) \neq \varnothing$ if and only if, for each $\phi \in B$, there exists a constant function in the closure C_ϕ of the set $\{\mu\phi : \mu \in P(G)\}$ in the topology of pointwise convergence on G;*

(ii) *if $\mathfrak{L}(B) \neq \varnothing$, then, for each $\phi \in B$,*

$$\{\alpha \in \mathbb{C} : \alpha 1 \in C_\phi\} = \{m(\phi) : m \in \mathfrak{L}(B)\}. \tag{1}$$

PROOF. (i) Suppose that $\mathfrak{L}(B) \neq \varnothing$. Let $m \in \mathfrak{L}(B)$ and $\phi \in B$. Then from the first equality of (2.5(v)), $m(\phi)1 = m\phi \in B$. Now from (0.1), we can find a net $\{\mu_\delta\}$ in $P(G)$ with $\hat{\mu}_\delta \to m$ weak* in B'. Using (2.5(vii)), $\mu_\delta\phi = \hat{\mu}_\delta\phi \to m\phi$ pointwise on G. Hence the constant function $m(\phi)1$ belongs to C_ϕ.

Conversely, suppose that there exists a constant function in C_ϕ for all $\phi \in B$. For each $\phi \in B$, let

$$D(\phi) = \{m \in \mathfrak{M}(B) : m\phi \in \mathbb{C}1\}$$

and

$$D = \bigcap \{D(\phi) : \phi \in B\}.$$

Obviously, $D \subset \mathfrak{L}(B)$, and so it is sufficient to show that $D \neq \varnothing$. Again, using (2.5(vii)), we see that each $D(\phi)$ is weak* compact in $\mathfrak{M}(B)$. So we need only show that the family $\{D(\phi) : \phi \in B\}$ has the finite intersection property. Let $\phi_1, \ldots, \phi_n \in B$. Now if $\{\mu_\delta\}$ is a net in $P(G)$ with $\hat{\mu}_\delta\phi_1$ converging pointwise to a constant function, then every weak* cluster point of $\{\hat{\mu}_\delta\}$ belongs to $D(\phi_1)$. So we can find $m_1 \in D(\phi_1)$. Now since B is left introverted, $m_1\phi_2 \in B$, and so, by the above, we can find $m_2 \in D(m_1\phi_2)$. Then if $m = m_2m_1 \in \mathfrak{M}(B)$,

$$m\phi_1 = m_1\phi_1 \in \mathbb{C}1, \qquad m\phi_2 = m_2(m_1\phi_2) \in \mathbb{C}1$$

so that $m \in D(\phi_1) \cap D(\phi_2)$. Applying this procedure as many times as required, we can find $m_1, \ldots, m_n \in \mathfrak{M}(B)$ such that if $m' = m_n \cdots m_2m_1$, then $m'\phi_i$ is constant $(1 \leq i \leq n)$. Thus $m' \in \bigcap_{i=1}^n D(\phi_i)$, and this concludes the proof of (i).

(ii) Suppose that $\mathfrak{L}(B) \neq \varnothing$. Let $\phi_0 \in B$, and suppose that $\alpha \in \mathbb{C}$ is such that $\alpha 1 \in C_{\phi_0}$. Now adapt the last part in the argument for (i) to show that the family $\{E(\phi) : \phi \in B\}$ has the finite intersection property, where

$$E(\phi) = \{n \in D(\phi) : n\phi_0 = \alpha 1\}.$$

This produces $m \in \mathfrak{L}(B)$ with $m(\phi_0) = \alpha$. The rest of the proof is obvious. □

Our next application of the "Arens product" idea relates invariant means to certain projection operators. Let S be a semigroup.

(2.14) Discussion. Invariant means are useful for producing projection operators on a Banach space. Indeed, if X is a right Banach S-space, then the map $F \to PF$ on X', given in (2.4), is, in a suitable sense, an invariant projection if P is a left invariant mean on $X(S)$. This, and its converse, are established in our next theorem.

An interesting application of this result to operator algebras is given in (2.35). Another application, dealing with the complementation of certain subalgebras of $L_\infty(G)$, is discussed in Problem 2-20.

Finally, we note that the result is related to the Mean Ergodic Theorem for amenable groups and semigroups in Chapter 5.

Let X be a right Banach S-space with $\|\xi s\| \leq \|\xi\|$ for all $\xi \in X$, $s \in S$. Let $T_sF = sF$ for $F \in X'$ $(s \in S)$. Of course, $T_s \in \mathbf{B}(X')$ and $\|T_s\| \leq 1$.

(2.15) THEOREM. *Suppose that* $1 \in X(S)$. *Then* (i) *and* (ii) *are equivalent.*

(i) $\mathfrak{L}(X(S)) \neq \emptyset$;

(ii) *there exists* $P \in \mathbf{B}(X')$ *such that*

(a) $P(X') = \mathfrak{J}_l(X)$, *and*

(b) *there exists a net* $\{A_\delta\}$ *in* $\mathrm{co}\{T_s : s \in S\}$ *such that* $A_\delta F \to P(F)$ *weak* for all* $F \in X'$.

If P *satisfies* (a) *and* (b) *of* (ii), *then* P *is a projection* (*that is,* $P^2 = P$) *and* $\|P\| \leq 1$.

PROOF. Let $B = X(S) \subset l_\infty(S)$. Suppose that (i) holds. Let $m \in \mathfrak{L}(B)$, and define

$$(1) \qquad P(F) = mF \qquad (F \in X').$$

Clearly, $P \in \mathbf{B}(X')$. Now for $s \in S$, $\xi \in X$, we have

$$T_s F(\xi) = (sF)(\xi) = F(\xi s) = \hat{s}(F\xi).$$

Now use (0.1) to approximate m by convex combinations of elements $\hat{s}$ and then use (2.5(viii)) to establish (ii)(b). Clearly, $P(X')$ is contained in $\mathfrak{J}_l(X)$ and $PF = F$ for $F \in \mathfrak{J}_l(X)$. This gives (i)(a) so that (i) implies (ii).

Conversely, suppose that (ii) holds. Let P and A_δ be as in (ii). Suppose that for $1 \leq i \leq n$, we have $\alpha_i \in \mathbb{C}$, $G_i \in X'$, and $\eta_i \in X$ such that

$$\sum_{i=1}^{n} \alpha_i G_i \eta_i = 0 \quad (\text{in } l_\infty(S)).$$

Observing that $\sum_{i=1}^{n} \alpha_i (T_s G_i)(\eta_i) = \sum_{i=1}^{n} \alpha_i (G_i \eta_i)(s) = 0$, we have, for each δ, $\sum_{i=1}^{n} \alpha_i (A_\delta G_i)(\eta_i) = 0$, and from (ii)(b), $\sum_{i=1}^{n} \alpha_i P(G_i)(\eta_i) = 0$. Hence we can define a linear functional m on $B = \mathrm{Span}\{F\xi : F \in X', \xi \in X\}$ by setting

$$(2) \qquad m(F\xi) = P(F)(\xi).$$

For $1 \leq j \leq k$, let $F_j \in X(S)'$, $\xi_j \in X$, and $\beta_j \in \mathbb{C}$. Since

$$\left| \sum_{j=1}^{k} \beta_j (A_\delta F_j)(\xi_j) \right| \leq \left\| \sum_{j=1}^{k} \beta_j F_j \xi_j \right\|_\infty ,$$

it follows that $\|m\| \leq 1$. Clearly, $m(1) = 1$. From (2) and (ii), (a) we have $m(\phi s) = m(\phi)$ for all $\phi \in B$. Thus $m \in \mathfrak{L}(B)$, and (ii) implies (i).

Now suppose that P satisfies (a) and (b) of (ii). From (a), $T_s(PF) = PF$, so that $A_\delta PF = PF$. Use (b) to deduce that $P(PF) = PF$, so that $P^2 = P$. Since $\|A_\delta\| \leq 1$, we have $\|P\| \leq 1$. □

(2.16) Discussion. We now continue with a discussion of some semigroup fixed-point theorems related to invariant means. Two typical theorems of this kind will be proved. Day's Fixed-Point Theorem ((1.14)) can be regarded as a discrete version of these theorems.

The procedure involved in proving all of these theorems is as follows. We are given a separately continuous semigroup S and are concerned with determining under what circumstances there exists an S-fixed-point in a compact, convex subset K of a locally convex space when S acts affinely and in a suitably continuous fashion on K. (We require an S-fixed-point in *every* such K.) The Banach space X of continuous, complex-valued, affine functions on K is a right invariant subspace of $l_\infty(K)$.

Let B be a left introverted subspace of $C(S)$. If we know that $X(S) \subset B$ and that $\mathfrak{L}(B) \neq \emptyset$, then it is easy to find an S-fixed-point in $\mathfrak{M}(X)$, viz., mp where $m \in \mathfrak{L}(B)$ and $p \in \mathfrak{M}(X)$. Since K can be identified canonically with $\mathfrak{M}(X)$, we have a fixed-point in K.

The problem, then, is to find a suitable B such that $X(S) \subset B$ whenever S acts in an appropriately continuous fashion on a K. A good help in resolving this is to find a B so that the natural left action of S on $\mathfrak{M}(B)$ satisfies the continuity condition. In the two cases with which we are concerned here, the B's associated with the joint and separate continuity conditions are $\mathrm{U_r}(S)$ and $\mathrm{WU_r}(S)$ respectively.

For other fixed-point theorems, see Problems 2-16, 2-17. References for Theorem (2.23) proved here are Furstenberg [**1**], Rickert [**2**], and Mitchell [**6**].

(2.17) Definitions. Let X_1 and X_2 be left S-sets. A mapping $\phi\colon X_1 \to X_2$ is called an *S-morphism* if $\phi(sx) = s\phi(x)$ for all $s \in S$ and $x \in X_1$. If $\mathscr{G}$ is a family of left S-sets, then an element $X_0 \in \mathscr{G}$ is called a *source for* $\mathscr{G}$ (or a *$\mathscr{G}$-source*) if for each $X \in \mathscr{G}$, there exists an S-morphism ϕ mapping X_0 into X. The significance of $\mathscr{G}$-sources for semigroup fixed-point theorems lies in the following trivial, but important, proposition.

(2.18) PROPOSITION. *An S-morphism carries an S-fixed-point into an S-fixed-point. Further, if a $\mathscr{G}$-source contains an S-fixed-point, then every $X \in \mathscr{G}$ contains an S-fixed-point.*

(2.19) Affine maps. Let K be a (nonempty) compact, convex subset of a locally convex space E. Recall that $A_f(K)$ and affine left S-sets were defined in (1.13). Clearly, $A_f(K)$ is a closed, linear subspace of $C(K)$ containing 1.

(2.20) PROPOSITION. *The mapping T, where Tk is the restriction of $\hat{k}$ to $A_f(K)$, is an affine homeomorphism from K onto $\mathfrak{M}(A_f(K))$.*

PROOF. Since $\{f|_K : f \in E'\} \subset A_f(K)$ and E is locally convex, it follows that T is one-to-one. The map T is obviously affine and continuous. The equality $T(K) = \mathfrak{M}(A_f(K))$ follows from (0.1), and compactness yields that T is a homeomorphism. □

(2.21) Notations. It is obvious that if K is an affine left S-set, then both $C(K)$, $A_f(K)$ are right invariant subspaces of $l_\infty(K)$. An important example of an affine left S-set is obtained as follows: if B is a right invariant subspace of $l_\infty(S)$ with $1 \in B$, then $\mathfrak{M}(B)$, with the relative weak* topology, is an affine left S-set with respect to the maps $m \to sm$. The essence of each of the fixed-point theorems to be proved is that for a suitable family $\mathscr{G}$ of affine left S-sets, there is a $\mathscr{G}$-source of the form $\mathfrak{M}(B)$ for some B. We now introduce the two families of affine left S-sets to be considered.

The family of affine left S-sets K for which the map $(s,k) \to sk$ is jointly [separately] continuous is denoted by $\mathscr{G}_{\mathrm{j}}$ [$\mathscr{G}_{\mathrm{s}}$].

Let $K_{\mathrm{j}} = \mathfrak{M}(\mathrm{U_r}(S))$ and $K_{\mathrm{s}} = \mathfrak{M}(\mathrm{WU_r}(S))$.

(2.22) PROPOSITION. *The affine left S-sets K_{j} and K_{s} are sources for $\mathscr{G}_{\mathrm{j}}$ and $\mathscr{G}_{\mathrm{s}}$ respectively.*

PROOF. We show first that $K_{\mathrm{j}} \in \mathscr{G}_{\mathrm{j}}$ and $K_{\mathrm{s}} \in \mathscr{G}_{\mathrm{s}}$. If $m_\delta \to m$ in K_{j}, $s_\delta \to s$ in S, and $\phi \in \mathrm{U_r}(S)$, then

$$|(s_\delta m_\delta - sm)(\phi)| \leq [\|\phi s_\delta - \phi s\| + |(m_\delta - m)(\phi s)|] \to 0$$

so that $K_{\mathrm{j}} \in \mathscr{G}_{\mathrm{j}}$. It is trivial to check that $K_{\mathrm{s}} \in \mathscr{G}_{\mathrm{s}}$.

It remains to show that K_{j} and K_{s} are sources for $\mathscr{G}_{\mathrm{j}}$ and $\mathscr{G}_{\mathrm{s}}$ respectively.

Let K be an affine left S-set. It is sufficient to show that for some $F_0 \in \mathfrak{M}(A_f(K))$ and all $\phi \in A_f(K)$, the function $F_0\phi \in l_\infty(S)$ is in the appropriate space of functions on S for appropriate K. For then, the map $P \to PF_0$, where $PF_0(\phi) = P(F_0\phi)$, is an S-morphism from K_{j} or K_{s} into K. Identifying F_0 with some $\hat{k}_0 \in K$ ((2.20)), we have to show that $\hat{k}_0\phi$ is in the appropriate space. Note that $\hat{k}_0\phi(s) = \phi(sk_0)$ $(s \in S)$ and that obviously $\hat{k}_0\phi$ belongs to $C(S)$ in both cases.

Suppose, first, that $K \in \mathscr{G}_{\mathrm{j}}$. Since the map $(s,k) \to sk$ is continuous and K is compact, it follows easily that the maps $s \to \phi s$ are norm continuous from S into $C(K)$, and as $\|(\hat{k}_0\phi)s - (\hat{k}_0\phi)t\| \leq \|\phi s - \phi t\|$, it follows that $\hat{k}_0\phi \in \mathrm{U_r}(S)$ as required.

Now suppose that $K \in \mathscr{G}_{\mathrm{s}}$. Let $s_\delta \to s$ in S and $F \in \mathfrak{M}(C(S))$. Let $k_1 \in K$ be such that $F\hat{k}_0 = \hat{k}_1$ in $\mathfrak{M}(A_f(K))$. Then

$$F((\hat{k}_0\phi)s_\delta) = (F\hat{k}_0)(\phi s_\delta) = \phi(s_\delta k_1) \to \phi(sk_1) = F((\hat{k}_0\phi)s)$$

so that ((0.1(ii))) $\hat{k}_0\phi \in \mathrm{WU_r}(S)$ as required. □

(2.23) THEOREM. *There exists an S-fixed-point in every member of $\mathscr{G}_{\mathrm{j}}$ ($\mathscr{G}_{\mathrm{s}}$) if and only if there exists a left invariant mean on $\mathrm{U_r}(S)$ ($\mathrm{WU_r}(S)$).*

PROOF. Use (2.22) and (2.18). □

(2.24) THEOREM. *The group G is amenable if and only if every affine left G-set K for which the map $(x,k) \to xk$ $(x \in G,\ k \in K)$ is jointly [separately] continuous contains a fixed-point for G.*

PROOF. The theorem follows from (1.10), with $B = \mathrm{U_r}(G)$ [$\mathrm{WU_r}(G)$], and (2.23). □

Let G be a locally compact group. Then $L_\infty(G)$ is an abelian C^*-algebra. Let $\Phi(G)$ be the carrier space of $L_\infty(G)$. Then the map $\phi \to \hat{\phi}$ is an isometric $*$-isomorphism from $L_\infty(G)$ onto $C(\Phi(G))$, and dualising this map and using the Riesz Representation Theorem, we obtain a linear isometry $F \to \hat{F}$ from $L_\infty(G)'$ onto the measure space $M(\Phi(G)) = C(\Phi(G))'$. Clearly

$$F(\phi) = \hat{F}(\hat{\phi}) = \int_{\Phi(G)} \hat{\phi}(p)\, d\hat{F}(p).$$

Similar considerations apply to the case of a semigroup S; in this case, $\Phi(S)$ is just βS, the Stone-Čech compactification of S, and, of course, $L_\infty(S) = l_\infty(S)$. One might hope that the Arens product on $l_\infty(S)'$ can be realized by some reasonable convolution formula involving $\Phi(S)$. Unfortunately, this is rarely the case.

Let Σ be either a locally compact group or a discrete semigroup.

Our next proposition collects some useful facts about invariant means when regarded as elements of $M(\Phi(\Sigma))$ as above. Note in (i) below that $\Phi(\Sigma)$, as a set of linear functionals on $L_\infty(\Sigma)$, is a subset of the Banach Σ-space $L_\infty(\Sigma)'$, and that in (ii), $\mathscr{B}(\Phi(\Sigma))$ is the Borel algebra of $\Phi(\Sigma)$. In (iii), Ext K is the set of extreme points of a compact convex subset K of a locally convex space.

(2.25) PROPOSITION. (i) *$\Phi(\Sigma)$ is a Σ-invariant subset of $L_\infty(\Sigma)'$ with the maps $P \to sP$, $P \to Ps$ continuous on $\Phi(\Sigma)$ for each $s \in \Sigma$. Further, with $C(\Phi(\Sigma))$ and $M(\Phi(\Sigma)) = C(\Phi(\Sigma))'$ given their natural Banach Σ-space structures, we have*

$$(1) \qquad (\phi s)^\wedge = \hat{\phi} s, \qquad (sF)^\wedge = s\hat{F}$$

for all $\phi \in L_\infty(\Sigma)$, $s \in \Sigma$, and $F \in L_\infty(\Sigma)'$.

(ii) *If $m \in \mathfrak{M}(\Sigma)$, then $m \in \mathfrak{L}(\Sigma)$ if and only if $\hat{m}(s^{-1}E) = \hat{m}(E)$ for all $s \in \Sigma$, $E \in \mathscr{B}(\Phi(\Sigma))$. Further, if $m \in \mathfrak{L}(\Sigma)$, then the support $\mathbf{S}(\hat{m})$ of $\hat{m}$ is a (compact) left invariant subset of $\Phi(\Sigma)$.*

(iii) *Let Σ be left amenable with the discrete topology, and let X be a closed, left invariant subset of $\Phi(\Sigma)$. Then there exists $m_0 \in \operatorname{Ext} \mathfrak{L}(\Sigma)$ with the support $\mathbf{S}(\hat{m}_0)$ of $\hat{m}_0$ contained in X.*

PROOF. (i) Let $P \in \Phi(\Sigma)$ and $s \in \Sigma$. Then for $\phi \in L_\infty(\Sigma)$,

$$(2) \qquad sP(\phi) = P(\phi s), \qquad (Ps)(\phi) = P(s\phi).$$

Since $(\phi s)(\psi s) = (\phi\psi)s$, $(s\phi)(s\psi) = s(\phi\psi)$ $(\phi, \psi \in L_\infty(\Sigma))$, it follows that sP, Ps are multiplicative on $L_\infty(\Sigma)$ and so belong to $\Phi(\Sigma)$. Thus $\Phi(\Sigma)$ is Σ-invariant.

Noting that the topology of $\Phi(\Sigma) \subset L_\infty(\Sigma)'$ is just the relative weak* topology, we see from (2) that the maps $P \to sP$, $P \to Ps$ are continuous for each $s \in \Sigma$.

It follows that we have a natural Banach Σ-space structure on $C(\Phi(\Sigma))$ given by

$$\psi s(P) = \psi(sP), \qquad (s\psi)(P) = \psi(Ps).$$

Of course, $M(\Phi(\Sigma))$ is given its dual structure. The equalities in (1) follow immediately from the first equality of (2) and the dual version.

(ii) Let $m \in \mathfrak{L}(\Sigma)$. We will show that

$$\hat{m}(s^{-1}E) = \hat{m}(E) \qquad (s \in \Sigma,\ E \in \mathscr{B}(\Phi(\Sigma))). \tag{3}$$

(I am grateful to John Pym for suggesting the following argument.)

Suppose that C is an open and closed subset of $\Phi(\Sigma)$. Then $\chi_C \in C(\Phi(\Sigma))$ and so we can find $\phi \in L_\infty(\Sigma)$ with $\hat{\phi} = \chi_C$. Using (i),

$$\hat{m}(s^{-1}C) = \hat{m}(\chi_C s) = m(\phi s) = m(\phi) = \hat{m}(C). \tag{4}$$

Now let U be an open subset of $\Phi(\Sigma)$ and

$$A = \{\chi_C \colon C \text{ is an open and closed subset of } \Phi(\Sigma), C \subset U\}.$$

Since $\Phi(\Sigma)$ is zero-dimensional, we have that A is upwards directed, and

$$\chi_U = \sup\{\chi_C \colon \chi_C \in A\}.$$

Also, the set $B = \{\chi_C s \colon \chi_C \in A\}$ is upwards directed, and $\sup B = \chi_{s^{-1}U}$. Further, every element of A, B is lower semicontinuous, and so using [**HR1**, (11.13)] and (4), we have

$$\hat{m}(s^{-1}U) = \sup\{\hat{m}(s^{-1}C) \colon \chi_C \in A\} = \sup\{\hat{m}(C) \colon \chi_C \in A\} = \hat{m}(U).$$

Since the family of open subsets of $\Phi(\Sigma)$ generates $\mathscr{B}(\Phi(\Sigma))$, we see, using the Monotone Class Lemma, that (3) holds.

Conversely, suppose that $m \in \mathfrak{M}(\Sigma)$ and (3) holds. Let $F \in \mathscr{M}(\Sigma)$ and let C be the open and closed subset of $\Phi(\Sigma)$ such that $\hat{\chi}_F = \chi_C$. From (3) with $E = C$, we obtain $m(\chi_F s) = m(\chi_F)$, and using the density of the span of the set of characteristic functions in $L_\infty(\Sigma)$, we obtain $m \in \mathfrak{L}(\Sigma)$.

Thus the first assertion of (ii) is established.

Now suppose that $m \in \mathfrak{L}(\Sigma)$ and let D be the support of $\hat{m}$. Let $p \in D$ and suppose that $sp \notin D$. Then we can find an open subset U of $\Phi(\Sigma)$ containing p and such that $U \cap D = \varnothing$. Then by the first assertion,

$$0 = \hat{m}(U) = \hat{m}(s^{-1}U) > 0$$

giving a contradiction. It follows that D is Σ-invariant.

(iii) Let $K = \mathrm{PM}(X)$, the set of probability measures in $M(X)$. Then K is convex. Now X is compact, and so K is weak* compact in $M(X)$. Also, in the obvious way, $M(X)$ can be regarded as a subspace of the Banach Σ-space $M(\Phi(\Sigma))$.

Let $\mu \in K$. We show that $s\mu \in K$. Indeed, if U is open in $\Phi(\Sigma)$ with $U \cap X = \varnothing$, and $u \in U$, then we can find $\phi \in L_\infty(\Sigma)$ with $0 \le \hat{\phi} \le 1$, $\hat{\phi}(u) = 1$, and $\hat{\phi}(X) = \{0\}$. Since X is left invariant, $\hat{\phi}s(p) = 0$ for $p \in X$, and so $s\mu(\hat{\phi}) = 0$. It follows that $\mathbf{S}(s\mu) \subset X$. Obviously $s\mu \in \mathrm{PM}(X) = K$.

Now K is an affine left Σ-set under the maps $\mu \to s\mu$. Since Σ is left amenable with the discrete topology, Day's Fixed-Point Theorem yields an element $\nu \in K$ with $s\nu = \nu$. Clearly, $\nu = \hat{n}$ for some $n \in \mathfrak{L}(\Sigma)$. Thus the convex set $\mathfrak{N}$, where

$$\mathfrak{N} = \{m \in \mathfrak{L}(\Sigma) \colon \mathbf{S}(\hat{m}) \subset X\},$$

is not empty. Use of the Tietze Extension Theorem shows that the weak* topology on the set $\hat{\mathfrak{N}} = \{\hat{m}: m \in \mathfrak{N}\} = \{\mu \in K: s\mu = \mu \text{ for all } s \in S\} \subset M(X)$ corresponds with the weak* topology on $\mathfrak{N}$, and since the former set is obviously weak* compact, so also is $\mathfrak{N}$.

By the Krein-Mil'man Theorem [**DS**, V.8.4], we can find $m_0 \in \operatorname{Ext}\mathfrak{N}$. We claim that $m_0 \in \operatorname{Ext}\mathfrak{L}(\Sigma)$. For if m_0 is a convex combination $\alpha m + (1-\alpha)m'$ with $0 < \alpha < 1$ and m, $m' \in \mathfrak{L}(\Sigma)$, then, since $\hat{m}_0 \geq \alpha\hat{m}$, $\hat{m}_0 \geq (1-\alpha)m'$, it follows that $\hat{m}, \hat{m}' \in K$. But in that case, m, $m' \in \mathfrak{N}$, so that $m_0 = m = m'$. Hence $m_0 \in \operatorname{Ext}\mathfrak{L}(\Sigma)$. □

We can say more about $\Phi(\Sigma)$ $(= \beta\Sigma)$ when Σ is a discrete semigroup, basically because $L_\infty(\Sigma)$ is introverted in that case.

Let S be a semigroup. Recall ((2.9)) that $\mathfrak{M}(S)$ is a semigroup in a natural way. Recall also that S is a dense subset of βS, an element $s \in S$ being identified with $\hat{s} \in \beta S$, where $\hat{s}(\phi) = \phi(s)$ $(\phi \in l_\infty(S))$. We thus have

$$S \subset \beta S \subset \mathfrak{M}(S).$$

(2.26) PROPOSITION. *The set βS is a subsemigroup of $\mathfrak{M}(S)$ containing S as a subsemigroup, and for each $q \in \beta S$, the map $p \to pq$ is continuous on βS.*

PROOF. Let p, $q \in \beta S$, ϕ_1, $\phi_2 \in l_\infty(S)$. Then for $s \in S$,

$$q((\phi_1\phi_2)s) = q((\phi_1 s)(\phi_2 s)) = q\phi_1(s)q\phi_2(s).$$

Thus $q(\phi_1\phi_2) = (q\phi_1)(q\phi_2)$, and so

$$pq(\phi_1\phi_2) = p(q(\phi_1\phi_2)) = pq(\phi_1)pq(\phi_2).$$

Thus $pq \in \beta S$, and βS is a subsemigroup of $\mathfrak{M}(S)$.

A routine check shows that S, with its given multiplication, is a subsemigroup of βS. The last assertion of the proposition follows from (2.5(viii)). □

We note that there is no conflict between sp, ps defined, on the one hand, as in (2.25(i)), and on the other, as semigroup products evaluated in βS. The two interpretations are obviously the same!

Note also that, in general, the maps $q \to pq$ are *not* continuous (cf. Problem 2-21).

We are now in a position to tackle the sixth problem in the theory of amenable semigroups posed in (1.15). The elegant solution is due to E. E. Granirer [**5**] after earlier work by Mitchell [**3**]. Mitchell introduced the following class of semigroups.

(2.27) DEFINITION. A semigroup S is called *extremely left amenable* (or ELA) if there exists a multiplicative left invariant mean on S. Similarly, S is called *extremely right amenable* (or ERA) if there exists a multiplicative right invariant mean on S.

So S is ELA if and only if there exists $m \in \mathfrak{L}(S)$ such that $\hat{m} = \delta_\alpha$ for some $\alpha \in \beta S$. The following lemma, which is of intrinsic interest, will enable us to characterise ELA semigroups.

(2.28) LEMMA. *Let S be a right cancellative semigroup and $x \in S$. If $x^n \neq x$ for all $n > 1$, then let N be any integer > 1; otherwise, let N be the smallest of the integers $n > 1$ for which $x^n = x$. Then there exists a family $\{A_i : 1 \leq i \leq N-1\}$ of subsets of S such that*

(i) *$A_i \cap A_j = \varnothing$ if $i \neq j$;*
(ii) *$S = \bigcup_{i=1}^{N-1} A_i$;*
(iii) *for each i, $xA_i \subset A_{i+1}$, where we take $A_N = A_1$.*

PROOF. Using Zorn's Lemma, we can find $(N-1)$ subsets $V_1, \ldots, V_{N-1}$ of S that are maximal with respect to the following properties:

$$V_i \cap V_j = \varnothing \quad \text{if } i \neq j, \qquad xV_i \subset V_{i+1} \qquad (1 \leq i \leq N-1) \tag{1}$$

where, of course, we take $V_N = V_1$. Let $V = \bigcup_{i=1}^{N-1} V_i$. It suffices to show that $V = S$, for then we can take $A_i = V_i$.

Suppose, on the contrary, that $V \neq S$. Let $b \in S \sim V$, and, for $1 \leq i \leq (N-1)$, $V_i(b) = \{x^{i+r(N-1)}b : r \geq 0\}$. We claim that

(a) $V_i(b) \cap V_j(b) = \varnothing$ if $i \neq j$;
(b) $xV_i(b) \subset V_{i+1}(b)$ $(1 \leq i \leq (N-1))$.

As (b) is obviously true, it remains to prove (a).

Suppose that $i \neq j$. If $x^n \neq x$ for all $n > 1$, then, as $i+r(N-1) \neq j+s(N-1)$ for all r, $s \geq 0$ and as S is right cancellative, we must have $x^{i+r(N-1)}b \neq x^{j+s(N-1)}b$, and (a) follows. If, however, $x^N = x$, then $xx^{N-1} = x$, so that

$$x^{i+r(N-1)}b = x^i b,$$

and as N is the smallest integer $n > 1$ with $x^n = x$ and S is right cancellative, we again see that (a) is true.

So (a) is true in general as required.

We establish the lemma by contradicting the maximality of the V_i's. There are two cases to be considered.

Suppose, first, that $x^n b \notin V$ for all n, and let $V_i' = V_i \cup V_i(b)$. Using (a), (b), and the fact that $V_i(b) \cap V = \varnothing$ for all i, we see that (1) is true with V_i replaced by V_i'. Since $V_i \neq V_i'$, the desired contradiction follows.

Now suppose that $x^m b \in V$ for some $m > 1$. Let m be the smallest such integer. For some unique j, $x^m b \in V_j$. If $j > 1$, set $V_i^* = V_i$ if $i \neq (j-1)$ and $V_{j-1}^* = V_{j-1} \cup \{x^{m-1}b\}$. If $j = 1$, set $V_i^* = V_i$ if $i \neq N$, and $V_N^* = V_N \cup \{x^{m-1}b\}$. Then $V_i \subset V_i^*$ for all i and $V_k \neq V_k^*$ for some k. The existence of the sets V_i^* contradicts the maximality of the V_i's. □

(2.29) THEOREM. *A left amenable semigroup S is ELA if and only if every nonempty, finite [two-point] subset of S has a right zero.*

PROOF. The following proof which deals with the "finite" version of the theorem trivially modifies to establish the "two-point" version.

Suppose that every finite subset of S has a right zero. Then for each $F \in \mathscr{F}(S)$, we can find $s_F \in S$ such that $Fs_F = \{s_F\}$. Now $\{s_F\}$ forms a net, where $s_F \leq s_E$ if $F \subset E$. By taking a suitable subnet, we can suppose that

$s_F \to p$ for some $p \in \beta S$. Since each of the maps $p \to sp$ is continuous and as $ss_F = s_F$ eventually for each $s \in S$, it follows that $Sp = \{p\}$. Define $m \in \mathfrak{M}(S)$ by setting $\hat{m} = \delta_p$. Using (1) of (2.25(i)), we see that $m \in \mathfrak{L}(S)$. Since m is trivially multiplicative, it follows that S is ELA.

Conversely, suppose that S is ELA, and let $\approx$ be the congruence of (1.24). Let $T = S/\approx$ and $Q\colon S \to T$ be the canonical homomorphism. Then if $p \in \mathfrak{L}(S)$ is multiplicative, so also is $r \in \mathfrak{L}(T)$, where

$$r(\phi) = p(\phi \circ Q) \qquad (\phi \in l_\infty(T)).$$

Hence T is also ELA.

We now show that we can assume $S = T$. For suppose that we know that every finite subset of T has a right zero, and let $F = \{s_1, \ldots, s_n\} \in \mathscr{F}(S)$. Let $a \in S$ be such that $Q(a)$ is a right zero for $Q(F)$. Then $s_i a \approx a$ for each i. Using the definition of $\approx$ and the fact that it is a right congruence on S, we can define, recursively, elements $b_1, \ldots, b_n$ in S such that

$$s_i a b_1 b_2 \cdots b_n = a b_1 b_2 \cdots b_n \qquad (1 \le i \le n).$$

Then $F(ab_1b_2\cdots b_n) = ab_1b_2\cdots b_n$ and every finite subset of S has a right zero.

The upshot is that we can suppose that $S = S/\approx$, that is, that S is right cancellative. We shall show that S is trivial. Suppose not. Then we can find $y_1, y_2 \in S$, $S, y_1 \ne y_2$. Suppose that $u^2 = u$ for all $u \in S$. Then for all $u, y \in S$, $yu^2 = yu$, so that $yu = y$. But then $y_1 S \cap y_2 S = \varnothing$ contradicting (1.23). So there exists $x \in S$ with $x^2 \ne x$. By (2.28), we can find $N > 2$ and subsets $A_1, \ldots, A_{N-1}$ of S such that (i), (ii), and (iii) of (2.28) hold. Let $m \in \mathfrak{L}(S)$ be multiplicative. Then for $1 \le i \le N-1$, we have, using (2.28(iii)), $m(A_{i+1}) = m(x^{-1}A_{i+1}) \ge m(A_i)$, and it follows that $m(A_i) = 1/(N-1)$ for all i. But $m(A_i)^2 = m(\chi_{A_i})^2 = m(\chi^2_{A_i}) = m(A_i)$ so that $m(A_i)$ is 1 or 0. This is a contradiction. So S is a singleton, and trivially every finite subset of S has a right zero. □

(2.30) COROLLARY. *The only right cancellative, ELA semigroup is the trivial singleton semigroup.*

A more general version of the ELA condition is the n-ELA condition: a semigroup S is *n-ELA* if there exists a finite, left ideal group G in βS with $|G| = n$. Thus ELA = 1-ELA. Problem 2-35 gives an algebraic characterization of n-ELA semigroups.

References

Berglund, Junghenn, and Milnes [**1**], Day [**2**], [**4**], [**5**], Dixmier [**1**], Følner [**1**], [**3**], Furstenberg [**1**], Granirer [**5**]-[**7**], Granirer and Lau [**1**], Lau [**5**], [**8**], de Leeuw and Glicksberg [**1**], [**S**], Mitchell [**2**], [**6**], Namioka [**3**], von Neumann [**1**], Rickert [**2**], Wong [**1**], [**4**].

Further Results

(2.31) On the L-space structure of $\mathfrak{J}_l(\Sigma)$. Let Σ be a semigroup or a locally compact group. Instead of considering the complex vector space $\mathfrak{J}_l(\Sigma)$, we consider $\mathfrak{J}_l(\Sigma, \mathbb{R})$, the real subspace of $\mathfrak{J}_l(\Sigma)$ spanned by $\mathfrak{L}(\Sigma)$. (In fact, use of (2.2) (or an easy direct argument) shows that $\mathfrak{J}_l(\Sigma)$ is the complexification of $\mathfrak{J}_l(\Sigma, \mathbb{R})$.)

We will show that $\mathfrak{J}_l(\Sigma, \mathbb{R})$ is an abstract L-space. We briefly recall the definition of such spaces. We state the axioms in the form most suitable for our purposes. A good account of this topic is given in [**Sch**]. A real Banach space X, with an ordering $\leq$, is said to be an *abstract L-space* if it has the following properties:

(i) $\leq$ is transitive, reflexive and antisymmetric;
(ii) $(X, \leq)$ is a vector lattice, that is, for x, y, $z \in X$ and $\lambda \geq 0$, we have
(a) if $x \leq y$, then $x + z \leq y + z$,
(b) if $x \leq y$, then $\lambda x \leq \lambda y$,
(c) x and 0 have a least upper bound $x \vee 0$;
(iii) if $x^+ = x \vee 0$, $x^- = (-x) \vee 0$, and $|x| = x^+ + x^-$, then
(a) $\|x\| = \||x|\|$,
(b) $\|y\| \leq \|x\|$ if $0 \leq y \leq x$,
(c) $\|x + y\| = \|x\| + \|y\|$ if x, $y \geq 0$.

Every abstract L-space is an L_1-space "in disguise," and (obviously) every L_1-space is an abstract L-space.

(1) (Namioka [**3**]). *The space $\mathfrak{J}_l(\Sigma, \mathbb{R})$ is an abstract L-space.*

[As in the proof of (2.2), we can identify $\mathfrak{J}_l(\Sigma, \mathbb{R})$ with a closed subspace of $M(Z, \mathbb{R})$, the space of real-valued measures in $M(Z)$, where $Z = \Phi(\Sigma)$. One readily checks that $M(Z, \mathbb{R})$ is an abstract L-space in the usual ordering. Give $\mathfrak{J}_l(\Sigma, \mathbb{R})$ the ordering it inherits as a subspace of $M(Z, \mathbb{R})$. Apart from (ii)(c) and (iii)(a) (which depends on (ii)(c) in order to make sense), all of the axioms above are obviously satisfied by $\mathfrak{J}_l(\Sigma, \mathbb{R})$. To prove (ii)(c), let $\nu \in \mathfrak{J}_l(\Sigma, \mathbb{R})$. From (2.2), ν^+, evaluated in $M(Z, \mathbb{R})$, belongs to $\mathfrak{J}_l(\Sigma, \mathbb{R})$. Since $\mathfrak{J}_l(\Sigma, \mathbb{R})$ is a subspace of $M(Z, \mathbb{R})$, ν^+ must also be the least upper bound $\nu \vee 0$ in $\mathfrak{J}_l(\Sigma, \mathbb{R})$. Finally, since (iii)(a) is true in $M(Z)$, it is also true in $\mathfrak{J}_l(\Sigma, \mathbb{R})$.]

(2.32) Invariant measures and the Translate Property. (von Neumann [**1**], Dixmier [**1**], Greenleaf [**2**], Rosenblatt [**1**], [**2**], [**4**], [**HR1**]). Let G be a group and X a left G-set with $ex = x$ for all $x \in X$. A natural and interesting question is *when does there exist an invariant, finitely additive, positive measure μ defined on the family $\mathscr{P}(X)$ of all subsets of X?* To make this question reasonable, we "normalise" μ by specifying a nonempty subset A of X and requiring $\mu(A) = 1$. Thus we say that an *invariant measure for* (G, X, A) is a finitely additive measure $\mu \colon \mathscr{P}(X) \to [0, \infty]$ such that $\mu(xB) = \mu(B)$ for all $x \in G$, $B \in \mathscr{P}(X)$, and $\mu(A) = 1$. A characterisation of when there exists an invariant measure for (G, X, A) (involving paradoxical decompositions) will be proved in (3.15).

When G is amenable, much more can be said about the existence of an invariant measure for (G, X, A). If (b) below holds, then (G, X, A) is said to have *the Translate Property*. Groups for which *every* triple (G, X, A) has the Translate Property will be briefly considered in (6.42).

(i) *Let G be an amenable group and (G, X, A) be as above. Then the following two statements are equivalent:*

(a) *there is an invariant measure for (G, X, A);*

(b) *if $n \geq 1$, $s_1, \dots, s_n \in G$, $\alpha_1, \dots, \alpha_n \in \mathbb{R}$ and*

$$\sum_{i=1}^{n} \alpha_i \chi_{s_i A} \geq 0, \tag{1}$$

then $\sum_{i=1}^{n} \alpha_i \geq 0$.

[Suppose that (a) holds. Let s_i and α_i be such that (1) holds. Let μ be an invariant measure for (G, X, A) and $B = \bigcup_{i=1}^{n} s_i A$. Note that $\mu(B) \geq \mu(A) = 1$. Let μ_B be the measure obtained by restricting μ to $\mathscr{P}(B)$ and $\nu_B = \mu_B / \mu(B)$. Then we can identify ν_B with a mean on X and

$$0 \leq \nu_B \left(\sum_{i=1}^{n} \alpha_i \chi_{s_i A} \right) = \sum_{i=1}^{n} \alpha_i \nu_B(s_i A) = \left(\sum_{i=1}^{n} \alpha_i \right) / \mu(B).$$

Thus (a) implies (b).

Conversely, suppose that (b) holds. Let $l_\infty(X, \mathbb{R})$ be the space of real-valued functions in $l_\infty(X)$ and C the linear span in $l_\infty(X, \mathbb{R})$ of the set $\{\chi_{sA} \colon s \in G\}$. Suppose that $\alpha_1, \dots, \alpha_n$, $\beta_1, \dots, \beta_m \in \mathbb{R}$ and $s_1, \dots, s_n$, $t_1, \dots, t_m \in G$ are such that

$$\sum_{i=1}^{n} \alpha_i \chi_{s_i A} = \sum_{j=1}^{m} \beta_j \chi_{t_j A}.$$

Then $(\sum_{i=1}^{n} \alpha_i \chi_{s_i A} - \sum_{j=1}^{m} \beta_j \chi_{t_j A}) = 0 \geq 0$, and so by (b), $\sum_{i=1}^{n} \alpha_i - \sum_{j=1}^{m} \beta_j \geq 0$. Similarly, $\sum_{j=1}^{m} \beta_j - \sum_{i=1}^{n} \alpha_i \geq 0$, and so $\sum_{i=1}^{n} \alpha_i = \sum_{j=1}^{m} \beta_j$. It follows that we can define a positive linear functional p on C by defining

$$p \left(\sum_{i=1}^{n} \alpha_i \chi_{s_i A} \right) = \sum_{i=1}^{n} \alpha_i.$$

It is readily checked that C is a right invariant subspace of $l_\infty(X, \mathbb{R})$ (under the usual G-action) and that p is right invariant.

Let us say that a subset E of X is *bounded* if it is contained in the union of a finite number of translates sA, and that $f \in l_\infty(X, \mathbb{R})$ has *bounded support* if it vanishes outside a bounded set. Let D be the subspace of such functions f, and note that $C \subset D$ and that D is a right invariant subspace of $l_\infty(X, \mathbb{R})$. Give the algebraic dual D^d of D its "weak*" topology—this topology is, of course, locally convex, and G acts affinely on D^d with the dual left action. Let $K = \{\beta \in D^d \colon \beta \text{ is positive}, \beta|_C = p\}$. Clearly, K is convex and G-invariant. Krein's Extension Theorem ([**Ri**, p. 227], Peressini [**S**, (2.8)]) gives that $K \neq \emptyset$. Finally, K is weak* compact—indeed, the usual proof of Alaoglu's Theorem

[**DS**, V.4.2] applies directly, noting that if $\beta \in K$ and if $f \in D$ vanishes outside a set of the form $E = \bigcup_{i=1}^{n} s_i A$, then $|\beta(f)| \le \beta(|f|) \le \|f\|_\infty \beta(\chi_E) \le n\|f\|_\infty$. Since G is amenable, Day's Fixed-Point Theorem applies to give a G-invariant functional $\alpha \in K$. Now define an invariant measure μ for (G, X, A) by setting $\mu(E) = \alpha(\chi_E)$ if E is bounded and $\mu(E) = \infty$ otherwise.]

(2.33) The nonintroversion of $C(G)$. Our main objective here is to prove the theorem that *if G is a locally compact group, then $C(G)$ is left introverted if and only if G is either discrete or compact.* The two main ingredients in the proof of this theorem are

(a) $\mathrm{WU_r}(G) = \mathrm{U_r}(G)$, and

(b) $\mathrm{U_r}(G) = C(G)$ if and only if G is either discrete or compact.

The result of (a) was initially proved by Mitchell [**6**] using a theorem of Ellis (Problem 2-24). Rao [**1**] proved a version of (a) for metric groups.

Following Milnes and Pym [**1**], we prove (a) (in both the locally compact and complete metric cases) using a remarkable theorem of Namioka dealing with points of joint continuity for a separately continuous function. Namioka's theorem requires the presence of a certain completeness condition satisfied by both locally compact and complete metric spaces. Some such completeness condition is required for the result to be true in view of the counterexample to the equality "$\mathrm{U_r}(G) = \mathrm{WU_r}(G)$" given in Milnes and Pym [**1**].

Namioka proves his theorem in his paper [**5**]. Another proof in the context of game theory is given by Christensen [**S**]. We will be content to state the theorem and use it to deal with the introversion question.

A topological space X is (Frolik [**1**]) called *strongly countably complete* or *s.c.c.* if there exists a sequence $\{\mathscr{A}_n\}$ of open coverings of X such that if $\{F_n\}$ is a decreasing sequence of nonempty closed subsets of X with F_n contained in a member of $\mathscr{A}_n$ for each n, then $\bigcap_{n=1}^{\infty} F_n \ne \varnothing$.

Every locally compact space Y is s.c.c.: for all n, take $\mathscr{A}_n = \{U : U$ is open in Y and contained in a compact subset of $Y\}$. Every complete metric space (Z, d) is also s.c.c. In this case, take $\mathscr{A}_n$ to be the family of all open balls of radius $1/n$. Both Y, Z are *regular* (in the sense that every neighbourhood of a point contains a closed neighbourhood).

Let X, Y, Z be topological spaces and $f \colon X \times Y \to Z$. Recall that f is said to be *separately continuous* if whenever $x_\delta \to x$ in X, $y_\delta \to y$ in Y and $x_0 \in X$, $y_0 \in Y$, then

$$f(x_\delta, y_0) \to f(x, y_0), \qquad f(x_0, y_\delta) \to f(x_0, y).$$

We say that f is *jointly continuous at* (x, y) if $f(x_\delta, y_\delta) \to f(x, y)$ whenever $x_\delta \to x$ in X and $y_\delta \to y$ in Y.

We can now state Namioka's theorem.

(i) (Namioka [**5**]). *Let X be a strongly countably complete, regular, topological space, Y a σ-compact, locally compact space, and (Z, d) a metric space. Let*

$f: X \times Y \to Z$ be separately continuous. Then there exists a dense G_δ-subset A of X such that f is jointly continuous at each point of $A \times Y$.

The following generalisation of Ellis's theorem is given by Namioka.

(ii) *Let G be a group that is also an s.c.c., regular topological space, and let Y be a left G-set that is also a locally compact Hausdorff space. Suppose that $ey = y$ for all $y \in Y$, and*

(a) *the map $(s, y) \to sy$ from $G \times Y$ into Y is separately continuous;*

(b) *the map $t \to ts$ from G into G is continuous for each $s \in G$.*

Then the map $(s, y) \to sy$ is continuous.

[Let Y_∞ be the one-point compactification of Y. We can make Y_∞ into a left G-set by defining

$$s\alpha = \begin{cases} s\alpha & \text{if } \alpha \in Y, \\ \infty & \text{if } \alpha = \infty. \end{cases}$$

A simple argument shows that condition (a) is true with Y replaced by Y_∞, and we may suppose that Y is compact. Since Y is now compact Hausdorff, its topology is determined by $C(Y)$. It therefore suffices to show that for each $\phi \in C(Y)$, the function $Q_\phi: G \times Y \to \mathbb{C}$ is continuous, where

$$Q_\phi(s, y) = \phi(sy).$$

Namioka's theorem obviously applies to the function Q_ϕ, and so there exists a dense, G_δ-subset A of G such that Q_ϕ is jointly continuous at each point of $A \times Y$. Let $s_0 \in A$ and $t_\delta \to t$ in G, $y_\delta \to y$ in Y. Then using the joint continuity of Q_ϕ at $(s_0, s_0^{-1}ty)$ and (a) and (b),

$$\begin{aligned} Q_\phi(t_\delta, y_\delta) = \phi((t_\delta t^{-1}s_0)(s_0^{-1}ty_\delta)) &= Q_\phi(t_\delta t^{-1}s_0, s_0^{-1}ty_\delta) \\ \to Q_\phi(s_0, s_0^{-1}ty) &= Q_\phi(t, y), \end{aligned}$$

so that Q_ϕ is jointly continuous on $G \times Y$.]

We note in passing that (ii) can be used to prove the interesting result of Ellis given in Problem 2-24.

We now proceed with our discussion of the left introversion question for $C(G)$. The next result, which also depends on (ii), implies that if G is either a locally compact or a complete metric topological group, then the maximal, left introverted subspace $\mathrm{WU_r}(G)$ of $C(G)$ is the same as the (more manageable) space $\mathrm{U_r}(G)$.

(iii) (Rao [**1**], Mitchell [**6**], Milnes [**1**], Milnes and Pym [**1**], Berglund, Junghenn, and Milnes [**1**]). *Let G be a separately continuous semigroup which is algebraically a group and topologically is s.c.c. and regular. Then*

$$\mathrm{WU_r}(G) = \mathrm{U_r}(G).$$

[We have to show that $\mathrm{WU_r}(G) \subset \mathrm{U_r}(G)$, the reverse inclusion being trivial. Let $\phi \in \mathrm{WU_r}(G)$ and let U be the closed unit ball of $C(G)'$ with the weak* topology. Let

$$Y = \{f\phi: f \in U\},$$

and give Y the topology of pointwise convergence on G. Now U is (weak*) compact and by (2.5(viii)), the map $f \to f\phi$ is continuous from U onto Y. Hence Y is a compact Hausdorff space. Clearly, Y is a left invariant subset of $C(G)$ ((2.5(iii))), and, since $\phi \in \mathrm{WU_r}(G)$, the maps $s \to sy$ are continuous from G into Y for all $y \in Y$. Clearly, the maps $y \to sy$ are continuous from Y into Y, so that the map $(s, y) \to sy$ from $G \times Y$ into Y is separately continuous. By (ii), the latter map is in fact jointly continuous. Define $F\colon G \times U \to \mathbb{C}$ by $F((s, f)) = s(f\phi)(e) = f\phi(s)$. Then F is continuous, and so

$$\|\phi s_\delta - \phi s\| = \sup_{f \in U} |F(s_\delta, f) - F(s, f)| \to 0.$$

Thus $\phi \in \mathrm{U_r}(G)$.]

Before proving the theorem concerning left introversion on $C(G)$, where G is locally compact, we need to know under what circumstances $\mathrm{U_r}(G) = C(G)$. The answer is given by Problem 1-3.

(iv) *Let G be a locally compact group. Then* $\mathrm{U_r}(G) = C(G)$ *if and only if G is either discrete or compact.*

(v) *Let G be a locally compact group. Then $C(G)$ is left introverted if and only if G is either discrete or compact.*

[The space $C(G)$ is left introverted if and only if $\mathrm{WU_r}(G) = C(G)$. Now use (iii) and (iv).]

(2.34) Left invariant means and non-Archimedean analysis. We discuss some of the results obtained by van Rooij [**1**] on the existence of left invariant means on certain subspaces of the space $l_\infty(\mathbb{P}, K)$ of bounded functions on $\mathbb{P}$ with values in a complete non-Archimedean field K. (See Schikhof [**1**] for the study of invariant means on a locally compact group in the non-Archimedean context.) A good account of non-Archimedean analysis is given in van Rooij [**2**], to which the reader is referred for further details. For convenience, we discuss briefly the definitions and facts from this theory that we need.

Throughout, $(K, |\cdot|)$ will be a non-Archimedean field. This means that K is a field and $|\cdot|$ (a *non-Archimedean valuation*) is a function from K into $\mathbb{R}$ such that

(a) $|x| \geq 0$ for all $x \in K$;
(b) $|x| = 0$ if and only if $x = 0$ $(x \in K)$;
(c) $|x + y| \leq \max(|x|, |y|)$ for all x, $y \in K$;
(d) $|xy| = |x||y|$ for all x, $y \in K$.

The non-Archimedean nature of $(K, |\cdot|)$ is expressed in (c), which is a particularly strong form of the triangular inequality. Clearly, (d) and (b) give that $|1| = 1$, where 1 is the unit of K, and then (c) gives that

$$|n1| \leq 1$$

for all $n \in \mathbb{P}$. Thus $(K, |\cdot|)$ does not satisfy a version of the well-known "Axiom of Archimedes."

Evidently, $|\cdot|$ determines a metric d on K, where

$$d(x,y) = |x-y|.$$

It easily follows from (c) that every ball in K is open and closed, so that K is zero-dimensional. (Not surprisingly, the zero-dimensional condition is often required in non-Archimedean analysis.) We say that $(K, |\cdot|)$ is *complete* if the metric d on K is complete.

The most familiar examples of complete, non-Archimedean fields are the fields $\mathbb{Q}_p$ of p-adic numbers. Here p is a prime, and $\mathbb{Q}_p$ can be thought of as the set of formal Laurent series

$$\sum_{n=N}^{\infty} a_n p^n$$

with $N \in \mathbb{Z}$ and $a_n \in \{0, 1, \ldots, (p-1)\}$. Addition and multiplication are the obvious formal processes, and if $\alpha \in \mathbb{Q}_p$, then we define

$$|\alpha|_p = \begin{cases} p^{-N} & \text{if } \alpha = \sum_{n=N}^{\infty} a_n p^n \text{ with } a_N \neq 0, \\ 0 & \text{if } \alpha = 0. \end{cases}$$

Then $(\mathbb{Q}_p, |\cdot|_p)$ is a complete, non-Archimedean field.

Let $(K, |\cdot|)$ be a complete, non-Archimedean field. Let Y be a vector space over K and $\|\cdot\|: Y \to [0, \infty)$. Then Y is said to be a *normed space* (over K) if the following properties hold:

(a) $\|\xi\| = 0$ if and only if $\xi = 0$ $(\xi \in Y)$;

(b) $\|\xi + \eta\| \leq \max(\|\xi\|, \|\eta\|)$ $(\xi, \eta \in Y)$,

(c) $\|\alpha\xi\| = |\alpha| \|\xi\|$ $(\alpha \in K, \xi \in Y)$.

Of course, Y is said to be a *Banach space* (over K) if Y is complete in the metric on Y induced by $\|\cdot\|$. The dual Y' of Y is the space of continuous, K-valued, linear functionals on Y and is a Banach space over K in the obvious way.

An example of such a Banach space is $l_\infty(X, K)$ where X is a nonempty set and

$$l_\infty(X, K) = \left\{ \phi \colon \phi \text{ maps } X \text{ into } K \text{ and } \sup_{x \in X} |\phi(x)| < \infty \right\}.$$

The norm $\|\cdot\|$ on $l_\infty(X, K)$ is, of course, given by $\|\phi\| = \sup_{x \in X} |\phi(x)|$. When S is a semigroup, $l_\infty(S, K)$ is a "Banach S-space" in the obvious way.

Let B be a K-linear subspace of $l_\infty(S, K)$ containing the constant function 1. The set $\mathfrak{M}(B)$ of means on B is defined as earlier. However, the analogue of (0.1(i)) does not make sense in the present context since the order properties, which are so important for real analysis, are absent in K. If B is, in addition, right invariant for S, then $\mathfrak{L}(B)$ is the set of left invariant means on S.

It is convenient to concentrate on the case where $S = \mathbb{P}$, the additive semigroup of positive integers.

The Hahn-Banach Theorem holds for every Banach space over K if and only if K is *spherically complete*, that is, every decreasing sequence of closed balls in K has nonempty intersection. Every $\mathbb{Q}_p$ is spherically complete.

Define $T \in \mathbf{B}(l_\infty(\mathbb{P}, K))$ by $T\phi(n) = \phi(n+1)$. Then for all $r \in \mathbb{P}$,

$$\phi(r) - \phi(r+n) = \sum_{i=0}^{n-1}(T^i\phi(r) - T^{i+1}\phi(r)) = \left(\sum_{i=0}^{n-1}\phi_i - T\left(\sum_{i=0}^{n-1}\phi_i\right)\right)(r),$$

where $\phi_i = T^i\phi$. Thus if B is invariant, then the set $\{\phi - T\phi \colon \phi \in B\}$ is a norm dense subspace of $D(B)$, the norm closure of $\mathrm{Span}\{(\phi - \phi n) \colon \phi \in B, n \in \mathbb{P}\}$. The proof of the next result is routine (cf. (2.1)).

(i) *Let K be spherically complete and B be an invariant subspace of $l_\infty(\mathbb{P}, K)$ containing* 1. *Then $\mathfrak{L}(B) \neq \varnothing$ if and only if*

$$\|1 - (\phi - T\phi)\| \geq 1$$

for all $\phi \in B$.

Define $j \in l_\infty(\mathbb{P}, K)$ by

$$j(n) = n1 \qquad (n \in \mathbb{P}).$$

Then $\|j\| = 1$. It is interesting that j plays such a central role in what follows.

(ii) $\mathfrak{L}(l_\infty(\mathbb{P}, K)) = \varnothing$.

[Suppose that $m \in \mathfrak{L}(l_\infty(\mathbb{P}, K))$. Then $m(T\phi) = m(\phi)$ for all $\phi \in l_\infty(\mathbb{P}, K)$. Thus as $Tj(n) = (n+1)1 = j(n) + 1$,

$$0 = m(Tj - j) = m(1) = 1$$

giving a contradiction.]

If Y is a linear subspace and ξ an element of a Banach space X over K, then we say that $\xi \perp Y$ if

$$\|\xi - \eta\| \geq \|\xi\|$$

for all $\eta \in Y$. The notion of orthogonality is, of course, familiar in the Hilbert space context.

Our next result characterises those subspaces B of $l_\infty(\mathbb{P}, K)$ for which $\mathfrak{L}(B)$ is *not* empty.

(iii) *Let K be spherically complete, and let B be an invariant subspace of $l_\infty(\mathbb{P}, K)$ which contains* 1. *Then $\mathfrak{L}(B) \neq \varnothing$ if and only if $j \perp B$.*

[Suppose that $\mathfrak{L}(B) \neq \varnothing$ and let $m \in \mathfrak{L}(B)$. Suppose that it is not true that $j \perp B$. Then there exists $\phi \in B$ such that $\|j - \phi\| < \|j\| = 1$. So there exists $\varepsilon > 0$ such that

$$|n1 - \phi(n)| < 1 - \varepsilon \qquad (n \in \mathbb{P}).$$

Thus

$$\begin{aligned} |1 - (\phi(n+1) - \phi(n))| &= |[(n+1)1 - \phi(n+1)] - [n1 - \phi(n)]| \\ &\leq \max(|(n+1)1 - \phi(n+1)|, |n1 - \phi(n)|) \\ &< 1 - \varepsilon. \end{aligned}$$

Thus $\|1 + (\phi - T\phi)\| < 1$ and (i) is contradicted (with $-\phi$ in place of ϕ). So $j \perp B$.

Conversely, suppose that $j \perp B$ and that $\mathfrak{L}(B) = \varnothing$. By (i), we can find $\psi \in B$ with $\|1 - (\psi - T\psi)\| < 1 - \eta$ for some $\eta > 0$. Then for all $n \geq 2$,

$$\begin{aligned}|(n-1)1 &+ \psi(n) - \psi(1)| \\ &= |[1 + \psi(2) - \psi(1)] + [1 + \psi(3) - \psi(2)] + \cdots + [1 + \psi(n) - \psi(n-1)]| \\ &\leq \max(|1 + \psi(2) - \psi(1)|, \ldots, |1 + \psi(n) - \psi(n-1)|) \\ &\leq \|1 - (\psi - T\psi)\| < 1 - \eta.\end{aligned}$$

Thus $\|j - 1 + \psi - \psi(1)1\| \leq \max(|1 - 1 + \psi(1) - \psi(1)|, 1 - \eta) = 1 - \eta$. Since $1 - \psi + \psi(1)1 \in B$, we have contradicted the fact that $j \perp B$. Hence $\mathfrak{L}(B) \neq \varnothing$.]

(iv) For an abelian semigroup S, van Rooij [**1**] also discusses invariant means on $l_\infty(S, K)$, on the non-Archimedean versions of $\mathrm{AP}(S)$ and $\mathrm{WP}(S)$, and also on the space $\mathrm{PC}(S, K)$ of functions in $l_\infty(S, K)$ with precompact range. In this, as also in the work of Schikhof [**1**] on invariant means on locally compact groups, the characteristic of the residue class field F of K plays an important role. The field F is the quotient of the subring $\{x \in K : |x| \leq 1\}$ of K by its maximal ideal $\{x \in K : |x| < 1\}$. (When $K = \mathbb{Q}_p$, the field F is just $\mathbb{Z}_p$.) Van Rooij characterises those semigroups S for which $\mathfrak{L}(S, K) \neq \varnothing$ in terms of the nonexistence of an element in S with "period" a multiple of p, where p is the characteristic of F.

Among other results, Schikhof shows that if G is an amenable, locally compact, zero-dimensional group and $p = 0$, then there exists a left invariant mean on the space $\mathrm{PC}(G, K)$ of functions in $C(G, K)$ with precompact range. Van Rooij [**2**, p. 346] asks if the converse to this result is true.

(2.35) Amenable von Neumann algebras. It would require a book on its own to do justice to the theory of amenable von Neumann algebras. We will be content to aim for the more modest objectives of discussing how the various formulations of amenability for von Neumann algebras are related to one another and to classical group amenability, and how earlier results of this chapter relate to some of the (easier) results in the theory. *In what follows, we will assume that the von Neumann algebras involved act on a separable Hilbert space.* (Many of the results discussed, however, are valid without the separability requirement.)

We discussed amenable C^*-algebras in (1.31). The latter class of C^*-algebras is defined in cohomological terms, that is, in terms of "many" derivations being inner. It is natural, then, to define amenable von Neumann algebras in similar terms. Since such algebras are in general very "large," it is reasonable to require extra ultraweak continuity conditions to "keep things under control." The relationship between amenability for C^*-algebras and amenability for von Neumann algebras is expressed in the deep theorem due to Connes [3] and Choi and Effros [1], [2] (cf. (1.31)): *A C^*-algebra B is amenable if and only if its second dual B'' is amenable as a von Neumann algebra.* (For the easier implication involved, cf. Problem 2-31.)

The amenability of a von Neumann algebra A is characterized by four other properties, the equivalences involved being far from obvious. These properties are *injectivity, Property P*, *semidiscreteness*, and *approximate finite dimensionality* (*AFD*).

The proof of the equivalence of the five formulations of amenability involves deep mathematics—in particular, it relies heavily on Connes's contributions to von Neumann algebras. A discussion of these matters is given in Connes [**S2**]. The present author has found the introduction to [**D1**] by E. C. Lance particularly helpful. The proof that amenable $\Rightarrow$ injective is given by Connes [**4**]—a more elementary proof is given by Bunce and Paschke [**S**]. The equivalence of injectivity and Property P follows from the work of Schwartz [**1**] and Tomiyama [**4**]. Effros and Lance [**1**] showed that semidiscrete $\Rightarrow$ injective; the converse follows from Connes [**3**] and Choi and Effros [**1**], the proof being substantially simplified by Wasserman [**1**] and Connes [**S1**]. The really difficult implication is that injective $\Rightarrow$ AFD. This was proved in Connes [**3**]; a new proof, in which many of the technicalities of Connes's proof are avoided, is given by Haagerup [**6**]. A short proof has been given by Pofa [S]. Johnson, Kadison, and Ringrose [**1**] showed that AFD $\Rightarrow$ amenable.

We will discuss each of the five characterizations of amenability for von Neumann algebras in turn. We then state some facts about the class of amenable von Neumann algebras. The section will conclude by discussing two related topics, viz., first the relationship between amenable G-actions and amenable von Neumann algebras, and second, how the inner amenability of a locally compact group G relates to the amenability of $VN(G)$.

(A) *Amenability.* Let A be a von Neumann algebra (acting on a separable Hilbert space $\mathfrak{H}$) and X a Banach A-module. The dual Banach A-module X' is called a *dual normal Banach A-module* if for each $\alpha \in X'$, the maps $a \to a\alpha$, $a \to \alpha a$ are ultraweak-weak* continuous from A to X'. A simple example of a dual Banach A-module is given by $X = A_*$, $X' = A$ under the obvious actions. The von Neumann algebra A is called *amenable* if every derivation from A to a dual normal Banach A-module is inner. There is a close connection between amenable C^*-algebras and amenable von Neumann algebras: *if a C^*-algebra $B \subset B(\mathfrak{H})$ is amenable, then the von Neumann algebra generated by B is amenable.* (This follows since B amenable $\Rightarrow B''$ amenable as a von Neumann algebra.)

We now discuss injectivity and illustrate how amenability and injectivity are related.

(B) *Injectivity.* The von Neumann algebra A is called *injective* if there exists a linear, norm one projection map P from $\mathbf{B}(\mathfrak{H})$ onto A. As it stands, this definition depends on the particular Hilbert space $\mathfrak{H}$ on which A is realised. However, it can be shown that if the injective property holds for one such $\mathfrak{H}$, then it holds for all of them (Hakeda and Tomiyama [**1**]). (It is possible to define such von Neumann algebras as the injective objects of a category.) Let A^c be the *commutant* of A: so $A^c = \{T \in \mathbf{B}(\mathfrak{h})\colon aT = Ta \text{ for all } a \in A\}$. If A is standard on $\mathfrak{H}$, then A is $*$-anti-isomorphic to its commutant A^c, using the

J-map of the Tomita-Takesaki theory. It follows that in the standard case, A is injective $\Leftrightarrow A^c$ is injective. Tomiyama [**4**] shows that the latter equivalence is true in the general case. Tomiyama [**4**] also shows that such a projection P is positive and satisfies $P(aTb) = aP(T)b$ for all a, $b \in A$, $T \in \mathbf{B}(\mathfrak{H})$. To relate amenability and injectivity, we need to show how the former gives rise to projection maps. The argument below shows how this can be done. This argument is taken from Bunce and Paschke [**S**], where it forms an important part of their proof that amenable $\Rightarrow$ injective. Note that the argument does not give immediately that A^c is injective, since the projection Q need not be of norm 1. However, the map Q can be "modified" using a polar decomposition argument to produce a norm one projection. The injectivity of A^c, and therefore of A, then follows.

Recall that for every von Neumann algebra B, there exists a unique Banach space B_* with $(B_*)' = B$. The space B_* is the *predual* of B, and when B acts on $\mathfrak{h}$, the elements of B_* can be realised in the form $\sum_{i=1}^{\infty} \omega_{f_i,g_i}$, where $f_i, g_i \in \mathfrak{h}$, $\sum_{i=1}^{\infty}(\|f_i\|^2 + \|g_i\|^2) < \infty$ and $\omega_{f,g}(T) = (Tf, g)$ for all $T \in B$, $f, g \in \mathfrak{h}$ ([D1], Part 1, Chapter 3). This realisation will be used in the following proof with $B = \mathbf{B}(\mathfrak{h})$.

Let A be amenable. Then there exists a bounded, linear projection Q from $\mathbf{B}(\mathfrak{H})$ onto A^c such that $Q(bTc) = bQ(T)c$ for all $T \in \mathbf{B}(\mathfrak{H})$, b, $c \in A^c$.

[Let $X = \mathbf{B}(\mathfrak{H}) \hat{\otimes} \mathbf{B}(\mathfrak{H})_*$. Then X is a Banach A-module under the actions: $a(T \otimes \phi) = T \otimes a\phi$, $(T \otimes \phi)b = T \otimes \phi b$ $(a, b \in A)$. (Here, of course, $R\phi(S) = \phi(SR)$, $\phi R(S) = \phi(RS)$ $(R, S \in \mathbf{B}(\mathfrak{H}))$.) Obviously, X' is a dual Banach A-module. We claim that X' is normal. To this end we can write $\phi \in \mathbf{B}(\mathfrak{H})_*$ in the form $\sum_{i=1}^{\infty} \omega_{f_i,g_i}$ as above. Note that $a\omega_{f,g} = \omega_{af,g}$, $\omega_{f,g}a = \omega_{f,a^*g}$. Now if $a_\delta \to a$ *strongly in a bounded subset of A, then $\sum_{i=1}^{\infty} \|(a_\delta - a)f_i\|^2 \to 0$, $\sum_{i=1}^{\infty} \|(a_\delta^* - a^*)g_i\|^2 \to 0$, and it follows that $\|a_\delta\phi - a\phi\| \to 0$, $\|\phi a_\delta - \phi a\| \to 0$. So if $\alpha \in X'$ and $\xi \in X$, then the functionals $a \to a\alpha(\xi)$, $a \to \alpha a(\xi)$ are *strongly continuous on bounded subsets of A, and so are ultraweakly continuous (cf. [**D1**, Part 1, Chapter 3, Theorem 1]). Thus X' is normal.

Now X' is canonically identified with $\mathbf{B}(\mathbf{B}(\mathfrak{H}))$, an element $\Phi \in \mathbf{B}(\mathbf{B}(\mathfrak{H}))$ corresponding to the functional $T \otimes \phi \to \phi(\Phi(T))$. With this identification, the A-actions on X' are given by $\Phi a(T) = \Phi(T)a$, $(a\Phi)(T) = a(\Phi(T))$. Let

$$Y = \{\Phi \in X' \colon \Phi(Tb) = \Phi(T)b, \Phi(bT) = b\Phi(T), \Phi(b) = 0 \quad \text{for all } T \in \mathbf{B}(\mathfrak{H}),\ b \in A^c\}.$$

One readily checks that Y is a closed submodule of X' and can be identified with $Z^\perp$ $(= (X/Z)')$, where Z is the closure of the submodule of X spanned by elements of the form $(Tb \otimes \phi - T \otimes b\phi)$, $(bT \otimes \phi - T \otimes \phi b)$, $b \otimes \phi$. So Y is a dual normal Banach A-module, and the derivation $D \colon A \to X'$, where $D(a)(T) = Ta - aT$, maps A into Y. Since A is amenable, there exists $\Psi \in Y$ with $D(a) = \Psi a - a\Psi$. The map Q that we require is given by $Q = (I - \Psi) \in \mathbf{B}(\mathbf{B}(\mathfrak{H}))$, where I is the identity on $\mathbf{B}(\mathfrak{H})$.]

There is a striking analogy between the above result and (2.15). Now (2.15) does involve an S-invariant mean, and one might wonder if the amenability of A can be formulated in such terms. The natural candidate for S is H, the unitary group $\mathbf{U}(A)$ of A, and we shall see in (C) that the amenability of A is equivalent to the existence of a left H-invariant mean on a space of functions on H. This brings us to Property P, the only one of the five formulations of amenability for A which explicitly involves H.

(C) *Property P.* The von Neumann algebra A is said to have *Property P* (Schwartz [**1**]) if for all $T \in \mathbf{B}(\mathfrak{H})$, the weak operator closure of $\mathrm{co}\{UTU^* : U \in H\}$ contains an element of the commutant A^c. (Property P is discussed in Sakai [**1**, (4.4)].) Again, Property P is intrinsic to A and does not depend on the Hilbert space $\mathfrak{H}$ on which A is realised (Hakeda [**1**]). We first show how Property P relates to injectivity.

For $U \in H$, define $\Phi_U \in \mathbf{B}(\mathbf{B}(\mathfrak{H}))$ by $\Phi_U(R) = URU^*$, and let $C = \mathrm{co}\{\Phi_U : U \in H\}$. Note that C forms a convex subsemigroup of the unit ball of $\mathbf{B}(\mathbf{B}(\mathfrak{H}))$ and that $C(T') = \{T'\}$ for all $T' \in A^c$.

(i) (Schwartz [**1**], Hakeda [**1**]). *Suppose that A has Property P. Then there exists a net $\{V_\delta\}$ in $\mathrm{co}\{\Phi_U : U \in H\}$ such that for each $T \in \mathbf{B}(\mathfrak{H})$, $\{V_\delta(T)\}$ converges in the weak operator topology to an element of A^c. Further, A^c is injective.*

[Let $\xi_1, \dots, \xi_n \in \mathfrak{H}$, $T_1, \dots, T_r \in \mathbf{B}(\mathfrak{H})$, and $\varepsilon > 0$. We claim that there exists $V \in C$ and $T'_1, \dots, T'_r \in A^c$ with $\|T'_j\| \le \|T_j\|$ $(1 \le j \le r)$ such that $\|V(T_j - T'_j)(\xi_i)\| < \varepsilon$ $(1 \le i \le n,\ 1 \le j \le r)$. Indeed, since A has Property P and the weak and strong operator closures of a bounded convex subset of $\mathbf{B}(\mathfrak{H})$ coincide, there exists $W_1 \in C$ such that $\|W_1(T_1 - T'_1)\xi_1\| < \varepsilon$. Arguing along the lines of the proof of (2.13(i)) produces $W_1, \dots, W_n \in C$ such that $V = W_n \cdots W_1$ has the desired property. Allowing the ξ_i, T_j, ε to vary, we see that the V's form a net $\{V'_\sigma\}$ in the natural way and that every ultraweak cluster point of $\{V'_\sigma(T)\}$ belongs to A^c for all $T \in \mathbf{B}(\mathfrak{H})$. Now use the ultraweak compactness of the unit ball of $\mathbf{B}(\mathfrak{H})$ together with Tychonoff's Theorem to produce the desired net $\{V_\delta\}$. A norm one projection P from $\mathbf{B}(\mathfrak{H})$ onto A^c is given by $P(T) = \lim V_\delta(T)$. (For use in (G), we note that $P(S_1TS_2) = S_1P(T)S_2$ for all S_1, $S_2 \in A^c$.)]

We now discuss how Property P fits in with (2.13) and (2.15) (cf. Lau [**8**], [**15**], Milnes [**3**], Yeadon [**1**]). From the viewpoint of (2.15), we should take $S = H = \mathbf{U}(A)$ and $X' = \mathbf{B}(\mathfrak{H})$ (so that $X = \mathbf{B}(\mathfrak{H})_*$). The dual left action of H on X' is given by $T \to U\,.\,T = UTU^*$ $(U \in H,\ T \in X')$. This left action is induced by the obvious right action of H on X: $\theta\,.\,U(T) = \theta(U\,.\,T)$ $(\theta \in X)$. It is easily checked that these actions are isometric. The space $X(H) \subset l_\infty(H)$ is spanned by functions of the form $T\theta$, where $T\theta(U) = T(\theta\,.\,U) = \theta(U\,.\,T)$. The space $\mathfrak{J}_l(X)$ is readily identified: indeed, $\mathfrak{J}_l(X) = \{T \in \mathbf{B}(\mathfrak{H}) : UT = TU \text{ for all } U \in H\} = A^c$ (since H spans A). We now show how (2.13) and (2.15) can be used to give an invariant mean formulation of Property P.

(ii) (de la Harpe [**2**]). *The von Neumann algebra A has Property P if and only if there exists a left invariant mean for H on $X(H)$.*

[Suppose that A has Property P. For the purpose of applying (2.13(i)), we take $G = H$ (as discrete) and $B = X(H)$. Note that B is left introverted ((2.7)) and $1 \in B$ (since $1 = I\theta$ for any state $\theta \in \mathbf{B}(\mathfrak{H})_*$, where I is the identity on $\mathfrak{H}$). Let $T_i \in X'$, $\theta_i \in X$ $(1 \leq i \leq n)$. With V_δ as in (i), $V_\delta(T_i) \to T_i'$ ultraweakly for some $T_i' \in A^c$, and noting that $U(T_i\theta_i) = \Phi_U(T_i)\theta_i$, we see that the constant function $(\sum_{i=1}^n T_i'(\theta_i))1$ belongs to the pointwise closure of $\text{co}\{U(\sum T_i\theta_i) \colon U \in H\}$. By (2.13(i)), $\mathfrak{L}(X(H)) \neq \emptyset$.

Conversely, suppose that $\mathfrak{L}(X(H)) \neq \emptyset$. Then $\mathfrak{J}_l(X) = A^c$, and A has Property P by (2.15).]

Suppose, now, that G is an amenable locally compact group, π is a continuous, unitary representation of G on a Hilbert space $\mathfrak{H}$, and B is the von Neumann algebra generated by $\pi(G)$. Noting that $\pi(G) \subset \mathbf{U}(B)$, we can define $X(G)$, in the obvious way. It is readily checked that $X(G) \subset C(G)$ and so carries a left invariant mean since G is amenable. It then follows as above that B has Property P, so that *the von Neumann algebra generated by a continuous, unitary representation of an amenable locally compact group has Property P.* In particular, if G is amenable, then the von Neumann algebra generated by the left regular representation π_2 of G ((1.31)) has Property P. The preceding result (ii) does not use any topology on H. However, H is a jointly continuous, complete, separable metric group in the strong operator topology, and $X(H)$ is a subspace of $\mathrm{U_r}(H)$ (Problem 2-26). De la Harpe shows, using the AFD property, that A *is amenable if and only if there exists a left invariant mean on* $\mathrm{U_r}(H)$ (Problem 2-27). This remarkable result shows that amenability for von Neumann algebras can be reformulated in familiar group terms and raises the speculation that the theories of amenable von Neumann algebras and amenable locally compact groups should be unifiable in an amenability theory for topological groups (cf. Paterson [**8**]).

A deep result related to that of de la Harpe above and involving the semigroup of isometries in A has been established by Haagerup [**3**]. This theorem uses Haagerup's extension of the Grothendieck-Pisier inequality and is used to show that nuclear $\Rightarrow$ amenable for C^*-algebras.

(D) *Semidiscreteness.* The semidiscrete property was introduced and studied by Effros and Lance [**1**]. The property can be defined using tensor products in a way analogous to the nuclear property for C^*-algebras ((1.31)). From an amenable group viewpoint, however, semidiscreteness can be regarded as a von Neumann algebra version of Problem 4-28. The appropriate definition of semidiscreteness involves the notion of *completely positive* maps. A map $T \in \mathbf{B}(A)$ is called completely positive if for each $n \geq 1$, the map $[a_{ij}] \to [T(a_{ij})]$ on $M_n \otimes A$ is positive. The von Neumann algebra A is called *semidiscrete* if there exists a net $\{T_\delta\}$ of weak* continuous, completely positive, unit preserving, finite rank operators in $\mathbf{B}(A)$ such that $T_\delta a \to a$ weak* in A for every $a \in A$. (This property is clearly reminiscent of Grothendieck's metric approximation property.) If G is an amenable locally compact group, then a certain net $\{f_\delta\}$ of continuous, positive definite functions with compact support gives rise to such a net $\{T_\delta\}$ on VN(G) so that VN(G) is semidiscrete (Problem 4-43).

(E) *Approximate finite dimensionality.* The AFD property is the most elementary and easily defined of the five formulations of amenability for A. The algebra A is called *approximately finite dimensional* (AFD) if A is generated by an increasing sequence of finite-dimensional *-subalgebras. (For an extension of this notion to the case where A cannot be realized on a separable Hilbert space, see Elliott [**1**], [**2**] and Elliott and Woods [**1**].) The AFD algebras form a very natural class of von Neumann algebras, and the determination of the AFD factors has been a major theme in operator algebras. This determination is completed in the work of Connes [**1**], [**3**] and Haagerup [**S3**]. Paradoxically, of the five formulations of von Neumann algebra amenability, the AFD formulation is the one which seems furthest away from group amenability. It is certainly not true that an amenable group is always an increasing union of finite subgroups! Note, however, that if A is AFD, then $\mathbf{U}(A)$ is generated (in the strong operator topology) by an increasing sequence of compact subgroups.

(F) *The class of amenable von Neumann algebras.* References for the following results are Arveson [**1**], Choda and Echigo [**S1**], [**S2**], Choi and Effros [**2**], Connes [**3**], [**S1**], Effros and Lance [**1**], Hakeda and Tomiyama [**1**], Schwartz [**1**], Tomiyama [**4**].

(a) Every Type 1 von Neumann algebra is amenable.

(b) The von Neumann algebra generated by an amenable C^*-algebra on a Hilbert space $\mathfrak{H}$ is amenable. In particular, every continuous, unitary representation of a locally compact group in the class $\mathscr{A}$ of (1.31) generates an amenable von Neumann algebra.

(c) The weak operator closure of the union of an upwards directed family of amenable von Neumann algebras on a Hilbert space is amenable. A similar result holds for the intersection of a downwards directed family of amenable von Neumann algebras.

(d) A von Neumann algebra on a Hilbert space is amenable if and only if its commutant is amenable.

(e) Let $\mathfrak{H}$ be a separable Hilbert space, X a standard Borel space with probability measure μ, and $x \to A_x$ a measurable field of von Neumann subalgebras of $\mathbf{B}(\mathfrak{H})$ [**D1**, Part 2]. Let $A = \int_X A_x \, d\mu(x)$. Then A is amenable if and only if A_x is amenable for μ-a.e. $x \in X$.

(f) If (A, ρ, G) is a von Neumann covariant system (Pedersen [**S**]), then the cross product von Neumann algebra $A \times_\rho G$ is amenable if A and G are amenable, while A is amenable if $A \times_\rho G$ is amenable (whether or not G is amenable).

(g) The tensor product of two von Neumann algebras is amenable if and only if each of the original von Neumann algebras is amenable.

(G) *Amenable G-spaces.* We saw in (f) above that a cross product $A \times_\rho G$ is amenable if A and G are amenable. Now the cross product construction is important for producing factors. In particular, the Murray–von Neumann construction [**D1**, Part 1, Chapter 9] associates with a free ergodic action of a group on an appropriate measure space a cross product factor M. An important question is *what conditions on the group and measure space will ensure that M is amenable*? There is a substantial literature on this theme, a good reference for which is Moore [**S**]. A related procedure involving ergodic equivalence relations produces the *Krieger factors*, and a similar question to that above arises for such factors. Zimmer [**1**]-[**7**] has introduced the notion of an amenable measured equivalence relation that answers these questions. Further developments have been made by Anantharaman-Delaroche [**1**]-[**3**]. See also Ocneanu [**1**], Hahn [**S**], and Bezuglyi and Golodets [**S**]. Connes, Feldman, and Weiss [**S**] formulate the above notion of amenability in terms of an operator that is invariant (in a suitable sense) under partial transformations. We shall be content to illustrate some of the ideas involved in this area by concentrating on the Murray–von Neumann construction and showing how the amenability of the resulting von Neumann algebra implies that the G-space involved is amenable. Our account is based on Zimmer [**3**]. We require some preliminaries.

Let X be a standard Borel space: this means that X is equipped with a σ-algebra $\mathscr{B}$ of subsets, where $\mathscr{B}$ is the Borel σ-algebra generated by the open sets for some complete, separable metric on X. Let G be a countable, discrete group acting invertibly and measurably on the right on X. We further suppose that μ is a probability measure on $(X, \mathscr{B})$ which is quasi-invariant in the sense that $\mu(Ea) = 0$ whenever $\mu(E) = 0$ ($E \in \mathscr{B}$, $a \in G$). For $a \in G$, we define the measure $a\mu$ by $a\mu(E) = \mu(Ea)$. By quasi-invariance, $a\mu \ll \mu$, and by the Radon-Nikodým Theorem and the countability of G, there is a Borel function $r\colon X \times G \to [0, \infty)$ such that $(d(a\mu)/d\mu)(x, a) = r(x, a)$. We can take $r(x, e) = 1$ for all $x \in X$. The quasi-invariance of μ yields a natural left Banach G-space structure on $L_\infty(X)$: $a\phi(x) = \phi(xa)$. Note that

$$\int \phi \, d(a\mu) = \int (a^{-1}\phi) \, d\mu = \int \phi(x) r(x, a) \, d\mu(x), \tag{1}$$

from which it follows that r satisfies the cocycle identity: $r(x, ab) = r(x, a)r(xa, b)$ a.e. x. There exist unitary representations U, V of G and isometric *-representations M, N of $L_\infty(X)$ on $\mathfrak{H} = L_2(X \times G) = L_2(X) \otimes l_2(G)$ given by

$$U_a f(x, b) = f(xa, ba) r(x, a)^{1/2}, \qquad V_a f(x, b) = f(x, a^{-1}b),$$

$$(M_\phi f)(x, b) = \phi(x) f(x, b), \qquad (N_\phi f)(x, b) = \phi(xb^{-1}) f(x, b)$$

$(a, b \in G, x \in X, \phi \in L_\infty(X), f \in \mathfrak{H})$. It is readily checked that

$$U_a M_\phi U_{a^{-1}} = M_{a\phi}, \qquad U_a F U_{a^{-1}} = aF \tag{2}$$

where $F \in L_\infty(X \times G)$ is identified as a multiplication operator on $\mathfrak{H}$ and $aF(x, b) = F(xa, ba)$. Note that M_ϕ "$=$" $\phi \otimes 1$ under this identification. The von Neumann algebra generated the operators V_a, N_ϕ is denoted by L, while that generated by the operators U_a, M_ϕ is denoted by R. If $J \in \mathbf{U}(\mathfrak{H})$ is given by $Jf(x,b) = f(xb^{-1}, b^{-1})r(x, b^{-1})^{1/2}$, then $J^2 = I$, $JV_aJ = U_a$, $JN_\phi J = M_\phi$, so that J implements a spatial isomorphism between L and R. Obviously, therefore, L has Property P $\Leftrightarrow$ R has Property P. It is easily checked that the elements of L commute with the elements of R, and indeed, by [**D1**, Part 1, Chapter 9, Exercise 1], R is the commutant L^c of L. (A tedious calculation shows that the representations considered by Dixmier are the "left action" versions of, and are equivalent to, those above.)

A simple, but important, example of the above situation occurs when X is a singleton $\{x_0\}$ and $\mu = \delta_{x_0}$. In this case, L, R are the von Neumann algebras generated by the left and right regular representations of G [**D2**, (13.1.6)].

We now introduce Zimmer's notion of an amenable G-space, motivation for which comes from Mackey's theory of virtual groups (Mackey [**S**]). Let E be a separable Banach space and $I(E)$ the group of isometric, invertible linear operators on E equipped with the strong operator topology. (So $T_\delta \to T$ in $I(E) \Leftrightarrow \|T_\delta\xi - T\xi\| \to 0$ for all $\xi \in E$.) With respect to its resulting Borel structure, the group operations in $I(E)$ are measurable. Since E is separable, the unit ball E'_1 of the dual space E' is a compact metric space in its weak* topology, and we give E'_1 the resulting Borel structure. Let $H(E'_1)$ be the group of weak* continuous homeomorphisms of E'_1. A cocycle $\gamma\colon X \times G \to I(E)$ is a Borel measurable map satisfying the cocycle identity: $\gamma(x, ab) = \gamma(x,a)\gamma(xa,b)$ $(x \in X, a, b \in G)$ for a.e. x. Every such cocycle γ induces a cocycle $\gamma^*\colon X \times G \to H(E')$ given by $\gamma^*(x,a) = (\gamma(x,a)^*)^{-1}|_{E'_1}$. A set-valued function $x \to A_x$ on X is called a *Borel field* (*of weak* compact, convex subsets* of E'_1) if every A_x is a weak* compact, convex subset of E'_1 and $\tilde{A} = \{(x,\alpha)\colon x \in X,\ \alpha \in A_x\}$ is a Borel subset of $X \times E'_1$. With γ as above, the field $x \to A_x$ is called γ-*invariant* if $\gamma^*(x,a)A_{xa} = A_x$ for a.e. x. A Borel function $\rho\colon X \to E'_1$ is called a γ-*invariant section* if for all a, $\gamma^*(x,a)\rho(xa) = \rho(x)$ a.e. Finally X is called an *amenable G-space* if for every such E, every cocycle $\gamma\colon X \times G \to I(E)$, and every γ-invariant Borel field $x \to A_x$ of weak* compact, convex subsets of E'_1, there exists a γ-invariant section ρ with $\rho(x) \in A_x$ a.e.

The amenable G-space property can be interpreted as a sophisticated extension of amenability fixed-point theorems. To illustrate this, consider the simple case (above) in which X is a singleton. Then the cocycle γ can be regarded as a mapping from G into $I(E)$, and the cocycle equality translates into an isometric G-action on E: $a\xi = \gamma(a)\xi$. A γ-invariant Borel field is just a G-invariant, weak* compact, convex subset K of E'_1, and a γ-invariant section is just a fixed-point of K. With Problem 2-14 in mind, we see that *a singleton is an amenable G-space* $\Leftrightarrow$ *G is amenable.* However note that in general, nonamenable groups can have amenable G-spaces (Zimmer [**6**]).

We now discuss the theorem of Zimmer alluded to above (Zimmer [**1**], [**3**]).
The G-space X is amenable if and only if the von Neumann algebra L has Property P.
[We shall give the proof of only one of the implications involved. Suppose that L (and therefore R) has Property P. We shall show that X is an amenable G-space.

Since L has Property P and $R = L^c$, there exists, by (C)(i), a norm one, linear, positive projection P from $\mathbf{B}(\mathfrak{H})$ onto R such that for all $T \in \mathbf{B}(\mathfrak{H})$, we have (a) $P(T)$ is in the weak operator closure of $\mathrm{co}\{UTU^*: U \in \mathbf{U}(L)\}$, and (b) $P(S_1TS_2) = S_1P(T)S_2$ for all S_1, $S_2 \in R$. We will use P to produce a norm one, linear, positive, unit-preserving map $\sigma: L_\infty(X \times G) \to L_\infty(X)$ such that (c) $\sigma(F(\phi \otimes 1)) = \sigma(F)\phi$ and (d) $\sigma(aF) = a\sigma(F)$ for all $F \in L_\infty(X \times G)$, $a \in G$, and $\phi \in L_\infty(X)$. The map σ can be produced by disintegration. The following argument produces σ directly and applies more generally.

Let D be the space of operators $T \in \mathbf{B}(\mathfrak{H})$ such that whenever $f_1, \ldots, f_n$, $g_1, \ldots, g_n \in L_2(X)$ are such that $\sum_{i=1}^n f_i\overline{g}_i = 0$, then $\sum_{i=1}^n (T(f_i \otimes e), g_i \otimes e) = 0$. Obviously, D is weak operator closed in $\mathbf{B}(\mathfrak{H})$. We claim that $R \subset D$. By (2), we need only show that each $U_aM_\phi \in D$. Indeed, if f_i, g_i are as above, then $U_aM_\phi(f_i \otimes e)(x, b) = \phi(xa)f_i(xa)\delta_e(ba)r(x, a)^{1/2}$, so that

$$\sum_{i=1}^n (U_aM_\phi(f_i \otimes e), g_i \otimes e) = \int \phi(xa)r(x,a)^{1/2}\delta_e(a) \sum_{i=1}^n f_i(xa)\overline{g_i(x)}\, d\mu(x) = 0,$$

and $U_aM_\phi \in D$. So $R \subset D$. It follows that for $T \in R$, if $h \in L_1(X)$ and we write $h = \sum_{j=1}^m f'_j\overline{g}'_j$, $f'_j, g'_j \in L_2(X)$, then the complex number $\sum_{j=1}^m (T(f'_j \otimes e), g'_j \otimes e)$ is independent of f'_j, g'_j. We define $\theta(T)(h)$ to be this common value. Clearly, $\theta(T)$ is a linear functional on $L_1(X)$ with (e) $\theta(T)(f\overline{g}) = (T(f \otimes e), g \otimes e)$. Writing $h = |h|\alpha$ where $\alpha: X \to \mathbb{T}$ is measurable, we have

$$\begin{aligned} |\theta(T)(h)| &= |(T(|h|^{1/2} \otimes e), |h|^{1/2}\overline{\alpha} \otimes e)| \\ &\le \|T\| \| |h|^{1/2} \|_2^2 = \|T\| \|h\|_1. \end{aligned}$$

So $\theta(T) \in L_\infty(X) = L_1(X)'$ and $\|\theta(T)\| \le \|T\|$. Define $\sigma(F) = \theta(P(F))$ $(F \in L_\infty(X \times G))$. It is easy to check that σ is a norm one, linear, positive, unit-preserving map from $L_\infty(X \times G)$ into $L_\infty(X)$. The equality (c) follows using (b), the fact that $\phi \otimes 1 = M_\phi \in R$, and (e). (Note that

$$(P(F)M_\phi(f \otimes e), g \otimes e) = (P(F)(\phi f \otimes e), g \otimes e).)$$

To prove (d), apply (2) and (e) to obtain

$$\sigma(aF)(f\overline{g}) = (P(F)U_{a^{-1}}(f \otimes e), U_{a^{-1}}(g \otimes e)).$$

Now note that

$$U_{a^{-1}}(k \otimes e) = (a^{-1}k)r_{a^{-1}}^{1/2} \otimes \delta_a = V_a((a^{-1}k)r_{a^{-1}}^{1/2} \otimes \delta_e),$$

where $r_{a^{-1}}(x) = r(x, a^{-1})$. Finally, use the fact that the unitary V_a commutes with $P(F)$ ($\in R$), together with (1), to obtain (d).

Now let γ be a cocycle and $x \to A_x$ a γ-invariant Borel field as in the definition of *amenable G-space.* We will use the map σ above to construct a γ-invariant section r. Since $X \times E_1'$ is a standard Borel space and $\tilde{A}$ is a Borel subset of $X \times E_1'$, a well-known result [**D1**, Appendix 5] shows that there exists a μ-measurable map $c: X \to E_1'$ such that $c(x) \in A_x$ a.e. Define $S: X \times G \to E_1'$ by $S(x,a) = \gamma^*(x, a^{-1})c(xa^{-1})$. Note that $S(x,a) \in A_x$ a.e. Using the measurability of γ and c, we can define, for $\xi \in E$, a function $\langle \xi, S\rangle \in L_\infty(X \times G)$ by $\langle \xi, S\rangle(x,a) = S(x,a)(\xi)$. Define $\Phi: E \to L_\infty(X)$ by $\Phi(\xi) = \sigma(\langle \xi, S\rangle)$. It is easily checked that $\|\langle \xi, S\rangle\|_\infty \le \|\xi\|$, and that Φ is linear with $\|\Phi\| \le 1$. Then Φ^*, restricted to $L_1(X)$ (or rather $L_1(X)^\wedge$), maps $L_1(X)$ into E', and applying [**DS**, V1.8.6] there exists a map $\rho: X \to E'$ such that for $\xi \in X$, the map $x \to \rho(x)(\xi)$ is in $L_\infty(X)$, $1 \ge \|\Phi\| = \text{ess sup}_{x\in X} \|\rho(x)\|$, and $\Phi(\xi)(x) = \rho(x)(\xi)$ a.e. x. In particular, $\rho(x) \in E_1'$ a.e. We claim that ρ is the required γ-invariant section so that X is G-amenable. We have to show (f) for each $a \in G$, $\gamma^*(x,a)\rho(x,a) = \rho(x)$ a.e. and (g) $\rho(x) \in A_x$ a.e.

We will prove (f) first. For any essentially bounded, measurable function $\psi: X \to E$, define $\langle \psi, S\rangle \in L_\infty(X \times G)$ by $\langle \psi, S\rangle(x,a) = S(x,a)(\psi(x))$. If $\xi \in X$, $\phi \in L_\infty(X)$, and $\psi(x) = \phi(x)\xi$, then using (c),

$$\begin{aligned}\sigma(\langle \psi, S\rangle)(x) &= \sigma(\langle \xi, S\rangle(\phi \otimes 1))(x) = \sigma(\langle \xi, S\rangle)(x)\phi(x)\\ &= \rho(x)(\psi(x)) \quad \text{a.e.}\end{aligned}$$

By approximating general ψ by linear combinations of functions of the form $x \to \phi(x)\xi$, we obtain that

$$\sigma(\langle \psi, S\rangle)(x) = \rho(x)(\psi(x)) \quad \text{a.e.} \tag{3}$$

Applying (3) with $\psi(x) = \gamma(xa^{-1}, a)^{-1}\xi$ and using (d),

$$\begin{aligned}\gamma^*(x,a)\rho(xa)(\xi) &= \rho(xa)(\gamma(x,a)^{-1}\xi) = \rho(xa)(\psi(xa))\\ &= \sigma(\langle \psi, S\rangle)(xa) = \sigma(a\langle \psi, S\rangle)(x).\end{aligned} \tag{4}$$

Now for $x \in X$, $b \in G$, we have, using the cocycle identity for γ,

$$\begin{aligned}a\langle \psi, S\rangle(x,b) &= \langle \psi, S\rangle(xa, ba) = S(xa, ba)(\psi(xa))\\ &= \gamma^*(xa, a^{-1}b^{-1})c(xb^{-1})(\gamma(x,a)^{-1}\xi)\\ &= c(xb^{-1})((\gamma(x,a)\gamma(xa, a^{-1}b^{-1}))^{-1}\xi)\\ &= c(xb^{-1})(\gamma(x, b^{-1})^{-1}\xi) = \langle \xi, S\rangle(x,b).\end{aligned}$$

So using (4), we obtain (f):

$$\gamma^*(x,a)\rho(xa)(\xi) = \sigma(\langle \xi, S\rangle)(x) = \rho(x)(\xi) \quad \text{a.e.}$$

We now turn to (g). If K is a weak* compact, convex subset of E' and $\alpha \in E' \sim K$, then α and K are separated in the sense that there exists $\xi \in E$ such that $\operatorname{Re}\hat{\xi}(\alpha) < \inf\{\operatorname{Re}\hat{\xi}(\beta) : \beta \in K\}$. From Problem 2-30, there exists a sequence

$\{b_n\}$ of Borel functions from X into E_1' such that for a.e. x, $\{b_n(x)\colon n \geq 1\}$ is weak* dense in A_x. Let Y be a countable dense subset of E, $\xi \in Y$, $q \in \mathbb{Q}$, and

$$X_0 = \left\{x \in X\colon \inf_{n\geq 1} \operatorname{Re}\hat{\xi}(b_n(x)) \geq q\right\}.$$

If, for some $x' \in X$, $\rho(x') \notin A_{x'}$, then we can separate $\rho(x')$ from $A_{x'}$ using some $\xi \in Y$, and noting that $\inf_{n\geq 1} \operatorname{Re}\hat{\xi}(b_n(x)) = \inf\{\hat{\xi}(\beta)\colon \beta \in A_x\}$ for a.e. $x \in X$, we will have $\rho(x) \in A_x$ a.e. provided we can show that $\{x \in X_0\colon \hat{\xi}(\rho(x)) \geq q\}$ has null complement in X_0. To this end, we can obviously suppose that $\mu(X_0) > 0$. Since $S(x,a) \in A_x$ a.e., we have for almost every $x \in X_0$, $\operatorname{Re}\langle \xi, S\rangle(x,a) = \operatorname{Re}\hat{\xi}(S(xa)) \geq q$. Using (c) and the fact that σ is positive and unit preserving, we have $\operatorname{Re}\sigma(\langle\xi, S\rangle\chi_{X_0} \otimes 1) \geq \sigma(q\chi_{X_0} \otimes 1) = q\chi_{X_0}$, so that for a.e. $x \in X_0$, $\operatorname{Re}\hat{\xi}(\rho(x)) = \operatorname{Re}\sigma(\langle\xi, S\rangle)(x) \geq q$. This gives (g).]

In the case where X is a singleton, the above result applies even when G is not countable, and becomes a well-known result of Schwartz [**1**]: *A discrete group G is amenable if and only if* VN(G) *has Property P.* (Schwartz extended this result in Schwartz [**S**].) It is natural to ask if the above result in Schwartz [**1**] is valid for general locally compact groups G. It follows from (C) that G amenable $\Rightarrow$ VN(G) has Property P. Since there exist nonamenable groups of Type 1 (for example, SL$(2,\mathbb{R})$), the converse is certainly not true. The quest for the missing ingredient that, together with Property P, will ensure that G is amenable leads to the study of a weaker kind of invariant mean on G that is particularly relevant for operator algebras. The means in question are inner invariant means.

(H) *Inner invariant means.* A mean m on $L_\infty(G)$ is called *inner invariant* if $m(x\phi x^{-1}) = m(\phi)$ for all $\phi \in L_\infty(G)$, $x \in G$. Inner amenability is definitely an L_∞-phenomenon—the mean δ_e is *always* inner invariant on $C(G)$, and since there exist familiar groups (such as SL$(2,\mathbb{R})$) that are not inner amenable, there is no hope of an inner invariant version of (1.10) holding! Every invariant mean is inner invariant, so that all amenable locally compact groups are inner amenable. In general, inner amenability is much weaker than amenability.

With our definition, every discrete group G is inner invariant with δ_e an inner invariant mean. E. Effros, whose paper [**1**] originated the study of inner invariant means, defines a discrete group to be inner amenable if there exists an inner invariant mean m on $l_\infty(G)$ with $m \neq \delta_e$. This fits in well with Property Γ of von Neumann algebras (Problem 2-33). However for locally compact groups in general, our definition of inner amenability seems preferable. We shall say that a discrete group G is *trivially inner amenable* if δ_e is the *only* inner invariant mean on $l_\infty(G)$.

There is now substantial literature on inner amenability—see, for example, Akemann and Walter [**3**], M. Choda [**S1**], [**S2**], H. and M. Choda [**S**], Bédos and de la Harpe [**S**], Lau and Paterson [**S2**], Losert and Rindler [**S2**], [**S3**], Paschke [**1**], Pier [**S**], and Yuan [**S**]. Some of these results are covered in Problems 2-32–2-34. Losert and Rindler [**S2**] (and Grosser, Losert, and Rindler [**S**]) show that a connected locally compact group G is inner amenable if and only if G is amenable. (The class of such groups is determined in (3.8).) More generally, amenability and inner amenability are equivalent for G in the class $\mathscr{A}$ of (1.31). Indeed, if $G \in \mathscr{A}$, then the amenability of $C^*(G)''$ implies that $\mathrm{VN}(G)$ is also amenable, and the result below applies. This result shows that inner amenability is the "missing ingredient" alluded to above.

(Lau and Paterson). *The locally compact group G is amenable if and only if* $\mathrm{VN}(G)$ *has Property P and G is inner amenable.*

[If G is amenable, then G is inner amenable, and by (C), $\mathrm{VN}(G)$ has Property P. Conversely, suppose that $\mathrm{VN}(G)$ has Property P and that G is inner amenable. Let $m \in \mathfrak{M}(G)$ be inner invariant. Then $xm = mx$ for all $x \in G$, and the usual (0.8) theme gives a net $\{f_\delta\}$ in $P(G)$ such that $\|x * f_\delta - f_\delta * x\|_1 \to 0$ for each $x \in G$. Let $g_\delta = f_\delta^{1/2}$ and note that $\|g_\delta\|_2 = 1$. Let π_r be the right regular representation of G on $L_2(G)$: so $\pi_r(x)g(t) = \Delta(x)^{1/2}g(tx)$ $(x, t \in G, g \in L_2(G))$. Then $(x * f_\delta)^{1/2} = \pi_2(x)g_\delta$, $(f_\delta * x)^{1/2} = \pi_r(x^{-1})g_\delta$, and (cf. (4.3(1)))

$$\|\pi_2(x)g_\delta - \pi_r(x^{-1})g_\delta\|_2 \le \|x * f_\delta - f_\delta * x\|_1^{1/2} \to 0.$$

Let ω_δ be the state on $\mathrm{VN}(G)$ given by $\omega_\delta(T) = (Tg_\delta, g_\delta)$. Let $H = \mathbf{U}(\mathrm{VN}(G)^{\mathrm{c}})$, $X = \mathbf{B}(\mathfrak{H})_*$, and let $X(H)$ be as in (C). Note that $\mathrm{VN}(G)^{\mathrm{c}}$ has Property P since $\mathrm{VN}(G)$ has Property P. (This is easy to see directly using the J-map of [**D1**, Part 1, Chapter 5].) For each δ, define $\Psi_\delta\colon L_\infty(G) \to X(H)$ by $\Psi_\delta(\phi)(U) = (UL_\phi U^* g_\delta, g_\delta)$, where L_ϕ is the multiplication operator on $L_2(G)$ associated with ϕ. By (C),(ii) there exists a left invariant mean m for H on $X(H)$. Then $n_\delta = m \circ \Psi_\delta \in \mathfrak{M}(G)$. We can suppose that $n_\delta \to n$ weak* for some $n \in \mathfrak{M}(G)$. We claim that n is a right invariant mean on $L_\infty(G)$ so that G is amenable. To this end, let $x \in G$, $\varepsilon > 0$, and let δ_0 be such that $\|\pi_2(x^{-1})g_\delta - \pi_r(x)g_\delta\|_2 < \varepsilon$ whenever $\delta \ge \delta_0$. For $\phi \in L_\infty(G)$, $U \in H$, we have $L_{\phi x^{-1}} = \pi_2(x)L_\phi\pi_2(x^{-1})$, and so, since U and $\pi_2(x)$ commute,

$$\begin{aligned}\Psi_\delta(\phi x^{-1})(U) &= (UL_{\phi x^{-1}}U^* g_\delta, g_\delta) = (U\pi_2(x)L_\phi\pi_2(x^{-1})U^* g_\delta, g_\delta)\\ &= (UL_\phi U^*\pi_2(x^{-1})g_\delta, \pi_2(x^{-1})g_\delta).\end{aligned}$$

Also $\Psi_\delta(\phi)\pi_r(x^{-1})(U) = (UL_\phi U^*\pi_r(x)g_\delta, \pi_r(x)g_\delta)$ and so by a simple triangular inequality argument, for $\delta \ge \delta_0$,

$$\begin{aligned}&|[\Psi_\delta(\phi x^{-1}) - \Psi_\delta(\phi)\pi_r(x^{-1})](U)|\\ &\qquad \le 2\|\phi\|\|\pi_2(x^{-1})g_\delta - \pi_r(x)g_\delta\|_2 \le 2\|\phi\|\varepsilon.\end{aligned}$$

So $\|\Psi_\delta(\phi x^{-1}) - \Psi_\delta(\phi)\pi_r(x^{-1})\|_\infty \to 0$, and since m is H-invariant and $\pi_r(G) \subset H$, we have $n(\phi x^{-1}) = n(\phi)$ as required.]

(2.36) Invariant means and almost periodicity. Let G be a locally compact group. If G is not amenable, then there is no left invariant mean on $\mathrm{U}(G)$. When we go to a smaller space, however, there is always the chance that left invariant means might exist. An interesting example of this possibility occurs with the space $\mathrm{WP}(G)$ of weakly almost periodic functions on G—this space *always* has a (unique) invariant mean. This result is a consequence of the Ryll-Nardzewski Fixed-Point Theorem, which goes as follows.

(THE RYLL-NARDZEWSKI THEOREM). *Let X be a locally convex space and K be a convex, weakly compact subset of X. Let Σ be a semigroup of weakly continuous, affine maps from K into K such that Σ is* distal *in the sense that whenever ξ, $\eta \in K$ with $\xi \neq \eta$, then $0 \notin \{(T\xi - T\eta): T \in \Sigma\}^-$. Then there exists $\xi_0 \in K$ such that $T\xi_0 = \xi_0$ for all $T \in \Sigma$.*

A recent short proof of this theorem is given by Namioka [**S**]. Let S be a separately continuous semigroup. References for the following theorem are de Leeuw and Glicksberg [**S**], Burckel [**S**], and Berglund, Junghenn, and Milnes [**1**]. Recall ((2.11)) that AP(S), WP(S) are invariant, left introverted, closed subspaces of $C(S)$.

(i) *The spaces* WP(S) *and* AP(S) *are C^*-subalgebras of $C(S)$.*

(ii) *The carrier space Y [Z] of* WP(S) [AP(S)] *is a separately [jointly] continuous semigroup under the Arens product* ((2.8)), *and if $\phi \in$* WP(S) [AP(S)], *then ϕS is relatively compact in the weak [norm] topology.*

(iii) *If G is a locally compact group, then*

$$\mathfrak{L}(\mathrm{WP}(G)) = \mathfrak{I}(\mathrm{WP}(G)) = \mathfrak{R}(\mathrm{WP}(G)),$$

and there is exactly one element in each. A similar result holds with AP(G) *in place of* WP(G).

[(i) Let ϕ, $\psi \in \mathrm{WP}(S)$. In order to show that $\phi\psi \in \mathrm{WP}(S)$, it is sufficient, by the Eberlein-Šmulian Theorem, to show that if $\{s_n\}$ is a sequence in S with $s_n\phi \to \phi'$ and $s_n\psi \to \psi'$ weakly in $C(S)$, then $s_n(\phi\psi) \to \phi'\psi'$ weakly. Let X be the carrier space of $C(S)$, and identify $C(S)$ with $C(X)$ and $C(S)'$ with $M(X)$. If $x \in X$, then

$$s_n(\phi\psi)(x) = (s_n\phi(x))(s_n\psi(x)) \to \phi'(x)\psi'(x) = (\phi'\psi')(x)$$

so that $s_n(\phi\psi) \to \phi'\psi'$ pointwise on X. Now apply Lebesgue's Dominated Convergence Theorem to obtain that $s_n(\phi\psi) \to \phi'\psi'$ weakly as required.

Since the product of two norm compact subsets of $C(S)$ is also norm compact, it follows that AP(S) is an algebra.

Trivially, both WP(S) and AP(S) are closed under complex conjugation so that they are C^*-algebras.

(ii) Let Y be the carrier space of the C^*-algebra WP(S). If p, $q \in Y$, then since $q\phi\psi = (q\phi)(q\psi)$ $(\phi, \psi \in \mathrm{WP}(S))$, it follows that $pq \in Y$. Thus Y is a semigroup. If $q_\delta \to q$ in Y and $\phi \in \mathrm{WP}(S)$, then $q_\delta\phi \to q\phi$ pointwise on S. But since the pointwise and weak topologies coincide on the (weakly compact) weak closure of co($S\phi$), it follows that $q_\delta\phi \to q\phi$ weakly. Hence the maps $q \to pq$ are continuous on Y. Of course, the maps $p \to pq$ are also continuous, so that Y is a separately continuous semigroup (the *weakly almost periodic compactification of S*). It now follows that for each $\phi \in \mathrm{WP}(S)$, the set $A = \{\hat{\phi}p: p \in Y\}$ is pointwise compact and bounded in $C(Y)$. Now apply a result of Grothendieck

[**DS**, IV.6.14] to obtain that A is weakly compact in $C(Y)$. Now for $s \in S$, we have $\hat{s} \in Y$, and so $(\phi S)^\wedge = \{\hat{\phi}\hat{s}: s \in S\} \subset A$. It follows that ϕS is relatively compact in the weak topology of $C(S)$ as required.

In the corresponding proof for AP(S), Z turns out to be a jointly continuous semigroup, and the relative norm compactness of ϕS follows easily. In this case, Z is the *almost periodic compactification of* S.

(iii) Let $\phi \in$ WP(G), and let C be the (compact) weak closure of co$(G\phi)$. For $x \in G$, define $L_x: C \to C$ by $L_x\psi = x\psi$. Let $\Sigma = \{L_x: x \in G\}$. Then Σ is a semigroup of weakly continuous, affine maps on C. Further, if $\psi_1, \psi_2 \in C$ with $\psi_1 \neq \psi_2$, then $\|L_x\psi_1 - L_x\psi_2\| = \|\psi_1 - \psi_2\|$, so that 0 does not belong to the norm closure of the set $\{T(\psi_1 - \psi_2): T \in \Sigma\}$. By the Ryll-Nardzewski Theorem, we can find $\phi_0 \in C$ with $x\phi_0 = \phi_0$ for all $x \in G$. Obviously, $\phi_0 \in \mathbb{C}1$. Since ϕ_0 is also in the pointwise closure ($= C$) of co$(G\phi)$ and WP(G) is left introverted, (2.13) applies (G discrete) to yield $\mathfrak{L}(\text{WP}(G)) \neq \varnothing$. Let $m \in \mathfrak{L}(\text{WP}(G))$ and suppose that $\alpha 1 \in D$, the weak closure of co(ϕG) in $C(G)$. (Interchanging C and D (using (ii)), we see that D contains a constant function.) Since $m(\text{co}(\phi G)) = \{m(\phi)\}$, it follows that $\alpha = m(\phi)$. Hence there exists exactly one constant function in D. Now D is weakly compact (using (ii)), and interchanging left and right, C contains exactly one constant function. Now apply (2.13(ii)) to give that $\mathfrak{L}(\text{WP}(G)) = \{m\}$. Further since $m(\phi)1$ is the only constant function in $C \cup D$, the "$\mathfrak{R}(B)$" version of (2.13(ii)) gives $\mathfrak{R}(\text{WP}(G)) = \{m\}$. The corresponding proofs for AP(G) are left to the reader.]

Kovacs and Szücs [**1**], generalising the notion of a finite von Neumann algebra, introduced the notion of a G-finite von Neumann algebra. If A is a von Neumann algebra and G is a group of $*$-automorphisms on A, then A is called *G-finite* if for every $a > 0$ in A, there exists a G-invariant element $\alpha \in A_*$ such that $\alpha(a) > 0$. Lau [**9**] investigates this topic using almost periodicity ideas. See also Størmer [**1**], Doplicher, Kastler, and Størmer [**1**], and Komlósi [**1**].

(2.37) On the range of a left invariant mean. Let S be a left amenable semigroup and $m \in \mathfrak{L}(S)$. The *range* R_m of m is the set $\{m(A): A \subset S\}$. Obviously, $R_m \subset [0,1]$. Granirer [**8**] essentially made the following conjecture: $R_m = [0,1]$ *for all* $m \in \mathfrak{L}(S)$ *if and only if* S *is not n-ELA for any* n. Granirer established the conjecture for all cases except when $S/\approx$ is an infinite, periodic group with the property that each infinite subgroup is *not* locally finite (where $\approx$ is as in (1.24)). Chou [**3**] established the conjecture when S is right cancellative. The truth of the conjecture in general follows using Chou's method.

(i) *Let* X *be a set,* $m \in \mathfrak{M}(X)$ *and suppose that the measure* $\hat{m}$ *on* βX *(defined as in* (2.24)) *is continuous, that is,* $\hat{m}(\{p\}) = 0$ *for all* $p \in \beta X$. *Then* $\{m(A): A \subset X\} = [0,1]$.

[By [**HR1**, (11.44)], if $E \in \mathscr{B}(\beta X)$, then

$$[0, \hat{m}(E)] = \{\hat{m}(F): F \in \mathscr{B}(\beta X), F \subset E\}. \tag{1}$$

Let $B \subset X$. Now βX is zero-dimensional. Further, every open and closed subset of βX is of the form A^-, where $A \subset S$ and A^- is the closure of A in βX.

It follows, using (1) and the regularity of $\hat{m}$, that

$$[0, m(B)] = \{m(A)\colon A \subset B\}^-. \tag{2}$$

Let $\alpha \in (0,1)$ and let $\{\alpha_n\}$ be a sequence in (0,1) such that $\alpha_n \to \alpha$, and for all n,

$$\alpha_{2n-1} > \alpha_{2n+1} > \alpha > \alpha_{2n+2} > \alpha_{2n}. \tag{3}$$

We construct recursively a sequence $\{A_n\}$ of subsets of X such that for all $n \geq 1$,

$$\alpha_{2n+1} < m(A_{2n-1}) < \alpha_{2n-1}, \alpha_{2n} < m(A_{2n}) < \alpha_{2n+2} \tag{4}$$

and

$$A_{2n-3} \supset A_{2n-1} \supset A_{2n} \supset A_{2n-2}. \tag{5}$$

Suppose then that $A_1, \ldots, A_{2k}$ $(k \geq 1)$ have been constructed so that (4) and (5) are valid whenever they make sense for $1 \leq n \leq k$. The following argument easily adapts to give the sets A_1, A_2 to start the recursion. From (4) and (3), we have

$$m(A_{2k-1}) > \alpha_{2k+1} > \alpha_{2k+3} > \alpha > \alpha_{2k+4} > \alpha_{2k+2} > m(A_{2k}). \tag{6}$$

Applying (2) with B replaced by $A_{2k-1} \sim A_{2k}$, and noting that, from (4) and (6), $0 < \alpha_{2k+3} - m(A_{2k}) < \alpha_{2k+1} - m(A_{2k}) < m(A_{2k-1} \sim A_{2k})$, we can find $C \subset A_{2k-1} \sim A_{2k}$ such that

$$\alpha_{2k+3} - m(A_{2k}) < m(C) < \alpha_{2k+1} - m(A_{2k}).$$

Setting $A_{2k+1} = A_{2k} \cup C$, we obtain

$$\alpha_{2k+3} < m(A_{2k+1}) < \alpha_{2k+1}.$$

Similarly, we can find $D \subset A_{2k+1} \sim A_{2k}$ such that $\alpha_{2k+2} - m(A_{2k}) < m(D) < \alpha_{2k+4} - m(A_{2k})$, and setting $A_{2k+2} = A_{2k} \cup D$, we obtain

$$\alpha_{2k+2} < m(A_{2k+2}) < \alpha_{2k+4}.$$

So (4) and (5) are valid whenever they make sense for $1 \leq n \leq k+1$, and the construction of the sets $\{A_n\}$ is completed.

Now let $A = \bigcap_{n=1}^{\infty} A_{2n-1}$. From (5), $A_{2n-1} \supset A \supset A_{2n}$ for all n, so that, using (4), $\alpha_{2n-1} > m(A) > \alpha_{2n}$, and since $\alpha_n \to \alpha$, $m(A) = \alpha$. This completes the proof.]

(ii) *The range R_m of every left invariant mean on a left amenable semigroup S is $[0,1]$ if and only if S is not n-ELA for any n.*

[Suppose that $R_m = [0,1]$ for all $m \in \mathfrak{L}(S)$. If S is n-ELA for some n, then we can find $m \in \mathfrak{L}(S)$ such that $\hat{m} = \frac{1}{n}\sum_{i=1}^{n} \delta_{p_i}$ for some $p_i \in \beta S$, and $R_m = \{r/n\colon 0 \leq r \leq n\}$. A contradiction results. Conversely suppose that S is not n-ELA for any n, and let $m \in \mathfrak{L}(S)$. By Problem 2-35, $\hat{m}$ is continuous. It follows from (i) that $R_m = [0,1]$.]

(iii) Granirer [**8**] showed that for "most" infinite, right cancellative, left amenable semigroups S, there exists a family $\{A(t): t \in [0,1]\}$ of subsets of S such that (a) $A(s) \subset A(t)$ if $s < t$, (b) $m(\chi_{A(t)}) = t$ for all $m \in \mathfrak{L}(S)$. (In particular, $\chi_{A(t)}$ is left almost convergent.) Further results in this direction are obtained by Snell [**S1**], [**S2**].

Problems 2

Throughout G is a locally compact group and S is a discrete semigroup.

1. Let $D(G)$ be as in (2.1). Show that G is amenable if and only if $D(G)$ is *not* norm dense in $L_\infty(G)$, and that if G is amenable, then

$$D(G) + \mathbb{C}1 = \{\phi \in L_\infty(G): m(\phi) = n(\phi) \text{ for all } m, n \in \mathfrak{L}(G)\}.$$

2. Let B be a left introverted subspace of $L_\infty(G)$ which is right invariant for $P(G)$, is closed under complex conjugation and contains 1. Prove that there exists a topologically left invariant mean on B if and only if whenever $\phi \in B$ is real-valued and μ, $\nu \in P(G)$, then

$$\inf_{x \in G} (\phi\mu - \phi\nu)(x) \leq 0.$$

3. Let $\mathfrak{J}_{lt}(G) = \{F \in L_\infty(G)': \mu F = F \text{ for all } \mu \in P(G)\}$. Prove that $\mathfrak{J}_{lt}(G)$ is an L-space and that $\mathfrak{L}_t(G)$ spans $\mathfrak{J}_{lt}(G)$ (cf. (2.31)).

4. Show that G is amenable if and only if whenever $n \geq 1$, $E_1, \ldots, E_n$ are Borel subsets of G, and $x_1, \ldots, x_n, y_1, \ldots, y_n \in G$, then

$$\operatorname*{ess\,inf}_{x \in G} \sum_{i=1}^{n} (\chi_{x_i^{-1}E_i} - \chi_{y_i^{-1}E_i})(x) < 1.$$

5. Show that G is *not* amenable if and only if there exists a real-valued function $\phi \in \mathrm{U_r}(G)$ and, for some n, elements r_i, s_i $(1 \leq i \leq n)$ in G such that

$$\sum_{i=1}^{n} (\phi r_i - \phi s_i) \geq 1.$$

6. Let A be a Banach algebra. Show that the second dual A'' of A is a Banach algebra under the following Arens product: if F_1, $F_2 \in A''$, $f \in A'$, and a, $b \in A$, we define $F_1F_2(f) = F_1(F_2 f)$, where $F_2 f \in A'$ is given by $F_2 f(a) = F_2(fa)$ with $fa(b) = f(ab)$. (There is another Arens product on A'' but this will not concern us.)

Show that the Arens product on $l_1(S)''$ coincides with the Arens product on $l_\infty(S)'$ obtained by regarding $l_\infty(S)$ as a left introverted subspace of itself.

Interpret the Arens product on the second dual of a Banach algebra A as a special case of the canonical product ((2.8)) on the dual of a left introverted subspace B of some $l_\infty(S)$.

7. A mean m on $L_\infty(G)$ is said to be *inversion invariant* if $m = m^*$, where $m^*(\phi) = m(\phi^*)$ and $\phi^*(x) = \phi(x^{-1})$. Show that if G is amenable then G admits an inversion invariant mean in $\mathfrak{I}(G)$.

8. Let $\mu \in M(G)$, F_1, $F_2 \in L_\infty(G)'$, and let $L_\infty(G)' = L_1(G)''$ be given its Arens product. Recall that $L_\infty(G)'$ is an $M(G)$-set ((1.1)). Prove that
(i) $\mu(F_1F_2) = (\mu F_1)F_2$, $(F_1\mu)F_2 = F_1(\mu F_2)$, $(F_1F_2)\mu = F_1(F_2\mu)$;
(ii) $\mathfrak{M}(G)$ is a subsemigroup of $L_\infty(G)'$;
(iii) if $Q \in \mathfrak{I}_{lt}(G)$ (see Problem 2-3), then $F_1Q = F_1(1)Q$.

9. Prove that $\mathrm{AP}(S) \subset \mathrm{U}(S)$, where S is a separately continuous semigroup.

10. Let S be an open, generating subsemigroup of G. Show that if $\mathfrak{L}(\mathrm{U}(S)) \neq \varnothing$, then there exists $m \in \mathfrak{L}_t(G)$ with $m(S) = 1$. Deduce that if S is an open subsemigroup of G and G is amenable, then $\mathfrak{L}(\mathrm{U}(S)) \neq \varnothing$ if and only if S is left reversible. (These are topological versions of (1.27) and (1.28).)

11. Let Y be a left Banach S-space with $\|sy\| \leq \|y\|$ for all $y \in Y$, let $s \in S$, and let $X = Y'$. Suppose that $1 \in X(S)$ and $\mathfrak{L}(X(S)) \neq \varnothing$. Suppose further that $y_0 \in Y$ and that the weak closure C_{y_0} of $\operatorname{co} Sy_0$ is weakly compact in Y. Show that there exists an S-fixed-point in C_{y_0}.

12. Let Y, X, y_0 and C_{y_0} be as in Problem 11. Show that if $m \in \mathfrak{M}(S)$, then $m\hat{y}_0 \in (C_{y_0})^\wedge$.

13. Let B be a left invariant, closed subspace of $l_\infty(S)$ containing 1. Let $\phi \in B$ be such that the weak closure C_ϕ of $\operatorname{co} S\phi$ is weakly compact in B. Deduce that $m\phi \in C_\phi$ $(\subset B)$ for all $m \in \mathfrak{M}(S)$.

14. Let G be σ-compact. Prove that G is amenable if and only if there exists a G-invariant probability measure on every (jointly continuous), separable, compact metric left G-space. Show also that G is amenable if and only if, whenever E is a separable right Banach G-space with jointly continuous, isometric action, then every G-invariant, weak* compact, convex subset of E' contains a fixed-point for G.

15. Let S be a separately continuous, locally compact Hausdorff semigroup and B a left introverted, closed subspace of $C(S)$ containing 1. Prove that B is right invariant for $\mathrm{PM}(S)$ (Problem 1-11) and that every left invariant mean on B is also left invariant for $\mathrm{PM}(S)$. (A version of this result for the case $B = \mathrm{U_r}(G)$ was proved in (1.8).)

16. Let S be separately continuous, and let $\mathscr{G}_\mathrm{e}$ be the family of sets $K \in \mathscr{G}_\mathrm{j}$ ((2.21)) for which the set of maps $k \to sk$ $(s \in S)$ is equicontinuous with respect to the (unique) uniformity for K which gives the topology of K (Kelley [**2**]).
(i) Prove that $\mathfrak{M}((\mathrm{AP}(S)) \in \mathscr{G}_\mathrm{e}$;
(ii) Prove that there exists an S-fixed-point in every member of $\mathscr{G}_\mathrm{e}$ if and only if $\mathfrak{L}(\mathrm{AP}(S)) \neq \varnothing$.

17. Let $(Z, \mathscr{B}, \nu)$ be a σ-finite measure space such that Z is a left G-set with G σ-compact and

(i) $ez = z$ for all $z \in Z$;

(ii) the map $(x, z) \to xz$ from $G \times Z$ into Z is a measurable transformation from $(G \times Z, \mathscr{M}(G) \times \mathscr{B})$ into $(Z, \mathscr{B})$;

(iii) $\nu(xE) = \nu(E)$ for all $x \in G$, $E \in \mathscr{B}$.

(a) Show that the map $f \to f . x$ $(x \in G)$, where $f . x(z) = f(xz)$, is an isometric right Banach space action of G on $L_1(Z)$, and that the dual left action of G on $L_\infty(Z)$ is given by $x . \phi(z) = \phi(x^{-1}z)$.

(b) Let K be a nonempty, weak* compact, convex, G-invariant subset of $L_\infty(Z)$. Prove that if G is amenable, then K contains a G-fixed-point.

18. Let K be a compact, convex subset of a Banach space X and S a left amenable semigroup of nonexpansive maps from K into K (so that $\|sx - sy\| \leq \|x - y\|$ for all $s \in S$, $x, y \in K$). Let M be a compact subset of X. Show

(i) if $|M| > 1$, then there exists $u \in \operatorname{co} M$ such that

$$\sup_{m \in M} \|u - m\| < \operatorname{diam} M;$$

(ii) there is an S-fixed point in K.

19. If S is a separately continuous, left reversible semigroup, show that there exists a left invariant mean on $\mathrm{AP}(S)$.

20. Let B be a norm closed, right invariant subspace of $L_\infty(G)$. The space B is called *invariantly complemented* if there exists a right invariant closed subspace C of $L_\infty(G)$ such that $L_\infty(G) = B \oplus C$ (vector space direct sum).

(i) Prove that B is invariantly complemented if and only if there exists a continuous, linear projection P from $L_\infty(G)$ onto B such that

$$P(\phi x) = P(\phi)x \qquad (\phi \in L_\infty(G),\ x \in G).$$

(ii) Now suppose that B is also a weak* closed, C^*-subalgebra of $L_\infty(G)$.

(a) Show that if $\mu \in P(G)$ and $\{\mu_\delta\}$ is a bounded approximate identity for $L_1(G)$ in $P(G)$, then for each $\phi \in B$, $\phi\mu_\delta \to \phi$, $\mu_\delta\phi \to \phi$ weak*, and $\phi\mu \in B$.

(b) Show that if $B \neq \{0\}$, then $1 \in B$.

(c) Show that if $B \neq \{0\}$, then $N = \{x \in G\colon x\phi = \phi \text{ for all } \phi \in B\}$ is a closed subgroup of G, and $B = \{\phi \in L_\infty(G)\colon x\phi = \phi \text{ for all } x \in N\}$.

(iii) Prove that G is amenable if and only if every right invariant, weak* closed, C^*-subalgebra of $L_\infty(G)$ is invariantly complemented.

21. Show that $\mathbb{Z} = \{p \in \beta\mathbb{Z}\colon \text{the map } q \to pq \text{ is continuous on } \beta\mathbb{Z}\}$.

22. Show that $\mathfrak{L}(G)$ is weakly compact in $L_\infty(G)'$ if and only if $\mathfrak{J}_l(G)$ is finite-dimensional.

23. Let G be amenable as discrete. Show that there exists a finitely additive positive measure μ on $\mathscr{P}(G)$ such that $\mu(xE) = \mu(E)$ for all $x \in G$, $E \subset G$ and the restriction of μ to $\mathscr{M}(G)$ coincides with λ.

24. Let H be a group which is also a locally compact Hausdorff space with separately continuous multiplication. Show that H is a locally compact group.

25. Let K be a complete, non-Archimedean field which is spherically complete. Show that

(a) there exists a maximal invariant subspace B_m of $l_\infty(\mathbb{P}, K)$ containing 1 and such that $\mathfrak{L}(B_m) \neq \varnothing$;

(b) if B_m is maximal in the sense of (a), then $\mathfrak{L}(B_m)$ is a singleton.

In Problems 26–28, A, $\mathfrak{H}$, X, and H are as in (2.35(C)).

26. Let H be given the strong operator topology.

(i) Show that H is a topological group.

(ii) Show that $X(H) \subset \mathrm{U_r}(H)$.

27. Let A be AFD. (So A contains an increasing sequence $\{M_n\}$ of finite-dimensional C^*-algebras with $\bigcup_{n=1}^{\infty} M_n$ strongly dense in A.) Show that $\mathfrak{L}(\mathrm{U_r}(H)) \neq \varnothing$. Deduce that AFD implies Property P. Show that if $A = \mathbf{B}(\mathfrak{H})$, then A is AFD yet $\mathfrak{L}(C(H)) = \varnothing$.

28. Prove that if A is abelian then A has Property P. More generally, show that if A is of Type 1, then A has Property P.

29. Prove that a Type 1, discrete group is amenable.

30. Let $x \to A_x$ be a Borel field of weak* compact, convex subsets of E_1' as in (2.35(G)). Show that there exists a sequence $\{b_n\}$ of Borel functions from X to E_1', such that $\{b_n(x) : n \geq 1\}^- = A_x$ a.e. (Hint: use the von Neumann Selection Theorem.)

31. Let A be a unital, nuclear C^*-algebra of operators on a Hilbert space $\mathfrak{H}$. Show that A^{c} is injective. Deduce that A'' is injective.

32. (i) Show that F_2 is trivially inner amenable.

(ii) Let π be the "inner regular" unitary representation of a discrete group G on $l_2(G)$: so $\pi(x)f(t) = f(x^{-1}tx)$. Let $C_\pi(G)$ be the C^*-algebra generated by $\pi(G)$ and P_e the orthogonal projection from $l_2(G)$ into $\mathbb{C}e$. Show that G is not trivially inner amenable $\Leftrightarrow P_e \notin C_\pi(G)$.

33. Let G be a discrete group. Prove that G is *not* trivially inner amenable if $\mathrm{VN}(G)$ has Murray and von Neumann's Property Γ, that is, given $T_1, \ldots, T_n \in \mathrm{VN}(G)$ and $\varepsilon > 0$, there exists $U \in \mathbf{U}(\mathrm{VN}(G))$ such that $\tau(U) = 0$ and $\|UT_jU^* - T_j\|_2 < \varepsilon$ $(1 \leq j \leq n)$, where $\tau(T) = (Te, e)$ and $\|T\|_2 = \tau(T^*T)^{1/2}$ $(T \in \mathbf{B}(l_2(G)))$.

34. Let G be a locally compact group, and let $\pi_\infty : G \to \mathbf{B}(L_\infty(G))$ be given by $\pi_\infty(x)\phi = x\phi x^{-1}$. Let A_∞^{c} be the commutant of $\pi_\infty(G)$ in $\mathbf{B}(L_\infty(G))$. Prove that G is inner amenable if and only if A_∞^{c} contains a nonzero compact operator.

35. Let S be a semigroup.

(i) Show that there exists $m \in \mathfrak{L}(S)$ with $\hat{m}$ not continuous if and only if βS contains a finite left ideal [group];

(ii) Prove that if S is left amenable, then S is n-ELA ((2.30)) if and only if n is the smallest positive integer with the property that whenever $F \in \mathscr{F}(S)$, then there exists $E \subset S$ with $|E| \leq n$ and $FE \subset E$;

(iii) Show that S is n-ELA if and only if S is left reversible and $S/\approx$ ((1.24)) is a group of order n;

(iv) Give examples of n-ELA semigroups.

36. Prove that S is ELA if

(i) S is a semilattice;

(ii) S is the set of transformations s on an infinite set X for which the set $\{x \in X : sx \neq x\}$ is finite;

(iii) S is the semigroup of all nonempty, countable subsets of a group G with multiplication given by $(A, B) \to AB$;

(iv) S is the Cartesian product of a family of ELA semigroups;

(v) $S = \mathbb{P} \times \mathbb{P}$ with multiplication given by

$$(m_1, n_1)(m_2, n_2) = (m_2, n_2) = (m_2, n_2)(m_1, n_1) \qquad (m_1 < m_2),$$
$$(m_1, n_1)(m_1, n_2) = (m_1, n_1 + n_2).$$

37. Prove that if S is ELA, then

$$\operatorname{Ext} \mathfrak{L}(S) = \{p \in \beta S : Sp = \{p\}\}.$$

38. Prove that a subset E of a left amenable semigroup S is left thick if and only if the closure $\hat{E}$ of E in βS contains a left ideal of βS.

39. (An amenability fixed subspace theorem.) Let X be a Banach space and H a closed linear subspace of X of codimension $n < \infty$. Suppose that $Y \subset X$ is such that $Y \cap (\xi + H)$ is a compact convex set for all $\xi \in X$ and that Y contains an n-dimensional linear subspace of X. Show that if $G \subset \mathbf{B}(X)$ is a group of invertible transformations on X that is amenable (as a discrete group) and that leaves Y invariant, then there exists an n-dimensional subspace L of X with $L \subset Y$ and $GL = L$.

CHAPTER 3

Free Groups and the Amenability of Lie Groups

(3.0) Introduction. The simplest example of a nonamenable group is the free group F_2 on two generators. In this chapter, we explore the extent to which nonamenability is related to the presence of F_2. We concentrate our discussion on groups that are either connected or "nearly" connected, leaving the discrete case (von Neumann's Conjecture) to (4.30)–(4.33). The main results of the chapter are the theorems (3.8) and (3.9). The result (3.8) characterises amenable, almost connected groups. (A locally compact group G is almost connected if G/G_e is compact.) The result (3.9) determines which connected Lie groups are amenable as discrete groups. These results are proved by a reduction process, the last steps of which involve considering the Lie groups $\mathrm{PSL}(2,\mathbb{R})$ and $\mathrm{SU}(2)$. The reduction process requires the introduction of the *radical* of a locally compact group ((3.7)), and uses an important theorem ((3.4)) on the identity component of the automorphism group of a compact group. The cases of $\mathrm{PSL}(2,\mathbb{R})$ and $\mathrm{SU}(2)$ are dealt with using (3.2): $\mathrm{PSL}(2,\mathbb{R})$ contains F_2 as a discrete subgroup, and $\mathrm{SU}(2)$ contains F_2 as a subgroup.

The following simple lemma is useful when discussing the existence of free subgroups in a group.

(3.1) LEMMA. *Let G and G' be locally compact groups and $\Phi\colon G \to G'$ a continuous homomorphism. If Φ is surjective and G' contains F_2 as a [discrete] subgroup, then so also does G. If G contains F_2 as a subgroup and* $\ker\Phi \subset Z(G)$, *the centre of G, then G' contains F_2 as a subgroup.*

PROOF. Suppose that Φ is surjective and that G' contains a subgroup H' that is a free group on two generators u', v'. Let $u, v \in G$ be such that $\Phi(u) = u'$, $\Phi(v) = v'$. Then the subgroup of G generated by u and v is free and, by the continuity of Φ, is discrete in G if H' is discrete in G'.

Now suppose that $\ker\Phi \subset Z(G)$ and that G contains a subgroup H which is free on two generators. Then

$$H \cap Z(G) \subset Z(H) = \{e\}$$

so that $\Phi\,|_H$ is one-to-one. Hence the subgroup $\Phi(H)$ of G' is isomorphic to F_2. $\square$

B references in the following are references to Appendix B. Recall (B19) that if $n \geq 1$ and $\mathbb{F}$ is either $\mathbb{R}$ or $\mathbb{C}$, then the Lie groups $\mathrm{SL}(n, \mathbb{F})$ and $\mathrm{SU}(n)$ are defined as follows:

$$\mathrm{SL}(n, \mathbb{F}) = \{A \in \mathrm{GL}(n, \mathbb{F})\colon \det A = 1\},$$
$$\mathrm{SU}(n) = \{A \in \mathrm{SL}(n, \mathbb{C})\colon A^*A = I = AA^*\},$$

where $\det A$ and A^* are the determinant and adjoint of A respectively. Of course, $\mathrm{SU}(n)$ is a compact Lie group.

We shall also be concerned with the group $\mathrm{PSL}(2, \mathbb{R})$ of linear fractional transformations

$$z \longrightarrow \frac{az + b}{cz + d}$$

defined on $\mathbb{C}^\infty$ such that $a, b, c, d \in \mathbb{R}$ and $(ad - bc) = 1$. It is elementary that $\mathrm{PSL}(2, \mathbb{R})$ is the group of linear fractional transformations preserving the upper half plane.

Now, in a natural way, $\mathrm{PSL}(2, \mathbb{R})$ can be identified with $\mathrm{SL}(2, \mathbb{R})/\{-I, I\}$. Indeed, the map Q, where

$$Q\left(\begin{bmatrix} a & b \\ c & d \end{bmatrix}\right)(z) = \frac{az + b}{cz + d},$$

is a homomorphism from $\mathrm{SL}(2, \mathbb{R})$ onto $\mathrm{PSL}(2, \mathbb{R})$. Of course $\mathrm{PSL}(2, \mathbb{R})$ is given the quotient topology. Under this topology $\mathrm{PSL}(2, \mathbb{R})$ becomes a Lie group. It is readily checked that for each $z \in \mathbb{C}^\infty$, the map $T \to T(z)$ from $\mathrm{PSL}(2, \mathbb{R})$ into $\mathbb{C}^\infty$ is continuous.

We shall use the elementary result (Rudin [**3**, Chapter 14]) that if C is a circle in the complex plane and $T \in \mathrm{PSL}(2, \mathbb{R})$, then $T(C)$ is either a line with ∞ included or is, again, a circle, and that in the latter case, T maps the interior of C either onto the interior or onto the exterior of $T(C)$ in $\mathbb{C}^\infty$.

As we shall see later, the question of which connected Lie groups are amenable [as discrete] reduces to the consideration of $\mathrm{PSL}(2, \mathbb{R})$ [$\mathrm{SU}(2)$]. The next result shows that F_2 is a subgroup (discrete in case (i)) of $\mathrm{PSL}(2, \mathbb{R})$ [$\mathrm{SU}(2)$], and we will be able to deduce the same result for every connected Lie group which is nonamenable [as discrete].

The proof of (i) was pointed out to me by Peter Waterman and probably goes back to Poincaré. A reference for (iii) is Dekker [**1**]; it can also be deduced from the result of Tits stated in (3.10).

(3.2) PROPOSITION. (i) $\mathrm{PSL}(2, \mathbb{R})$ *contains* F_2 *as a discrete subgroup.*
(ii) *The centre of* $\mathrm{PSL}(2, \mathbb{R})$ *is trivial.*
(iii) $\mathrm{SU}(2)$ *contains* F_2 *as a subgroup.*

PROOF. (i) Let C_1, C_1', C_2, C_2' be four circles in the complex plane such that each is exterior to all of the others, and, for $i = 1, 2$, let T_i be a linear fractional transformation in $\mathrm{PSL}(2, \mathbb{R})$ carrying C_i onto C_i' and the *interior* [*exterior*] of C_i onto the *exterior* [*interior*] of C_i'. For example, we could take C_1, C_1', C_2, C_2' to be the circles $|z+2|=1, |z-2|=1, |z+5|=1$, and $|z-5|=1$, respectively, and

$$T_1(z)=\frac{2z+3}{z+2}, \qquad T_2(z)=\frac{5z+24}{z+5}.$$

Obviously, if $z \in \mathbb{C}$ is exterior to both the circles C_i, C_i', then $T_i(z)$ is in the interior of C_i' and $T_i^{-1}(z)$ is in the interior of C_i. Further, $T_i(z)$ and $T_i^{-1}(z)$ are exterior to both circles C_j, C_j' where $j \neq i$. Let z_0 be exterior to all of the four circles, and let H be the subgroup of $\mathrm{PSL}(2, \mathbb{R})$ generated by T_1 and T_2. An easy induction argument on the length of words now shows that if $T_0 \in H$ corresponds to a nontrivial reduced word in $\{T_1, T_1^{-1}, T_2, T_2^{-1}\}$ starting with T_i $[T_i^{-1}]$, then $T_0(z_0)$ belongs to the interior of C_i' $[C_i]$, and so

$$|T_0(z_0)-z_0|>d>0,$$

where d is the distance of z_0 from $C_1 \cup C_1' \cup C_2 \cup C_2'$.

Since the function $T \to T(z_0)$ is continuous on $\mathrm{PSL}(2, \mathbb{R})$, it follows that H is a discrete subgroup of G. Clearly, H is isomorphic to F_2.

(ii) This fact is elementary. Suppose that $T \in Z(\mathrm{PSL}(2, \mathbb{R}))$. Let $Q\colon \mathrm{SL}(2, \mathbb{R}) \to \mathrm{PSL}(2, \mathbb{R})$ be the quotient map, and let $A \in \mathrm{SL}(2, \mathbb{R})$ be such that $Q(A)=T$. For all $B \in \mathrm{SL}(2, \mathbb{R})$, $ABA^{-1}B^{-1} \in \{I, -I\}$ so that

$$ABA^{-1} \in \mathbb{R}B \qquad (B \in \mathrm{SL}(2, \mathbb{R})). \tag{1}$$

By scaling, (1) is true for all $B \in E$, where

$$E=\{C \in M_2(\mathbb{R})\colon \det C>0\}.$$

Now E is an open subset of $M_2(\mathbb{R})$. By considering eigenvalues for the linear transformation $X \to AXA^{-1}$ $(X \in M_2(\mathbb{R}))$, we see that there exists $\lambda \in \mathbb{R}$ and an open, nonvoid subset U of $M_2(\mathbb{R})$ such that for all $X \in U$,

$$AXA^{-1}=\lambda X. \tag{2}$$

Since U spans $M_2(\mathbb{R})$, (2) is true for all $X \in M_2(\mathbb{R})$, and by putting $X=A$ in (2), we obtain $\lambda=1$, and

$$A \in Z(M_2(\mathbb{R})) \cap \mathrm{SL}(2, \mathbb{R})=\mathbb{R}I \cap \mathrm{SL}(2, \mathbb{R})=\{I, -I\}.$$

Thus T is trivial.

(iii) We first choose a certain angle θ and then an element $z \in \mathbb{T}$. We then define, in terms of θ, z, two matrices $A, B \in \mathrm{SU}(2)$ that generate a copy of F_2 in $\mathrm{SU}(2)$.

Choose $\theta \in \mathbb{R} \sim \{0\}$ so that π/θ is irrational. Let E be the set of all finite products in $\mathbb{R}$ of elements from the set

$$\{\cos r\theta\colon r \in \mathbb{Z}\} \cup \{\sin s\theta\colon s \in \mathbb{Z} \sim \{0\}\}.$$

Since π/θ is irrational, it follows that $0 \notin E$. Let S be the (countable) additive subsemigroup of $\mathbb{R}$ generated by E. Let F_E be the set of functions $f\colon \mathbb{T} \to \mathbb{C}$ given by a formula of the form

$$f(w)=\sum_{n=p}^{q} a_n w^n \qquad (w \in \mathbb{T}),$$

where $p \leq q$ in $\mathbb{Z}$ and every $a_n \in S$. The function f is said to be *nontrivial* if $a_n \neq 0$ for some $n \neq 0$.

For each nontrivial $f \in F_E$, the set $\{z \in \mathbb{T}: f(z) \in \{0,1\}\}$ is finite, and since S is countable and $\mathbb{T}$ uncountable, we can find $z \in \mathbb{T}$ such that for all nontrivial $f \in F_E$, $f(z) \notin \{0,1\}$.

Now let $A, B \in \mathrm{SU}(2)$ be given by

$$A = \begin{bmatrix} \cos\theta & \sin\theta \\ -\sin\theta & \cos\theta \end{bmatrix}, \qquad B = \begin{bmatrix} z & 0 \\ 0 & z^{-1} \end{bmatrix}.$$

Let L be the subgroup of $\mathrm{SU}(2)$ generated by A and B. It will be shown that L is free on the generators A, B.

We start by obtaining information about an element $T \in L$ of the form

$$A^{n_1}B^{m_1}\cdots A^{n_k}B^{m_k}, \tag{3}$$

where $k \geq 1$, and $n_i, m_i \in \mathbb{Z}$. First, an elementary calculation shows that

$$A^{n_1}B^{m_1} = \begin{bmatrix} z^{m_1}\cos n_1\theta & z^{-m_1}\sin n_1\theta \\ -z^{m_1}\sin n_1\theta & z^{-m_1}\cos n_1\theta \end{bmatrix}. \tag{4}$$

To find an expression for (3) for general k, we can proceed as follows. Let R be the set of k-tuples $r = (r_1, \ldots, r_k)$, where, for each i, r_i is either m_i or $-m_i$. Define

$$X = \{r \in R: r_k = m_k\}, \qquad Y = \{r \in R: r_k = -m_k\}.$$

For each $r \in R$, let $\alpha(r) = \sum_{i=1}^{k} r_i$. Note that different r's can give the same $\alpha(r)$. By either using (4) and induction on k, or the formula for the (i,j)th component of a matrix product, we can write

$$T = \begin{bmatrix} \sum_{r\in X} a_r z^{\alpha(r)} & \sum_{r\in Y} c_r z^{\alpha(r)} \\ \sum_{r\in X} b_r z^{\alpha(r)} & \sum_{r\in Y} d_r z^{\alpha(r)} \end{bmatrix} = \begin{bmatrix} p(z) & s(z) \\ q(z) & t(z) \end{bmatrix}, \tag{5}$$

where $a_r, b_r, c_r, d_r \in E \cup \{0\}$ and the functions p, q, s and t belong to F_E. Each of a_r, b_r, c_r, d_r is a product of terms from the set $\{\cos n_i\theta: 1 \leq i \leq k\} \cup \{\sin(\pm n_i\theta): 1 \leq i \leq k\}$ and so belongs to E if $n_i \neq 0$ for all i.

We now prove that L is free on A, B. Let w be a nontrivial reduced word in $\{A, B, A^{-1}, B^{-1}\}$ and T the element of L corresponding to w. It has to be shown that $T \neq I$. We can suppose that T is not of the form $A^{n_1}B^{m_1}$ since (4) and the choice of z, θ deal with the latter case.

This leaves four cases to be considered:

(a) $T = A^{n_1}B^{m_1}\cdots A^{n_k}B^{m_k}$,
(b) $T = A^{n_1}B^{m_1}\cdots B^{m_{k-1}}A^{n_k}$,
(c) $T = B^{m_1}A^{n_2}B^{m_2}\cdots A^{n_k}B^{m_k}$,
(d) $T = B^{m_1}A^{n_2}\cdots B^{m_{k-1}}A^{n_k}$,

where $k \geq 2$ and every n_i, m_i is nonzero.

Suppose that T is as in (a). Then $r_0 = (|m_1|, \ldots, |m_k|)$ belongs either to X or Y. Suppose that $r_0 \in X$. Then, referring to (5), the highest power of z occurring in the expansion of $q(z)$ (and $p(z)$) is $z^{\alpha(r_0)}$, and since every $m_i \neq 0$, r_0 is the only $r \in X$ with $\alpha(r) = \alpha(r_0)$. Hence the coefficient of $z^{\alpha(r_0)}$ in the expansion of $q(z)$ is a_{r_0}. Since $n_i \neq 0$ for all i, $a_{r_0} \neq 0$. Thus q is nontrivial in F_E, and by the choice of $z, q(z) \neq 0$. Hence $T \neq I$. A similar argument applies with $r_0 \in Y$, q being replaced by s. (With case (c) in mind, we have chosen q, s rather than p, t in this paragraph.)

Now suppose that T is as in (b) and that $T = I$. Then

$$A^{n_1}B^{m_1}\cdots A^{n_{k-1}}B^{m_{k-1}} = A^{-n_k} = \begin{bmatrix} \cos(-n_k\theta) & \sin(-n_k\theta) \\ -\sin(-n_k\theta) & \cos(-n_k\theta) \end{bmatrix}.$$

By the argument of the preceding paragraph, we can find $f \in F_E$, where f is the appropriate q or s, such that

(α) the highest power of z occurring in the expansion of $f(z)$ is z^d, where $d = \sum_{i=0}^{k-1} |m_i| > 0$;

(β) the coefficient of z^d in the above expansion belongs to E; and

(γ) $f(z) = \alpha \in E$.

Then $z \to (f(z) - \alpha)$ is nontrivial in F_E, and so $f(z) - \alpha \neq 0$. This is a contradiction. So again $T \neq I$.

Now suppose that T is as in (c) and that $T = I$. Then as in the preceding case,

$$A^{n_2}B^{m_2}\cdots A^{n_k}B^{m_k} = B^{-m_1} = \begin{bmatrix} z^{-m_1} & 0 \\ 0 & z^{m_1} \end{bmatrix}.$$

In this case we obtain an equation $f(z) = 0$ and a contradiction results as before.

Finally, case (d) reduces to case (a) or the $A^{n_1}B^{m_1}$-case by considering T^{-1}. $\square$

Note that the proof of (iii) actually provides uncountably many copies of F_2 in SU(2)—simply vary θ and z appropriately. We can also choose the generators A, B for F_2 arbitrarily close to I.

The Lie algebras of PSL$(2, \mathbb{R})$ and SL$(2, \mathbb{R})$ are the same, since PSL$(2, \mathbb{R})$ is the quotient of SL$(2, \mathbb{R})$ by the discrete subgroup $\{I, -I\}$. Recall (B19) that the Lie algebra of SL$(2, \mathbb{R})$ is the Lie subalgebra sl$(2, \mathbb{R})$ of $M_2(\mathbb{R})$, where

$$\mathrm{sl}(2, \mathbb{R}) = \{A \in M_2(\mathbb{R}) \colon \operatorname{tr} A = 0\}.$$

Further, the Lie algebra of SU(2) can be identified with the Lie algebra so$(3, \mathbb{R})$ of SO$(3, \mathbb{R})$, where

$$\mathrm{so}(3, \mathbb{R}) = \{A \in M_3(\mathbb{R}) \colon A' = -A\}.$$

Each of these Lie algebras has a natural basis such that the Lie products of the basis elements have particularly simple form. These Lie products, of course, determine all Lie products on the algebra in question.

In the case of sl$(2, \mathbb{R})$, we have the basis $\{H, E, F\}$ where

$$H = \begin{bmatrix} 1 & 0 \\ 0 & -1 \end{bmatrix}, \qquad E = \begin{bmatrix} 0 & 1 \\ 0 & 0 \end{bmatrix} \quad \text{and} \quad F = \begin{bmatrix} 0 & 0 \\ 1 & 0 \end{bmatrix}.$$

Then

$$(6) \qquad [H, E] = 2H, \qquad [H, F] = -2F, \qquad [E, F] = H.$$

Of course, every 3-dimensional, real Lie algebra with a basis satisfying the preceding equalities is isomorphic to $\mathrm{sl}(2,\mathbb{R})$.

A typical element of $\mathrm{so}(3,\mathbb{R})$ has the form

$$\begin{bmatrix} 0 & a & b \\ -a & 0 & c \\ -b & -c & 0 \end{bmatrix} \qquad (a,b,c \in \mathbb{R}).$$

Let $\{A, B, C\}$ be the basis of $\mathrm{so}(3,\mathbb{R})$ given by

$$A = \begin{bmatrix} 0 & 1 & 0 \\ -1 & 0 & 0 \\ 0 & 0 & 0 \end{bmatrix}, \quad B = \begin{bmatrix} 0 & 0 & 1 \\ 0 & 0 & 0 \\ -1 & 0 & 0 \end{bmatrix}, \quad \text{and} \quad C = \begin{bmatrix} 0 & 0 & 0 \\ 0 & 0 & -1 \\ 0 & 1 & 0 \end{bmatrix}.$$

Elementary matrix multiplication shows that

$$(7) \qquad [A,B] = C, \qquad [B,C] = A, \qquad [C,A] = B.$$

The latter equalities, of course, specify $\mathrm{so}(3,\mathbb{R})$ as a 3-dimensional, real Lie algebra.

(3.3) PROPOSITION. (i) *If $\mathfrak{k}$ is a real, semisimple, compact Lie algebra, then $\mathfrak{k}$ contains* $\mathrm{so}(3,\mathbb{R})$ *as a subalgebra.*

(ii) *If $\mathfrak{g}_0$ is a semisimple, real, noncompact Lie algebra, then $\mathfrak{g}_0$ contains* $\mathrm{sl}(2,\mathbb{R})$ *as a subalgebra.*

PROOF. (i) Let $\mathfrak{g}$ be the complexification of a real, semisimple compact Lie algebra $\mathfrak{k}$. Then $\mathfrak{g}$ is a complex, semisimple Lie algebra. Let Δ, H_α, and X_α be as in B53–B55. Then a compact real form $\mathfrak{g}_k$ of $\mathfrak{g}$ is obtained (B57) by setting

$$\mathfrak{g}_k = \sum_{\alpha\in\Delta} \mathbb{R}(iH_\alpha) + \sum_{\alpha\in\Delta} \mathbb{R}(X_\alpha - X_{-\alpha}) + \sum_{\alpha\in\Delta} \mathbb{R}(i(X_\alpha + X_{-\alpha})).$$

Let $\alpha \in \Delta$. Then (B55), $\alpha(H_\alpha) > 0$ and

$$(1) \quad [H_\alpha, X_\alpha] = \alpha(H_\alpha)X_\alpha, \quad [H_\alpha, X_{-\alpha}] = -\alpha(H_\alpha)X_{-\alpha}, \quad [X_\alpha, X_{-\alpha}] = H_\alpha.$$

Let

$$A' = (iH_\alpha)/\alpha(H_\alpha), \qquad B' = (X_\alpha - X_{-\alpha})/(2\alpha(H_\alpha))^{1/2},$$

and

$$C' = i(X_\alpha + X_{-\alpha})/(2\alpha(H_\alpha))^{1/2}.$$

Using (1), we have

$$(2) \qquad [A',B'] = C', \qquad [B',C'] = A', \qquad [C',A'] = B',$$

and comparing (7) of (3.2) with (2) above, there is an obvious isomorphism between $\mathrm{so}(3,\mathbb{R})$ and the subalgebra of $\mathfrak{g}_k$ spanned by $\{A', B', C'\}$.

Finally, since $\mathfrak{k}$ is trivially also a compact, real form of $\mathfrak{g}$, it follows from B58 that $\mathfrak{k}$ is isomorphic to $\mathfrak{g}_k$.

(ii) Let $\mathfrak{k}_0 + \mathfrak{p}_0$ be a Cartan decomposition of $\mathfrak{g}_0$, and $\mathfrak{k}_0 + \mathfrak{a}_0 + \mathfrak{n}_0$ the corresponding Iwasawa decomposition of $\mathfrak{g}_0$ (B58, B59). Since $\mathfrak{g}_0$ is noncompact, it follows from B60 that $\mathfrak{n}_0 \neq (0)$. Let $X \in \mathfrak{n}_0 \sim \{0\}$. By B60, $\operatorname{ad} X$ is a nonzero, nilpotent element of $L(\mathfrak{g}_0)$. From B63, $\mathrm{sl}(2,\mathbb{R})$ is a subalgebra of $\mathfrak{g}_0$ as required. □

The following theorem ((3.4)) due to Iwasawa [**1**] is of great intrinsic interest, and will be needed in establishing one of the main results of this chapter ((3.8)). Recall that the group $\mathrm{Aut}\, G$ of continuous automorphisms of G has a natural topology under which it becomes a topological group [**HR1**, §26]. Indeed, a base of neighbourhoods of the identity i of $\mathrm{Aut}\, G$ is provided by sets of the form

$$\{\alpha \in \mathrm{Aut}\, G\colon \alpha(c) \in Uc \text{ and } \alpha^{-1}(c) \in Uc \text{ for all } c \in C\}, \tag{3}$$

where C is a compact subset of G and U is a neighbourhood of e.

If K is a closed, normal subgroup of G, and $x \in G$, then we can define $\alpha_x^K \in \mathrm{Aut}\, K$ by setting $\alpha_x^K(k) = xkx^{-1}$. We write α_x for α_x^G. If $A \subset G$, then $I_A(K)$ is defined to be the set $\{\alpha_x^K : x \in A\}$. Thus the group of inner automorphisms $I(G)$ of G is just $I_G(G)$. The map $\Phi\colon G \to I(G)$, where $\Phi(x) = \alpha_x$, is a continuous homomorphism, $I(G)$ being given the relative topology.

If H is a topological group with identity e, then H_e is the identity component of H.

(3.4) THEOREM. *Let G be a compact group. Then the identity component $(\mathrm{Aut}\, G)_{\mathrm{i}}$ of $\mathrm{Aut}\, G$ is equal to $I_{G_e}(G)$.*

PROOF. The proof proceeds by considering three cases.

Case 1: *G is a connected Lie group.* The radical R of G is compact, connected, and solvable, and so is abelian (B51). It is therefore of the form $\mathbb{T}^n$ (B21). So $\mathrm{Aut}\, R$ is discrete (B21) and therefore has trivial identity component. Now as the property of being a maximal, connected solvable, normal subgroup of G is preserved under the action of an automorphism of G, we have $\alpha(R) = R$ for all $\alpha \in \mathrm{Aut}\, G$ (that is R is characteristic in G). It is easily shown that the map $\alpha \to \alpha\,|\,_R$ from $\mathrm{Aut}\, G$ into $\mathrm{Aut}\, R$ is continuous and so maps $(\mathrm{Aut}\, G)_{\mathrm{i}}$ into the (trivial) identity component of $\mathrm{Aut}\, R$. Hence $\alpha(r) = r$ for all $\alpha \in (\mathrm{Aut}\, G)_{\mathrm{i}}$, $r \in R$. Since G is connected and the map $\Phi\colon G \to I(G)$ is continuous, we have $\alpha_x \in (\mathrm{Aut}\, G)_{\mathrm{i}}$ for all $x \in G$, and so $\alpha_x(r) = r$ for all $x \in G$, $r \in R$. Hence $R \subset Z(G)$.

For $\beta \in \mathrm{Aut}\, G$, define $\beta' \in \mathrm{Aut}(G/R)$ by setting

$$\beta'(xR) = \beta(x)R \qquad (x \in G).$$

It is easily shown that the map $\beta \to \beta'$ from $\mathrm{Aut}\, G$ into $\mathrm{Aut}(G/R)$ is a continuous homomorphism. Let $\alpha \in (\mathrm{Aut}\, G)_{\mathrm{i}}$. Then α' belongs to the identity component of $\mathrm{Aut}\, G/R$, and since G/R is a connected, semisimple Lie group, it follows from B42 that α' is an inner automorphism. So we can find $x_0 \in G$, and, for each $x \in G$, an element $a_x \in R$ such that $\alpha(x) = (x_0xx_0^{-1})a_x$. Since $R \subset Z(G)$ and $\alpha(xy) = \alpha(x)\alpha(y)$ $(x, y \in G)$, we see that $a_{xy} = a_xa_y$ and $a_r = e$ if $r \in R$. Hence we can define a map $Q\colon G/R \to R$ by

$$Q(xR) = a_x.$$

Now Q is clearly a continuous homomorphism from the semisimple Lie group G/R into the solvable Lie group R, and so, by B42, the induced homomorphism between the Lie algebras of G/R and R is the zero map. Since G/R is connected, we can apply B11 and B18 to obtain $Q(xR) = e$ for all $x \in G$. Hence $a_x = e$ for all $x \in G$, and $\alpha = \alpha_{x_0}$. Thus $(\operatorname{Aut} G)_{\mathrm{i}} \subset I(G)$. The reverse inclusion is obvious, so that $(\operatorname{Aut} G)_{\mathrm{i}} = I(G)$ $(= I_{G_e}(G))$ as required.

Case 2: G is a Lie group. Define subsets A, N of G by

$$A = \{\alpha \in \operatorname{Aut} G\colon \alpha(x) \in xG_e \text{ for all } x \in G\},$$
$$N = \{\alpha \in A\colon \alpha(y) = y \text{ for all } y \in G_e\}.$$

It is easily checked that A is a subgroup of $\operatorname{Aut} G$. Now A is open and therefore closed in $\operatorname{Aut} G$. Indeed, since G is a Lie group, G_e is open in G. Further, $xG_e = G_e x$ for all $x \in G$ and since G_e is a characteristic subgroup of G, $\alpha(x) \in xG_e$ if and only if $\alpha^{-1}(x) \in xG_e$. By taking $U = G_e$ and $C = G$ in (3) of (3.3), we see that A is open as required. Obviously, N is a closed subgroup of $\operatorname{Aut} G$ and, by definition, $N \subset A$. If $y \in G_e$, then $yxy^{-1} = x[x^{-1}yx]y^{-1} \in xG_e$, and so $I_{G_e}(G) \subset A$.

We claim that it is sufficient to prove

$$|N/(N \cap I_{G_e}(G))| < \infty. \tag{1}$$

For suppose that (1) holds and let $\Psi\colon \operatorname{Aut} G \to \operatorname{Aut} G_e$ be the continuous homomorphism given by $\Psi(\alpha) = \alpha\,|\,_{G_e}$. Clearly $A \cap \ker \Psi = N$. Let $H = \Psi^{-1}(I(G_e)) \cap A$. Then H is a closed subgroup of $\operatorname{Aut} G$. Since $\Psi(I_{G_e}(G)) = I(G_e)$, we have $NI_{G_e}(G) = H$, and using (1) and an isomorphism theorem for groups, we have

$$\begin{aligned} |H/I_{G_e}(G)| &= |(NI_{G_e}(G))/I_{G_e}(G)| \\ &= |N/(N \cap I_{G_e}(G))| < \infty. \end{aligned} \tag{2}$$

Now $\operatorname{Aut} G_e$ is a Lie group (B20) so that $(\operatorname{Aut} G_e)_{\mathrm{i}}$ is open in $\operatorname{Aut} G_e$. By Case 1, $(\operatorname{Aut} G_e)_{\mathrm{i}} = I(G_e)$, and it follows that H is open in A. As the map $x \to \alpha_x$ from G into $\operatorname{Aut} G$ is continuous, and as G_e is compact and $\operatorname{Aut} G$ is Hausdorff, the set $I_{G_e}(G) = \{\alpha_x\colon x \in G_e\}$ is compact and so closed in $\operatorname{Aut} G$ and H. So every left $I_{G_e}(G)$-coset is also closed in H. By (2), H is the disjoint union of a finite number of left $I_{G_e}(G)$-cosets, each of which is closed (and therefore open) in H. So $I_{G_e}(G)$ is open in H, which, in turn, is open in A, which, in turn, is open in $\operatorname{Aut} G$. Hence $I_{G_e}(G)$ is a connected, open and closed subgroup of $\operatorname{Aut} G$, so that $I_{G_e}(G) = (\operatorname{Aut} G)_{\mathrm{i}}$ as required. (The openness of $I_{G_e}(G)$ in $\operatorname{Aut} G$ will be used in Case 3.)

It remains to establish (1).

Let $\alpha \in N$. Since $N \subset A$, we can write $\alpha(x) = xu(x)$, where $u\colon G \to G_e$. If $y \in G_e$ and $x \in G$, then since $\alpha \in N$ and G_e is normal in G, we have

$$\begin{aligned} xyu(x) &= (xyx^{-1})(xu(x)) = \alpha(xyx^{-1})\alpha(x) \\ &= \alpha(xyx^{-1}x) = \alpha(xy) = xu(x)y \end{aligned}$$

so that $u(x) \in Z(G_e)$. Further, $xyu(x) = xu(x)y = \alpha(xy) = xyu(xy)$, so that $u(x) = u(xy)$ and u is constant on each left G_e-coset. Also, if $x_1, x_2 \in G$, then

$$u(x_1x_2) = x_2^{-1}x_1^{-1}\alpha(x_1x_2) = x_2^{-1}x_1^{-1}x_1u(x_1)x_2u(x_2)$$
$$= (x_2^{-1}u(x_1)x_2)u(x_2).$$

Since G is compact and G_e is open, it follows that $|G/G_e| = n < \infty$. Let $\{x_1, \dots, x_n\}$ be a left transversal of G with respect to G_e, and let $v = u(x_1)u(x_2)\cdots u(x_n) \in Z(G_e)$. It is easy to check that for each $x \in G$, $\{x_1x, \dots, x_nx\}$ is also a left transversal of G with respect to G_e. Using the earlier results of this paragraph, we have, for each $x \in G$,

$$\begin{aligned} v &= u(x_1x)u(x_2x)\cdots u(x_nx) \\ &= [(x^{-1}u(x_1)x)u(x)][(x^{-1}u(x_2)x)u(x)]\cdots[(x^{-1}u(x_n)x)u(x)] \\ &= (x^{-1}vx)(u(x))^n. \end{aligned} \tag{3}$$

Now $Z(G_e)$ is an abelian, compact Lie group, and so $Z_e = Z(G_e)_e = \mathbb{T}^m$ for some m. Clearly Z_e is a divisible subgroup of $Z(G_e)$, and so there is a subgroup Z_1 of $Z(G_e)$ such that $Z(G_e) = Z_e \times Z_1$ [**HR1**, (A8)]. The subgroup Z_1 is finite since $Z(G_e)$ is covered by a finite number of translates of its identity component Z_e. Let $p = |Z_1|$ and $Z_2 = \{x \in Z(G_e)\colon x^p = e\}$. Then Z_2 is a finite, characteristic subgroup of $Z(G_e)$, and since $Z_1 \subset Z_2$, we have $Z(G_e)/Z_2 = \mathbb{T}^q$ for some q. So we can find $w \in Z(G_e)$ such that $w^nv^{-1} \in Z_2$. Clearly $\alpha_w \in N \cap I_{G_e}(G)$. Now define $u'\colon G \to G_e$ by $\alpha_w\alpha(x) = xu'(x)$. Then $\alpha_w\alpha \in N$, and as in the case of u, we have that $u'(G) \subset Z(G_e)$ and u' is constant on each left G_e-coset. Using (3) and the fact that Z_2 is characteristic, we have

$$(u'(x))^n = [(x^{-1}wx)u(x)w^{-1}]^n = (x^{-1}wx)^nu(x)^nw^{-n}$$
$$= x^{-1}w^nxx^{-1}v^{-1}xvw^{-n} = x^{-1}(w^nv^{-1})x(w^nv^{-1})^{-1} \in Z_2.$$

So u' can be regarded as a function from the finite group G/G_e into the finite group $\{z \in Z(G_e)\colon z^{np} = e\}$. Now the number of such functions u' is finite. This establishes (1). For u' determines an element of the right coset $(N \cap I_{G_e}(G))\alpha$, and every right $(N \cap I_{G_e}(G))$-coset in N is so determined. So $|N/(N \cap I_{G_e}(G))| = |N\backslash(N \cap I_{G_e}(G))| < \infty$.

Case 3: G is a compact group. Let $R(G)$ be the class of finite-dimensional, continuous, unitary representations of G. If $\pi \in R(G)$ and $\alpha \in \operatorname{Aut} G$, then $\pi\alpha \in R(G)$ is defined by $\pi\alpha(x) = \pi(\alpha(x))$. Let

$$A_\pi = \{\alpha \in \operatorname{Aut} G\colon \pi\alpha \text{ is equivalent to } \pi\}.$$

It follows from [**HR2**, (27.32)] that $\alpha \in A_\pi$ if and only if $\chi_\alpha = \chi_{\pi\alpha}$, where χ_σ is the character of a representation $\sigma \in R(G)$.

We will show first that A_π is an open (and closed) subgroup of $\operatorname{Aut} G$. By composing and inverting intertwining operators, we see that A_π is a subgroup of $\operatorname{Aut} G$. Suppose that $\beta \in (\operatorname{Aut} G) \sim A_\pi$. Since $\pi\beta$ is not equivalent to π, we can find σ in the dual $\hat{G}$ of G such that $n_\sigma \neq m_\sigma$, where n_σ $[m_\sigma]$ is the number of times σ occurs in π $[\pi\beta]$. Using [**HR2**, (27.30)], we have

$$\begin{aligned}\|\chi_\pi - \chi_{\pi\beta}\|_\infty^2 &\geq \int |\chi_\pi - \chi_{\pi\beta}|^2 d\lambda_G \\ &= (\chi_\pi - \chi_{\pi\beta}, \chi_\pi - \chi_{\pi\beta}) \geq (n_\sigma - m_\sigma)^2 \geq 1.\end{aligned}$$

So the set

$$W = \{\alpha \in \operatorname{Aut} G\colon \|\chi_\pi - \chi_{\pi\alpha}\|_\infty < 1\}$$

is contained in A_π. Now note that for $x \in G$, $\chi_{\pi\alpha}(x) = \operatorname{tr}\pi(\alpha(x)) = \chi_\pi(\alpha(x))$. Since G is compact, the continuous function χ_π is uniformly continuous, and we can find a neighbourhood U of e in G such that $|\chi_\pi(a) - \chi_\pi(b)| < \frac{1}{2}$ whenever $ab^{-1} \in U$. So the neighbourhood $\{\beta \in \operatorname{Aut} G\colon \beta(x) \in Ux \text{ for all } x \in G\}$ of i is contained in $W \subset A_\pi$, and hence A_π is an open (and therefore closed) subgroup of $\operatorname{Aut} G$.

Now if $\alpha \in A_\pi$, then $\alpha(x) \in \ker\pi$ if and only if $x \in \ker\pi$. So for each $\alpha \in A_\pi$, we can define an (algebraic) automorphism $\Phi_\pi(\alpha)$ of $H = \pi(G)$ by setting

$$\Phi_\pi(\alpha)(\pi(z)) = \pi(\alpha(z)) \qquad (z \in G).$$

It is routine to check that $\Phi_\pi(\alpha)$ is continuous and that $\Phi_\pi\colon A_\pi \to \operatorname{Aut} H$ is continuous. (The compactness of G entails that H can be identified with the quotient group $G/(\ker\pi)$.) If $x \in G$, then $\pi(x)$ is an intertwining operator for π and $\pi\alpha_x$, so that $\alpha_x \in A_\pi$. Since $\Phi_\pi(\alpha_x) = \alpha_{\pi(x)}$, we have $I_{H_{\pi(e)}}(H) \subset \Phi_\pi(A_\pi)$. Now H is a compact Lie group, and from Case 2, $I_{H_{\pi(e)}}(H)$ is an open subgroup of $\operatorname{Aut} H$. Hence $A'_\pi = \Phi_\pi^{-1}(I_{H_{\pi(e)}}(H))$ is an open and closed subgroup of A_π, and so of $\operatorname{Aut} G$. Let $X = \bigcap\{A'_\pi\colon \pi \in R(G)\}$. Then $I_{G_e}(G) \subset (\operatorname{Aut} G)_{\mathrm{i}} \subset X$, and Case 3 will be proved once we have established that $X \subset I_{G_e}(G)$.

So let $\beta \in X$. For each $\pi \in R(G)$, let $B_\pi = \{x \in G_e\colon \Phi_\pi(\alpha_x) = \Phi_\pi(\beta)\}$. Now $\pi\colon G \to \pi(G) = H$ $(= G/(\ker\pi))$ is an open mapping and, using [**HR1**, (7.12)], $\pi(G_e) = H_{\pi(e)}$. As $\beta \in A'_\pi$, $\Phi_\pi(\beta) = \alpha_h$ for some $h \in H_{\pi(e)}$, and if $x \in G_e$ is such that $\pi(x) = h$, then $x \in B_\pi$. So B_π is a nonempty, compact subset of G. If $\pi_1, \ldots, \pi_n \in R(G)$, then consideration of $\pi = \pi_1 \oplus \cdots \oplus \pi_n$ shows that $\bigcap_{i=1}^n B_{\pi_i} \neq \varnothing$. Hence we can find $x \in \bigcap\{B_\pi\colon \pi \in R(G)\}$. Since $R(G)$ separates the points of G, we have $\beta = \alpha_x$. This concludes the proof. □

(3.5) COROLLARY. *Let G be a connected, locally compact group, K a compact, normal subgroup of G, and $H = \{x \in G\colon xk = kx \text{ for all } k \in K\}$. Then $G = K_eH$.*

PROOF. Let $\Psi\colon G \to \operatorname{Aut} K$ be given by $\Psi(x)(k) = xkx^{-1}$. Then, since Ψ is a continuous homomorphism, $\Psi(G)$ is a connected subgroup of $\operatorname{Aut} K$ and hence by (3.4), is contained in $I_{K_e}(K)$. So if $x \in G$, then there exists $u \in K_e$ such that $xkx^{-1} = uku^{-1}$ for all $k \in K$. Hence $u^{-1}x \in H$, and $x = u(u^{-1}x) \in K_eH$. □

References for the next result are Yosida [**1**] and Goto [**1**].

(3.6) PROPOSITION. *Let G be a semisimple, Lie group and H a connected, semisimple, Lie subgroup of G. Then H is closed in G.*

PROOF. Suppose first that H is a Lie subgroup of $\mathrm{GL}(W)$, where W is a finite dimensional vector space over $\mathbb{R}$. Suppose that $H^- \neq H$. We will derive a contradiction. Note B14 that H^- is a connected Lie subgroup of $\mathrm{GL}(W)$ with the relative topology (B16). By B15, the Lie algebra $\mathfrak{h}$ of H is a proper subalgebra of the Lie algebra $\mathfrak{k}$ of H^-. Let $Y \in \mathfrak{k} \sim \mathfrak{h}$ and $t \in \mathbb{R}$. Find a sequence $\{x_n\}$ in H with $x_n \to \exp(tY)$ in $\mathrm{GL}(W)$. Since the Lie group topology of H^- is the relative topology induced by $\mathrm{GL}(W)$, and as Ad (on H^-) is continuous, $\mathrm{Ad}\, x_n(\mathfrak{h}) \subset \mathfrak{h}$, and $\exp(t \,\mathrm{ad}\, Y) = \mathrm{Ad}(\exp(tY))$ (B12), we have $\exp(t\,\mathrm{ad}\, Y)(\mathfrak{h}) \subset \mathfrak{h}$, so that $\mathrm{ad}\, Y(\mathfrak{h}) \subset \mathfrak{h}$. Since $\mathfrak{h}$ is semisimple, the derivation $\mathrm{ad}\, Y \mid_{\mathfrak{h}}$ is inner (B42). So by subtracting an element of $\mathfrak{h}$ from Y, we can suppose that $\mathrm{ad}\, Y(\mathfrak{h}) = \{0\}$. Identifying $\mathfrak{k}$ in the canonical way with a Lie subalgebra of $L(W)$, the space of linear transformations of W, we see that Y commutes with the elements of $\mathfrak{h}$.

Let $W_{\mathbb{C}}[\mathfrak{h}_{\mathbb{C}}]$ be the complexifications of W and $\mathfrak{h}$, and identify the complexification $(L(W))_{\mathbb{C}}$ in the obvious way with the complex Lie algebra $L(W_{\mathbb{C}})$. If $T \in L(W)$, let $T_{\mathbb{C}}$ be the element of $L(W_{\mathbb{C}})$ corresponding to T. So $T_{\mathbb{C}}(v+iw) = T(v) + iT(w)$ $(v, w \in W)$. Of course, $Y_{\mathbb{C}}$ commutes with all the elements of $\mathfrak{h}_{\mathbb{C}}$.

Since $\mathfrak{h}_{\mathbb{C}}$ is semisimple, the $\mathfrak{h}_{\mathbb{C}}$-module $W_{\mathbb{C}}$ can be written as a direct sum

$$W_{\mathbb{C}} = \bigoplus_{i=1}^{k} W_i,$$

where W_i is an $\mathfrak{h}_{\mathbb{C}}$-invariant, irreducible subspace of W (B41).

Let $i \in [1; k]$. Since $(\exp X)_{\mathbb{C}}(W_i) \subset W_i$ for all $X \in \mathfrak{h}$ and $\exp \mathfrak{h}$ generates H, it follows that $x_{\mathbb{C}}(W_i) \subset W_i$ for all $x \in H$. An argument similar to the one above involving $\{x_n\}$ shows that $Y_{\mathbb{C}}(W_i) \subset W_i$. By Schur's Lemma, $Y_i = (Y_{\mathbb{C}})\mid_{W_i} = \lambda_i I_i$, where $\lambda_i \in \mathbb{C}$ and I_i is the identity of $L(W_i)$.

We now show that $\lambda_i = 0$. To this end, let $\mathfrak{h}_i = \{A\mid_{W_i} : A \in \mathfrak{h}_{\mathbb{C}}\}$. Then the Lie algebra $\mathfrak{h}_i$ is semisimple, and so $[\mathfrak{h}_i, \mathfrak{h}_i] = \mathfrak{h}_i$. Hence $\mathrm{tr}\, T = 0$ for all $T \in \mathfrak{h}_i$. This implies that for all such T, $\det(\exp T) = \exp(\mathrm{tr}\, T) = 1$. Thus for all $X \in \mathfrak{h}$, $\det(\exp(X_{\mathbb{C}})) = 1$, and it follows that $\det((x_{\mathbb{C}})\mid_{W_i}) = 1$ for all $x \in H$. As earlier, we obtain $\det(\exp t(Y_{\mathbb{C}})) = 1$ for all $t \in \mathbb{R}$, giving that $\lambda_i = (1/n_i)\mathrm{tr}\, Y_{\mathbb{C}} = 0$ as required, where $n_i = \dim W_i$.

Hence $Y = 0$ (since $Y_{\mathbb{C}} = 0$) and the resulting contradiction establishes $H^- = H$.

Now let H and G be as in the statement of the proposition and let $\mathfrak{h}$ and $\mathfrak{g}$ be the Lie algebras of H and G respectively. Let $\mathrm{Ad}\colon G \to \mathrm{GL}(\mathfrak{g})$ be the adjoint representation. By the first part of the proof, $\mathrm{Ad}\, H$ is a closed subgroup of $\mathrm{GL}(\mathfrak{g})$. Let $H' = (\mathrm{Ad})^{-1}(\mathrm{Ad}\, H)$. Then H' is closed in G, and since $\ker \mathrm{Ad} \subset Z(G)$, we have $H' = HZ(G)$. Since G is semisimple, $Z(G)$ is discrete in G. Now $Z(G)$ is countable (G is σ-compact) and H is connected, so that $H \times Z(G)$ is σ-compact. Hence the canonical homomorphism from $H \times Z(G)$ onto H' is open by [**HR1**, (5.29)], and it follows that $H = (H')_e$. So H is closed in H', and hence also in G. □

We now generalize the well-known fact that there exists, in every Lie group G, a maximal, connected, normal, solvable subgroup, called the *radical* of G (cf. B35).

(3.7) PROPOSITION. *Let G be a locally compact group. Then G contains a connected, closed, normal, solvable subgroup $r(G)$ which contains every connected, normal, solvable subgroup of G. Further*

(i) *$r(G/r(G))$ is trivial;*

(ii) *if G is connected, $r(G)$ is trivial and L is a compact, normal subgroup of G, then $r(G/L)$ is trivial.*

PROOF. Let $\mathscr{S}$ be the family of connected, normal, solvable subgroups of G, and S_0 the subgroup of G generated by $\bigcup\mathscr{S}$. Then $\bigcup\mathscr{S}$ is connected and invariant under the inner automorphisms of G, and so (Appendix A, (*)), S_0 is a connected, normal subgroup of G. If we can show that S_0 is solvable, then it will follow that S_0^- is solvable, and the first assertion of the proposition will have been established with $r(G) = S_0^-$ ($= S_0$). To this end, we can suppose that G is connected, and we can find a compact, normal subgroup K of G such that G/K is a Lie group (B6). Let $Q\colon G \to G/K$ be the quotient map. If $S \in \mathscr{S}$, then $Q(S)$ is a connected, normal, solvable subgroup of G/K and so is contained in the radical R of G/K. Hence $Q(S_0) \subset R$. For any group H, let $\{D_1(H), D_2(H), \dots\}$ be the derived series of H; so $D_1(H) = H$ and $D_{i+1}(H)$ is the commutator subgroup of $D_i(H)$. Since R is solvable, we can find r so that $D_r(Q(S_0)) = \{e\}$. So $D_r(S_0) \subset K$. Now if $S_1, \dots, S_m$ belong to $\mathscr{S}$, then so does the subgroup of G generated by $\bigcup_{i=1}^m S_i$ (Appendix A, (*)). It follows that $D_r(S_0)$ is the union of the upwards-directed family of subgroups $\{D_r(S)\colon S \in \mathscr{S}\}$. Now for each $S \in \mathscr{S}$, the compact subgroup $D_r(S)^-$ is solvable, and so abelian (B51). Hence $D_r(S_0)$ is abelian, and S_0 is solvable as required.

The result (i) follows by observing that if $\Phi\colon G \to G/r(G)$ is the quotient map and T is a connected, normal, solvable subgroup of $G/r(G)$, then $\Phi^{-1}(T)$ is a connected, normal, solvable subgroup of G.

Now suppose that G is connected and $r(G)$ and L are as in (ii). Let $H = \{h \in G\colon hl = lh \text{ for all } l \in L\}$. By (3.5), $G = L_eH$. Let $\Psi\colon G \to G/L$ be the quotient map, $R = r(G/L)$, and $R_1 = \Psi^{-1}(R) \cap H$. Since $G = LH$, we have $\Psi(H) = G/L$, and it follows that $R_1L = \Psi^{-1}(R)$.

Using the equality $R_1 \cap L = H \cap L$,

$$R_1/(H \cap L) = R_1/(R_1 \cap L) = (R_1L)/L = \Psi^{-1}(R)/L = R. \tag{1}$$

Clearly, R_1 is a closed subgroup of G. We claim that R_1 is a normal subgroup of G. To prove this, it is sufficient to show that H is a normal subgroup of G. This follows from the equalities below, where $x \in G, h \in H$, and $l \in L$:

$$(xhx^{-1})l = xh(x^{-1}lx)x^{-1} = x(x^{-1}lx)hx^{-1} = l(xhx^{-1}).$$

Now $H \cap L$ is solvable (being abelian) and R is solvable. It follows from (1) that R_1 is also solvable. Hence $(R_1)_e \subset r(G) = \{e\}$, and R_1 is totally disconnected.

It follows from [**HR1**, (5.29)] and [**HR1**, (7.12)] that $R_1/(H \cap L) = R = \{e\}$. So (ii) is established. □

Consonant with the terminology used in the theory of Lie groups, we shall call the subgroup $r(G)$ above the *radical* of G. The group G is called *semisimple* if $r(G) = \{e\}$. The group G is called *almost connected* if G/G_e is compact. The main section of this chapter culminates in the two theorems below (Rickert [**1**], [**2**], Balcerzyk and Mycielski [**S**]).

(3.8) THEOREM. *The following three statements are equivalent for an almost connected, locally compact group G:*

(i) *G is amenable;*

(ii) *G does not contain F_2 as a discrete subgroup;*

(iii) *$G/r(G)$ is compact.*

PROOF. It follows from (1.12) that (i) implies (ii) and from (1.13) and Problem 1–6 that (iii) implies (i). Suppose that (ii) holds. It remains to show that (iii) holds. Since G/G_e is compact, it is sufficient to show that $H = G_e/r(G)$ is compact.

Since H is connected, we can find a compact normal subgroup K of H such that H/K is a Lie group. The compactness of H will follow once we have shown that H/K is compact. Suppose, on the contrary, that H/K is not compact. Now H is semisimple by (3.7(i)) and so therefore also is H/K by (3.7(ii)). Note also that H/K is connected. Let $\mathfrak{h}$ be the Lie algebra of H/K. If $\mathfrak{h}$ were a compact Lie algebra, then, by B52, the group H/K would be compact. Hence $\mathfrak{h}$ is semisimple and noncompact, and so by (3.3(ii)), contains a subalgebra $\mathfrak{l}$ isomorphic to $\mathrm{sl}(2, \mathbb{R})$. Let L be the connected, Lie subgroup of H/K with Lie algebra $\mathfrak{l}$. Then L is semisimple (since its Lie algebra is), and so, by (3.6), is closed in H/K. Let W be the simply connected covering group of $\mathrm{PSL}(2, \mathbb{R})$ (and L). Then (B6) $\mathrm{PSL}(2, \mathbb{R}) = W/A_1$ and $L = W/A_2$, where A_i is a subgroup of $Z(W)$. Since $\mathrm{PSL}(2, \mathbb{R})$ has trivial centre ((3.2(ii))), it follows that $A_1 = Z(W) \supset A_2$, and there exists a continuous homomorphism from L onto $\mathrm{PSL}(2, \mathbb{R})$. Hence by (3.2(i)) and (3.1), L contains F_2 as a discrete subgroup. Since L is closed in H/K, its topology coincides with the relative topology induced by H/K (B16), and it follows that the group H/K also contains F_2 as a discrete subgroup. Now use (3.1) twice to contradict (ii). So H/K is compact as required. □

(3.9) THEOREM. *The following three statements are equivalent for a connected Lie group G:*

(i) *G is amenable as a discrete group;*

(ii) *G does not contain F_2 as a subgroup;*

(iii) *G is solvable.*

PROOF. Clearly, (iii) implies (i) and (i) implies (ii). Suppose that (ii) holds. Then G does not contain F_2 as a discrete subgroup, and so by (3.8), $G/r(G)$ is compact. Suppose that $G/r(G)$ is nontrivial. Then by (3.3(i)), we can find a Lie subgroup H of $G/r(G)$ whose Lie algebra is $\mathrm{so}(3, \mathbb{R})$. Now $\mathrm{SU}(2)$ is the simply connected Lie group with Lie algebra so$(3, \mathbb{R})$ (B19), and so there exists a continuous homomorphism $Q\colon \mathrm{SU}(2) \to H$ with $\ker Q \subset Z(\mathrm{SU}(2))$. By (3.2(iii)) and (3.1), H, and therefore $G/r(G)$, and therefore G, contains F_2 as a subgroup. This contradiction of (ii) yields that $G/r(G)$ is trivial. Hence $G = r(G)$ is solvable, and (ii) implies (iii). □

(3.10) A THEOREM OF TITS. *Let V be a finite-dimensional vector space over a field $\mathbb{F}$ and G a subgroup of* $\mathrm{GL}(V)$. *Then G is amenable if and only if it does not contain the free group F_2 on two generators as a subgroup. Now suppose that G is amenable. Then*

(a) *if $\mathbb{F}$ has characteristic zero, then G contains a normal, solvable subgroup of finite index;*

(b) *if $\mathbb{F}$ has nonzero characteristic, then G contains a normal, solvable subgroup H with G/H locally finite.*

This remarkable theorem is proved by Tits [**2**]. A Lie group version of this result is proved above.

References

Dekker [**1**], van Dijk [**1**], Goto [**2**], Greenleaf [**2**], Iwasawa [**1**], Milnor [**4**], Moore [**2**], von Neumann [**1**], [**R**], Chapter 8, Rickert [**1**], [**2**], Takenouchi [**1**], Tits [**2**], Wagon [**S2**], and Yosida [**1**].

Further Results

(3.11) A group in NF **but not in** EG. The classes NF and EG were introduced in (0.16). We will show here that EG $\neq$ NF, basing our account on Chou [**14**].

R. I. Grigorchuk [**S1**] has proved the stronger result that EG $\neq$ AG.

The proof uses the existence of a periodic group A that is not locally finite. (So every element of A has finite order, and there exists a finite subset of A that generates an infinite subgroup.) The existence of such a group A is well known, and settles the Burnside problem negatively. The existence of A follows from Golod [**1**] and Golod and Shafarevitch [**1**]. A short, lucid account of the construction of A is given in Chapter 8 of Herstein [**2**]. See also Grigorchuk [**1**], who realises such a group in terms of measure preserving transformations of $[0, 1]$.

The group A is in NF *but not in* EG. [Obviously, $A \in$ NF. To prove that $A \notin$ EG, it is sufficient to show that every periodic group in EG is locally finite. To this end, we define, by transfinite recursion, for each ordinal γ, a class EG_γ of groups as follows:

(i) EG_0 is the union of the class of finite groups and the class of abelian groups;

(ii) if γ is a limit ordinal, then $\mathrm{EG}_\gamma = \bigcup\{\mathrm{EG}_\beta \colon \beta < \gamma\}$;

(iii) if γ has a predecessor β, then $G \in \mathrm{EG}_\gamma$ if and only if either G is an extension of H by K with $H, K \in \mathrm{EG}_\beta$ or G is the union of an upwards-directed family of members of EG_β.

Let $\mathrm{EG}' = \bigcup_\gamma \mathrm{EG}_\gamma$. We show first that $\mathrm{EG}' = \mathrm{EG}$. (The reason for working with EG' rather than with EG directly is that it is easier to check in EG' rather than in EG that every periodic group is locally finite, since the two processes in (0.16) of taking subgroups and forming factor groups need not be considered.)

Using transfinite induction, it is easily checked that $\mathrm{EG}_\gamma \subset \mathrm{EG}$ for all γ. Hence $\mathrm{EG}' \subset \mathrm{EG}$.

We now prove the reverse inclusion. Firstly, as $\mathrm{EG}_0 \subset \mathrm{EG}'$, it follows that EG' contains all finite and all abelian groups. It is an easy consequence of (iii) that EG' is closed under the process of forming group extensions. Now let $\{H_\delta\}$ $(\delta \in \Delta)$ be a net in EG' with $H_{\delta_1} \subset H_{\delta_2}$ if $\delta_1 \leq \delta_2$. For each δ, let $H_\delta \in \mathrm{EG}_{\alpha(\delta)}$ and let $\gamma = (\bigcup_{\delta\in\Delta} \alpha(\delta)) + 1$. Then γ is an ordinal, and from (iii), $\bigcup_{\delta\in\Delta} H_\delta \in \mathrm{EG}_\gamma \subset \mathrm{EG}'$. Thus EG' is closed under the process of forming upwards directed unions. It remains to prove that EG' is closed under the process (I) of taking subgroups and (II) of forming factor groups. For then EG' is closed under all four processes of (0.16) and so must contain EG.

It is clearly sufficient to show that, for each ordinal γ, EG_γ is closed under (I) and (II). This is proved by transfinite induction.

The result is obvious when $\gamma = 0$. Now suppose that α is an ordinal, and that EG_β is closed under (I) and (II) for all $\beta < \alpha$. Let $G \in \mathrm{EG}_\alpha$, and let B be a subgroup of G, C a normal subgroup of G, and $D = G/C$. We have to show that $B, D \in \mathrm{EG}_\alpha$. If α is a limit ordinal, then $G \in \mathrm{EG}_\beta$ for some $\beta < \alpha$, and it follows that B, D belong to $\mathrm{EG}_\beta \subset \mathrm{EG}_\alpha$. Suppose then that α has a predecessor ε. Then either (a) there are groups L, M in EG_ε with L a normal subgroup of G and $G/L = M$ or (b) G is the union of an upwards directed family $\{G_\tau\}$ of groups in EG_ε. Suppose that (a) holds. Then $L \cap B$ and $B/(L\cap B) = LB/L$ belong to EG_ε as they are subgroups of L and M respectively. Hence $B \in \mathrm{EG}_\alpha$. Now $(CL)/C = L/(L \cap C) \in \mathrm{EG}_\varepsilon$, and so

$$D/(CL/C) = G/CL = (G/L)/(CL/L) = M/(CL/L) \in \mathrm{EG}_\varepsilon.$$

Hence $D \in \mathrm{EG}_\alpha$ as required. Now suppose that (b) holds. Then

$$B = \bigcup_\tau (G_\tau \cap B) \in \mathrm{EG}_\alpha \quad \text{and} \quad D = \bigcup_\tau (CG_\tau/C) \in \mathrm{EG}_\alpha.$$

So EG_α is closed under (I) and (II) and it follows that $\mathrm{EG}' = \mathrm{EG}$. To show that $A \notin \mathrm{EG}$, it is sufficient to show that every periodic group in every EG_γ is locally finite. This is proved (yet again!) by transfinite induction.

The desired result is easy when $\gamma = 0$. Suppose that α is an ordinal and that the result is true for all $\gamma < \alpha$. Let G be a periodic group in EG_α. If α is a limit ordinal, then $G \in \mathrm{EG}_\gamma$ for some $\gamma < \alpha$ and so is locally finite. Suppose, then, that α has a predecessor γ. There are two cases to be considered.

Suppose that $H, K \in \mathrm{EG}_\gamma$ with H a normal subgroup of G and $G/H = K$. Since G is periodic, so also are H and K. By assumption, both H and K are locally finite. By Problem 3–10, G is also locally finite.

The other case is that in which G is the union of an upwards directed family $\{G_\tau\}$ of groups in EG_γ. Since every finite subset of G is contained in some G_τ, it follows that G is locally finite.

This completes the proof.]

(3.12) Three problems considered by Banach. In his paper [**1**], Banach considers three questions, all of which involve invariance of finitely additive measures. A commentary on Banach's paper is given by Mycielski [**2**]. Mycielski [**6**] and Wagon [**S2**] are useful references for the discussion below.

The first problem originates with Lebesgue [**1**]. Essentially, Lebesgue asked if the monotone convergence theorem, which was a property of his integral on $\mathbb{R}$, could be dispensed with. Now the latter property is equivalent to countable additivity, and thus Lebesgue's question can be formulated: *if μ is a finitely additive, positive, translation invariant measure on $\mathscr{M}_b(\mathbb{R})$, the family of bounded Lebesgue measurable sets, such that $\mu([0,1]) = 1$, does $\mu = \lambda_1$, where λ_1 is Lebesgue measure (restricted to $\mathscr{M}_b(\mathbb{R})$).*

Banach showed that the answer to this question is in the negative. He constructed a finitely additive, positive, translation invariant measure μ on $\mathscr{P}_b(\mathbb{R})$, the family of bounded subsets of $\mathbb{R}$, such that

(i) $\mu(A) < \infty$ for every bounded subset A of $\mathbb{R}$ (so that μ gives rise, in the obvious way, to an element μ_A of $l_\infty(A)'$ for each such A);

(ii) $\mu_{[a,b]}(\phi) = \int_a^b \phi(x)\,dx$ for every Riemann integrable function ϕ on an interval $[a,b]$;

(iii) there exists a Lebesgue integrable function ψ on an interval $[c,d]$ such that

$$\mu_{[c,d]}(\psi) \neq \int_c^d \psi(x)\,d\lambda_1(x).$$

(A similar result to this for the circle group $\mathbb{T}$, with continuous functions in place of Riemann integrable functions, follows from (7.17).)

The other two questions considered by Banach involve isometry invariance rather than translation invariance. (Indeed, in terms of the physical world, invariance under isometries (rigid motions) is more natural than just invariance under translations.) Let G_n be the group of isometries of $\mathbb{R}^n$. Then G_n is a semidirect product $\mathbb{R}^n \times_\rho \mathrm{O}(n,\mathbb{R})$. Further, Lebesgue measure λ_n on $\mathbb{R}^n$ is G_n-invariant on $\mathscr{M}(\mathbb{R}^n)$. (Indeed, the equality

$$\lambda_n(TA) = \lambda_n(A)$$

follows from the "change of variables" formula for n-variables, since the Jacobian of the function $x \to Tx$ is the determinant of an element of $\mathrm{O}(n,\mathbb{R})$ and so has modulus 1.)

The second problem considered by Banach arose out of results proved by Hausdorff [**2**]. Hausdorff raised the following question, which originated with

Lebesgue and which Banach calls "*le problème large de la mesure*". It is often referred to as "*the problem of measure.*" The question is (Hausdorff [**2**, p. 401]) *does there exist an invariant measure for* $(G_n, \mathbb{R}^n, [0,1]^n)$? So ((2.32)) we are asking if there exists a G_n-invariant, finitely additive, positive measure μ on $\mathscr{P}(\mathbb{R}^n)$ such that $\mu([0,1]^n) = 1$.

Hausdorff showed (ibid., p. 469ff.) that there is no such measure when $n \geq 3$. The crucial step of his argument was his construction of a partition $\{P, S_1, S_2, S_3\}$ of $\mathbb{S}^2 = \{(x_1, x_2, x_3) \in \mathbb{R}^3 \colon \sum_{i=1}^3 x_i^2 = 1\} \subset \mathbb{R}^3$ and elements $\phi, \psi \in \mathrm{SO}(3, \mathbb{R})$ such that P is countable and

$$\phi(S_1) = S_2 \cup S_3, \qquad \psi(S_1) = S_2, \qquad \psi^2(S_1) = S_3.$$

This initiated a line of development leading to the Banach-Tarski Theorem of 1924 (Banach and Tarski [**1**]). (This theorem embodies the Banach-Tarski Paradox.) A short account of the theorem is given in (3.16).

This leaves the cases $n = 1, 2$ to be dealt with. Banach showed in these cases that there does exist an invariant measure for $(G_n, \mathbb{R}^n, [0,1]^n)$. This follows from (2.32(i)). Indeed, in the latter, we can take $X = \mathbb{R}^n, A = [0,1]^n$, and (b) easily follows by applying λ_n to the left-hand side of (1). Since G_1 and G_2 are amenable as discrete (Problem 0–25), the desired result follows.

Von Neumann [**1**] realised that the case $n = 1, 2$ could be generalised to groups carrying a finitely additive, invariant positive measure of total mass one, that is, to amenable groups. For example, if H is an amenable subgroup of G_n for any n, then there exists an invariant measure for $(H, \mathbb{R}^n, [0,1]^n)$. Von Neumann also showed that versions of the Banach-Tarski Theorem can be obtained in two dimensions by considering, instead of the amenable group G_2, the (nonamenable) group A_2 of all affine transformations of $\mathbb{R}^2$ that preserve area. (The nonamenability of A_2 follows from the inclusions ((3.2)): $F_2 \subset \mathrm{SL}(2, \mathbb{R}) \subset A_2$, where the fact that $T \in \mathrm{SL}(2, \mathbb{R})$ is area preserving follows by using the change of variables formula for multiple integrals and the equality $\det T = 1$.) Von Neumann also realised the importance of F_2 in the problem of measure: G_n is not amenable if and only if G_n contains F_2 as a subgroup. (For more in this direction see Wagon [**2**], [**S2**].) Indeed, if $n \geq 3$, we can regard $\mathrm{SO}(3, \mathbb{R})$ as a subgroup of G_n by allowing the elements of $\mathrm{SO}(3, \mathbb{R})$ to act on the first three coordinates, and then (3.9) gives F_2 as a subgroup of $\mathrm{SO}(3, \mathbb{R})$. The reader is referred to S. Wagon's monograph [**S2**] on the Banach-Tarski Paradox for a comprehensive account of Banach's second problem and its concomitants.

The general problem of measure, considered in (2.32), is concerned, of course, with the existence of an invariant measure μ for a triple (G, X, A). We showed in (2.32) that if G is amenable, then such a measure μ exists if and only if (G, X, A) has the Translate Property.

Tarski proved a remarkable theorem (which we shall call *Tarski's Theorem*) in which it is shown that a measure μ exists if and only if A admits a "paradoxical decomposition." (One might hope for such a characterisation, given the "paradoxes" inherent in Hausdorff's partition of $\mathbb{S}^2$ mentioned earlier and in the Banach-Tarski Theorem.)

The set A is said to *admit a paradoxical decomposition* if there exists a partition $A_1, \ldots, A_m, B_1, \ldots, B_n$ of A and elements $x_1, \ldots, x_m, y_1, \ldots, y_n$ of G such that $\{x_1A_1, \ldots, x_mA_m\}$ is a partition of A and $\{y_1B_1, \ldots, y_nB_n\}$ is a partition of A.

If such a decomposition exists, then there cannot exist an invariant measure for (G, X, A). For if μ were such a measure, then

$$1 = \mu(A) = \sum_{i=1}^{m} \mu(A_i) + \sum_{j=1}^{n} \mu(B_j) = \sum_{i=1}^{m} \mu(x_iA_i) + \sum_{j=1}^{n} \mu(y_jB_j)$$
$$= \mu\left(\bigcup_{i=1}^{m} x_iA_i\right) + \mu\left(\bigcup_{j=1}^{n} y_jB_j\right) = 2\ (!).$$

This is the easy implication in Tarski's Theorem. The converse implication is much more difficult, and is established, using (3.13) and (3.14), in (3.15).

It remains to mention briefly the third problem considered by Banach. This was raised by Lebesgue and Ruziewicz and is often referred to as the *Banach-Ruziewicz Problem.*

Recall that λ_n is Lebesgue measure on $\mathbb{R}^n$ and that $\mathscr{M}_b(\mathbb{R}^n)$ is the algebra of bounded Lebesgue measurable sets. The Banach-Ruziewicz Problem then is *does there exist a finitely additive, G_n-invariant, positive measure μ on $\mathscr{M}_b(\mathbb{R}^n)$ such that $\mu([0,1]^n) = 1$ and $\mu \neq \lambda_n$.* (There is a version of this problem for the n-sphere $\mathbb{S}^n$.)

Banach showed that the answer to this question is affirmative if $n = 1$ or 2, but left the cases $n \geq 3$ unanswered.

The problem for $n \geq 3$ was only solved in the 1980s and is much the deepest of the three problems considered by Banach.

The remarkable solution to the Banach-Ruziewicz Problem (for the n-sphere) will be considered in Chapter 4.

(3.13) Two results on equidecomposability. We shall state two straightforward results on equidecomposability that are needed to prove Tarski's Theorem on paradoxical decompositions ((3.15)). A third equidecomposability result is also needed for Tarski's Theorem. This result is deeper than the other two, and depends on König's Theorem which will be stated and proved in (3.14).

Let G be a group and X a left G-set with $ex = x$ for all $x \in X$. Following Wagon [**3**] we say that subsets A and B of X are *G-equidecomposable* (or simply *equidecomposable*) if there exist partitions $\{A_1, \ldots, A_m\}$ and $\{B_1, \ldots, B_m\}$ of A and B respectively and elements $s_1, \ldots, s_m$ of G such that $s_iA_i = B_i$ for $1 \leq i \leq m$. If A and B are equidecomposable, then we write $A \cong B$. We say $A \precsim B$ if $A \cong C$ for some subset C of B. The two following results appear in Banach and Tarski [**1**]. See also Stromberg [**1**].

The proofs of these two results are set as Problem 3–13.

(i) $\cong$ *is an equivalence relation on $\mathscr{P}(X)$.*

(ii) *Let* $A, B \in \mathscr{P}(X)$ *be such that* $A \precsim B$ *and* $B \precsim A$. *Then* $A \cong B$.

(3.14) König's Theorem and division. Let G and X be as in (3.13). The following result will be called (Tarski [**2**]) *the Division Theorem* (for equidecomposable sets), since it is the equidecomposable version of the trivial arithmetical fact: *if* $a, b \in \mathbb{R}$, $r \in \mathbb{P}$, *and* $ra = rb$, *then* $a = b$ (which, of course, is obtained by *dividing* by r)!

(i) (The Division Theorem). *Let* $A_1, \ldots, A_r, B_1, \ldots, B_r \in \mathscr{P}(X)$ *and be such that*

(a) $A_i \cap A_j = \varnothing = B_i \cap B_j$ *whenever* $i \neq j$;

(b) $A_i \cong A_1$ and $B_i \cong B_1$ *for all* i;

(c) $(\bigcup_{i=1}^r A_i) \cong (\bigcup_{i=1}^r B_i)$.

Then $A_1 \cong B_1$.

This result was proved by Banach and Tarski in the case where n is a power of 2. The general case depends on a Theorem due to König [**1**] and is proved in (iii) below. König's proof of his Theorem is rather involved; the modern approach is substantially simpler and uses the combinatorial result known as the "Marriage Theorem." Our account is based on the treatment of the Marriage Theorem by Bondy and Murty [**5**] and uses ideas in König [**1**]. We have also been greatly influenced by Wagon's treatment of König's Theorem (Wagon [**S2**, Theorem 8.11]).

To ease the exposition, it is convenient to express König's Theorem in graph-theoretic terms.

Let X, Y, and E be disjoint sets. We think of X and Y as sets of vertices (points) and E as a set of edges joining points of X to points of Y. Let $F \colon E \to X \times Y$. If $F(e) = (x, y)$, then we say that e *joins* x to y (or y to x), and x, y are the *ends* of e. Note that x, y can be the ends of more than one edge. A *path* in (X, Y, E) is a finite sequence of edges $(e_1, \ldots, e_n)$ together with a finite sequence $(a_0, \ldots, a_n)$ of vertices with a_i an end-point of e_i and e_{i+1} $(1 \leq i \leq n-1)$ and a_0, a_n end-points of e_1, e_n. We say that the path *joins* a_0 to a_n. The diagram below illustrates a path joining a_0 to a_6.

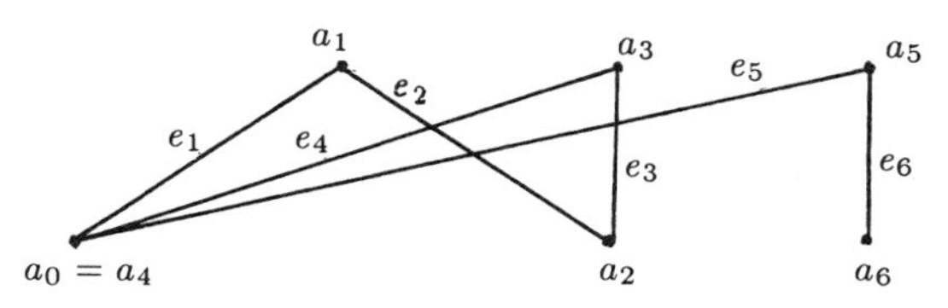

Let $k \in \mathbb{P}$. We say that (X, Y, E) is a *k-regular bipartite graph* if each $x \in X$ is an end of exactly k edges and each $y \in Y$ is an end of exactly k edges.

A *matching* of subsets A of X and B of Y is a subset F of E such that every $a \in A$ $[b \in B]$ is an end of exactly one $f \in F$ and the ends of every $f \in F$ are in $A \cup B$. When X and Y are finite, König's Theorem below implies the colourful classical "Marriage Theorem": if every girl in a village knows exactly k boys and every boy knows exactly k girls, then the boys and girls can be married off in such a way that each boy marries a girl he knows.

In the following proof, the reader will probably find it helpful to have a sketch of a bipartite graph in front of himself/herself.

(ii) (*König's Theorem*) *Let (X, Y, E) be a k-regular bipartite graph. Then there is a matching of X and Y.*

Let $G' = X \cup Y$. For $S \subset G'$, let $N(S)$ be the set of vertices in G' joined by an edge to a point of S. There are two cases to consider.

Case 1: *G is finite.* By k-regularity, $k|X| = |E| = k|Y|$ so that $|X| = |Y| = n$ for some $n \in \mathbb{P}$.

(a) *$|N(S)| \geq |S|$ for all $S \subset X$.* The number of edges joining vertices in S to vertices in $N(S)$ is $k|S|$. So $\sum_{t \in N(S)} n_t = k|S|$, where n_t is the number of edges joining $t \in N(S)$ to members of S. If $|N(S)| < |S|$, then $n_t > k$ for some t, which is impossible.

(b) *There exists a matching of X and Y.* Let A, B be subsets of X, Y admitting a matching F with $|F|$ maximum possible. Suppose that $A \neq X$ and let $x \in X \sim A$. Consider the set Z of points $z \in G'$ for which there is a path $(e_1, \ldots, e_n)$ with the properties: (1) the path joins x to z, (2) the e_i are alternately in F and $E \sim F$. If $x, a_1, \ldots, z$ are the vertices associated with such a path, then all of the X-vertices $x, a_2, a_4, \ldots$ lie in A (except for x) and all of the Y-vertices $a_1, a_3, \ldots$ lie in B except possibly for z. Suppose that $z \in Y$. We show that by the maximality of F, the vertex z lies, in fact, in B. Indeed $n = 2k+1$ is odd, and if $z \notin B$ and we take $F' = (F \sim \{e_2, e_4, \ldots, e_{2k}\}) \cup \{e_1, e_3, \ldots, e_n\}$ then F' is a matching of $A \cup \{x\}$ with $B \cup \{z\}$ and $|F'| = |F| + 1 > |F|$, contradicting the maximality of F. So $T = Z \cap Y \subset B$. Let $S = Z \cap X$. It is easy to check that $T = N(S)$ and that $|T| = |S| - 1$. Then $|N(S)| < |S|$, contradicting (a). So $A = X, B = Y$ as required.

Case 2: *G' is infinite.* Clearly, being joined by a path gives an equivalence relation on G', and by k-regularity, each equivalence class is a countable, k-regular bipartite graph. If we can find a matching for each such class, then a matching for G' follows. So we can assume that G' is denumerable and every pair of vertices is path-connected.

Let $(x, y) \in X \times Y$ and define subsets Z_i of G' by: $Z_1 = \{x, y\} \cup N(\{x, y\})$, $Z_{i+1} = N(Z_i)$. Clearly, by connectedness, $G' = \bigcup_{i=1}^{\infty} Z_i$. Let $X_i = X \cap Z_i$, $Y_i = Y \cap Z_i$. Note that $Z_i \subset Z_{i+1}$, so that $X_i \subset X_{i+1}$, $Y_i \subset Y_{i+1}$. Let $j \geq 1$. By adding new vertices to X_j, Y_j and new edges, we obtain a finite k-regular bipartite graph (X'_j, Y'_j, E'_j). [We illustrate the case $j = 1, k = 2$ below.

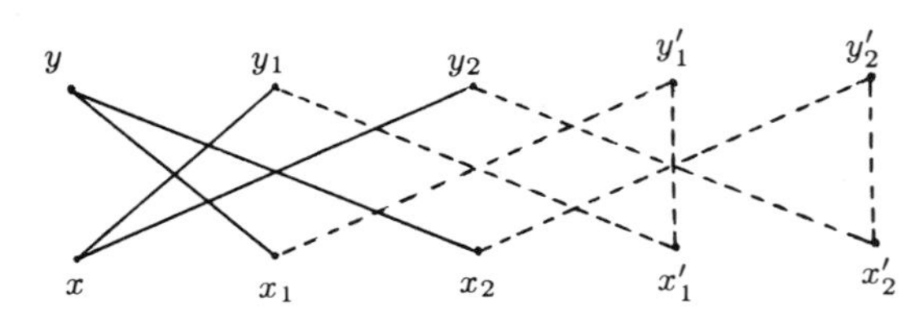

Here, $N(x) = \{y_1, y_2\}, N(y) = \{x_1, x_2\}$, and points x'_1, x'_2, y'_1, y'_2 are added on to X and Y. The new edges introduced are dotted.] By Case 1, there exists a matching of X'_j and Y'_j. For $i < j$, this matching induces a matching of X_i with a subset of Y.

An *i-matching* is a matching of X_i with a subset of Y. The above shows that there is an i-matching for all i. If $i < j$ and M_j is a j-matching, then M_j induces an i-matching $M_j \mid_{X_i}$. We now use "König's Infinity Lemma" to obtain a matching of X and Y. Since the set of 1-matchings is finite, there exist r-matchings M_r^1 $(r \geq 1)$ such that $M_r^1 \mid_{X_1} = M_1^1$. Let $M_1 = M_1^1$. Similarly, there exist r-matchings M_r^2 $(r \geq 2)$ such that $M_2^2 \mid_{X_1} = M_1$ and $M_r^2 \mid_{X_2} = M_2^2$. Let $M_2 = M_2^2$. An obvious induction argument produces a sequence $\{M_i\}$ of i-matchings with $M_{i+1} \mid_{X_i} = M_i$. Then $M = \bigcup_{i=1}^{\infty} M_i$ is a matching for X and Y.]

(iii) (*Proof of the Division Theorem.*) Let G, X, r, A_i, B_i be as in (i). Let $\phi_i \colon A_i \to A_1$, $\psi_i \colon B_i \to B_1$, and $\phi \colon \bigcup_{i=1}^{n} A_i \to \bigcup_{i=1}^{n} B_i$ be the equidecomposability bijections associated with (b) and (c) of (i). We take ϕ_1, ψ_1 to be the identity maps. We define an r-regular bipartite graph (X_1, Y_1, E_1) as follows. Let $X_1 = A_1 \times \{0\}$, $Y_1 = B_1 \times \{1\}$. (The reason for the "$\times\{0\}$" and "$\times\{1\}$" is that of ensuring disjointness for X_1 and Y_1.) Assign an edge joining $(a, 0) \in X_1$ to $(b, 1) \in Y_1$ for each pair i, j such that $\phi\phi_i^{-1}(a) \in B_j$ and $\psi_j\phi\phi_i^{-1}(a) = b$. (The diagram below will help clarify the situation.

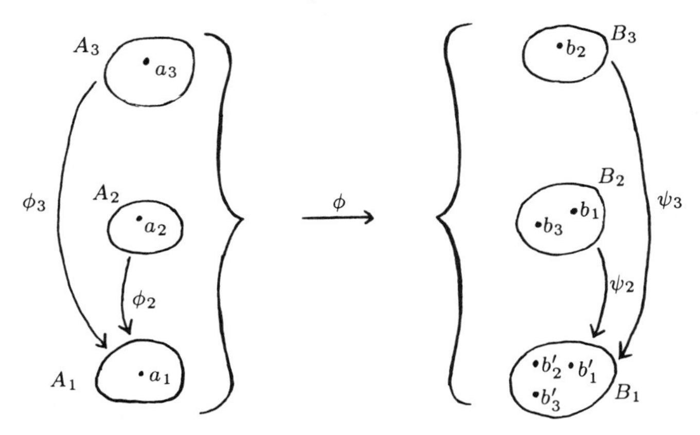

Here, $r = 3$, $a_1 = a$, $a_2 = \phi_2^{-1}(a)$, $a_3 = \phi_3^{-1}(a)$, and $b_i = \phi(a_i)$. Also $b'_1 = \psi_2(b_1)$, $b'_2 = \psi_3(b_2)$ and $b'_3 = \psi_2(b_3)$. Note that ϕ need not map A_i into B_i, and that the b'_i need not be distinct. We assign edges joining $(a_1, 0)$ to $(b'_1, 1), (b'_2, 1)$ and $(b'_3, 1)$.) By König's Theorem, there exists a matching M for X_1 and Y_1. For $a \in A_1$, let $\alpha(a) \in B_1$ be such that $(\alpha(a), 1)$ is the other end of the edge in M containing $(a, 0)$. Then $\alpha(a) = \psi_j\phi\phi_i^{-1}(a)$ for some i, j. Since ψ_j, ϕ and ϕ_i are given by G-translates, $\alpha(a)$ is of the form $y_{j,m}s_kx_{i,l}^{-1}a$. There are only a finite number of possibilities for the $y_{j,m}$, s_k and $x_{i,l}$ in G, and we obtain a partition $K_1, \ldots, K_q$ of A_1 and elements z_p, each of the form $y_{j,m}s_kx_{i,l}^{-1}$, such that $\{z_1K_1, \ldots, z_qK_q\}$ is a partition of B_1. So $A_1 \cong B_1$.]

(3.15) Tarski's Theorem on amenability and paradoxical decompositions. (Tarski [**2**], [**4**]). Let (G, X, A) be as in (2.32). Tarski's Theorem is concerned with the equivalence between the existence of an invariant measure for (G, X, A) and the nonexistence of a paradoxical decomposition of A. Tarski's beautiful proof depends on showing that such a measure μ exists if and only if there exists a certain homomorphism on a semigroup Σ associated with (G, X, A) and then using the Division Theorem of (3.14) to complete the proof of his theorem.

(Is there a short proof of Tarski's Theorem? This question seems open.) Let $\mathscr{D} = \{B \subset X$: there exist $n \in \mathbb{P}$ and $s_1, \ldots, s_n \in G$ such that $B \subset \bigcup_{i=1}^n s_i A\}$. The elements of $\mathscr{D}$ were called *bounded* in (2.32). From the last sentence of (2.32), we see that there exists an invariant measure on (G, X, A) if and only if there exists a positive, G-invariant, finitely additive measure ν on $\mathscr{D}$ with $\nu(A) = 1$.

Reference to the definition of a paradoxical decomposition of A (given in (3.12)) shows that 3 partitions of A are involved. To cope with this, it is convenient to have disjoint copies of A available. For this reason, we introduce $Y = X \times \mathbb{P}$, where X is identified with $X \times \{1\}$. Let $\Pi(\mathbb{P})$ be the group of permutations of $\mathbb{P}$ (that is, the group of bijective maps from $\mathbb{P}$ onto $\mathbb{P}$). Let $H = G \times \Pi(\mathbb{P})$. Then H acts on Y by means of the formula:

$$(s,p)(x,n) = (sx, p(n)) \qquad ((s,p) \in H,\ (x,n) \in Y).$$

Let I be the identity permutation on $\mathbb{P}$. Then (e, I) is the identity of H and $(e, I)y = y$ for all $y \in Y$. Let

$$\mathscr{N} = \{C \in \mathscr{P}\{(Y): C \subset B \times F \text{ for some } B \in \mathscr{D}, F \in \mathscr{F}(\mathbb{P})\}.$$

Clearly, $\mathscr{N}$ is H-invariant and is a ring of subsets of Y.

(i) *There exists an invariant measure for (G, X, A) if and only if there exists an H-invariant, positive, finitely additive measure ν on $\mathscr{N}$ with $\nu(A) = 1$.*

[Every $N \in \mathscr{N}$ can be written uniquely in the form

$$\bigcup_{i=1}^{r} C_i \times \{j_i\}$$

where $1 \leq j_1 < j_2 < \cdots < j_r, C_i \in \mathscr{D}$ and each $C_i \neq \varnothing$. Note that if $(s,p) \in H$, then

$$(s,p)(C) = \bigcup_{i=1}^{r} sC_i \times \{k_i\}$$

where the k_i are distinct. Every invariant measure ν for $\mathscr{D}$ gives rise to an appropriate invariant measure ξ on $\mathscr{N}$, where

$$\xi(N) = \sum_{i=1}^{r} \nu(C_i).$$

Conversely, given an invariant measure ξ on $\mathscr{N}$, we obtain an invariant measure ν on $\mathscr{D} \subset \mathscr{N}$ by restriction and by identifying each $s \in G$ with $(s, I) \in H$.]

Let us write $N_1 \cong N_2$ in $\mathscr{N}$ whenever N_1 and N_2 are H-equidecomposable. By (3.13(i)), $\cong$ is an equivalence relation on $\mathscr{N}$. Let $\Sigma = \mathscr{N}/\cong$, and $\tau: \mathscr{N} \to \Sigma$ be the canonical surjection. Let $\alpha = \tau(A)$.

We make Σ into an abelian semigroup as follows. Let $N_1, N_2 \in \mathscr{N}$. There exists $h \in H$ such that $h(N_1) \cap N_2 = \varnothing$. Define $N_1 +_h N_2 = h(N_1) \cup N_2$.

One readily checks that if $N_1 \cong N_1'$, $N_2 \cong N_2'$, and $k \in H$ is such that $k(N_1') \cap N_2' = \varnothing$, then $N_1 +_h N_2 \cong N_1' +_k N_2'$. Hence it makes sense to define an element $\tau(N_1) + \tau(N_2)$ by setting it equal to $\tau(N_1 +_h N_2)$. It is readily checked that Σ is an abelian semigroup under the map $(\gamma, \tau) \to (\gamma + \tau)$.

The next result reduces the problem of measure for (G, X, A) to that of the existence of a certain homomorphism from Σ into the (additive) semigroup $[0, \infty)$.

(ii) *There exists an invariant measure for (G, X, A) if and only if there exists a homomorphism $f: \Sigma \to [0, \infty)$ such that $f(\alpha) = 1$.*

[Suppose that there exists an invariant measure ν for (H, Y, A). Since $\nu(N_1) = \nu(N_2)$ if $N_1 \cong N_2$ in $\mathscr{N}$, it follows that we can define a map $f: \Sigma \to [0, \infty)$ by

$$f(\tau(N)) = \nu(N) \qquad (N \in \mathscr{N}).$$

If $N_1, N_2 \in \mathscr{N}$ with $N_1 \cap N_2 = \varnothing$, then $\nu(N_1 \cup N_2) = \nu(N_1) + \nu(N_2)$. Referring to the definition of addition in Σ, we see that f is a homomorphism. Since $\nu(A) = 1$, it follows that $f(\alpha) = 1$.

Conversely, suppose that $g: \Sigma \to [0, \infty)$ is a homomorphism with $g(\alpha) = 1$. For $N \in \mathscr{N}$, define $\xi(N) = g(\tau(N))$. Then $\xi(A) = 1$ and since, for $h \in H$ and $N \in \mathscr{N}, N \cong hN$, it follows that $\xi(hN) = \xi(N)$. If $N_1, N_2 \in \mathscr{N}$ with $N_1 \cap N_2 = \varnothing$, then $\tau(N_1 \cup N_2) = \tau(N_1) + \tau(N_2)$, so that $\xi(N_1 \cup N_2) = \xi(N_1) + \xi(N_2)$. Obviously $\xi(\varnothing) = 0$, so that ξ is an H-invariant, positive, finitely additive measure on $\mathscr{N}$ with $\xi(A) = 1$.

Now use (i) to complete the proof.]

The following extension result for homomorphisms on an abelian semigroup S will be required. We shall say $s \le t$ in S for $s, t \in S$ if either $s = t$ or there exists $w \in S$ such that $s + w = t$.

For related extension results, see Ross [**1**], [**2**] and Comfort and Hill [**1**].

(iii) *Let T be a subsemigroup of the abelian semigroup S and $F: T \to [0, \infty)$ an (additive) homomorphism. Let $e \in T$ be such that $F(e) = 1$ and, for each $s \in S$, there exists $n \in \mathbf{P}$ such that $s \le ne$. Then there exists a homomorphism $g: S \to [0, \infty)$ with $g\,|_T = F$ if and only if $F(s) \le F(t)$ whenever $s, t \in T$ and $s \le t$ in S.*

[If F can be extended to a homomorphism $g: S \to [0, \infty)$, then obviously $g(s) \le g(t)$ whenever $s, t \in T$ and $s \le t$.

Conversely, suppose that $F(s) \le F(t)$ whenever $s, t \in T$ and $s \le t$. Let $\mathscr{B}$ be the set of all pairs (F', B') where B' is a subsemigroup of S, F' is homomorphism from B' into $[0, \infty)$, $T \subset B'$, $F'\,|_T = F$, and $F'(s) \le F'(t)$ whenever $s, t \in B'$ and $s \le t$ (in S). Clearly, $(F, T) \in \mathscr{B}$, so that $\mathscr{B} \ne \varnothing$. Then $\mathscr{B}$ becomes partially ordered under $\le$, where $(F', B') \le (F'', B'')$ if $B' \subset B''$ and $F''\,|_{B'} =$

F'. Let $\mathscr{A}$ be a linearly ordered subset of $\mathscr{B}$, and let $S_1 = \bigcup\{B'\colon (F', B') \in \mathscr{A}\}$. Define $F_1\colon S_1 \to [0, \infty)$ by $F_1(s) = F'(s)$, where $(F', B') \in \mathscr{A}$, and $s \in B'$. Clearly (F_1, S_1) is an upper bound for $\mathscr{A}$ in $\mathscr{B}$. Thus Zorn's Lemma applies to give a maximal element (g, W) of $\mathscr{B}$. We need to show that $W = S$.

Suppose, on the contrary, that $W \neq S$, and let $a \in S \sim W$. Let

$$W' = \{w + na\colon w \in W,\ n \in \mathbb{P}\} \cup \mathbb{P}_a.$$

Obviously, W' is a subsemigroup of S. Let $x \in W'$ and define

$$g'(x) = \sup\{(g(w_1) - g(w_2))/n\colon n \in \mathbb{P}, w_1, w_2 \in W, \text{ and } w_1 \leq w_2 + nx\}. \tag{1}$$

Since for $w \in W$, $w \leq w + 1x$, $g'(x)$ is well defined and ≥ 0. Further, $e \in T \subset W$ and, for some $k \in \mathbb{P}$, $x \leq ke$, so that if $w_1 \leq w_2 + nx$, then $w_1, (w_2 + nke) \in W$ and $w_1 \leq (w_2 + nke)$ in S. Since $(g, W) \in \mathscr{B}$,

$$(g(w_1) - g(w_2))/n \leq g(ke) = F(ke) = k.$$

Thus $g'(x) < \infty$, and $g'\colon W' \to [0, \infty)$.

We will contradict the maximality of (g, W) by showing that $(g', W') \in \mathscr{B}$. To this end, it is sufficient to show that if $x = w + ka$ ($w \in W$, $k \in \mathbb{N}$) and $n \in \mathbb{P}$, then

$$g'(x) = g(w) + kg'(a), g'(na) = ng'(a). \tag{2}$$

Suppose that $x = w + ka$ as above and let $w_1, w_2 \in W$ and $m \in \mathbb{P}$ be such that

$$w_1 \leq w_2 + ma. \tag{3}$$

Then

$$\begin{aligned} mw + kw_1 \leq mw + kw_2 + kma &= kw_2 + m(w + ka) \\ &= kw_2 + mx. \end{aligned}$$

Referring to (1), we obtain

$$\begin{aligned} g'(x) &\geq [g(mw + kw_1) - g(kw_2)]/m \\ &= g(w) + k[(g(w_1) - g(w_2))/m]. \end{aligned} \tag{4}$$

Taking the supremum (over all choices of w_1, w_2, m satisfying (3)) of the right-hand side of (4) and using (1) with a in place of x, we have

$$g'(x) \geq g(w) + kg'(a). \tag{5}$$

Conversely, suppose that $v_1, v_2 \in W$ and $n \in \mathbb{P}$ are such that

$$v_1 \leq v_2 + nx. \tag{6}$$

Then $v_1 \leq (v_2 + nw) + nka$, and so

$$\begin{aligned} g'(a) &\geq [g(v_1) - g(v_2 + nw)]/(nk) \\ &= \tfrac{1}{k}[(g(v_1) - g(v_2))/n] - \tfrac{1}{k}g(w). \end{aligned} \tag{7}$$

Taking the supremum of the right-hand side of (7) over all choices of v_1, v_2, n satisfying (6), we obtain

$$g'(a) \geq \tfrac{1}{k} g'(x) - \tfrac{1}{k} g(w).$$

So

$$g'(x) \leq g(w) + kg'(a). \tag{8}$$

The first equality of (2) now follows from (5) and (8), and the second equality of (2) from an easier version of the above proof. The resulting contradiction gives $W = S$. This completes the proof.]

We now continue with the proof of Tarski's Theorem.

(iv) *There exists an invariant measure for (G, X, A) if and only if $k\alpha \neq l\alpha$ whenever $k, l \in \mathbb{P}$ with $k \neq l$.*

[Suppose that there exists an invariant measure for (G, X, A). By (ii), there exists a homomorphism $f\colon \Sigma \to [0, \infty)$ such that $f(\alpha) = 1$. If $k, l \in \mathbb{P}$ with $k \neq l$, then $f(k\alpha) = k$, $f(l\alpha) = l$ so that $k\alpha \neq l\alpha$.

Conversely, suppose that $k\alpha \neq l\alpha$ whenever $k, l \in \mathbb{P}$ with $k \neq l$. Let $T = \{n\alpha\colon n \in \mathbb{P}\}$. Then T is a subsemigroup of Σ. If $\beta \in T$, then there exists exactly one $n \in \mathbb{P}$ with $\beta = n\alpha$. Define $F\colon T \to [0, \infty)$ by $F(\beta) = n$. Obviously F is a homomorphism, and $F(\alpha) = 1$. We now show that the hypotheses of (iii) are satisfied with S, e replaced by Σ, α.

Let $\gamma \in \Sigma$. We require to show that there exists $\delta \in \Sigma$ and $r \in \mathbb{P}$ such that $\gamma + \delta = r\alpha$. To this end, let $C \in \mathscr{N}$ be such that $\tau(C) = \gamma$. We can write C as a disjoint, finite union of sets $B_m \times \{m\}$, where $B_m \in \mathscr{D}$, and since $\tau(C) = \sum \tau(B_m)$, we can suppose that $C \in \mathscr{D}$. So $C \subset \bigcup_{i=1}^r s_i A$ for some $s_i \in G$. Write C as a disjoint union $\bigcup_{i=1}^r C_i'$, with $C_i' \subset s_i A$ and find $h_i \in H$ such that $h_i(s_i A) \cap h_j(s_j A) = \varnothing$ if $i \neq j$. Let $B_i' = h_i(s_i A) \sim h_i(C_i')$ and $B' = \bigcup B_i'$. Then $\gamma + \tau(B') = r\alpha$ and we can take $\delta = \tau(B')$.

We now show that if $s, t \in T$ with $s \leq t$ in Σ, then $F(s) \leq F(t)$. Suppose, then, that $s = n\alpha$ and $t = m\alpha$ with $s \leq t$ in Σ. If $n \leq m$, there is nothing to prove. Suppose that $n > m$. Let $C_n, C_m \in \mathscr{N}$ with $\tau(C_n) = n\alpha$, $\tau(C_m) = m\alpha$. Since $n\alpha = m\alpha + (n - m)\alpha$, it follows that $C_m \precsim C_n$ (where we use the notation of (3.13) with respect to the action of H on Y). Since $s \leq t$, we can find $\beta \in \Sigma$ with $n\alpha + \beta = m\alpha$, and it follows that $C_n \precsim C_m$. By (3.13(ii)), we have $C_n \cong C_m$ and $n\alpha = m\alpha$. But by hypothesis, $n\alpha \neq m\alpha$ since $n \neq m$, and a contradiction results. So $n \leq m$, and $F(s) \leq F(t)$.

Thus (iii) applies to give a homomorphism $g\colon \Sigma \to [0, \infty)$ with $g(\alpha) = 1$. Hence, by (ii), there exists an invariant measure for (G, X, A).]

(v) *There exists an invariant measure for (G, X, A) if and only if $\alpha \neq 2\alpha$.*

[By (iv) we need only establish that if $\alpha \neq 2\alpha$, then $k\alpha \neq l\alpha$ whenever $k \neq l$ in $\mathbb{P}$. Suppose, then, that $\alpha \neq 2\alpha$ and that there exist $k, l \in \mathbb{P}$ with $k \neq l$ such that $k\alpha = l\alpha$. We will derive a contradiction. Without loss of generality, we can suppose that $k < l$. Let $l = k + r$ with $r \geq 1$. Then by adding on $r\alpha$ at each stage,

$$k\alpha = (k+r)\alpha, \qquad (k+r)\alpha = (k+2r)\alpha, \ldots$$

and we obtain $k\alpha = (k+pr)\alpha$ for all $p \geq 1$. Putting $p = k$, we obtain $k\alpha = k[(1+r)\alpha]$. Translating this equality into the context of $\mathscr{N}$, we see that the Division Theorem of (3.14) applies to give $\alpha = (1+r)\alpha$. If $r = 1$, then $\alpha = 2\alpha$ and the desired contradiction follows. If $r > 1$, then by adding on $(r-1)\alpha$ to both sides of the equality $\alpha = (1+r)\alpha$, it follows that $r(\alpha) = r(2\alpha)$, and again using the Division Theorem, we obtain $\alpha = 2\alpha$ and the desired contradiction.]

(vi) (Tarski's Theorem). *There exists an invariant measure for* (G, X, A) *if and only if* A *does not admit a paradoxical decomposition* ((3.12)) (*with respect to* G).

[If there exists an invariant measure for (G, X, A), then, by (3.12), A does not admit a paradoxical decomposition.

Conversely, suppose that G does not admit a paradoxical decomposition. By (v) it is sufficient to show that $\alpha \neq 2\alpha$. Suppose, on the contrary, that $\alpha = 2\alpha$. Then $A \cong A \cup (A \times \{2\})$ in $\mathscr{N}$ with respect to H. (Recall that A is identified with $A \times \{1\}$.) So there exists a partition $A_1, \ldots, A_m, B_1, \ldots, B_n$ of A and elements $\{(s_1, p_1), \ldots, (s_m, p_m), (t_1, q_1), \ldots, (t_n, q_n)\}$ of H such that $p_i(1) = 1$, $q_i(1) = 2$, and $\{(s_1, p_1)A_1, \ldots, (s_m, p_m)A_m, (t_1, q_1)B_1, \ldots, (t_n, q_n)B_n\}$ is a partition of $A \cup (A \times \{2\})$. It follows that $\{s_1A_1, \ldots, s_mA_m\}$ and $\{t_1B_1, \ldots, t_nB_n\}$ are partitions of A so that A admits a paradoxical decomposition. This contradiction establishes the Theorem.]

Tarski's Theorem, applied with (G, G, G) in place of (G, X, A) yields the following remarkable characterisation of amenable groups.

(vii) *The group* G *is not amenable if and only if there exists a partition* $\{A_1, \ldots, A_m, B_1, \ldots, B_n\}$ *of* G *and elements* $x_1, \ldots, x_m, y_1, \ldots, y_n$ *of* G *such that* $\{x_1A_1, \ldots, x_mA_m\}$ *and* $\{y_1B_1, \ldots, y_nB_n\}$ *are partitions of* G.

Is there a similar characterisation of amenability for discrete semigroups? See Problem 3–15 for the locally compact group version.

Rosenblatt [**11**] discusses paradoxical decompositions in the context of Boolean algebras. A version of Tarski's Theorem involving *countably* additive measures is given by Chuaqui [**1**].

(3.16) The Banach-Tarski Theorem. The Banach-Tarski Paradox (Theorem) was established in the paper of Banach and Tarski [**1**]. We will be content to base our account on the short proof in Wagon [**S2**]. A good account of the classical proof, which uses the fact that $\mathbb{Z}_2 * \mathbb{Z}_3$ is a subgroup of $\mathrm{SO}(3, \mathbb{R})$, is given by Stromberg [**1**].

(*The Banach-Tarski Theorem*). *Let* A *and* B *be bounded subsets of* $\mathbb{R}$ *each having non-empty interior. Then* $A \cong B$ (*with respect to the action of* G_3 *on* $\mathbb{R}^3$).

The "Paradox" associated with this Theorem can be expressed in the following dramatic form: *a billiard ball can be chopped into a finite number of pieces and the pieces fitted together to form a life-size statue of Banach*! The flaw in the "paradox" is that the "real" world is being confused with its geometrical model. (Observe, for example, that at least one of the "pieces" has to be *nonmeasurable* (since Lebesgue measure preserves volumes)!)

The key to the proof of the Theorem is that $SO(3, \mathbb{R})$ contains a subgroup H isomorphic to F_2. This is an easy consequence of (3.2(iii))—see Problem 3–5. Each element of $H \sim \{I\}$ is a rotation A about an axis l_A through 0, and l_A intersects the sphere $\mathbb{S}^2$ in two points α_A, β_A. Let D be the (countable) set of such points α_A, β_A. Then H acts freely on $\mathbb{S}^2 \sim D$—the free condition means that $A = I$ whenever $Ax = x$ for some $x \in \mathbb{S}^2 \sim D$. By Tarski's Theorem, H admits a paradoxical decomposition (p.d.)—in fact, it is easy to write down an explicit p.d. for F_2 acting on itself (Problem 3–12). Let $T \subset \mathbb{S}^2 \sim D$ be such that $|T \cap Hx| = 1$ for every $x \in \mathbb{S}^2 \sim D$. (The existence of T uses the Axiom of Choice.) Then $\mathbb{S}^2 \sim D$ admits a p.d. for H—indeed, if A'_i, B'_j implement a p.d. for H, then A'_iT, B'_jT implement an $\mathbb{S}^2 \sim D$-p.d.

We would like to have $\mathbb{S}^2$ itself having an $SO(3, \mathbb{R})$-p.d.—the problem is that of dealing with D. To this end, since D is countable, there exists a line l through 0 intersecting $\mathbb{S}^2$ at points of $\mathbb{S}^2 \sim D$. Further, there exists an angle θ such that the rotation ρ through θ with axis l is such that the sets $\{\rho^n(D)\colon n \geq 0\}$ are pairwise-disjoint. Let $\overline{D} = \bigcup_{n=0}^{\infty} \rho^n(D)$. Then $\rho(\overline{D}) \cup D = \overline{D}$, and it readily follows that $\mathbb{S}^2 \cong \mathbb{S}^2 \sim D$ (for $SO(3, \mathbb{R})$); and that $\mathbb{S}^2$ has an $SO(3, \mathbb{R})$-p.d. If A_i, B_j implement a p.d. for $\mathbb{S}^2$ (with respect to $SO(3, \mathbb{R})$), then A'_i, B'_j implement a p.d. for $B \sim \{0\}$, where B is the unit ball of $\mathbb{R}^3$, and for $L \subset \mathbb{S}^2$, we define $L' = \{\alpha x\colon \alpha \in (0, 1],\ x \in L\}$. Arguing as above with ρ replaced by a suitable rotation whose axis does not pass through 0, we can readily show that B has a p.d. for G_3. It follows that B is equidecomposable with the union of a finite number $n > 1$ of disjoint balls of radius 1, and an application of (3.13(ii)) yields the Banach-Tarski Theorem (cf. Problem 3–14).

R. M. Robinson [**S**] has shown that there exists a p.d. of B into five sets and that this is best possible. Robinson's work was extended by Dekker and de Groot [**1**]. A full discussion of this work is given in Chapter 4 of Wagon [**S2**].

Problems 3

1. Prove that F_2 is isomorphic to a subgroup of a Cartesian product $\prod_{\alpha \in A} G_\alpha$ of finite groups. Deduce that the Cartesian product of a family of amenable groups need not be amenable.

2. Show that there exists a compact group that is not amenable as discrete and yet contains a dense subgroup that is amenable as discrete.

3. Show that if G is the free product $*_{\alpha\in A}\mathbb{Z}_{n_\alpha}$ $(n_\alpha > 1)$, then $G \not\supset F_2$ if and only if either $|A| = 2$ and $n_\alpha = 2$ for all $\alpha \in A$ or $|A| = 1$. (Hint: use the Kurosh Subgroup Theorem.)

4. Let G be a group presented by a set $\{x_i : i \in I\}$ of generators and defining relations $x_j^{n_j} = e$ $(j \in J)$ where $n_j \geq 2$ and $J \subset I$. Show that G is amenable if and only if either $|I| = 2 = |J|$ and $n_j = 2$ $(j \in J)$ or $|I| = 1$. Show that von Neumann's Conjecture is true for the class of such groups G. (Hint: use Problem 3–3.)

5. Assuming (3.2(iii)), give an easy proof that $\mathrm{SO}(3,\mathbb{R})$ contains F_2 as a subgroup. Give an example of elements A, B which generate F_2 in $\mathrm{SO}(3,\mathbb{R})$.

6. Check Iwasawa's Theorem (3.4) directly when G is $\mathbb{T}^n$ $(n \geq 1)$.

7. Let us say that a finite-dimensional Lie algebra $\mathfrak{g}$ over $\mathbb{R}$ is amenable if its Levi-Malcev Decomposition $\mathfrak{s} \oplus \mathfrak{r}$ is such that $\mathfrak{s}$ is a compact Lie algebra (B48). Show that a connected Lie group G is amenable if and only if its Lie algebra is amenable.

8. Let G be a connected Lie group with $\mathfrak{g}$ as Lie algebra. The group G is said to be of *Type R* if for every $X \in \mathfrak{g}$, the eigenvalues of $\operatorname{ad} X$ are purely imaginary (that is, $\mathrm{Sp}(\operatorname{ad} X) \subset i\mathbb{R}$). Give examples of Type R groups. Is the "$ax + b$" group of Type R? Show that G is amenable if G is of Type R. (For more on the Type R theme, see Chapter 6.)

9. Show that every connected Lie group that, as a discrete group, has polynomial growth ((0.12)), is solvable.

10. Let G be a discrete group and $H \lhd G$. Show that G is locally finite if and only if both H and G/H are locally finite. (This was used in (3.11).)

11. Consider the Lebesgue problem as stated in (3.12). Why is there no point in replacing $\mathscr{M}_b(\mathbb{R})$ by $\mathscr{M}(\mathbb{R})$?

12. Give a paradoxical decomposition for F_2 acting on itself.

13. Prove ((3.13(i),(ii)) that $\cong$ is an equivalence relation on $\mathscr{P}(X)$ and that the "Cantor-Bernstein Theorem" holds for $\cong$.

14. Let (G, X, A) be as in (3.12), and suppose that $B \subset A$ is such that A is contained in the union of a finite number of G-translates of B. Show that if A admits a paradoxical decomposition (p.d.) then so does B and that $A \cong B$. Deduce that every subset of $\mathbb{S}^n$ with nonempty interior admits an $\mathrm{SO}(n+1,\mathbb{R})$-p.d. and that all such subsets are $\mathrm{SO}(n+1,\mathbb{R})$-equidecomposable. (Use 3–18 below.)

15. (i) Let G be a locally compact group. Show that if an open subgroup H of G admits a *Borel* p.d. then so also does G. Show that G admits a Borel p.d. if a quotient group of G does.

(ii) Prove that a locally compact group G is not amenable if and only if it admits a *Borel* paradoxical decomposition.

16. Let G be a nonamenable locally compact group. Show that if $\alpha\colon L_\infty(G) \to \mathbb{C}$ is a translation-invariant linear functional, then $\alpha(1) = 0$.

17. Fill in the details of the Banach-Tarski Theorem ((3.16)).

18. Show that if $n \geq 3$, $\mathbb{S}^n$ has a paradoxical decomposition under the action of $\mathrm{SO}(n+1, \mathbb{R})$. Deduce that the unit ball in $\mathbb{R}^n$ has a paradoxical decomposition for G_n.

CHAPTER 4

Følner Conditions

(4.0) Introduction. In this chapter, we examine in detail the growth conditions on an amenable locally compact group G known as the *Følner conditions.* These were briefly discussed in the introductory chapter. The starting place for these conditions is the Reiter-type condition which was proved in (0.8): *there exists a net* $\{f_\delta\}$ *in* $P(G)$ *such that* $\|x * f_\delta - f_\delta\|_1 \to 0$ *for all* $x \in G$. Our first objective ((4.4)) is to show that the f_δ can be taken to be such that the convergence to 0 is uniform on compacta. Our next objective is to replace the f_δ's by scaled characteristic functions $\chi_K/\lambda(K)$.

This leads to the remarkable condition of (4.10), which roughly asserts that G is amenable if and only if whenever $C \in \mathscr{C}(G)$, then we can find nonnull $K \in \mathscr{C}(G)$ that, relative to its size, is "almost invariant" under left translation by elements of C. The deepest of the conditions considered here is (4.13). In (4.16), we show that when G is amenable, we can form useful averages of functions in $L_\infty(G)$; these averages are analogues of the averages $(\sum_{r=-n}^{n} \phi(r))/(2n+1)$ in the case where $G = \mathbb{Z}$.

We complete the main section by discussing three important applications of the Reiter-type conditions that characterise the amenability of G. The first ((4.18)) determines the precise distance of $\mathrm{co}\{f * x \colon x \in G\}$ from 0 for $f \in L_1(G)$: the distance is just $|\int f \, d\lambda|$. The second and third are related to the relevance of amenability for the representation theory of G.

Our second application ((4.19)) shows that $\|f\|_1 = \|\pi_2(f)\|$ for all $f \geq 0$ in $L_1(G)$, where π_2 is the left regular representation of G. The result (4.20(ii)) gives a converse: G is amenable if there exists a self-adjoint $f \geq 0$ whose support generates a dense subgroup of G and for which $\|f\|_1 = \|\pi_2(f)\|$. (This will be important in our discussion of von Neumann's conjecture in (4.32).) The third application ((4.21)) is the "weak containment" property, which can be neatly expressed as asserting that $C_l^*(G) = C^*(G)$, that is, the representation theories of G and the (accessible) C^*-algebra $C_l^*(G)$ coincide. (See Chapter 0.)

Let G be a locally compact group. Our first result is proved in exactly the same way as (0.8), taking m to be in $\mathfrak{L}_t(G)$ rather than $\mathfrak{L}(G)$. (Recall (1.10).)

(4.1) PROPOSITION. *The group G is amenable if and only if there exists a net $\{f_\delta\}$ in $P(G)$ such that $\|\mu * f_\delta - f_\delta\|_1 \to 0$ for all $\mu \in P(G)$.*

(4.2) Notes. Examination of the proof of (0.8) shows that if $m \in \mathfrak{L}_t(G)$, we can actually ensure that $\hat{f}_\delta \to m$ weak*.

Obvious modifications of the proof, using $m \in \mathfrak{I}_t(G)$, yield that G is amenable if and only if there is a net $\{f_\delta\}$ in $P(G)$ such that for all $\mu \in P(G)$, we have $\|\mu * f_\delta - f_\delta\|_1 \to 0$ and $\|f_\delta * \mu - f_\delta\|_1 \to 0$. The natural right-handed versions of (4.1) (for example, involving $\|f_\delta * \mu - f_\delta\|_1$ rather than $\|\mu * f_\delta - f_\delta\|_1$) hold.

For $1 \leq p < \infty$ and G a locally compact group, we define

$$P_p(G) = \{f \in L_p(G)\colon f \geq 0,\ \|f\|_p = 1\}.$$

Note that $P_1(G) = P(G)$. We recall [**HR1**, §20] that $L_p(G)$ becomes a left $M(G)$-space with action given by

$$\mu * f(x) = \int f(y^{-1}x)\,d\mu(y)$$

and that $\|\mu * f\|_p \leq \|\mu\|\|f\|_p$ for all $\mu \in M(G)$, $f \in L_p(G)$.

(4.3) PROPOSITION. *Let $\{g_\delta\}$ be a net in $P_p(G)$ such that $\|t * g_\delta - g_\delta\|_p \to 0$ $(t \in G)$ uniformly on compacta. Then $\|\mu * g_\delta - g_\delta\|_p \to 0$ for all $\mu \in \mathrm{PM}(G)$.*

PROOF. Since $\|\mu * g\|_p \leq \|\mu\|\|g\|_p$ $(g \in L_p(G))$ and $\|g_\delta\|_p = 1$ for all δ, we can suppose that μ has compact support C. For any $g \in L_p(G)$ and $k \in L_q(G)$, where $p^{-1} + q^{-1} = 1$, we have, using the theorems of Fubini and Hölder (which are easily seen to apply) and the fact that $\mu \in \mathrm{PM}(G)$,

$$\begin{aligned}\left|\int (\mu * g - g)(x)k(x)\,d\lambda(x)\right| &= \left|\int_C d\mu(t)\int (t * g - g)(x)k(x)\,d\lambda(x)\right| \\ &\leq \int_C \|t * g - g\|_p\|k\|_q\,d\mu(t) \\ &\leq \|k\|_q \sup_{t\in C}\|t * g - g\|_p.\end{aligned}$$

Hence

$$\begin{aligned}\|\mu * g - g\|_p &= \sup\left\{\left|\int (\mu * g - g)(x)k(x)\,d\lambda(x)\right| : k \in L_q(G),\ \|k\|_q \leq 1\right\} \\ &\leq \sup_{t\in C}\|t * g - g\|_p.\end{aligned}$$

Since $\sup_{t\in C}\|t * g_\delta - g_\delta\|_p \to 0$, the required result follows. □

Our next result will require the use of the following inequalities: if $f, g \in P(G)$ and $p \in [1, \infty)$, then

$$(\|f^{1/p} - g^{1/p}\|_p)^p \leq \|f - g\|_1 \leq p2^{p-1}\|f^{1/p} - g^{1/p}\|_p. \tag{1}$$

This is trivial for $p = 1$. One can prove (1) for $p > 1$ by first establishing, by elementary differentiation, that

$$|b - a|^p \leq |b^p - a^p| \leq p|b - a|(a + b)^{p-1} \tag{2}$$

for all $a, b \geq 0$, and then using (2), with $b = f(x)^{1/p}$, $a = g(x)^{1/p}$, and Hölder's inequality.

References for the result below are Reiter [**1**], [**3**], [**14**], [**R**], Dieudonné [**2**], Stegeman [**1**], Hulanicki [**4**], and Day [**8**]. The characterisation of amenability given in (4.4(ii)) is often called *Reiter's condition* or *Property P_p*.

(4.4) THEOREM. *Let G be a locally compact group and $1 \leq p < \infty$. Then the following statements are equivalent:*

(i) *G is amenable;*

(ii) *if $C \in \mathscr{C}(G)$ and $\varepsilon > 0$, then there exists $f \in P_p(G)$ such that*

$$\|x * f - f\|_p < \varepsilon \qquad (x \in C); \tag{1}$$

(iii) *there exists a net $\{g_\delta\}$ in $P_p(G)$ such that $\|\mu * g_\delta - g_\delta\|_p \to 0$ for all $\mu \in \mathrm{PM}(G)$.*

PROOF. We prove that (i) implies (ii). Suppose that (i) holds. Let $C \in \mathscr{C}(G)$ and $\varepsilon > 0$. Let $g \in P(G)$ and $\eta > 0$. Using the continuity of the map $x \to x * g$ from G to $L_1(G)$, we can find an open, relatively compact neighbourhood U of e such that

$$\|u * g - g\|_1 < \eta \qquad (u \in U). \tag{2}$$

Let $x_1, \ldots, x_n$ in G be such that $x_1 = e$ and $C \subset \bigcup_{i=1}^n x_i U$. Since $x_i * g \in P(G)$ for each i and as G is amenable, we can find, by (4.1), an element $g_0 \in P(G)$ such that

$$\|x_i * g * g_0 - g_0\|_1 < \eta \qquad (1 \leq i \leq n). \tag{3}$$

Set $h = g * g_0$. Let $c \in C$. Then $c = x_i u$ for some i and some $u \in U$. So

$$\begin{aligned}\|c * h - h\|_1 &\leq \|x_i * u * g * g_0 - x_i * g * g_0\|_1 + \|x_i * g * g_0 - g * g_0\|_1 \\ &\leq \|u * g - g\|_1 + \|x_i * g * g_0 - g_0\|_1 + \|g * g_0 - g_0\|_1 < 3\eta\end{aligned}$$

by (2), (3), and the fact that e is an x_i. Using (4.3(1)) and noting that $(c*h)^{1/p} = c * h^{1/p}$, we have

$$\|c * h^{1/p} - h^{1/p}\|_p < (3\eta)^{1/p}. \tag{4}$$

Taking $f = h^{1/p}$ and $(3\eta)^{1/p} < \varepsilon$, we have that (i) implies (ii).

Now suppose that (ii) holds. Then there exists a net $\{g_\delta\}$ in $P_p(G)$ such that $\|t * g_\delta - g_\delta\|_p \to 0$ uniformly on compacta, and using (4.3), (ii) implies (iii).

Finally, suppose that (iii) holds. Then a fortiori, $\|x * g_\delta - g_\delta\|_p \to 0$ for all $x \in G$. It follows from (4.3(1)) that $\|x * g_\delta^p - g_\delta^p\|_1 \to 0$ for all $x \in G$ and G is amenable ((0.8)). □

(4.5) Notes. Obvious modifications of the preceding proof give other characterisations of amenability involving a right-handed version of (ii) of (4.4). Indeed, let G be amenable, $C \in \mathscr{C}(G)$, and $\varepsilon > 0$. From (4.4), there exists $f \in P(G)$ such that $\|x * f - f\|_1 < \varepsilon$ $(x \in C^{-1})$. Since the involution $k \to k^\sim$ on $L_1(G)$ is isometric, we have

$$\|f^\sim * c - f^\sim\|_1 < \varepsilon \qquad (c \in C). \tag{1}$$

Note that for $g \in L_p(G)$ $(1 \leq p < \infty)$ and $\mu \in M(G)$ with compact support, we have

$$g * \mu(x) = \int g(xy^{-1})\Delta(y^{-1})\, d\mu(y),$$

and $g * \mu \in L_p(G)$ with

$$\|g * \mu\|_p \leq \|g\|_p \int \Delta(y)^{-1/q} d|\mu|(y) \qquad [\mathbf{HR1}, \S 20].$$

In particular, $g * c(x) = g * \delta_c(x) = g(xc^{-1})\Delta(c^{-1})$.

Due to the complications caused by the modular function Δ, care has to be taken in obtaining the analogue of (4.4(ii)) when $p > 1$. Indeed, for $h \in P(G)$,

$$\begin{aligned}(h * c)^{1/p}(y) &= [h(yc^{-1})\Delta(c^{-1})]^{1/p} \\ &= \Delta(c)^{1/q}[h(yc^{-1})^{1/p}\Delta(c^{-1})] \\ &= \Delta(c)^{1/q}(h^{1/p} * c)(y).\end{aligned}$$

Taking $h = f^{\sim}$ and using (1) and (4.3(1)), we readily obtain the result: *G is amenable if and only if, given $C \in \mathscr{C}(G)$ and $\varepsilon > 0$, there exists $h \in P_p(G)$ such that $\|\Delta(x)^{1/q} h * x - h\|_p < \varepsilon$ $(x \in C)$* (cf. Skudlarek [**1**]).

One easily checks that if C, ε, and f are as in (4.4(ii)), if $p = 1$, and if $g \in P(G)$ is such that $\|g * x - g\|_1 < \varepsilon$ for all $x \in C$, then $k = f * g$ belongs to $P(G)$ and satisfies the two-sided condition

$$(2) \qquad \|x * k - k\|_1 < \varepsilon, \quad \|k * x - k\|_1 < \varepsilon \qquad (x \in C).$$

The amenability of G is obviously equivalent to such a two-sided condition. By taking $g = f^{\sim}$ as above and replacing C in (4.4(ii)) by $C \cup C^{-1}$, we can take k in (2) to be of the form $f * f^{\sim}$ with $f \in P(G)$.

Note also that in the preceding results of this chapter, we can replace $P(G)$ $[P_p(G)]$ by $P(G) \cap C_c(G)$ $[P_p(G) \cap C_c(G)]$ since the latter set is norm dense in the former. Here, $C_c(G)$ is the space of functions in $C(G)$ vanishing at infinity.

Our next aim is to improve the characterisation (4.4(ii)) of amenability in the case $p = 1$. It will be shown ((4.10)) that the function f of (4.4(ii)) can be taken to be of the form $\chi_K/\lambda(K)$ for some nonnull $K \in \mathscr{C}(G)$. Recall that for sets A, B, the symmetric difference $A \triangle B$ is defined to be $(A \sim B) \cup (B \sim A)$. Note that

$$|\chi_A - \chi_B| = \chi_{A \sim B} + \chi_{B \sim A} = \chi_{A \triangle B}.$$

Since

$$\|x * f - f\|_1 = \|\chi_{xK} - \chi_K\|_1/\lambda(K) = \lambda(xK \triangle K)/\lambda(K)$$

for such a function f, it follows that the inequality (4.4(1)) becomes:

$$\lambda(xK \triangle K)/\lambda(K) < \varepsilon \qquad (x \in C).$$

We start by establishing some simple, but useful, results on symmetric differences in G and Haar measure.

(4.6) LEMMA. *Let $C, K \in \mathscr{C}(G)$.*

(i) *If $x \in G$, then $\lambda(xK \cap K) = \lambda(K) - \frac{1}{2}\lambda(xK \triangle K)$.*

(ii) *If $x \in C$ and $C_1 = C \cup C^{-1}$, then*

$$\lambda(xK \triangle K) \leq 2\lambda(C_1 K \triangle K).$$

(iii) *If $C \neq \varnothing$, $C_2 \in \mathscr{C}_e(G)$, and $C \subset C_2$, then*

$$\lambda(CK \triangle K) \leq 2\lambda(C_2K \triangle K).$$

PROOF. (i)

$$\begin{aligned} 2\lambda(xK \cap K) &= [\lambda(xK) - \lambda(xK \sim K)] + [\lambda(K) - \lambda(K \sim xK)] \\ &= 2\lambda(K) - \lambda(xK \triangle K). \end{aligned}$$

(ii) If $x \in C$, then

$$\lambda(xK \triangle K) = \lambda(xK \sim K) + \lambda(x^{-1}K \sim K) \leq 2\lambda(C_1K \triangle K).$$

(iii) We have $\lambda(CK) \geq \lambda(cK) = \lambda(K)$ for any $c \in C$. So

$$\begin{aligned} \lambda(CK \triangle K) &= \lambda(CK \sim K) + \lambda(K \sim CK) \\ &= (\lambda(CK \sim K) + \lambda(K) - \lambda(K \cap CK)) \\ &\leq \lambda(CK \sim K) + \lambda(CK) - \lambda(CK \cap K) \\ &= 2\lambda(CK \sim K) \leq 2\lambda(C_2K \sim K) = 2\lambda(C_2K \triangle K) \end{aligned}$$

(since $C_2K \supset K$ as $e \in C_2$). □

References for (4.7)–(4.13) are Namioka [**1**], Emerson and Greenleaf [**1**], Følner [**2**], and Frey [**1**].

(4.7) LEMMA. *Let G be amenable, $C \in \mathscr{C}(G)$ nonnull, and $\varepsilon, \delta > 0$. Then there exists nonnull $K \in \mathscr{C}(G)$ and a compact subset N of C such that $\lambda(N) < \delta$, and*

$$\lambda(xK \triangle K)/\lambda(K) < \varepsilon \qquad (x \in C \sim N). \tag{1}$$

PROOF. Let $\eta > 0$. By (4.4(ii)), we can find $f \in P(G)$ such that $\|x*f - f\|_1 < \eta$ for all $x \in C$. Then we can find a sequence $\{g_n\}$ of simple, positive functions in $L_1(G)$ such that $\|g_n - f\|_1 \to 0$. Now if $g \in L_1(G)$ is a simple function $\sum_{i=1}^N \gamma_i \chi_{E_i}$, where $\gamma_i > 0$ for all i, $E_i \cap E_j = \varnothing$ if $i \neq j$, and $0 < \lambda(E_i) < \infty$ for each i, then for each i, we can find (by inner regularity of λ) an element D_i of $\mathscr{C}(G)$ with $D_i \subset E_i$ and $\lambda(E_i \sim D_i)$ as small as desired. It follows that we can suppose that each g_n is of the form $\sum_{i=1}^N \gamma_i \chi_{D_i}$ with $\gamma_i > 0$, $D_i \in \mathscr{C}(G)$, $\lambda(D_i) > 0$, and $D_i \cap D_j = \varnothing$ for $i \neq j$. Since $g_n \to f$ in $L_1(G)$, it follows that $\|g_n\|_1 \to \|f\|_1$, and we can suppose that $\|g_n\|_1 = 1$, that is, that $g_n \in P(G)$. Finally, since

$$\|(x * g_n - g_n) - (x * f - f)\|_1 \to 0$$

it follows that we can find $g \in P(G)$ with

$$\|x * g - g\|_1 < \eta \qquad (x \in C) \tag{2}$$

such that $g = \sum_{i=1}^n \alpha_i \chi_{C_i}$, where $C_i \in \mathscr{C}(G)$ are such that $\lambda(C_i) > 0$, $C_i \cap C_j = \varnothing$ if $i \neq j$, and the order of summation is arranged so that $\alpha_1 > \alpha_2 > \cdots >$

$\alpha_n > 0$. Let $B_r = \bigcup_{i=1}^r C_i$ $(1 \le r \le n)$. Thus $B_r \subset B_s$ if $r < s$, $\lambda(B_r) > 0$ for all r, and $\chi_{C_r} = (\chi_{B_r} - \chi_{B_{r-1}})$. So, with $B_0 = \varnothing$, we have

$$\begin{aligned} g &= \sum_{i=1}^n \alpha_i \chi_{C_i} = \sum_{i=1}^n \alpha_i (\chi_{B_i} - \chi_{B_{i-1}}) \\ &= \alpha_n \chi_{B_n} + \sum_{r=1}^{n-1} (\alpha_r - \alpha_{r+1}) \chi_{B_r} = \sum_{r=1}^n \beta_r (\chi_{B_r}/\lambda(B_r)). \end{aligned}$$

Note that

$$\beta_r > 0 \qquad (1 \le r \le n), \qquad \sum_{r=1}^n \beta_r = 1. \tag{3}$$

For $x \in G$ and $1 \le r \le n$,

$$(x * \chi_{B_r} - \chi_{B_r}) = (\chi_{xB_r} - \chi_{B_r}) = (\chi_{xB_r \sim B_r} - \chi_{B_r \sim xB_r}).$$

Further, if $1 \le r, s \le n$, then $(xB_r \sim B_r) \cap (B_s \sim xB_s) = \varnothing$ since either $B_r \subset B_s$ or $B_s \subset B_r$. Hence

$$\begin{aligned} \|x * g - g\|_1 &= \left\| \sum_{r=1}^n (\beta_r/\lambda(B_r))(\chi_{xB_r \sim B_r} - \chi_{B_r \sim xB_r}) \right\|_1 \\ &= \left\| \left(\sum_{r=1}^n (\beta_r/\lambda(B_r)) \chi_{xB_r \sim B_r} \right) - \left(\sum_{r=1}^n (\beta_r/\lambda(B_r)) \chi_{B_r \sim xB_r} \right) \right\|_1 \\ &= \left\| \sum_{r=1}^n (\beta_r/\lambda(B_r)) \chi_{xB_r \sim B_r} \right\|_1 + \left\| \sum_{r=1}^n (\beta_r/\lambda(B_r)) \chi_{B_r \sim xB_r} \right\|_1 \\ &= \sum_{r=1}^n (\beta_r/\lambda(B_r)) \lambda(xB_r \,\triangle\, B_r). \end{aligned} \tag{4}$$

Integrating the continuous function $x \to \|x * g - g\|_1$ over C and using (2) and (4), we obtain

$$\sum_{r=1}^n \beta_r \int_C (\lambda(xB_r \,\triangle\, B_r)/\lambda(B_r))\, d\lambda(x) < \eta \lambda(C).$$

Using (3), we can find s such that

$$\int_C (\lambda(xB_s \,\triangle\, B_s)/\lambda(B_s))\, d\lambda(x) < \eta \lambda(C). \tag{5}$$

Then (1) is satisfied with $K = B_s$, $\eta = (\varepsilon\delta)/\lambda(C)$, and $N = \{x \in C: \lambda(xB_s \,\triangle\, B_s)/\lambda(B_s) \ge \varepsilon\}$. □

(4.8) Notes. If G is unimodular, then, by (4.5), we can suppose f in the above proof is such that $f^{\sim} = f$. Modifying the proof and replacing g by $\frac{1}{2}(g + g^{\sim})$, we can further suppose that $g^{\sim} = g$. Then the sets C_i are symmetric and we conclude *if G is unimodular, then the set K of* (4.7) *can be taken to be symmetric.*

The analogue of (4.7) for a discrete left amenable semigroup S is false in general ((4.22)). (In such an analogue, the sets C and K are finite and we can take $N = \varnothing$.) Following the argument of (4.7) in this case and using the semigroup version of (4.1) instead of (4.4(ii)), we obtain the results of the first paragraph of the above proof for S. If S is left cancellative, then the rest of the argument is valid. But if S is not left cancellative, then $x * \chi_{B_r}$ may not equal χ_{xB_r} so that the step "$x * \chi_{B_r} - \chi_{B_r} = \chi_{xB_r} - \chi_{B_r}$" in the argument fails.

Recalling the formula for the convolution in $l_1(S)$ ((0.18)), we see that if $x \in S$ and $E \in \mathscr{F}(S)$, we have $x * \chi_E(s) = 0$ if $s \notin xE$, and that

$$x * \chi_E(s) = |\{e \in E: xe = s\}| \geq 1$$

if $s \in xE$.

Thus, replacing E by B_r, it follows that $x * \chi_{B_r} \geq \chi_{xB_r}$. We still have $(xB_r \sim B_r) \cap (B_s \sim xB_s) = \varnothing$ if $s \neq r$. So if $A = \bigcup_{r=1}^n (xB_r \sim B_r)$, then

$$(\chi_{x*B_r} - \chi_{B_r})|_A = (\chi_{x*B_r} - \chi_{B_r \cap xB_r} - \chi_{B_r \sim xB_r})|_A \geq (\chi_{xB_r \sim B_r})|_A,$$

and so

$$\begin{aligned} \|x * g - g\|_1 &\geq \left\| \sum_{r=1}^n (\beta_r/\lambda(B_r))(x * \chi_{B_r} - \chi_{B_r})|_A \right\|_1 \\ &\geq \sum_{r=1}^n (\beta_r/\lambda(B_r))\lambda(xB_r \sim B_r). \end{aligned}$$

Recalling that, in this case, λ is counting measure and taking $\delta = \frac{1}{2}(!)$, the rest of the argument of (4.7) applies mutatis mutandis, and we have the following result.

(4.9) PROPOSITION. *Let S be a left amenable semigroup, C a finite subset of S, and $\varepsilon > 0$. Then there exists a nonempty, finite subset K of S such that*

$$|xK \sim K|/|K| < \varepsilon \qquad (x \in C).$$

See (4.22) for further discussion of Følner conditions for left amenable semigroups. The next result is the fundamental Følner condition for a locally compact group. See Problem 4-10 for a symmetric version.

(4.10) THEOREM. *A locally compact group G is amenable if and only if whenever $\varepsilon > 0$ and $C \in \mathscr{C}(G)$, then there exists a nonnull, compact subset K of G such that*

$$(1) \qquad \lambda(xK \,\triangle\, K)/\lambda(K) < \varepsilon \qquad (x \in C).$$

PROOF. If, given ε and C, we can find nonnull $K \in \mathscr{C}(G)$ such that (1) is satisfied, then G is amenable ((4.5)).

Conversely, suppose that G is amenable, $\varepsilon > 0$, and $C \in \mathscr{C}(G)$. Replacing C by a larger compact subset of G if necessary, we can suppose that $\lambda(C) > 0$. Let $D = C \cup C^2$. Applying (4.7) with $\delta = \frac{1}{2}\lambda(C)$, we can find nonnull $K \in \mathscr{C}(G)$

and a Borel set $N \subset D$ such that $\lambda(N) < \delta$ and $\lambda(xK \triangle K)/\lambda(K) < \frac{1}{2}\varepsilon$ for all $x \in D \sim N$. Now if $x, y \in D \sim N$, then

$$\begin{aligned}\lambda(xy^{-1}K \triangle K) &= \lambda(y^{-1}K \triangle x^{-1}K) \leq \lambda(y^{-1}K \triangle K) + \lambda(x^{-1}K \triangle K) \\ &= \lambda(yK \triangle K) + \lambda(xK \triangle K) < \varepsilon\lambda(K).\end{aligned} \tag{2}$$

Further, if $x \in C$, then $xC \subset D$ and

$$xD \cap D \subset [x(D \sim N) \cap (D \sim N)] \cup [x(D \sim N) \cap N] \cup xN,$$

so that

$$\begin{aligned}2\delta = \lambda(C) = \lambda(xC) &= \lambda(xC \cap D) \leq \lambda(xD \cap D) \\ &\leq \lambda(x(D \sim N) \cap (D \sim N)) + \lambda(N) + \lambda(xN) \\ &< \lambda(x(D \sim N) \cap (D \sim N)) + 2\delta.\end{aligned}$$

Hence $\lambda(x(D \sim N) \cap (D \sim N)) > 0$ so that $x \in (D \sim N)(D \sim N)^{-1}$. Thus $C \subset (D \sim N)(D \sim N)^{-1}$, and (1) follows using (2). □

Note that by (4.6(i)), (1) is equivalent to the inequality

$$\lambda(xK \cap K)/\lambda(K) > 1 - \varepsilon/2. \tag{3}$$

Our next objective is to prove that amenable locally compact groups satisfy the deep Følner condition of (4.13). We require two rather technical lemmas. References for the first are [**R**, Chapter 8] and Milnes and Bondar [**1**]. Let G be a locally compact group.

(4.11) LEMMA. *Let V be a relatively compact subset of G with nonempty interior. Then there exists a subset T of G and an integer $R > 0$ such that*

(i) *$G = \bigcup\{tV : t \in T\}$, and*

(ii) *for each $x \in G$, we have $|\{t \in T : x \in tV\}| \leq R$.*

PROOF. Suppose first that V is a compact, symmetric neighbourhood of e. If G is compact, then trivially we can find a finite subset T of G such that (i) and (ii) are satisfied. Suppose, then, that G is not compact.

Suppose, next, that $G = \bigcup_{n=1}^{\infty} V^n$. Let $Z \in \mathscr{C}_e(G)$ be such that Z is symmetric and $Z^2 \subset V$. We construct recursively sequences $\{t_i\}$ in G and $\{N_i\}$ in $\mathbb{P}$ such that for all n

(a) $N_n \leq \lambda(V^n Z)/\lambda(Z)$;

(b) $t_j \notin \bigcup_{i=1}^{j-1} t_i V$ $(j > 1)$;

(c) $t_j \in V^k$ if $1 \leq j \leq N_{k+1}$;

(d) $V^n \subset \bigcup\{t_i V : 1 \leq i \leq N_{n+1}\}$;

(e) $N_n \leq N_{n+1}$.

Take $N_1 = 1$ and $t_1 = e$. Let $m \in \mathbb{P}$ and suppose that N_n $(1 \leq n \leq m)$ and t_j $(1 \leq j \leq N_m)$ have been constructed so that (a)–(e) are valid whenever they make sense. If $V^m \subset \bigcup_{i=1}^{N_m} t_i V$, then we take $N_{m+1} = N_m$. Suppose that $V^m \not\subset \bigcup_{i=1}^{N_m} t_i V$. Define recursively a (finite) sequence $t_{N_m+1}, t_{N_m+2}, \ldots$ with $t_i \in V^m \sim [\bigcup_{k=1}^{i-1} t_k V]$ for $i \geq N_m + 1$. If t_i is defined with $i \geq N_m + 1$, then, using (b), $t_j Z \cap t_k Z = \varnothing$ $(1 \leq j < k \leq i)$, and, using (c), $\bigcup_{k=1}^{i} t_k Z \subset V^m Z$, so that $i \leq \lambda(V^m Z)/\lambda(Z)$. So the sequence $t_{N_m+1}, \ldots$ has a last member $t_{N_{m+1}}$. One readily checks that this completes the recursion process.

From (d), $G = \bigcup_{i=1}^{\infty} t_i V$. If $x \in G$ and $x \in t_i V$ for some i, then $t_i \in xV^{-1} = xV$ and $t_i Z \subset xVZ$. Since $t_i Z \cap t_j Z = \varnothing$ if $i \neq j$, we have $|\{i: x \in t_i V\}| \leq \lambda(VZ)/\lambda(Z)$. So the lemma is established in this case, with $T = \{t_i : i \in \mathbb{P}\}$ and R any integer $\geq \lambda(VZ)/\lambda(Z)$.

We now deal with the case where $H = \bigcup_{n=1}^{\infty} V^n$ is no longer assumed to be G. By the earlier part of the proof, we can find a subset S of H and an integer $R > 0$ such that (i) and (ii) are satisfied with G, T replaced by H, S. Let A be a left transversal of G with respect to the subgroup H. Then the theorem is established by taking $T = \{as\colon a \in A,\ s \in S\}$.

Finally, we now deal with the case where V is no longer assumed to be a symmetric compact neighbourhood of e. Let $W \in \mathscr{C}_e(G)$ be symmetric. Find a subset T' of G and an integer $R' > 0$ such that (i) and (ii) are satisfied with T, R, V replaced by T', R', W. Since V^- is compact with $V^0 \neq \varnothing$ and as $W \in \mathscr{C}_e(G)$, we can find $n, m \in \mathbb{P}$ and elements x_i $(1 \leq i \leq n)$ and y_i $(1 \leq i \leq m)$ in G such that

$$W \subset \bigcup_{i=1}^{n} x_i V \subset \bigcup_{i=1}^{m} W y_i.$$

One readily checks that (i) and (ii) are satisfied with $T = \{t' x_i\colon 1 \leq i \leq n,\ t' \in T'\}$ and $R = nmR'$. □

Now let $C \in \mathscr{C}_e(G)$ and $C' = CC^{-1}C$. Let T and R be as in (4.11) with $V = C$. Let $M = \max\{\Delta(x)\colon x \in C^{-1}\}$, where Δ is the modular function of G, and $b = \frac{1}{2}\lambda(C')$. The next lemma is analogous to (4.7). This result and (4.13) are due to Emerson and Greenleaf [**1**]. Leptin [**1**], [**2**], [**5**] studied the quantity $\mathbf{I}(G)$ of (4.14).

(4.12) LEMMA. *Let $\eta \in (0, (2\lambda(C^{-1}))/(M\lambda(C')))$ and suppose that $K' \in \mathscr{C}(G)$ is nonnull and is such that*

$$\lambda(xK' \,\Delta\, K')/\lambda(K') < \eta \qquad (x \in C'). \tag{1}$$

Let $\delta \in (\eta b, \lambda(C^{-1})/M)$ and define

$$K'(\delta) = \{x \in K'\colon \lambda(C' \sim K'x^{-1}) \leq \delta\}.$$

Then we have

(i) *$K'(\delta)$ is compact and $\lambda(K'(\delta))/\lambda(K') \geq 1 - (\eta b)/\delta$;*

(ii) *$\lambda(CK'(\delta))/\lambda(K') \leq 1 + [\delta MR/(\lambda(C^{-1}) - \delta M)]$.*

PROOF. Since the map $x \to \lambda(C' \sim K'x^{-1}) = \lambda(C') - \lambda(C' \cap K'x^{-1})$ is continuous (Problem 4-5), the set $K'(\delta)$ is obviously compact. Let $E(\delta) = K' \sim K'(\delta)$. (So $E(\delta) = \{x \in K'\colon \lambda(C' \sim K'x^{-1}) > \delta\}$.) Using Fubini's

Theorem,

$$\begin{aligned}
\delta\lambda(E(\delta)) &\leq \int_{E(\delta)} \lambda(C' \sim K't^{-1})\,d\lambda(t) \\
&= \int_{E(\delta)} d\lambda(t) \int_{C'} \chi_{C'\sim K't^{-1}}(x)\,d\lambda(x) \qquad (2)\\
&= \int_{C'} \lambda(E(\delta) \sim x^{-1}K')\,d\lambda(x).
\end{aligned}$$

Now for $x \in C'$,

$$\begin{aligned}
\lambda(E(\delta) \sim x^{-1}K') \leq \lambda(xK' \sim K') &= \lambda(xK') - \lambda(xK' \cap K') \\
&= \tfrac{1}{2}\lambda(xK' \,\Delta\, K') < \tfrac{1}{2}\eta\lambda(K')
\end{aligned}$$

using (4.6(i)) and (1). Using (2), we have $\delta\lambda(E(\delta)) \leq \frac{1}{2}\eta\lambda(C')\lambda(K')$, and so

$$\begin{aligned}
\lambda(K'(\delta)) = \lambda(K') - \lambda(E(\delta)) &\geq \lambda(K')(1 - (\tfrac{1}{2}\eta\lambda(C')/\delta)) \\
&= \lambda(K')(1 - (\eta b)/\delta).
\end{aligned}$$

The inequality of (i) follows. We now turn to (ii).

Using (4.11(i)), $G = \bigcup\{tC\colon t \in T\} = \bigcup\{C^{-1}t^{-1}\colon t \in T\}$. Letting $I(\delta) = ((K'(\delta))^{-1}C^{-1}) \cap T$ and noting that for $t \in T$, $K'(\delta) \cap C^{-1}t^{-1} \neq \emptyset$ if and only if $t \in I(\delta)$, we have $K'(\delta) \subset \bigcup\{C^{-1}t^{-1}\colon t \in I(\delta)\}$, so that

$$CK'(\delta) \subset \bigcup\{CC^{-1}t^{-1}\colon t \in I(\delta)\}. \qquad (3)$$

For each $t \in I(\delta)$, choose $y_t \in C^{-1}t^{-1} \cap K'(\delta)$. Recalling that $C' = CC^{-1}C$, noting that $t^{-1} = cy_t$ for some $c \in C$, and using (3) and the definition of $K'(\delta)$, we have[1]

$$\begin{aligned}
\lambda(CK'(\delta)) &\leq \lambda(K') + \lambda\left(\bigcup\{CC^{-1}t^{-1} \sim K'\colon t \in I(\delta)\}\right) \\
&\leq \lambda(K') + \sum_{t\in I(\delta)} \lambda(C'y_t \sim K') \\
&\leq \lambda(K') + \sum_{t\in I(\delta)} \Delta(y_t)\lambda(C' \sim K'y_t^{-1}) \\
&\leq \lambda(K') + \delta M \sum_{t\in I(\delta)} \Delta(t^{-1}).
\end{aligned}$$

Clearly the required result (ii) will follow once we have established

$$\sum_{t\in I(\delta)} \Delta(t^{-1}) \leq R\lambda(K')/(\lambda(C^{-1}) - \delta M). \qquad (4)$$

Now if $t_1, \ldots, t_m \in T$, then, by (4.11(ii)), each $x \in G$ is contained in at most R of the sets $C^{-1}t_i^{-1}$, so that $\sum_{i=1}^{m} \chi_{C^{-1}t_i^{-1}} \leq R$. Hence

$$R\lambda(K') \geq \int_{K'} \left(\sum_{i=1}^{m} \chi_{C^{-1}t_i^{-1}}(x)\right) d\lambda(x) = \sum_{i=1}^{m} \lambda(C^{-1}t_i^{-1} \cap K').$$

[1] It is straightforward to show using (4.11(ii)) that if $D \in \mathscr{C}(G)$, then $D \cap T$ is finite. It therefore follows that $I(\delta)$ is finite.

Hence

$$\begin{aligned}
R\lambda(K') &\geq \sum_{t\in I(\delta)} \lambda(C^{-1}t^{-1}\cap K') \\
&= \sum_{t\in I(\delta)} [\lambda(C^{-1}t^{-1}) - \lambda(C^{-1}t^{-1} \sim K')] \\
&\geq \sum_{t\in I(\delta)} [\lambda(C^{-1}t^{-1}) - \lambda(C^{-1}Cy_t \sim K')] \\
&\geq \sum_{t\in I(\delta)} [\lambda(C^{-1}t^{-1}) - \lambda(C'y_t \sim K')] \\
&\geq \sum_{t\in I(\delta)} [\lambda(C^{-1})\,\Delta\,(t^{-1}) - \delta M\Delta(t^{-1})] \\
&= (\lambda(C^{-1}) - \delta M) \sum_{t\in I(\delta)} \Delta(t^{-1}).
\end{aligned}$$

The inequality (4) now follows. □

(4.13) THEOREM. *The following statements are equivalent:*

(i) *G is amenable;*

(ii) *if $C \in \mathscr{C}(G)$ with $C \neq \varnothing$ and $\varepsilon > 0$, then there exists nonnull $K \in \mathscr{C}(G)$ such that*

$$\lambda(CK \,\Delta\, K)/\lambda(K) < \varepsilon;$$

(iii) *if $C, L \in \mathscr{C}(G)$ with $C \neq \varnothing$ and $\varepsilon > 0$, then there exists nonnull $K \in \mathscr{C}(G)$ with $L \subset K$ such that*

$$\lambda(CK \,\Delta\, K)/\lambda(K) < \varepsilon. \tag{1}$$

PROOF. Trivially, (iii) implies (ii). That (ii) implies (i) follows from (4.6(ii)) and (4.10).

We now prove that (i) implies (iii). Suppose, then, that G is amenable, and let $C, L \in \mathscr{C}(G)$ with $C \neq \varnothing$, and $\varepsilon > 0$. By (4.6(iii)), we can suppose that $C \in \mathscr{C}_e(G)$. Let $\varepsilon > 0$ and b, M, and R be as in (4.12). Choose $\delta > 0$ so that $\delta < \lambda(C^{-1})/M$ and $\delta MR/(\lambda(C^{-1}) - \delta M) < \frac{1}{2}\varepsilon$. Now choose $\eta > 0$ so that $\eta b < \delta$ and $[1 - (\eta b)/\delta]^{-1}(1 + \frac{1}{2}\varepsilon) < 1 + \varepsilon$. Then with K' as in (4.12), we have

$$\begin{aligned}
\frac{\lambda(CK'(\delta) \,\Delta\, K'(\delta))}{\lambda(K'(\delta))} &= \frac{\lambda(CK'(\delta)) - \lambda(K'(\delta))}{\lambda(K'(\delta))} \\
&= \frac{\lambda(CK'(\delta))}{\lambda(K')} \cdot \frac{\lambda(K')}{\lambda(K'(\delta))} - 1 \\
&\leq \frac{1 + (\delta MR/(\lambda(C^{-1}) - \delta M))}{1 - (\eta b)/\delta} - 1 \\
&< \varepsilon.
\end{aligned}$$

So given $\varepsilon > 0$, there exists nonnull $K_\varepsilon \in \mathscr{C}(G)$ such that

$$\lambda(CK_\varepsilon \,\Delta\, K_\varepsilon)/\lambda(K_\varepsilon) < \varepsilon. \tag{2}$$

It remains to be shown that K_ε can be chosen so that $L \subset K_\varepsilon$. Again we can suppose that $e \in C$. If G is compact then we can take $K_\varepsilon = G$. Suppose that G is not compact. We construct by recursion a sequence $\{K_r\}$ in $\mathscr{C}(G)$ such that for all r

(a) $\lambda(K_{r+1}) \geq 2\lambda(K_r) > 0$;

(b) $\lambda(CK_r \triangle K_r) < \frac{1}{2}\varepsilon\lambda(K_r)$.

By (2), we choose K_1 so that (b) is satisfied. Suppose that $K_1, \ldots, K_n$ have been constructed so that (a) and (b) are satisfied whenever they make sense. Let $A = \{x \in G\colon \Delta(x) \geq 1\}$. Since G is not compact and $G = A \cup A^{-1}$, the set A is also not compact, and we can find $y_n \in A \sim K_n^{-1}C^{-1}CK_n$. Let $K_{n+1} = K_n \cup K_n y_n$. Now $K_n y_n \cap K_n = \varnothing$, and so

$$\lambda(K_{n+1}) = \lambda(K_n) + \lambda(K_n y_n) = (1 + \Delta(y_n))\lambda(K_n) \geq 2\lambda(K_n).$$

Further, $CK_n \cap CK_n y_n = \varnothing$, and so

$$\begin{aligned}\lambda(CK_{n+1} \triangle K_{n+1}) &= \lambda([CK_n \cup CK_n y_n] \sim [K_n \cup K_n y_n]) \\ &\leq \lambda(CK_n \sim K_n) + \lambda(CK_n y_n \sim K_n y_n) \\ &< \tfrac{1}{2}\varepsilon\lambda(K_n) + \Delta(y_n)\tfrac{1}{2}\varepsilon\lambda(K_n) = \tfrac{1}{2}\varepsilon\lambda(K_{n+1}).\end{aligned}$$

This completes the construction of the sets K_n. Note that from (a), $\lambda(K_n) \to \infty$. Let $L_n = K_n \cup L$. Then $CL_n \triangle L_n = CL_n \sim L_n \subseteq (CK_n \triangle K_n) \cup CL$. So

$$\lambda(CL_n \triangle L_n)/\lambda(L_n) \leq [\lambda(CK_n \triangle K_n) + \lambda(CL)]/\lambda(K_n),$$

and using (b), we can take the set K of (iii) to be L_n for large enough n. □

(4.14) COROLLARY. *Let*

$$\mathbf{I}(G) = \sup_{C \in \mathscr{C}(G)} \{\inf\{\lambda(CK)/\lambda(K)\colon K \in \mathscr{C}(G),\ \lambda(K) > 0\}\}.$$

Then $\mathbf{I}(G) = 1$ *if and only if* G *is amenable.*

PROOF. We can obviously restrict attention to the case $C \neq \varnothing$ in the definition of $\mathbf{I}(G)$. Since $\lambda(CK)/\lambda(K) \geq 1$, we have $\mathbf{I}(G) = 1$ if and only if

$$\inf\{\lambda(CK)/\lambda(K)\colon K \subset \mathscr{C}(G),\ \lambda(K) > 0\} = 1 \tag{1}$$

for each C.

Suppose that G is amenable and let C ($\neq \varnothing$) be in $\mathscr{C}(G)$. Find a set $C_1 \in \mathscr{C}_e(G)$ with $C \subset C_1$. By (4.13), we can find a sequence $\{K_n\}$ of nonnull elements of $\mathscr{C}(G)$ with $\lambda(C_1K_n \triangle K_n)/\lambda(K_n) \to 0$. Since

$$\lambda(C_1K_n \triangle K_n) = \lambda(C_1K_n) - \lambda(K_n),$$

it follows that $\lambda(C_1K_n)/\lambda(K_n) \to 1$. As

$$1 \leq \lambda(CK_n)/\lambda(K_n) \leq \lambda(C_1K_n)/\lambda(K_n),$$

we see that $\lambda(CK_n)/\lambda(K_n) \to 1$, and (1) follows.

Conversely, if (1) is true with C replaced by C_1, then we can find a sequence $\{K'_n\}$ of nonnull elements of $\mathscr{C}(G)$ with $\lambda(C_1K'_n)/\lambda(K'_n) \to 1$. Now use (4.6(iii)) again to deduce that

$$\lambda(CK_n \triangle K_n)/\lambda(K_n) \leq 2[\lambda(C_1K_n) - \lambda(K_n)]/\lambda(K_n) \to 0$$

so that G is amenable by (4.13). □

It is easy to prove that if G is not amenable, then $\mathbf{I}(G) = \infty$ (Problem 4-11). An application of (4.13) to transference theory is given in Problem 4-44.

(4.15) DEFINITION. A net $\{K_\delta\}$ $(\delta \in \Delta)$ of nonnull, compact subsets of G is called a *summing net for* G if the following conditions are satisfied:

(1) $K_\delta \subset K_\sigma$ if $\delta \leq \sigma$;
(2) $G = \bigcup\{K_\delta^0 \colon \delta \in \Delta\}$;
(3) $\lambda(xK_\delta \triangle K_\delta)/\lambda(K_\delta) \to 0$ uniformly on compacta.

Of special interest is the case where Δ is the sequence of positive integers n, in which case $\{K_n\}$ is called a *summing sequence.* It is easily checked that $\{[-n, n]\}$ and $\{[-n; n]\}$ are summing sequences for $\mathbb{R}$ and $\mathbb{Z}$ respectively. References for the result below are Emerson [**2**] and Bondar and Milnes [**1**].

(4.16) THEOREM. *A locally compact group G is amenable if and only if there exists a summing net for G. If G is σ-compact, then G is amenable if and only if there exists a summing sequence for G.*

PROOF. Let G be locally compact. If there exists a summing net for G, then G is amenable by (4.10).

Conversely, suppose that G is amenable. Let $\{H_\gamma \colon \gamma \in \Gamma\}$ be the set of σ-compact, open subgroups of G. Every compact subset of G is contained in some H_γ. Further ((1.12)) each H_γ is amenable. For each γ, find a sequence $\{C_n^\gamma\}$ in $\mathscr{C}_e(H_\gamma)$ such that $C_n^\gamma \subset (C_{n+1}^\gamma)^0$, $(C_n^\gamma)^{-1} = C_n^\gamma$, and $H_\gamma = \bigcup_{n=1}^\infty C_n^\gamma$. Let $\Delta = \mathbb{P} \times \Gamma$. For each $\delta = (n, \gamma)$ in Δ, find, using (4.13) and (4.6(ii)), a set $K_\delta \in \mathscr{C}(H_\gamma)$ such that $K_\delta \supset C_n^\gamma$ and

$$\lambda(xK_\delta \triangle K_\delta) < \lambda(K_\delta)/n \qquad (x \in C_n^\gamma).$$

The set Δ is made into a net by defining $(n, \gamma) \leq (n', \gamma')$ if $n \leq n'$, $H_\gamma \subset H_{\gamma'}$, $C_n^\gamma \subset C_{n'}^{\gamma'}$, and $K_{(n,\gamma)} \subset K_{(n',\gamma')}$. To prove this, let $\delta_1 = (n_1, \gamma_1)$, $\delta_2 = (n_2, \gamma_2)$ be elements of Δ. For $i = 1, 2$, let $\mathscr{C}_i$ be a countable family of symmetric, compact subsets of G, such that $H_{\gamma_i} = \bigcup \mathscr{C}_i$. Define

$$H_\gamma = \bigcup\{A_1 \cdots A_n \colon n \geq 1,\ A_j \in \mathscr{C}_1 \cup \mathscr{C}_2\}.$$

Then H_γ is a σ-compact, open subgroup of G containing $H_{\gamma_1} \cup H_{\gamma_2}$. Now find $N \geq n_1, n_2$ such that

$$C_N^\gamma \supset C_{n_1}^{\gamma_1} \cup C_{n_2}^{\gamma_2} \cup K_{\delta_1} \cup K_{\delta_2}.$$

Then $(N, \gamma) \geq (n_1, \gamma_1)$, (n_2, γ_2), so that Δ is a net as required.

It is obvious that (1) and (2) of (4.15) are satisfied. We check (4.15(3)). Let $C \in \mathscr{C}(G)$ and $\varepsilon > 0$. Find $\delta_0 = (n_0, \gamma_0)$ such that $C \subset C_{n_0}^{\gamma_0}$, and $1/n_0 < \varepsilon$. Then for $x \in C$ and $\delta \geq (n_0, \gamma_0)$, we have

$$\lambda(xK_\delta \bigtriangleup K_\delta)/\lambda(K_\delta) < \varepsilon$$

and (4.15(3)) is established.

Obvious modifications of the above argument deal with the case of a summing sequence for σ-compact G. □

The next result shows that all topologically left invariant means on G can be obtained from a given summing net. Let G be an amenable locally compact group and $\{K_\delta\}$ a summing net for G. For each δ, let $\mu_\delta = \chi_{K_\delta}/\lambda(K_\delta) \in P(G)$. For each net $\{x_\delta\}$ in G, let $\mathfrak{L}(\{x_\delta\})$ be the set of weak* cluster points of the net $\{(\mu_\delta * x_\delta)^\wedge\}$ of means in $\mathfrak{M}(G)$. (One readily checks that $\mu_\delta * x_\delta = \chi_{K_\delta x_\delta}/\lambda(K_\delta x_\delta)$.) Let $\mathfrak{L} = \bigcup\{\mathfrak{L}(\{x_\delta\})\colon \{x_\delta\}$ is a net in $G\}$. The theorem below is due to Chou [**4**].

(4.17) THEOREM. *The set $\mathfrak{L}_t(G)$ is the weak* closure $\overline{\text{co}}\,\mathfrak{L}$ of $\mathfrak{L}$.*

PROOF. Let $\{x_\delta\}$ be a net in G and m be a weak* cluster point of $\{(\mu_\delta * x_\delta)^\wedge\}$. Then $m \in \mathfrak{M}(G)$. By (4.15(3)) and (4.3), for $\mu \in P(G)$, $\|\mu * \mu_\delta * x_\delta - \mu_\delta * x_\delta\|_1 \leq \|\mu * \mu_\delta - \mu_\delta\|_1 \to 0$. Hence $\mu m = m$ and $m \in \mathfrak{L}_t(G)$. So $\overline{\text{co}}\,\mathfrak{L} \subset \mathfrak{L}_t(G)$.

Conversely, let $m_1 \in \mathfrak{L}_t(G)$. Suppose that $m_1 \notin \overline{\text{co}}\,\mathfrak{L}$. Separating m_1 and $\overline{\text{co}}\,\mathfrak{L}$, we can find a real-valued function $\psi \in L_\infty(G)$ and positive numbers c, ε such that

$$(1) \qquad .m(\psi) \leq c - \varepsilon < c \leq m_1(\psi) \qquad (m \in \overline{\text{co}}\,\mathfrak{L}).$$

For each δ, let $M_\delta = \sup_{x \in G} \psi\mu_\delta(x)$. Fix $n \in \mathbb{P}$ and for each δ, choose $x_\delta(n) \in G$ such that

$$(2) \qquad [M_\delta - \psi\mu_\delta(x_\delta(n))] < n^{-1}.$$

Let m_2 be a weak* cluster point of the net $\{(\mu_\delta * x_\delta(n))^\wedge\}$ in $\mathfrak{M}(G)$. Then $m_2 \in \mathfrak{L}$. Let $\eta > 0$. Then we can find σ such that

$$(3) \qquad |m_2(\psi) - \psi\mu_\sigma(x_\sigma(n)| < \eta,$$

where we have used the equality: $(\mu * x)^\wedge(\psi) = \psi\mu(x)$ ((1.1)). So using (1), (2), and (3),

$$\begin{aligned} m_2(\psi) + \varepsilon \leq m_1(\psi) = m_1(\psi\mu_\sigma) \leq M_\sigma &\leq \psi\mu_\sigma(x_\sigma(n)) + n^{-1} \\ &< m_2(\psi) + \eta + n^{-1}. \end{aligned}$$

By choosing n large enough and then η small enough, we contradict (1).

So $m_1 \in \overline{\text{co}}\,\mathfrak{L}$ and the proof is complete. □

We conclude this chapter with some applications of Reiter and Følner conditions. The first is due to Glicksberg [**2**] and Reiter [**14**], [**R**].

For $f \in L_1(G)$, let $C_f = \text{co}\{f * x\colon x \in G\}$ and

$$d(0, C_f) = \inf\{\|g\|_1\colon g \in C_f\}.$$

(4.18) THEOREM. *The locally compact group G is amenable if and only if, for all $f \in L_1(G)$,*

$$d(0, C_f) = \left|\int f\,d\lambda\right|. \tag{1}$$

PROOF. Suppose that G is amenable. Since $\|f * x\|_1 \geq |\chi_G(f * x)| = |\int f\,d\lambda|$, we have

$$d(0, C_f) \geq \left|\int f\,d\lambda\right|. \tag{2}$$

To show equality in (2), we assume first that $\int f\,d\lambda = 0$. Suppose that $d(0, C_f) > \delta > 0$. Separating the convex sets C_f and $\{g \in L_1(G) : \|g\|_1 \leq \delta\}$ [**DS**, V.2.8], we can find $\phi_\delta \in L_1(G)' = L_\infty(G)$ and $\gamma \in \mathbb{R}$ such that

$$\operatorname{Re}\phi_\delta(g) \leq \gamma \leq \operatorname{Re}\phi_\delta(h) \tag{3}$$

for all $h \in C_f$ and $g \in L_1(G)$ with $\|g\|_1 \leq \delta$. We can suppose that $\|\phi_\delta\| = 1$. Since $\sup\{\operatorname{Re}\phi_\delta(g) : \|g\|_1 = \delta\} = \delta$, we can take $\gamma = \delta$ in (3). Let $D = \{\phi \in L_\infty(G) : \|\phi\| = 1, \operatorname{Re}\phi(h) \geq d(0, C_f) \text{ for all } h \in C_f\}$. By taking a weak* cluster point of $\{\phi_\delta\}$, we see that D is not empty. Clearly, D is a weak* compact subset of $L_\infty(G)$. If $\psi \in L_\infty(G)$ and $\operatorname{Re}\psi(h) \geq d(0, C_f)$ for all $h \in C_f$, then $\|\psi\| \geq 1$. It follows that D is convex. Since $h * x \in C_f$ if $h \in C_f$, $x \in G$, we have that D is left invariant for G, and since each map $(h, x) \to h * x$ is norm continuous (cf. (1.2)), it follows that the map $(x, \phi) \to x\phi$ is continuous on D. Applying the fixed-point theorem (2.24), we can find $\phi \in D$ with $x\phi = \phi$ for all $x \in G$. Hence ϕ is constant (cf. (1.7)), and so $= \alpha 1$ where $\alpha \in \mathbb{T}$. So

$$d(0, C_f) \leq \operatorname{Re}\phi(f) \leq \left|\alpha \int f\,d\lambda\right| = 0$$

giving a contradiction. So (1) is proved for functions f with $\int f\,d\lambda = 0$.

Now suppose that f is general, and let $h \in P(G)$, $a = \int f\,d\lambda$. Applying the preceding result to $(f - ah)$, we obtain

(4)

$$0 = \inf\left\{\left\|\sum_{i=1}^{n} \alpha_i (f * x_i - ah * x_i)\right\|_1 : n \geq 1,\ \alpha_i \geq 0,\ \sum_{i=1}^{n} \alpha_i = 1,\ x_i \in G\right\}.$$

Now

$$\begin{aligned}
&\left\|\sum_{i=1}^{n} \alpha_i (f * x_i - ah * x_i)\right\|_1 \\
&\quad \geq \left\|\sum_{i=1}^{n} \alpha_i f * x_i\right\|_1 - |a| \int \left|\left(\sum_{i=1}^{n} \alpha_i h * x_i\right)\right| d\lambda \\
&\quad = \left\|\sum_{i=1}^{n} \alpha_i f * x_i\right\|_1 - |a| \quad (\text{since } h \in P(G)) \\
&\quad \geq 0,
\end{aligned}$$

and (1) follows from (2) and (4).

Conversely, suppose that (1) holds for all $f \in L_1(G)$. Let $\varepsilon > 0$, $x_1, \ldots, x_n \in G$, $f \in P(G)$. Since $d(0, C_{x_1 * f - f}) = 0$, there exists a convex combination g_1 of $f * y$'s such that $\|x_1 * g_1 - g_1\|_1 < \varepsilon$. Similarly, there exists a convex combination g_2 of $g_1 * y$'s such that $\|x_2 * g_2 - g_2\|_1 < \varepsilon$. Since $\|x_1 * g_1 - g_1\|_1 < \varepsilon$, we also have $\|x_1 * g_2 - g_2\|_1 < \varepsilon$. Proceeding in this way, we produce $g_n \in P(G)$ such that

$$\|x_i * g_n - g_n\|_1 < \varepsilon \qquad (1 \leq i \leq n)$$

and G is amenable ((0.8)). □

Our next result concerns the norms of operators associated with the left regular representation of $M(G)$ on $L_p(G)$ $(1 \leq p < \infty)$. We prove a more general result which is the key to the "L_p-conjecture" for amenable locally compact groups (Problem 4-17). References for the result are Dieudonné **[2]**, **[R]** and Leptin **[1]**.

(4.19) THEOREM. *Let μ be a positive, regular Borel measure on an amenable locally compact group G such that $\mu * L_p(G) \subset L_p(G)$. Define $T\mu\colon L_p(G) \to L_p(G)$ by $T\mu(f) = \mu * f$. Then $T\mu \in \mathbf{B}(L_p(G))$, $\mu \in M(G)$, and $\|T\mu\| = \|\mu\|$.*

PROOF. Of course, the assertion that $\mu * L_p(G) \subset L_p(G)$ means that if $f \in L_p(G)$, then the function $y \to f(y^{-1}x)$ is in $L_1(\mu)$ for μ-a.e. $x \in G$, and that the function $\mu * f$ given by

$$\mu * f(x) = \int_G f(y^{-1}x)\, d\mu(y)$$

belongs to $L_p(G)$. We show first that $T\mu$ is bounded.

Let μ^* be the measure on $\mathscr{B}(G)$ given by $\mu^*(E) = \mu(E^{-1})$. It is easily checked that μ^* is a positive, regular Borel measure on G, and that if $k \in L_1(\mu)$, then the function k^*, where $k^*(x) = k(x^{-1})$, belongs to $L_1(\mu^*)$, and

$$\int k\, d\mu = \int k^*\, d\mu^*. \tag{1}$$

Let $f \in L_p(G)$, $g \in L_q(G)$. We write $(f, g) = \int fg\, d\lambda$. Since

$$\int d\lambda(x) \int |f(y^{-1}x)g(x)|\, d\mu(y) = (\mu * |f|, |g|) < \infty,$$

Fubini's Theorem applies, and we have, using (1),

$$\begin{aligned}
(\mu * f, g) &= \int d\mu(y) \int f(y^{-1}x)g(x)\, d\lambda(x) \\
&= \int d\mu(y) \int f(x)g(yx)\, d\lambda(x) \\
&= \int d\lambda(x) \int f(x)g(yx)\, d\mu(y) \\
&= \int d\lambda(x) \int f(x)g(y^{-1}x)\, d\mu^*(y).
\end{aligned} \tag{2}$$

Define $h\colon G \to \mathbb{C}$ by $h(x) = \int g(y^{-1}x)\, d\mu^*(y)$ whenever the latter integral exists, and $h(x) = 0$ otherwise. Then from (2), $fh \in L_1(G)$. Since this is true for all

$f \in L_p(G)$, it follows that $h \in L_q(G)$. So $S\mu(g) = h$ belongs to $L_q(G)$. From (2),

$$(T\mu(f), g) = (f, S\mu(g))$$

so that $S\mu$ is an adjoint for $T\mu$. By the Closed-Graph Theorem, $T\mu$ is bounded. It remains to show that μ is bounded and $\|\mu\| = \|T\mu\|$.

Let $C \in \mathscr{C}(G)$ and $\varepsilon \in (0,1)$. By (4.10(3)), we can find nonnull $K \in \mathscr{C}(G)$ such that

$$(3) \qquad \lambda(cK \cap K) \geq (1-\varepsilon)\lambda(K) \qquad (c \in C).$$

Let $f = \lambda(K)^{-1/p}\chi_K$ and $g = \lambda(K)^{-1/q}\chi_K$. Then $f \in L_p(G)$, $g \in L_q(G)$, and $\|f\|_p = 1 = \|g\|_q$. Recalling that the function $y \to \lambda(yK \cap K)$ is measurable (indeed continuous) and that $p^{-1}+q^{-1} = 1$, and using (3) and Fubini's Theorem,

$$\begin{aligned}
\mu(C)\lambda(K)(1-\varepsilon) &\leq \mu(C) \inf_{c \in C} \lambda(cK \cap K) \\
&\leq \int_C \lambda(yK \cap K)\, d\mu(y) \\
&= \int_C d\mu(y) \int \chi_K(y^{-1}x)\chi_K(x)\, d\lambda(x) \\
&= \lambda(K) \int_C d\mu(y) \int f(y^{-1}x)g(x)\, d\lambda(x) \\
&= \lambda(K) \int (\mu * f)(x) g(x)\, d\lambda(x) \\
&= \lambda(K)(\mu * f, g) \\
&\leq \lambda(K)\|T\mu\|\|f\|_p\|g\|_q = \lambda(K)\|T\mu\|.
\end{aligned}$$

Hence $\mu(C) \leq \|T\mu\|/(1-\varepsilon)$, and letting $\varepsilon \to 0$ and using the inner regularity of μ, it follows that μ is bounded and $\|\mu\| \leq \|T\mu\|$. The reverse inequality follows since $\mu \in M(G)$. □

It follows from the preceding result that the amenability of G implies that for all $\mu \in M(G)$, $\mu \geq 0$, we have $\|\pi_p(\mu)\| = \|\mu\| = \mu(G)$, where $\pi_p(\mu)\colon L_p(G) \to L_p(G)$ is given by $\pi_p(\mu)f = \mu * f$. The converse is also true. It is sometimes impractical to use this as a test for amenability, and our next result is a great improvement in this direction: under very modest assumptions, we need only check that $\|\pi_p(\mu)\| = \|\mu\|$ for *one* measure μ. This will prove useful in our discussion of von Neumann's Conjecture. References for the following result are Kesten [**1**], Derriennic and Guivarc'h [**1**] and Berg and Christensen [**2**]. See also Day [**6**], [**8**], Dieudonné [**2**], Eymard [**1**], Gilbert [**1**], and Leptin [**1**], [**5**]. We deal with the case $p = 2$, the extension to general p being left to Problem 4-25.

Let $\mu \in M(G)$, $\mu \geq 0$. Recall that the *support* $\mathbf{S}(\mu)$ of μ is the set of elements $x \in G$ such that $\mu(U) > 0$ whenever U is an open neighbourhood of x in G. If the property $\|\pi_2(\mu)\| = \|\mu\|$ is to give us the amenability of G, it is reasonable to require that $\mathbf{S}(\mu)$—the set on which μ "lives"—should be algebraically "large" in G. To make this precise, we could require that the subgroup K_μ of G generated by $\mathbf{S}(\mu)$ is dense in G. (Such a measure μ is sometimes called *aperiodic* or *adapted.*) This is the condition used in (ii) below. In (i), we require the subgroup H_μ generated by $\mathbf{S}(\mu)^{-1}\mathbf{S}(\mu)$ to be dense in G. (Such a measure μ is sometimes called *strictly aperiodic.*) If $\nu = \mu^\sim * \mu$, where $^\sim$ is the natural involution on $m(G)$ ((1.1)), then $\mathbf{S}(\nu) = (\mathbf{S}(\mu)^{-1}\mathbf{S}(\mu))^-$, so that H_μ is dense in G if and only if K_ν is dense in G. Notice that in general H_μ^- will be a proper subgroup of K_μ^- (so that requiring $H_\mu^- = G$ is stronger than requiring $K_\mu^- = G$). For example, if G is the free group on two generators x, y and $\mu = \delta_x + \delta_y$, then $K_\mu = G$ while H_μ is the (abelian) subgroup of G generated by $y^{-1}x$. The results (i) and (ii) are obviously similar in kind: is there a theorem which unifies (i) and (ii)?

The author is grateful to J. M. Rosenblatt for pointing out an error in an earlier version of (4.20), and for suggesting the elegant proof of (ii).

(**4.20**) THEOREM. *Let $\mu \in M(G)$, $\mu \geq 0$, and H_μ, K_μ be as above. Then:*
(i) *G is amenable if $H_\mu^- = G$ and $\|\pi_2(\mu)\| = \mu(G)$;*
(ii) *G is amenable if $K_\mu^- = G$, $\mu^\sim = \mu$, and $\|\pi_2(\mu)\| = \mu(G)$.*

PROOF. (i) Suppose that $H_\mu^- = G$ and $\|\pi_2(\mu)\| = \mu(G)$. We can suppose that $\mu(G) = 1$. Let $\nu = \mu^\sim * \mu$. Then $\pi_2(\nu) \geq 0$, and we have

$$1 = \|\pi_2(\nu)\| = \sup\{(\pi_2(\nu)f, f): f \in L_2(G),\ \|f\|_2 = 1\}.$$

Since $\nu \geq 0$, there exists a sequence $\{f_n\}$ in $C_c(G)$ with $f_n \geq 0$, $\|f_n\|_2 = 1$ such that $(\pi_2(\nu)f_n, f_n) \to 1$. Let $g_n(x) = (\pi_2(x)f_n, f_n)$. (In the notation below, $g_n = f_n * f_n^\dagger$.) Then $0 \leq g_n \leq 1$ and $g_n \in C_c(G)$. Further,

$$(\pi_2(\nu)f_n, f_n) = \int g_n(x)\,d\nu(x) \to 1.$$

Since $\nu(G) = 1$ and $0 \leq g_n \leq 1$, we have $g_n \to 1$ pointwise ν-a.e. Let $L = \{x \in G: g_n(x) \to 1\}$. Then $L^- \supset \mathbf{S}(\nu)$. Since for $x \in G$,

$$\text{(1)} \qquad \|x * f_n - f_n\|_2^2 = (x * f_n - f_n, x * f_n - f_n) = [2\|f_n\|_2^2 - 2g_n(x)],$$

it follows that $L = \{x \in G: \|x * f_n - f_n\|_2 \to 0\}$. Since

$$\|x * f_n - f_2\|_2 = \|x^{-1} * f_n - f_n\|$$

and

$$\|xy * f_n - f_n\|_2 \leq \|y * f_n - f_n\|_2 + \|x * f_n - f_n\|_2,$$

we see that L is a subgroup of G with $L^- \supset \mathbf{S}(\nu)$. Since $\mathbf{S}(\nu)$ generates the dense subgroup H_μ of G, it follows that L is dense in G.

Now if we knew that L was all of G, then a "$p = 2$" Reiter-type condition would immediately give G amenable. Unfortunately, we only know that $L^- = G$. To overcome this difficulty, let $h_n = f_n^2 \in \mathrm{P}(G)$ $(n \geq 1)$. Then ((4.3(1))) $\|x * h_n - h_n\|_1 \to 0$ for each $x \in L$. Any weak* cluster point m of $\{\hat{h}_n\}$ in $\mathfrak{M}(\mathrm{U_r}(G))$ then satisfies: $xm = m$ for all $x \in L$. Since the left action of G on $\mathfrak{M}(\mathrm{U_r}(G))$ is jointly continuous for the weak* topology ((2.22)), we obtain $xm = m$ for $x \in L^- = G$. Hence $m \in \mathfrak{L}(\mathrm{U_r}(G))$ and G is amenable ((1.10)).

(ii) Suppose that $K_\mu^- = G$, $\mu^\sim = \mu$, and $\|\pi_2(\mu)\| = \mu(G)$. We can also suppose that $\mu(G) = 1$. Let $A = \mathbf{S}(\mu)$ $(= \mathbf{S}(\mu)^{-1})$ and $N = H_\mu^-$, the closure of the subgroup generated by A^2. Since A generates K_μ, $K_\mu^- = G$, and $aA^2a^{-1} \subset N$ for all $a \in A$, it follows that N is normal in G. Let $a_0 \in A$. We claim that $G = N \cup a_0 N$. Indeed, if $x \in G \sim N$, then since $K_\mu^- = G$, there exists a net $x_\delta \to x$ in G where each x_δ is the product of an *odd* number of elements of A. Then $x_\delta = a_0(a_0^{-1}x_n) \in a_0 N$, and so $x \in a_0 N$. Thus G/N is of order ≤ 2 and N is open in G. Clearly, the amenability of G will follow from that of N. We will now show, using (i), that N is amenable.

Consider the measure $a_0\mu$. Clearly, $\mathbf{S}(a_0\mu) \subset A^2 \subset N$. Further, if $a, b \in \mathbf{S}(\mu)$, then $a^{-1}b = (a_0 a)^{-1}(a_0 b) \in \mathbf{S}(a_0\mu)^{-1}\mathbf{S}(a_0\mu)$, so that the latter set generates a dense subgroup of N. By (i), with G, μ replaced by N, $a_0\mu$, we will have N amenable if we can show that $\|\pi_2^N(a_0\mu)\| = 1$, where π_2^N is the left regular representation of N on $L_2(N)$.

If $N = G$, then $\pi_2^N(a_0\mu) = \pi_2(a_0\mu) = \pi_2(a_0)\pi_2(\mu)$, and we obviously have $\|\pi_2^N(a_0\mu)\| = 1$ as required. Suppose then that $N \neq G$, so that G is the disjoint union $N \cup Na_0$. Then $\lambda_G|_N$ is a left Haar measure on N, and it is easily checked that $L_2(G)$ can be identified with $L_2(N) \oplus L_2(N)$, and under this identification, $\pi_2(a_0\mu) = \pi_2^N(a_0\mu) \oplus \pi_2^N(a_0\mu)$. Then $1 = \|\pi_2(a_0\mu)\| = \|\pi_2^N(a_0\mu)\|$ as required. □

We conclude this section by proving the "weak containment" characterisation of amenability. (For "weak containment" see (4.29).) Recall that the *reduced* C^*-*algebra* of G is $C_l^*(G)$, the C^*-algebra generated by $\pi_2(L_1(G))$. The universal C^*-algebra $C^*(G)$ can be described as the completion of the Banach $*$-algebra $L_1(G)$ under its largest C^*-norm $\|\cdot\|$ $(\leq \|\cdot\|_1)$ [**D2**, (13.9)]. Since $\|f\| \geq \|\pi_2(f)\|$ $(f \in L_1(G))$, we have a canonical homomorphism Q_2 from $C^*(G)$ onto $C_l^*(G)$. We shall say that $C^*(G) = C_l^*(G)$ if Q_2 is an isomorphism. We discussed in Chapter 0 the significance for representation theory of the equality $C^*(G) = C_l^*(G)$.

Before proving the theorem, we briefly discuss some fundamental facts about representations of and states on $L_1(G)$ and $C^*(G)$. References for these facts are [**HR2**] and [**D2**]. A continuous function $\phi\colon G \to \mathbb{C}$ is called *positive definite* if, whenever $x_1, \ldots, x_n \in G$ and $\alpha_1, \ldots, \alpha_n \in \mathbb{C}$, then $\sum_{i,j} \alpha_i \bar{\alpha}_j \phi(x_i x_j^{-1}) \geq 0$. Such functions are precisely the matrix coefficients arising from continuous unitary representations of G. Thus, if π is such a representation on a Hilbert space $\mathfrak{H}$ and $\xi \in \mathfrak{H}$, then the function $x \to (\pi(x)\xi, \xi)$ is positive definite. Conversely, every positive definite function ϕ can be so realised. Every such ϕ satisfies $\|\phi\|_\infty = \phi(e)$, and the set of states on $L_1(G)$ (or $C^*(G)$) can be identified with those ϕ for which $\phi(e) = 1$. If ϕ, ψ are positive definite on G, then so also is $\phi\psi$.

The ϕ's arising from π_2 are rather special. Indeed, for such a ϕ, there exists $f \in L_2(G)$ such that for all $x \in G$,

$$\phi(x) = (\pi_2(x)f, f) = \int_G f(x^{-1}y)\overline{f(y)}\, d\lambda(y) = \bar{f} * f^*(x) = g * g^\dagger(x)$$

where $g = \bar{f} \in L_2(G)$ and $g^\dagger(x) = \overline{g(x^{-1})}$ $(= \overline{f^*(x)})$.

We require an important result due to Godement [**D2**, (13.8.6)]: *if* $\phi \in C_c(G)$ *is positive definite, then there exists* $g \in C_c(G)$ *with* $\phi = g * g^\dagger$.

References for the theorem below are Godement [**1**], [**D2**], Greenleaf [**2**], Hulanicki [**2**], and [**R**]. It is sometimes called the "weak containment" theorem for amenable groups.

(4.21) THEOREM. *The locally compact group* G *is amenable if and only if* $C^*(G) = C_l^*(G)$.

PROOF. Suppose that G is amenable. Let F be a state on $C^*(G)$. Then $F|_{L_1(G)}$ is a state on $L_1(G)$, and there exists a (continuous) positive definite function ϕ on G with $\phi(e) = 1$ and

$$F(f) = \int_G f\phi \, d\lambda \qquad (f \in L_1(G)).$$

Let $f \in L_1(G)$. By (4.19) and the proof of (4.20)(i) with $\mu = |f| + \delta_e$, we can find a sequence $\{f_n\}$ in $C_c(G)$ with $f_n \geq 0$, $\|f_n\|_2 = 1$ such that $f_n * f_n^\dagger \to 1$ pointwise $\mu^\sim * \mu$ a.e. and hence λ-a.e. on the support of f ($\subset \mathbf{S}(\mu^\sim * \mu)$). Now $(f_n * f_n^\dagger)\phi$ is positive definite and of compact support, and so by Godement's Theorem, there exists $k_n \in C_c(G)$ with $\|k_n\|_2 = 1$ such that $(f_n * f_n^\dagger)\phi = k_n * k_n^\dagger$. So $k_n * k_n^\dagger \to \phi$ a.e. on the support of f, and by the Dominated Convergence Theorem,

$$|F(f)| = \lim \left| \int_G f(k_n * k_n^\dagger) \, d\lambda \right| = \lim |(\pi_2(f)k_n, k_n)| \leq \|\pi_2(f)\|.$$

So for $g \in L_1(G)$,

$$\begin{aligned}\|g\|^2 = \|g * g^\sim\| &= \sup\{F(g * g^\sim) \colon F \text{ is a state on } C^*(G)\} \\ &\leq \|\pi_2(g * g^\sim)\| = \|\pi_2(g)\|^2 \leq \|g\|^2,\end{aligned}$$

giving $C^*(G) = C_l^*(G)$.

Conversely, suppose that $C^*(G) = C_l^*(G)$. Let $1(f) = \int f \, d\lambda$ ($f \in L_1(G)$). Then 1 is a continuous positive functional on $C_l^*(G)$. A simple separation argument ([D2, 3.4.4]) shows that we can approximate 1 weak* by convex combinations of functionals of the form $f \to (\pi_2(f)g, g)$ ($g \in C_c(G)$, $\|g\|_2 = 1$). Godement's result gives that any such combination is also of the form $f \to (\pi_2(f)g, g)$. The upshot is that we can find a net $\{g_\delta\}$ in $C_c(G)$, $\|g_\delta\|_2 = 1$, $g_\delta \geq 0$ such that $g_\delta * g_\delta^\dagger \to 1$ pointwise on G. (Indeed, the convergence is uniform on compacta by [**D2**, (13.5.2)].) Now use (1) of (4.20) to obtain a Reiter condition for G and hence the amenability of G. □

It easily follows from the above proof that G *is amenable if and only if* 1 *is weakly contained in* π_2. (See Problem 4-28.) M. Bekka [**S**] has recently proved that G *is amenable if and only if* 1 *is weakly contained in* $\pi \otimes \bar{\pi}$ *for every continuous, unitary representation* π *of* G. (Here, $\bar{\pi}$ is the conjugate representation of π ([**D2**, 13.1.5]).) While on the "tensor product" theme, we note that *if* G *is discrete, then* $\pi_2 \otimes \pi$ *is a multiple of* π_2 *for every unitary representation* π *of* G. The latter result is useful in K-theory (Cuntz [**S2**]).

The weak containment property can be formulated for inverse semigroups. An interesting, open question is that of characterising those inverse semigroups with this property. The papers Paterson [**1**], Duncan and Paterson [**1**], [**S**] investigate this question for Clifford semigroups, the class of inverse semigroups which are unions of groups.

References

Berg and Christensen [**2**], Bondar and Milnes [**1**], Chou [**4**], Day [**2**], [**3**], [**4**], [**6**], Dieudonné [**2**], del Junco and Rosenblatt [**1**], Emerson [**2**], Emerson and Greenleaf [**1**], Følner [**2**], Frey [**1**], Glicksberg [**2**], Godement [**1**], Greenleaf [**4**], Greenleaf and Emerson [**1**], Hulanicki [**4**], Jerison [**2**], Kesten [**1**], Leptin [**1**], [**2**], [**5**], Milnes and Bondar [**1**], Namioka [**1**], Raimi [**3**], and Reiter [**1**], [**13**], [**14**].

Further Results

(4.22) Semigroups and Følner conditions. The results of this paragraph are due to Argabright and Wilde [**1**]. Let us say that a semigroup S satisfies (FC) ("Følner condition") [(SFC) ("strong Følner condition")] if for every $C \in \mathscr{F}(S)$ and $\varepsilon > 0$, there exists nonvoid $K \in \mathscr{F}(S)$ such that $|xK \sim K|/|K| < \varepsilon$ $[|K \sim xK|/|K| < \varepsilon]$ for all $x \in C$. Note that we obtain a condition equivalent to (SFC) if $K \sim xK$ is replaced by $xK \Delta K$ in the statement of the (SFC) condition. [Indeed, $|K \sim xK| = |K| - |K \cap xK|$ and $|xK \sim K| = |xK| - |K \cap xK|$, so that $|K \sim xK| - |xK \sim K| = |K| - |xK| \geq 0$. Thus $2|K \sim xK| \geq |xK \,\Delta\, K| \geq |K \sim xK|$, and this establishes the desired equivalence. It also follows that (SFC) implies (FC) (thereby justifying the "S" in (SFC)).] Now according to (4.9), S satisfies (FC) if S is left amenable. The converse to the latter statement is easily seen to be false, since, with $K = S$, every finite S, left amenable or not ((1.19)), satisfies (FC)! However, *S is left amenable if S satisfies (SFC).* Argabright and Wilde also show that *every left amenable semigroup that is either left cancellative or finite satisfies (SFC).* (See Problem 4-32.) They also show that every abelian semigroup satisfies (SFC).

More information about (SFC) is given in Rajagopalan and Ramakrishnan [**1**]. However, Klawe [**2**] shows that *left amenability is equivalent to (SFC) if and only if Sorenson's Conjecture is true.* Since the latter conjecture *is* false ((1.29)), it follows that there exists a left amenable semigroup which does not satisfy (SFC). Indeed, *every right cancellative, left amenable semigroup which is not left cancellative does not satisfy (SFC).* [An example of such a semigroup is the semigroup $F \times_\rho \mathbb{P}$ of (1.29(iv)). Let S be right cancellative, left amenable, and not left cancellative. We can find $r, s, t \in S$ such that $rs = rt$ and $s \neq t$. Suppose that S satisfies (SFC). Let $\varepsilon = 1/8$. Then we can find nonvoid $K \in \mathscr{F}(S)$ such that $|K \sim xK| < \varepsilon|K|$ for $x \in \{r, s, t\}$. Thus for $x \in \{r, s, t\}$, $|xK \cap K| > (1 - \varepsilon)|K|$, and so $|\{k \in K : xk \in K\}| > (1-\varepsilon)|K|$. Hence if $U = \{k \in K : sk \in K, tk \in K\}$, then $|U| > (1-2\varepsilon)|K|$. Let $W = sU \cup tU$. (So $W \subset K$.) If $k \in U$, then $sk \neq tk$, yet $r(sk) = r(tk)$. Thus if $\sim$ is the equivalence relation on W defined by $w \sim w'$ whenever $rw = rw'$, then $|rW| = |W/\sim|$, and as each equivalence class contains at least two elements, we have $|W| \geq |rW| \cdot 2$. It follows that $|rW| \leq \frac{1}{2}|W|$. Now

$$\begin{aligned} |W| \geq |sU| \geq |sK| - |s(K \sim U)| &> (1-\varepsilon)|K| - |K \sim U| \\ &> (1-\varepsilon)|K| - 2\varepsilon|K| = (1-3\varepsilon)|K| \end{aligned}$$

So

$$(1-\varepsilon)|K| < |rK| \leq |rW| + |r(K \sim W)| < \tfrac{1}{2}|K| + 3\varepsilon|K|.$$

Hence $(1-\varepsilon) < \frac{1}{2} + 3\varepsilon$ and we have contradicted the choice of ε.]

Recently, long-standing problems posed by Namioka [**1**] on Følner conditions for semigroups have been solved by Z. Yang [**S2**].

Rosenblatt ([**1**], [**2**]) discusses a Følner type condition in connection with the general problem of measure and the translate property ((2.32)).

(4.23) The groups G for which $\mathfrak{L}_t(G) = \mathfrak{R}_t(G)$. (Paterson [**3**], Milnes [**4**], Kotzmann and Rindler [**1**].) The following conjecture is made by Paterson: *if G is an amenable locally compact group, then $\mathfrak{L}_t(G) = \mathfrak{R}_t(G)$ if and only if $G \in [\mathrm{FC}]^-$.* We say (Palmer [**1**]) that a locally compact G is an $[\mathrm{FC}]^-$ *group* (or $G \in [\mathrm{FC}]^-$) if the closure of every conjugate class of G is compact. (Such groups are amenable and unimodular (Problem 6-7, (6.9)).

The condition that $\mathfrak{L}_t(G) = \mathfrak{R}_t(G)$ is easily seen to be equivalent to the condition *every topologically left invariant mean on G is actually topologically invariant.*

The conjecture is obviously true when G is either abelian or compact. (Abelian and compact groups belong to $[\mathrm{FC}]^-$.)

The conjecture will be shown to be true when G is σ-compact or discrete ((iii)). For general G, the conjecture seems to be open.

(i) *Let $G \in [\mathrm{FC}]^-$, and be either σ-compact or discrete. Then $\mathfrak{L}_t(G) = \mathfrak{R}_t(G)$.* [Since $m \in \mathfrak{L}_t(G)$ if and only if $m^* \in \mathfrak{R}_t(G)$, it is sufficient to show that $\mathfrak{L}_t(G) \subset \mathfrak{R}_t(G)$. (For m^*, see Problem 2-7.)

Let $\{K_\delta\}$ $(\delta \in \Delta)$ be a summing net for G with $K_\delta^{-1} = K_\delta$ for all δ (Problem 4-10). If G is σ-compact, take ((4.15)) $\Delta = \mathbb{P}$ (so that $\{K_\delta\}$ is a summing sequence). Let $\{x_\delta\}$ be a net in G and μ_δ, $\mathfrak{L}(\{x_\delta\})$ be as in the paragraph preceding (4.17). By (4.17), it suffices to show that $\mathfrak{L}(\{x_\delta\}) \subset \mathfrak{R}_t(G)$. Since G is unimodular, $\lambda(K_\delta x_\delta) = \lambda(K_\delta)$ and $\mu_\delta * x_\delta = \chi_{K_\delta x_\delta}/\lambda(K_\delta)$. If $t \in G$, then

$$\begin{aligned} \|\mu_\delta * x_\delta * t - \mu_\delta * x_\delta\| &= \lambda(K_\delta x_\delta t \,\Delta\, K_\delta x_\delta)/\lambda(K_\delta) \\ &= \lambda(K_\delta x_\delta t x_\delta^{-1} \,\Delta\, K_\delta)/\lambda(K_\delta) = \lambda([K_\delta x_\delta t x_\delta^{-1} \,\Delta\, K_\delta]^{-1})/\lambda(K_\delta) \\ &= \lambda(x_\delta t^{-1} x_\delta^{-1} K_\delta \,\Delta\, K_\delta)/\lambda(K_\delta), \end{aligned}$$

where we have used the symmetry of the K_δ and the fact that G is unimodular. Since $G \in [\mathrm{FC}]^-$, we can find $C \in \mathscr{C}(G)$ such that $x_\delta t^{-1} x_\delta^{-1} \in C$ for all δ. Hence

$$\lambda(x_\delta t^{-1} x_\delta^{-1} K_\delta \,\Delta\, K_\delta)/\lambda(K_\delta) \to 0$$

by (4.15(3)). It follows from the above that

$$(1) \qquad \lambda(K_\delta x_\delta t \,\triangle\, K_\delta x_\delta)/\lambda(K_\delta) \to 0.$$

Using Problem 4-8 and the equality

$$\lambda(K_\delta x_\delta t \,\triangle\, K_\delta x_\delta)/\lambda(K_\delta) = \lambda(t^{-1}(x_\delta^{-1}K_\delta^{-1}) \,\triangle\, x_\delta^{-1}K_\delta^{-1})/\lambda(x_\delta^{-1}K_\delta^{-1})$$

to cope with the σ-compact case, we see that the convergence in (1) is uniform (in t) on compacta. Referring to the earlier equalities, it follows that

$$\|\mu_\delta * x_\delta * t - \mu_\delta * x_\delta\| \to 0$$

uniformly (in t) on compacta, and thus the right-handed version of (4.3), with $p = 1$, gives

$$\|\mu_\delta * x_\delta * \nu - \mu_\delta * x_\delta\| \to 0$$

for each $\nu \in P(G)$. Hence if $m \in \mathfrak{L}(\{x_\delta\})$ (that is, m is a weak* cluster point of $\{(\mu_\delta * x_\delta)^\wedge\}$), then $m\nu = m$, and $m \in \mathfrak{R}_t(G)$ as required.]

(ii) *Let G be an amenable, locally compact, σ-compact group. Let $\{K_n\}$ be a sequence in $\mathscr{C}(G)$. Suppose that $x \in G$ is such that $(\mathrm{Cl}(x))^-$ is not compact, where $\mathrm{Cl}(x)$ is the conjugate class of x. Then there exists a sequence $\{x_n\}$ in G such that if $A = \bigcup_{n=1}^\infty K_n x_n$, then $x \notin A^{-1}A$.*

[The sequence $\{x_n\}$ is constructed recursively. Let $n \in \mathbb{P}$ and suppose that the elements x_s $(s < n)$ have been constructed so that if $A_s = \bigcup_{r=1}^s K_r x_r$, then $x \notin A_s^{-1}A_s$. If $y \in G$, then

$$(2) \qquad \begin{aligned}&(A_{n-1} \cup K_n y)^{-1}(A_{n-1} \cup K_n y)\\ &\quad = A_{n-1}^{-1}A_{n-1} \cup A_{n-1}^{-1}K_n y \cup y^{-1}K_n^{-1}A_{n-1} \cup y^{-1}K_n^{-1}K_n y.\end{aligned}$$

Let $D = \{y \in G \colon yxy^{-1} \notin K_n^{-1}K_n\}$. If D is relatively compact, then, as $\mathrm{Cl}(x) \subset DxD^{-1} \cup K_n^{-1}K_n$, we would have $\mathrm{Cl}(x)$ relatively compact, and so a contradiction. Hence D is not relatively compact, and we can find $y \in D$ such that $x \notin A_{n-1}^{-1}K_n y \cup y^{-1}K_n^{-1}A_{n-1}$. Setting $x_n = y$ and referring to (2), we have $x \notin A_n^{-1}A_n$. This completes the construction of $\{x_n\}$.]

(iii) *Let G be an amenable, locally compact group which is either σ-compact or discrete. Then $\mathfrak{L}_t(G) = \mathfrak{R}_t(G)$ if and only if $G \in [\mathrm{FC}]^-$.*

[By (i), if $G \in [\mathrm{FC}]^-$, then $\mathfrak{L}_t(G) = \mathfrak{R}_t(G)$.

Conversely, suppose that $\mathfrak{L}_t(G) = \mathfrak{R}_t(G)$. We deal with the σ-compact and discrete cases separately.

Suppose that G is σ-compact and not in $[\mathrm{FC}]^-$. Let $\{K_n\}$ be a summing sequence for G and $\mu_n = \chi_{K_n}/\lambda(K_n)$ as usual. Since $G \notin [\mathrm{FC}]^-$, we can find $x \in G$ with $(\mathrm{Cl}(x))^-$ not compact. Let $\{x_n\}$ and A be as in (ii), and $m \in \mathfrak{L}(\{x_n\})$. For each n, $K_n x_n \subset A$, and noting that $\chi_A \in L_\infty(G)$, we have $1 \geq (\mu_n * x_n)^\wedge(\chi_A) = \hat{\mu}_n(\chi_{Ax_n^{-1}}) \geq \hat{\mu}_n(\chi_{K_n}) = 1$. Hence $m(\chi_A) = 1$. Further,

$$(\mu_n * x_n)^\wedge(x^{-1}\chi_A) = (\mu_n * x_n)^\wedge(\chi_{Ax}) = \lambda(K_n x_n)^{-1}\lambda(Ax \cap K_n x_n) = 0$$

since $Ax \cap K_n x_n \subset Ax \cap A = \varnothing$ by (ii). So $m(x^{-1}\chi_A) = 0$ and $m \in \mathfrak{L}_t(G) \sim \mathfrak{R}(G) \subset \mathfrak{L}_t(G) \sim \mathfrak{R}_t(G)$. The resulting contradiction establishes the desired result in the σ-compact case.

Suppose, then, that G is discrete and not in $[\mathrm{FC}]^-$ ($= [FC]$). Then we can find $x \in G$ and a countable subset B of G with the set $\{bxb^{-1} : b \in B\}$ infinite. Let H be the countable (and so σ-compact!) subgroup of G generated by $B \cup \{x\}$, and note that $H \notin [\mathrm{FC}]$. We shall derive a contradiction by showing that $\mathfrak{L}(H) = \mathfrak{R}(H)$.

We will use the map $\beta : \mathfrak{L}(H) \to \mathfrak{L}(G)$ of Problem 1-8. So $\beta(p)(\psi) = m_0(\psi_p)$, where $m_0 \in \mathfrak{L}(G)$ is fixed, and $\psi_p(x) = p((\psi x))|_H)$ for $p \in \mathfrak{L}(H)$, $\psi \in l_\infty(G)$, and $x \in G$. Let T be a transversal for the left cosets of H in G. For $\phi \in l_\infty(H)$, define ${}_T\phi \in l_\infty(G)$ by

$$ {}_T\phi(th) = \phi(h) \qquad (t \in T,\ h \in H). $$

Let $m \in \mathfrak{L}(H)$. We will eventually show that $m \in \mathfrak{R}(H)$. For $t_0 \in T$, $h_0 \in H$, we have

$$ ({}_T\phi)_m(t_0h_0) = m((({}_T\phi)t_0h_0)|_H) = m(\phi h_0) = m(\phi), $$

and so $\beta(m)({}_T\phi) = m(\phi)$. Since $\beta(m) \in \mathfrak{L}(G)$, and, by hypothesis, $\mathfrak{L}(G) = \mathfrak{R}(G)$, it follows that for $h_1 \in H$,

$$ \beta(m)(h_1({}_T\phi)) = \beta(m)({}_T\phi) = m(\phi). \tag{3} $$

But if $x = t_0h_0$ with $t_0 \in T$, $h_0 \in H$, then

$$ h_1({}_T\phi)(x) = {}_T\phi(xh_1) = {}_T\phi(t_0(h_0h_1)) = \phi(h_0h_1) = h_1\phi(h_0) $$

so that $h_1({}_T\phi) = {}_T(h_1\phi)$. Hence from (3),

$$ m(h_1\phi) = \beta(m)(h_1({}_T\phi)) = m(\phi) $$

and $m \in \mathfrak{R}(H)$. Thus $\mathfrak{L}(H) = \mathfrak{R}(H)$ giving a contradiction.]

(4.24) Amenability and statistics. The paper by Bondar and Milnes [**1**] is a useful source of information about the applications of amenability to mathematical statistics. Two typical such applications are discussed in detail in (4.25)–(4.26). See also Problem 4-19.

The first, and most well-known, of these applications is the *Hunt-Stein Theorem* ((4.25)). This theorem was proved during the Second World War by G. A. Hunt and C. Stein and appeared in the unpublished paper Hunt and Stein [**1**]. The usual proof of the Hunt-Stein Theorem involves the existence of an asymptotically invariant sequence of probability measures on a σ-compact locally compact group G. A sequence $\{\mu_n\}$ of probability measures in $M(G)$ is called *asymptotically right invariant* (cf. Wesler [**1**]) if $(\mu_n(Bx) - \mu_n(B)) \to 0$ for all $B \in \mathscr{B}(G)$ and $x \in G$. If $\{\mu_n\}$ is such a sequence, then, since each $\phi \in C(G)$ is the uniform limit of a sequence in $\mathrm{Span}\{\chi_B : B \in \mathscr{B}(G)\}$, $\hat{\mu}_n(x^{-1}\phi - \phi) \to 0$ for all $\phi \in C(G)$, and any weak* cluster of $\{\hat{\mu}_n\}$, where $\hat{\mu}_n$ is regarded as an element of $C(G)'$, is in $\mathfrak{R}(C(G))$. Also, if G is amenable, then Property P_1 ((4.4(ii))) yields a sequence $\{\nu_n\}$ in $P(G)$ with $\|x * \nu_n - \nu_n\|_1 \to 0$ for all $x \in G$, and it follows that $\{\nu_n^{\sim}\}$ is asymptotically right invariant. Consequently, there exists an asymptotically right invariant sequence for G if and only if G is amenable.

However, as with Bondar and Milnes, we have preferred to use a fixed-point theorem in place of such a sequence when proving the Hunt-Stein Theorem. (See Bondar and Milnes [**1**, p. 111] for a converse to the Hunt-Stein Theorem.)

Applications of amenability to the theory of symmetric random walks on groups are given in Kesten [**1**], [**2**] and C. Stone [**1**].

An application of summing sequences to informational futures is given in Pickel [**1**].

Emerson [**5**], [**6**] makes interesting use of summing sequences in his development of ideas in abstract probability due to J. C. Kieffer.

(4.25) The Hunt-Stein Theorem. (Hunt and Stein [**1**], Peisakoff [**1**], Wesler [**1**], Lehman [**1**].) Following Wesler [**1**], we shall discuss the theorem in the language of testing problems. Schmetterer's book [**S**] provides a useful source for the theory of hypothesis testing.

Let $(Z, \mathscr{B}, \nu)$ be a measure space with ν σ-finite, Ω a set, G a σ-compact, locally compact group, and P a mapping from Ω into the set of probability measures on $(Z, \mathscr{B})$. We set $P_\omega = P(\omega)$ $(\omega \in \Omega)$. We require that both Z and Ω be left G-sets, with e acting as the identity transformation on both Z and Ω. We also require

(1) the map $(x, z) \to xz$ from $(G \times Z, \mathscr{M}(G) \times \mathscr{B})$ into $(Z, \mathscr{B})$ is measurable;

(2) $\nu(xE) = \nu(E)$ for all $x \in G$, $E \in \mathscr{B}$;

(3) $P_\omega \ll \nu$ for all ω, and $P_{x\omega}(xE) = P_\omega(E)$ for all $x \in G$, $\omega \in \Omega$, and $E \in \mathscr{B}$. (In particular, $P_\omega \in L_1(Z)$ $(= L_1(Z, \nu))$ so that $\hat{P}_\omega \in L_\infty(Z)'$.) A *test* is a measurable function $\phi \colon Z \to [0,1]$. (So every test is in $L_\infty(Z)$.) A test ϕ is called *invariant* if, for all $x \in G$, $\phi(xz) = \phi(z)$ ν-a.e.

Let $\Omega_0, \Omega_1 \in \mathscr{B}$ be G-invariant and be such that Ω is the disjoint union $\Omega_0 \cup \Omega_1$. For each $\gamma \in [0,1]$, let

$$T_\gamma = \{\phi \colon \phi \text{ is a test and } \hat{P}_\omega(\phi) \le \gamma \text{ for all } \omega \in \Omega_0\}.$$

Note that $T_\gamma \ne \varnothing$ since $0 \in T_\gamma$. It is routine to check that T_γ is weak* compact in $L_\infty(Z)$. A *minimax test for* T_γ is an element $\phi \in T_\gamma$ for which

$$\sup_{\psi \in T_\gamma} \inf_{\omega \in \Omega_1} \hat{P}_\omega(\psi) = \inf_{\omega \in \Omega_1} \hat{P}_\omega(\phi).$$

The Hunt-Stein Theorem is then *if G is amenable, then there exists an invariant, minimax test in T_γ for all $\gamma \in [0,1]$*. The proof falls into two parts. The first part (a) is the result of Problem 2-17 for measurable actions. Note that $L_\infty(Z)$ is a left Banach G-space, where we define $x.\psi(z) = \psi(x^{-1}z)$ $(x \in G$, $z \in Z$, $\psi \in L_\infty(Z))$.

(a) *Let K be a nonempty, weak* compact, convex, G-invariant subset of $L_\infty(Z)$, and let G be amenable. Then K contains a G-fixed-point.*

(b) *Let G be amenable and $\gamma \in [0,1]$. Then there exists an invariant, minimax test in T_γ.*

[Since $x.\chi_E = \chi_{xE}$, the equality of (3) gives $\hat{P}_{x\omega}(x\chi_E) = \hat{P}_\omega(\chi_E)$, and it follows that $\hat{P}_{x\omega}(x.\phi) = \hat{P}_\omega(\phi)$ for all $\phi \in L_\infty(Z)$. Hence T_γ is G-invariant. Define $F\colon T_\gamma \to [0,1]$ by

$$F(\phi) = \inf_{\omega \in \Omega_1} \hat{P}_\omega(\phi).$$

Since each of the functions $\phi \to \hat{P}_\omega(\phi)$ is (weak*) continuous on T_γ, it follows that F is upper semicontinuous on T_γ. Since T_γ is weak* compact in $L_\infty(Z)$, it follows that A_γ, where

$$A_\gamma = \left\{ \phi \in T_\gamma \colon F(\phi) = \sup_{\psi \in T_\gamma} F(\psi) \right\},$$

is nonvoid and weak* compact in $L_\infty(Z)$. It is easy to check that $F(x.\phi) = F(\phi)$ and $F(\alpha\phi + (1-\alpha)\psi) \geq \alpha F(\phi) + (1-\alpha)F(\psi)$ for all $x \in G$, $\phi, \psi \in T_\gamma$, and $\alpha \in [0,1]$, and so A_γ is convex and G-invariant. Now apply (a) to obtain an invariant, minimax test for T_γ.]

(4.26) Amenability and translation experiments. The definitions below are based on Le Cam [**1**], [**2**]. In the discussion below, we could have taken A and B to be abstract L-spaces. However, every abstract L-space can be realised as some $L_1(X,\mu)$, so that there is no loss of generality in the concrete approach adopted below.

Let X be a locally compact, Hausdorff space and μ_0 a positive, regular, Borel measure on X. Let $A = L_1(X)$ $(= L_1(X,\mu_0))$, and $P(A) = P(X,\mu_0)$. Thus

$$P(A) = \{\nu \in L_1(X) \colon \nu \geq 0,\ \nu(X) = 1\}.$$

An *experiment* for A (over Θ) is a function $\mathbf{P}\colon \Theta \to P(A)$, where Θ is some nonempty set.

Classically we can think of Θ as a set of possible theories about a certain physical system on which we perform an experiment. We take X to be the set of possible results of the experiment, and the probability that, on the basis of a theory θ, the result belongs to a measurable subset E of X (an "event") is just $\mathbf{P}(\theta)(E)$.

Now suppose that Y is a locally compact, Hausdorff space, ν_0 is a positive, regular, Borel measure on Y, and $B = L_1(Y)$ $(= L_1(Y,\nu_0))$. Suppose that we have an experiment $\mathbf{Q}$ for B over the same set (of "theories") Θ. Both experiments $\mathbf{P}$, $\mathbf{Q}$ will give us information about Θ.

In order to compare the amounts of information provided by experiments over Θ, Le Cam introduced a "distance" function $\delta(\mathbf{P},\mathbf{Q})$. For motivation, observe that to effect the comparison, we need to "lift" the information from one space into the other. This "lifting" is achieved by means of "transitions." A *transition* from A to B is a linear mapping $T\colon A \to B$ such that

(a) $T\mu \geq 0$ if $\mu \geq 0$ in A;

(b) $\|T\mu\| = \|\mu\|$ if $\mu \geq 0$ in A.

The set of transitions from A into B is denoted by $\mathbf{T}(A,B)$. (Such operators are called *stochastic* in [**Sch**, p. 191].)

Let $T \in \mathbf{T}(A, B)$. We claim that $T \in \mathbf{B}(A, B)$ with $\|T\| = 1$. [This is well known (cf. [**Sch**, Chapter II, (5.3)]) and can be proved as follows. If $\mu \in A = L_1(X)$, then, using the Jordan decomposition, we can write $\mu = \mu_1 - \mu_2 + i(\mu_3 - \mu_4)$ with $\mu_j \geq 0$ in A and $\|\mu_j\| \leq \|\mu\|$. Applying (b), we see that $\|T\mu\| \leq 4\|\mu\|$ so that $T \in \mathbf{B}(A, B)$. Now consider the adjoint $T^*: L_\infty(Y) \to L_\infty(X)$. Since $\phi \geq 0$ in $L_\infty(X)$ if $\phi(\mu) \geq 0$ for all $\mu \geq 0$ in $L_1(X)$, it follows that $T^*\psi \geq 0$ whenever $\psi \geq 0$ in $L_\infty(Y)$. Further, $T^*1(\mu) = \mu(X)$ (by (b)) for all $\mu \in L_1(Y)$ so that $T^*1 = 1$. Evaluating $(T^*\psi)^\wedge$ at points of the maximal ideal space of $L_\infty(X)$ gives states on $L_\infty(Y)$, and it follows that T^*, and therefore T, is norm decreasing.] Clearly, $\mathbf{T}(A, B)$ is a convex subset of $\mathbf{B}(A, B)$.

A measure of the amount of information yielded by $\mathbf{P}$ relative to $\mathbf{Q}$ is provided by the real number $\delta(\mathbf{P}, \mathbf{Q})$, where we define $\delta(\mathbf{P}, \mathbf{Q})$ by

$$\delta(\mathbf{P}, \mathbf{Q}) = \inf\left\{\sup_{\theta \in \Theta} \|\mathbf{Q}(\theta) - T(\mathbf{P}(\theta))\| : T \in \mathbf{T}(A, B)\right\}.$$

In general, the computation of $\delta(\mathbf{P}, \mathbf{Q})$ is difficult. However, the situation is better when we consider translation experiments. (Such an experiment is just a variant of a "translation family" introduced in Problem 4-19.)

Let G be a locally compact group. A *translation experiment* (for G) is an experiment $\mathbf{P}$ for $L_1(G)$ over G of the form *for some* $\mu \in P(G)$, $\mathbf{P}(x) = x\mu$ *for all* $x \in G$. We will write $\boldsymbol{\mu}$ in place of $\mathbf{P}$. Writing $\mathbf{T}(L_1(G))$ in place of $\mathbf{T}(L_1(G), L_1(G))$ we have, for $\mu, \nu \in P(G)$,

$$\begin{aligned} \delta(\boldsymbol{\mu}, \boldsymbol{\nu}) &= \inf\left\{\sup_{x \in G} \|x\nu - T(x\mu)\| : T \in \mathbf{T}(L_1(G))\right\} \\ &= \inf\left\{\sup_{x \in G} \|\nu - x^{-1}T(x\mu)\| : T \in \mathbf{T}(L_1(G))\right\}. \end{aligned} \tag{1}$$

The next result shows that when G is amenable, then, for $\mu, \nu \in P(G)$, $\delta(\boldsymbol{\mu}, \boldsymbol{\nu})$ is much more manageable. For this result, see Boll [**1**], Torgersen [**1**], and Paterson [**7**]. What can be said when G is not amenable? What can be said, even, for $G = F_2$? Paterson [**7**] gives a result corresponding to the one below for left amenable semigroups.

Let G be amenable and $\mu, \nu \in P(G)$. Then

$$\delta(\boldsymbol{\mu}, \boldsymbol{\nu}) = \inf\{\|\nu - \mu\xi\| : \xi \in P(G)\}. \tag{2}$$

[The essential ideas of the proof of (2) are as follows. The space $\mathbf{B}(L_1(G))$ is a left Banach G-space, where we define

$$(xT)(\xi) = xT(x^{-1}\xi) \qquad (\xi \in L_1(G)), \tag{3}$$

and $\mathbf{T}(L_1(G))$ is a convex subset of $\mathbf{B}(L_1(G))$. If, somehow, we could "find" a weak* topology on $\mathbf{B}(L_1(G))$, we might hope to have $\mathbf{T}(L_1(G))$ weak* compact and to use an amenability fixed-point theorem to obtain a fixed-point T_0 in $\mathbf{T}(L_1(G))$. Hopefully, with equation (1) in mind, we can choose T_0 so that $\sup_{x \in G} \|\nu - x^{-1}T_0(x\mu)\|$ is close to $\delta(\boldsymbol{\mu}, \boldsymbol{\nu})$. Since T_0 is a fixed-point for G, we see from (3) that $T_0(x\xi) = xT_0(\xi)$ $(\xi \in L_1(G))$, that is, T_0 is a multiplier of $L_1(G)$. Thus by Wendel's Theorem [**HR2**, (35.5)], we can find $\eta \in M(G)$ such that $T_0(\mu) = \mu\eta$. Hopefully, we can take $\eta \in P(G)$ and we would have

$$\sup_{x \in G} \|\nu - x^{-1}T_0(x\mu)\| = \|\nu - \mu\eta\|$$

and so the right-hand side of (2) is less than or equal to its left-hand side. The reverse inclusion is easy.

This crude argument falls down in two crucial aspects. First, if we want a weak* topology, we ought to replace

$$\mathbf{B}(L_1(G)) = \mathbf{B}(L_1(G), L_1(G)) \quad \text{by} \quad \mathbf{B}(L_1(G), M(G))$$

since the latter space has the natural weak* topology, which it inherits when identified with the dual space $(L_1(G)\hat{\otimes}C_0(G))'$. Even with this replacement and $\mathbf{T}(L_1(G))$ replaced by $\mathbf{T}(L_1(G), M(G))$, there is a second problem, since the lack of an identity in $C_0(G)'$ in general, jeopardises the weak* compactness of $\mathrm{PM}(G)$ and hence of $\mathbf{T}(L_1(G), M(G))$.

We will suppose that G is not compact, since the latter case is easier.

To overcome the above difficulties, we replace $C_0(G)$ by $C(G_\infty)$, where G_∞ is the one-point compactification of G. Then $C(G_\infty)$ can be identified with the space $C_0(G) \oplus \mathbb{C}1$ of functions in $C(G)$ that tend to a limit as $x \to \infty$. Clearly, $C(G_\infty)$ is a G-invariant subspace of $C(G)$. The G-action on $C(G_\infty)$ can also be obtained in the canonical way from the natural action of G on G_∞—define $x\infty = \infty = \infty x$ $(x \in G)$.

We will identify $M(G)$ in the obvious way with the following subspace of $M(G_\infty)$:

$$\{\mu \in M(G_\infty) \colon \mu(\{\infty\}) = 0\}.$$

Clearly

$$M(G_\infty) = M(G) \oplus \mathbb{C}\delta_\infty, \tag{4}$$

where δ_∞ is the point-mass at ∞, and the right-hand side of (4) is a Banach space direct sum.

We now proceed to the proof of (2). Let $\varepsilon > 0$, and define K_ε as follows:

$$K_\varepsilon = \{T \in \mathbf{T}(L_1(G), M(G_\infty)) \colon \|x\nu - T(x\mu)\| \leq \delta(\boldsymbol{\mu}, \boldsymbol{\nu}) + \varepsilon \text{ for all } x \in G\}.$$

Since

$$\mathbf{T}(L_1(G)) \subset \mathbf{T}(L_1(G), M(G_\infty)),$$

K_ε is not empty. Clearly K_ε is convex in $X = \mathbf{B}(L_1(G), M(G_\infty))$. Now X is canonically identified with $(L_1(G)\hat{\otimes}C(G_\infty))'$, an operator $T \in X$ corresponding to the linear functional

$$\xi \otimes \phi \to (T\xi)^\wedge(\phi).$$

We now show that K_ε is weak* compact in X. Let $T_\delta \to T$ weak* in X with $T_\delta \in K_\varepsilon$ for all δ. Then for $\mu \geq 0$ in $L_1(G)$ and $\phi \geq 0$ in $C(G_\infty)$, we have $(T\mu)^\wedge(\phi) = \lim_\delta (T_\delta\mu)^\wedge(\phi) \geq 0$.

So $T\mu \geq 0$, and

$$\|T\mu\| = (T\mu)^\wedge(1) = \lim_\delta (T_\delta \mu)^\wedge(1) = \|\mu\|.$$

So $T \in \mathbf{T}(L_1(G), M(G_\infty))$ and K_ε is weak* closed in X. Since K_ε is also bounded in X, it is weak* compact.

In order to apply an amenability fixed-point theorem, we need an action of G on K_ε. In fact, we can make X into a left Banach G-space (with K_ε G-invariant) as follows. The space $L_1(G)\hat{\otimes}C(G_\infty)$ is a right Banach G-space under the action:

$$\xi \otimes \phi \to x^{-1}\xi \otimes \phi x \qquad (x \in G),$$

where we regard $C(G_\infty)$ as a G-invariant subspace of $C(G)$ as commented earlier. Dualising this action, we obtain that X is a left Banach G-space. One readily checks that this action is given by

$$(5) \qquad (xT)(\xi) = x[T(x^{-1}\xi)] \qquad (x \in G).$$

Note that $\|yT\| = \|T\|$ for all $y \in G$, $T \in X$.

We claim that $C(G_\infty) \subset \mathrm{U}(G)$. Indeed, $C(G_\infty) = C_0(G) \oplus \mathbb{C}1$, and since $C_c(G)$ is norm dense in $C_0(G)$ and is contained in $\mathrm{U}(G)$, it follows that $C_0(G)$, and so $C(G_\infty)$, is contained in $\mathrm{U}(G)$ as claimed.

If $x_\delta \to x$ in G and $T_\delta \to T$ weak* in a bounded subset of X then, since, for each $\phi \in C(G_\infty)$, $\xi \in L_1(G)$, the maps $\phi \to \phi t$, $t \to t^{-1}\xi$ $(t \in G)$ are norm continuous, it follows using (5) that

$$(x_\delta T_\delta)(\xi)(\phi) = T_\delta(x_\delta^{-1}\xi)(\phi x_\delta) \to (xT)(\xi)(\phi)$$

so that the action of G on every bounded subset of X is jointly continuous for the weak* topology.

If $T \in K_\varepsilon$ and $x, y \in G$, then

$$\|x\nu - (yT)(x\mu)\| = \|(y^{-1}x)\xi - T((y^{-1}x)\mu)\| \leq \delta(\boldsymbol{\mu}, \boldsymbol{\nu}) + \varepsilon,$$

and it easily follows that K_ε is G-invariant.

Applying the fixed-point theorem ((2.24)), we can find $T \in K_\varepsilon$ with $xT = T$ for all $x \in G$. From (5), we see that T is a right multiplier in the sense that

$$(6) \qquad T(x\xi) = x[T(\xi)]$$

for all $x \in G$, $\xi \in L_1(G)$.

Define $T' \in \mathbf{B}(L_1(G), M(G))$ and $p \in L_1(G)'$ by

$$(7) \qquad T(\xi) = T'(\xi) + p(\xi)\delta_\infty,$$

where we use the direct sum of (4).

Now $x\delta_\infty = \delta_\infty$ for all x, $M(G)$ is G-invariant, and (6) holds. So

$$(8) \qquad T'(x\xi) = xT'(\xi), \qquad p(x\xi) = p(\xi).$$

Our objective now is to show that (9) below holds.

We observe that p, being left invariant, is represented by a constant function c in $L_\infty(G)$. So $p(\xi) = c$ for all $\xi \in P(G)$. Since $T \in \mathbf{T}(L_1(G), M(G_\infty))$, we must have $0 \le c \le 1$. If $c = 1$, then $T' = 0$ and $T(\xi) = \delta_\infty$, giving

$$2 = \|x\nu - T(x\mu)\| \le \delta(\boldsymbol{\mu}, \boldsymbol{\nu}) + \varepsilon.$$

Now the inequality (9) below is trivial if $\delta(\boldsymbol{\mu}, \boldsymbol{\nu}) = 2$. We therefore suppose that $\delta(\boldsymbol{\mu}, \boldsymbol{\nu}) < 2$. Clearly, for small enough ε, the case $c = 1$ cannot occur. We can therefore suppose that $0 \le c < 1$. Let $T'' = (1-c)^{-1}T'$. We claim that $T'' \in K_\varepsilon \cap \mathbf{T}(L_1(G), M(G))$.

Clearly, $T''(x\xi) = xT''(\xi)$ and $T'' \in \mathbf{T}(L_1(G), M(G))$. Also

$$\begin{aligned}\|x\nu - T(x\mu)\| &= \|x\nu - T'(x\mu)\| + c = \|x\nu - (1-c)T''(x\mu)\| + c\|T''(x\mu)\| \\ &\ge \|x\nu - [(1-c)T''(x\mu) + cT''(x\mu)]\| = \|x\nu - T''(x\mu)\|.\end{aligned}$$

So $T'' \in K_\varepsilon \cap \mathbf{T}(L_1(G), M(G))$ as asserted.

Let $\{\nu_\delta\}$ be a bounded, approximate identity in $P(G)$ and let $T'_\delta\colon L_1(G) \to L_1(G)$ be given by

$$T'_\delta(\xi) = T''(\xi)\nu_\delta.$$

Then $T'_\delta \in \mathbf{T}(L_1(G))$ and $T'_\delta(x\xi) = xT'_\delta(\xi)$ for all $x \in G$, $\xi \in L_1(G)$. By Wendel's Theorem, there exists $\nu'_\delta \in \mathrm{PM}(G)$ such that

$$T'_\delta(\xi) = \xi\nu'_\delta \qquad (\xi \in L_1(G)).$$

Then $\nu'_\delta\nu_\delta \in P(G)$ and

$$\begin{aligned}\|\nu - \mu\nu'_\delta\nu_\delta\| &= \|\nu - T''(\mu)\nu_\delta^2\| \\ &\le \|\nu - \nu\nu_\delta^2\| + \|\nu - T''(\mu)\| \\ &\le \|\nu - \nu\nu_\delta^2\| + \delta(\boldsymbol{\mu}, \boldsymbol{\nu}) + \varepsilon\end{aligned}$$

since $T'' \in K_\varepsilon$. Since $\|\nu - \nu\nu_\delta^2\| \to 0$, it follows that

$$\inf\{\|\nu - \mu\xi\| \colon \xi \in P(G)\} \le \delta(\boldsymbol{\mu}, \boldsymbol{\nu}). \tag{9}$$

To prove the reverse inequality, let $\xi \in P(G)$ and define $T_\xi \in \mathbf{T}(L_1(G))$ by $T_\xi(\eta) = \eta\xi$. Then

$$\delta(\boldsymbol{\mu}, \boldsymbol{\nu}) \le \sup_{x\in G} \|x\nu - T_\xi(x\mu)\| = \|\nu - \mu\xi\|.$$

This completes the proof.]

(4.27) Uniqueness of invariant means. The burden of Chapter 7 is that, normally, if there exists *one* invariant mean on a semigroup or group, then there are *many* other such means. Much less seems to be known about the corresponding question for a group acting on a set. Perhaps the most outstanding particular case of this question which has been studied is the *Banach-Ruziewicz Problem* ((3.12)). The latter problem is concerned with whether or not normalised Lebesgue measure λ_n is the *unique* finitely additive, positive, normalised measure on the ring $\mathscr{M}_b(\mathbb{R}^n)$ of bounded, measurable subsets of $\mathbb{R}^n$, which is invariant under isometries. Banach showed that λ_n is *not* unique when $n = 1, 2$.

For $n \geq 3$, the problem had to wait till the 1980s, when Margulis [**3**] demonstrated the *uniqueness* of λ_n $(n \geq 3)$. We will discuss the solution of the $\mathbb{S}^n$ version of the problem in (4.28) and (4.29).

An essential ingredient in the solution of the Banach-Ruziewicz Problem is the uniqueness theorem below. This result is stated in del Junco and Rosenblatt [**1**], and a more general version is proved by Rindler [**S11**]. (Other relevant references are Losert and Rindler [**2**] and Schmidt [**1**], [**2**].) The proof we give involves careful adjusting of Namioka's proof ((4.7)).

Let $(X, \mathscr{B}, \mu)$ be a measure space with μ a probability measure. Let G be a group and X a left G-set, such that $ex = x$ for all $x \in X$, and, for each $s \in G$, the map $x \to sx$ from X to X is measurable. We also suppose that μ is G-invariant in the sense that $\mu(sA) = \mu(A)$ for all $s \in G$, $A \in \mathscr{B}$.

The group G is said to be *ergodic* (on X) if whenever $A \in \mathscr{B}$ and

$$\mu(sA \triangle A) = 0$$

for all $s \in G$, then $\mu(A)$ is 0 or 1. (Sometimes, either of the expressions "metrically ergodic" or "metrically transitive" is used where we have used "ergodic.") The set X is said *to admit arbitrarily small, almost invariant sets* if there exists a sequence $\{A_n\}$ of nonnull, measurable subsets of X such that $\mu(A_n) \to 0$ and

$$\mu(sA_n \triangle A_n)/\mu(A_n) \to 0 \tag{1}$$

for all $s \in G$.

As in the case where $X = G$, the space $L_\infty(X)$ is a right Banach G-space under the maps $\phi \to \phi s$. The set of means on $L_\infty(X)$ is denoted by $\mathfrak{M}(X)$, and the set of G-invariant means in $\mathfrak{M}(X)$ is denoted by $\mathfrak{L}(X)$. Obviously, $\hat{\mu} \in \mathfrak{L}(X)$. We now state and prove the uniqueness theorem alluded to above.

(i) *Let G be countable and ergodic on X. Then $\mathfrak{L}(X) \neq \{\hat{\mu}\}$ if and only if there exist arbitrarily small, almost invariant sets in X.*

[Suppose that X admits arbitrarily small, almost invariant sets, and let $\{A_n\}$ be a sequence as in (1). We can suppose that $\mu(A_n) < 2^{-n-1}$. Let $C = X \sim (\bigcup_{n=1}^\infty A_n)$. Then $C \in \mathscr{B}$ and $\mu(C) > \frac{1}{2}$. Define $m_n \in \mathfrak{M}(X)$ by

$$m_n = (\chi_{A_n}/\mu(A_n))^\wedge,$$

and m be a weak* cluster point of the sequence $\{m_n\}$. Let $\phi \in L_\infty(X)$ and $x \in G$. Then using the invariance of μ,

$$\begin{aligned} |m_n(\phi s - \phi)| &= \mu(A_n)^{-1} \left| \int \chi_{A_n s^{-1}}(st)\phi(st)\, d\mu(t) - \int \chi_{A_n}(t)\phi(t)\, d\mu(t) \right| \\ &= \mu(A_n)^{-1} \left| \int [\chi_{A_n s^{-1}}(t) - \chi_{A_n}(t)]\phi(t)\, d\mu(t) \right| \\ &\leq \mu(A_n)^{-1} \mu(sA_n \triangle A_n)\|\phi\|, \end{aligned}$$

and it follows from (1) that $m \in \mathfrak{L}(X)$. Now $m_n(C) = 0$ since $A_n \cap C = \varnothing$, so that $m(C) = 0$. But $\mu(C) \geq \frac{1}{2}$ so that $\hat{\mu} \neq m$. Hence $\mathfrak{L}(X) \neq \{\hat{\mu}\}$.

For the converse, it is sufficient to show that if X does not admit arbitrarily small, almost invariant sets, then $\mathfrak{L}(X) = \{\hat{\mu}\}$. Suppose, then, that X does not admit arbitrarily small, almost invariant sets. Then using the countability of G, there exist $\varepsilon, k > 0$ and a finite subset F of G such that if $A \in \mathscr{B}$ is nonnull and $\mu(sA \,\triangle\, A)/\mu(A) < \varepsilon$ for all $s \in F$, then $\mu(A) \geq k$.

Let us say that a sequence $\{B_n\}$ of nonnull, measurable subsets of X is a *Følner sequence* if $\mu(sB_n \,\triangle\, B_n)/\mu(B_n) \to 0$ for all $s \in G$. If $\{B_n\}$ is a Følner sequence, then, from the argument of the first paragraph of the present proof, every weak* cluster point of the sequence $\{(\chi_{B_n}/\mu(B_n))^\wedge\}$ belongs to $\mathfrak{L}(X)$. Let $m \in \mathfrak{L}(X)$. We need to prove that $m = \hat{\mu}$. The proof of this falls into two parts (a) and (b). In the first, we suppose that m is such a weak* cluster point, and in the second, we deal with the general case.

(a) Suppose that $m \in \mathfrak{L}(X)$ is a weak* cluster point of $\{p_n\}$, where $p_n = (\chi_{B_n}/\mu(B_n))^\wedge$, and $\{B_n\}$ is a Følner sequence. We can suppose that

$$\mu(sB_n \,\triangle\, B_n)/\mu(B_n) < \varepsilon$$

for all $s \in F$ and all n, so that we have $\mu(B_n) \geq k$ for all n. For $C \in \mathscr{B}$, we have

$$p_n(C) = \left(\int_C \chi_{B_n}\, d\mu\right) / (\mu(B_n)) = \mu(C \cap B_n)/\mu(B_n)$$

and so $p_n \leq k^{-1}\mu$. Hence $m \leq k^{-1}\mu$, and since μ is countably additive and finite, $m(E_n) \to 0$ for every decreasing sequence $\{E_n\}$ in $\mathscr{B}$ with empty intersection. So m is countably additive on $\mathscr{B}$. Since μ is a finite measure, we can apply the Radon-Nikodým Theorem to obtain a function $f \geq 0$ in $L_1(\mu)$ such that $dm = f\, d\mu$.

Let $A = \{x \in X \colon f(x) < 1\}$. Let $s \in G$, and suppose that $\mu(sA \sim A) > 0$. Let $B = A \sim s^{-1}A = s^{-1}(sA \sim A)$. Then $\mu(B) > 0$ and $f(b) < 1$, $f(sb) \geq 1$ for all $b \in B$. Since $m(B) = m(sB)$, we have

$$\int_B f\, d\mu = \int_{sB} f\, d\mu.$$

So

$$\mu(B) > \int_B f\, d\mu = \int_{sB} f\, d\mu \geq \mu(sB) = \mu(B)$$

giving a contradiction. So $\mu(sA \sim A) = 0$ for all $s \in G$, and $\mu(A \sim sA) = \mu(s^{-1}A \sim A) = 0$ also. Thus $\mu(sA \,\triangle\, A) = 0$ for all $s \in G$, and since G is ergodic, $\mu(A) = 0$ or 1. If $\mu(A) = 1$, then

$$m(X) = \int_A f\, d\mu < \mu(A) = 1,$$

and we contradict the fact that m is a mean. Thus $\mu(A) = 0$. A similar argument shows that

$$m(\{x \in X \colon f(x) > 1\}) = 0$$

so that $f = 1$ μ-almost everywhere, and $m = \hat{\mu}$ as required.

We note that, from the above, $\hat{\mu}$ is the only weak* cluster point of $\{p_n\}$, so that $p_n \to \hat{\mu}$ weak*.

(b) Now suppose that m is a general point in $\mathfrak{L}(X)$. Suppose also that $m \neq \hat{\mu}$. We will eventually derive a contradiction.

Let $\alpha = \sup\{\beta \geq 0 \colon m \geq \beta\mu \text{ on } \mathscr{B}\}$. Then $(m - \alpha\mu) \geq 0$. Since $\|m - \alpha\hat{\mu}\| = (m - \alpha\hat{\mu})(1) = 1 - \alpha$ and $m \neq \hat{\mu}$, it follows that $\alpha < 1$. By subtracting $\alpha\hat{\mu}$ from m and dividing by $(1 - \alpha)$ we can suppose that if $\beta > 0$ then it is *not* true that $\beta\mu \leq m$. For each $p \in \mathbb{P}$, it follows, by putting $\beta = 1/p$, that there exists nonnull $C_p \in \mathscr{B}$ such that

$$\mu(C_p) > pm(C_p). \tag{2}$$

We now adapt ideas in (0.8) and (4.2). We can find a net $\{g_\delta\}$ in $P(X)$ such that $\hat{g}_\delta \to m$ weak*. (Here, of course, $P(X)$ is the set of probability measures in $L_1(X)$.) For each p, $p\hat{g}_\delta(C_p) \to pm(C_p) < \mu(C_p)$ by (2). Using the countability of G, we can construct a sequence $\{f_n\}$ in $P(X)$ with

$$\|s * f_n - f_n\|_1 \to 0 \qquad (s \in G) \tag{3}$$

and

$$\mu(C_p) > p\hat{f}_n(C_p) \qquad (n \geq p). \tag{4}$$

(Of course, $s * f(x) = f(s^{-1}x)$.)

Our next task is to modify the proof of (4.7). We will assume that G is denumerable, the case where G is finite being easier.

Enumerate $G = \{x_r \colon r \geq 1\}$. Let $n \geq 1$. By choosing f_r for suitably large r and then approximating f_r by simple functions, we can find a simple function

$$g_n = \sum_{r=1}^{N_n} \beta_r^n (\chi_{B_r^n}/\mu(B_r^n)) \qquad \left(\beta_r^n \geq 0,\ \sum_{r=1}^{N_n} \beta_r^n = 1\right)$$

such that $B_r^n \subset B_s^n$ if $1 \leq r < s \leq N_n$, $\mu(B_r^n) > 0$ for $1 \leq r \leq N_n$, and

$$\|x_i * g_n - g_n\|_1 < n^{-2} \quad (1 \leq i \leq n), \qquad \mu(C_p) > p\hat{g}_n(C_p) \quad (n \geq p). \tag{5}$$

Arguing as in (4.7), with summation over $\{x_1, \ldots, x_n\}$ replacing integration over C,

$$\sum_{r=1}^{N_n} \beta_r^n \sum_{i=1}^{n} (\mu(x_i B_r^n \triangle B_r^n)/\mu(B_r^n)) < n \cdot n^{-2} = n^{-1}.$$

Let $U_n = \{r \colon \sum_{i=1}^n (\mu(x_i B_r^n \triangle B_r^n)/\mu(B_r^n)) < 2n^{-1}\}$, and $V_n = \{1, \ldots, N_n\} \sim U_n$. Then

$$n^{-1} > \sum_{r \in V_n} \beta_r^n \sum_{i=1}^{n} (\mu(x_i B_r^n \triangle B_r^n)/\mu(B_r^n)) \geq \left(\sum_{r \in V_n} \beta_r^n\right) 2n^{-1},$$

so that

$$\sum_{r \in V_n} \beta_r^n < \tfrac{1}{2}.$$

In particular, since $\sum_{r=1}^{N_n} \beta_r^n = 1$, it follows that $U_n \neq \varnothing$.

Summarising, we thus have, for each n, a nonvoid subset U_n of $[1; N_n]$ and for each $r \in U_n$, a nonnull set $B_r^n \in \mathscr{B}$ and a number $\beta_r^n \geq 0$ such that

$$\mu(x_i B_r^n \triangle B_r^n)/\mu(B_r^n) < 2n^{-1} \qquad (1 \leq i \leq n,\ r \in U_n), \tag{6}$$

$$1 \geq \sum_{r \in U_n} \beta_r^n > \tfrac{1}{2}, \tag{7}$$

$$0 \leq \sum_{r \in U_n} \beta_r^n (\chi_{B_r^n}/\mu(B_r^n)) \leq g_n. \tag{8}$$

By choosing a suitable subsequence, we can suppose that $\sum_{r \in U_n} \beta_r^n \to t$ where, from (7), $\frac{1}{2} \leq t \leq 1$.

We now claim that for each $C \in \mathscr{B}$,

$$\max_{r \in U_n} |(\hat{\chi}_{B_r^n}(C)/\mu(B_r^n)) - \mu(C)| \to 0 \tag{9}$$

as $n \to \infty$. For if not, we could, using (6), construct a Følner sequence $\{A_m\}$, where each A_m is some B_r^n, with a weak* cluster point n for which $n(C) \neq \mu(C)$, and the conclusion of (a) is contradicted. This establishes (9) and it follows from (7) and (9) that

$$\left[\sum_{r \in U_n} \beta_r^n (\chi_{B_r^n}/\mu(B_r^n))^\wedge(C) - \left(\sum_{r \in U_n} \beta_r^n\right) \mu(C)\right] \to 0.$$

Recalling that $\sum_{r \in U_n} \beta_r^n \to t$, we obtain

$$\sum_{r \in U_n} \beta_r^n (\chi_{B_r^n}/\mu(B_r^n))^\wedge \to t\hat{\mu}$$

weak*. But this is impossible, since a consequence of (8) and (5) is that for every $p \in \mathbb{P}$,

$$\lim_{n \to \infty} \sum_{r \in U_n} \beta_r^n (\chi_{B_r^n}/\mu(B_r^n))^\wedge(C_p) \leq p^{-1}\mu(C_p)$$

and so $t \leq p^{-1}$ for all p. But $t \geq \frac{1}{2}$. This contradiction establishes (i).]

A related question is the following. Let G be a compact group. A function $\phi \in L_\infty(G)$ is said to have *a unique left invariant mean* if $m(\phi) = \hat{\lambda}(\phi)$ for all $m \in \mathfrak{L}(G)$. Since $\mathfrak{L}(C(G)) = \{\hat{\lambda}\}$, it is trivial that every $\phi \in C(G)$ has a unique left invariant mean.

A larger set of functions in $L_\infty(G)$ with unique left invariant mean is the set of *Riemann measurable* functions on G (Problem 4-34). See Talagrand [**S2**], [**S4**].

A sufficient condition for a function $\phi \in L_\infty(G)$ to have a unique left invariant mean is the following: given $\varepsilon > 0$, there exist $n \geq 1$ and elements $x_1, \ldots, x_n \in G$ such that

$$\left\| n^{-1}\left(\sum_{i=1}^{n} \phi x_i\right) - \left(\int \phi \, d\lambda\right) 1 \right\|_\infty < \varepsilon. \tag{10}$$

(By applying $m \in \mathfrak{L}(G)$ to the L_∞-function on the left-hand side of (10), the condition is obviously sufficient.) Talagrand [**5**] asserts that this condition can be used to construct for many compact groups G, functions ϕ which are not Riemann-measurable and yet have a unique left invariant mean.

Rubel and Shields [**1**] raised the following question: suppose that $\phi \in L_\infty(G)$ and that $\phi\psi$ has a unique left invariant mean for all $\psi \in C(G)$. Does it follow that ϕ is Riemann-measurable?

Talagrand shows that this is not the case for many compact groups, using (10) in his proof.

See Rosenblatt and Yang [**S**] for recent work on functions with unique left invariant mean.

(4.28) The Banach-Ruziewicz Problem: the $\mathbb{S}^n$-version. In the sequel we shall often abbreviate "the Banach-Ruziewicz Problem" to "the B-R Problem". Recall ((3.12), (4.27)) that this problem (for $\mathbb{R}^n$) is: *does there exist a finitely additive, G_n-invariant, positive measure μ on $\mathscr{M}_b(\mathbb{R}^n)$ such that $\mu_n([0,1]^n) = 1$ and $\mu \neq \lambda_n$?*

With the uniqueness result (i) of (4.27) in mind, it is desirable to change the problem to one in which we have a probability measure in place of the infinite measure λ_n. This is achieved by studying the corresponding problem on the (compact) set $\mathbb{S}^{n-1}$. Recall that $\mathbb{S}^{n-1}$ is invariant under the action of the orthogonal group $\mathrm{O}(n) \subset G_n$. The required probability measure μ_{n-1} on $\mathbb{S}^{n-1}$ is obtained from λ_n as follows.

For each subset E of $\mathbb{S}^{n-1}$, define $E' \subset \mathbb{R}^n$ by

$$E' = \{\alpha x \colon x \in E,\ 0 < \alpha \leq 1\}.$$

(The correspondence $E \to E'$ was used earlier in our discussion of the Banach-Tarski Theorem ((3.16)).) If E is a Borel subset of $\mathbb{S}^{n-1}$, then the set E' is a Borel subset of $\mathbb{R}^n$. (This is clearly true when E is closed, since then $E' \cup \{0\}$ is closed in $\mathbb{R}^n$. The Borel measurability of E' in general follows using the Monotone Class Lemma.)

It is easily checked that we obtain a probability measure μ_{n-1}, defined on the Borel subsets of $\mathbb{S}^{n-1}$, by defining

$$\mu_{n-1}(E) = \lambda_n(E')/\lambda_n(B_n)$$

where $B_n = \{x \in \mathbb{R}^n \colon \sum_{i=1}^n x_i^2 \leq 1\}$. It is left as a pleasant, measure-theoretic exercise for the reader to check that μ_{n-1} is a regular Borel measure on $\mathbb{S}^{n-1}$ and that if E is a μ_{n-1}-measurable subset of $\mathbb{S}^{n-1}$, then E' is λ_n-measurable in $\mathbb{R}^n$. Let $\mathscr{M}(\mathbb{S}^{n-1})$ be the algebra of μ_{n-1}-measurable sets.

Clearly, if $T \in \mathrm{O}(n)$, then $(T(E))' = T(E')$, and since λ_n is $\mathrm{O}(n)$-invariant, it follows that μ_{n-1} is also $\mathrm{O}(n)$-invariant. For convenience, we now replace n by $(n+1)$ (so that λ_n, μ_{n-1}, $\mathbb{S}^{n-1}$ are replaced by λ_{n+1}, μ_n, and $\mathbb{S}^n$ respectively). The B-R problem for $\mathbb{S}^n$ is the following: *is μ_n the only $\mathrm{O}(n+1)$-invariant, finitely additive, positive measure on $\mathscr{M}(\mathbb{S}^n)$ of total mass 1?*

The solution to the B-R *problem for* $\mathbb{S}^n$ *is negative if* $n = 1$ *and affirmative for* $n \geq 2$. This solution is discussed in (4.29).

In order to allow the use of functional analytic techniques, it is convenient to formulate the B-R problem for $\mathbb{S}^n$ in terms of invariant means on $L_\infty(\mathbb{S}^n)$ $(= L_\infty(\mathbb{S}^n, \mu_n))$. Let $H_n = \mathrm{O}(n+1)$. Then as in (4.27), $L_\infty(\mathbb{S}^n)$ is a right Banach H_n-space. The set of left invariant means on $L_\infty(\mathbb{S}^n)$ for H_n is denoted by $\mathfrak{L}(\mathbb{S}^n)$.

Clearly $\hat{\mu}_n \in \mathfrak{L}(\mathbb{S}^n)$, and every $m \in \mathfrak{L}(\mathbb{S}^n)$ is associated with an H_n-invariant, finitely additive, positive measure on $\mathscr{B}(\mathbb{S}^n)$ of total mass 1. The following result shows that the B-R problem for $\mathbb{S}^n$ can be reformulated for $n \geq 2$:

$$(*) \qquad \textit{does } \mathfrak{L}(\mathbb{S}^n) = \{\hat{\mu}_n\}?$$

The result is due to Tarski; the $\mathbb{R}^n$-version is discussed in Rosenblatt [**14**] and Wagon [**3**].

Let $n \geq 2$ *and* μ *be an* H_n*-invariant, finitely additive, positive measure on* $\mathscr{M}(\mathbb{S}^n)$ *of total mass* 1. *Then every* μ_n*-null subset of* $\mathbb{S}^n$ *is also* μ*-null* (*so that* μ *can be identified with an element of* $\mathfrak{L}(\mathbb{S}^n)$). [We will deal with the case $n = 2$, the proof for general $n \geq 2$ being similar. (Note that $\mathbb{S}^n$ has an H_n-paradoxical decomposition for $n \geq 2$ by Problem 3-18.) As in the proof of Tarski's Theorem ((3.15)), let $Y = \mathbb{S}^2 \times \mathbb{P}$ and $H = H_2 \times \Pi(\mathbb{P})$. Identify $\mathbb{S}^2$ with $\mathbb{S}^2 \times \{1\}$. Then H acts on Y. Let $m \in \mathbb{P}$. Since $\mathbb{S}^2$ has a paradoxical decomposition,

$$(1) \qquad \mathbb{S}^2 \cong \bigcup_{r=1}^{m} (\mathbb{S}^2 \times \{r\}).$$

For such an m, "slice" the sphere into parts $A_0, \ldots, A_{m-1}$ by the planes

$$y\cos(2r\pi/m) - x\sin(2r\pi/m) = 0 \qquad (0 \leq r \leq (m-1)).$$

(The process is rather like separating a skinned orange into its divisions.) More precisely, let

$$A_r = \Big\{ \Big((\cos\phi)\sqrt{1-z^2},\ (\sin\phi)\sqrt{1-z^2}, z \Big) : \\ -1 < z < 1,\ 2r\pi/m < \phi \leq 2(r+1)\pi/m \Big\}.$$

Then $\mathbb{S}^2 \sim \{(0,0,1),(0,0,-1)\}$ is the disjoint union of the A_r's. Further, $A_r \cong A_0$ with respect to H_2 since rotation about the z-axis through an angle of $2r\pi/m$ carries A_0 onto A_r. Since singletons are μ_2-null and μ_2 is H_2-invariant, it follows that $\mu_2(A_r) = (\mu_2(\mathbb{S}^2))/m = 1/m$. The measure μ also vanishes on singletons (since any point can be rotated to infinitely many other positions and $\mu(\mathbb{S}^2) = 1$.) Hence $\mu(A_r) = 1/m$ as well. Finally, as in the proof of the Banach-Tarski Theorem ((3.16)), $\mathbb{S}^2 \sim (\{(0,0,1),(0,0,-1)\}) \cong \mathbb{S}^2$ with respect to H_2. Hence $\bigcup_{r=0}^{m-1} A_r \cong \mathbb{S}^2$ with respect to H_2, and using (1) and the Division Theorem ((3.14)) (for the action of H on Y), we obtain

$$(2) \qquad A_0 \cong \mathbb{S}^2.$$

Now let $E \in \mathscr{B}(\mathbb{S}^2)$ be μ_2-null. From (2), we can find a partition $E_1, \ldots, E_k$ of E and elements $x_1, \ldots, x_k \in H_n$ such that $x_iE_i \cap x_jE_j = \emptyset$ if $i \neq j$ and $\bigcup_{i=1}^k x_iE_i \subset A_0$. Normally, one cannot relate μ to the sets E_i since these sets will not, in general, belong to $\mathscr{M}(\mathbb{S}^2)$. However, in the present case, E_i is a subset of a μ_2-null set, and so itself is μ_2-null (and, in particular, belongs to $\mathscr{M}(\mathbb{S}^2)$).

Thus

$$\mu(E) = \mu\left(\bigcup_{i=1}^k E_i\right) = \mu\left(\bigcup_{i=1}^k x_iE_i\right) \leq \mu(A_0) = m^{-1}.$$

Since this is true for all $m \in \mathbb{P}$, we have $\mu(E) = 0$ as required.]

(4.29) The Banach-Ruziewicz Problem and the Kazhdan Property (T). We now examine the B-R Problem for $\mathbb{S}^n$, in the form of (*) of (4.28). The notations of (4.28) will be used without comment.

We start by dealing with the case $n = 1$. A result which will be proved later is used in the proof. A useful fact is that $\mathbb{S}^1$ is the circle group $\mathbb{T}$, and so is a compact, abelian group (and hence amenable as a discrete group).

(i) (Banach [**1**]). *The answer to the B-R Problem when $n = 1$ is negative, that is, $\mathfrak{L}(\mathbb{S}^1) \neq \{\hat{\mu}_1\}$.* [The following result follows from (7.19). *Let G be a non-discrete compact group that is amenable as a discrete group. Then there exists an element $m \in \mathfrak{L}(G)$ such that the probability measures $\hat{m}$, $\hat{\lambda}_G$ on the maximal ideal space $\Phi(G)$ of $L_\infty(G)$ have disjoint supports.*

Let $G = \mathbb{T}$ $(= \mathbb{S}^1)$. Clearly, multiplication by an element of $\mathbb{S}^1$ corresponds to a rotation $(\in H_1 = \mathrm{O}(2))$, and the uniqueness of Haar measure gives $\mu_1 = \lambda_G$.

Now H_1 is generated by the group R_1 of rotations of $\mathbb{S}^1$ and the reflection T, where $T(w) = \bar{w}$. Let $m \in \mathfrak{L}(\mathbb{T})$ be such that $\hat{m}$ and $(\hat{\mu}_1)\hat{}$ have disjoint supports. Then

(1)
for each $\alpha > 0$, it is *not* the case that $\alpha\hat{m} \leq (\hat{\mu}_1)\hat{}$ in $L_\infty(G)'(= M(\Phi(G)))$.

Define $T(m) \in \mathfrak{M}(\mathbb{S}^1)$ by $T(m)(\phi) = m(\phi T)$, where $\phi T(w) = \phi(T(w))$ for all $w \in \mathbb{S}^1$, $\phi \in L_\infty(\mathbb{S}^1)$. Now for $\phi \in L_\infty(\mathbb{S}^1)$, we have $T(m)(\phi) = m(\tilde{\phi})$, where $\tilde{\phi}(w) = \phi(\bar{w})$, and if $z \in \mathbb{S}^1$, then

$$T(m)(\phi z) = m(\widetilde{(\phi z)}) = m(\tilde{\phi}\bar{z}) = m(\tilde{\phi}) = T(m)(\phi)$$

so that $T(m) \in \mathfrak{L}(\mathbb{T})$.

Let $n = \frac{1}{2}(m + T(m)) \in \mathfrak{L}(\mathbb{T})$. Since $T(n) = n$, we have $n \in \mathfrak{L}(\mathbb{S}^1)$. If $n = \hat{\mu}_1$, then $\frac{1}{2}\hat{m} \leq \hat{n} = (\hat{\mu}_1)\hat{}$, and this contradicts (1). So $n \neq \hat{\mu}_1$ and $\mathfrak{L}(\mathbb{S}^1) \neq \{\hat{\mu}_1\}$.]

When $n > 1$, the answer to the B-R Problem is affirmative, that is, $\mathfrak{L}(\mathbb{S}^n) = \{\hat{\mu}_n\}$. This is a uniqueness theorem, and so it is not surprising that the uniqueness result (4.27(i)) is used in the proof. Obviously, the latter result cannot be applied directly—for a start, H_n is uncountable! However, it turns out that

(4.27(i)) can be applied to a certain subgroup of $\mathrm{O}(n+1)$, using the Property (T) of Kazhdan (Kajdan). This Property, which we will now briefly discuss, has been found useful for other questions (such as that of the finiteness of the number of faces of a fundamental polyhedron).

(ii) *Kazhdan's Property* (*T*). The Property was introduced and studied by Kazhdan [**1**]. Detailed proofs of Kazhdan's results are given in the indispensable paper of Delaroche and Kirillov [**1**]. (Unfortunately, the volume in the Séminaire Bourbaki series containing this paper seems to be out of print.)

We start by discussing briefly some basic ideas from the representation theory of locally compact groups. (See, for example, [**D2**].)

Let G be a *separable*, locally compact group. The class of all continuous, unitary representations of G is denoted by $\operatorname{Rep} G$. For $\pi \in \operatorname{Rep} G$, $\mathfrak{H}_\pi$ stands for the Hilbert space of the representation π. We will denote by $G^{\#}$ the set of equivalence classes of representations $\pi \in \operatorname{Rep} G$ for which $\mathfrak{H}_\pi$ is a separable, Hilbert space (that is, $\dim \mathfrak{H}_\pi \leq \aleph_0$). The dual of G is denoted by $\hat{G}$. Thus $\hat{G}$ is the set of equivalence classes of representations $\pi \in \operatorname{Rep} G$ that are irreducible. Since G is separable, we have that $\hat{G} \subset G^{\#}$.

The notion of *weak containment* of representations is important in discussing the topology on $\hat{G}$. To define this notion, let us say that a positive definite function $\phi \in C(G)$, is *associated* with $\pi \in \operatorname{Rep} G$ if there exists $\xi \in \mathfrak{H}_\pi$ such that for all $x \in G$,

$$\phi(x) = (\pi(x)\xi, \xi).$$

We write $\phi = \phi_\xi$. If $S \subset \operatorname{Rep} G$, we say that π is *weakly contained* in S if, given $\xi \in \mathfrak{H}_\pi$, there exists a net $\{\phi_\xi\}$, where each ϕ_ξ is a sum of positive definite functions associated with elements of S, such that $\phi_\xi \to \phi_\xi$ uniformly on compacta. The set $\hat{G}$ is topologised by requiring $S \subset \hat{G}$ to be *closed* if and only if $\pi \in S$ whenever $\pi \in \hat{G}$ is weakly contained in S. Of course, if we identify $\hat{G}$ with $C^*(G)^\wedge$, then weak containment is formulated in terms of weak* convergence of positive functionals.

It is tempting to use the weak containment property as above to define a topology on $G^{\#}$. Unfortunately, we cannot define a topology in this way: as Fell [**3**] points out, if $\pi, \pi' \in G^{\#}$ are such that neither is weakly contained in the other, then $\pi \oplus \pi'$ is weakly contained in $S = \{\pi, \pi'\}$ but not in $\{\pi\}$ or in $\{\pi'\}$. Instead, (Fell [**3**], Delaroche and Kirillov [**1**]) we define a base of sets $V(\varepsilon, C, \xi_1, \ldots, \xi_n, \pi)$ for a topology on $G^{\#}$, where $\varepsilon > 0$, $C \in \mathscr{C}(G)$, $\pi \in G^{\#}$ and $\xi_i \in \mathfrak{H}_\pi$, $\|\xi_i\| = 1$, and where $\pi' \in G^{\#}$ is in $V(\varepsilon, C, \xi_1, \ldots, \xi_n, \pi)$ if and only if there exist $\eta_i \in \mathfrak{H}_{\pi'}$, $\|\eta_i\| = 1$ $(1 \leq i \leq n)$ such that

$$|(\pi'(x)\eta_i, \eta_j) - (\pi(x)\xi_i, \xi_j)| \leq \varepsilon \qquad (1 \leq i, j \leq n,\ x \in C).$$

The separability of G ensures that $G^{\#}$ is "first countable", that is, each point in $G^{\#}$ has a countable base of neighbourhoods. The relative topology inherited by $\hat{G}$ from $G^{\#}$ coincides with the "weak containment" topology above.

The following simple result will be useful. Let π_0 be the trivial representation $x \to 1$ of G on $\mathbb{C}$. Of course, $\pi_0 \in \hat{G}$.

(a) *Let* $\pi \in G^{\#}$. *Then* $\pi_n \to \pi_0$ *in* $G^{\#}$, *where* $\pi_n = \pi$ *for all* $n \geq 1$, *if, given* $\varepsilon > 0$ *and* $C \in \mathscr{C}(G)$, *there exists* $\xi \in \mathfrak{H}_\pi$ *with* $\|\xi\| = 1$ *such that*

$$\|\pi(x)\xi - \xi\| < \varepsilon \qquad (x \in C). \tag{2}$$

[If (2) holds, then, for each $\varepsilon > 0$, $C \in \mathscr{C}(G)$, we can find $\xi_\varepsilon \in \mathfrak{H}_\pi$ with $\|\xi_\varepsilon\| = 1$ and $\|\pi(x)\xi_\varepsilon - \xi_\varepsilon\| < \varepsilon$ for all $x \in C$. But then

$$|(\pi(x)\xi_\varepsilon, \xi_\varepsilon) - 1| = |((\pi(x) - I)\xi_\varepsilon, \xi_\varepsilon)| \leq \|\pi(x)\xi_\varepsilon - \xi_\varepsilon\| < \varepsilon,$$

and since 1 is the only function associated with π_0, it follows from the definition of the $G^{\#}$-topology that $\pi_n \to \pi_0$.]

A much stronger notion than that of "weak containment" is that of "containment." If $\pi, \pi^1 \in \operatorname{Rep} G$, then we say that π^1 is *contained* in π if there exists a $\pi(G)$-invariant, closed subspace $\mathfrak{H}^1$ of $\mathfrak{H}_\pi$ such that the representation $x \to \pi(x)|_{\mathfrak{H}^1}$ is unitarily equivalent to π^1. It is almost trivial that if π^1 is contained in π, then π^1 is weakly contained in π.

(b) *Definition of Property (T). The group G is said to have* Property (T) *if, whenever* $\pi_n \to \pi_0$ *in* $G^{\#}$, *then* π_n *contains* π_0 *eventually.*

We now discuss some facts about groups with Property (T). These facts are due to Kazhdan [**1**], and detailed proofs are given in Delaroche and Kirillov [**1**]. We shall not reproduce the full proofs here. However, for the convenience of the reader acquainted with the representation theory of locally compact groups, the easier proofs will be sketched.

If G has Property (T), then $\{\pi_0\}$ *is an open subset of* $\hat{G}$. [To see this, suppose that G has Property (T) and that $\{\pi_0\}$ is *not* an open subset of $\hat{G}$. Then using the first countability of $\hat{G}$, there exists a sequence $\pi_n \to \pi_0$ in $\hat{G}$ with $\pi_n \neq \pi_0$. Since $\pi_n \to \pi_0$ in $G^{\#}$, Property (T) gives that π_n contains π_0 eventually, and so $= \pi_0$ eventually by irreducibility. This is a contradiction.]

The converse to the preceding result is also true (using disintegration theory) so that *G has Property (T) if and only if* $\{\pi_0\}$ *is an open subset of* $\hat{G}$.

Obviously, *if G is compact, then G has Property (T)*, since in that case, $\hat{G}$ is discrete. Also, *if G is abelian, then G has Property (T) if and only if G is compact*, since the dual of an abelian, locally compact group is discrete if and only if the group is compact.

Now if H is a closed, normal subgroup of G, then the canonical map from $(G/H)^\wedge$ into $\hat{G}$ is continuous, so that *G/H has Property (T) if G has Property (T)*. Combining the last two assertions, we obtain that *if G has Property (T) and H is the closure of the commutator subgroup of G, then G/H is compact.* This easily implies that *groups with Property (T) are unimodular.*

We now come to the two results involving Property (T) that are of particular importance for solving the B-R Problem.

Recall from (1.11) that if H is a closed, unimodular subgroup of G, then there is a canonical measure $\lambda_{G/H}$ on G/H associated with the function $q(x) =$

$\Delta_G(x)^{-1}$. The subgroup H is said to have *cofinite volume* if $\lambda_{G/H}(G/H) < \infty$. If H has cofinite volume, then $\lambda_{G/H}$ is G-invariant.

(c) *Let H be a closed, unimodular subgroup of G with cofinite volume. If G has Property* (*T*), *then H also has Property* (*T*). [From the Mackey theory, each $\pi' \in H^{\#}$ induces a representation $\phi(\pi') \in G^{\#}$. Further, ϕ is continuous. Let π'_0 be the trivial representation of H. Then the finiteness of $\lambda_{G/H}(G/H)$ yields that $\phi(\pi'_0)$ contains π_0. (Indeed, the space of $\phi(\pi'_0)$ is just $L_2(G/H)$, and every constant function in $L_2(G/H)$ is fixed under the action of $\phi(\pi'_0)(G)$.) So if $\pi'_n \to \pi'_0$ in $H^{\#}$, then, using the definition of the $G^{\#}$-topology, $\phi(\pi'_n) \to \pi_0$ in $G^{\#}$. Since G has Property (T), $\phi(\pi'_n)$ contains π_0 eventually. By looking at the formula relating π'_n to $\phi(\pi'_n)$, we see that π'_n contains π'_0 eventually. Hence H has Property (T).]

The second result on Property (T) that we shall need in the solution of the B-R Problem lies considerably deeper and involves the theory of algebraic groups. We shall not attempt to sketch this theory. References for the theory are Borel [**1**], Borel and Tits [**1**], and Bruhat and Tits [**1**]. An excellent, introductory account is given in the book by Humphreys [**1**]. We will be content to state the theorem and then comment briefly on it.

(d) *Let G be an algebraic, semisimple, Lie group over a nondiscrete, locally compact field $\mathbb{F}$, and be such that its simple components are of rank ≥ 2. Then G has property* (*T*). The *rank* of a simple algebraic group is defined to be the dimension of a maximal torus of G. The basic examples of a group G with rank ≥ 2 are $\mathrm{SL}(3,\mathbb{F})$ and $\mathrm{Sp}(4,\mathbb{F})$, and the proof of (d) in general reduces to the consideration of these two cases.

The next two results enable us to use the uniqueness result of (4.27(i)) to obtain uniqueness results for *uncountable* groups through the mediation of countable subgroups with Property (T). These results are stated in Margulis [**2**]. See also Rosenblatt [**14**] and Sullivan [**1**].

(iii) *Let $(X,\mathscr{B},\mu)$ be a probability measure space and G a group of invertible, measure-preserving transformations of X* (*as in* (4.27)). *Let M be a countable subgroup of G that is ergodic on X and* (*as a discrete group*) *has Property* (*T*). *Then $\hat{\mu}$ is the only G-invariant mean on $L_\infty(X)$.*

[Let $\pi\colon G \to \mathbf{B}(L_2(X))$ be given by

$$(\pi(s)f)(x) = f(s^{-1}x) \qquad (s \in G,\ f \in L_2(X),\ x \in X).$$

It is routine to check that π is a unitary representation of the discrete group G on $L_2(X)$.

Since every G-invariant mean on $L_\infty(X)$ is also M-invariant, it is sufficient to show that

$$\{\hat{\mu}\} = \mathfrak{L}_M(X) \quad (= \{m \in \mathfrak{M}(X)\colon sm = m \text{ for all } s \in M\}).$$

Applying (4.27(i)) (with M in place of G), it is sufficient to show that there do *not* exist arbitrarily small, almost invariant sets in X for the action of M.

Suppose that there *do* exist such sets. Then we can find a sequence $\{A_n\}$ of nonnull, measurable subsets of X such that $\mu(A_n) \to 0$ and

$$(3) \qquad \mu(xA_n \triangle A_n)/\mu(A_n) \to 0 \qquad (x \in M).$$

Let

$$L_2^0(X) = \left\{g \in L_2(X) \colon \int g\,d\mu = 0\right\}.$$

(Note that since $\mu(X) < \infty$, $L_2(X) \subset L_1(X)$ and the preceding integral makes sense.) Clearly, $L_2^0(X)$ is a closed subspace of $L_2(X)$ that is invariant under $\pi(G)$.

Let $f_n = (\chi_{A_n} - \mu(A_n)1) \in L_2^0(X)$. Then for $s \in M$,

$$(4) \qquad \begin{aligned} \|\pi(s)f_n - f_n\|_2^2 &= \int |\chi_{A_n}(s^{-1}x) - \chi_{A_n}(x)|^2\,d\mu(x) \\ &= \int \chi_{sA_n \triangle A_n}(x)\,d\mu(x) = \mu(sA_n \triangle A_n). \end{aligned}$$

Also

$$(5) \qquad \begin{aligned} \|f_n\|_2^2 &= \int [\chi_{A_n} - 2\mu(A_n)\chi_{A_n} + [\mu(A_n)]^2 1]\,d\mu \\ &= \mu(A_n) - 2[\mu(A_n)]^2 + [\mu(A_n)]^2 = \mu(A_n)[1 - \mu(A_n)]. \end{aligned}$$

Since $\mu(A_n) \to 0$, we can suppose that $\mu(A_n) < 1$. Let $g_n = f_n/\|f_n\|_2$. From (3), (4), and (5), it follows that for each $s \in M$,

$$(6) \qquad \|\pi(s)g_n - g_n\|_2^2 = \mu(sA_n \triangle A_n)/[\mu(A_n)(1 - \mu(A_n))] \to 0.$$

Let $\mathfrak{H}$ be the closed, $\pi(M)$-invariant subspace of $L_2^0(X)$ generated by $\{g_n \colon n \in \mathbb{P}\}$. Let $\pi'(s) = \pi(s)|_{\mathfrak{H}}$ for $s \in M$. Then $\pi' \in M^{\#}$. It follows from (6) and (ii), (a) that $\pi_n \to \pi_0$ in $G^{\#}$, where $\pi_n = \pi'$ for all $n \geq 1$. Since M has Property (T), π_0 is actually contained in π'. Hence there exists $f \in L_2^0(X)$ with $\|f\|_2 = 1$ and

$$\pi(s)f = f \qquad (s \in M).$$

Now $f \in L_1(X)$, and the positive and negative parts of $\operatorname{Re} f$ and $\operatorname{Im} f$ are also $\pi(M)$-invariant. Also M is ergodic on X and arguing as in the proof of (4.27(i)), we see that each of these parts is constant almost everywhere. Hence f is constant almost everywhere, and since $\int f\,d\mu = 0$, we obtain $f = 0$ almost everywhere. This contradicts the fact that $\|f\|_2 = 1$, and the desired result follows.]

(iv) *Let G be a compact group, M a subgroup of G, X a locally compact, Hausdorff space that is a left G-set, and let $\mu \in \mathrm{PM}(X)$ be such that*

(a) *for each $s \in G$, the map $x \to sx$ is measurable and measure preserving with respect to μ;*

(b) *for each $x \in X$, the map $s \to sx$ is continuous from G into X;*

(c) *G acts transitively on X (that is, if $x, y \in X$, then there exists $s \in G$ such that $sx = y$);*

(d) *M has Property (T) (as a discrete group), and is a countable, dense subgroup of G.*
Then $\hat{\mu}$ is the only G-invariant mean on $L_\infty(X, \mu)$.

[The idea of the proof is to identify X with a quotient G/L, where L is a closed subgroup of G. This enables us to identify μ with $\lambda_{G/L}$, and the result becomes a domestic matter for G.

Let $x_0 \in X$ and L be the stabiliser of x_0; thus

$$L = \{s \in G\colon sx_0 = x_0\}.$$

Using (b), L is a closed subgroup of G, and using (c), the map $sL \to sx_0$ is a bijection from G/L onto X. The latter map is continuous, since the quotient map $Q\colon G \to G/L$ is open and (b) holds. It follows that X is compact (since G is), and since X is also Hausdorff, X and G/L are homeomorphic. Thus we can (and shall) identify X and G/L as left G-sets.

We now claim that $\mu = \lambda_{G/L}$, where λ_G, λ_L, and $\lambda_{G/L}$ are connected by Weil's formula ((1.11(1))). For let μ' be the regular, Borel measure on G given by

$$\hat{\mu}'(\phi) = \int_{G/L} d\mu(xL) \int_L \phi(xl)\, d\lambda_L(l) \tag{7}$$

for $\phi \in C(G)$. Clearly, by (a), μ' is a left Haar measure on G with total mass 1, and so coincides with λ_G. Now (7) is valid with ϕ replaced by χ_U, where U is open in G, and the Monotone Class Lemma yields the validity of (7) for $\phi = \chi_A$ with $A \in \mathscr{B}(G)$. Let $E \in \mathscr{B}(G/L)$. With $\phi = \chi_{Q^{-1}(E)}$ in (7) and using Weil's formula, we obtain $\mu = \lambda_{G/L}$ as required.

We now show that the action of G on X is ergodic and that for each $A \in \mathscr{B}(X)$, the map $s \to \mu(sA \,\triangle\, A)$ is continuous on G. Indeed, with $\lambda = \lambda_G$,

$$\mu(sA \,\triangle\, A) = \lambda(Q^{-1}(sA \,\triangle\, A)) = \lambda(sQ^{-1}(A) \,\triangle\, Q^{-1}(A)),$$

and the desired results follow by (1.2) and the ergodicity of the action of G on G. (For the latter fact, note that if $E \in \mathscr{B}(G)$ with $\lambda(E) > 0$ and $\lambda(sE \,\triangle\, E) = 0$ for all $s \in G$, then the measure $A \to \lambda(E \cap A)$ is a Haar measure on G.)

Finally it will be shown that M is ergodic on X. The fact that $\hat{\mu}$ is the only G-invariant mean on $L_\infty(X, \mu)$ will then follow from (iii).

Suppose that $A \in \mathscr{B}(X)$ is such that $\mu(tA \,\triangle\, A) = 0$ for all $t \in M$. Since the map $s \to \mu(sA \,\triangle\, A)$ is continuous on G and M is dense in G, it follows that $\mu(sA \,\triangle\, A) = 0$ for all $s \in G$. Thus $\mu(A) = 0$ or 1, and so M is ergodic on X as required.]

The preceding results give a criterion for determining when the answer to the B-R Problem is affirmative. Recall that $H_n = \mathrm{O}(n+1)$.

(v) *Let $n \in \mathbb{P}$ and suppose that there exists an algebraic, simple Lie group G' over $\mathbb{R}$ with rank ≥ 2 and a countable group M such that*

(a) *M is a subgroup of both G' and H_n;*

(b) *M is a discrete subgroup of G' and of cofinite volume in G';*

(c) *M is dense in H_n.*

Then $\mathfrak{L}(\mathbb{S}^n) = \{\hat{\mu}_n\}$.

[From (ii)(d), G' has Property (T). Using (ii)(c), M also has Property (T). It is now sufficient to check that the conditions of (iv) are satisfied with $G = H_n$, $X = \mathbb{S}^n$, and $\mu = \mu_n$. Conditions (a) and (b) of (iv) are obviously satisfied. From the above, M has Property (T). The latter fact, together with (v)(c), gives (iv)(d).

It remains to show G acts transitively on $\mathbb{S}^n$. Let p, q be distinct points of $\mathbb{S}^n$, and 0 be the origin of $\mathbb{R}^{n+1}$. Then the orthogonal transformation

$$x \to x - \left(\frac{2(p-q).x}{\|p-q\|^2}\right)(p-q)$$

reflects points of $\mathbb{R}^{n+1}$ in the hyperplane through 0 perpendicular to PQ and carries p into q. This establishes (iv)(c).]

(vi) *Sketch of Sullivan's solution to the B-R Problem for $n > 3$.* This solution is given in Sullivan [**1**]. Margulis [**2**] gives another solution. (Drinfeld [**S**] has settled the remaining cases $n = 2, 3$ so that *for all $n \geq 2$, $\mathfrak{L}(\mathbb{S}^n) = \{\hat{\mu}_n\}$*.)

The solution is obtained by applying (v). The group G' is taken to be $\mathrm{O}(n-1, 2)$, where $\mathrm{O}(n-1, 2)$ is the group of $(n+1) \times (n+1)$ invertible, real matrices preserving the quadratic form

$$x_0^2 + x_1^2 + \cdots + x_{n-2}^2 - \sqrt{2}x_{n-1}^2 - \sqrt{2}x_n^2.$$

Then $\mathrm{O}(n-1, 2)$ is an algebraic, simple, Lie group over $\mathbb{R}$ with rank ≥ 2.

Now let M be the set of elements of $\mathrm{O}(n-1, 2)$ with matrix coefficients of the form $(n + m\sqrt{2})$ $(n, m \in \mathbb{Z})$. Results from the theories of arithmetic and algebraic groups give that M is a discrete subgroup of cofinite volume in G'.

It remains to identify M with a dense subgroup of $\mathrm{O}(n+1)$. For $A \in M$, let $\tilde{A}$ be the matrix obtained by replacing each matrix entry $(n + m\sqrt{2})$ by $(n - m\sqrt{2})$ and let $\tilde{M} = \{\tilde{A} \colon A \in M\}$. Then $\tilde{M}$ is isomorphic to M in the obvious way, and since the map $(r + s\sqrt{2}) \to (r - s\sqrt{2})$ is an automorphism of $\mathbb{Q}(\sqrt{2})$, the elements of $\tilde{M}$ preserve the quadratic form

$$x_0^2 + x_1^2 + \cdots + x_{n-2}^2 + \sqrt{2}x_{n-1}^2 + \sqrt{2}x_n^2.$$

It is elementary that we can find an invertible $(n+1) \times (n+1)$ matrix C such that, for all $B \in \tilde{M}$, the matrix CBC^{-1} preserves the quadratic form

$$x_0^2 + x_1^2 + \cdots + x_n^2.$$

Thus the map $A \to C\tilde{A}C^{-1}$ enables us to identify M with a subgroup of H_n. Again using the theory of algebraic groups, it can be shown that M is a dense subgroup of H_n.

Thus (v) applies.

(vii) *Other related results.* For our first such result, let $\mathbb{T}^n = \mathbb{T} \times \cdots \times \mathbb{T}$ be the n-dimensional torus. Then $\mathbb{T}^n$ is a compact, abelian group. Let $H = \operatorname{Aut} \mathbb{T}^n$. Since $\mathbb{T}^n$ is compact, it follows that each element of H preserves the

Haar measure μ on $\mathbb{T}^n$. By B21, H contains $\mathrm{SL}(n, \mathbb{Z})$ as a subgroup. Let $n \geq 3$. We can apply (ii)(d) to obtain that $\mathrm{SL}(n, \mathbb{R})$ has Property (T). The discrete subgroup $\mathrm{SL}(n, \mathbb{Z})$ of $\mathrm{SL}(n, \mathbb{R})$ has cofinite volume in $\mathrm{SL}(n, \mathbb{R})$ (cf. Humphreys [**2**, Chapter 3]) and so, by (ii)(c), also has Property (T). Finally, the action of $\mathrm{SL}(n, \mathbb{Z})$ on $\mathbb{T}^n$ is ergodic, and we obtain from (iii) the following result (Losert and Rindler [**1**], Rosenblatt [**13**], Schmidt [**1**], and Margulis [**2**]): *$\hat{\mu}$ is the unique H-invariant mean on $L_\infty(\mathbb{T}^n, \mu)$.*

Another application, pointed out by Rosenblatt, concerns the uniqueness of left invariant means on a compact group G. Of course, $\mathfrak{L}_t(G) = \{\lambda_G\}$, but what can one say about $\mathfrak{L}(G)$? If G is finite, then, obviously, $\mathfrak{L}(G) = \{\lambda_G\}$. However, if G is infinite and amenable as a discrete group, then $\mathfrak{L}(G)$ is infinite ((7.19)). The natural question is *can it ever happen that $\mathfrak{L}(G) = \{\lambda_G\}$ with G infinite?*

The answer to this question is affirmative. Indeed, let $G = \mathrm{O}(n+1)$, where $n > 3$. From (vi), G contains a countable, dense subgroup with Property (T). Now apply (iv) with $X = G$, $\mu = \lambda_G$ to obtain that $\mathfrak{L}(G) = \{\lambda_G\}$.

(4.30) Von Neumann's Conjecture. The most celebrated conjecture in the history of amenability is *von Neumann's* (or *Day's*) *Conjecture* ((0.16)): *a (discrete) group is not amenable if and only if it contains F_2 as a subgroup.* Evidence in favour of the conjecture is provided by (3.9), Problem 3-4, and the groups G_n.

However, A. Yu. Ol′shanskii ([**2**], [**4**]) has constructed a group G_0 that is not amenable and does not contain F_2 as a subgroup, so that the conjecture is false.

Our discussion of this remarkable piece of mathematics falls into two parts, in (4.31) and (4.32) respectively.

The actual construction of the group G_0 giving the counterexample involves a long and intricate argument in combinatorial group theory, and we shall not attempt to describe the details. A brief discussion of the group is given in (4.33).

From the construction of G_0, it is clear that G_0 does not contain F_2 as a subgroup. How are we going to show that G_0 is not amenable? Now the group G_0 is finitely generated—indeed, it is generated by two elements—and so what is needed is a criterion for deciding when a group G on k generators $u_1, \ldots, u_k$ is amenable. We can obviously take $k > 1$.

Given such a group G, there is a canonical homomorphism $Q \colon F_k \to G$, where F_k is the free group on k generators and Q carries the canonical generators of F_k onto the G-generators $\{u_1, \ldots, u_k\}$. Let $N = \ker Q$, so that $G = F_k/N$. One would expect that the "size" of N (in a suitable sense) is critical in determining whether or not G is amenable. If, for example, N is too "large" (e.g., if N contains all commutators, so that G is abelian), we will have an amenable group G that is too special. If, however, N is too "small," then G will be too "close" to F_k to allow it to be amenable.

To make these intuitions precise, define, for each $n \in \mathbb{P}$, an integer $\gamma_n \geq 0$, where γ_n is the number of elements in N of length n (in the generators of

F_k). Suppose that N is nontrivial. We will show that $\gamma_{2n} > 0$ eventually, and $\lim_{n\to\infty}(\gamma_{2n})^{1/2n} = \gamma$ exists. It is easy to show that $\gamma \geq (2k-1)^{1/2}$.

The following deep theorem, due to Grigorchuk (and Cohen), is proved in (4.32):

$$(*) \quad G \textit{ is amenable if and only if } \lim_{n\to\infty}(\gamma_{2n})^{1/2n} = (2k-1).$$

$(*)$ now provides a criterion for establishing the nonamenability of G_0. For the detailed construction of G_0 enables one to estimate $(\gamma_{2n})^{1/2n}$ for G_0 and to show that $\lim_{n\to\infty}(\gamma_{2n})^{1/2n} = \gamma$ is *not* $2 \cdot 2 - 1$ $(= 3)$!

The proof of $(*)$ proceeds by a surprising route. Recall that $\pi_2 \colon G \to \mathbf{B}(l_2(G))$ is the left regular representation of G. It follows from (4.20(ii)) that if $g \in l_1(G)$ is ≥ 0, self-adjoint, and is such that its support generates G, then G is amenable if and only if $\|\pi_2(g)\| = \|g\|_1$.

In order to use this result, we have to pick a suitable $g \in l_1(G)$ and calculate $\|\pi_2(g)\|$ in terms of γ. We choose

$$(1) \qquad g = u_k^{-1} + u_{k-1}^{-1} + \cdots + u_1^{-1} + u_1 + \cdots + u_{k-1} + u_k.$$

To prove $(*)$ it will be shown that

$$(2) \qquad \|\pi_2(g)\| = \gamma + (2k-1)/\gamma.$$

This easily gives $(*)$, since for $\gamma \geq \sqrt{2k-1} > 1$, $\gamma + (2k-1)/\gamma = 2k$ $(= \|g\|_1)$ if and only if $\gamma = (2k-1)$.

The proof of (2) is somewhat involved. An expression for $\|\pi_2(f)\|$, where $f = f^\sim$ in G, will be obtained in (4.31). This expression involves the natural trace on the C^*-algebra $C_l^*(G)$ generated by $\pi_2(G)$, and facilitates the calculation of $\|\pi_2(f)\|$. We will also require the value of the norm of $\pi_2(h) \in \mathbf{B}(l_2(F_k))$, where $h \in l_1(F_k)$ is defined as in (1). Finally, a power series argument, described in (4.32), gives the desired equality (2).

(4.31) On the norms of operators in $C_l^*(G)$. Recall that $\pi_2 \colon G \to \mathbf{B}(l_2(G))$ is the left regular representation of a (discrete) group G and that $C_l^*(G)$ is the C^*-subalgebra generated by $\pi_2(G)$. The map π_2 extends to a norm-decreasing, *-representation of the Banach *-algebra $l_1(G)$. This extension will also be denoted by π_2.

For $T \in C_l^*(G)$, define

$$\operatorname{tr} T = (T(e), e).$$

Here, of course, e, and in general, any element of $l_1(G)$, is identified with an element of $l_2(G)$ in the obvious way.

In the language of operator algebras [**D2**, (6.1.1)], the restriction of tr to the positive part of $C_l^*(G)$ is a "finite trace". It is easy to show that tr is a state on $C_l^*(G)$, and $\operatorname{tr}(TS) = \operatorname{tr}(ST)$ for all $T, S \in C_l^*(G)$.

Clearly, $\operatorname{tr} T$ is easier to calculate, in general, than $\|T\|$. The next result shows that we can use tr to calculate the norms of operators $\pi_2(f)$, where $f = f^\sim \in l_1(G)$. For this result, see Kesten [**2**].

(i) *Let $f \in l_1(G)$ with $f^\sim = f \neq 0$. Then*

$$(1) \qquad \|\pi_2(f)\| = \lim_{n\to\infty} [\operatorname{tr}(\pi_2(f)^{2n})]^{1/(2n)}.$$

[Let A be the commutative C^*-algebra generated by $T = \pi_2(f)$ and I, and let X be the carrier space of A. Now it is routine that $\operatorname{Sp}_A(T) = \operatorname{Sp}(T)$, the spectrum of T in $\mathbf{B}(l_2(G))$, and since T generates A, the map $f \to f(T)$ from X onto $\operatorname{Sp}(T)$ is a homeomorphism. We thus can, and shall, identify X with $\operatorname{Sp}(T) \subset \mathbb{R}$. So $\hat{T}(t) = t$ for all $t \in X$.

The linear functional $\hat{S} \to \operatorname{tr} S$ is a state on $C(X)$, and so, by the Riesz Representation Theorem, there exists $\mu \in \operatorname{PM}(X)$ such that

$$\operatorname{tr} S = \int_X \hat{S}\, d\mu$$

for all $S \in A$.

We now show that X is the support $\mathbf{S}(\mu)$ of μ. For suppose otherwise. Then $X \sim \mathbf{S}(\mu)$ is a nonempty, open subset of X, and so there exists non-zero $W \in A$ with $W \geq 0$ and $\hat{W}(t) = 0$ for all $t \in \mathbf{S}(\mu)$. Let $V = W^{1/2}$. Then

$$0 = \operatorname{tr} W = (W(e), e) = \|V(e)\|^2$$

so that $V(e) = 0$. Recall ((2.35(H))) that π_r is the right regular representation of G and that $\pi_r(l_1(G))$ is in the commutant of $\operatorname{VN}(G)$, and so of $C_l^*(G)$. Hence, for all $x \in G$,

$$(2) \qquad V(x) = V\pi_r(x)(e) = \pi_r(x)(V(e)) = 0.$$

It follows that $V(l_2(G)) = \{0\}$, so that V and W are 0. This contradiction establishes that $\mathbf{S}(\mu) = X$.

We now prove (1). For all $n \geq 1$,

$$(3) \qquad \operatorname{tr}(T^{2n}) = \int_X t^{2n}\, d\mu(t).$$

Let $M = \|T^2\|$ $(= \|\pi_2(f * f^\sim)\| = \|\pi_2(f)\|^2)$. Note that $M > 0$ since $f \neq 0$ and π_2 is one-to-one on $l_1(G)$. Let $\varepsilon \in (0, M)$ and define

$$U_\varepsilon = \{t \in X : t^2 > M - \varepsilon\}.$$

Then U_ε is an open, nonempty subset of X, and since $\mathbf{S}(\mu) = X$ we have $\mu(U_\varepsilon) = k_\varepsilon > 0$. So, using (3),

$$\begin{aligned}
\|\pi_2(f)\| &= \lim_{n\to\infty} \|T^{2n}\|^{1/2n} \geq \limsup_{n\to\infty} (\operatorname{tr}(T^{2n}))^{1/2n} \\
&\geq \liminf_{n\to\infty} (\operatorname{tr}(T^{2n}))^{1/2n} \\
&= \liminf_{n\to\infty} \left[\int_{U_\varepsilon} t^{2n}\, d\mu(t) + \int_{X\sim U_\varepsilon} t^{2n}\, d\mu(t)\right]^{1/2n} \\
&\geq \liminf_{n\to\infty} \left(\int_{U_\varepsilon} t^{2n}\, d\mu(t)\right)^{1/2n} \geq \liminf_{n\to\infty} ((M-\varepsilon)^n k_\varepsilon)^{1/2n} \\
&= (M-\varepsilon)^{1/2} = \|\pi_2(f)\|(1 - \varepsilon/M)^{1/2}.
\end{aligned}$$

The equality (1) now follows by letting $\varepsilon \to 0$.]

In (4.32), we will require the value of an operator norm associated with the left regular representation of a free group. (Indeed this value will be used to obtain the norm of the corresponding operator for *any* finitely generated group.)

Let F_k be the free group on k generators $x_1, \ldots, x_k$. To any element $x \in F_k$, we can assign a "length" $l(x)$, where $l(x)$ is the length of the reduced word in $\{x_k^{-1}, \ldots, x_1^{-1}, x_1, \ldots, x_k\}$ corresponding to x. Let E_n be the set of elements in F_k of length n. For example, $E_0 = \{e\}$ and $E_1 = \{x_k^{-1}, \ldots, x_1^{-1}, x_1, \ldots, x_k\}$.

For each $n \geq 0$, define $h_n \in l_1(F_k)$ by

$$h_n = \sum_{x \in E_n} x.$$

So $h_0 = e$ and $h_1 = (x_k^{-1} + \cdots + x_1^{-1} + x_1 + \cdots + x_k)$. Set $h = h_1$.

Our next result shows that each power h^n of h can be expressed as a linear combination

$$\sum_{i=0}^{n} a_{i,n} h_i$$

with each $a_{i,n}$ an integer, and obtains a number of simple, useful facts about the $a_{i,n}$. These integers $a_{i,n}$ will also be used in (4.32).

(ii) (Cohen [**1**]). *There exist integers $a_{i,n} \geq 0$ $(0 \leq i \leq n,\ n \geq 0)$ such that for each n,*

$$h^n = \sum_{i=0}^{n} a_{i,n} h_i. \tag{4}$$

The integers $a_{i,n}$ are uniquely determined by (4), and the following properties hold:

(a) *for all n, $a_{n,n} = 1$;*
(b) *for all n, $a_{i,n} = 0$ if $(n-i)$ is odd;*
(c) *for all $n \geq 1$, $a_{0,n+1} = 2ka_{1,n}$;*
(d) *for $i \geq 1$ and $n \geq (i+1)$,*

$$a_{i,n+1} = a_{i-1,n} + (2k-1)a_{i+1,n};$$

(e) *if $\gamma \in (0,1)$, then, for all $i \leq n$,*

$$a_{i,n} \leq \gamma^i(\gamma^{-1} + (2k-1)\gamma)^n.$$

[We start by establishing the following equalities:

$$h * h_n = h_{n+1} + (2k-1)h_{n-1} \qquad (n \geq 2), \tag{5}$$

$$h^2 = h_2 + 2kh_0. \tag{6}$$

The proof of (6) is very easy. To prove (5), let $n \geq 2$. For $y \in E_1$, let A_y $[B_y]$ be the set of elements in F_k of length n $[n-1]$ not beginning with y^{-1} $[y]$. Then

$$yh_n = \sum_{x \in A_y} yx + \sum_{x' \in B_y} x'. \tag{7}$$

Summing the equalities (7) with y ranging over E_1 gives (5).

The equality (4) is trivially true for $n = 0$. If $n \geq 1$ and we have

$$h^n = \sum_{i=0}^{n} a_{i,n} h_i,$$

where $a_{i,n} \in \mathbb{N}$, then, using (5) and (6),

$$\begin{aligned} h^{n+1} &= \sum_{i=0}^{n} a_{i,n} h * h_i \\ &= a_{0,n} h + a_{1,n}(h_2 + 2kh_0) + \sum_{i=2}^{n} a_{i,n}(h_{i+1} + (2k-1)h_{i-1}) \end{aligned} \tag{8}$$

and we can write

$$h^{n+1} = \sum_{i=0}^{n+1} a_{i,n+1} h_i,$$

where $a_{i,n+1} \in \mathbb{N}$. Thus, by induction, (4) is established for all n. By considering the length of the group elements involved in each h_n, it is evident that (4) determines uniquely the integers $a_{i,n}$.

Now (8) gives $a_{n+1,n+1} = a_{n,n}$, $a_{0,n+1} = 2ka_{1,n}$, $a_{n,n+1} = a_{n=1,n}$, and $a_{i,n+1} = a_{i-1,n} + (2k-1)a_{i+1,n}$ $(i \geq 1,\ n \geq (i+1))$. The results of (a), (c), and (d) now follow. As for (b), note that if $(n+1) - i$ is odd, then so also are $n - (i-1)$ and $n - (i+1)$. Now use (d) and induction to yield (b).

We now prove (e). Let $\gamma \in (0,1)$ and $\beta = [\gamma^{-1} + (2k-1)\gamma]$. We have to show that for each $n \geq 0$,

$$a_{i,n} \leq \gamma^i \beta^n \qquad (0 \leq i \leq n). \tag{9}$$

The proof is, again, by induction. If $n = 0$, then

$$1 = a_{0,0} = \gamma^0 \beta^0$$

and if $n = 1$, then

$$a_{0,1} = 0 \leq \gamma^0 \beta^1, \qquad 1 = a_{1,1} = \gamma \cdot \frac{1}{\gamma} \leq \gamma^1 \beta^1.$$

So (9) is true for $n = 0, 1$. Now suppose that (9) is true for some $n \geq 1$. Then since $0 < \gamma < 1$, we have, using (c),

$$a_{0,n+1} = 2ka_{1,n} \leq 2k\gamma^1 \beta^n < (\gamma^{-1} + (2k-1)\gamma)\beta^n = \gamma^0 \beta^{n+1}.$$

If $1 \leq i \leq (n-1)$, then

$$\begin{aligned} a_{i,n+1} &= a_{i-1,n} + (2k-1)a_{i+1,n} \\ &\leq \gamma^{i-1}\beta^n + (2k-1)\gamma^{i+1}\beta^n \\ &= \gamma^i \beta^n (\gamma^{-1} + (2k-1)\gamma) = \gamma^i \beta^{n+1}. \end{aligned}$$

Finally, since $a_{n,n+1} = 0$ and $a_{n+1,n+1} = 1$, we see that (9) is true with n replaced by $(n+1)$. This completes the proof.]

Cohen comments that there does not seem to be a conveniently computable expression for the $a_{i,n}$.

Our next result entails that

$$\|\pi_2(h)\| = 2(2k-1)^{1/2},$$

where π_2 is the left regular representation of F_k. This result is due to Kesten [**2**], and also follows from Akemann and Ostrand [**1**]. The elegant proof given here is due to Cohen [**1**].

$$\text{(iii) } \operatorname{Sp}(\pi_2(h)) = [-2(2k-1)^{1/2}, 2(2k-1)^{1/2}]. \tag{10}$$

[Let $T = \pi_2(h)$ and A be the C^*-subalgebra of $C_l^*(F_k)$ generated by T and I. Using (4), we see that A is the norm closure of $\operatorname{Span}\{\pi_2(h_n)\colon n \geq 0\}$. Define $\Phi\colon A \to l_2(F_k)$ by $\Phi(W) = We$. If $\alpha_0, \ldots, \alpha_N \in \mathbb{C}$, then

$$\Phi\left(\sum_{n=0}^{N} \alpha_n \pi_2(h_n)\right) = \sum_{n=0}^{N} \alpha_n h_n.$$

Since Φ is continuous and $\{h_n\}$ is obviously an orthogonal sequence in $l_2(F_k)$, it follows that for each $W \in A$, $\Phi(W)$ is of the form $\sum_{n=0}^{\infty} \beta_n h_n$, where

$$\sum_{n=0}^{\infty} |\beta_n|^2 \|h_n\|_2^2 < \infty.$$

We now turn to the proof of the equality (10).

Let $\alpha \in \mathbb{R}$ be such that $|\alpha| < 2(2k-1)^{1/2}$. We shall show that $\alpha \in \operatorname{Sp}(T)$. Suppose, on the contrary, that $\alpha \notin \operatorname{Sp}(T)$. Since $\operatorname{Sp}(T) = \operatorname{Sp}_A(T)$, there exists $S \in A$ such that $(T - \alpha I)S = I$. Let

$$\Phi(S) = \sum_{n=0}^{\infty} \beta_n h_n$$

as above. Now if $W \in A$ is a polynomial in T with real coefficients, then one checks directly that the coefficients of h_n in the expression for $\Phi(W)$ are all real. Since S is the norm limit of a sequence of such polynomials W, it follows that $\beta_n \in \mathbb{R}$ for all n. Now

$$e = Ie = (T-\alpha)(Se) = (T-\alpha)\left(\sum_{n=0}^{\infty} \beta_n h_n\right).$$

Hence using (5) and (6),

$$\begin{aligned} e &= \sum_{n=0}^{\infty} \beta_n h * h_n - \alpha \sum_{n=0}^{\infty} \beta_n h_n \\ &= \beta_0 h + \beta_1(h_2 + 2ke) + \sum_{n=2}^{\infty} \beta_n(h_{n+1} + (2k-1)h_{n-1}) - \alpha \sum_{n=0}^{\infty} \beta_n h_n. \end{aligned} \tag{11}$$

Since $\{h_n\}$ is an orthogonal sequence, we can equate the coefficients of each h_n on both sides of (11), obtaining

$$1 = (r+1)\beta_1 - \alpha\beta_0, \tag{12}$$

$$0 = \beta_{n-1} + r\beta_{n+1} - \alpha\beta_n \qquad (n \geq 1), \tag{13}$$

where $r = 2k - 1$.

Thus

$$\beta_1 = \frac{\alpha}{r+1}\beta_0 + \frac{1}{r+1}, \qquad \beta_{n+1} = \frac{\alpha}{r}\beta_n - \frac{\beta_{n-1}}{r} \quad (n \geq 1). \tag{14}$$

Clearly $\{\beta_n\}$ is determined by β_0, and direct checking shows that the solution to (14) is of the form:

$$\beta_n = pr^{-(n/2)}\sin(n\theta + q) \tag{15}$$

where $p, q \in \mathbb{R}$ satisfy (17) below, and the angle θ satisfies the equalities

$$\cos\theta = \frac{\alpha}{2\sqrt{r}}, \qquad \sin\theta = \frac{(4r - \alpha^2)^{1/2}}{2\sqrt{r}}. \tag{16}$$

Note that since $|\alpha| < 2r^{1/2}$, it follows that $4r - \alpha^2 > 0$. It also follows that we can take $0 < \theta < \pi$. The numbers p, q have to satisfy the equalities

$$\beta_0 = p\sin q, \qquad pr^{-1/2}\sin(\theta + q) = \frac{\alpha}{r+1}\beta_0 + \frac{1}{r+1}. \tag{17}$$

Now from the definition of h_n in (i),

$$\|h_n\|_2^2 = |E_n|.$$

Elementary combinatorics shows that $|E_n| = 2k(2k-1)^{n-1}$, so that

$$\begin{aligned}\|\beta_n h_n\|_2^2 &= p^2 r^{-n}\sin^2(n\theta + q)(r+1)r^{n-1} \\ &= p^2 r^{-1}(r+1)\sin^2(n\theta + q).\end{aligned}$$

Since $0 < \theta < \pi$, $\{\sin^2(n\theta + q)\}$ does not converge. Hence $\sum_{n=0}^{\infty} \|\beta_n h_n\|^2$ diverges. This is impossible since $\sum_{n=0}^{\infty} \|\beta_n h_n\|^2 = \|\Phi(S)\|^2 < \infty$.

We thus obtain that $(-2(2k-1)^{1/2}, 2(2k-1)^{1/2}) \subset \mathrm{Sp}(T)$, and since $\mathrm{Sp}(T)$ is closed, it follows that

$$[-2(2k-1)^{1/2}, 2(2k-1)^{1/2}] \subset \mathrm{Sp}(T). \tag{18}$$

Let b be the spectral radius of T. The equality (10) will follow from (18) once we have shown that $b \leq 2(2k-1)^{1/2}$. This is proved using (i).

For each n, we see from (4) that

$$\mathrm{tr}(T^{2n}) = \mathrm{tr}\left(\sum_{i=0}^{2n} a_{i,2n}\pi_2(h_i)\right) = a_{0,2n}$$

since $\pi_2(h_0) = I$ and $(\pi_2(h_i)e, e) = (h_i, e) = 0$ if $i \neq 0$. Using (i) and recalling that T is hermitian,

$$b = \|T\| = \lim_{n\to\infty}(a_{0,2n})^{1/2n}. \tag{19}$$

Let $\gamma \in (0,1)$. From (ii) (e),

$$(a_{0,2n})^{1/2n} \leq \gamma^{-1} + (2k-1)\gamma.$$

Put $\gamma = (2k-1)^{-1/2}$ and use (19) to obtain that $b \leq 2(2k-1)^{1/2}$ as required.]

The preceding proof also established the following result.

(iv) $2(2k-1)^{1/2} = \lim_{n\to\infty}(a_{0,2n})^{1/2n}$.

(4.32) An amenability criterion for finitely-generated groups. This remarkable criterion is due to Grigorchuk [**2**]. We shall follow the proof of Cohen [**2**]. The author is grateful to Joel Cohen for very helpful correspondence on this result.

Let G be a group generated by the finite set $\{u_1,\ldots,u_k\}$, where $k > 1$. (Clearly, there is no point in looking at the case $k = 1$, since every abelian group is amenable!) Let F_k be the free group on k generators $\{x_1,\ldots,x_k\}$. Let $Q\colon F_k \to G$ be the canonical homomorphism from F_k onto G determined by the equalities $Q(x_i) = u_i$ $(1 \leq i \leq k)$.

As in (4.31(i)), $l(x)$ is the length of the reduced word in $\{x_k^{-1},\ldots,x_1^{-1}, x_1,\ldots,x_k\}$ corresponding to $x \in F_k$. For $n \geq 0$, let

$$E_n = \{x \in F_k\colon l(x) = n\}.$$

Let $N = \ker Q$, and $N_n = N \cap E_n$. Let $\gamma_n = |N_n|$. Obviously, $N = \bigcup_{n=1}^{\infty} N_n$.

Now, of course, some of the N_n might be empty. Indeed, if $N = \{e\}$ (so that $F_k = G$), then $N_n = \varnothing$ for all $n \geq 1$. In our next result, we shall show that if $N \neq \{e\}$, then $\gamma_{2n} \neq 0$ eventually, and that $\gamma = \lim_{n\to\infty}(\gamma_{2n})^{1/2n}$ exists. The number γ can be regarded as a measure of "how far away" G is from F_k.

(i) *Suppose that $N \neq \{e\}$. Then there exists $n_0 \in \mathbb{P}$ such that $\gamma_{2n} > 0$ for all $n \geq n_0$, and*

$$\gamma = \lim_{n\to\infty}(\gamma_{2n})^{1/2n}$$

exists, and belongs to the interval $[(2k-1)^{1/2}, (2k-1)]$.

[The following two inequalities will be proved first for all $m, n > 0$:

$$(1) \qquad (2k-2)(2k-1)^{n-1}\gamma_m \leq \gamma_{m+2n},$$

$$(2) \qquad \gamma_n\gamma_m \leq \gamma_{m+n+2}.$$

Let $A_k = \{x_k^{-1},\ldots,x_1^{-1},x_1,\ldots,x_k\}$ and $x \in N_m$. Write x in reduced form $w_1\cdots w_m$ with $w_i \in A_k$. Let B_n be the set of elements $y \in F_k$ which can be written in reduced form $v_1\cdots v_n$ with $v_n \neq w_1^{-1}$, $w_m \neq v_n$. Then

$$(3) \qquad yxy^{-1} = v_1\cdots v_n w_1\cdots w_m v_n^{-1}\cdots v_1^{-1}$$

is in reduced form, so that $l(yxy^{-1}) = (2n+m)$. We now count the number of elements in B_n. First, $|A_k| = 2k$. In the expression $v_1\cdots v_n$ for y, there are $(2k-2)$ choices for v_n and $(2k-1)$ choices for every other v_i. Hence $|B_n| = (2k-2)(2k-1)^{n-1}$. Since N is normal in F_k, it follows that $yxy^{-1} \in N_{2n+m}$.

It is clear from (3) that a different choice of (x, y) $(x \in N_m, y \in B_n)$ gives rise to a different element yxy^{-1}. Thus

$$\gamma_{2n+m} \geq (2k-2)(2k-1)^{n-1}|N_m| \tag{4}$$

giving (1).

We now turn to (2). Let $x' \in N_m$ and $y' \in N_n$ have reduced forms $w_1' \cdots w_m'$ and $v_1' \cdots v_n'$ respectively. If $u \in A_k \sim \{(w_m')^{-1}, (v_1')^{-1}, v_n'\}$, then

$$x'(uy'u^{-1}) = w_1' \cdots w_m' u v_1' \cdots v_n' u^{-1}$$

is in reduced form, and so belongs to N_{m+n+2}. By considering all possible choices of x', y' we obtain (2).

We now show that for some n_0, $\gamma_{2n} > 0$ for all $n \geq n_0$. To prove this, note first that $\gamma_m > 0$ for some $m \in \mathbb{P}$ since $N \neq \{e\}$. It follows from (2) that $\gamma_{2(m+1)} > 0$, so that we can take m to be even. Let $m = 2p$ $(p \in \mathbb{P})$. Using (1), $\gamma_{2r} > 0$ for all $r \geq p$, so that we can take $n_0 = p$.

It will now be shown that $\{(\gamma_{2n})^{1/2n}\}$ converges. In this argument, it will always be assumed that $n, m > n_0$.

Let $a_n = \log \gamma_{2n} \geq 0$. Since $\log[(\gamma_{2n})^{1/2n}] = a_n/2n$, it is sufficient to show that $\{a_n/n\}$ converges. From (2)

$$a_n + a_m \leq a_{n+m+1}. \tag{5}$$

Let $b_n = a_{n-1}$. Clearly, $\{a_n/n\}$ converges if $\{b_n/n\}$ converges. We will prove the latter. From (5),

$$b_n + b_m \leq b_{n+m}. \tag{6}$$

Thus $2b_m \leq b_{2m}$, and, by induction,

$$rb_m \leq b_{rm} \quad (r \geq 1). \tag{7}$$

From (6) we also have that $\{b_n\}$ is increasing. Now

$$\gamma_n = |N_n| \leq |E_n| = 2k(2k-1)^{n-1},$$

so that

$$b_n = a_{n-1} = \log \gamma_{2n-2} \leq \log[2k(2k-1)^{2n-3}].$$

Since $2k < (2k-1)^3$,

$$b_n \leq \log(2k) + (2n-3)\log(2k-1) < 2n \log(2k-1). \tag{8}$$

Thus the sequence $\{b_n/n\}$ is bounded above.

Let $\alpha = \limsup_{n\to\infty}(b_n/n)$ and $\beta = \liminf_{n\to\infty}(b_n/n)$. So $\alpha \geq \beta$. Let $\eta > 0$ and $p \in \mathbb{P}$. Then we can find m such that $b_m/m > (\alpha - \eta)$, and then find $n \geq pm$ with $b_n/n < (\beta + \eta)$. Write $n = rm + q$ where $r \in \mathbb{P}$ and $0 \leq q < m$. Note that $r \geq p$. Using (7) and the fact that $\{b_n\}$ is increasing, we have:

$$\begin{aligned}\alpha - \eta < b_m/m = rb_m/rm &\leq b_{rm}/rm \leq b_n/rm \\ &= (b_n/n)(n/m)(1/r) = (b_n/n)(r + q/m)(1/r) \\ &< (b_n/n)((r+1)/r) < (\beta + \eta)(1 + r^{-1}) \leq (\beta + \eta)(1 + p^{-1}).\end{aligned}$$

Letting $p \to \infty$ and then $\eta \to 0$ gives $\alpha \le \beta$. Hence $\alpha = \beta$ and $\{b_n/n\}$ converges. Thus $\gamma = \lim_{n\to\infty}((\gamma_{2n})^{1/2n})$ exists.

It remains to be shown that $(2k-1)^{1/2} \le \gamma \le (2k-1)$. From (1),

$$\gamma_{2(n+n_0)} \ge (2k-2)(2k-1)^{n-1}\gamma_{2n_0} \ge (2k-2)(2k-1)^{n-1},$$

so that

$$\gamma = \lim_{n\to\infty}[(\gamma_{2(n+n_0)})^{1/(2(n+n_0))}] \ge (2k-1)^{1/2}.$$

Finally, using (8),

$$\gamma = \lim_{n\to\infty} \exp(a_n/(2n)) = \lim_{n\to\infty} \exp(b_n/(2n)) \le (2k-1).]$$

It is easy to show that, for any group G with k generators $u_1, \ldots, u_k$, one of (a), (b), (c) below holds:

(a) $\gamma_n = 0$ for all $n > 0$ (that is, $N = \{e\}$);

(b) $\gamma_{2n-1} = 0$ for all $n \ge 1$, and, for some n_0, $\gamma_{2n} > 0$ for all $n \ge n_0$;

(c) for some n_1, $\gamma_n > 0$ for all $n \ge n_1$.

An easier version of the above argument shows that in case (c),

$$\lim_{n\to\infty} (\gamma_n)^{1/n}$$

exists.

Our objective is to prove the following criterion for the amenability of G (with $N \ne \{e\}$):

G is amenable if and only if $\gamma = (2k-1)$.

This is proved in (vii) below.

Let

$$g = u_k^{-1} + \cdots + u_1^{-1} + u_1 + \cdots + u_k.$$

Obviously, $g \ge 0$ in $l_1(G)$, $g^\sim = g$, and $\mathbf{S}(g)$ generates G. Further, $\|g\|_1 = 2k$. We will calculate $\|\pi_2(g)\|$ and then apply (4.20) to establish the criterion.

Let $a_{i,n}$ be the integers introduced in (4.31(ii)).

(ii) $\|\pi_2(g)\| = \lim_{n\to\infty}(\sum_{i=0}^{2n} a_{i,2n}\gamma_i)^{1/2n}$. [Let $T = \pi_2(g)$. By (4.31(4)),

$$\operatorname{tr}(T^{2n}) = \sum_{i=0}^{2n} a_{i,2n} \operatorname{tr}(\pi_2(Q(h_i))), \tag{9}$$

where Q is extended, in the obvious way, to a linear map, also denoted Q, from $l_1(F_k)$ onto $l_1(G)$. Of course, $Q(h) = g$. Now, from the definition of h_i,

$$\begin{aligned} \operatorname{tr}(\pi_2(Q(h_i))) &= \sum_{x\in E_i} (\pi_2(Q(x))e, e) \\ &= |\{x \in E_i : Q(x) = e\}| \\ &= |N_i| = \gamma_i. \end{aligned} \tag{10}$$

Now use (9), (10) and (4.31(i)).]

For large even i, $(\gamma_i)^{1/i}$ is close to γ. This suggests replacing $\sum_{i=0}^{2n} a_{i,2n}\gamma_i$ in (ii) by $\sum_{i=0}^{2n} a_{i,2n}\gamma^i$. Motivated by the formula for the radius of convergence of a power series, we are led to consider the power series in $z \in \mathbb{C}$ for each fixed $s > 0$:

$$\sum_{n=0}^{\infty} a_n(s) z^n \tag{11}$$

where $a_n(s) = \sum_{i=0}^{n} a_{i,n} s^i$. We shall also have to consider the power series:

$$\sum_{n=0}^{\infty} a_{0,n} z^n. \tag{12}$$

Let ρ_s and ρ be the radii of convergence of the power series in (11) and (12) respectively. Since $a_{0,n} \leq a_n(s)$, it is obvious that $\rho_s \leq \rho$. Let

$$f_s(z) = \sum_{n=0}^{\infty} a_n(s) z^n, \ (|z|) < \rho_s) \qquad g(z) = \sum_{n=0}^{\infty} a_{0,n} z^n \ (|z| < \rho).$$

(iii) (a) *For* $n \geq 1$,

$$a_{n+1}(s) = (2k)^{-1} a_{0,n+1} - ((2k-1)/s) a_{0,n} + (s + ((2k-1)/s)) a_n(s). \tag{13}$$

(b) If $|z| < \rho_s$, then

$$2k[1 - (s + (2k-1)/s)z] f_s(z) = (2k-1) + [1 - (2k(2k-1)z/s)] g(z). \tag{14}$$

[(a) Let $n \geq 1$ and $r = (2k-1)$. Then using (d), (a), (b) and (c) of (4.31(ii)):

$$\begin{aligned}
a_{n+1}(s) &= a_{0,n+1} + a_{1,n+1}s + \cdots + a_{n+1,n+1}s^{n+1} \\
&= a_{0,n+1} + (a_{0,n} + ra_{2,n})s + (a_{1,n} + ra_{3,n})s^2 + (a_{2,n} + ra_{4,n})s^3 \\
&\quad + \cdots + (a_{n-4,n} + ra_{n-2,n})s^{n-3} + (a_{n-3,n} + ra_{n-1,n})s^{n-2} \\
&\quad + (a_{n-2,n} + ra_{n,n})s^{n-1} + a_{n,n+1}s^n + a_{n+1,n+1}s^{n+1} \\
&= a_{0,n+1} + a_{0,n}s + a_{1,n}s^2 + (r/s + s)a_{2,n}s^2 + (r/s + s)a_{3,n}s^3 \\
&\quad + \cdots + (r/s + s)a_{n-2,n}s^{n-2} + rs^{n-1} + s^{n+1} \\
&= a_{0,n+1} + a_{0,n}s + a_{1,n}s^2 + (s + r/s)\left(\sum_{i=2}^{n} a_{i,n}s^i\right) \\
&= a_{0,n+1} + a_{0,n}s + a_{1,n}s^2 \\
&\quad - (s + r/s)(a_{0,n} + a_{1,n}s) + (s + r/s)a_n(s) \\
&= a_{0,n+1}[1 + s^2(2k)^{-1} - (s + r/s)s(2k)^{-1}] \\
&\quad + a_{0,n}(s - s - r/s) + (s + r/s)a_n(s) \\
&= a_{0,n+1}(1 - (2k-1)/(2k)) - (r/s)a_{0,n} + (s + r/s)a_n(s) \\
&= (2k)^{-1}a_{0,n+1} - ((2k-1)/s)a_{0,n} + (s + ((2k-1)/s))a_n(s).
\end{aligned}$$

(b) Using (13),

$$\begin{aligned}
f_s(z) &= a_0(s) + a_1(s)z + \sum_{n=1}^{\infty}[(2k)^{-1}a_{0,n+1} - ((2k-1)/s)a_{0,n} \\
&\qquad + (s + (2k-1)/s)a_n(s)]z^{n+1} \\
&= 1 + sz + (2k)^{-1}[g(z) - 1] - ((2k-1)/s)z[g(z) - 1] \\
&\quad + z(s + (2k-1)/s)[f_s(z) - 1].
\end{aligned}$$

So

$$\begin{aligned}2k[1-(s+(2k-1)/s)z]f_s(z) \\ &= 2k\{1+sz+g(z)[(2k)^{-1}-((2k-1)/s)z] \\ &\quad -(2k)^{-1}+((2k-1)/s)z-z(s+(2k-1)/s)\} \\ &= 2k\{1-(2k)^{-1}+(2k)^{-1}g(z)[1-(2k(2k-1)/s)z]\} \\ &= (2k-1)+[1-(2k(2k-1)z/s)]g(z).]\end{aligned}$$

We now prove an elementary result on real sequences.

(iv) *Let* $\{a_n\}$ *and* $\{b_n\}$ *be sequences with* $a_n > 0$ *for all* n, *and* $L, K \in \mathbb{R}$ *with* $0 \le L < K$, $|b_n| < L^n$ *eventually,* $\limsup_{n\to\infty} a_n^{1/n} = K$, *and*

$$a_{n+1} = b_n + Ka_n. \tag{15}$$

Then $a_n^{1/n} \to K$. [Dividing both sides of (15) by K^{n+1} gives:

$$a_{n+1}/K^{n+1} = b_n/K^{n+1} + a_n/K^n.$$

Replacing a_n, b_n by a_n/K^n and b_n/K^{n+1} respectively, we can suppose that $L < K = 1$. Now for all n,

$$a_{n+1} - a_n = b_n. \tag{16}$$

Since $|b_n| < L^n$ eventually and $L < 1$, the series $\sum_{n=0}^{\infty} |b_n|$ converges. Let $\varepsilon > 0$ be such that $L < (1-\varepsilon)$. Let $N_1 \in \mathbb{P}$ be such that both:

$$|b_n| < L^n \quad (n \ge N_1), \qquad \frac{[L/(1-\varepsilon)]^n}{1-L} < 1 \ (n \ge N_1). \tag{17}$$

Now find $N \ge N_1$ such that $a_N^{1/N} \ge 1-\varepsilon$. Let $m \in \mathbb{P}$. Adding up the equalities (16) for $N \le n \le (m+N)$, we have

$$a_{m+N+1} - a_N = \sum_{n=N}^{m+N} b_n.$$

So, using the first inequality of (17),

$$a_{m+N+1} \ge a_N - \sum_{r=N}^{m+N} L^n \ge (1-\varepsilon)^N - L^N/[1-L].$$

Using the second inequality of (17),

$$\begin{aligned}&\liminf_{m\to\infty}(a_{m+N+1})^{1/(m+N+1)} \\ &\quad \ge \lim_{m\to\infty}[(1-\varepsilon)^N - L^N/[1-L]]^{1/(m+N+1)} = 1.\end{aligned}$$

So $\liminf_{n\to\infty} a_n^{1/n} = 1 = \limsup_{n\to\infty} a_n^{1/n}$, and $a_n^{1/n} \to 1$.]

(v) *Let* $s > 0$. *Then* $\rho_s = (s + (2k-1)/s)^{-1}$.

[Recall that ρ_s $[\rho]$ is the radius of convergence of the power series in (11) [(12)] giving $f_s(z)$ $[g(z)]$, and that $\rho_s \le \rho$. Now $a_{0,n} = 0$ for odd n by (4.31(ii)(b)), and from (4.31(iv)),

$$\rho = \left[\limsup_{n\to\infty}(a_{0,2n})^{1/2n}\right]^{-1} = \left[\lim_{n\to\infty}(a_{0,2n})^{1/2n}\right]^{-1} = \tfrac{1}{2}(2k-1)^{-1/2}.$$

For $x \in (0,\infty)$, let $r_x = (x + (2k-1)/x)^{-1}$. By elementary differentiation, the function $x \to r_x$ has a unique maxiumum on $(0,\infty)$ at $x = (2k-1)^{1/2}$, so that $r_x \leq r_{(2k-1)^{1/2}} = \rho$ for all x. If $|z| < r_s$, then, from (14), the power series obtained by multiplying out

$$(2k)^{-1}\sum_{n=0}^{\infty}[(s+(2k-1)/s)z]^n\left\{(2k-1)+[1-(2k(2k-1)z/s)]\left(\sum_{n=0}^{\infty}a_{0,n}z^n\right)\right\}$$

has to coincide with the power series in (11) for $f_s(z)$. This means that the radius of convergence ρ_s of the power series in (11) is $\geq r_s$. So

$$r_s \leq \rho_s \leq \rho.$$

We now show that either $\rho_s = r_s$ or $\rho_s = \rho$ $(= \frac{1}{2}(2k-1)^{-1/2})$.

Suppose that $\rho_s > r_s$. Substituting $z = r_s$ in (14), we see that

$$(2k-1) + [1 - (2k(2k-1)r_s/s)]g(r_s) = 0.$$

Then the function

$$\left(\frac{2k}{r_s}\right)^{-1}(r_s - z)^{-1}[(2k-1) + [1 - (2k(2k-1)z/s)]g(z)]$$

extends to an analytic function on the disc $\{z \in \mathbb{C}\colon |z| < \rho\}$ using the theorem on removable singularities and, by (14), is given by the power series expansion of (11), thus coinciding with f_s. So if $\rho_s > r_s$, then $\rho_s = \rho$. So either $\rho_s = r_s$ or $\rho_s = \rho$. We shall now show that $\rho_s = r_s$.

To this end, let

$$A = \{x \in (0,\infty) \sim \{(2k-1)^{1/2}\}\colon (2k-1) + [1 - (2k(2k-1)r_x/x)]g(r_x) = 0\}.$$

Since for $x \in A$, $r_x < r_{(2k-1)^{1/2}} = \rho$, we see that $g(r_x)$ makes sense.

Let $x \in A$. We claim that x is isolated in A. To prove this we suppose that $x > (2k-1)^{1/2}$. (A more complicated version of the argument applies to the case where $x < (2k-1)^{1/2}$ letting $x \to 0$ rather than ∞ in (18).) We can find an open, connected set U in $\mathbb{C}$ containing $((2k-1)^{1/2},\infty)$ such that the following function $h\colon U \to \mathbb{C}$ is analytic:

$$h(z) = (2k-1) + \left[1 - \frac{2k(2k-1)}{z^2 + (2k-1)}\right] g\left(\frac{z}{z^2+(2k-1)}\right).$$

(The function $h(z)$ is obtained by replacing z, s in the right-hand side of (14) by $(z + (2k-1)/z)^{-1}$, z.) Suppose that x is not isolated in A. Then x is a nonisolated zero of h in U, so that h is identically 0 on U. So

$$0 = \lim_{x\to\infty} h(x) = (2k-1) + g(0) = 2k \tag{18}$$

giving a contradiction. So the elements of A are isolated.

If $s \notin A$, then ρ_s cannot be $> r_s$; for then substituting $z = r_s$ in (14) gives $h(s) = 0$. Since ρ_s is either r_s or ρ, we have $\rho_s = r_s$.

If $s = (2k-1)^{1/2}$, then $r_s = (2(2k-1)^{1/2})^{-1} = \rho$, so that $\rho_s = r_s = \rho$. Suppose, finally, that $s \in A \sim \{(2k-1)^{1/2}\}$. Then we can find $x \notin A$ with $x \neq (2k-1)^{1/2}$, $x < s$. Since $\sum_{i=0}^n a_{i,n}x^i \leq \sum_{i=0}^n a_{i,n}s^i$ and $x \notin A$, we have

$$\rho_s \leq \rho_x = r_x < \rho.$$

So $\rho_s = r_s$ as required.]

After the "classical analysis" excursion above, we are now in a position to evaluate $\|\pi_2(g)\|$. Recall that $g \in l_1(G)$ and γ are defined in (i).

(vi) $\|\pi_2(g)\| = (\gamma + (2k-1)/\gamma)$.
[Let $s \in (0,\infty) \sim \{(2k-1)^{1/2}\}$, and $r_s = (s + (2k-1)/s)^{-1}$. From (13),

$$a_{n+1}(s) = b_n + r_s^{-1}a_n(s),$$

where

$$b_n = [(2k)^{-1}a_{0,n+1} - ((2k-1)/s)a_{0,n}].$$

Since $s \neq (2k-1)^{1/2}$, $r_s^{-1} > 2(2k-1)^{1/2}$. Let $L \in (2(2k-1)^{1/2}, r_s^{-1})$. From (4.31(iv)) and the fact that $a_{0,r} = 0$ if r is odd,

$$|b_n| < L^n$$

eventually. Now using (v) and the "lim sup" formula for the radius of convergence of a power series, we have

$$\limsup_{n\to\infty} a_n(s)^{1/n} = \rho_s^{-1} = r_s^{-1}.$$

From (iv),

$$a_n(s)^{1/n} \to r_s^{-1} = s + (2k-1)/s.$$

Now let $\varepsilon > 0$ and choose $\eta \in (0,\varepsilon)$ such that both $\gamma - \eta, \gamma + \eta \in (0,\infty) \sim \{(2k-1)^{1/2}\}$ and $1 < (\gamma - \eta)$. Since $\gamma_{2n}^{1/2n} \to \gamma$, we can find $N \in \mathbb{P}$ such that

$$(\gamma-\eta)^{2j} < \gamma_{2j} < (\gamma+\eta)^{2j} \qquad (j \geq N).$$

Recalling that $a_{i,2n} = 0$ whenever i is odd, we have for $n > N$,

$$A + \sum_{i=2N}^{2n} a_{i,2n}(\gamma-\eta)^i < \sum_{i=0}^{2n} a_{i,2n}\gamma_i < A + \sum_{i=2N}^{2n} a_{i,2n}(\gamma+\eta)^i,$$

where $A = \sum_{i=0}^{2N-1} a_{i,2n}\gamma_i$. So we can find $B, C \in \mathbb{R}$ such that for all $n > N$,

$$(B + a_{2n}(\gamma-\eta))^{1/2n} < \left(\sum_{i=0}^{2n} a_{i,2n}\gamma_i\right)^{1/2n} < (C + a_{2n}(\gamma+\eta))^{1/2n},$$

that is,

$$(a_{2n}(\gamma-\eta))^{1/2n}\left(1 + \frac{B}{a_{2n}(\gamma-\eta)}\right)^{1/2n} < \left(\sum_{i=0}^{2n} a_{i,2n}\gamma_i\right)^{1/2n}$$
$$< (a_{2n}(\gamma+\eta))^{1/2n}\left(1 + \frac{C}{a_{2n}(\gamma+\eta)}\right)^{1/2n}.$$

Since $1 < \gamma - \eta$, $\gamma + \eta$, it follows that $a_{2n}(\gamma - \eta)$, $a_{2n}(\gamma + \eta) \to \infty$ as $n \to \infty$. So

$$r_{\gamma-\eta}^{-1} \leq \lim_{n\to\infty} \left(\sum_{i=0}^{2n} a_{i,2n}\gamma_i \right)^{1/2n} \leq r_{\gamma+\eta}^{-1}.$$

Now let $\eta \to 0$ and use the continuity of the function $s \to r_s^{-1}$ to obtain

$$\lim_{n\to\infty} \left(\sum_{i=0}^{2n} a_{i,2n}\gamma_i \right)^{1/2n} = (\gamma + (2k-1)/\gamma).$$

Finally, the desired result follows from (ii).]

We can now prove the amenability criterion for G for which the above has been the preparation.

(vii) *The group G is amenable if and only if $\gamma = 2k - 1$.*

[By (4.20(ii)), G is amenable if and only if

$$\|\pi_2(g)\| = \|g\|_1 = 2k.$$

Now, by (vi), $\|\pi_2(g)\| = \gamma + (2k-1)/\gamma$. Hence G is amenable if and only if $\gamma + (2k-1)/\gamma = 2k$, that is, $\gamma^2 - 2k\gamma + (2k-1) = 0$, that is, $(\gamma - (2k-1))(\gamma - 1) = 0$. Since $\gamma \geq (2k-1)^{1/2} > 1$, it follows that G is amenable if and only if $\gamma = (2k-1)$.]

(4.33) Ol′shanskii's counterexample to von Neumann's Conjecture. The result of (4.32) suggests an approach for resolving von Neumann's Conjecture negatively. Suppose that we have a finitely generated group G_0 that does not contain F_2 as a subgroup, and that, we suspect, is nonamenable. If we have detailed information about the relations satisfied by the generators $u_1, \ldots, u_k$ of G_0, then we might hope to be able to estimate γ_{2n} and hence $\gamma = \lim_{n\to\infty}(\gamma_{2n})^{1/2n}$. If $\gamma \neq (2k-1)$, then G_0 is, indeed, a counterexample to the Conjecture.

Ol′shanskii [**2**] constructed a group G_0 for which the above procedure can be carried out. Here are some of the properties possessed by G_0.

(a) G_0 is generated by two elements $\{a, b\}$ and there exist equality and conjugacy algorithms for G_0;

(b) G_0 is a simple, infinite group;

(c) every proper subgroup of G_0 is infinite cyclic;

(d) G_0 is Noetherian, that is, there are no infinite, ascending chains of subgroups;

(e) root extraction is unique, that is, if $x, y \in G_0$ and $x^n = y^n$ for some $n \in \mathbb{P}$, then $x = y$.

The construction of this group requires a formidable induction argument involving 80 lemmas and a large number of parameters. It relies on the well-known work by Adian and Novikov on the Burnside problem (concerning the existence of a finitely generated, infinite group H such that for some $n \in \mathbb{P}$, $x^n = e$ for all $x \in H$). It also involves the use of geometrical methods, originating in a result of van Kampen, and embodied in the so-called "small cancellation theory" (Lyndon and Schupp [**1**]). We shall not attempt to describe the argument.

The details of the construction enable Ol′shanskii [**4**] to estimate the γ_n for G_0 with respect to the generators $\{a, b\}$ of (a). He shows that the parameters defining G_0 can be chosen so that

$$\gamma_n \le 3^{(3n/4)}.$$

It then follows that

$$\gamma = \lim_{n\to\infty} (\gamma_{2n})^{1/2n} \le 3^{3/4} < 2 \cdot 2 - 1,$$

so that G_0 is not amenable.

Ol′shanskii comments that an example of a periodic, nonamenable group is given in his paper [**3**].

Is there an easier counterexample to von Neumann's Conjecture? What about the Golod-Shafarevitch groups ((3.11))?

(4.34) Amenability and the Fourier algebra of a locally compact group. Duality theory for a locally compact group G can be investigated in two different ways. The first way is that of representation theory in which one tries to determine $\hat{G}$ and the Plancherel measure (for G of Type 1). The latter approach uses extensively the theory of induced representations, and when combined with Lie theory, has produced the remarkable representation theory of Lie groups. (For a related approach in the non-Type 1 case, see Pukanzsky [**N1**].)

Another approach to duality, developed by P. Eymard [**N1**], is to study certain Banach algebras of continuous functions on G, which, in a sense, contain duality information about G. These algebras are the *Fourier algebra* $A(G)$ and the *Fourier-Stieltjes algebra* $B(G)$ of G. For abelian G, $A(G)$ and $B(G)$ are just $L_1(\hat{G})$ and $M(\hat{G})$ "in disguise." Indeed in this case, the Fourier and Fourier-Stieltjes transforms, applied to $L_1(\hat{G})$ and $M(\hat{G})$, produce two subspaces $A(G)$, $B(G)$ of $C_0(G)$ and $C(G)$ respectively. The convolution product on $L_1(\hat{G})$, $M(\hat{G})$ goes over to the pointwise product of functions on G, and $A(G)$, $B(G)$ become commutative Banach algebras under the $L_1(\hat{G})$, $M(\hat{G})$ norms.

To obtain versions of $A(G)$, $B(G)$ for G nonabelian, we need to reformulate $A(G)$, $B(G)$ above in a way that only uses concepts present for general G. These concepts are $C^*(G)$ (the enveloping C^*-algebra of G) and $\mathrm{VN}(G)$, the von Neumann algebra generated by the left regular representation π_2 of G. To see how this is possible, let us suppose, once again, that G is abelian. Then $L_1(\hat{G})$ is the predual of $L_\infty(\hat{G})$. Now the Plancherel Theorem asserts that the map $f \to \hat{f}$ is a linear isometry from $L_2(G)$ onto $L_2(\hat{G})$, and hence induces a $*$-isomorphism Φ from $\mathbf{B}(L_2(\hat{G}))$ onto $\mathbf{B}(L_2(G))$. Regard $L_\infty(\hat{G})$ as a $*$-algebra of multiplication operators on $L_2(\hat{G})$. What is $\Phi(L_\infty(\hat{G}))$? To answer this, simply note that $\Phi(\hat{x}) = \pi_2(x)$. Since $\{\hat{x} : x \in G\}$ generates $L_\infty(\hat{G})$, it follows that $\Phi(L_\infty(\hat{G})) = \mathrm{VN}(G)$. This and the observations that $C^*(G) \cong C_0(\hat{G})$ and that $C_0(\hat{G})' = M(\hat{G})$ give a way of defining $A(G)$ and $B(G)$ for general G.

So let G be an arbitrary locally compact group. We define $A(G) = \mathrm{VN}(G)_*$, the predual of $\mathrm{VN}(G)$, and $B(G) = C^*(G)'$. Now the positive functionals on $C^*(G)$ correspond to continuous positive definite functions on G [**D2**, (13.4)], and since every element of $C^*(G)'$ is a linear combination of positive functionals, it follows that $B(G)$ can be identified with $\mathrm{Span}\{\phi\colon \phi$ is continuous and positive definite on $G\} \subset C(G)$. Turning to $A(G) = \mathrm{VN}(G)_*$, we observe that $C_l^*(G)$ is strongly dense in $\mathrm{VN}(G)$, and so by the Kaplansky Density Theorem [**D1**, Part 1, Chapter 3, Theorem 3], $A(G)$ can be regarded as a closed subspace of $C_l^*(G)'$. Since $C_l^*(G)$ is a homomorphic image of $C^*(G)$, $A(G)$ can be regarded as a closed subspace of $C^*(G)'$ and hence as a closed subspace of $B(G)$. The context will make clear if $B(G)$ (or $A(G)$) is being regarded as a space of linear functionals or as a subspace of $C(G)$.

The natural question now is can we identify those functions in $B(G) \subset C(G)$ that belong to $A(G)$? To answer this, we recall ((2.35(B))) that the predual of $\mathrm{VN}(G)$ consists of all those functionals α of the form $\sum_{i=1}^{\infty} \omega_{h_i,k_i}$, where $h_i, k_i \in L_2(G)$, $\sum_{i=1}^{\infty} \|h_i\|_2^2 < \infty$, $\sum_{i=1}^{\infty} \|k_i\|_2^2 < \infty$, and $\omega_{h_i,k_i}(T) = (Th_i, k_i)$ $(T \in \mathrm{VN}(G))$. When α is identified with a function $\phi \in C(G)$, we have $\phi(x) = \alpha(\pi_2(x))$ (since $\omega_{h_i,k_i}(\pi_2(f)) = \int f(x)(\pi_2(x)h_i, k_i)\, d\lambda(x)$ for $f \in L_1(G)$), and so it follows that $\phi(x) = \sum_{i=1}^{\infty}(\pi_2(x)h_i, k_i)$. The following useful formulae are easily checked:

$$f * g^\dagger(x) = (f, \pi_2(x)g), \quad f * g^*(x) = (f, \pi_2(x)\bar{g}) \qquad (f, g \in L_2(G), x \in G),$$

where, as in (4.20), $g^*(y) = g(y^{-1})$ $(y \in G)$, $g^\dagger = \overline{g^*}$. So $(\pi_2(x)h_i, k_i) = \bar{k}_i * h_i^*(x)$, and $A(G)$ *is the set of all functions of the form* $\sum_{i=1}^{\infty} f_i * g_i^*$, *where* $f_i, g_i \in L_2(G)$ *and* $\sum_{i=1}^{\infty} \|f_i\|_2^2 < \infty$, $\sum_{i=1}^{\infty} \|g_i\|_2^2 < \infty$. (In fact Eymard [**N1**] shows that $A(G) = \{f * g^\dagger\colon f, g \in L_2(G)\}$.)

We will use $\|\cdot\|$ to denote the norm of $B(G)$ or $A(G)$; the $C(G)$-sup-norm will be denoted by $\|\cdot\|_\infty$.

Now if $f \in L_1(G)$ and $\|\cdot\|_*$ is the $C^*(G)$-norm on $L_1(G)$, then for $\phi \in B(G)$, $|\phi(f)| \le \|\phi\|\|f\|_* \le \|\phi\|\|f\|_1$, so that $\|\phi\|_\infty \le \|\phi\|$. By considering functions $f * g^\dagger$, where f, g have suitably chosen compact supports, and using (i) below, we see that the Stone-Weierstrass Theorem applies to give that $C_C(G) \cap A(G)$ *is* $\|\cdot\|_\infty$ *dense in* $C_0(G)$. So $A(G) \subset C_0(G)$.)

(i) (Eymard [**N1**]) (a) $B(G)$ *is a Banach algebra with pointwise product, and* $A(G)$ *is an ideal in* $B(G)$.

(b) *The map* $f \otimes g \to f * g^*$ *extends to a continuous linear map* $\Phi\colon L_2(G)\hat{\otimes}L_2(G) \to C(G)$, *and* $(L_2(G)\hat{\otimes}L_2(G))/\ker\Phi$ *is isometrically isomorphic with* $A(G)$.

[(a) Let $\phi \in B(G)$. We show first that if π is the universal representation of $C^*(G)$ on the Hilbert space $\mathfrak{H}$, then there exist $\xi, \eta \in \mathfrak{H}$ such that $\phi(a) = (\pi(a)\xi, \eta)$ $(a \in C^*(G))$, $\|\phi\| = \|\xi\|\|\eta\|$. To this end, since $\phi \in (C^*(G)'')_*$, the polar decomposition theory applies [**D2**, 12.2.4]: so there exists a partial isometry u in $C^*(G)''$ and a positive functional $p \in C^*(G)'$ such that $\|\phi\| = \|p\|$ and $g(\phi) = (ug)(p)$ for all $g \in C^*(G)''$.

Now $C^*(G)''$ can be identified with the weak operator closure of $\pi(C^*(G))$, and there exists $\xi \in \mathfrak{H}$ such that $p(a) = (\pi(a)\xi, \xi)$ $(a \in C^*(G))$ [**D2**, 12.1.3]. Then $\phi(a) = (\pi(a)\xi, \eta)$, where $\eta = u^*\xi$. Further $\|\xi\|^2 = \|p\| = \|\phi\| \leq \|\xi\|\|\eta\| \leq \|\xi\|^2$, so that $\|\phi\| = \|\xi\|\|\eta\|$ as required.

Now let $\psi \in B(G)$ and $\xi_1, \eta_1 \in \mathfrak{H}$ be such that $\psi(x) = (\pi(x)\xi_1, \eta_1)$ and $\|\psi\| = \|\xi_1\|\|\eta_1\|$. Then $(\phi\psi)(a) = ((\bar{\pi} \otimes \pi)(a)(\xi \otimes \xi_1), \eta \otimes \eta_2)$, and so $\|\phi\psi\| \leq \|\xi \otimes \xi_1\|\|\eta \otimes \eta_1\| = (\|\xi\|\|\eta\|)(\|\xi_1\|\|\eta_1\|) = \|\phi\|\|\psi\|$. So $B(G)$ is a Banach algebra.

We now show that $A(G)$ is an ideal in $B(G)$. Let $h \in A(G)$ and $k \in B(G)$. We show that $hk \in A(G)$. We can suppose that h and k are positive definite. If $h \in C_c(G)$, then $hk \in A(G)$ by Godement's Theorem ((4.20)). It is therefore sufficient to show that $C_c(G) \cap A(G)$ is dense in $A(G)$. Let $p = \sum_{i=1}^{\infty} f_i * g_i^* \in A(G)$ as above. Note that as a member of $\mathrm{VN}(G)_*$, $p = \sum_{i=1}^{\infty} \omega_{g_i, \bar{f}_i}$, and also that, since $|\omega_{g_i,\bar{f}_i}(T)| \leq \|T\|\|g_i\|_2\|f_i\|_2$ $(T \in \mathrm{VN}(G))$, we have $\|\omega_{g_i,\bar{f}_i}\| \leq \|g_i\|_2\|f_i\|_2$. Then the Cauchy-Schwarz inequality gives

(1)
$$\left\|p - \sum_{i=1}^{N} f_i * g_i^*\right\| = \left\|\sum_{i=N+1}^{\infty} \omega_{g_i,\bar{f}_i}\right\| \leq \left(\sum_{i=N+1}^{\infty} \|g_i\|^2\right)^{1/2} \left(\sum_{i=N+1}^{\infty} \|f_i\|^2\right)^{1/2},$$

while if $F_i, G_i \in C_c(G)$, then by the triangular inequality,

$$\|f_i * g_i^* - F_i * G_i^*\| \leq [\|f_i - F_i\|_2\|g_i\|_2 + \|F_i\|_2\|g_i - G_i\|_2].$$

It follows that $C_c(G) \cap A(G)$ is dense in $A(G)$ as required.

(b) Regard $\mathrm{VN}(G)_* \subset \mathrm{VN}(G)'$, $\mathbf{B}(L_2(G))_* \subset \mathbf{B}(L_2(G))'$. The restriction map $j \colon \mathbf{B}(L_2(G))_* \to \mathrm{VN}(G)_*$ is clearly norm continuous and surjective. Now $\mathbf{B}(L_2(G))_*$ is identified with the Banach space $L_2(G)\hat{\otimes}L_2(G)$ of trace class operators, the duality being given by $(T, f \otimes g) = (Tg, \bar{f})$ $(T \in \mathbf{B}(L_2(G))$, $f, g \in L_2(G))$. (Use the result $(Y\hat{\otimes}X)' = B(X, Y')$ with $X = Y = Y' = L_2(G)$ the map $g \to \alpha_g$ identifies $L_2(G)$ with $L_2(G)'$, where $\alpha_g(f) = (g, \bar{f})$. So $\mathrm{VN}(G)_* = (L_2(G)\hat{\otimes}L_2(G))/\ker j$. Now $\mathrm{VN}(G)_*$ is identified with $A(G) \subset C(G)$ by the map $\omega_{h,k} \to \bar{k} * h^*$, and composing j with this map gives the map Φ of the statement of (b).]

As we shall see, (b) above gives a useful device for estimating norms in $A(G)$. A simple illustration of this is the inequality $\|f * g^*\| \leq \|f\|_2\|g\|_2$.

The following two results explore the consequences of the amenability of G for $A(G)$ and $B(G)$.

(ii) (Leptin [**4**]). *The Banach algebra $A(G)$ has a bounded approximate identity if and only if G is amenable.*

[Suppose that G is amenable. For each $C \in \mathscr{C}_e(G)$ and each $\varepsilon > 0$, let $K = K(C, \varepsilon)$ be such that the inequality of (4.13(ii)) holds. Let

$$\alpha_{C,\varepsilon} = (\lambda(CK)\lambda(K))^{1/2},$$

and $u_{C,\varepsilon} = (\chi_{CK} * \chi_K^*)/\alpha_{C,\varepsilon} \in A(G)$. The set of pairs (C, ε) is a net under the natural ordering (C increasing, ε decreasing). We claim that the net $\{u_{C,\varepsilon}\}$ is a bounded approximate identity for $A(G)$.

First, $\|u_{C,\varepsilon}\| \leq (\lambda(CK)\lambda(K))^{-1/2}\|\chi_{CK}\|_2\|\chi_K\|_2 = 1$. Second, let $f \in A(G) \cap C_c(G)$ and let f vanish outside some $C \in \mathscr{C}_e(G)$. Let ε, K be as above. Then

$$(u_{C,\varepsilon}f - f)(x) = f(x)(u_{C,\varepsilon} - 1)(x).$$

Now direct calculation shows that $u_{C,\varepsilon}(x) = (\lambda(xK \cap CK))/\alpha_{C,\varepsilon}$, and obviously, $\lambda(xK \cap CK) = \lambda(K)$ if $x \in C$. Since f vanishes outside C, we have $u_{C,\varepsilon}f - f = (\lambda(K)/(\lambda(CK)\lambda(K))^{1/2} - 1)f$. So

$$\|u_{C,\varepsilon}f - f\| = |(\lambda(K)/\lambda(CK))^{1/2} - 1|\|f\| \to 0$$

by (4.13(ii)). From the proof of (i)(a), $A(G) \cap C_c(G)$ is norm dense in $A(G)$. So $\{u_{C,\varepsilon}\}$ is a bounded approximate identity for $A(G)$.

Conversely suppose that there exists a bounded approximate identity $\{v_\delta\}$ in $A(G)$. Let $M = \sup_\delta \|v_\delta\|$. Using (4.20) and Problem 1-7, the amenability of G will follow once we have shown that $\|\mu\| = \|\pi_2(\mu)\|$ for all $\mu \geq 0$ with compact support in $M(G)$. To this end, we first show that given $K \in \mathscr{C}(G)$ and $\varepsilon > 0$, there exists $\phi \in A(G)$ such that

$$(2) \qquad |\phi(x) - 1| < \varepsilon, \qquad (x \in K), \qquad \|\phi\| \leq M.$$

To prove this, let $W = \{f|_K : f \in A(G)\}$. Using the Stone-Weierstrass Theorem (as in the proof that $A(G) \cap C_c(G)$ is $\|\cdot\|_\infty$-dense in $C_0(G)$), we see that W is dense in $C(K)$. So given $\eta \in (0, 1)$, there exists $h \in A(G)$ with $|h(x) - 1| < \eta$ for all $x \in K$. Now for some σ, $\|hv_\sigma - h\|_\infty \leq \|hv_\sigma - h\| < \eta$. If $x \in K$, then

$$|v_\sigma(x) - 1| = |h(x)v_\sigma(x) - h(x)|/|h(x)| \leq \eta/(1 - \eta).$$

For (2), choose η so that $\eta/(1 - \eta) = \varepsilon$ and then take $\phi = v_\sigma$.

Now let $p \in A(G)$. Suppose that $p = \sum_{i=1}^\infty f_i * g_i^*$ ($f_i, g_i \in L_2(G)$, $\sum_{i=1}^\infty \|f_i\|_2^2 < \infty$, $\sum_{i=1}^\infty \|g_i\|_2^2 < \infty$). Let $\mu \in M(G)$, $\mu \geq 0$. Then using (1),

$$\begin{aligned} \left|\int p\,d\mu\right| &= \left|\sum_{i=1}^\infty \int (f_i * g_i^*)(x)\,d\mu(x)\right| = \left|\sum_{i=1}^\infty \int (f_i, \pi_2(x)\bar{g}_i)\,d\mu(x)\right| \\ &= \left|\sum_{i=1}^\infty (f_i, \pi_2(\mu)\bar{g}_i)\right| \leq \|\pi_2(\mu)\| \sum_{i=1}^\infty \|f_i\|_2\|g_i\|_2. \end{aligned}$$

Taking the inf of $\sum_{i=1}^\infty \|f_i\|_2\|g_i\|_2$ over all such representations $\sum_{i=1}^\infty f_i * g_i^*$ for p and using (i)(b), we obtain that $|\int p\,d\mu| \leq \|\pi_2(\mu)\|\|p\|$ ($p \in A(G)$). Let μ be as above with compact support K and ϕ be as in (2). Then applying the preceding inequality with $p = \phi$,

$$\|\mu\| = \mu(K) \leq \left|\int \phi\,d\mu\right| + \varepsilon\|\mu\| \leq \|\pi_2(\mu)\|\|\phi\| + \varepsilon\|\mu\| \leq M\|\pi_2(\mu)\| + \varepsilon\|\mu\|.$$

Letting $\varepsilon \to 0$, we have $\|\mu\| \leq M\|\pi_2(\mu)\|$. Now $\|\mu^n\| = \|\mu\|^n$ ($n \geq 1$), and replacing μ by μ^n in the preceding inequality, taking nth roots, and then letting

$n \to \infty$ gives $\|\mu\| \leq \|\pi_2(\mu)\|$. It is elementary that $\|\pi_2(\mu)\| \leq \|\mu\|$, and so $\|\pi_2(\mu)\| = \|\mu\|$ as required.]

From the above, when G is amenable, $A(G)$ is a commutative Banach algebra with bounded approximate identity $\{u_\delta\}$ with $\|u_\delta\| = 1$ for all δ. We will now determine the multiplier algebra $\Delta(A(G))$ in that case. Multiplier algebras were briefly discussed in (1.30). Since $A(G)$ is commutative and has an approximate identity, one readily checks that if $(T_1, T_2) \in \Delta(A(G))$, then $T_2 = T_1$ so that $\Delta(A(G))$ can be identified with the *left multiplier algebra* $\{T \in \mathbf{B}(A(G)) \colon T(fg) = (Tf)g \text{ for all } f, g \in A(G)\}$.

(iii) (Cowling [**3**], Herz [**6**], McKennon [**S**], Renaud [**2**]). *Let G be amenable. Then the multiplier algebra $\Delta(A(G))$ of $A(G)$ is canonically isometrically isomorphic to $B(G)$.*

[Since $A(G)$ is an ideal in the Banach algebra $B(G)$, the map $\phi \to L_\phi$, where $L_\phi(f) = \phi f$ $(f \in A(G))$, is a norm-decreasing homomorphism from $B(G)$ into $\Delta(A(G))$. It remains to show that this map is isometric and onto. To this end, let $T \in \Delta(A(G))$. We claim that there exists $\psi \in C(G)$ such that $Tu_\delta \to \psi$ uniformly on compacta. Indeed, from (2) in the proof of (ii), if $C \in \mathscr{C}(G)$ and $\varepsilon > 0$, then there exists $h \in A(G)$ such that $|h(x) - 1| < \varepsilon$ $(x \in C)$ and $\|h\| \leq 1$. Then if $x \in C$ and $\delta, \sigma \in \Delta$,

$$\begin{aligned} |Tu_\delta(x) - Tu_\sigma(x)| &\leq \|(Tu_\delta)h - (Tu_\sigma)h\|_\infty + 2\varepsilon\|T\| \\ &\leq \|(Th)u_\delta - (Th)u_\sigma\| + 2\varepsilon\|T\|, \end{aligned}$$

and so $\{Tu_\delta\}$ converges uniformly on compacta to some $\psi \in C(G)$. If $f \in A(G)$, then $\{(Tu_\delta)f\}$ converges to Tf in the $A(G)$-norm, and hence also in the topology of uniform convergence on compacta. So $Tf = \phi f$. It remains to show that $\phi \in B(G)$ and $\|T\| \geq \|\phi\|$. Let $f \in C_c(G)$ have compact support C, and let h be as above. Let $\|f\|_*$ be the $C^*(G)$-norm of f. Then $\|\phi h\| = \|Th\| \leq \|T\|$, and

$$\begin{aligned} \left|\int_G f\phi\, d\lambda\right| &\leq \left|\int_G f\phi h\, d\lambda\right| + \left|\int_C f\phi(1-h)\, d\lambda\right| \\ &\leq \|f\|_*\|\phi h\| + \varepsilon\|f\|_1\|\phi\|_\infty \leq \|f\|_*\|T\| + \varepsilon\|f\|_1\|\phi\|_\infty. \end{aligned}$$

Letting $\varepsilon \to 0$ and using the density of $C_c(G)$ in $C^*(G)$ we see that $\phi \in C^*(G)' = B(G)$ and $\|\phi\| \leq \|T\|$. This completes the proof.]

The converse to (iii) is also true (Losert [**S6**]).

Herz introduced and studied L_p-versions of $A(G)$, $B(G)$ called $A_p(G)$, $B_p(G)$. In the latter terminology, $A(G) = A_2(G)$ and $B(G) = B_2(G)$, and much of the $A(G)$-theory extends to the $A_p(G)$ case. An account of $A_p(G)$ when G is abelian appears in Larsen [**S**].

There is a large literature on $A(G)$, $B(G)$—indeed, more generally, on $A_p(G)$, $B_p(G)$. This includes papers by Akemann and Walter, Arsac, Božejko, Carling, Cecchini, Cowling, Derighetti, Eymard, Figà-Talamanca and Picardello, Flory, Forrest, Gilbert, Granirer, Haagerup, Herz, Kugler, Lai, Lai and Chen, Lau, Leinert, Losert, de Michele and Soardi, Nebbia, Renaud, and Walter.

Problems 4

Unless otherwise specified, G is a locally compact group and S is a discrete semigroup.

1. Let Σ be either a σ-compact, amenable, locally compact group or a countable, left amenable semigroup. Prove that there exists a sequence $\{f_n\}$ in $P(\Sigma)$ such that $\|x * f_n - f_n\|_1 \to 0$ for all $x \in \Sigma$.

2. Let G be a locally compact group and

$$D(G) = \left\{ \phi \in L_\infty(G) \colon \inf_{\mu \in P(G)} \|\phi\mu\|_\infty = 0 \right\}.$$

Prove

(i) (a) $D(G)$ is a closed subset of $L_\infty(G)$;
(b) if $\phi \in L_\infty(G, \mathbb{R})$ and $\operatorname{ess\,inf}_{y \in G} \phi(y) > 0$, then $\phi \notin D(G)$;
(c) if $\phi \in D(G)$ and $\alpha \in \mathbb{C}$, then $\alpha\phi \in D(G)$;
(d) $(\phi x - \phi) \in D(G)$ for all $\phi \in L_\infty(G)$, $x \in G$.
(ii) G is amenable if and only if $D(G)$ is a subspace of $L_\infty(G)$.

Give another proof that G is amenable if G is abelian.

3. The group G is said to be *contractible* if for every $U \in \mathscr{C}_e(G)$ and $F \in \mathscr{F}(G)$, there exists $\alpha \in \operatorname{Aut} G$ such that $\alpha(F) \subset U$. Prove that every contractible group is amenable. Give examples of contractible groups and of noncontractible amenable groups.

4. Suppose that G is not amenable and let $p \in (1, \infty)$.
(i) Show that there exist $x_1, \ldots, x_n \in G$ and $c_1, \ldots, c_n \geq 0$, $\sum_{i=1}^n c_i = 1$, such that $\|\pi_p(n^{-1} \sum_{i=1}^n c_i \delta_{x_i})\| < 1$, where π_p is defined in (4.19).
(ii) Prove that there does not exist a nonzero, left invariant linear functional on $L_p(G)$.

5. Show that if $C, D \in \mathscr{C}(G)$ then the map $x \to \lambda(Cx \cap D)$ is continuous. (This was used in (4.12).)

6. Let $p \in (1, \infty)$ and $p^{-1} + q^{-1} = 1$. Prove that G is amenable if and only if, given $\varepsilon > 0$ and $C \in \mathscr{C}(G)$, there exists $f \in P_p(G)$ such that

$$\|x * f - f\|_p < \varepsilon, \quad \|\Delta(x)^{1/q} f * x - f\|_p < \varepsilon \qquad (x \in C).$$

7. Give an example of a summing sequence $\{K_n\}$ in $\mathbb{R}$ for which $\lambda(C + K_n \mathbin{\triangle} K_n)/\lambda(K_n) \not\to 0$ for all $C \in \mathscr{C}(\mathbb{R})$.

8. Let $\{K_n\}$ be a sequence of nonnull, compact subsets of G such that $\lambda(xK_n \mathbin{\triangle} K_n)/\lambda(K_n) \to 0$ for all $x \in G$. Prove that $\lambda(xK_n \mathbin{\triangle} K_n)/\lambda(K_n) \to 0$ uniformly on compacta. Show also that the smallest closed subgroup of G containing $\bigcup_{n=1}^\infty K_n$ is G itself.

9. Suppose that $U \in \mathscr{B}(G)$ and $\{V_n\}$ is a sequence in $\mathscr{B}(G)$ such that $\lambda(U) < \infty$, $0 < \lambda(V_n) < \infty$ for each n, and $\lambda(uV_n \triangle V_n)/\lambda(V_n) \to 0$ for all $u \in U$. Prove that

$$\liminf_{n\to\infty} \lambda(V_n^{-1}) \geq \lambda(U).$$

Deduce that if G is noncompact and unimodular and $\{K_n\}$ is a sequence of nonnull, compact subsets of G such that $\lambda(xK_n \triangle K_n)/\lambda(K_n) \to 0$ for all $x \in G$, then $\lambda(K_n) \to \infty$.

10. Show that G is amenable if and only if, given $C, L \in \mathscr{C}(G)$ and $\varepsilon > 0$, then there exists a nonnull symmetric set $K \in \mathscr{C}(G)$ such that $L \subset K$ and $\lambda(xK \triangle K)/\lambda(K) < \varepsilon$ for all $x \in C$. (Hint: distinguish between the unimodular and nonunimodular cases.) Deduce that every amenable G admits a symmetric summing net.

11. Prove that G is not amenable if and only if $\mathbf{I}(G) = \infty$ ((4.14)).

12. Suppose that G is amenable. Show that if nonnull sets $C_1, C_2 \in \mathscr{C}(G)$ and $\varepsilon > 0$, then there exist nonnull sets $K_1, K_2 \in \mathscr{C}(G)$ such that

$$\lambda(C_1K_1 \triangle C_2K_2)/(\lambda(K_1) + \lambda(K_2)) < \varepsilon.$$

13. Prove that G is amenable if and only if there exists a net $\{E_\delta\}$ of closed subsets of G with $0 < \lambda(E_\delta) < \infty$ for all δ such that for all $C \in \mathscr{C}(G)$, $\lambda(\bigcap_{c\in C}(cE_\delta))/\lambda(E_\delta) \to 1$.

14. Let G be amenable. A function $\phi \in \mathrm{U}(G)$ is called *left almost convergent* if $\{m(\phi): m \in \mathfrak{L}(\mathrm{U_r}(G))\}$ is a singleton. The set of such functions is denoted by $\mathrm{AC}_l(G)$.

(i) Show that $\mathrm{AC}_l(G)$ is a closed, right invariant subspace of $\mathrm{U_r}(G)$ containing $W(G)$.

(ii) Show that if $\{K_\delta\}$ is a summing net for G and $\mu_\delta = \chi_{K_\delta}/\lambda(K_\delta)$, then $\phi \in \mathrm{AC}_l(G)$ if and only if $\|\phi\mu_\delta - \alpha 1\|_\infty \to 0$ for some $\alpha \in \mathbb{C}$.

(iii) Show that $\mathrm{AC}_l(G)$ admits a unique left invariant mean m, and

$$m(\phi) = \lim_\delta \lambda(K_\delta)^{-1} \int_{K_\delta} \phi(t)\, d\lambda(t),$$

where $\{K_\delta\}$ is as in (ii).

15. We can define the notion of almost convergence for a semigroup S as in Problem 4-14. We also defined almost convergence for sequences in Problem 0-13. Show that the two notions for $S = \mathbb{P}$ are equivalent.

16. Let G be amenable. A function $\phi \in \mathrm{U_r}(G)$ is called *Bohr almost periodic* ($\phi \in \mathrm{BAP}(G)$) if given $\varepsilon > 0$, there exists $C \in \mathscr{C}(G)$ such that for all $x \in G$, there exists $c_x \in C$ such that

$$\|x\phi - c_x\phi\|_\infty < \varepsilon. \qquad (1)$$

Show that

$$\mathrm{AP}(G) = \mathrm{U}(G) \cap \mathrm{BAP}(G) \subset \mathrm{BAP}(G) \subset \mathrm{AC}_l(G).$$

17. (L_p-conjecture for amenable groups). Let us say that for $p > 1$, $L_p(G)$ is closed under convolution if for $f, g \in L_p(G)$, the function $f * g$, where

$$f * g(x) = \int f(y) g(y^{-1}x)\, d\lambda(y)$$

is well defined and belongs to $L_p(G)$. Show that if G is amenable and $p \in (1, \infty)$, then $L_p(G)$ is closed under convolution if and only if G is compact.

18. For $\phi \in L_\infty(G \times G)$ and $\nu \in L_1(G)$ define functions $\nu \circ_1 \phi$, $\nu \circ_2 \phi$ on G by

$$\nu \circ_1 \phi(s) = \int \phi(xs, s)\, d\nu(x), \qquad \nu \circ_2 \phi(s) = \int \phi(s, xs)\, d\nu(x).$$

Show that $\nu \circ_1 \phi, \nu \circ_2 \phi \in L_\infty(G)$. Let $m \in \mathfrak{M}(G)$. Show that $m \in \mathfrak{L}_t(G)$ if and only if $m(\nu \circ_2 \phi) = m(\nu^\sim \circ_1 \phi)$ for all $\nu \in P(G)$, $\phi \in L_\infty(G \times G)$.

19. Let $\mu \in P(G)$ and $\mathbf{P}\colon G \to P(G)$ be the "translation family" given by $\mathbf{P}(s) = \mu s$. Let $m \in \mathfrak{M}(G)$. A *posterior* for m (relative to $\mathbf{P}$) is a map $\mathbf{Q}\colon G \to P(G)$ such that for every bounded measurable function $\phi\colon G \times G \to \mathbb{C}$, we have

$$m(\theta \to \mathbf{P}^\wedge(\theta)({}_\theta\phi)) = m_1(x \to \mathbf{Q}^\wedge(x)(\phi_x)),$$

where $m_1 \in \mathfrak{M}(G)$ is given by $m_1(E) = m(\theta \to \mathbf{P}(\theta)(E))$, ${}_\theta\phi(y) = \phi(\theta, y)$, $\phi_x(y) = \phi(y, x)$ and it is assumed that the functions $x \to \mathbf{Q}^\wedge(x)(\phi_x)$ are in $L_\infty(G)$. (Show that $\theta \to \mathbf{P}^\wedge(\theta)({}_\theta\phi)$ is in $L_\infty(G)$ for each ϕ.) Show that if G is amenable and $m \in \mathfrak{L}_t(G)$, then the translation family $\mathbf{Q}$ associated with $\mu^\sim$ is a posterior for m relative to $\mathbf{P}$.

20. Let H be a closed normal, amenable subgroup of G and $Q_H\colon L_1(G) \to L_1(G/H)$ be the canonical $*$-homomorphism of (1.11). Show that $\|Q_H f\|_1 = d(0, C_f(H))$ for all $f \in L_1(G)$, where $C_f(H) = \mathrm{co}\{f * x\colon x \in H\}$ and $d(0, C_f(H)) = \inf\{\|g\|_1\colon g \in C_f(H)\}$.

Prove also that G is amenable if and only if for all $f \in L_1(G)$, $|\int f\, d\lambda|$ equals the distance of $\mathrm{co}\{x * f\colon x \in G\}$ from the zero function.

21. Let H be a closed, normal subgroup of G. Suppose that G is amenable and that $\{K_\delta\}$ is a summing net for H. Show that if $f \in L_1(G)$, then

$$\begin{aligned}(1)\qquad &\int_{G/H} d\lambda_{G/H}(xH) \left| \int_H f(xh)\, d\lambda_H(h) \right| \\ &= \lim_\delta \lambda_H(K_\delta)^{-1} \int_G d\lambda(x) \left| \int_{K_\delta} f * h(x)\, d\lambda_H(h) \right|.\end{aligned}$$

Deduce the classical summation formula: if $f \in L_1(\mathbb{R})$, then

$$\int_0^1 \left| \sum_{n=-\infty}^{\infty} f(x+n) \right| dx = \lim_{N \to \infty} \int_{-\infty}^{\infty} \left| \frac{1}{2N+1} \sum_{n=-N}^{N} f(x+n) \right| dx.$$

22. The group G is said to be *uniformly distributed* (u.d.) if there exists a sequence $\{x_r\}$ in G such that for all $f \in L_1(G)$,

$$\lim_{n\to\infty} \left\| \frac{1}{n} \sum_{r=1}^{n} f * x_r \right\|_1 = \left| \int f \, d\lambda \right| .$$

Such a sequence $\{x_r\}$ is said to be u.d. in G.

(i) Show that if G is u.d., then G is amenable.

(ii) Show that G is amenable and separable if and only if G is u.d.

23. Let G be a compact group. Let $\{x_n\}$ be a sequence in G and suppose that for all $\phi \in C(G)$,

$$n^{-1} \sum_{r=1}^{n} \phi(x_r) \to \int \phi \, d\lambda. \tag{1}$$

Show that $\{x_r\}$ is u.d. for G. (Problem 4-22).

Suppose that $G = \mathbb{R}/\mathbb{Z}$ ($= [0,1)$ with mod 1 addition). Prove Weyl's Criterion: *$\{x_r\}$ satisfies* (1) *if and only if* $n^{-1} \sum_{r=1}^{n} \exp(2\pi i k x_r) \to 0$ *for all* $k \in \mathbb{Z} \sim \{0\}$. Give an example of such a sequence $\{x_r\}$.

24. Prove that G is amenable if and only if $1 \in \operatorname{Sp} \pi_2(\mu)$ for all $\mu \in P(G)$.

25. Prove that if $1 < p < \infty$, then (4.20(i)) is true with $\pi_2(\mu)$ replaced by $\pi_p(\mu)$.

26. Check directly that $1 \in \operatorname{Sp} \pi_2(\mu)$ for all $\mu \in P(G)$ when G is abelian (Problem 4-24).

27. Give a direct proof of the weak containment equality $C^*(G) = C_l^*(G)$ when G is abelian and when G is compact.

28. Show that G is amenable if and only if there exists a net $\{\phi_\delta\}$ in $C_c(G)$ such that $\phi_\delta * \phi_\delta^\dagger \to 1$ pointwise on G. Show also that G is amenable if and only if the map $f \to \int f \, d\lambda$ is continuous for the $C_l^*(G)$-norm on $L_1(G)$.

29. Let G be separable and of Type 1. Prove that G is amenable if and only if the support of the Plancherel measure on $\hat{G}$ is the whole of $\hat{G}$. Illustrate this result when G is (a) abelian, (b) the Heisenberg group of real 3×3 matrices of the form

$$\begin{bmatrix} 1 & x_3 & x_1 \\ 0 & 1 & x_2 \\ 0 & 0 & 1 \end{bmatrix} \qquad (x_i \in \mathbb{R}).$$

30. Let G be a locally compact group containing a nonamenable open, proper subgroup H. Let $S_l(G)$ and $S(G)$ be the sets of states on $C_l^*(G)$ and $C^*(G)$ respectively. We can regard $S_l(G) \subset S(G)$. Let 1 be regarded as a state on $C^*(G)$: $1(f) = \int f \, d\lambda$ $(f \in L_1(G))$. Show that

$$S_l(G) \subsetneqq \operatorname{co}(S_l(G) \cup 1) \subsetneqq S(G).$$

31. Show that a separable, amenable locally compact group G has Property (T) ((4.29)) if and only if G is compact.

32. Show that a semigroup S is left amenable if S satisfies (SFC) ((4.22)). Show also that if S is left amenable and is either left cancellative or finite, then S satisfies (SFC).

33. Let S be a left amenable semigroup and T_0 be a countable subset of S. Show that there exists a countable left amenable subsemigroup T of S with $T_0 \subset T$.

34. Let G be compact. A function $\phi \in L_\infty(G)$ is called *Riemann measurable* if there exists a null set $E \subset G$ such that ϕ is continuous at each point of $G \sim E$. Show that every Riemann measurable function on G has a unique left invariant mean.

35. We proved in (1.12) that a closed subgroup H of an amenable locally compact group G is itself amenable. Give a quick proof of this result using Problem 4-28 and Godement's Theorem ((4.20)).

36. Let G be an amenable locally compact group. When does there exist a right summing net for G, that is, a net $\{K_\delta\}$ satisfying (4.15) with $K_\delta x$ in place of xK_δ?

37. Prove that if G is amenable, then $C_l^*(G)$ is not a simple C^*-algebra.

38. Let $L^0(G) = \{f \in L_1(G): \int f\,d\lambda = 0\}$ be the "augmentation ideal" of $L_1(G)$. Prove that $L^0(G)$ has a bounded approximate identity if and only if G is amenable.

39. Show that $A(G) = B(G)$ if and only if G is compact.

40. For a Banach $*$-algebra A, let $\mathrm{Prim}_* A$ be the set of ideals of A that are kernels of nonzero, topologically irreducible $*$-representations of A on a Hilbert space. (So if A is C^*-algebra then $\mathrm{Prim}\, A = \mathrm{Prim}_* A$).

(i) Show that the hull kernel topology for $\mathrm{Prim}_*(A)$ makes sense. (For the rest of the question, $\mathrm{Prim}_* A$ is given this topology for $A = L_1(G)$, $C^*(G)$.)

(ii) Show that the map $\Psi_G\colon \mathrm{Prim}\, C^*(G) \to \mathscr{P}(L_1(G))$, where $\Psi_G(I) = I \cap L_1(G)$, is a continuous bijection onto $\mathrm{Prim}_* L_1(G)$.

(iii) Let $[\Psi]$ be the class of locally compact groups G for which Ψ_G is a homeomorphism. Prove that G is amenable if $G \in [\Psi]$. [Note: Problem 6-25 asserts that $G \in [\Psi]$ if G has polynomial growth.]

41. Assume that G is separable. Show that G is amenable if and only if it possesses the following "weak Frobenius" property: *whenever N is a closed normal subgroup of G, $\pi \in \hat{G}$, $\sigma = \pi|_N$, and U^σ is the representation of G obtained by inducing σ to G, then U^σ weakly contains π.*

42. Let $P(A(G))$ be the semigroup of positive definite functions $\phi \in A(G)$ with $\phi(e) = 1$.

(i) Let $f, g \in C_c(G)$ with g vanishing outside $C \in \mathscr{C}(G)$ and $\|f\|_2 = 1 = \|g\|_2$. Let Φ be as in (4.34(i)(b)). Show that

$$\Phi\left(\int (tf)g \otimes \overline{(tf)g}\, d\lambda(t)\right) = (f * f^\dagger)(g * g^\dagger).$$

Deduce that

$$\|(f * f^\dagger)(g * g^\dagger) - g * g^\dagger)\| \leq 2 \sup_{x \in C} \|x^{-1} * f - f\|_2.$$

(ii) Prove that there exists a net $\{\phi_\delta\}$ in $P(A(G))$ such that $\|\phi\phi_\delta - \phi_\delta\| \to 0$ for all $\phi \in P(A(G))$.

(iii) Deduce that there exists a state m on $\mathrm{VN}(G)$ that is "topologically invariant" in the sense that $m(\phi T) = m(T)$ for all $\phi \in P(A(G))$, $T \in \mathrm{VN}(G)$. [Here, $\mathrm{VN}(G) = A(G)'$ and $\phi T(\psi) = T(\phi\psi)$ for all $\psi \in A(G)$.]

43. Prove that if G is an amenable, discrete group, then $\mathrm{VN}(G)$ is semidiscrete ((2.35)).

44. Let G be a σ-compact, amenable, locally compact group, $1 \leq p < \infty$ and $k \in L_1(G)$ have compact support C. Let $T_k \in \mathbf{B}(L_p(G))$ be the convolution operator

$$T_k F(v) = \int_G k(u) F(uv)\, d\lambda(u).$$

Now let (X, Σ, μ) be a σ-finite measure space and $R\colon G \to \mathbf{B}(L_p(X))$ be a strongly continuous antirepresentation of G on $L_p(X)$ such that the map $(u, x) \to R_u f(x)$ is $\lambda \times \mu$-measurable from $G \times X$ into $\mathbb{C}$ for all $f \in L_p(X)$, and such that $R_e = I$ and $\sup_{u \in G} \|R_u\| = M < \infty$. "Transfer" T_k to an operator $R_k \in \mathbf{B}(L_p(X))$, where

$$R_k f(x) = \int_G k(u) R_u f(x)\, d\lambda(u).$$

Show that $\|R_k\| \leq M^2 \|T_k\|$. [Hint: use (4.13).]

CHAPTER 5

Ergodic Theorems for Amenable Locally Compact Groups

(5.0) Introduction. The classical Mean Ergodic and Pointwise Ergodic Theorems are concerned with the convergence (in the appropriate topology) of the sequence of Cesàro means $\{n^{-1}\sum_{r=0}^{n-1}T^r\}$, where T is a bounded linear operator on a suitable Banach space. Our primary concern in this chapter is to prove versions of these theorems for an amenable locally compact group G. (We shall also establish ((5.2)) a Mean Ergodic Theorem for amenable semigroups.) Both theorems involve a (suitably measurable) representation $x \to T_x$ of G on a Banach space X. The Cesàro means referred to above are replaced by "operator averages" of the form

$$\lambda(K_\delta)^{-1}\int_{K_\delta} T_{x^{-1}}\,d\lambda(x),$$

where $\{K_\delta\}$ is a summing net for G. The Mean Ergodic Theorem is established in this general context ((5.7)). Not surprisingly, the Pointwise Ergodic Theorem requires care, but it will be shown ((5.20)) that if G is, in addition, connected, then we can construct a summing sequence for G so that a natural version of the Pointwise Ergodic Theorem holds. The proof uses the Dunford-Schwartz-Zygmund Theorem and Lie group arguments.

Other aspects of the relationship between amenability and ergodic theory are mentioned in Problems 5.

(5.1) Preliminaries. Let Σ be either a locally compact group or a discrete semigroup. If Σ is discrete, then, of course, $L_1(\Sigma) = l_1(\Sigma)$.

Let $\mu \to S_\mu$ be an antirepresentation of the Banach algebra $L_1(\Sigma)$ on a Banach space X. Thus $\mu \to S_\mu$ is a norm continuous, linear map from $L_1(\Sigma)$ into $\mathbf{B}(X)$ such that $S_{\mu\nu} = S_\nu S_\mu$ for all $\mu, \nu \in L_1(\Sigma)$. Let $M > 0$ be such that $||S_\mu|| \leq M||\mu||$ for all $\mu \in L_1(\Sigma)$.

The antirepresentation $\mu \to S_\mu$ is said to be *ergodic* if there exists a net $\{B_\delta\}$ in $\mathbf{B}(X)$ such that

(a) for all $\mu \in P(\Sigma)$,

$$\lim_\delta B_\delta(I - S_\mu) = 0,$$

the convergence being in the strong operator topology of $\mathbf{B}(X)$;

(b) for each $\xi \in X$ and each δ, $B_\delta(\xi) \in C_\xi$, the norm closure of the convex set $\{S_\mu(\xi): \mu \in P(\Sigma)\}$.

The notion of an *ergodic representation* of Σ is defined similarly.

We now define three important subspaces X_f, X_0, X_Σ of X:

$$X_f = \{\xi \in X: S_\mu(\xi) = \xi \text{ for all } \mu \in P(\Sigma)\},$$
$$X_0 = [\mathrm{Span}\{(I - S_\mu)\xi: \xi \in X, \ \mu \in P(\Sigma)\}]^-,$$
$$X_\Sigma = X_f + X_0.$$

By definition, X_0 is a closed subspace of X. It is obvious that X_f is also a closed subspace of X. Note that, since $P(\Sigma)$ spans $L_1(\Sigma)$, an element ξ is in X_f if and only if, for all $\nu \in L_1(\Sigma)$,

$$S_\nu(\xi) = \nu(\Sigma)\xi.$$

The justification of the use of the adjective "ergodic" above lies in the following version of the Mean Ergodic Theorem. References for (5.2), (5.3) are Eberlein [**1**] and Day [**9**].

(5.2) THEOREM. *Let $\nu \to S_\nu$ be ergodic with $\{B_\delta\}$ satisfying* (a) *and* (b) *of* (5.1). *Then*

(i) $||B_\delta|| \leq M$ *for all δ, and if $\xi \in X_\Sigma$, then $\{B_\delta(\xi)\}$ converges in norm to an element of $X_f \cap C_\xi$ and* $\lim B_\delta(\xi) = 0$ *if $\xi \in X_0$;*

(ii) *X_Σ is invariant under every S_ν ($\nu \in L_1(\Sigma)$) and every B_δ and is the vector space direct sum $X_f \oplus X_0$; further, $C_\xi \subset X_\Sigma$ for all $\xi \in X_\Sigma$;*

(iii) *the net $\{B'_\delta\}$, where B'_δ is the restriction of B_δ to X_Σ, converges in the strong operator topology of $\mathbf{B}(X_\Sigma)$ to the projection $P: X_\Sigma \to X_f$ associated with the direct sum decomposition of* (ii);

(iv) *for $\xi \in X_\Sigma$, $\{P\xi\} = X_f \cap C_\xi$.*

PROOF. (i) For $\xi \in X$, $\mu \in P(\Sigma)$, we have $||S_\mu(\xi)|| \leq M||\xi||$, and using (5.1(b)), we see that $||B_\delta|| \leq M$ for all δ. Now let $\xi \in X_\Sigma$. Write $\xi = \xi_f + \xi_0$, where $\xi_f \in X_f$, $\xi_0 \in X_0$. By definition, $S_\mu(\xi_f) = \xi_f$ for all $\mu \in P(\Sigma)$, so that $C_{\xi_f} = \{\xi_f\}$. Since each $B_\delta(\xi_f) \in C_{\xi_f} = \{\xi_f\}$, it follows that $B_\delta(\xi_f) = \xi_f$ for all δ.

Now by (5.1(a)), if $\eta \in X$ and $\mu \in P(\Sigma)$, we have

$$||B_\delta(I - S_\mu)(\eta)|| \to 0.$$

So $||B_\delta(\varsigma)|| \to 0$ for all $\varsigma \in \mathrm{Span}\{(I - S_\mu)(\eta): \eta \in X\}$. Now X_0 is the norm closure of this span, and since $||B_\delta|| \leq M$ for all δ, it follows that $||B_\delta(\varsigma)|| \to 0$ for all $\varsigma \in X_0$. In particular, $||B_\delta(\xi_0)|| \to 0$.

Hence $B_\delta(\xi) = [B_\delta(\xi_f) + B_\delta(\xi_0)] \to \xi_f \in X_f$. Since $B_\delta(\xi) \in C_\xi$ and C_ξ is norm closed in X, it follows that $\xi_f \in X_f \cap C_\xi$. This establishes (i).

(ii) Let $\mu \in P(\Sigma)$. For $\mu' \in P(\Sigma)$ and $\eta \in X$,

$$S_\mu(I - S_{\mu'})(\eta) = [(I - S_{\mu'\mu})(\eta) - (I - S_\mu)(\eta)] \in X_0,$$

and it follows that $S_\mu(X_0) \subset X_0$.

Now let $\xi = \xi_f + \xi_0 \in X_\Sigma$ as in (i). Then

$$S_\mu(\xi) = S_\mu(\xi_f) + S_\mu(\xi_0) = \xi_f + S_\mu(\xi_0) \in X_f + X_0 = X_\Sigma,$$

so that X_Σ is invariant under each S_μ. Since $P(\Sigma)$ spans $L_1(\Sigma)$, X_Σ is invariant under every S_ν $(\nu \in L_1(\Sigma))$.

We now prove that $C_\xi \subset X_\Sigma$. Indeed, recalling that X_0 is closed in X,

$$\begin{aligned} C_\xi - \xi_f &= \{[\xi_f + S_\mu(\xi_0)] - \xi_f \colon \mu \in P(\Sigma)\}^- \\ &= \{S_\mu(\xi_0) \colon \mu \in P(\Sigma)\}^- \subset X_0 \subset X_\Sigma \end{aligned}$$

so that $C_\xi \subset X_\Sigma$. Thus $B_\delta(\xi) \in C_\xi \subset X_\Sigma$, and X_Σ is B_δ-invariant for all δ.

Finally, if $\xi \in X_f \cap X_0$, then, using (i), $B_\delta(\xi) \to \xi$ on the one hand (since $\xi \in X_f$), and on the other hand, $B_\delta(\xi) \to 0$ (since $\xi \in X_0$). Thus $\xi = 0$, and the asserted direct sum decomposition follows.

(iii) This follows as in the preceding paragraph.

(iv) Let $\xi \in X_\Sigma$ and $\eta \in X_f \cap C_\xi$. By approximating η by elements of $\{S_\mu(\xi) \colon \mu \in P(\Sigma)\}$ and using the fact that X_0 is closed in X, we have $\xi - \eta \in X_0$. Hence

$$0 = \lim_\delta B_\delta(\xi - \eta) = P(\xi) - P(\eta) = P(\xi) - \eta$$

so that $P(\xi) = \eta$. By (i), $P(\xi) \in X_f \cap C_\xi$, and so (iv) is established. □

There are two obvious deficiencies in the preceding result. In the first place, it does not give that X_Σ is closed in X. (Is there a counterexample available?) Secondly, it does not provide information about when X_Σ is "large" in X.

To remedy these deficiencies, we can impose an additional condition on the $\{B_\delta\}$:

$$\text{(c)} \qquad \lim_\delta (I - S_\mu) B_\delta = 0,$$

the convergence being in the strong operator topology of $\mathbf{B}(X)$.

The weak compactness condition of (ii) below is obviously satisfied if X is reflexive.

(5.3) PROPOSITION. *Suppose that the net* $\{B_\delta\}$ *of* (5.2) *satisfies the additional property* (c) *above. Then*

(i) X_Σ *is closed in* X;

(ii) *if* C_ξ *is weakly compact for all* $\xi \in X$, *then* $X_\Sigma = X$.

PROOF. Let $\xi \in X$ and suppose that $B_\delta \xi \to \eta$ *weakly* in X for some $\eta \in X$. Then for $\mu \in P(\Sigma)$, $S_\mu(\eta)$ is, using (c), the weak limit η of the net

$$\{-(I - S_\mu)B_\delta(\xi) + B_\delta(\xi)\}.$$

So $\eta \in X_f$. Now C_ξ is weakly closed in X, since it is convex and norm closed in X. Since $B_\delta(\xi) \in C_\xi$ for all δ, it follows that $\eta \in C_\xi \cap X_f$. Further,

$$(\xi - \eta) \in \{(I - S_\mu)\xi \colon \mu \in P(\Sigma)\}^- \subset X_0.$$

So $\xi \in X_\Sigma$.

Using (5.2(i)), X_Σ is the closed subspace

$$\{\xi \in X\colon \{B_\delta \xi\} \text{ is norm convergent in } X\}$$

of X. This establishes (i).

Suppose, now, that C_ξ is weakly compact in X for all $\xi \in X$. Let $\varsigma \in X$. Then we can find a subnet $\{B_{\alpha(\sigma)}\}$ $(\sigma \in \Delta')$ of $\{B_\delta\}$ such that $\{B_{\alpha(\sigma)}(\varsigma)\}$ is weakly convergent in C_ς. Replace $\{B_\delta\}$ by $\{B_{\alpha(\sigma)}\}$ in the first paragraph of the present proof, and observe that X_Σ depends only on the antirepresentation $\nu \to S_\nu$. Then $\varsigma \in X_\Sigma$. So $X = X_\Sigma$, giving (ii). □

We now show that when Σ is left amenable, then a net $\{\nu_\delta\}$ in $P(\Sigma)$ satisfying a Følner-type condition of (4.1) gives rise to a net $\{B_\delta\}$ satisfying (a) and (b) of (5.1).

(5.4) PROPOSITION. (i) *Let Σ be left amenable and* ((4.1)) $\{\nu_\delta\}$ *be a net in* $P(\Sigma)$ *such that*

$$\|\nu_\delta - \mu * \nu_\delta\| \to 0 \tag{1}$$

for all $\mu \in P(\Sigma)$. Let $\nu \to S_\nu$ be an antirepresentation of $L_1(\Sigma)$ on a Banach space X, and let $B_\delta = S_{\nu_\delta}$. Then $\{B_\delta\}$ satisfies (a) *and* (b) *of* (5.1) (*so that $\nu \to S_\nu$ is ergodic*).

(ii) *Σ is left amenable if and only if every antirepresentation of $L_1(\Sigma)$ on a Banach space is ergodic.*

PROOF. (i) For $\mu \in P(\Sigma)$,

$$\|B_\delta(I - S_\mu)\| = \|S_{\nu_\delta} - S_{\mu*\nu_\delta}\| \le M\|\nu_\delta - \mu * \nu_\delta\| \to 0$$

so that (a) is satisfied. Trivially, (b) is satisfied.

(ii) Suppose that Σ is left amenable. By (i), every antirepresentation of $L_1(\Sigma)$ on a Banach space is ergodic.

Conversely, suppose that every antirepresentation of $L_1(\Sigma)$ on a Banach space is ergodic. Let $\nu \to S_\nu$ be the antirepresentation of $L_1(\Sigma)$ on $L_\infty(\Sigma)$ given by

$$S_\nu(\phi) = \phi\nu \qquad (\phi \in L_\infty(\Sigma)).$$

By hypothesis, $\nu \to S_\nu$ is ergodic. Clearly $1\mu = 1$ for all $\mu \in P(\Sigma)$, so that, in the notation of (5.2), $1 \in (L_\infty(\Sigma))_f$. By (5.2(ii)),

$$1 \notin (L_\infty(\Sigma))_0.$$

Hence (cf. (2.1) and (2.2)), $\mathfrak{L}_t(\Sigma) \neq \emptyset$. So, appealing to (1.10) in the nondiscrete case, we see that Σ is left amenable. □

(5.5) Discussion. If Σ is amenable, then, from (4.2), the net $\{\nu_\delta\}$ of (5.4(i)) can be chosen so that we also have

$$\|\nu_\delta - \nu_\delta * \mu\| \to 0$$

for all $\mu \in P(\Sigma)$. In this case, with B_δ as in (5.4(i)), the third condition (c) of (5.2) is also satisfied.

When Σ is discrete, then an antirepresentation $\mu \to S_\mu$ of Σ determines, and is determined by, the antirepresentation $x \to S_x$ $(= S_{\delta_x})$ of Σ. In this case, the preceding results can be reformulated in terms of $x \to S_x$ in the obvious way. (For example, C_ξ is defined as the norm closure of $\mathrm{co}\{S_x(\xi)\colon x \in \Sigma\}$ in X.)

We now concentrate on the Mean and Pointwise Ergodic Theorems for a locally compact group. In this case, of course, the relationship between antirepresentations of $L_1(\Sigma)$ and antirepresentations of Σ is more complicated.

Let G be a locally compact group and $x \to S_x$ an antirepresentation of G on a Banach space X. Thus $S_x \in \mathbf{B}(X)$ and $S_{xy} = S_y S_x$ for all $x, y \in G$. Further, for some $M > 0$, $\|S_x\| \leq M$ for all $x \in G$. No measurability conditions on $x \to S_x$ are assumed at the moment.

Of course, X is a right Banach G-space (G regarded as discrete), where the right action is given by

$$\xi x = S_x(\xi) \qquad (\xi \in X,\ x \in G).$$

As in (2.4), for $\xi \in X$, $f \in X'$, we define $f\xi \in l_\infty(G)$ by

$$f\xi(x) = f(\xi x) \qquad (x \in G).$$

We define the antirepresentation $x \to S_x$ to be *weakly measurable* if $f\xi \in L_\infty(G)$ for all $f \in X'$, $\xi \in X$. So in the notation of (2.4), the antirepresentation is weakly measurable if and only if $X(G) \subset L_\infty(G)$. Weak measurability for a *representation* of G is similarly defined.

Our next result shows that, under certain conditions, the antirepresentation $x \to S_x$ can be "integrated up" to give an antirepresentation of the Banach algebra $L_1(G)$ on X. A familiar, special case of this result occurs in the theory of unitary representations of locally compact groups. (Compare, also, (1.30).)

For $\xi \in X$, let D_ξ be the norm closure of $\mathrm{co}\{S_x\xi\colon x \in G\}$ in X.

(5.6) PROPOSITION. *Let the antirepresentation $x \to S_x$ be weakly measurable on G, and suppose that each D_ξ ($\xi \in X$) is weakly compact in X. Then there exists an antirepresentation $\nu \to A_\nu$ of the Banach algebra $L_1(G)$ on X such that*

(i) $\|A_\nu\| \leq M\|\nu\|$ *for all* $\nu \in L_1(G)$;

(ii) *if* $f \in X'$, $\nu \in L_1(G)$, *and* $\xi \in X$, *then*

$$f(A_\nu(\xi)) = \int_G f(S_x\xi)\, d\nu(x) \qquad (= \hat{\nu}(f\xi));$$

(iii) $A_\nu S_y = A_{y*\nu}$ *and* $S_y A_\nu = A_{\nu * y}$ *for all* $y \in G$, $\nu \in L_1(G)$;

(iv) *if* $\mu \in P(G)$ *and* $\xi \in X$, *then* $A_\mu(\xi) \in D_\xi$.

PROOF. Recall ((2.7)) that $X(G)$ is left introverted and is a Banach algebra with multiplication given by

$$pq(\phi) = p(q\phi) \qquad (p, q \in X(G)',\ \phi \in X(G)), \tag{1}$$

where

$$q\phi(x) = q(\phi x) \qquad (x \in G). \tag{2}$$

Further, again from (2.7), there is a representation $p \to \Phi_p$ of the Banach algebra $X(G)'$ on X', where

$$(3) \qquad \Phi_p(f) = pf$$

with $pf(\xi) = p(f\xi)$.

Recalling that $X(G) \subset L_\infty(G)$, we see that the map $\mu \to \hat{\mu}$ (strictly $\hat{\mu}|_{X(G)}$) from $L_1(G)$ into $X(G)'$ is a norm decreasing homomorphism. Let $B_\mu = \Phi_{\hat{\mu}}$. Then $\mu \to B_\mu$ is a representation of $L_1(G)$ on X' with $||B_\mu|| \le M||\mu||$ for all $\mu \in L_1(G)$. Hence $\mu \to B_\mu^*$ is an antirepresentation of $L_1(\Sigma)$ on X'', and $||B_\mu^*|| \le M||\mu||$ for all $\mu \in L_1(\Sigma)$.

Let $\mu \in P(G)$. Then we can find a mean $m \in \mathfrak{M}(l_\infty(G))$ such that $m|_{X(G)} = \hat{\mu}$. Let $\xi \in X$. Then D_ξ is weakly compact in X, and by the right Banach G-space version of Problem 2-12, $\hat{\xi}m \in (D_\xi)^\wedge$. (Of course, for $F \in X''$, $p \in \mathfrak{M}(l_\infty(G))$, $f \in X'$, we have $Fp(f) = F(pf)$.) But for $f \in X'$, using (3),

$$\hat{\xi}m(f) = \hat{\xi}(mf) = m(f\xi) = \hat{\mu}(f\xi) = \hat{\mu}f(\xi) = (B_\mu^*\hat{\xi})(f),$$

so that $B_\mu^*(\hat{\xi}) \in \hat{D}_\xi$. Since $P(G)$ spans $L_1(G)$, $B_\nu^*(\hat{\xi}) \in \hat{X}$ for all $\nu \in L_1(G)$. Define $A_\nu(\xi) \in X$ by $(A_\nu(\xi))^\wedge = B_\nu^*(\hat{\xi})$.

Clearly $\nu \to A_\nu$ is an antirepresentation of $L_1(G)$ satisfying (i) and (iv). As for (ii),

$$\begin{aligned} f(A_\nu(\xi)) &= \hat{\xi}(B_\nu f) = \hat{\nu}(f\xi) \\ &= \int_G f\xi(x)\, d\nu(x) = \int_G f(S_x\xi)\, d\nu(x). \end{aligned}$$

The first equality of (iii) is a consequence of the following argument, which uses (ii):

$$\begin{aligned} f(A_{y*\nu}(\xi)) &= (y*\nu)^\wedge(f\xi) = \hat{\nu}((f\xi)y) = \hat{\nu}(f\xi y) \\ &= f(A_\nu(\xi y)) = f(A_\nu S_y(\xi)). \end{aligned}$$

The second equality of (iii) is proved similarly. □

In the circumstances of (5.6), we shall sometimes write

$$A_\nu = \int_G S_x\, d\nu(x).$$

If E is a measurable subset of G and $d\nu = \chi_E\, d\lambda$, then we write

$$\int_E S_x\, d\lambda(x) = \int_G S_x\, d\nu.$$

Let $x \to T_x$ be a weakly measurable *representation* of G on a Banach space X. Let $S_x = T_{x^{-1}}$ for all $x \in G$. Then $x \to S_x$ is a weakly measurable antirepresentation of G on X. Suppose that, in the notation of (5.5), D_ξ is weakly compact for all $\xi \in X$. Then by (5.6), we can form the antirepresentation $\nu \to A_\nu$ of $L_1(G)$ on X.

Now let $\{K_\delta\}$ be a summing net for G. Then it follows from (4.15(3)) and (4.3) that if $\mu_\delta = \chi_{K_\delta}/\lambda(K_\delta)$, then $\{\mu_\delta\}$ satisfies (1) of (5.4).

The following result can be referred to as "the Mean Ergodic Theorem" for amenable, locally compact groups (Greenleaf [**4**]).

(5.7) THEOREM. *Let G be an amenable, locally compact group and $\{K_\delta\}$ a summing net for G. Let $x \to T_x$ be a weakly measurable* representation *of G on a Banach space X such that $D_\xi = (\mathrm{co}\{T_x\xi\colon x \in G\})^-$ is weakly compact for all $\xi \in X$. For each δ, let $A_\delta \in \mathbf{B}(X)$ be given by* ((5.6)):

$$A_\delta = \lambda(K_\delta)^{-1} \int_{K_\delta} T_{x^{-1}}\, d\lambda(x) \tag{1}$$

and define

$$X_f = \{\xi \in X\colon T_x\xi = \xi \text{ for all } x \in G\},$$
$$X_0 = [\mathrm{Span}\{(I - T_x)\xi\colon \xi \in X,\ x \in G\}]^-.$$

Then

(i) *X is the direct sum $X_f \oplus X_0$ of X_f and X_0;*

(ii) *if $P\colon X \to X_f$ is the projection onto X_f associated with the direct sum decomposition of* (i), *then $P \in \mathbf{B}(X)$, and*

(a) *$\{P\xi\} = X_f \cap D_\xi$ for all $\xi \in X$;*

(b) *$A_\delta \to P$ in the strong operator topology of $\mathbf{B}(X)$.*

PROOF. Applying (5.6) with $S_x = T_{x^{-1}}$, we obtain an antirepresentation $\nu \to A_\nu$ of G on X with the properties (i)–(iv) of (5.6). By (5.6(iv)), for each $\xi \in X$,

$$C_\xi = \{A_\mu(\xi)\colon \mu \in P(\Sigma)\}^- \subset D_\xi.$$

Since D_ξ is weakly compact and C_ξ is norm, and therefore weakly, closed in D_ξ, it follows that C_ξ is also weakly compact. From (5.5) and (5.3), we see that $X = Y_f + Y_0$, where

$$Y_f = \{\xi \in X\colon A_\mu(\xi) = \xi \text{ for all } \mu \in P(G)\},$$
$$Y_0 = [\mathrm{Span}\{(I - A_\mu)(\xi)\colon \xi \in X,\ \mu \in P(G)\}]^-.$$

Applying (5.2) to the antirepresentation $\nu \to A_\nu$ with $B_\delta = A_{\mu_\delta} = A_\delta$, the proof will be complete once we have shown

$$Y_f = X_f, \qquad Y_0 = X_0. \tag{2}$$

For then (i) and (ii)(b) follow immediately, and (ii)(a) follows by a simple adaptation of the proof of (5.2(iv)). If $\xi \in Y_f$, $x \in G$, and $\mu \in P(G)$, then, by (5.6(iii)),

$$T_x(\xi) = S_{x^{-1}}A_\mu(\xi) = A_{\mu * x^{-1}}(\xi) = \xi$$

since $\mu * x^{-1} \in P(G)$. So $Y_f \subset X_f$. Conversely, if $\xi \in X_f$, then, by (5.6(ii)), for all $\nu \in P(G)$, $f \in X'$,

$$f(A_\nu(\xi)) = f(\xi)$$

so that $\xi \in Y_f$. Thus $X_f = Y_f$.

We now turn to the proof that $X_0 = Y_0$. Let $\xi \in X$ and $x \in G$. Then

$$\begin{aligned} \|A_\delta(I - T_x)\xi\| &= \|A_{(\mu_\delta - x^{-1} * \mu_\delta)}(\xi)\| \\ &\leq M\|\mu_\delta - x^{-1} * \mu_\delta\|_1\|\xi\| \\ &= M(\lambda(xK_\delta \,\triangle\, K_\delta)/\lambda(K_\delta))\|\xi\| \to 0. \end{aligned}$$

It follows from (5.2(iii)) that $P((I-T_x)\xi)=0$, so that $(I-T_x)\xi \in Y_0$. Hence $X_0 \subset Y_0$.

Conversely, let $\mu \in P(G)$, $\xi \in X$. By (0.1), we can find a net $\{a_\sigma\}$, where each a_σ is a convex combination of elements $\hat{x}$ $(x \in G)$ in $(X(G)+\mathbb{C}1)'$, with $a_\sigma \to \hat{\mu}$ weak*. But then, using (5.6(i)), for $f \in X'$,

$$[f(\xi)-a_\sigma(f\xi)] \to [f(\xi)-\hat{\mu}(f\xi)] = f[(I-A_\mu)\xi].$$

Now $[f(\xi)-a_\sigma(f\xi)]$ is of the form $f(K_\sigma\xi)$ with $K_\sigma\xi \in X_0$ and K_σ independent of f. Clearly $K_\sigma\xi \to (I-A_\mu)\xi$ weakly. Thus $(I-A_\mu)\xi$ is in the weak closure X_0 of X_0. Hence $Y_0 \subset X_0$.

So $X_0 = Y_0$ and (2) is established. □

Before turning to the Pointwise Ergodic Theorem, we require some preliminary results on vector-valued measurability and integration, which will be used, without comment, in the sequel. A good reference for these results is Hille and Phillips [**1**, pp. 71–85]. See also [**DS**] for a more extensive treatment and Diestel and Uhl [**S**].

(5.8) Preliminaries. Let X be a Banach space and $(N, \mathfrak{N}, \mu)$ be a complete, σ-finite, measure space. A function $g\colon N \to X$ is said to be a *measurable, countably valued function* if g is of the form $\sum_{n=1}^{\infty} \chi_{E_n}\xi_n$, where $\{\xi_n\}$ is a sequence in X and $\{E_n\}$ is a disjoint sequence in $\mathfrak{N}$.

A function $f\colon N \to X$ is called *μ-measurable* if there exists a sequence $\{f_n\}$ of measurable, countably valued functions from N into X such that $||f_n(x)-f(x)|| \to 0$ almost everywhere on N. It is obvious that if $f\colon N \to X$ is μ-measurable, then the real-valued function $x \to ||f(x)||$ is measurable.

The following useful result characterises μ-measurable functions: *a function $f\colon N \to X$ is μ-measurable if and only if f is weakly measurable* (*that is, the complex-valued function $x \to F(f(x))$ is measurable for all $F \in X'$*) *and there exists a null subset D' of N such that $f(N \sim D')$ is norm separable.*

A function $f\colon N \to X$ is said to be *μ-integrable* (or *Bochner integrable*) if there exists a sequence $\{f_n\}$ of measurable, countably valued functions from N into X such that $\int ||f_n(x)||\,d\mu(x) < \infty$ for all n, and

$$||f_n(x)-f(x)|| \to 0 \quad \text{almost everywhere}, \qquad \int ||f(x)-f_n(x)||\,d\mu(x) \to 0.$$

A necessary and sufficient condition for $f\colon N \to X$ to be μ-integrable is that f *be μ-measurable and the function $x \to ||f(x)||$ be in $L_1(N)$*. The space $B(N;X;\mu)$ of equivalence classes of μ-integrable functions from N to X is a Banach space, the norm being given by

$$||f|| = \int_N ||f(x)||\,d\mu(x).$$

The integral for functions in $B(N;X;\mu)$ is defined first in the obvious way on the (dense) subspace of countably valued functions in $B(N;X;\mu)$ and then extended to the whole of $B(N;X;\mu)$ by continuity. Further, if $f \in B(N;X;\mu)$, then

$$\left\| \int f(x)\, d\mu(x) \right\| \leq \int ||f(x)||\, d\mu(x),$$

and if $T \in \mathbf{B}(X)$, then $T \circ f \in B(N;X;\mu)$, and

$$T\left(\int f(x)\, d\mu(x) \right) = \int T(f(x))\, d\mu(x).$$

Also, the natural analogue of Fubini's Theorem holds for a μ-integrable function of several variables. We note that if $g\colon N \to X$ is μ-measurable, bounded, and vanishes outside a measurable set of finite measure, then $g \in B(N;X;\mu)$.

For the remainder of this section, the group G will be assumed to be σ-compact. A representation $x \to T_x$ of G on X is said to be *strongly measurable* if each of the maps $x \to T_x\xi$ ($\xi \in X$) is λ-measurable (for the measure space $(G, \mathscr{M}(G), \lambda)$, where $\mathscr{M}(G)$ is the σ-algebra of λ-measurable sets). Similarly, a representation $t \to R_t$ of the additive semigroup $[0,\infty)$ on X is said to be *strongly measurable* if each of the maps $t \to R_t\xi$ ($\xi \in X$) is μ-measurable, where μ is the restriction of Lebesgue measure to $[0,\infty)$. The next result is well known but for the sake of completeness, a proof is given. (The corresponding result for representations of $[0,\infty)$ is given in [**DS**, VIII.1.3].)

(5.9) PROPOSITION. *A representation $x \to T_x$ of G on X is strongly measurable if and only if each of the maps $x \to T_x\xi$ ($\xi \in X$) is continuous.*

PROOF. Suppose that $x \to T_x$ is strongly measurable, and for $\eta \in X$, let $F_\eta(x) = T_x\eta$ ($x \in G$). Then F_η is λ-measurable for all $\eta \in X$. Let $\xi \in X$. We have to show that F_ξ is continuous. Since each T_{x_0} is continuous ($x_0 \in G$), it is sufficient to show that F_ξ is continuous at e. Fix $C \in \mathscr{C}_e(G)$ and $x_0 \in G$. Then $F_\xi|_C$, $F_{T_{x_0}\xi}|_C$ belong to $B(G;X;\lambda)$, and so, with $M = \sup_{x\in G} ||T_x||$,

$$\begin{aligned} \lambda(C)||(T_{x_0} - T_e)\xi|| &= \left\| \int_C T_{x^{-1}}(T_{xx_0} - T_x)\xi\, d\lambda(x) \right\| \\ &\leq M \int_C ||(T_{xx_0} - T_x)\xi||\, d\lambda(x). \end{aligned} \tag{1}$$

Fix $D \in \mathscr{C}(G)$ such that $C \subset D^0$. Now let $\varepsilon > 0$. First approximate $F_\xi|_C$ by a finitely valued function $g \in B(G;X;\lambda)$ supported in C and then, using the regularity of λ and Urysohn's Lemma, approximate g by continuous functions on D to obtain a continuous function $f \in B(G;X;\lambda)$ with support in D such that

$$\int_D ||F_\xi(x) - f(x)||\, d\lambda(x) < \varepsilon. \tag{2}$$

Since f is uniformly continuous on D, we can find a neighbourhood U of e in G such that $CU \subset D$ and $||f(xx_0) - f(x)|| < \varepsilon$ for all $x \in C$, $x_0 \in U$. Using (1) and (2), we have, for $x_0 \in U$,

$$\begin{aligned}||(T_{x_0} - T_e)\xi|| &\leq (M/\lambda(C)) \left[\int_G \chi_{Cx_0}(xx_0)||F_\xi(xx_0) - f(xx_0)||\,d\lambda(x)\right. \\ &\qquad \left. + \int_C ||f(xx_0) - f(x)||\,d\lambda(x) + \int_C ||f(x) - F_\xi(x)||\,d\lambda(x)\right] \\ &\leq (M/\lambda(C))[M_1 + \lambda(C) + 1]\varepsilon\end{aligned}$$

where $M_1 = \sup\{\Delta(y^{-1}) \colon y \in D\}$. The continuity of F_ξ now follows.

Conversely, suppose that (in the above notation) every F_ξ is continuous. Then for every ξ, $F_\xi(G)$ is σ-compact (since G is) and so is norm separable in X. Since F_ξ is trivially weakly measurable, we see that F_ξ is λ-measurable, so that $x \to T_x$ is strongly measurable. □

(5.10) COROLLARY. *Let X be a Banach space, $\xi \in X$, G_i $(1 \leq i \leq n)$ be a σ-compact, locally compact group, D be a compact subset of $G_1 \times \cdots \times G_n$ and, for each i, T_i be a strongly measurable representation of G_i on X. Then the function $(t_1, \ldots, t_n) \to \chi_D(t_1, \ldots, t_n)T_1(t_1)\cdots T_n(t_n)\xi$ from $G_1 \times \cdots \times G_n$ into X is $\lambda_{G_1} \times \cdots \times \lambda_{G_n}$-integrable.*

PROOF. Let $F(t_1, \ldots, t_n) = T_1(t_1)\cdots T_n(t_n)\xi$. It follows, using (5.9) and the joint continuity of multiplication on a bounded subset of $\mathbf{B}(X)$ for the strong operator topology, that F is continuous, and so $\lambda_{G_1} \times \cdots \times \lambda_{G_n}$-measurable. Now use the fact that D has finite measure. □

(5.11) The Dunford-Schwartz-Zygmund Theorem. This theorem, stated below, will provide a basis for the Pointwise Ergodic Theorem for connected amenable locally compact groups. The theorem is proved in [**DS**, VIII.7.10].

Let $(Z, \mathscr{Z}, \mu)$ be a measure space. Denote the operator norm on $\mathbf{B}(L_1(Z))$ by $||\cdot||_1$. Let $W = L_1(Z) \cap L_\infty(Z)$. Then W is a dense subspace of $L_p(Z)$ for $1 \leq p < \infty$. (Indeed, if $\phi \in W$, and $1 < p < \infty$, then

$$\int |f|^p\,d\mu = \int |f|\,|f|^{p-1}\,d\mu \leq ||f||_\infty^{p-1}||f||_1 < \infty$$

so that $\phi \in L_p(Z)$. Consideration of simple functions shows that W is dense in $L_p(Z)$.) For a measurable function $f \colon Z \to \mathbb{C}$ define $||f||_p \leq \infty$ in the obvious way. For $T \in \mathbf{B}(L_1(Z))$ and $p \in [1, \infty]$, let

$$||T||_p = \sup\{||Tf||_p \colon f \in W,\ ||f||_p = 1\}$$

(where the sup of an unbounded subset of $[0, \infty]$ is taken to be ∞). Note that if $p \in [1, \infty)$ and $||T||_p < \infty$, then $T|_W$ extends uniquely to an element of $\mathbf{B}(L_p(Z))$ of norm $||T||_p$. Of particular importance is the case where $||T||_1 \leq 1$ and $||T||_\infty \leq 1$. In this case, $||T||_p \leq 1$ for all $p \in [1, \infty]$ by the Riesz-Thorin Convexity Theorem [**HR2**, (E18)], so that T can be regarded as a bounded linear operator on $L_p(Z)$ $(p \in [1, \infty))$ with norm $||T||_p \leq 1$.

The Dunford-Schwartz-Zygmund Theorem is the following result.

Let $(Z, \mathscr{Z}, \mu)$ *be a measure space,* $n \in \mathbb{P}$, *and let* $t \to T_i(t)$ $(1 \leq i \leq n)$ *be strongly measurable representations of* $[0, \infty)$ *on* $L_1(Z)$ *such that* $\|T_i(t)\|_1 \leq 1$, $\|T_i(t)\|_\infty \leq 1$ *for* $1 \leq i \leq n$ *and* $t \geq 0$. *Let* $p \in (1, \infty)$ *and for* $R = (R_1, \ldots, R_n) \in (0, \infty)^n$ *define* $A(R) \in \mathbf{B}(L_p(Z))$ *by*

$$A(R)g = (R_1 \cdots R_n)^{-1} \int_0^{R_n} \cdots \int_0^{R_1} T_1(t_1) \cdots T_n(t_n) g \, dt_1 \cdots dt_n$$

for $g \in L_p(Z)$. *If* $f \in L_p(Z)$ $(1 < p < \infty)$, *then there exist functions* f', f_1 *in* $L_p(Z)$ *such that*

(i) $A(R)f \to f'$ μ*-a.e. as* $R_1, \ldots, R_n \to \infty$;
(ii) $A(R)f \to f'$ *in* $L_p(Z)$ *as* $R_1, \ldots, R_n \to \infty$;
(iii) $|A(R)f| \leq |f_1|$ μ*-a.e. for all* $R \in (0, \infty)^n$.

It is convenient to have a version of this theorem involving representations of $\mathbb{R}$.

(5.12) PROPOSITION. *Let* $(Z, \mathscr{Z}, \mu)$ *be a measure space,* K *a compact group, and* $t \to T_i(t)$ $(1 \leq i \leq n)$ *and* $k \to T_k$ *strongly measurable representations of* $\mathbb{R}$ *and* K *respectively on* $L_1(Z)$ *such that* $\|T_i(t)\|_1 \leq 1$, $\|T_i(t)\|_\infty \leq 1$ $(1 \leq i \leq n$, $t \in \mathbb{R})$, *and* $\|T_k\|_1 \leq 1$, $\|T_k\|_\infty \leq 1$ $(k \in K)$. *Let* $p \in (1, \infty)$ *and for* $R = (R_1, \ldots, R_n) \in (0, \infty)^n$, *define* $A'(R) \in \mathbf{B}(L_p(Z))$ *by*

$$(1) \quad A'(R)g = (2^n R_1 \cdots R_n)^{-1} \times \int_K \int_{-R_n}^{R_n} \cdots \int_{-R_1}^{R_1} T_1(t_1) \cdots T_n(t_n) T_{k^{-1}} g \, dt_1 \cdots dt_n \, d\lambda_K(k).$$

Then (i), (ii), *and* (iii) *of* (5.11) *hold with* $A(R)$ *replaced by* $A'(R)$.

PROOF. The integral in (1) exists by (5.10). Using Fubini's Theorem and replacing g by $\int_K T_{k^{-1}} g \, d\lambda_K(k)$, we may delete the terms "$\int_K$", "$T_{k^{-1}}$", and "$\lambda_K(k)$" in (1). Now write $[-R_i, R_i] = [-R_i, 0] \cup [0, R_i]$ and expand the expression for $A'(R)g$ into a sum of 2^n repeated integrals I, the range of each t_i in each I being either $[-R_i, 0]$ or $[0, R_i]$. Whenever the range of t_i in some I is $[-R_i, 0]$, change the variable t_i to $s_i = -t_i$. (The appropriate "change of variables" formula is obviously valid for countably valued, integrable functions, and its truth for general integrable functions follows by a routine approximation argument.) Now apply the Dunford-Schwartz-Zygmund Theorem to each I. □

The proof of the Pointwise Ergodic Theorem (5.20) (due to Greenleaf and Emerson [**1**]) involves first a reduction to the Lie group case and then to a situation in which we are dealing with a group that is obtained by forming a finite sequence of semidirect products of copies of $\mathbb{R}$, a compact group K entering at the last stage, so that (5.12) can be applied.

In the following two lemmas, G is a semidirect product $N \times_\rho H$. Recall (Appendix A) that $\rho(h)(n) = hnh^{-1}$ $(= (e, h)(n, e)(e, h^{-1}))$ in G $(n \in N$, $h \in H)$. Define $j\colon N \times H \to G$ by $j(n, h) = (hnh^{-1}, h)$. Observe that $j(n, h) = (e, h)(n, e)$, or simply hn. The group H is identified with G/N in the obvious way, and the Haar measures λ_H, λ_N and $\lambda_G(=\lambda)$ are chosen so that Weil's formula holds.

(5.13) LEMMA. *The map j is a homeomorphism from $N \times H$ onto G, and for all $f \in L_1(G)$, $E \in \mathscr{B}(G)$, we have*

$$\int_G f(x)\,d\lambda(x) = \iint\limits_{N\times H} f(j(n,h))\,d\lambda_H(h)\,d\lambda_N(n), \qquad \lambda_N \times \lambda_H(j^{-1}(E)) = \lambda(E).$$

PROOF. Trivially, j is a homeomorphism with $j^{-1}(n,h) = (h^{-1}nh, h)$. If $f \in L_1(G)$, then by Weil's formula ((1.11)),

$$\begin{aligned} \int f(x)\,d\lambda(x) &= \int_H d\lambda_H(h) \int_N f(hn)\,d\lambda_N(n) \\ &= \iint\limits_{N\times H} f(j(n,h))\,d(\lambda_N \times \lambda_H)(n,h). \end{aligned} \tag{1}$$

Now (1) is also true when f is replaced by χ_U for some open subset U of G. But then $\lambda_N \times \lambda_H(j^{-1}(U)) = \lambda(U)$, and the second equality of the lemma follows by outer regularity. □

A more general result than the above holds. Indeed, let $N_1, \ldots, N_m$ be locally compact groups, and let G be obtained by forming semidirect products of N_i: thus

$$G = (\cdots((N_1 \times_{\rho_1} N_2) \times_{\rho_2} N_3) \times_{\rho_3} \cdots) \times_{\rho_m} N_m.$$

Let λ_{N_i} be a left Haar measure on N_i and λ_G the left Haar measure on G determined by the λ_{N_i}'s through Weil's formula. Let $j\colon N_1 \times \cdots \times N_m \to G$ be given by

$$j(x_1, \ldots, x_m) = x_m x_{m-1} \cdots x_1. \tag{2}$$

The result alluded to above is as follows. *The map j is a homeomorphism from $N_1 \times \cdots \times N_m$ onto G, and for all $f \in L_1(G)$, $E \in \mathscr{B}(G)$,*

$$\begin{gathered} \int f(x)\,d\lambda_G(x) = \int f(j(x_1,\ldots,x_m))\,d\lambda_{N_1}(x_1)\cdots d\lambda_{N_m}(x_m), \\ \lambda_{N_1} \times \cdots \times \lambda_{N_m}(j^{-1}(E)) = \lambda_G(E). \end{gathered} \tag{3}$$

Using the above lemma and Fubini's Theorem, this result easily follows by induction.

We now introduce some notation. Let G be a semidirect product $N \times_\rho H$ as in the above lemma. For $x \in G$, write $x = (x_N, x_H)$, where $x_N \in N$, $x_H \in H$. If $A \subset G$, then we set

$$A_N = \{x_N : x \in A\}, \qquad A_H = \{x_H : x \in A\}.$$

The set A_N [A_H] is sometimes denoted $(A)_N$ [$(A)_H$].

Note that, given $C \in \mathscr{C}(G)$, the existence of sets K_1, K_2 in the next lemma is assured by (4.13). The lemma enables us to find, for given $C \in \mathscr{C}(G)$, a set $K \in \mathscr{C}(G)$ satisfying the inequality of (4.13(ii)), given that the corresponding problem for N and H can be solved.

(5.14) LEMMA. *Let $G = N \times_\rho H$ be amenable, $\varepsilon > 0$, and $C \in \mathscr{C}_e(G)$. Let $K_1 \in \mathscr{C}(H)$ be nonnull and such that*

$$\lambda_H(C_H K_1 \triangle K_1)/\lambda_H(K_1) < \varepsilon. \tag{1}$$

Define $C_N^* \in \mathscr{C}(N)$ by

$$C_N^* = \{k_1^{-1} c_H^{-1} c_N c_H k_1 \colon c \in C,\ k_1 \in K_1\},$$

and let $K_2 \in \mathscr{C}(N)$ be nonnull and such that

$$\lambda_N(C_N^* K_2 \triangle K_2)/\lambda_N(K_2) < \varepsilon. \tag{2}$$

Then if $K = j(K_2 \times K_1)$ $(= K_1 K_2)$, we have

$$\lambda(CK \triangle K)/\lambda(K) < \varepsilon(2 + \varepsilon).$$

PROOF. Let $c \in C$, $k_1 \in K_1$, $k_2 \in K_2$, and $k = j(k_2, k_1)$. Then

$$\begin{aligned} ck &= (c_N, c_H)(k_1 k_2 k_1^{-1}, k_1) = (c_N c_H k_1 k_2 k_1^{-1} c_H^{-1}, c_H k_1) \\ &= (c_H k_1 (k_1^{-1} c_H^{-1} c_N c_H k_1 k_2)(c_H k_1)^{-1}, c_H k_1). \end{aligned}$$

Hence $CK \subset j(C_N^* K_2 \times C_H K_1)$, and using (5.13), together with the elementary set inclusion

$$(A \times B) \sim (C \times D) \subset (A \times (B \sim D)) \cup ((A \sim C) \times D)$$

and (1) and (2), we have

$$\begin{aligned} \lambda(CK \triangle K) &\leq \lambda(j(C_N^* K_2 \times C_H K_1) \sim j(K_2 \times K_1)) \\ &= \lambda_N \times \lambda_H((C_N^* K_2 \times C_H K_1) \sim (K_2 \times K_1)) \\ &\leq \lambda_N(C_N^* K_2)\lambda_H(C_H K_1 \sim K_1) + \lambda(C_N^* K_2 \sim K_2)\lambda_H(K_1) \\ &= [(1+\varepsilon)\varepsilon + \varepsilon]\lambda_N(K_2)\lambda_H(K_1) = \varepsilon(2+\varepsilon)\lambda(K). \quad \square \end{aligned}$$

We now embark on the reduction procedure described in the second last paragraph of (5.12). The B references in the sequel refer to Appendix B.

(5.15) PROPOSITION. *Let G be a connected Lie group. Then there exists a maximal, compact, normal subgroup $K(G)$ of G. The subgroup $K(G)$ contains every compact, normal subgroup of G, and if S is a closed, connected, normal subgroup of G, then $K(S)$ is a normal subgroup of G contained in $K(G)$.*

PROOF. Let $\mathscr{A}$ be the set of maximal compact subgroups of G. By B62, $\mathscr{A}$ is not empty. If $K \in \mathscr{A}$ and K' is a compact normal subgroup of G, then $K'K$ is a compact subgroup containing K, so that $K' \subset K'K = K$. Observing that $xKx^{-1} \in \mathscr{A}$ whenever $x \in G$, $K \in \mathscr{A}$, it readily follows that $\bigcap \mathscr{A}$ is the maximal, compact, normal subgroup of G. Let S be a closed, connected, normal subgroup of G. Then $xK(S)x^{-1}$ is a normal subgroup of S for all $x \in G$, and the maximality of $K(S)$ yields that $xK(S)x^{-1} \subset K(S)$. Hence, $K(S)$ is a closed, normal subgroup of G and is thus contained in $K(G)$. $\square$

(5.16) PROPOSITION. *Let G be a connected, amenable Lie group such that $K(G)$ is trivial. Then there exists $n \in \mathbb{P}$, closed subgroups $L_1, \dots, L_n$ of G each of which is isomorphic to $\mathbb{R}$, and a compact subgroup K of G such that*

(i) *the map $j\colon L_1 \times \cdots \times L_n \times K \to G$, where $j(l_1, \dots, l_n, k) = kl_n \cdots l_1$ is a homeomorphism onto G (as in (5.13(2)));*

(ii) *the subsets $H_i = j(L_1 \times \cdots \times L_i)$ $(1 \le i \le n)$ are closed subgroups of G;*

(iii) *with $H_{n+1} = G$ and $L_{n+1} = K$, we have that each H_{i+1} $(1 \le i \le n)$ is a semidirect product $H_i \times_{\rho_i} L_{i+1}$. In particular, H_i is a normal subgroup of H_{i+1}.*

PROOF. Let N be the nil-radical of G. (So N is the maximal (closed), connected, nilpotent, normal subgroup of G (B35).) We show first that N is simply connected.

Let $(\tilde{N}, \pi)$ be the simply connected, covering group of N (B6). Since $\tilde{N}$ is nilpotent and simply connected, its centre $Z(\tilde{N})$ is simply connected (B34, 44) and hence of the form R^m for some $m > 0$.

Since $\ker \pi$ is a discrete subgroup of $Z(\tilde{N}) = \mathbb{R}^m$, we can find $r \ge 0$ and vectors $e_1, \dots, e_r$ in $\mathbb{R}^m$, independent over $\mathbb{Z}$, such that

$$\ker \pi = \left\{ \sum_{i=1}^{r} n_i e_i \colon n_i \in \mathbb{Z} \right\}.$$

It follows that $Z(N) \supset Z(\tilde{N})/\ker \pi = \mathbb{R}^{m-r} \times \mathbb{T}^r$, and since, by (5.15), $K(N) \subset K(G) = \{e\}$, we have $r = 0$. Hence π is an isomorphism and $N = \tilde{N}$ is simply connected.

Let R be the radical of G, and $\mathfrak{r}$ and $\mathfrak{n}$ be the Lie algebras of R and N respectively. By B37, $[\mathfrak{r}, \mathfrak{r}] \subset \mathfrak{n}$, so that $\mathfrak{r}/\mathfrak{n}$ is abelian. Hence R/N is abelian. Thus $R/N = \mathbb{R}^p \times \mathbb{T}^q$ for certain integers $p, q \ge 0$. Let $Q\colon R \to R/N$ be the canonical quotient map and $S = Q^{-1}(\mathbb{R}^p)$. Then S is obviously closed in G and is simply connected since both N and S/N are simply connected (B45).

We claim that S is a normal subgroup of G. To this end, let $(\tilde{G}, \pi')$ be the simply connected covering group of G. Since S is simply connected, $\tilde{S} \subset \tilde{G}$ is isomorphic under π' to S. Let $D\tilde{G}$ be the commutator subgroup of $\tilde{G}$. Then, by B24, $\mathscr{D}\mathfrak{g}$ is the Lie algebra of $D\tilde{G}$, and since $\tilde{G}$ is simply connected, $D\tilde{G}$ is a closed, connected subgroup of $\tilde{G}$ (B17). Clearly, $\tilde{R}$ can be identified with the radical of $\tilde{G}$, and $\tilde{R} \supset \tilde{S}$. Now the Lie algebra of $D\tilde{G} \cap \tilde{R}$ is $\mathscr{D}\mathfrak{g} \cap \mathfrak{r}$ (B15), and by B36, $D\tilde{G} \cap \tilde{R} \subset \tilde{N} = N$. So if $x \in \tilde{G}$ and $s \in \tilde{S}$, then

$$xsx^{-1}s^{-1} = (xsx^{-1})s^{-1} \in D\tilde{G} \cap \tilde{R} \subset \tilde{N}.$$

Thus $xsx^{-1} \in \tilde{N}\tilde{S} = \tilde{S}$. Thus $\tilde{S}$ is a normal subgroup of $\tilde{G}$, and it follows that $S = \pi'(\tilde{S})$ is a normal subgroup of $G = \pi'(\tilde{G})$.

Clearly, R/S is compact, since it is isomorphic to $\mathbb{T}^q$. Further, since G is amenable, it follows from (3.8) that G/R is compact. So G/S is also compact. By B47, G is a semidirect product $S \times_\rho K$ with K a compact group. By B46, there exists a basis $\{X_1, \dots, X_n\}$ for the Lie algebra $\mathfrak{s}$ of S such that

(α) the map $\Psi\colon \mathfrak{s} \to S$, where

$$\Psi\left(\sum_{i=1}^{n} \alpha_i X_i\right) = \exp(\alpha_1 X_1) \cdots \exp(\alpha_n X_n) \qquad (\alpha_i \in \mathbb{R})$$

is a homeomorphism from $\mathfrak{s}$ onto S;

(β) if $\mathfrak{s}_i = \operatorname{Span}\{X_1, \ldots, X_i\}$ $(1 \le i \le n)$, then $\Psi(\mathfrak{s}_i)$ is a closed subgroup H_i of G;

(γ) H_{i+1} is a semidirect product $H_i \times_{\rho_i} L_{i+1}$ $(1 \le i \le (n-1))$, where $L_{i+1} = \exp(\mathbb{R}X_{i+1}) \cong \mathbb{R}$.

This gives (ii) and (iii) of the Proposition, recalling that $H_{n+1} = G = S \times_\rho K = H_n \times_\rho K$. Note now that G is formed by a sequence of semidirect products

$$(\cdots(L_1 \times_{\rho_1} L_2) \times_{\rho_2} \cdots \times_{\rho_{n-1}} L_n) \times_\rho K.$$

Finally, (i) follows by (5.13) and the equalities

$$\begin{aligned} \jmath(\exp(\alpha_1 X_1), \ldots, \exp(\alpha_n X_n), k) &= k \exp(\alpha_n X_n) \cdots \exp(\alpha_1 X_1) \\ &= \jmath\left(\left[\Psi\left(-\sum_{i=1}^{n} \alpha_i X_i\right)\right]^{-1}, k\right), \end{aligned}$$

with an obvious abuse of the $\jmath$-notation. □

(5.17) Definition. Let G be a σ-compact, amenable, locally compact group. A summing sequence $\{K_n\}$ for G is said to be *admissible for the Pointwise Ergodic Theorem* if whenever $(Z, \mathcal{Z}, \mu)$ is a measure space, $x \to T_x$ is a strongly measurable representation of G on $L_1(Z)$ such that $||T_x||_1 \le 1$, $||T_x||_\infty \le 1$ for all x ((5.11)), $p \in (1, \infty)$, $f \in L_p(Z)$, and, for each n, $A_n \in \mathbf{B}(L_p(Z))$ is defined by ((5.7(1))

$$A_n = \lambda(K_n)^{-1} \int_{K_n} T_{x^{-1}}\, d\lambda(x), \tag{1}$$

then

(a) there exists $f' \in L_p(Z)$ such that $T_x f' = f'$ for all $x \in G$, and $A_n f \to f'$ both in $L_p(Z)$ and pointwise μ-a.e.;

(b) there exists $f_1 \in L_p(Z)$ such that $|A_n f| \le |f_1|$ μ-a.e. for all n.

(5.18) Note. If $\{K_n\}$ is *any* summing sequence for G and $\{A_n\}$, $x \to T_x$ are as above, then the conclusions of the Mean Ergodic Theorem (5.7) hold with $\{A_n\}$ in place of $\{A_\delta\}$ and $L_p(Z)$ $(p \in (1, \infty))$ in place of X. (Of course, $L_p(Z)$ is reflexive.) So there exists f' in $L_p(Z)$ such that $T_x f' = f'$ $(x \in G)$ and $A_n f \to f'$ in $L_p(Z)$. (Indeed, in the notation of (5.7), $f' = Pf$.) However, the other two conditions in (a) and (b) of (5.17) are *not* always satisfied (Problem 5-4). As we shall see, the next two results give a procedure for constructing summing sequences that *are* admissible for the Pointwise Ergodic Theorem when G is a connected, amenable, Lie group (cf. Problem 5-3).

(5.19) LEMMA. *Let G be a connected, amenable Lie group such that $K(G)$ is trivial, and let $L_1, \ldots, L_n$, K, j be as in* (5.16). *Then there exist sequences $\{R_i(m)\}$ $(1 \leq i \leq n)$ with $0 < R_i(m) < \infty$ for all i, m and $R_i(m) \to \infty$ as $m \to \infty$ for each i, and such that if*

$$K'_m = j([-R_1(m), R_1(m)] \times \cdots \times [-R_n(m), R_n(m)] \times K), \tag{1}$$

then $\lambda(CK'_m \triangle K'_m)/\lambda(K'_m) \to 0$ for all $C \in \mathscr{C}(G)$.

PROOF. We see from (5.16) that G is obtained by forming a sequence of n group semidirect products involving $\mathbb{R}$ or K. It is clearly sufficient to show that if G is a semidirect product of (σ-compact) locally compact groups N, H where H is $\mathbb{R}$ [$\mathbf{K}$] and if $\{L'_m\}$ is a sequence of nonnull, compact subsets of N such that

$$\lambda(C'L'_m \triangle L'_m)/\lambda(L'_m) \to 0 \qquad (C' \in \mathscr{C}_e(N)), \tag{2}$$

then there is a sequence $\{X_m\}$, where $X_m = [-R_m, R_m]$, $R_m > 0$ [$X_m = K$] for all m and a subsequence $\{L'_{\alpha(m)}\}$ of $\{L'_m\}$ such that if $K''_m = j(L'_{\alpha(m)} \times X_m) = X_m L'_{\alpha(m)}$, then

$$\lambda(CK''_m \triangle K''_m)/\lambda(K''_m) \to 0 \qquad (C \in \mathscr{C}_e(G)). \tag{3}$$

Let G, N, H, $\{L'_m\}$ be as above.

Suppose first that $H = \mathbb{R}$. Find a sequence $\{C_m\}$ in $\mathscr{C}_e(G)$ such that $C_m \subset C^0_{m+1}$ for all m and $G = \bigcup_{m=1}^{\infty} C_m$. Observing that for $D \in \mathscr{C}(\mathbb{R})$,

$$\lambda_H((D + [-p, p]) \triangle [-p, p])/\lambda_H([-p, p]) \to 0$$

as $p \to \infty$, we can find a sequence $\{R_m\}$ in $(0, \infty)$ such that $R_m \to \infty$ and, using the notation of (5.14),

$$\lambda_H(((C_m)_H + [-R_m, R_m]) \triangle [-R_m, R_m])/\lambda_H([-R_m, R_m]) \to 0. \tag{4}$$

We define $(C_m)^*_N$ as in (5.14), with C, K_1 replaced by C_m, $[-R_m, R_m]$.

Now choose a subsequence $\{L'_{\alpha(m)}\}$ of $\{L'_m\}$ such that

$$\lambda_N(((C_m)^*_N L'_{\alpha(m)}) \triangle L'_{\alpha(m)})/\lambda_N(L'_{\alpha(m)}) \to 0. \tag{5}$$

The required result (3) now follows using (4), (5), (5.14), and (4.13).

Suppose now that $H = K$. Then (4) is valid with $[-R_m, R_m]$ replaced by K and (5) follows, yielding the required result. □

We can now state and prove the Pointwise Ergodic Theorem for amenable groups.

(5.20) THEOREM. *Let G be a connected, amenable, locally compact group. Then there exists a summing sequence $\{K_m\}$ for G which is admissible for the Pointwise Ergodic Theorem.*

PROOF. Find (B6) a compact, normal subgroup K_0 of G such that G/K_0 is a Lie group, and let $G' = G/K_0$ and $Q\colon G \to G'$ be the quotient map. Replacing

K_0 by $Q^{-1}(K(G'))$, we can suppose that $K(G') = \{e\}$. Applying (5.19) with G' in place of G, we can find $R_i(m)$, K, and K'_m such that

(1) $$\lambda_{G'}(CK'_m \triangle K'_m)/\lambda_{G'}(K'_m) \to 0 \qquad (C \in \mathscr{C}(G')).$$

Let $K_m = Q^{-1}(K'_m)$. A simple application of Weil's formula yields that

$$\lambda(Q^{-1}(D)) = \lambda_{G'}(D)$$

for all $D \in \mathscr{C}(G')$, and the latter equality together with (4.6(ii)) enables us to deduce that $\{K_m\}$ is a summing sequence for G. By (5.18), the first two assertions of (5.17(a)) follow. The rest of the proof is devoted to showing that $\{K_m\}$ is admissible for the Pointwise Ergodic Theorem.

Let $x \to T_x$ be a strongly measurable representation of G on some $L_1(Z)$ such that $||T_x||_1 \leq 1$, $||T_x||_\infty \leq 1$ for all $x \in G$. Using (5.9) and Mazur's Theorem, we see that for each $f \in L_1(Z)$, the set $(\mathrm{co}\{T_k f\colon k \in K_0\})^-$ is norm, and hence weakly, compact in $L_1(Z)$. Further the representation $k \to T_k$ is strongly measurable (using (5.9)) and so, a fortiori, the antirepresentation $k \to T_{k^{-1}}$ is weakly measurable. Hence the Mean Ergodic Theorem (5.7) applies to the antirepresentation $k \to T_{k^{-1}}$ of the compact group K_0. So if

(2) $$P_0 = \int_{K_0} T_{k^{-1}}\, d\lambda_{K_0}(k),$$

then $P_0 \in \mathbf{B}(L_1(Z))$ is the projection from $L_1(Z)$ onto the subspace $L_1^f(Z) = \{f \in L_1(Z)\colon T_k f = f \text{ for all } k \in K_0\}$. Clearly $||P_0||_1 \leq 1$, and if $f \in W$ ((5.11)), then the function $k \to T_{k^{-1}} f$ is λ_{K_0}-integrable and

$$||P_0 f||_\infty \leq \int ||T_{k^{-1}} f||_\infty\, d\lambda_{K_0}(k) \leq \int ||f||_\infty\, d\lambda_{K_0}(k) = ||f||_\infty.$$

Hence $||P_0||_1 \leq 1$, $||P_0||_\infty \leq 1$. If $x \in G$, $k \in K_0$, then $T_x T_k P_0 = T_x P_0$, and if $F \in L_1^f(Z)$, then $T_k(T_x F) = T_x T_{x^{-1}kx} F = T_x F$, so that $T_x F \in L_1^f(Z)$. It follows that we can define a representation $a \to S_a$ of G' on $L_1(Z)$ by

$$S_{xK_0} = T_x P_0 \qquad (x \in G).$$

Using (5.9), the representation $a \to S_a$ is strongly measurable, and since $||P_0||_1 \leq 1$, $||P_0||_\infty \leq 1$, $||T_x||_1 \leq 1$, and $||T_x||_\infty \leq 1$ for all $x \in G$, we have $||S_a||_1 \leq 1$, $||S_a||_\infty \leq 1$ for all $a \in G'$.

Now if $x \in G$ and $\phi\colon \mathscr{B}(K_0) \to \mathscr{B}(K_0)$ is given by the formula $\phi(E) = xEx^{-1}$, then the uniqueness of normalised Haar measure on K_0 yields that $\lambda_{K_0}(\phi^{-1}(E)) = \lambda_{K_0}(E)$. Thus

$$\int_{K_0} \chi_E(xkx^{-1})\, d\lambda_{K_0}(k) = \int_{K_0} \chi_E(k)\, d\lambda_{K_0}(k),$$

and it follows that

(3) $$\int_{K_0} f(k)\, d\lambda_{K_0}(k) = \int_{K_0} f(xkx^{-1})\, d\lambda_{K_0}(k)$$

for $f \in L_1(K_0)$.

Let $1 < p < \infty$, and for $m \in \mathbb{P}$, $g \in L_p(Z)$, define $A_m, A'_m \in \mathbf{B}(L_p(Z))$ by

$$A_m g = (\lambda(K_m))^{-1} \int_{K_m} T_{x^{-1}} g \, d\lambda(x),$$
$$A'_m g = (\lambda_{G'}(K'_m))^{-1} \int_{K'_m} S_{a^{-1}} g \, d\lambda_{G'}(a).$$

Recalling that $K_m = Q^{-1}(K'_m)$ and using Weil's formula again, (3), (2), and the weak measurability of the representations involved, we have, for $f \in (L_p(Z))'$,

$$\begin{aligned} f(A_m g) &= \int_{K'_m} d\lambda_{G'}(xK_0) \int_{K_0} f(T_{x^{-1}} T_{xk^{-1}x^{-1}} g) \, d\lambda_{K_0}(k) \\ &= \int_{K'_m} d\lambda_{G'}(xK_0) \int_{K_0} f(T_{x^{-1}} T_{k^{-1}} g) \, d\lambda_{K_0}(k) \\ &= \int_{K'_m} f(T_{x^{-1}} P_0 g) \, d\lambda_{G'}(xK_0) = f\left(\int_{K'_m} S_{a^{-1}} g \, d\lambda_{G'}(a) \right). \end{aligned}$$

Thus $A_m = A'_m$ for all m.

Let $E_m = [-R_1(m), R_1(m)] \times \cdots \times [-R_n(m), R_n(m)] \times K$, so that $K'_m = j(E_m)$. Then using the first equality of (5.13(3)),

$$\begin{aligned} f(A_m g) &= f(A'_m g) \\ &= (\lambda(K_m))^{-1} \int_{G'} \chi_{j(E_m)}(a) f(S_{a^{-1}} g) \, d\lambda_{G'}(a) \\ &= (\lambda(K_m))^{-1} \int_K \int_{\mathbb{R}} \cdots \int_{\mathbb{R}} \chi_{j(E_m)}(j(t_1, \ldots, k)) f(S_{(j(t_1,\ldots,t_n,k))^{-1}} g) \\ &\qquad\qquad dt_1 \cdots dt_n \, d\lambda_K(k) \end{aligned}$$
$$= (\lambda(K_m))^{-1} f\left(\int_K \int_{-R_n(m)}^{R_n(m)} \cdots \int_{-R_1(m)}^{R_1(m)} S_{-t_1} \cdots S_{-t_n} S_{k^{-1}} g \, dt_1 \cdots dt_n \, d\lambda_K(k) \right).$$

Now $\lambda(K_m) = \lambda'(K'_m) = 2^n R_1(m) \cdots R_n(m)$, using the second equality of (5.13(3)). The proof is completed by applying (5.12). □

References

Alaoglu and Birkoff [**1**], Aribaud [**1**], [**2**], Chatard [**1**], Conze and Dang Ngoc [**1**], [**2**], Day [**1**], [**2**], [**4**], [**9**], Eberlein [**1**], Greenleaf [**4**], Greenleaf and Emerson [**1**], Greenleaf, Moskowitz, and Rothschild [**1**], R. T. Moore [**1**], Peck [**1**], Renaud [**1**], and Witz [**1**].

Further Results

(5.21) Other ergodic theorems. We now state another Mean and Pointwise Ergodic Theorem for amenable groups. References for this theorem are Calderon [**1**], [**2**] Templeman [**1**], Chatard [**1**], Bewley [**1**], and Emerson [**4**].

Let G be σ-compact and amenable. Suppose that there exists a summing sequence $\{K_n\}$ such that

$$\sup_{n\in\mathbb{P}}(\lambda(K_n^{-1}K_n)/\lambda(K_n)) < \infty. \tag{1}$$

Let $(Z,\mathscr{Z},\mu)$ be a σ-finite measure space such that Z is a left G-set, the map $(x,z)\to xz$ is measurable from $G\times Z$ into Z, and μ is G-invariant. Let $1 < p < \infty$, $f\in L_p(Z)$, and $x\to T_x$ be the representation of G on $L_p(Z)$ given by $T_xf(z) = f(x^{-1}z)$. If $\{A_n\}$ is defined as in (5.17(1)), then the sequence $\{A_nf\}$ converges both in the L_p-norm and pointwise μ-a.e. to a G-invariant function in $L_p(Z)$.

An obviously important question, in connection with the preceding theorem, is the following. *Which groups G admit summing sequences $\{K_n\}$ satisfying* (1)? On the positive side, Auslander and Brezin [**1**] prove that every simply connected, nilpotent Lie group admits a summing sequence satisfying (1), and they use this fact to obtain results on uniform distribution on nilmanifolds and almost periodicity on nilpotent groups. We shall see in Problem 6-15 that if G is abelian or if G is discrete nilpotent, then G admits a summing sequence $\{K_n\}$ satisfying (1).

On the negative side, Greenleaf and Emerson [**1**] assert that the condition (1) and the summing sequence condition (4.15(3)) "are actually incompatible in many amenable Lie groups." In particular, there are difficulties for nonunimodular groups. In this connection it is very easy to see (Chatard [**1**]) that if the summing sequence $\{K_n\}$ satisfies the related condition

$$\sup_{n\in\mathbb{P}}(\lambda(K_nK_n^{-1})/\lambda(K_n)) < \infty,$$

then G *is* unimodular. Is it true that if $\{K_n\}$ satisfies (1), then G is unimodular?

We now turn to another problem in ergodic theory—the existence of equivalent G-invariant measures.

Let $(Z,\mathscr{Z},\mu)$ be a measure space, and G a group of invertible, measurable transformations of Z. In view of the importance of measure-preserving transformations, the following question is of interest: *under what conditions can one find a (positive) measure ν on $(Z,\mathscr{Z})$ which is G-invariant and (in a suitable sense), is equivalent to μ?* A complete solution to this question when $(Z,\mathscr{Z},\mu)$ is σ-finite and nonatomic and ν is finite is given in Hajian and Ito [**1**]: *if μ is σ-finite and nonatomic, then there exists a finite, G-invariant measure ν on $(Z,\mathscr{Z})$ equivalent to μ if and only if there does not exist $E\in\mathscr{Z}$ with $\mu(E)>0$ which is weakly wandering under G.* (Here, a set $W\in\mathscr{Z}$ is called *weakly wandering* (under G) if there exists a sequence $\{x_n\}$ in G such that $\mu(x_iW\cap x_jW) = 0$ if $i\neq j$.)

Less is known about the above question when μ [ν] is no longer required to be σ-finite [finite]. For results in this general case see Friedman [**1**] (with its references), Millet and Sucheston [**1**], and Rosenblatt [**3**]. We now state a result due to Rosenblatt [**12**] that solves the question when G is amenable. It is not known if the result is true without the hypothesis of amenability. (It is of interest that the amenable case of the Hajian-Ito result was proved in the (earlier) work of Natarajan [**1**], [**2**]. See also Granirer [**12**].)

We require a stronger version of equivalence of measures. Recall that if $(Z, \mathscr{Z}, \mu)$ and $(Z, \mathscr{Z}, \nu)$ are measure spaces, then μ and ν are said to be *equivalent* if they have the same null sets. In order to cope with sets of infinite measure in the non-σ-finite context, we say that μ and ν are *strongly equivalent* if μ and ν have the same null sets and the same sets of infinite measure, that is, if $E \in \mathscr{Z}$, then $\mu(E) = 0$ if and only if $\nu(E) = 0$, and $\mu(E) = \infty$ if and only if $\nu(E) = \infty$. Rosenblatt's result is the following.

Let G be amenable and μ be nonatomic. Then (i) *and* (ii) *are equivalent:*

(i) *there exists a G-invariant measure ν on $(Z, \mathscr{Z})$ that is strongly equivalent to μ;*

(ii) *if $E \in \mathscr{Z}$, then $\inf_{x \in G} \mu(xE) > 0$ if $\mu(E) > 0$ and $\sup_{x \in G} \mu(xE) < \infty$ if $\mu(E) < \infty$.*

Problems 5

Throughout, G is a locally compact group.

1. Let G, $\{K_\delta\}$ be as in (5.7) and $x \to T_x$ be a weakly measurable representation of G on a reflexive Banach space X. Show that the conclusions of (5.7) hold.

2. What does (5.7) say when $X = \mathrm{WP}(G)$, the space of weakly almost periodic functions on G under the natural left G-action? Identify X_f, X_0, and P.

3. By following the proofs of (5.19) and (5.20), construct a summing sequence for the "$ax + b$"-group that is admissible for the Pointwise Ergodic Theorem.

4. Let $Z = [0, 1)$ and μ be the restriction of Lebesgue measure to $[0, 1)$. Let α be irrational and $n \to T_n$ the representation of $\mathbb{Z}$ on $L_1(Z)$ given by

$$T_n g(x) = g(n.x)$$

where $n.x$ is the unique element of $[0, 1) \cap (x + n\alpha + \mathbb{Z})$. Let $a_n = \sum_{r=1}^{n-1} r^{-1}$.

(i) Show that $||T_n||_1 \leq 1$, $||T_n||_\infty \leq 1$.

(ii) Show that if $m > 0$ and $x \in [0, 1)$ then

$$A_{x,m} = \{r \in \mathbb{Z} \colon r.x \in [m^{-1}, 2m^{-1})\}$$

is infinite.

(iii) Define inductively a sequence $\{K_m\}$ in $\mathscr{C}(\mathbb{Z})$ by setting $K_1 = [-1; 1]$ and, for $m \geq 2$, $K_m = [-(m!); m!] \cup K_{m-1} \cup D_m$, where $D_m \subset A_{a_n,m}$ is such that $(m!)m^{-1/4} - 1 \leq |D_m| < (m!)m^{-1/4}$. Show that $\{K_m\}$ is a summing sequence for $\mathbb{Z}$.

(iv) Show that $\{K_m\}$ is not admissible for the Pointwise Ergodic Theorem. [Hint: show that $\{A_n f\}$ does not converge pointwise, where $f \in L_p(Z)$ $(1 < p < 2)$ is given by $f(x) = x^{-1/2}$.]

5. A sequence $\{\mu_n\}$ of probability measures in $M(G)$ is called *ergodic* if, whenever π is a continuous unitary representation of G on a Hilbert space $\mathfrak{H}$,

then $\pi(\mu_n) \to P^\pi$, in the weak operator topology of $\mathbf{B}(\mathfrak{H})$, where P^π is the orthogonal projection of $\mathfrak{H}$ onto the subspace $\mathfrak{H}_f = \{\xi \in \mathfrak{H}: \pi(x)\xi = \xi$ for all $x \in G\}$.

(i) Show that if $\{K_n\}$ is a summing sequence for G and $\mu_n = \chi_{K_n}/\lambda(K_n)$, then $\{\mu_n\}$ is ergodic.

(ii) For each $\pi \in \hat{G}$ let B_π be the set of functions on G of the form $x \to (\pi(x)\xi, \eta)$ for some $\xi, \eta \in \mathfrak{H}_\pi$, the Hilbert space of π. Let $B_I(G)$ be the closed subspace of $C(G)$ spanned by $\bigcup\{B_\pi : \pi \in \hat{G}\}$. Prove that if G is abelian, then $B_I(G) = \mathrm{AP}(G)$. Show that $B_I(G) \subset \mathrm{WP}(G)$.

(iii) Prove that the following are equivalent for separable G:

(a) $\{\mu_n\}$ is an ergodic sequence;

(b) $\pi(\mu_n) \to 0$ in the weak operator topology for all $\pi \in \hat{G} \sim \{1\}$;

(c) $\hat{\mu}_n \to m$ weak* on $B_I(G)$, where m is the (unique) invariant mean on $B_I(G)$ (obtained by restricting the invariant mean on $\mathrm{WP}(G)$);

(iv) Show that if G is abelian, then $\{\mu_n\}$ is ergodic if and only if the sequence $\{\mu_n\}$, regarded as a sequence of probability measures on the Bohr compactification βG of G, converges weak* to the Haar measure on βG.

(v) If H is the Heisenberg group (Problem 4-29) show that

$$B_I(H) = (\mathrm{AP}_0(\mathbb{R})\check{\otimes}C_0(\mathbb{R}^2)) \oplus (\mathbb{C} \otimes \mathrm{AP}(\mathbb{R}^2))$$

where $\mathrm{AP}_0(\mathbb{R}) = \{\phi \in \mathrm{AP}(\mathbb{R}): m(\phi) = 0$, m the unique invariant mean on $\mathrm{AP}(\mathbb{R})\}$ and $\check{}$ denotes the injective tensor product completion.

6. Let S be an amenable semigroup of bounded linear operators on some $C(X)$ (X compact Hausdorff). Suppose that

(i) $(\mathrm{co}\{T\phi: T \in S\})^-$ contains an S-invariant function for each $\phi \in C(X)$;

(ii) each $T \in S$ is *Markov* (in the sense that $T(1) = 1$ and $T\phi \geq 0$ whenever $\phi \geq 0$ in $C(X)$).

Show that there exists a positive projection $P \in \mathbf{B}(C(X))$ that maps onto the fixed-point subspace of $C(X)$ and is such that $PT = TP = P$ for all $T \in S$.

7. (Sine's Mean Ergodic Theorem). Let X be a Banach space and $T \in \mathbf{B}(X)$, $||T|| \leq 1$. Show that the sequence $\{(n+1)^{-1}\sum_{r=0}^n T^r\}$ converges in the strong operator topology if and only if the fixed points of T separate the fixed points of $T^* \in B(X')$.

8. (An ergodic mixing theorem). Let S be a left amenable semigroup with a continuous left action on a compact Hausdorff space X. (So $C(X)$ is a right Banach S-space and $l_\infty(S)'$ has its "Arens product" left action on$M(X) = C(X)'$ as in (2.5), (2.6).) Let Y be the carrier space of $C(X)''$ so that $M(X)'' = M(Y)$. For $m \in \mathfrak{L}(S)$ define $P_m: M(X) \to M(X)$ by $P_m\nu = m\nu$. Let $Q_m = P_m^{**}: M(Y) \to M(Y)$.

(i) Show that Q_m is a positive, linear, weak* continuous projection with $||Q_m(\xi)|| = ||\xi||$ for $\xi \geq 0$ in $M(Y)$. Show also that $F = Q_m(M(Y))$ is weak* closed in $M(Y)$, and is the same set for all $m \in \mathfrak{L}(S)$.

(ii) Show that if $\xi \in F$, $\eta \in M(Y)$, and $0 \leq \eta \leq \xi$, then $Q_m(\eta) = Q_n(\eta)$ for all $m, n \in \mathfrak{L}(S)$.

(iii) Show that if $\mu \in \mathrm{PM}(X)$ is S-invariant (Problem 1-11) and $\phi, \psi \in C(X)$, then the function $s \to \int_X \phi(x)\psi(sx)\, d\mu(x)$ is left almost convergent on S (Problem 4-15).

CHAPTER 6

Locally Compact Groups of Polynomial Growth

(6.0) Introduction. In this chapter, we shall investigate the properties of a remarkable class of amenable, locally compact groups. Consider the following properties that such a group G might possess:

(i) *G has polynomial growth*, that is, if $C \in \mathscr{C}_e(G)$, then there exists a real polynomial p such that for all $n \geq 1$, $\lambda(C^n) \leq p(n)$. (Note that we can take $p(n)$ to be of the form kn^r for some $k > 0$, $r \in \mathbb{N}$.)

(ii) *G is exponentially bounded*, that is, if $C \in \mathscr{C}_e(G)$, then $\lambda(C^n)^{1/n} \to 1$ as $n \to \infty$.

(iii) *G is of Type R*, that is, there exists a compact, normal subgroup K of G such that G/K is a Lie group with $\mathrm{Sp}(\mathrm{ad}\, X) \subset i\mathbb{R}$ for all X in the Lie algebra of G/K.

(iv) *G does not contain a free, uniformly discrete semigroup in two generators*, that is, there is no subsemigroup T of G which is a copy of FS_2 (Problem 0-28) and is such that for some $U \in \mathscr{C}_e(G)$, $sU \cap tU = \varnothing$ whenever s, $t \in T$, $s \neq t$.

(v) *$\mathfrak{L}(U(S)) \neq \varnothing$ for every open subsemigroup S of G* (where $U(S)$ is the space of bounded, uniformly continuous, complex-valued functions on S).

We shall prove the theorem ((6.39)) that when G is connected, the first four of the above conditions are all equivalent, and that each implies (v). (All five properties are equivalent if G is assumed to be a connected, solvable, Lie group.)

The class of groups satisfying the "growth" conditions (i) and (ii) is large and includes all nilpotent and many solvable groups. (Complete information about growth in discrete nilpotent groups is given in (6.17).)

Every group G of polynomial growth is amenable, and the amenability properties of such a group are strikingly sharp. Consider, for example, the property (iv). There are amenable groups (such as the "$ax + b$" group) which contain FS_2 as a subsemigroup. Which amenable (discrete) groups do *not* contain FS_2 as a subsemigroup? The equivalence of (i) and (iv) shows that in the connected case the answer to the topological version of this question is *groups of polynomial growth.* Property (v) is concerned with those locally compact groups whose open subsemigroups are (in a suitable sense) left amenable. Further, a group G of polynomial growth satisfies a very strong Følner condition: (6.8) asserts that if $C \in \mathscr{C}_e(G)$, then the set K of the Følner condition (4.13(ii)) can be taken to be one of the sets C^N (see also (6.43)). A nonzero fixed-point theorem is discussed briefly in (6.44). Groups of polynomial growth have a particularly strong "weak containment" property ((4.20))—see Problem 6-25.

The scope of the theory presented in this chapter is wide. There are, for example, applications to Riemannian manifolds, the representation theory of connected, solvable Lie groups, the theory of Poisson spaces, a $C^{(r)}$-functional calculus for $L_1(G)$, and the much studied symmetry problem for $L_1(G)$.

We shall be primarily concerned with the study of polynomial growth for connected locally compact groups. The corresponding theory for discrete groups is discussed in (6.40). However, the polynomial growth of discrete nilpotent groups is examined in detail in (6.10)–(6.17).

We start by establishing some elementary results on polynomial growth and exponential boundedness. Throughout the chapter, G will be a locally compact group with left Haar measure λ.

For the next result, see Milnor [**1**], Guivarc'h [**1**].

(6.1) PROPOSITION. *Let $C \in \mathscr{C}_e(G)$ be such that $\bigcup_{n=1}^{\infty} C^n = G$. Then $\lim_{n\to\infty} \lambda(C^n)^{1/n}$ exists and is greater than or equal to* 1.

PROOF. For $m, n \geq 1$,

$$
\begin{aligned}
\lambda(C^{m+1})\lambda(C^{-1}C^n) &= \int_G \chi_{C^{m+1}}(z)\lambda(zC^{-1}C^n)\, d\lambda(z) \\
&= \int_G \chi_{C^{m+1}}(z)\, d\lambda(z) \int_G \chi_{C^{-1}C^n}(z^{-1}x)\, d\lambda(x) \\
&= \int_G d\lambda(x) \int_G \chi_{C^{m+1}}(z)\chi_{C^{-1}C^n}(z^{-1}x)\, d\lambda(z) \\
&\geq \int_{C^{m+n}} \lambda(C^{m+1} \cap xC^{-n}C)\, d\lambda(x).
\end{aligned} \tag{1}
$$

Now if $x \in C^{m+n}$, then $x = c_1c_2\cdots c_{m+n}$ ($c_i \in C$) and

$$C^{m+1} \cap xC^{-n}C \supset c_1 \cdots c_m C \cap (c_1 \cdots c_{m+n})(c_{m+n}^{-1} \cdots c_{m+1}^{-1})C = (c_1 \cdots c_m)C.$$

Hence $\lambda(C^{m+1} \cap xC^{-n}C) \geq \lambda(C)$, and from (1),

$$\lambda(C^{m+1})\lambda(C^{-1}C^n) \geq \lambda(C^{m+n})\lambda(C). \tag{2}$$

Since $C \in \mathscr{C}_e(G)$ and $\bigcup_{n=1}^{\infty} C^n = G$, there exists $p \geq 1$ such that $C^{-1} \subset C^p$. So from (2),

$$\lambda(C^{m+n}) \leq (\lambda(C))^{-1}\lambda(C^{m+1})\lambda(C^{n+p}). \tag{3}$$

Substituting $n = 2$ in (3) yields

$$\lambda(C^{m+2}) \leq (\lambda(C))^{-1}\lambda(C^{m+1})\lambda(C^{2+p}) \tag{4}$$

and substituting $n = p + 1$, $m = r - 1$ ($r \geq 2$) in (3) yields

$$\lambda(C^{r+p}) \leq (\lambda(C))^{-1}\lambda(C^r)\lambda(C^{2p+1}). \tag{5}$$

Using (3), (4), and (5), we have for $m,\ n \geq 2$,

$$\begin{aligned}\lambda(C^{m+n}) &\leq (\lambda(C))^{-1}\lambda(C^{m+1})\lambda(C^{n+p}) \leq (\lambda(C))^{-2}\lambda(C^m)\lambda(C^{2+p})\lambda(C^{n+p})\\ &\leq [(\lambda(C))^{-3}\lambda(C^{2+p})\lambda(C^{2p+1})]\lambda(C^m)\lambda(C^n).\end{aligned}$$

Setting $\gamma(m) = \log(\lambda(C^m))$, we see that there exists $\alpha \in \mathbb{R}$ such that

$$\gamma(m+n) \leq \alpha + \gamma(m) + \gamma(n) \qquad (m, n \geq 2).$$

To establish the existence of $\lim_{n\to\infty} \lambda(C^n)^{1/n}$, it is sufficient to show that $\{\gamma(n)/n\}$ converges. Let $\delta(n) = \gamma(n+1)$. Then since $\lambda(C^r) \leq \lambda(C^{r+1})$,

(6) $\delta(m+n) = \gamma(m+n+1) \leq \gamma(m+n+2) \leq \alpha+\delta(m)+\delta(n) \qquad (m, n \geq 1).$

It is clearly sufficient to show that $\{\delta(n)/n\}$ converges (c.f. [**DS**], VIII.1.4).

Let $N \geq 1$, and for $n \geq 1$, write $n = q_n N + r_n$, where $q_n \in \mathbb{N}$ and $0 \leq r_n < N$. From (6) with $n \geq N$,

$$\begin{aligned}\delta(n) &\leq \alpha + \delta(q_n N) + \delta(r_n) \leq \alpha + [(q_n - 1)\alpha + q_n\delta(N)] + \delta(r_n)\\ &= q_n\alpha + q_n\delta(N) + \delta(r_n).\end{aligned}$$

So $\delta(n)/n \leq (q_n/n)[\alpha+\delta(N)]+\delta(r_n)/n$, and since $q_n/n \to 1/N$ and $\delta(r_n)/n \to 0$ as $n \to \infty$, we have

$$\limsup_{n\to\infty}(\delta(n)/n) \leq \delta(N)/N + \alpha/N.$$

So

$$\limsup_{n\to\infty}(\delta(n)/n) \leq \liminf_{n\to\infty}(\delta(n)/n).$$

It follows that $\{\delta(n)/n\}$, and hence $\{\lambda(C^n)^{1/n}\}$, converge. Since $\lambda(C^n)^{1/n} \geq \lambda(C)^{1/n}$, we have $\lim_{n\to\infty} \lambda(C^n)^{1/n} \geq 1$. □

Note that if G is exponentially bounded and $D \in \mathscr{C}(G)$ is nonnull, then, by putting D inside some $C \in \mathscr{C}_e(G)$, it readily follows that $\lim_{n\to\infty}(\lambda(D^n))^{1/n}$ exists and equals 1.

If there exists a nonnull, compact subset C of G such that $\lim_{n\to\infty} \lambda(C^n)^{1/n} > 1$, then G is said to have *exponential growth*. It is easy to see that either G is exponentially bounded or has exponential growth. (Exponential boundedness and polynomial growth are defined in (6.0).) The proof of the next proposition is trivial.

(6.2) PROPOSITION. *Every compact group has polynomial growth.*

(6.3) PROPOSITION. *Suppose that C_1, $C_2 \in \mathscr{C}_e(G)$ are such that*

$$\bigcup_{n=1}^{\infty} C_1^n = G = \bigcup_{n=1}^{\infty} C_2^n.$$

Let $r \in \mathbb{N}$. Then $\{\lambda(C_1^n)/n^r\}$ is bounded if and only if $\{\lambda(C_2^n)/n^r\}$ is bounded.

PROOF. Find $p, q \in \mathbb{P}$ such that $C_1 \subset C_2^p$, $C_2 \subset C_1^q$. Then $\lambda(C_1^n) \leq \lambda(C_2^{nq})$, $\lambda(C_2^n) \leq \lambda(C_1^{nq})$, and the desired result follows. □

The preceding proposition justifies the following definition.

(6.4) DEFINITION. *The* degree *of a compactly generated, locally compact group G of polynomial growth is the smallest integer $r \geq 0$ such that for each $C \in \mathscr{C}_e(G)$ with $\bigcup_{n=1}^{\infty} C^n = G$, there exists $k_C > 0$ such that for all n, $\lambda(C^n) \leq k_C n^r$.*

(6.5) PROPOSITION. *If G is connected, then G is exponentially bounded [has polynomial growth] if and only if there exists $C_0 \in \mathscr{C}_e(G)$ such that $\lambda(C_0^n)^{1/n} \to 1$ as $n \to \infty$ [for some polynomial p, $\lambda(C_0^n) \leq p(n)$ for all $n \geq 1$].*

PROOF. The proof of the case involving polynomial growth follows from (6.3). The other case is dealt with by a simple modification of the proof of (6.3). □

(6.6) PROPOSITION. *If G has polynomial growth, then G is exponentially bounded.*

PROOF. Let $C \in \mathscr{C}_e(G)$ and G have polynomial growth. Then for some $k > 0$ and $r \in \mathbb{P}$, $\lambda(C^n) \leq kn^r$, and so

$$1 = \lim_{n\to\infty} (\lambda(C))^{1/n} \leq \lim_{n\to\infty} (\lambda(C^n))^{1/n} \leq \lim_{n\to\infty} (kn^r)^{1/n} = 1. \quad \square$$

(6.7) PROPOSITION. *Let G be exponentially bounded [have polynomial growth] as a discrete group. Then G is exponentially bounded [has polynomial growth].*

PROOF. Let $C \in \mathscr{C}_e(G)$. A simple compactness argument shows that there exists a finite subset F of G such that $C^2 \subset FC$. The desired result follows from the inequalities

$$\lambda(C^n) \leq \lambda(F^{n-1}C) \leq |F^{n-1}|\lambda(C). \quad \square$$

The preceding proof is due to Hulanicki [**6**], who attributes it to Greenleaf and Emerson. An improvement of the next result for certain groups is given in (6.43). A reference for the next result is Glasner [**S**].

(6.8) PROPOSITION. *Let G be exponentially bounded and $C \in \mathscr{C}_e(G)$, $\varepsilon > 0$. Then there exists arbitrarily large n such that $\lambda(C^{n+1}\Delta C^n)/\lambda(C^n) < \varepsilon$. Further, G is amenable.*

PROOF. From an elementary result on convergent sequences (Rudin [**5**, p. 68]),

$$1 \leq \liminf_{n\to\infty}(\lambda(C^{n+1})/\lambda(C^n)) \leq \liminf_{n\to\infty}(\lambda(C^n)^{1/n}) = 1.$$

Now observe that $\lambda(C^{n+1})/\lambda(C^n) = 1 + (\lambda(C \cdot C^n \,\Delta\, C^n)/\lambda(C^n))$, and use (4.13). □

(6.9) PROPOSITION. *Every exponentially bounded group is unimodular.*

PROOF. Let G be exponentially bounded and $x \in G$, $C \in \mathscr{C}_e(G)$ with $x \in C$. Then for all $n \geq 1$,

$$\lambda(C)\Delta(x)^{n-1} = \lambda(Cx^{n-1}) \leq \lambda(C^n),$$

where Δ is the modular function on G, and taking nth roots and letting $n \to \infty$, we have $\Delta(x) \leq 1$. Hence G is unimodular. □

Our next aim is to show that every nilpotent group has polynomial growth, and to determine the degree ((6.4)) of a finitely generated nilpotent group. The arguments involved in this determination can be regarded as an apotheosis of induction! References for the results are Wolf [**1**], Bass [**1**], Tits [**4**], Guivarc'h [**1**], and Hulanicki [**3**].

Our account is an exegesis of Tits [**4**].

(6.10) Preliminaries. *Throughout* (6.10)–(6.17), *G will be a finitely generated nilpotent group.* For $x, y \in G$, we define the commutator $[x, y]$ to be $x^{-1}y^{-1}xy$. The *lower central series*

$$G = G_1 \rhd G_2 \rhd \cdots \rhd G_c \rhd G_{c+1} = \{e\}$$

is defined by setting $G_{i+1} = [G_i, G]$, the subgroup generated by elements of the form $[x, y]$, $x \in G_i$, $y \in G$. So in the notation of Appendix A, $G_i = C^iG$. When $G = \{e\}$ we take $c = 0$. Otherwise, we require $G_c \neq \{e\}$. So c is the *class* of G. Whenever convenient in the sequel, we can suppose that $c > 0$. Recall that $[G_i, G_j] \subset G_{i+j}$.

A generating subset for G is a subset E of G such that $\bigcup_{r=1}^{\infty}(E \cup E^{-1})^r = G$, and in such a case, E is said to generate G. We choose a finite, generating subset E for G so that for each i $(1 \leq i \leq c+1)$, $E_i = E \cap G_i$ generates G_i. (We take $E_{c+1} = \varnothing$. It follows from Appendix A that each G_i is finitely generated.) Let $F_i = E \sim E_{i+1}$ $(0 \leq i \leq c)$. Since $E_{i+1} \subset E_i$, it follows that $F_i \subset F_{i+1}$. Note that $F_0 = \varnothing$.

For each word $w = (w_1, \ldots, w_n)$ in $E \cup E^{-1}$, let $\alpha(w) = w_1 w_2 \cdots w_n \in G$ and $n_i(w)$ be the number of w_j's in $F_i \cup F_i^{-1}$. Observe that every element of G is of the form $\alpha(w)$, and that $n_i(w) \leq n_{i+1}(w)$. For $r = (r_1, \ldots, r_c)$, where $r_i \geq 0$, let

$$A_r^E = \{\alpha(w) \colon n_i(w) \leq r_i \ (1 \leq i \leq c)\}.$$

For each $k \in \mathbb{P}$, define $A^E(k)$ to be A_r^E, where $r_i = k^i$. The following arguments depend on using $A^E(k)$ rather than the related, more natural, set $(E \cup E^{-1})^k$.

If N is a finitely generated abelian group, then, from the structure theorem for such groups, we can write $H = \mathbb{Z}^k \oplus F$, where $k \in \mathbb{N}$ and F is a finite (abelian) group. The integer k is an invariant of H and is called the *rank* of H. The rank of H is denoted by $r'(H)$. Further, if L is a normal subgroup of H, then [**M**, Chapter 5] L is also finitely generated, and we have

$$r'(H) = r'(L) + r'(H/L). \tag{1}$$

In particular, each of the quotient groups G_i/G_{i+1} is a finitely generated, abelian group. Let $d_i = r'(G_i/G_{i+1})$ and $d = \sum_{i=1}^{c} id_i$. We shall prove in (6.17) that *d is the degree of G.*

(6.11) LEMMA. (i) *For each $k \in \mathbb{P}$, $(E \cup E^{-1})^k \subset A^E(k)$.*

(ii) *Let E' be a finite, generating subset of G with $E_i' = E' \cap G_i$ generating G_i $(1 \leq i \leq c)$ and $E_{c+1}' = \varnothing$. Let $F_i', n_i'(w'), \alpha'(w'), A_r^{E'}$ be defined with respect to*

E' in the same way as $F_i, n_i(w), \alpha(w), A_r^E$ were defined with respect to E. Then there exist $a, b > 0$ such that for all $r = (r_1, \ldots, r_c)$, we have

$$(1) \qquad |A_{ar}^{E'}| \le |A_r^E| \le |A_{br}^{E'}|$$

where $ar = (ar_1, \ldots, ar_c)$, $br = (br_1, \ldots, br_c)$.

PROOF. (i) Let $x \in (E \cup E^{-1})^k$. Then $x = \alpha(w)$, where $w = \{w_1, \ldots, w_r\}$ is a word in $E \cup E^{-1}$. Clearly, $n_i(w) \le k \le k^i$, so that (i) follows.

(ii) For $1 \le i \le c$, let $H_i = (F_i \sim F_{i-1}) \cup (F_i \sim F_{i-1})^{-1}$ and $H_i' = (F_i' \sim F_{i-1}') \cup (F_i' \sim F_{i-1}')^{-1}$. Using a simple recursion argument, the fact that the canonical image of each of the (finite) sets H_i' $(= (E_i' \sim E_{i+1}') \cup (E_i' \sim E_{i+1}')^{-1})$ generates the group G_i/G_{i+1}, and arguing as in the proof of (6.3), we can find finite subsets C_i of G_i and integers $m_i > 0$ such that

$$(2) \qquad C_1 = \varnothing, \qquad H_i \cup C_i \subset (H_i')^{m_i} C_{i+1}.$$

Let $w = (w_1, \ldots, w_n)$ be a word in $E \cup E^{-1}$ and $x = \alpha(w)$. We construct a word w' in $E' \cup (E')^{-1}$ with $x = \alpha'(w')$ as follows. Let $p_i(w)$ be the number of w_j in H_i. Replace each of the $p_1(w)$ elements in the expansion $x = w_1 w_2 \cdots w_n$ by an element of $(H_1')^{m_1} C_2$. This produces a new product expression for x, involving $m_1 p_1(w)$ elements from H_1', $p_1(w)$ elements from C_2, and $p_i(w)$ elements from H_i $(2 \le i \le c)$. The same procedure enables us to replace the elements of $H_2 \cup C_2$ by elements of $(H_2')^{m_2} C_3$. Repeating this process, we obtain a word w' in $E' \cup (E')^{-1}$ with $x = \alpha(w')$ and, in an obvious notation,

$$\begin{aligned} p_1'(w') &= m_1 p_1(w), \\ p_2'(w') &= m_2(p_2(w) + p_1(w)), \\ &\ \vdots \qquad \vdots \\ p_c'(w') &= m_c(p_c(w) + \cdots + p_1(w)). \end{aligned}$$

Let $m = \max\{m_i \colon 1 \le i \le c\}$. Then

$$n_i'(w') = \sum_{r=1}^{i} p_r'(w') \le mc \left(\sum_{r=1}^{i} p_r(w) \right) = mcn_i(w).$$

It follows that $|A_r^E| \le |A_{br}^{E'}|$, where $b = mc$. Reversing the roles of E and E' gives (1). $\square$

(6.12) LEMMA. *Let E be such that*

(i) $|F_1|$ is smallest possible;

(ii) $[u, v] \in E$ whenever $u, v \in E \cup E^{-1}$.

Let $y \in F_1$ and $x \in A^E(k)$. Then there exists $s \in \mathbb{Z}$ with $|s| \le k$ and a word w, with $w_i \in (E \sim \{y\}) \cup (E \sim \{y\})^{-1}$ and $n_i(w) \le (c+1)k^i - |s|$ $(1 \le i \le c)$, such that

$$(1) \qquad x = y^s \alpha(w).$$

PROOF. Noting that every commutator $[u, v]$ in G is contained in $G_2 \subset G \sim F_1$, it is easily checked that E can be chosen so that (i) and (ii) are satisfied.

Note that, using (i), $y^{-1} \notin E$, so that the word w in (1) does not involve y or y^{-1}.

Let $x = \alpha(w')$, where $w' = (w'_1, \ldots, w'_N)$ is a word in $E \cup E^{-1}$ and $n_i(w') \leq k^i$. Let $w'_{i_1}, \ldots, w'_{i_q}$ $(i_1 < i_2 < \cdots < i_q)$ be those w_j's which are either y or y^{-1}. Using the equality

$$(2) \qquad w'_j w'_{i_l} = w'_{i_l} w'_j [w'_j, w'_{i_l}]$$

we bring the elements $w'_{i_1}, \ldots, w'_{i_q}$ to the left, one at a time, in the expression for x. Note that by (ii), $[w'_j, w'_{i_l}]$ in (2) belongs to E. When $w'_{i_1}, \ldots, w'_{i_p}$ $(1 \leq p \leq q)$ have been brought to the left, we have

$$x = w'_{i_1} \cdots w'_{i_p} \alpha(w^p),$$

where w^p is some word in $E \cup E^{-1}$. Take $w^0 = w'$. We now prove by induction on p that for all $i \leq c$,

$$(3) \qquad n_i(w^p) \leq \sum_{j=0}^{p} \binom{p}{j} n_{i-j},$$

where $n_l = n_l(w^0)$ $(l \geq 1)$ and $n_l = 0$ for $l < 1$. The result is obviously true when $p = 0$. Suppose that it is true for $p = m$, $0 \leq m < q$. Applying (2) with $l = (m+1)$, we see that if $w'_j \in F_i \cup F_i^{-1}$, then w'_j still appears on the right-hand side of (2), while a new element $[w'_j, w'_{i_l}]$ appears. Now if $w'_j \in E_k$, then $z = [w'_j, w'_{i_l}] \in E_{k+1}$. Hence if $k \geq i$, then $z \notin F_i$. So if $z \in F_i$, then $w'_j \in F_{i-1}$. Recalling that $n_{i-1}(w^m) = 0$ if $i \leq 1$, we have

$$\begin{aligned} n_i(w^{m+1}) \leq n_i(w^m) + n_{i-1}(w^m) &\leq \sum_{j=0}^{m} \binom{m}{j} n_{i-j} + \sum_{j=0}^{m} \binom{m}{j} n_{i-1-j} \\ &= n_i + \sum_{j=1}^{m} \left(\binom{m}{j} + \binom{m}{j-1} \right) n_{i-j} + n_{i-(m+1)} \\ &= \sum_{j=0}^{m+1} \binom{m+1}{j} n_{i-j}. \end{aligned}$$

Thus (3) is established.

Putting $p = q$ in (3) and setting $w = w^q$, we have $x = y^s \alpha(w)$, where $|s| \leq q \leq n_1 \leq k$ (since $x \in A^E(k)$). Clearly $w_i \in (E \sim \{y\}) \cup (E \sim \{y\})^{-1}$. We now use (3) with $p = q$. If $i \leq q$, then $n_i(w) \leq \sum_{j=0}^{i-1} \binom{q}{j} n_{i-j} \leq \sum_{j=0}^{i-1} q^j k^{i-j} \leq \sum_{j=0}^{i-1} k^j k^{i-j} = ik^i = (i+1)k^i - k^i \leq (c+1)k^i - |s|$. If $i > q$, then $n_i(w) \leq \sum_{j=0}^{q} \binom{q}{j} n_{i-j} \leq (q+1)k^i \leq ik^i \leq (c+1)k^i - |s|$. This completes the proof. □

(6.13) LEMMA. *Let H be a subgroup of G of class l and such that $H \supset G_2$. Let $\{H_i\}$ be the lower central series for H and $s_i = r'(H_i/H_{i+1})$. Then, in the notation of* (6.10),

$$(1) \qquad \sum_{i=1}^{l} i s_i \leq d - r'(G/H).$$

PROOF. Note that H is normal in G since G/G_2 is abelian. Since $H \subset G$, we have $H_i \subset G_i$. Hence $l \leq c$. We can obviously replace l by c in (1), since $s_i = 0$ for $i > l$.

We establish (1) by induction on c. The result is trivial if $c = 0$. Suppose that $N \in \mathbb{P}$ and the result is true for groups of class $< N$. Suppose that $c = N$ and let $G' = G/G_c$. Applying the induction hypothesis to G', we have

$$\sum_{i=1}^{c-1} id_i = \sum_{i=1}^{c-1} ir'(G'_i/G'_{i+1}) \geq \sum_{i=1}^{c-1} is'_i + r'(G/H), \tag{2}$$

where $s'_i = r'((H_iG_c)/(H_{i+1}G_c))$. Now

$$(H_iG_c)/(H_{i+1}G_c) = (H_{i+1}G_c)H_i/(H_{i+1}G_c) = H_i/(H_i \cap (H_{i+1}G_c)).$$

For $1 \leq i \leq (c-1)$, we have, using (6.10(1)),

$$\begin{aligned} s_i &= r'(H_i/(H_i \cap (H_{i+1}G_c))) + r'((H_i \cap (H_{i+1}G_c))/H_{i+1}) \\ &= s'_i + r'(H_{i+1}(H_i \cap G_c)/H_{i+1}) \\ &= s'_i + r'((H_i \cap G_c)/(H_{i+1} \cap G_c)). \end{aligned}$$

So using (2) and (6.10(1)) again,

$$\begin{aligned} \sum_{i=1}^{c} is_i + r'(G/H) &= \sum_{i=1}^{c-1} is'_i + r'(G/H) \\ &\quad + \sum_{i=1}^{c-1} ir'((H_i \cap G_c)/(H_{i+1} \cap G_c)) + cr'(H_c) \\ &\leq \sum_{i=1}^{c-1} id_i + \sum_{i=1}^{c} r'(H_i \cap G_c) \\ &\leq \sum_{i=1}^{c-1} id_i + cr'(G_c) = d. \quad \square \end{aligned}$$

(6.14) LEMMA. *There exists $k_0 > 0$ such that, for all $k \geq 1$, $|A^E(k)| \leq k_0(k^d)$.*

PROOF. The proof involves an induction argument within an induction argument. The result is trivally true if c (the class of G) is 0. Let $p \geq 0$ and suppose that the result is true for all groups of class $\leq p$. Suppose that $c = (p+1)$. By (6.11(1)), E can be chosen as in (6.12).

Let $\mathscr{S}$ be the set of subgroups H of G generated by a subset of the form $F_H \cup E_2$, where $F_H \subset F_1$. For such a subgroup H, let $m(H) = |F_H|$. Note that $H \supset G_2$, and that G_2 and G belong to $\mathscr{S}$. We now prove, by induction on $m(H)$, that the result is true for all $H \in \mathscr{S}$ (and in particular, for G itself).

When $m(H) = 0$, then $H = G_2$, and since G_2 is of class $< (p+1)$, the result is true for G_2. Now let $0 \leq m < m(G)$, and suppose that the result is true for all H such that $m(H) \leq m$. Let $H \in \mathscr{S}$ be such that $m(H) = (m+1)$ and let $\Delta = F_H \cup E_2$. Let $y \in F_H$ and H' be the subgroup of G generated by

$\Delta' = (F_H \sim \{y\}) \cup E_2$. Then $H' \in \mathcal{S}$ and $m(H') \leq m$. By hypothesis and (6.13) (with H, H' in place of G, H), there exists $k_1 > 0$ such that for $k \geq 1$,

$$|A^{\Delta'}(k)| \leq k_1 k^{(d_H - r(H/H'))}, \tag{1}$$

where $d_H = \sum_{i=1}^{l} ir'(H_i/H_{i+1})$ with l the class of H. There are now two cases to be considered.

(i) Suppose that H/H' is infinite. Then $r'(H/H') = 1$, and from (1)

$$|A^{\Delta'}(k)| \leq k_1 k^{(d_H - 1)}. \tag{2}$$

By (6.12), every element x of $A^{\Delta}(k)$ is of the form $y^s x'$, where $x' = \alpha(w)$ with $w_i \in \Delta' \cup (\Delta')^{-1}$, $n_i(w) \leq (l+1)k^i - |s| \leq ((l+1)k)^i$, and $|s| \leq k$. Since there are $(2k+1)$ possibilities for s and as $w \in A^{\Delta'}((l+1)k)$, we have, using (2),

$$|A^{\Delta}(k)| \leq k_1(2k+1)((l+1)k)^{d_H - 1} \leq k_2(k^{d_H})$$

for some k_2 independent of k.

(ii) Suppose now that $|H/H'| = t < \infty$. So $r'(H/H') = 0$. Then there exists a smallest positive integer p, with $1 < p \leq t$, such that $y^p \in H'$. Using (6.11(1)), we can suppose that (1) is true with Δ' replaced by $\Delta'' = \Delta' \cup \{y^p\}$. Let $x \in A^{\Delta}(k)$ and write $x = y^s x'$, $x' = \alpha(w)$ as in (6.12). Now set $s = np + s_1$, where $n \in \mathbb{Z}$ and $0 \leq s_1 < p \leq t$. So

$$x = y^{s_1}((y^p)^n x'). \tag{3}$$

Clearly, $(y^p)^n x' = \alpha(v)$, where $v = (v_1, \dots, v_q)$ is a word in $\Delta'' \cup (\Delta'')^{-1}$, and $n_i(v) \leq (l+1)k^i - |s| + |n| \leq (l+1)k^i$, since $(|s| - |n|) \geq 0$. Hence $\alpha(v) \in A^{\Delta''}((l+1)k)$, and from (1), there exists $k' > 0$ such that

$$|A^{\Delta''}((l+1)k)| \leq k'(k^{d_H}) \qquad (k \geq 1).$$

Using (3), $|A^{\Delta}(k)| \leq pk'(k^{d_H})$, and the proof is complete. □

For $z \in G$, let $l_{G,E}(z)$ be the smallest integer s such that for some $w = (w_1, \dots, w_s)$, $w_i \in E \cup E^{-1}$, we have $z = \alpha(w)$.

(6.15) LEMMA. *If $z \in G_c$, then there exists $k' > 0$ such that*

$$l_{G,E}(z^n) \leq k'(n^{1/c}) \qquad (n \geq 1). \tag{1}$$

PROOF. The proof proceeds by induction on c. The case $c = 1$ is almost trivial since then G is abelian. Suppose that $p \geq 1$ and the result is true for all groups H of class p. Let G have class $c = (p+1)$. As $[G, G_p] = G_{p+1}$, we can write

$$z = [x_1, y_1]^{\varepsilon_1}[x_2, y_2]^{\varepsilon_2} \cdots [x_r, y_r]^{\varepsilon_r},$$

where $x_i \in G$, $y_i \in G_p$, and $\varepsilon_i \in \{-1, 1\}$. Let $z_i = [x_i, y_i]^{\varepsilon_i}$. Since G_{p+1} is abelian, we have $z^n = z_1^n z_2^n \cdots z_r^n$. Obviously, $l_{G,E}(z_i^n) = l_{G,E}(z_i^{-n})$, and $l_{G,E}(z_1^n \cdots z_r^n) \leq \sum_{i=1}^{r} l_{G,E}(z_i^n)$. Thus to establish (1), we can suppose $z = [x, y]$ for some $x \in G$, $y \in G_p$. Now it is readily checked that if $x \in (E \cup E^{-1})^m$, then

$$l_{G,E}(u) \leq m l_{G,E \cup \{x\}}(u) \qquad (u \in G).$$

Since E generates G, we can suppose that $x \in E$. For each $n \in \mathbb{P}$, find $n_1 \in \mathbb{P}$ such that

$$(2) \qquad n^{1/(p+1)} < n_1 \leq n^{1/(p+1)} + 1$$

and write $n = a_1 n_1^p + a_2$, where $a_1, a_2 \in \mathbb{N}$, $a_2 < n_1^p$. It follows from (2) that $n < n_1^{(p+1)} = n_1 \cdot n_1^p$, so that $a_1 < n_1$. Since $n \geq (n_1 - 1)^{p+1} = n_1^{p+1}(1 - (n_1)^{-1})^{p+1}$ and $n_1 \to \infty$ as $n \to \infty$, we see that $a_1 > 0$ eventually. For the purposes of (1), we can suppose that $a_1 > 0$. Now $H = G/G_{p+1}$ is nilpotent of class p, so that by hypothesis, there exists $k_1 > 0$ (independent of n) such that if y' $[E']$ is the image of y $[E]$ in H, then

$$l_{H,E'}((y')^{n_1^p}) \leq k_1 n_1,$$

$$l_{H,E'}((y')^{a_2}) \leq k_1 (a_2)^{1/p} \leq k_1 n_1.$$

So there exist u_1, $u_2 \in G_{p+1}$, y_1, $y_2 \in G_p$ such that $y^{n_1^p} = u_1 y_1$, $y^{a_2} = u_2 y_2$, and

$$(3) \qquad l_{G,E}(y_1) \leq k_1 n_1, \qquad l_{G,E}(y_2) < k_1 n_1.$$

Now for $v \in G$, $[v, y] \in G_{p+1} \subset Z(G)$, and so, for all $n \geq 1$,

$$\begin{aligned}
[x^n, y][x, y] &= (x^{-n} y^{-1} x^n y)(x^{-1} y^{-1} x y) = x^{-n}(y^{-1} x^n y x^{-n})(x^{n-1} y^{-1} x y) \\
&= x^{-n}(x^{n-1} y^{-1} x y)(y^{-1} x^n y x^{-n}) = x^{-1} y^{-1} x^{n+1} y x^{-n} \\
&= x^{-1}(y^{-1} x^{n+1} y x^{-n-1}) x = x^{-1} x (y^{-1} x^{n+1} y x^{-n-1}) \\
&= y^{-1}(x^{n+1} y x^{-n-1} y^{-1}) y = x^{n+1} [y^{-1}, x^{n+1}] x^{-n-1} \\
&= y [x^{n+1}, y] y^{-1} = [x^{n+1}, y].
\end{aligned}$$

Thus for all $m \geq 1$, $[x, y]^m = [x^m, y]$. Similarly, we also have $[x, y]^m = [x, y^m]$. More generally, it readily follows that $[x, y]^{m_1 m_2} = [x^{m_1}, y^{m_2}]$ for all $m_1, m_2 \geq 1$.

Recalling that $y^{n_1^p} = u_1 y_1$, $y^{a_2} = u_2 y_2$ for $u_i \in G_{p+1}$, we have, for $a_2 > 0$,

$$\begin{aligned}
z^n = [x, y]^n &= ([x, y]^{a_1})^{n_1^p} [x, y]^{a_2} = [x^{a_1}, y^{n_1^p}][x, y^{a_2}] \\
&= [x^{a_1}, u_1 y_1][x, u_2 y_2] = [x^{a_1}, y_1][x, y_2]
\end{aligned}$$

since $u_1, u_2 \in Z(G)$. Of course, if $a_2 = 0$, then $z^n = [x^{a_1}, y_1]$. Hence, using (3) and (2),

$$\begin{aligned}
l_{G,E}(z^n) &\leq (2a_1 + 2k_1 n_1) + (2 + 2k_1 n_1) \\
&= 2a_1 + 4k_1 n_1 + 2 \\
&\leq 2(1 + 2k_1) n_1 + 2 \\
&\leq 2(1 + 2k_1)(n^{1/(p+1)} + 1) + 2.
\end{aligned}$$

The induction step readily follows. □

The following lemma is left as an exercise to the reader (Problem 6-4).

(6.16) LEMMA. *For $m, r \in \mathbb{P}$, let*

$$A_{m,r} = \left\{ (n_1, \ldots, n_m) \colon n_q \in \mathbb{N},\ \sum_{q=1}^{m} n_q = r \right\}.$$

Then $|A_{m,r}| = \binom{r+m-1}{m-1}$ $\left(= \binom{r+m-1}{r}\right)$.

(6.17) THEOREM. *The nilpotent group G has polynomial growth with degree d.*

PROOF. By (6.14) and (6.11(i)), there exists $k_1 > 0$ such that $|E^k| \le k_1(k^d)$ $(k \ge 1)$. It remains to prove that there exists $k_2 > 0$ such that

$$(1) \qquad |E^k| \ge k_2(k^d) \qquad (k \ge 1).$$

The proof proceeds by induction on c. Trivially, the result is true when $c = 0$. Suppose that the result is true for all groups H of class $m \ge 0$. Let $c = (m+1)$. Let $d' = d - cd_c$, $G' = G/G_c$, $Q\colon G \to G'$ be the quotient map, and $E' = Q(E)$. Then G' is of class m and $r'(G'_i/G'_{i+1}) = r'(G_i/G_{i+1})$ $(1 \le i \le m)$. By hypothesis, there exists $k_3 > 0$ such that

$$(2) \qquad |(E')^k| \ge k_3(k^{d'}) \qquad (k \ge 1).$$

Let $p = d_c$ and $G_c = \mathbb{R}^p \times H$, where H is a finite group, and let $z_1, \ldots, z_p$ be the standard basis for $\mathbb{R}^p$. We can find ((6.15)) $k_4 > 0$ such that

$$(3) \qquad l_{G,E}(z_i^n) \le k_4 n^{1/c} \qquad (n \ge 1, 1 \le i \le p).$$

If, in the notation of (6.16), $(n_1, \ldots, n_p) \in A_{p,n}$ and $T = \{i \colon n_i \ne 0\}$, then, for $c > 1$,

$$l_{G,E}(z_1^{n_1} \cdots z_p^{n_p}) \le \sum_{i \in T} l_{G,E}(z_i^{n_i}) \le k_4 \sum_{i \in T} (n_i^{1/c} \cdot 1)$$
$$\le k_4 \left(\sum_{i \in T} n_i \right)^{1/c} \left(\sum_{i \in T} 1^q \right)^{1/q},$$

where $c^{-1} + q^{-1} = 1$ and Hölder's inequality has been used. So for $c \ge 1$, there exists $k' > 0$ such that

$$(4) \qquad l_{G,E}(z_1^{n_1} \cdots z_p^{n_p}) \le k' n^{1/c} \qquad (n \ge 1).$$

For convenience, we now suppose that $e \in E$. From (4) and (6.16), if $m \ge k' n^{1/c}$, then

$$(5) \qquad |G_c \cap E^m| \ge |\{z_1, \ldots, z_p\}^n| = |A_{p,n}| = \binom{n+p-1}{p} \ge k_5 n^p,$$

where $k_5 > 0$ is independent of n. Replacing n by n^c in (5), we have

$$(6) \qquad |G_c \cap E^m| \ge k_5 n^{cp} \qquad (n \ge 1, m \ge k'n).$$

From (2) and (6), if $N \ge (1 + k')n$, then

$$|E^N| \ge |(E')^n||G_c \cap E^{N-n}| \ge k_3 k_5 n^d.$$

The inequality (1) now readily follows. $\square$

(6.18) COROLLARY. *Every locally compact nilpotent group has polynomial growth. In particular, every abelian, locally compact group has polynomial growth.*

PROOF. Use (6.7). □

(6.19) COROLLARY. *Every abelian, locally compact group has polynomial growth.*

The next three results discuss polynomial growth for subgroups and quotient groups of a locally compact group G.

(6.20) PROPOSITION. *Let H be a closed, normal subgroup of G. Then*

(i) *if H is compact and G/H has polynomial growth, then G has polynomial growth;*

(ii) *if H is discrete and G has polynomial growth, then G/H has polynomial growth.*

PROOF. Let $\pi\colon G \to G/H$ be the quotient map and $V \in \mathscr{C}_e(G)$. By Weil's formula,

$$\text{(1)}\qquad \begin{aligned} \lambda_G(V) &= \int_{G/H} d\lambda_{G/H}(xH) \int_H \chi_V(xh)\, d\lambda_H(h) \\ &= \int_{\pi(V)} \lambda_H(x^{-1}V \cap H)\, d\lambda_{G/H}(xH). \end{aligned}$$

So if H is compact, then $\lambda_G(V) \leq \lambda_H(H)\lambda_{G/H}(\pi(V)) = \lambda_{G/H}(\pi(V))$, and (i) follows.

Now suppose that H is discrete, so that λ_H can be taken to be counting measure. If $x \in VH$, then $\lambda_H(x^{-1}V \cap H) \geq 1$, and so from (1), $\lambda_G(V) \geq \lambda_{G/H}(\pi(V))$. Thus (ii) follows, noting that every element of $\mathscr{C}_e(G/H)$ is of the form $\pi(V)$. □

The next result is due to Hulanicki [**6**].

(6.21) PROPOSITION. *Let H be a closed, normal subgroup of G. Let H have polynomial growth and be such that either G/H is finite or G is separable with G/H compact. Then G has polynomial growth.*

PROOF. Let $Q\colon G \to G/H$ be the quotient map. Since G/H is compact and Q is open, we can find $D \in \mathscr{C}(G)$ with $Q(D) = G/H$. Using Appendix C to deal with the separable case, we can find a Borel subset B of D such that $Q|_B$ is one-to-one and $Q(B) = G/H$. Then $C = B^-$ is compact. Replacing B by $b_0^{-1}B$, where $b_0 \in B$ is such that $Q(b_0)$ is the identity of G/H, we can suppose that $e \in B$.

Let $A \in \mathscr{C}_e(G)$ and $A_0 = H \cap C^{-1}A$. Then $A_0 \in \mathscr{C}_e(H)$, and since $G = BH$,

$$\text{(1)}\qquad A \subset BA_0.$$

We define the following subsets D, D_0, and A_1 of H as follows:

$$D = C^{-1}C^2 \cap H, \qquad D_0 = \bigcup_{x \in C} (x^{-1}A_0x), \qquad A_1 = DD_0.$$

Clearly D_0, and hence A_1, belong to $\mathscr{C}_e(H)$.

Since, by hypothesis, H has polynomial growth, the proposition will be established once we have shown that for all n,

$$(2) \qquad \lambda(A^n) \leq \lambda_H(A_1^n).$$

Since G/H is compact, we can scale λ so that $\lambda_{G/H}(G/H) = 1$. Now if we know that

$$(3) \qquad (BA_0)^n \subset B(A_1)^n,$$

then, using (1) and Weil's formula, we would have (cf. (6.20))

$$\begin{aligned} \lambda(A^n) &\leq \lambda((BA_0)^n) \leq \lambda(B(A_1)^n) \\ &= \int_{G/H} \lambda_H(x^{-1}B(A_1)^n \cap H)\, d\lambda_{G/H}(xH), \end{aligned}$$

and since G/H is compact, and, since, using the properties of B given in the first paragraph of the present proof, for each $x \in G$, $x^{-1}B(A_1)^n \cap H = h_x A_1^n$ for some $h_x \in H$, the inequality (2) would follow. So it suffices to prove (3). We will be content to give the proof of (3) for $n = 3$, the proof for general n being similar.

Let $b_1, b_2, b_3 \in B$ and $a_1, a_2, a_3 \in A_0$. Find $b_4, b_5 \in B$ and $h, h' \in H$ such that

$$b_2 b_3 = b_4 h, \qquad b_1 b_2 b_3 = b_1 b_4 h = b_5 h' h.$$

Then both h, h' belong to $B^{-1}(B^2) \cap H \subset D$. So

$$\begin{aligned} (b_1a_1)(b_2a_2)(b_3a_3) &= (b_1b_2b_3)[(b_2b_3)^{-1}a_1(b_2b_3)][b_3^{-1}a_2b_3]a_3 \\ &= b_5h'h(h^{-1}b_4^{-1}a_1b_4h)(b_3^{-1}a_2b_3)a_3 \\ &= b_5(h'b_4^{-1}a_1b_4)(hb_3^{-1}a_2b_3)(eea_3e^{-1}) \in B(A_1)^3. \quad \square \end{aligned}$$

(6.22) PROPOSITION. (i) *Let G be a discrete group containing a nilpotent subgroup of finite index. Then G has polynomial growth.*

(ii) *Let G be a semidirect product of the form $H \times_\rho K$, where H has polynomial growth and K is compact. Then G has polynomial growth.*

PROOF. (i) It is an elementary fact of group theory that if N is a subgroup of G of finite index, then the normal subgroup $N' = \bigcap\{xNx^{-1} : x \in G\} \subset N$ is also of finite index. Now apply (6.17) and (6.21).

(ii) Use (6.21). (The proof of the latter applies without the separability hypothesis by taking $B = K$.) $\square$

The converse to (6.22(i)) is true when G is finitely generated. This result is much harder to prove than (6.22(i)) and is briefly discussed in (6.40).

A locally compact group of the form $H \times_\rho K$, with H abelian and K compact, is called a *motion group*. (An example of such a group is the isometry group $\mathbb{R}^n \times_\rho \mathrm{O}(n, \mathbb{R})$ of $\mathbb{R}^n$.) It follows from (6.22) and (6.19) that every motion group has polynomial growth.

For the remainder of this section, it will be assumed that G is connected. References for the work below are Jenkins **[6]**, **[9]** and Auslander and Moore **[1]**.

We now examine the properties of Type R groups ((6.0(iii)). The first major result to be proved is ((6.30)) that if G is of Type R, then G has polynomial growth. We start by stating an elementary result from linear algebra; the reader is invited to provide the easy proof. Of course, for a linear transformation T on a complex, finite-dimensional vector space, the spectrum $\mathrm{Sp}(T)$ of T is just the set of eigenvalues of T.

(6.23) LEMMA. *Let W be a linear subspace of a complex, finite-dimensional vector space V, and let $T \in L(V)$ be such that $T(W) \subset W$. Let $T_W = T|_W$ and $T_{V/W} \in L(V/W)$ be given by $T_{V/W}(\xi + W) = T\xi + W$. Then*

$$\mathrm{Sp}(T) = \mathrm{Sp}(T_W) \cup \mathrm{Sp}(T_{V/W}).$$

All B references in the following are references to Appendix B.

A real Lie algebra $\mathfrak{g}$ is said to be of *Type R* if $\mathrm{Sp}(\operatorname{ad} X) \subset i\mathbb{R}$ for all $X \in \mathfrak{g}$.

(6.24) PROPOSITION. *Let G be of Type R, and let L be a compact, normal subgroup of G such that G/L is a Lie group. Then* $\mathrm{Sp}(\operatorname{ad} X) \subset i\mathbb{R}$ *for all X in the Lie algebra $\mathfrak{g}$ of G/L.*

PROOF. By definition, there exists a compact, normal subgroup K of G as in (6.0(iii)). In particular, the Lie algebra $\mathfrak{g}_K$ of G/K is of Type R. Now the Lie algebra of $G/(LK)$ is canonically identified with a quotient of $\mathfrak{g}_K$, and so ((6.23)) is also of Type R. We can thus suppose that L is trivial. In particular, G is now a Lie group, and since the Type R condition is formulated in Lie algebra terms, we can also take K to be connected.

Let $\mathfrak{g}$ and $\mathfrak{k}$ be the Lie algebras of G and K respectively, and let $X \in \mathfrak{g}$. Since G/K is of Type R and has Type R Lie algebra $\mathfrak{g}/\mathfrak{k}$ (B18), it is sufficient to show ((6.23)) that $\mathrm{Sp}(D) \subset i\mathbb{R}$, where $D = (\operatorname{ad} X)_{\mathfrak{k}}$. Now the radical $r(K)$ of K is a connected, compact, normal subgroup of G, and so by B51, B21, $r(K) \subset Z(G)$. Let $\mathfrak{r}$ be the radical of $\mathfrak{k}$. Then $[X, Y] = 0$ for $y \in \mathfrak{r}$, so that $D(\mathfrak{r}) = \{0\}$.

Thus $\mathrm{Sp}(D_{\mathfrak{r}}) = \{0\} \subset i\mathbb{R}$. It remains to show that if $\mathfrak{s} = \mathfrak{k}/\mathfrak{r}$ and $D' = D_{\mathfrak{s}}$, then $\mathrm{Sp}(D') \subset i\mathbb{R}$. Note that $\mathfrak{s}$ is a semisimple, compact Lie algebra. Let B be the Killing form on $\mathfrak{s}$. Then $-B$ is an inner product on $\mathfrak{s}$ with respect to which D is skew symmetric (B49, B39). Hence $\mathrm{Sp}(D') \subset i\mathbb{R}$ and the result follows. □

(6.25) COROLLARY. *If G is a Lie group, then G is of Type R if and only if* $\mathrm{Sp}(\operatorname{ad} X) \subset i\mathbb{R}$ *for all X in the Lie algebra of G.*

The next result relates the possession of purely imaginary eigenvalues by $T \in L(\mathbb{R}^r)$ with the polynomial growth of the norm of e^{tT} $(t \in \mathbb{R})$.

(6.26) PROPOSITION. *Let $r \in \mathbb{P}$ and $T \in L(\mathbb{R}^r)$ be such that* $\mathrm{Sp}(T) \subset i\mathbb{R}$. *Let $\|\cdot\|$ be a norm on $L(\mathbb{R}^r)$. Then there exists an (even) polynomial p such that $\|e^{tT}\| \le p(t)$ for all $t \in \mathbb{R}$.*

PROOF. Extend $\|\cdot\|$ in any way to a norm, also denoted $\|\cdot\|$, on $L(\mathbb{C}^r)$. In the obvious way, we have $T \in L(\mathbb{C}^r)$. Let S and N be respectively the semisimple and nilpotent components of T in $L(\mathbb{C}^r)$ (B31). So $T = S + N$, $SN = NS$, $N^r = 0$, and for some basis for $\mathbb{C}^r$, S can be realised as a diagonal matrix with entries in $i\mathbb{R}$. Using the equivalence of norms on $L(\mathbb{C}^r)$, we can find $C > 0$ such that $\|e^{tS}\| \leq C$ for all $t \in \mathbb{R}$. Then for some $K > 0$,

$$\|e^{tT}\| = \|e^{tS} \cdot e^{tN}\| \leq K\|e^{tS}\|\|e^{tN}\| \leq KC\sum_{s=0}^{r} |t|^s(\|N^s\|/s!) \leq p(t),$$

where p is some even polynomial. □

(6.27) PROPOSITION. *Every simply connected, Type R, solvable Lie group has polynomial growth.*

PROOF. Let G be a simply connected, Type R, solvable Lie group with Lie algebra $\mathfrak{g}$. The proof proceeds by induction on the vector space dimension $\dim \mathfrak{g}$ of $\mathfrak{g}$. If $\dim \mathfrak{g} = 1$, then $G = \mathbb{R}$, and the result is obviously true.

Let $n > 1$ with $\dim \mathfrak{g} = n$ and suppose that the result is true for every group of dimension less than n. Then (B46) we can write G as a semidirect product $H \times_\rho L$, where $L = \mathbb{R}$. Let $\mathfrak{h}$, $\mathfrak{l}$ be the Lie algebras of H and L respectively. If $X \in \mathfrak{h}$, then $\operatorname{Sp}(\operatorname{ad}_\mathfrak{h}(X)) \subset \operatorname{Sp}(\operatorname{ad}_\mathfrak{g}(X)) \subset i\mathbb{R}$, so that H is of Type R and hence, by hypothesis, has polynomial growth. Let $X_0 \in \mathfrak{l} \sim \{0\}$. Then $\mathfrak{l} = \mathbb{R}X_0$. Note that $\mathfrak{h}$ is an ideal of $\mathfrak{g}$ and $\mathfrak{l}$ a Lie subalgebra of $\mathfrak{g}$.

Let $h(t) = \exp(tX_0)$ $(t \in \mathbb{R})$ and $\operatorname{ad} X_0 = \operatorname{ad}_\mathfrak{g} X_0$. Now $\operatorname{Ad}_G(h(t)) = \exp(t \operatorname{ad}(X_0))$ (B12) and $\operatorname{Sp}((\operatorname{ad} X_0)_\mathfrak{h}) \subset i\mathbb{R}$. Identify $\mathfrak{h}$ as a linear space with $\mathbb{R}^{n-1}$ and let $\|\cdot\|_2$ be the Euclidean norm on $\mathfrak{h}$. Applying (6.26) with $T = (\operatorname{ad} X_0)_\mathfrak{h}$ and $r = (n-1)$, there exists a nonzero, even polynomial p, which we can take to have nonnegative integer coefficients, such that

$$\|(\operatorname{Ad}_G(h(t)))_\mathfrak{h}\| \leq p(t) \qquad (t \in \mathbb{R}), \tag{1}$$

where $\|\cdot\|$ is the operator norm on $L(\mathfrak{h})$ corresponding to $\|\cdot\|_2$. Let $W = \{X \in \mathfrak{h} \colon \|X\|_2 \leq 1\}$, $V = \exp W$, and $C = h([-1,1])V$. Then $V \in \mathscr{C}_e(H)$, and using B46, B9(4), it follows that $C \in \mathscr{C}_e(G)$. It remains to find a polynomial q such that $\lambda(C^n) \leq q(n)$ $(n \geq 1)$ ((6.5)).

Let $t_1, t_2 \in [-1,1]$, $X_1, X_2 \in W$, and $v_i = \exp(X_i)$ $(i = 1,2)$. Then

$$\begin{aligned} h(t_1)v_1h(t_2)v_2 &= h(t_1+t_2)[h(-t_2)v_1h(t_2)]v_2 \\ &= h(t_1+t_2)\exp(\operatorname{Ad}_G h(-t_2)(X_1))\exp X_2, \end{aligned}$$

and using (1), we have $C^2 \subset h([-2,2])\exp(p(1)W)\exp W$. Now $p(1) \in \mathbb{P}$ and if $Z \in p(1)W$, then $\exp Z = (\exp(Z/p(1)))^{p(1)} \in (\exp W)^{p(1)}$. Hence

$$C^2 \subset h([-2,2])(\exp W)^{1+p(1)} = h([-2,2])V^{1+p(1)}.$$

An easy induction argument shows that if $a_r = 1 + \sum_{i=1}^{r-1} p(i)$, then

$$C^r \subset h([-r,r])V^{a_r} \qquad (r \geq 1). \tag{2}$$

Since H has polynomial growth, we can find a polynomial P with coefficients in $\mathbb{N}$ such that $\lambda_H(V^r) \leq P(r)$ $(r \geq 1)$. Choose λ_L so that the isomorphism $h\colon \mathbb{R} \to L$ takes Lebesgue measure on $\mathbb{R}$ into λ_L. Then using (2) and (5.13), with $j\colon H \times L \to G$ given by $j(h, l) = lh$, we have

$$\begin{aligned}\lambda(C^r) &\leq \lambda_H \times \lambda_L(j^{-1}(j(V^{a_r} \times h([-r, r])))) \\ &= 2r\lambda_H(V^{a_r}) \leq Q(r),\end{aligned}$$

where $Q(x) = 2xP(1 + (x-1)p(x-1))$. □

(6.28) PROPOSITION. *Let G be a Type R, Lie group and H a closed, connected, normal subgroup of G. Then both H and G/H are of Type R.*

PROOF. Both H and G/H are Lie groups and the Lie algebra $\mathfrak{h}$ of H is an ideal of the Lie algebra $\mathfrak{g}$ of G, while $\mathfrak{g}/\mathfrak{h}$ is the Lie algebra of G/H. Now use (6.25) and (6.23). □

Recall ((3.7)) that $r(G)$ is the radical of G.

(6.29) PROPOSITION. *Let G be a Type R Lie group. Then $G/r(G)$ is compact.*

PROOF. By (6.28), the (Lie) group $H = G/r(G)$ is of Type R. Suppose that $G/r(G)$ is not compact. Then the (semisimple) Lie algebra $\mathfrak{h}$ of H is not compact. So ((3.3(ii))) $\mathfrak{h}$ contains $\mathrm{sl}(2, \mathbb{R})$ as a subalgebra. Let

$$X = \begin{bmatrix} 1 & 1 \\ \cdot & -1 \end{bmatrix}, \qquad Y = \begin{bmatrix} 2 & 1 \\ -4 & -2 \end{bmatrix}.$$

Then $X, Y \in \mathrm{sl}(2, \mathbb{R})$, and $\operatorname{ad} X(Y) = XY - YX = -2Y$, and a contradiction of the fact that H is of Type R results. □

(6.30) PROPOSITION. *Every (connected) locally compact, Type R group has polynomial growth.*

PROOF. Suppose that G is of Type R. Appealing to (6.20(i)), we can suppose that G is a Lie group. By (6.29) and (6.21), we can further suppose that G is solvable. Let $\tilde{G}$ be the simply connected covering group of G and $\pi\colon \tilde{G} \to G$ the canonical homomorphism. As $\ker \pi$ is a discrete subgroup of $\tilde{G}$, it follows from (6.27) and (6.20(ii)) that G has polynomial growth. □

Our next objective (6.32) reduces the study of polynomial growth for G to the solvable case.

(6.31) LEMMA. *Let W be a finite-dimensional vector space over $\mathbb{R}$ and $\mathfrak{k}$ a compact, semisimple subalgebra of* $\mathrm{gl}(W)$. *Then* $\mathrm{Sp}(X) \subset i\mathbb{R}$ *for all* $X \in \mathfrak{k}$.

PROOF. Let K be the connected Lie subgroup of $\mathrm{GL}(W)$ whose Lie algebra is $\mathfrak{k}$. Then K is compact (B52) in $\mathrm{GL}(W)$ and so the eigenvalues of its elements have modulus 1. Since $e^X \in K$ for $X \in \mathfrak{k}$, the lemma is established. □

(6.32) PROPOSITION. *Let G be a Lie group. Then G is of Type R if and only if $r(G)$ is of Type R and $G/r(G)$ is compact.*

PROOF. If G is of Type R, then the required implication follows from (6.28) and (6.29).

Conversely, suppose that $r(G)$ is of Type R and $G/r(G)$ is compact. Let $\mathfrak{g}$, $\mathfrak{r}$, and $\mathfrak{k}$ be the Lie algebras of G, $r(G)$, and $G/r(G)$ respectively. Observe that $\mathfrak{k} = \mathfrak{g}/\mathfrak{r}$ and is a compact, semisimple Lie algebra. Let $X \in \mathfrak{g}$. We have to show that $\mathrm{Sp}(\mathrm{ad}\, X) \subset i\mathbb{R}$. Now $(\mathrm{ad}\, X)_{\mathfrak{k}}$ is a derivation on $\mathfrak{k}$ and so belongs to $\mathrm{ad}\,\mathfrak{k}$ (B42). By (6.31), $\mathrm{Sp}((\mathrm{ad}\, X)_{\mathfrak{k}}) \subset i\mathbb{R}$. So ((6.23)) it is sufficient to show that $\mathrm{Sp}((\mathrm{ad}\, X)_{\mathfrak{r}})$ is contained in $i\mathbb{R}$.

Let $\mathfrak{n}$ be the nil-radical of $\mathfrak{g}$. Then $\mathrm{ad}\, X(\mathfrak{r}) \subset \mathfrak{n}$ (B36), and so it is sufficient to prove that $\mathrm{Sp}(\mathrm{ad}\, X)_{\mathfrak{n}}) \subset i\mathbb{R}$. Let $\{\mathfrak{n}^{(j)}\}$ be the lower central series for $\mathfrak{n}$ and r the class of $\mathfrak{n}$. So $\mathfrak{n}^{(1)} = \mathfrak{n}$, $\mathfrak{n}^{(j+1)} = [\mathfrak{n}, \mathfrak{n}^{(j)}]$ $(1 \le j \le r-1)$, $\mathfrak{n}^{(r)} \neq \{0\}$, and $\mathfrak{m}^{(r+1)} = \{0\}$. Let $T(j) = (\mathrm{ad}\, X)_{\mathfrak{n}^{(j)}}$. We will show, by downwards induction on j, that $\mathrm{Sp}(T(j)) \subset i\mathbb{R}$ for all j. The result is trivial if $j = r+1$. Suppose that $1 < j \le r+1$ and that $\mathrm{Sp}(T(j)) \subset i\mathbb{R}$. Now by the Levi-Malcev Theorem (B43) we can find a Lie subalgebra $\mathfrak{k}'$ of $\mathfrak{g}$ isomorphic to $\mathfrak{k}$ such that $\mathfrak{g}$ is the vector space direct sum $\mathfrak{r} \oplus \mathfrak{k}'$. Further, the map $Y \to (\mathrm{ad}\, Y)_{\mathfrak{n}^{(j-1)}/\mathfrak{n}^{(j)}}$ is a Lie homomorphism from $\mathfrak{g}$ into $\mathrm{gl}(\mathfrak{n}^{(j-1)}/\mathfrak{n}^{(j)})$. Since $\mathfrak{r}$ is of Type R and $\mathfrak{k}'$ is compact, we can write $T(j-1)_{\mathfrak{n}^{(j-1)}/\mathfrak{n}^{(j)}} = A + B$, where $A, B \in \mathrm{gl}(\mathfrak{n}^{(j-1)}/\mathfrak{n}^{(j)})$, $\mathrm{Sp}(A) \subset i\mathbb{R}$, and B belongs to a compact, semisimple Lie subalgebra of $\mathrm{gl}(\mathfrak{n}^{(j-1)}/\mathfrak{n}^{(j)})$. Further, since $[\mathfrak{g}, \mathfrak{r}] \subset \mathfrak{n}$ and $\mathrm{ad}\, W(\mathfrak{n}^{(j-1)}) \subset \mathfrak{n}^{(j)} (W \in \mathfrak{n})$, we see that $[A, B] = 0$ so that $AB = BA$. Also by (6.31), $\mathrm{Sp}(B) \subset i\mathbb{R}$. Hence $\mathrm{Sp}(T(j-1)_{\mathfrak{n}^{(j-1)}\mathfrak{n}^{(j)}}/{=}\mathrm{Sp}(A + B) \subset \mathrm{Sp}(A) + \mathrm{Sp}(B) \subset i\mathbb{R}$. Since, by hypothesis, $\mathrm{Sp}(T(j-1)_{\mathfrak{n}^{(j)}}) = \mathrm{Sp}(T(j)) \subset i\mathbb{R}$, we have $\mathrm{Sp}(T(j-1)) \subset i\mathbb{R}$. This completes the induction argument, and $\mathrm{Sp}(\mathrm{ad}\, X) = \mathrm{Sp}(T(1)) \subset i\mathbb{R}$ as required. □

The preceding result, in a sense, reduces the study of Type R Lie groups to the solvable case. We now investigate the Type R condition for solvable Lie algebras. We start by discussing some results from the theory of nilpotent Lie algebras. We use the results and terminology of B31f. Our objective is (6.37).

(6.33) Preliminaries. Let V be a finite-dimensional vector space over $\mathbb{R}$ and $\mathfrak{l}$ a nilpotent Lie subalgebra of $L(V)$. In the natural way, we can regard $\mathfrak{l}_{\mathbb{C}}$ as a nilpotent Lie subalgebra of $L(V_{\mathbb{C}})$. Let Ψ' be the set of weights of $\mathfrak{l}_{\mathbb{C}}$ with respect to $V_{\mathbb{C}}$, and, for each ψ' in Ψ', let $V_{\psi'}$ be the weight space associated with ψ'. Then $V_{\mathbb{C}} = \bigoplus_{\psi' \in \Psi'} V_{\psi'}$. Let $\Psi = \{\psi'|_{\mathfrak{l}} \colon \psi' \in \Psi'\}$. Let us call the elements of Ψ the *weights* of $\mathfrak{l}$ (with respect to V). A simple exercise in complexifications shows that $\overline{\psi} \in \Psi$ if $\psi \in \Psi$. Let V_ψ be $V_{\psi'}$, where $\psi'|_{\mathfrak{l}} = \psi$.

There is a basis for $V_{\mathbb{C}}$ such that the matrix of each $T \in \mathfrak{l}$ relative to this basis is of the form

$$\begin{bmatrix} B_1 & & & \\ & B_2 & & 0 \\ & & \ddots & \\ & 0 & & B_n \end{bmatrix},$$

where B_i is of the form

$$\begin{bmatrix} \psi_i(T) & & * \\ & \ddots & \\ 0 & & \psi_i(T) \end{bmatrix}$$

with $\psi_1, \dots, \psi_n$ the weights of $\mathfrak{l}$.

Now let $\mathfrak{g}$ be a real, finite-dimensional, Lie algebra and $X \in \mathfrak{g}$. Applying the results of the preceding paragraph with $\mathfrak{l} = \mathbb{R}(\operatorname{ad} X)$, $W = \mathfrak{g}$, we see that, in the obvious way, $\Psi = \operatorname{Sp}(\operatorname{ad} X)$, and $\mathfrak{g}_{\mathbb{C}}$ is the direct sum of weight spaces $\mathfrak{g}_\mu^X$ ($\mu \in \operatorname{Sp}(\operatorname{ad} X)$). Further the subspace $[\mathfrak{g}_{\mu_1}^X, \mathfrak{g}_{\mu_2}^X]$ is contained in $\mathfrak{g}_{(\mu_1+\mu_2)}^X$ if $(\mu_1 + \mu_2) \in \operatorname{Sp}(\operatorname{ad} X)$ and is $\{0\}$ otherwise.

For any Lie algebra $\mathfrak{h}$, let $\{\mathfrak{h}^{(i)}\}$ be the lower central series for $\mathfrak{h}$ (B23).

In the next three results, $\mathfrak{g}$ is a solvable, real, finite-dimensional Lie algebra, and $\mathfrak{n}$ is its nil-radical.

(6.34) LEMMA. *Let $\mathfrak{h}$ be a subalgebra of $\mathfrak{g}$. If $\mathfrak{h} + \mathfrak{n}^{(2)} = \mathfrak{g}$, then $\mathfrak{h} = \mathfrak{g}$.*

PROOF. Suppose that $\mathfrak{h}+\mathfrak{n}^{(2)} = \mathfrak{g}$. It is sufficient to show that $\mathfrak{n} \subset \mathfrak{h}$ (for then $\mathfrak{h}+\mathfrak{n}^{(2)} \subset \mathfrak{h}$). Let $r \in \mathbb{P}$ and $Z_1, \dots, Z_r \in \mathfrak{n}$. Write $Z_i = X_i + Y_i$, where $X_i \in \mathfrak{h} \cap \mathfrak{n}$ and $Y_i \in \mathfrak{n}^{(2)}$. Substituting for Z_i in $[Z_1, [Z_2, [\dots, Z_r]] \dots]$ and using the fact that (by Jacobi's identity) $[\mathfrak{n}^{(2)}, \mathfrak{n}^{(p)}] \subset \mathfrak{n}^{(p+2)}$, we see that $\mathfrak{n}^{(r)} \subset \mathfrak{h} + \mathfrak{n}^{(r+1)}$. Since $\mathfrak{n}^{(m)} = \{0\}$ for some m, it follows that $\mathfrak{n} \subset \mathfrak{h}$. □

(6.35) COROLLARY. *Let $\mathfrak{k}$ be an ideal of $\mathfrak{g}$ contained in $\mathfrak{n}^{(2)}$ such that $\mathfrak{g}/\mathfrak{k}$ is of Type R. Then $\mathfrak{g}$ is also of Type R.*

PROOF. Let $X \in \mathfrak{g}$ and $E = \operatorname{Sp}(\operatorname{ad} X) \cap i\mathbb{R}$. Obviously, if μ, $\mu' \in E$ and $(\mu + \mu') \in \operatorname{Sp}(\operatorname{ad} X)$, then $(\mu + \mu') \in E$. So $\mathfrak{h} = \bigoplus_{\mu \in E} \mathfrak{g}_\mu^X$ is a subalgebra of $\mathfrak{g}_{\mathbb{C}}$. Let $\mu \in E' = \operatorname{Sp}(\operatorname{ad} X) \sim E$. Then, since $\mathfrak{g}/\mathfrak{k}$ is of Type R, μ is not an eigenvalue of $\operatorname{ad}(X + \mathfrak{k})$. However, if $\xi \in \mathfrak{g}_{\mathbb{C}}$ is such that for some n,

$$(\operatorname{ad} X - \mu)^n(\xi) = 0,$$

then it follows that $(\operatorname{ad}(X + \mathfrak{k}) - \mu)^n(\xi + \mathfrak{k}_{\mathbb{C}}) = 0$ in $(\mathfrak{g}/\mathfrak{k})_{\mathbb{C}}$ $(= \mathfrak{g}_{\mathbb{C}}/\mathfrak{k}_{\mathbb{C}})$. Hence $\xi \in \mathfrak{k}_{\mathbb{C}}$ and

$$\mathfrak{g}_{\mathbb{C}} = \Big(\bigoplus_{\mu \in E} \mathfrak{g}_\mu^X\Big) \oplus \Big(\bigoplus_{\mu \in E'} \mathfrak{g}_\mu^X\Big) \subset \mathfrak{h} \oplus \mathfrak{k}_{\mathbb{C}} \subset \mathfrak{h} + \mathfrak{n}_{\mathbb{C}}^{(2)} \subset \mathfrak{g}_{\mathbb{C}}.$$

So $\mathfrak{h}+\mathfrak{n}_{\mathbb{C}}^{(2)} = \mathfrak{g}_{\mathbb{C}}$. By (6.34) (applied to $\mathfrak{g}_{\mathbb{C}}$), we have $\mathfrak{h} = \mathfrak{g}_{\mathbb{C}}$, so that $E = \operatorname{Sp}(\operatorname{ad} X)$ and $\mathfrak{g}$ is of Type R. □

The solvable Lie algebras of (i), (ii), and (iii) below are the "building blocks" (in a suitable sense) of solvable Lie algebras which are *not* of Type R.

(i) The algebra $\mathfrak{s}_2$ is the two-dimensional Lie algebra for which there is a basis $\{X_1, X_2\}$ such that $[X_1, X_2] = X_2$.

(ii) Let $\sigma \in \mathbb{R} \sim \{0\}$. Then $\mathfrak{s}_3^\sigma$ is the three-dimensional Lie algebra for which there is a basis $\{X_1, X_2, X_3\}$ such that

$$[X_3, X_1] = \sigma X_1 - X_2,$$
$$[X_3, X_2] = X_1 + \sigma X_2,$$
$$[X_1, X_2] = 0.$$

(iii) The algebra $\mathfrak{s}_4$ is the four-dimensional Lie algebra for which there is a basis $\{X_1, X_2, X_3, X_4\}$ such that

$$[X_1, X_2] = 0 = [X_3, X_4],$$
$$[X_1, X_3] = X_2 = -[X_2, X_4],$$
$$[X_1, X_4] = -X_1 = [X_2, X_3].$$

The simply connected Lie groups yielding these Lie algebras are given after the following result, which is due to Auslander and Moore [**1**].

(6.36) PROPOSITION. *If $\mathfrak{g}$ is not of Type R, then it has (at least) one of the algebras $\mathfrak{s}_2$, $\mathfrak{s}_3^\sigma$, $\mathfrak{s}_4$ as a homomorphic image.*

PROOF. Suppose that $\mathfrak{g}$ is not of Type R. We shall construct Lie algebras $\mathfrak{g}_0 = \mathfrak{g}$, $\mathfrak{g}_1$, $\mathfrak{g}_2$, $\mathfrak{g}_3$, and $\mathfrak{g}_4$, each $\mathfrak{g}_{i+1}$ being a quotient algebra of $\mathfrak{g}_i$, every $\mathfrak{g}_i$ not of Type R, and $\mathfrak{g}_4 \in \{\mathfrak{s}_2, \mathfrak{s}_3^\sigma, \mathfrak{s}_4\}$. For each i, set $\mathfrak{m}_i = \mathfrak{g}_i^{(2)}$, and let $\mathfrak{n}_i$ be the nil-radical of $\mathfrak{g}_i$. Let $\mathfrak{a}_i = (\mathfrak{m}_i)_{\mathbb{C}}$, and $\mathfrak{l}_i = \{(\operatorname{ad} X)_{\mathfrak{a}_i} : X \in \mathfrak{g}_i\}$. Note that $\mathfrak{m}_i \subset \mathfrak{n}_i$ (B37).

(i) *Let* $\mathfrak{g}_1 = \mathfrak{g}/\mathfrak{m}_0^{(2)}$. Since $\mathfrak{m}_0 \subset \mathfrak{n}_0$, we have $\mathfrak{m}_0^{(2)} \subset \mathfrak{n}_0^{(2)}$. By (6.35), $\mathfrak{g}_1$ is not of Type R. Trivially, $\mathfrak{m}_1^{(2)} = \{0\}$. If $Y, Z \in \mathfrak{g}_1$ and $A \in \mathfrak{a}_1$, then by Jacobi's identity, we have

$$(1) \qquad [Y,[Z,A]] - [Z,[Y,A]] = [Y,[Z,A]] + [Z,[A,Y]] + [A,[Y,Z]] = 0$$

since $\mathfrak{m}_1^{(2)} = \{0\}$ and $A, [Y,Z] \in \mathfrak{a}_1$. Hence $\mathfrak{l}_1$ is an abelian (and so nilpotent) Lie subalgebra of $\operatorname{gl}(\mathfrak{a}_1)$. Let Ψ be the set of weights of $\mathfrak{l}_1$. If $Y \in \mathfrak{g}_1$, $\lambda \in \operatorname{Sp}(\operatorname{ad} Y) \sim \{0\}$, and Z is an eigenvector for $\operatorname{ad} Y$ associated with λ in $(\mathfrak{g}_1)_{\mathbb{C}}$, then

$$\lambda Z = \operatorname{ad} Y(Z) = [Y, Z]$$

so that $Z = \lambda^{-1}[Y,Z] \in \mathfrak{a}_1$, and $\lambda \in \operatorname{Sp}((\operatorname{ad} Y)_{\mathfrak{a}_1})$. Hence, since $\mathfrak{g}_1$ is not of Type R, there exists $\alpha \in \Psi$ such that $\alpha(\mathfrak{l}_1)$ is not contained in $i\mathbb{R}$. With W_ψ the weight space for $\psi \in \Psi$, let

$$\mathfrak{b} = \bigoplus\{W_\psi : \psi \in \Psi \sim \{\alpha, \overline{\alpha}\}\}$$

and $\mathfrak{a}_1' = \mathfrak{b} \cap \mathfrak{g}_1$. Then $\mathfrak{a}_1'$ is an ideal in $\mathfrak{g}_1$, and $(\mathfrak{a}_1')_{\mathbb{C}} = \mathfrak{b}$.

(ii) *Let* $\mathfrak{g}_2 = \mathfrak{g}_1/\mathfrak{a}_1'$. Then $\mathfrak{a}_2 = \mathfrak{a}_1/\mathfrak{b}$. Clearly $\mathfrak{a}_2$ can be identified with W_α if $\alpha = \overline{\alpha}$, and with $W_\alpha \oplus W_{\overline{\alpha}}$ if $\alpha \neq \overline{\alpha}$. Suppose that $\alpha = \overline{\alpha}$, and (B32) let $\{e_1, \ldots, e_s\}$ be a basis for $\mathfrak{m}_2$ such that for all $X \in \mathfrak{g}_2$, $(\operatorname{ad} X)_{\mathfrak{m}_2}$ is upper triangular with all diagonal entries equal to $\alpha(X)$ for this basis. Let $\mathfrak{q}_2$ be the span of $\{e_1, \ldots, e_{s-1}\}$. Note that $\mathfrak{q}_2$ is an ideal in $\mathfrak{g}_2$. Suppose now that $\alpha \neq \overline{\alpha}$.

This time, we choose $\{f_1, \dots, f_r\}$ to be a basis for W_α such that relative to this basis, $(\operatorname{ad} X)_{W_\alpha}$ is upper triangular for all $X \in \mathfrak{g}_2$. Note that $\{\overline{f}_1, \dots, \overline{f}_r\}$ is a basis for $W_{\overline{\alpha}}$. Then $\mathfrak{c} = \operatorname{Span}\{f_1, \dots, f_{r-1}\}$ is an ideal of $(\mathfrak{g}_2)_{\mathbb{C}}$. Let $\mathfrak{q}_2$ be the ideal of $\mathfrak{g}_2$ spanned by $\operatorname{Re} \mathfrak{c}$ (in an obvious notation). Clearly $\mathfrak{c} + \overline{\mathfrak{c}}$ is the complex span of $\mathfrak{q}_2$ in $(\mathfrak{g}_2)_{\mathbb{C}}$.

(iii) *Let* $\mathfrak{g}_3 = \mathfrak{g}_2/\mathfrak{q}_2$. Then $\mathfrak{g}_3$ is not of Type R (since $\alpha, \overline{\alpha}$ induce nonimaginary weights of $\mathfrak{l}_3$). We prove that $\mathfrak{n}_3^{(2)} = \{0\}$. If $\alpha = \overline{\alpha}$, then $\mathfrak{a}_3 = \mathbb{C}X$ for some $X \in \mathfrak{m}_3$, so that $\mathfrak{m}_3$ is one-dimensional. Now $\mathfrak{n}_3^{(2)} \subset \mathfrak{m}_3$. If $\mathfrak{n}_3^{(2)} = \mathfrak{m}_3$, then trivially, $\mathfrak{g}_3/\mathfrak{n}_3^{(2)}$ is abelian and so of Type R, contradicting (6.35). It follows that $\mathfrak{n}_3^{(2)} = \{0\}$ in this case.

Suppose, then, that $\alpha \neq \overline{\alpha}$. Then $\mathfrak{a}_3 = \mathbb{C}Z \oplus \mathbb{C}\overline{Z}$, where Z is the image of f_r in $(\mathfrak{g}_3)_{\mathbb{C}}$. So $\operatorname{Re} Z$ and $\operatorname{Im} Z$ are linearly independent over $\mathbb{R}$. (For otherwise, $\{f_r, \overline{f}_r\}$ would be linearly dependent over $\mathbb{C}$.) Hence $\mathfrak{m}_3$ is two-dimensional. By the argument at the end of the preceding paragraph, $\mathfrak{n}_3^{(2)} \neq \mathfrak{m}_3$. So if $\mathfrak{n}_3^{(2)} \neq \{0\}$, then it is one-dimensional. If X is a nonzero element of $\mathfrak{n}_3^{(2)}$, then there is a *real* weight γ of $\mathfrak{l}_3$. This is impossible. Hence $\mathfrak{n}_3^{(2)} = \{0\}$ in both cases. Now let $\mathfrak{h} = \{(\operatorname{ad} Y)_{(\mathfrak{n}_3)_{\mathbb{C}}} : Y \in \mathfrak{g}_3\}$. Arguing as in (i), $\mathfrak{h}$ is an abelian Lie subalgebra of $\mathrm{gl}((\mathfrak{n}_3)_{\mathbb{C}})$. If β is a nonzero weight of $\mathfrak{h}$, then, from the definition of "weight space" and with $W = (\mathfrak{n}_3)_{\mathbb{C}}$, we have $W_\beta \subset \mathfrak{a}_3$. So $(\mathfrak{n}_3)_{\mathbb{C}} = \mathfrak{a}_3 \oplus W_0$, where W_0 is the weight space associated with 0 if 0 is a weight of $\mathfrak{h}$, and is $\{0\}$ otherwise. Let $\mathfrak{k}_0 = W_0 \cap \mathfrak{g}_3$, and note that $W_0 = (\mathfrak{k}_0)_{\mathbb{C}}$.

(iv) *Let* $\mathfrak{g}_4 = \mathfrak{g}_3/\mathfrak{k}_0$. Let $\mathfrak{p}$ be the inverse image of $\mathfrak{n}_4$ in $\mathfrak{g}_3$. Then $\mathfrak{p}$ is a nilpotent ideal in $\mathfrak{g}_3$ by B29. Since $\mathfrak{p}$ obviously contains $\mathfrak{n}_3$, we have $\mathfrak{p} = \mathfrak{n}_3$, and $\mathfrak{n}_4 = \mathfrak{n}_3/\mathfrak{k}_0 = \mathfrak{m}_4$. But $\mathfrak{m}_3$ has dimension one or two. So $\mathfrak{m}_4 = \mathfrak{n}_4$ has dimension one or two. Also, $\mathfrak{n}_4^{(2)} = \{0\}$. Write $\mathfrak{n} = \mathfrak{n}_4$, and, abusing notation, write $\mathfrak{g} = \mathfrak{g}_4$.

Suppose, first, that $\mathfrak{n}$ is one-dimensional. If $X \in \mathfrak{g}$ is such that $[X, \mathfrak{n}] = \{0\}$, then $(\mathbb{R}X + \mathfrak{n})$ is a nilpotent ideal in $\mathfrak{g}$, and so $X \in \mathfrak{n}$. It follows that $\mathfrak{n}$ is the kernel of the map $X \to (\operatorname{ad} X)_{\mathfrak{n}}$ $(X \in \mathfrak{g})$, and since $\mathfrak{n}$ is one-dimensional and $\mathfrak{g} \neq \mathfrak{n}$ (as $\mathfrak{g}$ is not of Type R), we must have $\mathfrak{g}$ two-dimensional. Let $\{X_1, X_2\}$ be a basis for $\mathfrak{g}$ with $X_2 \in \mathfrak{n}$. Since $\mathfrak{g} \neq \mathfrak{n}$, $[X_1, X_2] = kX_2$ where $k \neq 0$. Scaling X_1, it follows that $\mathfrak{g} = \mathfrak{s}_2$.

Suppose, then, that $\mathfrak{n}$ is two-dimensional. Then consideration of the weights $\delta, \overline{\delta}$ of $\mathfrak{l}_4$ and their relationship to $\alpha, \overline{\alpha}$ shows that there are two cases to be considered. Let $\delta_1 = \operatorname{Re}\delta$, $\delta_2 = \operatorname{Im}\delta$, and $\{X_1, X_2\}$ be a basis for $\mathfrak{n}$ such that $[X, X_1 + iX_2] = \delta(X)(X_1 + iX_2)$ $(X \in \mathfrak{g})$.

In the first of the cases, δ_1 and δ_2 are linearly dependent. Since both δ_1, δ_2 are nonzero, there exists $\sigma \in \mathbb{R} \sim \{0\}$ such that $\delta_1 = \sigma\delta_2$. Choose $X_3 \in \mathfrak{g}$ such that $\delta_2(X_3) = 1$. Then with respect to the basis $\{X_1, X_2\}$,

$$(\operatorname{ad} X)_{\mathfrak{n}} = \delta_2(X) \begin{bmatrix} \sigma & 1 \\ -1 & \sigma \end{bmatrix} \qquad (X \in \mathfrak{g}). \tag{2}$$

So $\operatorname{ad}(X - \delta_2(X)X_3)(\mathfrak{n}) = \{0\}$ for all $X \in \mathfrak{g}$, and, as in the one-dimensional case, $\mathfrak{n} = \{X \in \mathfrak{g} \colon [X, \mathfrak{n}] = \{0\}\}$. So $\{X_1, X_2, X_3\}$ is a basis for $\mathfrak{g}$, and using (2) and the fact that $\mathfrak{n}^{(2)} = \{0\}$, we have $\mathfrak{g} = \mathfrak{s}_3^\sigma$.

In the second case, we suppose that δ_1 and δ_2 are linearly independent. Choose $X_3, X_4 \in \mathfrak{g}$ such that $\delta_1(X_3) = 0$, $\delta_2(X_3) = 1$, $\delta_1(X_4) = 1$, and $\delta_2(X_4) = 0$. Then with respect to the basis $\{X_1, X_2\}$ of $\mathfrak{n}$,

$$(\mathrm{ad}\, X_3)_{\mathfrak{n}} = \begin{bmatrix} 0 & 1 \\ -1 & 0 \end{bmatrix}, \qquad (\mathrm{ad}\, X_4) = \begin{bmatrix} 1 & 0 \\ 0 & 1 \end{bmatrix}. \tag{3}$$

As above,

$$\mathfrak{n} = \{X \in \mathfrak{g} \colon (\mathrm{ad}\, X)_{\mathfrak{n}} = 0\}. \tag{4}$$

Replacing X_3 by $(X_3 + [X_3, X_4])$ and using (3), we can suppose that $[X_3, X_4] = 0$. Further, by (3) and (4), for all $X \in \mathfrak{g}$,

$$(X - (\delta_1(X)X_4 + \delta_2(X)X_3)) \in \mathfrak{n}$$

so that $\{X_1, X_2, X_3, X_4\}$ is a basis for $\mathfrak{g}$. It readily follows that $\mathfrak{g} = \mathfrak{s}_4$. □

The simply connected Lie groups with Lie algebras $\mathfrak{s}_2$, $\mathfrak{s}_3^\sigma$, and $\mathfrak{s}_4$ respectively are the groups S_2, S_3^σ, and S_4 defined below:

(i) S_2 is the "$ax + b$" group; so S_2 is $\mathbb{R}^2$ with multiplication given by

$$(s, t)(s', t') = (s + e^t s', t + t');$$

(ii) S_3^σ is the semidirect product $\mathbb{R}^2 \times_\rho \mathbb{R}$, with multiplication given by

$$(\alpha, t)(\alpha', t') = (\alpha + A_\sigma(t)\alpha', t + t'),$$

where

$$A_\sigma(t) = e^{\sigma t} \begin{bmatrix} \cos t & \sin t \\ -\sin t & \cos t \end{bmatrix};$$

(iii) S_4 is the semidirect product $\mathbb{R}^2 \times_\rho \mathbb{R}^2$, with multiplication given by

$$(\alpha, \beta)(\alpha', \beta') = (\alpha + B(\beta)\alpha', \beta + \beta'),$$

where

$$B(\beta) = e^t \begin{bmatrix} \cos s & \sin s \\ -\sin s & \cos s \end{bmatrix} \qquad (\beta = (s, t)).$$

[To check this, note that each of these groups is of the form $G \times_\rho H$ with G, $H \in \{\mathbb{R}, \mathbb{R}^2\}$, and, by B22, the Lie algebra $\mathfrak{s}$ of $G \times_\rho H$ is, in an obvious notation, $\mathfrak{g} \times_{\rho^*} \mathfrak{h}$, where $\rho^*(X) = (\mathrm{ad}\, X)_{\mathfrak{g}}$ $(X \in \mathfrak{h})$. For example, in the case of S_4, we have $G = H = \mathbb{R}^2$ and $\rho = B$. Since $\exp \beta = \beta$ $(\beta \in \mathfrak{g})$ and $\mathrm{Ad}(\exp(u\beta)) = e^{u\, \mathrm{ad}\, \beta}$, we have, for $\beta = (s, t)$,

$$\rho^*(\beta) = \frac{d}{du}\left(e^{ut}\begin{bmatrix} \cos us & \sin us \\ -\sin us & \cos us \end{bmatrix}\right)_{u=0} = \begin{bmatrix} t & s \\ -s & t \end{bmatrix}.$$

The multiplication in $\mathfrak{s} = \mathbb{R}^2 \times_{\rho^*} \mathbb{R}^2$ is given by B5:

$$\begin{aligned}[(\alpha, \beta), (\alpha', \beta')] &= ([\alpha, \alpha'] + \rho^*(\beta)(\alpha') - \rho^*(\beta')(\alpha), [\beta, \beta']) \\ &= (\rho^*(\beta)(\alpha') - \rho^*(\beta')(\alpha), 0)\end{aligned}$$

and it readily follows that $\mathfrak{s} = \mathfrak{s}_4$, with $\{-X_1, -X_2, X_3, X_4\}$ the standard basis for $\mathfrak{s}$.]

(6.37) PROPOSITION. *Let G be a simply connected, solvable Lie group that is not of Type R. Then there is a continuous homomorphism from G onto (at least) one of the groups S_2, S_3^σ, S_4.*

PROOF. Use (6.36), B11, and B7. □

The next two results involve the (so far neglected) conditions (iv) and (v) of (6.0). The results (6.38), (6.39) are due to Jenkins [**6**], [**9**].

(6.38) LEMMA. (i) *Let G be one of the groups S_2, S_3^σ, S_4. Then G contains an open subsemigroup T with disjoint right ideals I, J.*

(ii) *Let G be a solvable Lie group with the property that there exists a left invariant mean on $U(S)$ for every open subsemigroup S of G. Then G is of Type R.*

PROOF. (i) It is elementary to check that the map $((a,b),(s,t)) \to (a+ib, t-is)$ is an isomorphism from S_4 onto the complex version $\mathbb{C} \times_\rho \mathbb{C}$ of S_2, multiplication in $\mathbb{C} \times_\rho \mathbb{C}$ being given by

$$(w,z)(w',z') = (w + e^z w', z + z').$$

We thus identify S_4 with $\mathbb{C} \times_\rho \mathbb{C}$.

Suppose that G is S_2 or S_4. For $j \in \{-1, 1\}$, define

$$U_j = \{(w,z) \in G\colon \operatorname{Re} z < -2, |w - j| < e^{-2}\}.$$

Let T be the (open) subsemigroup of G generated by $U_1 \cup U_{-1}$ and let $I = U_1 T$, $J = U_{-1}T$. Clearly, I and J are right ideals of T. We show that $I \cap J = \varnothing$. To this end, let $n \geq 1$ and $(w_i, z_i) \in U_1 \cup U_{-1}$ $(1 \leq i \leq n)$. Let $(w,z) = (w_1,z_1)(w_2,z_2)\cdots(w_n,z_n)$. Then

$$\begin{aligned}(w,z) &= (w_1 + e^{z_1}w_2, z_1 + z_2)(w_3,z_3)\cdots(w_n,z_n)\\ &= \left(w_1 + e^{z_1}w_2 + \cdots + e^{\sum_{i=1}^{n-1} z_i}w_n, \sum_{i=1}^{n} z_i\right).\end{aligned}$$

If $(w_1,z_1) \in U_1$, then (w,z) is a typical element of I, and since $|e^{z_i}| = e^{\operatorname{Re} z_i} < e^{-2}$ and $|w_i| \leq 1 + e^{-2}$, we have

$$\begin{aligned}|w - 1| &< |w_1 - 1| + (1 + e^{-2})\sum_{r=1}^{n-1}(e^{-2})^r\\ &< e^{-2} + ((1+e^{-2})e^{-2})/(1-e^{-2}) = \frac{2e^{-2}}{1-e^{-2}} < 1.\end{aligned}$$

Similarly, if $(w_1,z_1) \in U_{-1}$, then $|w+1| < 1$. Hence $I \cap J = \varnothing$ as required.

Suppose, now, that $G = S_3^\sigma$ for some $\sigma \neq 0$. Then G is identified, in the natural way, with the semidirect product $\mathbb{C} \times_\rho \mathbb{R}$, where

$$(w,t)(w',t') = (w + e^{(\sigma - i)t}w', t + t').$$

We shall assume that $\sigma > 0$, the argument for the case $\sigma < 0$ being very similar.

Let $l, k > 0$. (The numbers l, k will be specified later.) For $j \in \{-1, 1\}$, let

$$V_j = \{(w,t)\colon t < -l, |w - j| < k\}.$$

Let T be the (open) subsemigroup of G generated by $V_1 \cup V_{-1}$. Arguing as in the S_2, S_4 cases, we see that if $j \in \{-1, 1\}$ and $(w,t) \in V_jT$, then $|w - j| < (k + e^{-l\sigma})/(1 - e^{-l\sigma})$. Now choose $k = 1/2$. Then for large enough l, we can take $I = V_1T$, $J = V_{-1}T$.

(ii) Suppose that G is not of Type R. Let $\tilde{G}$ be the simply connected covering group of G and $Q\colon \tilde{G} \to G$ the canonical homomorphism. Since G and $\tilde{G}$ have the same Lie algebra, $\tilde{G}$ is also not of Type R. Let Q' be a continuous homomorphism from $\tilde{G}$ onto one of the groups S_2, S_3^{σ}, S_4 ((6.37)). Let T be an open subsemigroup of $Q'(\tilde{G})$ such that (i) applies, and let $T' = (Q')^{-1}(T)$. Then T' is an open subsemigroup of $\tilde{G}$ and there exist disjoint, right ideals I', J' of T'. Now G is amenable (since it is solvable), and, by hypothesis, $\mathfrak{L}(U(Q(T'))) \neq \varnothing$. Using Problem 2-10, $Q(I') \cap Q(J') \neq \varnothing$. Hence we can find $x \in I'$, $y \in J'$ such that $x^{-1}y \in \ker Q \subset Z(\tilde{G})$, the centre of $\tilde{G}$. Hence $xy = x^2(x^{-1}y) = x(x^{-1}y)x = yx$, and as I', J' are right ideals of $\tilde{G}$, $xy \in I' \cap J'$, yielding a contradiction. So G is of Type R. □

We now state and prove the theorem for which much of this chapter has been a preparation.

(6.39) THEOREM. *Let G be a connected, locally compact group. Then the statements* (i), (ii), (iii), *and* (iv) *below are equivalent, and each implies* (v). *If G is, in addition, a solvable Lie group, then all five statements below are equivalent.*

(i) *G has polynomial growth.*

(ii) *G is exponentially bounded.*

(iii) *G is of Type R.*

(iv) *G does not contain a free, uniformly discrete semigroup in two generators* ((6.0(iv))).

(v) *If S is an open subsemigroup of G, then $\mathfrak{L}(U(S)) \neq \varnothing$.*

PROOF. It is proved in (6.30) that (iii) implies (i), and in (6.6) that (i) implies (ii). Further, if G is a solvable Lie group, then (6.38(ii)) yields that (v) implies (iii). So to establish the theorem it is sufficient to show that (ii) implies (iv), (iv) implies (v), and (iv) implies (iii).

Suppose that (ii) is true, and that G contains a free, uniformly discrete semigroup T in two generators a, b. Let $U \in \mathscr{C}_e(G)$ be such that $sU \cap tU = \varnothing$ whenever s, $t \in T$ with $s \neq t$. Since G is connected and U is compact, there exists $k \in \mathbb{P}$ such that $aU \cup bU \subset U^k$. For each n, let

$$T_n = \{s_1 s_2 \cdots s_n\colon s_i \in \{a, b\}, 1 \leq i \leq n\}.$$

Then $|T_n| = 2^n$, and we have $\lambda(U^{nk}) \geq \lambda(T_nU) = 2^n\lambda(U)$, so that

$$\lim_{n\to\infty} \lambda(U^n)^{1/n} \geq 2^{1/k} > 1,$$

thus contradicting (ii). So (ii) implies (iv).

Now assume that (iv) holds. Then, a fortiori, G does not contain F_2 as a discrete subgroup, so that by (3.8), G is amenable. Suppose that there exists an open subsemigroup S of G such that $\mathfrak{L}(U(S)) = \varnothing$. By Problem 2-10, there exist disjoint, open, right ideals I, J of S. Let $a \in I$, $b \in J$, and let T be the subsemigroup of S generated by $\{a, b\}$. Since I and J are open in G, we can find $C \in \mathscr{C}_e(G)$ such that $aC \subset I$, $bC \subset J$. A simple argument shows that T is a free, uniformly discrete semigroup in G in the two generators a, b, with $sC \cap tC = \varnothing$ whenever s, t in T are different. This contradicts (iv), so that (iv) implies (v).

Still assuming (iv), let K be a compact normal subgroup of G such that G/K is a Lie group. If G/K contains a free, uniformly discrete semigroup in two generators, then so does G (cf. (3.1)), and (iv) is contradicted. So G/K does not contain a free, uniformly discrete semigroup in two generators. One easily shows that such a semigroup cannot be contained in any closed subgroup of G/K, in particular, in the radical H of G/K. Since (iv) implies (v), we see, using (6.38(ii)), that H is of Type R. Now G/K is amenable (since G is). Applying (3.8) and (6.32), G/K is of Type R, and hence so also is G. Thus (iv) implies (iii). □

References

Adelson-Welsky and Sreider [**1**], Dixmier [**4**], Guivarc'h [**1**], Jenkins [**6**], [**9**], Milnor [**1**], Hulanicki [**3**], [**6**], Greenleaf [**2**], Wolf [**1**], Bass [**1**], Tits [**4**], and Auslander and Moore [**1**].

Further Results

(6.40) Discrete groups and polynomial growth. Theorem (6.17) asserts that every nilpotent group has polynomial growth, and, in the finitely generated case, gives the degree. What can be said about finitely generated, solvable groups? The question is examined by Wolf [**1**] and Milnor [**2**]. (See also Bass [**1**].)

A group G is said to be *polycyclic* if there exists a normal series

$$G = A_1 \triangleright A_2 \triangleright \cdots \triangleright A_r = \{e\}$$

such that every quotient A_i/A_{i+1} is cyclic. The following theorem follows from the work of Wolf and Milnor.

Let G be a finitely generated, solvable group. Then

(a) *if G has polynomial growth, then it is polycyclic.*

(b) *The following three statements are equivalent:*

(i) *G has polynomial growth;*

(ii) *G contains a nilpotent subgroup N of finite index;*

(iii) *G is exponentially bounded.*

It follows using (6.21(2)) that the degrees of G and N are equal. The latter, of course, is given by (6.17).

A natural question (raised by Milnor, Wolf and Bass) is the following: *is the preceding theorem true if G is no longer required to be solvable*? Major progress on this problem has been made by M. Gromov [**1**], who proved the deep theorem that *a finitely generated group has polynomial growth if and only if it contains a nilpotent subgroup of finite index*. An account of the proof of this theorem is given in the useful paper of Tits [**4**].

Milnor [**2**] essentially raised the following natural question: *is every exponentially bounded group of polynomial growth*? R. I. Grigorchuk [**S1**] has shown that this is not the case. Property (iv) of (6.39) for a discrete group G is investigated by Rosenblatt [**1**], [**4**]. (Note that for discrete amenable groups, properties (iv) and (v) of (6.39) are equivalent (Problem 1-23).) Rosenblatt shows that if G is solvable, then G has polynomial growth if and only if G does not contain FS_2 as a subsemigroup. Rosenblatt defines a group G to be *supramenable* if, in the notation of (2.32), there exists an invariant measure for every triple (G, X, A). By considering (G, G, G), we see that every supramenable group is amenable. Further, by (2.32), G is supramenable if and only if every triple (G, X, A) satisfies the translate property. He also shows that every exponentially bounded group is supramenable (cf. (6.42)), and that no group containing FS_2 as a subsemigroup is supramenable. A natural conjecture is that *an amenable group G is supramenable if and only if G does not contain FS_2 as a subsemigroup*. This conjecture is established when G is solvable.

(6.41) Polynomial growth and fundamental groups. It is interesting to record that groups of polynomial growth arise, somewhat unexpectedly, in the work of Milnor on the relationship between curvature in a Riemannian manifold and the fundamental group of the manifold. The book by Bishop and Crittenden [**1**] is a good reference for the differential geometry required.

Let (M, d) be a complete, connected, Riemannian manifold that is assumed to have nonnegative mean curvature. Let $(\tilde{M}, \tilde{d})$ be the simply connected covering Riemannian manifold of M with covering map $p\colon \tilde{M} \to M$. Then $(\tilde{M}, \tilde{d})$ is also complete, and by relating the Riemannian connections on M and $\tilde{M}$ in the natural way, we see that $\tilde{M}$ also has nonnegative mean curvature. An important "comparison" theorem is the following: if ω_n is the volume of the unit ball in $\mathbb{R}^n$, then

$$\mu(B_x(r)) \leq \omega_n r^n, \tag{1}$$

where, for $x \in M$, $r \geq 0$, $B_x(r)$ is the ball with centre x and radius r in $\tilde{M}$, μ is the volume measure on $\tilde{M}$, and n is the dimension of M.

Recall (B3) that $\pi_1(M)$ can be realised as the group of covering transformations of $\tilde{M}$. Also each $T \in \pi_1(M)$ is isometric on $\tilde{M}$, that is,

$$\tilde{d}(Ta, Tb) = \tilde{d}(a, b) \qquad (a, b \in \tilde{M}). \tag{2}$$

The following beautiful result is proved in Milnor [**1**].

Let (M, d) be a complete, n-dimensional, Riemannian manifold with nonnegative mean curvature. Let F be a finite subset of $G = \pi_1(M)$. Then there exists $C > 0$ such that

$$|F|^r \leq Cr^n \qquad (r \in \mathbb{P}).$$

In particular, G has polynomial growth.

[Let $x_0 \in \tilde{M}$ and $k = \max_{T \in F} \tilde{d}(x_0, Tx_0)$. Using (2) and the triangular inequality, we have

$$F^r(x_0) \subset B_{x_0}(kr) \qquad (r \in \mathbb{P}). \tag{3}$$

It follows from (3) that for any $\varepsilon > 0$,

$$F^r(B_{x_0}(\varepsilon)) \subset B_{x_0}(kr + \varepsilon) \qquad (r \in \mathbb{P}). \tag{4}$$

Find an open neighbourhood U of $p(x_0)$ such that $p^{-1}(U)$ is a disjoint union of open sets V_α $(\alpha \in A)$, with $p|_{V_\alpha}$ a diffeomorphism from V_α onto U and such that if T, $S \in G$ with $T \neq S$, then $T(V_\alpha) \cap S(V_\alpha) = \varnothing$ for all α. (See B3.) Hence we can find $\varepsilon > 0$ such that the sets $T(B_{x_0}(\varepsilon))$ $(T \in G)$ are pairwise disjoint. Noting that $T(B_{x_0}(\varepsilon)) = B_{Tx_0}(\varepsilon)$, it follows using (1) and (4) that

$$|F|^r \mu(B_{x_0}(\varepsilon)) = \mu(F^r(B_{x_0}(\varepsilon))) \leq \mu(B_{x_0}(kr + \varepsilon)) \leq \omega_n(kr + \varepsilon)^n$$

and the desired result follows.]

Milnor also showed that if a Riemannian manifold M is compact and is such that all of its sectional curvatures are less than 0, then $\pi_1(M)$ has exponential growth ((6.1)). He conjectures that a similar conclusion is true if some of the sectional curvatures are allowed to be zero. P. Eberlein [**1**] and Chen [**3**] have made some progress in this direction.

J. A. Wolf [**1**] showed that if M is compact with every sectional curvature nonpositive, then $\pi_1(M)$ contains a nilpotent subgroup of finite index if and only if M is flat (that is, every sectional curvature is 0). S. T. Yau [**1**] shows that, with the same conditions on M, if $\pi_1(M)$ is solvable, then M is flat. Other papers in this area are Preissman [**1**], Byers [**1**], Myers [**1**], Chen [**2**], [**1**], Wood [**1**], Gromoll and Wolf [**1**], and Lawson and Yau [**1**].

The study of the relationship between a manifold M and the polynomial growth or amenability of its fundamental group is continued in the paper of Hirsch and Thurston [**1**]. Let $\mathscr{C}$ be the smallest class of groups that contains all amenable groups, is such that the free product $G * H \in \mathscr{C}$ whenever G, $H \in \mathscr{C}$, and contains every group K with a subgroup $L \in \mathscr{C}$ of finite index in K. They show that if M is a compact, Riemannian manifold with negative sectional curvatures and $\pi_1(M) \in \mathscr{C}$, then the Euler characteristic $\chi(M)$ of M is 0. Chen [**2**] shows that if $\pi_1(M) \in \mathscr{C}$, then $\pi_1(M)$ is actually free, and any amenable subgroup of $\pi_1(M)$ is infinite cyclic.

Milnor [**4**] studies the question: *which groups occur as fundamental groups of complete affinely flat manifolds?* He shows that every torsion free group with a polycyclic subgroup of finite index is such a fundamental group. The converse seems to be unresolved.

A discrete group G is said to have *subexponential growth* if there exists a finite, generating set F for G, with $F^{-1} = F$, such that

$$\liminf_{n\to\infty}(|F^{n+1}|/|F^n|) = 1.$$

This is a Følner-type condition, and is closely related to conditions considered in (6.8) and (6.43). Indeed, it follows from (6.8) that every finitely generated, exponentially bounded group has subexponential growth. There is a close connection between "subexponential growth" of volume in a Riemannian covering space and subexponential growth of a related fundamental group (Plante [**1**], [**2**], [**3**]).

(6.42) Exponentially bounded groups and the Translate Property (Rosenblatt [**1**], [**4**]; Jenkins [**9**], [**10**], [**11**]). Recall ((2.32)) that if G is discrete, then G has the Translate Property if whenever $x_1, \ldots, x_n \in G$, $\alpha_1, \ldots, \alpha_n \in \mathbb{R}$, and $\sum_{i=1}^n \alpha_i \chi_{x_i A} \geq 0$, then $\sum_{i=1}^n \alpha_i \geq 0$. We note that if $\mu = \sum_{i=1}^n \alpha_i \delta_{x_i^{-1}} \in M(G)$, then $\sum_{i=1}^n \alpha_i \chi_{x_i A} = \chi_A \mu$, and that $\sum_{i=1}^n \alpha_i = (\sum_{i=1}^n \alpha_i \delta_{x_i^{-1}})(G) = \mu(G)$. This motivates the version of the Translate Property for general locally compact groups in (i) below. The proof of (i) is left as Problem 6-13.

(i) *Let G be an exponentially bounded, locally compact group. Then G has the* Generalised Translate Property, *that is, if $\mu \in M(G)$ has compact support and $\phi \in L_\infty(G)$ is nonzero such that $\phi \geq 0$, $\phi\mu \geq 0$, then $\mu(G) \geq 0$.*

(ii) *Let G be an amenable locally compact group having the Generalised Translate Property. Then $\mathfrak{L}(U(S)) \neq \varnothing$ for every open subsemigroup S of G.*

[Suppose that there is an open subsemigroup S of G with $\mathfrak{L}(U(S)) = \varnothing$. By Problem 2-10, we can find open, disjoint, right ideals I, J of S. Note that S, I, J are not $\{e\}$. Let $a \in I \sim \{e\}$ and $b \in J \sim \{e\}$. Choose $C \in \mathscr{C}_e(G)$ such that $C = C^{-1}$, $CaC \subset I$, and $CbC \subset J$. Let $\phi = \chi_S$ and let $\nu \in L_1(G)$ be given by

$$\nu = \chi_{aC} - \chi_{abC} - \chi_{a^2C}.$$

Let $\mu = \nu^\sim$. We will contradict the Translate Property by showing that $\mu(G) < 0$ while $\phi\mu \geq 0$.

First, $\mu(G) = \nu(G) = (\lambda(aC) - \lambda(abC) - \lambda(a^2C)) = -\lambda(C) < 0$. Second, let $t \in G$. Then

$$\begin{aligned} \phi\mu(t) &= \mu(St^{-1}) = \nu(tS^{-1}) \\ &= (\lambda(tS^{-1} \cap aC) - \lambda(tS^{-1} \cap abC) - \lambda(tS^{-1} \cap a^2C)). \end{aligned} \tag{1}$$

There are three cases to consider. Suppose that $tS^{-1} \cap abC = \varnothing = tS^{-1} \cap a^2C$. Then, trivially, $\phi\mu(t) \geq 0$. Suppose now that $tS^{-1} \cap abC \neq \varnothing$. Then $t = abcs$ for some $c \in C$, $s \in S$. If $c_1 \in C$, then, noting that $sS^{-1} \supset s(Ss)^{-1} = S^{-1}$ and $c_1^{-1}bc \in CbC \subset J \subset S$,

$$\begin{aligned} tS^{-1} &= abcsS^{-1} \supset abcS^{-1} = ac_1(c_1^{-1}bc)S^{-1} \\ &\supset \{ac_1(c_1^{-1}bc)(c_1^{-1}bc)^{-1}\} = \{ac_1\}. \end{aligned}$$

Hence $\lambda(tS^{-1} \cap aC) = \lambda(aC) = \lambda(C)$. Now if $z \in b^{-1}a^{-1}tS^{-1} \cap a^{-2}tS^{-1} \cap C$, then $t \in a(bzS \cap azS) \subset a(bCS \cap aCS) \subset a(I \cap J)$. This is impossible since $I \cap J = \varnothing$. Hence

$$\begin{aligned}\lambda(tS^{-1} \cap abC) + \lambda(tS^{-1} \cap a^2C) &= \lambda(b^{-1}a^{-1}tS^{-1} \cap C) + \lambda(a^{-2}tS^{-1} \cap C) \\ &\leq \lambda(C) \leq \lambda(tS^{-1} \cap aC),\end{aligned}$$

and from (1), $\phi\mu(t) \geq 0$. A similar argument deals with the case where we suppose that $tS^{-1} \cap a^2C \neq \varnothing$. Hence $\phi\mu \geq 0$ as required.]

(6.43) A Følner condition for exponentially bounded groups (Jenkins [**9**]). If G is amenable, $C \in \mathscr{C}(G)$, $C \neq \varnothing$, and $\varepsilon > 0$, then it follows from (4.13(ii)) that there exists nonnull $K \in \mathscr{C}(G)$ such that $\lambda(CK \,\triangle\, K)/\lambda(K) < \varepsilon$. In general, it is not clear how to construct such a set K. We will show that if G is an exponentially bounded, connected, solvable Lie group and $C^0 \neq \varnothing$, then very strong information about K is available: *K can be taken to be one of the sets C^n!* It seems plausible that, in general, if G is a connected, locally compact group, then G is exponentially bounded if and only if it satisfies the Følner condition (1) below. (This result is asserted in Jenkins [**9**], but there are gaps in the proof.) The Følner condition we will be considering is: *given $C \in \mathscr{C}(G)$ with $C^0 \neq \varnothing$ and $\varepsilon > 0$, there exists $N \in \mathbb{P}$ such that*

$$\lambda(C^{N+1} \,\triangle\, C^N)/\lambda(C^N) < \varepsilon. \tag{1}$$

(i) *Let G be exponentially bounded and $C \in \mathscr{C}(G)$ with C nonnull. Suppose that for some $p \in \mathbb{P}$, there exists $x \in C$ such that $x^{p+1} \in C^p$. Then given $\varepsilon > 0$, there exists $N \in \mathbb{P}$ such that the inequality* (1) *is true.*

[Let $r \geq 1$. Since $x^{p+1} \in C^{p+1} \cap C^p$, we have $C^{r+p+1} \cap C^{r+p+2} \supset x^{p+1}C^{r+1}$ so that

$$\begin{aligned}&\lambda(C^{r+p+1} \,\triangle\, C^{r+p+2})/\lambda(C^{r+p+1}) \\ &\quad\leq \lambda(C^{r+p+1} \,\triangle\, C^{r+p+2})/\lambda(C^{r+1}) \\ &\quad\leq [\lambda(C^{r+p+1}) + \lambda(C^{r+p+2}) - 2\lambda(C^{r+p+1} \cap C^{r+p+2})]/\lambda(C^{r+1}) \\ &\quad\leq 2[\lambda(C^{r+p+2}) - \lambda(C^{r+1})]/\lambda(C^{r+1}) \\ &\quad= 2[(\lambda(C^{r+p+2})/\lambda(C^{r+1})) - 1].\end{aligned} \tag{2}$$

Let $\alpha_r = \lambda(C^r)$. Since G is exponentially bounded, $\alpha_r^{1/r} \to 1$ as $r \to \infty$, and

$$\begin{aligned}1 &\leq \liminf_{r\to\infty}(\alpha_{r+p+2}/\alpha_{r+1}) \\ &= \liminf_{r\to\infty}(\alpha_{r+p+2}\alpha_{r+p+1}\cdots\alpha_{r+2})/(\alpha_{r+p+1}\alpha_{r+p}\cdots\alpha_{r+1}) \\ &\leq \lim_{r\to\infty}(\alpha_{r+p+1}\cdots\alpha_{r+1})^{1/r} = 1.\end{aligned}$$

The required result now follows.]

Note that by taking $C \in \mathscr{C}_e(G)$ and $x = e$ we obtain (6.8).

In the discrete case, it is hard for (1) to be satisfied for all C. Indeed, if we take $G = \mathbb{Z}$ and C to be a finite, nonempty, set of *odd* integers, then, for all n, $|C^{n+1} \triangle C^n|/|C^n| \geq 2$! We therefore concentrate on the case when G is connected.

(ii) *Let G be a connected, solvable Lie group. Then G is exponentially bounded if and only if* (1) *is true.*

[Suppose that G is exponentially bounded, and let $C \in \mathscr{C}(G)$ with $C^0 \neq \varnothing$. Let $\mathfrak{g}$ be the Lie algebra of G. By Appendix E, $\exp(\mathfrak{g})$ is dense in G. Hence we can find $X \in \mathfrak{g}$ and $r > 1$, $r \in \mathbb{P}$ such that both $\exp X$ and $\exp((1 - r^{-1})X)$ are in C^0. Let $x = \exp((1 - r^{-1})X)$. Then $x^r = (\exp X)^{r-1} \in C^{r-1}$. Then the inequality of (1) follows using (i).

Conversely, suppose that (1) is true. Let S be an open subsemigroup of G. By (6.39), it is sufficient to show that $\mathfrak{L}(U(S)) \neq \varnothing$. Suppose that $\mathfrak{L}(U(S)) = \varnothing$. By Problem 2-10, there exist disjoint, open, right ideals I, J of S. Let $C_1 \in \mathscr{C}(I)$ and $D_1 \in \mathscr{C}(J)$ with $C_1^0 \neq \varnothing$, $D_1^0 \neq \varnothing$. Let $C = C_1 \cup D_1$, and write $C^n = C_n \cup D_n$, where $C_n = C^n \cap I$, $D_n = C^n \cap J$. Then $C_n, D_n \neq \varnothing$, and if $c \in C_1$, then $C_{n+1} \supset (cC_n) \cup (cD_n)$, so that $\lambda(C_{n+1}) \geq \lambda(C_n) + \lambda(D_n)$. Similarly, $\lambda(D_{n+1}) \geq \lambda(C_n) + \lambda(D_n)$, so that $\lambda(C^{n+1}) \geq 2(\lambda(C_n) + \lambda(D_n)) = 2\lambda(C^n)$. So $\lambda(C^{n+1} \triangle C^n)/\lambda(C^n) \geq 1$ for all n, and (1) is contradicted.]

(iii) Emerson and Greenleaf [**2**] obtain sharp estimates on the growth of $\lambda(U^n)$, where U is an open, relatively compact subset of a locally compact abelian group G. They show that there exists a constant $A > 0$ and $k \in \mathbb{N}$ such that

$$\lambda(U^n) = An^k + O(n^{k-1} \log n) \qquad (n \to \infty). \tag{3}$$

It immediately follows that $\lambda(U^{n+1})/\lambda(U^n) \to 1$ as $n \to \infty$. It also follows from their work that if $C \in \mathscr{C}(G)$ has nonempty interior, then $\lambda(C^{n+1})/\lambda(C^n) \to 1$ as $n \to \infty$. Similar issues for a nilpotent, simply connected Lie group are treated in Porada [**1**].

(6.44) Polynomial growth and the nonzero fixed-point property (Jenkins [**12**]). Recall ((2.24)) that the amenability of a locally compact group G is equivalent to the existence of a fixed-point in every affine left G-set K on which G acts in a jointly continuous manner. Now such a set K is, by definition, a compact convex subset of some locally convex space Z. One can thus enquire what properties of an amenable, locally compact group G are necessary for there to exist a *nonzero* fixed-point in every such set K? (Of course, by the above fixed-point theorem, such a question is relevant only when $0 \in K$.) Particular examples show that we need to qualify the question as follows.

The group G is said to have the *nonzero fixed-point property* if whenever Z is a locally convex space, $x \to T_x$ is a homomorphism from G into the algebra of linear operators on Z, with T_e the identity operator, such that the map $(x, \xi) \to T_x(\xi)$ is continuous from $G \times Z$ into Z, and $K \neq \{0\}$ is a compact convex subset of Z, with $T_x(K) = K$ for all $x \in G$, for which there exists $\alpha \in Z'$ such that for all $\xi \in K \sim \{0\}$, $\alpha(\xi) > 0$ and $\xi/(\sup_{x \in G} \alpha(x\xi)) \in K$, then K has a nonzero fixed-point for G.

Jenkins proves that if G *is either a connected, locally compact group or a discrete, finitely generated, solvable group, then* G *has the nonzero fixed-point property if and only if* G *has polynomial growth.*

(6.45) The representation theory of solvable Lie groups and the Type R condition. Recall that a locally compact group G is liminal (CCR) if $\pi(f)$ is a compact operator whenever $f \in L_1(G)$ and π is an irreducible, unitary representation of G. Every liminal group is of Type 1 (postliminal, GCR). If G is nilpotent, then G is liminal (Dixmier [**3**], Kirillov [**1**], Fell [**2**]).

Not every Type R group is liminal. Indeed the Mautner group (Problem 6-8) is not even of Type 1. C. C. Moore, in Auslander and Moore [**1**], proved the following remarkable theorem. *Let* G *be a Type* 1, *simply connected, solvable Lie group. Then* G *is liminal if and only if* G *is of Type* R. [We shall be content to outline the proof of the easier of the two implications involved.

Let G be a liminal, simply connected, solvable Lie group. Suppose that G is not of Type R. Then we can find a closed, normal subgroup H of G such that G/H is one of the groups S_2, S_3^σ, S_4. Since the map $Q_H\colon L_1(G) \to L_1(G/H)$ of (1.11) is a continuous *-epimorphism, it follows that G/H is also liminal. A contradiction is obtained by showing that none of the groups S_2, S_3^σ, S_4 is liminal. (This is well known for S_2 (Nelson and Stinespring [**1**]).) Each of the groups is a semidirect product of the form $\mathbb{R}^n \times_\rho \mathbb{R}^m$, so that a well-known theorem of Mackey (Warner [**1**, Vol. 1, pp. 439–440]), together with general results of Fell on induced representations, can be used to establish the contradiction (Problem 6-10).]

See p. 129 of the above memoir of Auslander and Moore for a discussion of the importance of the Type R condition for applying the "Mackey machine."

The question of characterising Type 1, simply connected, solvable Lie groups is investigated by Auslander and Konstant [**1**]. They develop a theory in the spirit of Kirillov's work on nilpotent Lie groups. In the work of Pukanszky [**2**], [**1**] it turns out that, for a locally compact group G, the primitive ideal space $\mathrm{Prim}(G)$, rather than the dual space $\hat{G}$, is the appropriate object of study. He proves that *if* G *is a simply connected, solvable Lie group, then* $\mathrm{Prim}(G)$ *is a* T_1*-topological space* (*or equivalently, every element of* $\mathrm{Prim}(G)$ *is maximal in* $C^*(G)$) *if and only if* G *is of Type R.* (Note that this generalises Moore's result above since if G is of Type 1, then $\mathrm{Prim}(G)$ can be identified with $\hat{G}$ and $\hat{G}$ is a T_1-topological space if and only if G is liminal.) Moore and Rosenberg [**2**] and Pukanszky [**4**] have generalised this theorem to cover the case of almost connected, locally compact groups.

(6.46) Poisson spaces, amenability and the Type R condition. Let G be a locally compact group and $\mu \in \mathrm{PM}(G)$. We define H_μ, the space of μ-harmonic functions on G, by

$$H_\mu = \{\phi \in \mathrm{U_r}(G) \colon \mu\phi = \phi\}.$$

Recall that $\mathrm{U_r}(G)$ is the space of bounded, right uniformly continuous, complex-valued functions on G, and that for $\psi \in \mathrm{U_r}(G)$,

$$\mu\psi(x) = \int_G \psi(xy)\,d\mu(y) \qquad (x \in G).$$

Clearly, $1 \in H_\mu$, and $\phi x_0 \in H_\mu$ if $\phi \in H_\mu$, $x_0 \in G$, so that H_μ is a right invariant subspace of $\mathrm{U_r}(G)$. Obviously, H_μ is closed in $\mathrm{U_r}(G)$.

We now formulate the notions of *Poisson space* and *Poisson kernel* of μ. A *Poisson space* for μ is a pair (X, ν) such that

(i) X is a compact, Hausdorff space;

(ii) X is a left G-set with $e\xi = \xi$ for all $\xi \in X$, and the map $(x, \xi) \to x\xi$ is continuous from $G \times X$ into X;

(iii) $\nu \in \mathrm{PM}(X)$, and the map $T_\nu \colon C(X) \to \mathrm{U_r}(G)$, where

$$T_\nu f(x) = \int_X f(x\xi)\,d\nu(\xi) \tag{1}$$

is an isometry from $C(X)$ onto H_μ. It is immediate from (1) that T_ν is "equivariant", that is, $(T_\nu f)x_0 = T_\nu(fx_0)$ for all $f \in C(X)$, $x_0 \in G$.

The measure ν is the *Poisson kernel* of the space. Sometimes we just refer to X as a Poisson space, reference to ν being left implicit. If (X, ν), (Y, ν') are Poisson spaces of μ, then $T_{\nu'}^{-1}T_\nu$ is a linear isometry from $C(X)$ onto $C(Y)$ that preserves the action of G, and the Banach-Stone Theorem [**DS**, (V.8.8)] then shows that (X, ν) and (Y, ν') can be identified. So provided there exists a Poisson space for μ, then we can talk of *the* Poisson space for μ.

The above notions were essentially introduced and studied in the important paper of Furstenberg [**1**].

We now establish the existence of a Poisson space in general. This was proved, using probabilistic techniques, by Furstenberg when G is separable. (See Furstenberg [**1**], [**3**], [**4**], Cartier [**1**], Zimmer [**6**], and Azencott [**1**].) We will follow the proof in Paterson [**6**]; this proof does not require G to be separable and even applies to locally compact semigroups.

(i) *There exists a Poisson space for every* $\mu \in \mathrm{PM}(G)$. [Let $\mu \in \mathrm{PM}(G)$. Let L be the weak* closure of

$$\mathrm{co}\{\widehat{\mu^n} \colon n \geq 1\}$$

in $\mathrm{U_r}(G)'$. Now $\mathrm{U_r}(G)$ is left introverted ((2.11)), and by (1.4) and Problem 2-15 (Solution), $\phi\nu \in \mathrm{U_r}(G)$ and $\hat{\nu}(p\phi) = p(\phi\nu)$ for all $\nu \in \mathrm{PM}(G)$, $p \in \mathrm{U_r}(G)'$ and $\phi \in \mathrm{U_r}(G)$. Hence if $p_\delta \to p$ weak* in $\mathrm{U_r}(G)'$, then $\hat{\nu}p_\delta \to \hat{\nu}p$ weak* in the Banach algebra $\mathrm{U_r}(G)'$. For $n \geq 1$, we thus have a weak* continuous affine map T_n on L given by $T_n(p) = \mu^n p$. Applying Day's Fixed-Point Theorem (or even the Markov-Kakutani Theorem) we can find $p_0 \in L$ such that $\mu p_0 = p_0$. If $\hat{\nu}_\sigma \to p_0$ weak* with $\hat{\nu}_\sigma \in \mathrm{co}\{\widehat{\mu^n} \colon n \in \mathbb{P}\}$, then $\hat{\nu}_\sigma p_0 = p_0$ for all σ, and since $\hat{\nu}_\sigma p_0 \to p_0 p_0$ weak*, it follows that $(p_0)^2 = p_0$. We can therefore define a positive, norm one, unit preserving, linear projection P on $\mathrm{U_r}(G)$ by setting $P\phi = p_0\phi$. We claim that $P(\mathrm{U_r}(G)) = H_\mu$. Indeed, if $\phi \in \mathrm{U_r}(G)$ and $p_0\phi = \phi$, then $\mu\phi = (\hat{\mu}p_0)\phi = p_0\phi = \phi$, so that $\phi \in H_\mu$. Conversely, if $\phi \in H_\mu$ and $\hat{\nu}_\sigma \to p_0$ as above, then $\phi = \hat{\nu}_\sigma\phi \to p_0\phi$ pointwise, so that $P\phi = \phi$.

By Problem 6-16 H_μ is a commutative, unital C^*-algebra with 1 as the identity and multiplication and involution given by

$$\phi \times \psi = P(\phi\psi), \quad \phi^* = \overline{\phi} \qquad (\phi, \psi \in H_\mu).$$

Let X be the maximal ideal space of H_μ. As the map $(\phi, x) \to \phi x$ is jointly continuous from $H_\mu \times G$ into H_μ (since $H_\mu \subset \mathrm{U_r}(G)$), it follows, after dualising, that X is a left G-space with $e\xi = \xi$ for all $\xi \in X$, and that the map $(x, \xi) \to x\xi$ is continuous. Let $T\colon H_\mu \to C(X)$ be the Gelfand transform, and define $\nu \in \mathrm{PM}(X)$ by setting $\nu(f) = T^{-1}(f)(e)$ $(f \in C(X))$. Define T_ν as in (1). Then for $\phi \in H_\mu$, $x \in G$,

$$\begin{aligned} T_\nu\hat{\phi}(x) &= \hat{\nu}(\hat{\phi}x) = \hat{\nu}(\widehat{\phi x}) = T^{-1}(\widehat{\phi x})(e) \\ &= \phi x(e) = \phi(x) = T^{-1}\hat{\phi}(x). \end{aligned}$$

Thus $T_\nu^{-1} = T$ so that T_ν is a linear isometry. Thus (X, ν) is a Poisson space of μ.]

(ii) Furstenberg investigates the deeper issue of determining the Poisson spaces that can arise when G is a connected, semisimple, Lie group with finite centre. An excellent account of this, and of Poisson spaces in general, is given in Azencott [**1**].

Let G be a connected, semisimple, Lie group with finite centre, and let KAN be an Iwasawa decomposition of G (B61). Then K is a maximal, compact subgroup of G and $S = AN$ is a closed, solvable subgroup of G. Let M be the centraliser of A in K; so

$$M = \{x \in K\colon xa = ax \text{ for all } a \in A\}.$$

Then $xSx^{-1} \subset S$ for all $x \in M$, and it follows that $H(G) = MS$ is a closed subgroup of G. The group $H(G)$ is called a *minimal, parabolic subgroup* of G and is of importance in the representation theory of semisimple Lie groups. Clearly, S is a normal subgroup of $H(G)$ and M is compact, so that $H(G)$ is amenable.

The work of Furstenberg, together with a result of C. C. Moore, shows that there are only a finite number of possibilities for the Poisson spaces X_μ as μ ranges over $P(G)$: such an X_μ has to be of the form $G/(P_\mu S)$, where P_μ is a (compact) subgroup of G such that

$$M_e \subset P_\mu \subset M.$$

(Of course, there is only a finite number of such subgroups since M/M_e is finite.) The subgroup P_μ depends only on the semigroup generated by the support of μ. In particular, G acts transitively on every Poisson space.

We will not give the proofs of these remarkable results. However, we note that the amenability of $H(G)$ is used in the proof in an application of the fixed-point theorem (2.24) and that $G/H(G)$ is the "maximal boundary" of G.

It is convenient at this point to mention some results of C. C. Moore [**2**] on the maximal (closed) amenable subgroups of G. Two examples of such subgroups are the maximal compact subgroup K and the minimal parabolic subgroup $H(G)$. Moore proves that all such subgroups are obtained as follows. Let $X = G/H(G)$. For each $\mu \in \mathrm{PM}(X)$, let

$$G_\mu = \{x \in G \colon x\mu = \mu\},$$

where, of course, the action of G on $M(X)$ is that induced by the action of G on the homogeneous space X. Moore proves, using techniques from the theory of algebraic groups, that *G_μ is always an amenable subgroup of G.* (This result is also proved by Guivarc'h [**2**].) Of particular importance are those subgroups of the form G_μ, where μ is in the weak* closure E of the set of measures $\nu \in \mathrm{PM}(X)$ that are invariant under the actions of some maximal compact subgroup of G and the maximal, *normal*, amenable subgroup of G. It turns out that, with a certain connectivity condition assumed, the set $\mathscr{A}$ of maximal amenable subgroups of G is precisely the set $\{G_\mu \colon \mu \in E\}$. The set E is G-invariant, and by examining the orbits of G in E, the elements of $\mathscr{A}$ can be written down explicitly and the cardinality of the set of conjugate classes of $\mathscr{A}$ determined.

We now proceed with our discussion of Poisson spaces.

Let G be a locally compact group. A measure $\mu \in \mathrm{PM}(G)$ is said to be *spread out* (étalée) if for some $n \in \mathbb{P}$, μ^n is not singular with respect to Haar measure λ. The group G is said to be of *Type* (T) if whenever μ is spread out on G, then G acts transitively on the Poisson space X_μ for μ. (So X_μ can be identified with some compact quotient space G/H, where H is a closed subgroup of G.) Azencott [**1**] generalises Furstenberg's results above to the case of a spread out measure on a group of Type (T).

Paterson [**6**] shows that if S is a compact, jointly continuous semigroup and $\mu \subset \mathrm{PM}(S)$ is such that its support $\mathbf{S}(\mu)$ generates a dense subsemigroup of S, then the Poisson space of μ can be identified with X, where the kernel of S is a "Rees product" $X \times G \times Y$.

(iii) What can be said about H_μ, X_μ, and the amenability of G? The space X_μ is said to be *trivial* if X_μ is a singleton, or equivalently $H_\mu = \mathbb{C}1$. *If there exists $\mu \in \mathrm{PM}(G)$ such that X_μ is trivial then G is amenable.* [Suppose that $H_\mu = \mathbb{C}1$ for some μ, and let $P\colon \mathrm{U_r}(G) \to H_\mu$ be the projection in the proof of (i). Then $m \in \mathfrak{L}(\mathrm{U_r}(G))$, where $P\phi = m(\phi)1$, and G is amenable.] Furstenberg [**3**, p. 213] gives a simple example of a measure μ on the (amenable) "$ax+b$" group with X_μ *nontrivial.*

However, *if G is a locally compact abelian group and $\mu \in \mathrm{PM}(G)$ is such that $\mathbf{S}(\mu)$ generates a dense subgroup of G, then X_μ is trivial.* An elegant proof of this has been given by Choquet and Deny, and appears in Furstenberg [**4**]. See also Revuz [**1**, Chapter 5, §1]. A related result for nilpotent groups is given in Furstenberg [**3**].

As we shall see in (6.47), if μ is "recurrent," then X_μ is trivial.

If G is σ-compact, then there is an important criterion for determining whether or not X_μ is trivial. This involves the augmentation ideal $L^0(G) = \{f \in L_1(G)\colon \int f\,d\lambda = 0\}$ of $L_1(G)$. (Recall (Problem 4-38) that G is amenable if and only if $L^0(G)$ has a bounded approximate identity.) We consider two related conditions for $\mu \in \mathrm{PM}(G)$:

(A) for all $f \in L^0(G)$, $N^{-1}\|f * \sum_{n=1}^N \mu^n\|_1 \to 0$ as $N \to \infty$,

(B) for all $f \in L^0(G)$, $\|f * \mu^n\|_1 \to 0$ as $n \to \infty$.

It is elementary that (B) $\Rightarrow$ (A). The criterion alluded to above is (cf. Rosenblatt [**S1**]): *X_μ is trivial if* (B) *holds.* [Suppose that (B) holds and let $\phi \in H_\mu$, so that $\mu^n\phi = \phi$ for all n. Then for $f \in L^0(G)$,

$$\|f * \phi\|_\infty = \lim_{N\to\infty} N^{-1}\left\|\left(f * \sum_{n=1}^N \mu^n\right) * \phi\right\|_\infty \leq \lim_{N\to\infty}\left(N^{-1}\left\|f * \sum_{n=1}^N \mu^n\right\|_1\right)\|\phi\|_\infty = 0.$$

So $f * \phi = 0$. If $g \in L_1(G)$ and $x \in G$, then putting $f = (g * x - g)$ gives $(x\phi - \phi)(L_1(G)g) = \{0\}$. Since $L_1(G)^2 = L_1(G)$, we have $x\phi = \phi$ so that $\phi \in \mathbb{C}1$.] The converse to this result also holds.

Rosenblatt [**S1**] shows that *if G is σ-compact and amenable, then there exists $\mu = \mu^\sim \in P(G)$ with support equal to G such that* (B) *holds.* Since (B) $\Rightarrow$ (A), this gives a positive answer to a question of Furstenberg [**3**]: *a σ-compact, locally compact group G is amenable if and only if there exists $\mu = \mu^\sim \in P(G)$ with $\mathbf{S}(\mu) = G$ and X_μ trivial.* An earlier proof of the last result (when G is discrete) was given by Kaimanovich and Vershik, and their paper [**S**] contains a useful detailed discussion of how amenability relates to boundaries and entropy. See also Birgé and Raugi [**1**].

(iv) Polynomial growth, in its Type R form, turns out to be relevant to the problem of determining when a locally compact group G is of Type (T). The following theorem is the result of work by Azencott [**1**], Brown and Guivarc'h [**1**], and Moore and Rosenberg [**1**]. *Let G be a connected Lie group with radical S and $\mathfrak{g}$ and $\mathfrak{s}$ their respective Lie algebras. If G/S has finite centre, then G is of Type (T) if and only if* $\mathrm{Sp}(\mathrm{ad}(X)_\mathfrak{s}) \subset i\mathbb{R}$ *for all $X \in \mathfrak{g}$.* This result, combined with Azencott [**1**, Théorème V.3], yields that *the following three statements are equivalent for a connected, solvable Lie group G*:

(a) *G has polynomial growth*;

(b) *G is of Type (T)*;

(c) *X_μ is finite for every spread out measure $\mu \in \mathrm{PM}(G)$.*

(6.47) Random walks and amenability. Let G be a *separable*, locally compact group and $\mu \in \mathrm{PM}(G)$. Let $\Omega' = G^{\mathbb{P}}$, $\Omega = G^{\mathbb{P}}$. Let $\mu_i = \mu$ ($i \in \mathbb{N}$), and let P be $\Omega' = G^{\mathbb{P}}$, the infinite product $\prod_{i=1}^\infty \mu_i$ on Ω'. Then P is a probability measure on Ω'. For each i, let $X_i\colon \Omega \to G$ be the canonical projection onto the

ith coordinate space G_i $(= G)$. For $x \in G$, let $T_x: \Omega' \to \Omega$ be given by

$$T_x(x_1, x_2, \dots) = (x, xx_1, xx_1x_2, \dots)$$

and P_x be the Ω-probability measure $P \circ T_x^{-1}$. Then the triple $\{\Omega, \{P_x\}, \{X_n\}\}$ is a *random walk of law* μ *on* G. Now suppose that *the subgroup of* G *generated by* $\mathbf{S}(\mu)$ *is dense in* G. Consider the following two properties:

(i) $P_e(\{w \in \Omega: X_n(w) \in C$ for infinitely many $n\}) = 0$ for every $C \in \mathscr{C}(G)$;

(ii) $P_e(\{w \in \Omega: X_n(w) \in U$ for infinitely many $n\}) = 1$ for every open neighborhood U of e.

If (i) [(ii)] holds, then μ is said to be *transient* [*recurrent*]. It can be shown that μ is either transient or recurrent. Furstenberg [**3**] shows that *if* G *is not amenable, then* μ *is necessarily transient.* (This result is also a consequence of (6.46(iii)) and the Dacunha-Castelle et al. result below.) On the other hand (Dacunha-Castelle et al. [**1**, p. 286]) μ *is recurrent if and only if* $H_\mu = \mathbb{C}1$ (*so that* G *is amenable*) ((6.46)). However, there exist amenable groups with transient measures μ (cf. (6.4(iii))). Indeed, it is proved in Brunel, Crepel et al. [**1**] that *if* G *is not unimodular, then every* μ *is transient.* Thus, every μ on the "$ax + b$" group is transient. Even more striking, if $G = \mathbb{R}^d$ or $\mathbb{Z}^d$ with $d \geq 3$, then *every* μ *is transient* (Revuz [**1**, p. 100]). It is shown in Dacunha-Castelle et al. [**1**, p. 294] that *if there exists a recurrent* μ *on* G, *where* G *is a connected, Lie group, then* G *has polynomial growth.*

The measure $\overline{\mu}$, where $\overline{\mu} = \sum_{n=0}^{\infty} \mu^n$ is defined on $\mathscr{B}(G)$, is of importance. It can be shown (Revuz [**1**, p. 89]) that μ *is recurrent if and only if* $\overline{\mu}(U) = \infty$ *for every open neighborhood* U *of* e. If μ is transient, then $\overline{\mu}$ is a Radon measure in the sense that $\overline{\mu}(C) < \infty$ for all $C \in \mathscr{C}(G)$ and $\overline{\mu} \geq 0$.

Suppose that μ is transient. The *vague topology* on the set $R(G)$ of Radon measures on G is the weakest topology for which the maps $\nu \to \int f\,d\nu$ $(f \in C_c(G))$ are continuous. Let $K = \{\delta_x * \overline{\mu}: x \in G\}$. The transience of μ entails that K is relatively compact in the vague topology. Consider the closure $\overline{K}$ of K in $R(G)$. It can be shown (ibid, p. 140) that either $\overline{K} \sim K = \{0\}$ or there exists $c > 0$ such that $\overline{K} \sim K = \{0, c\lambda\}$, where $\lambda = \lambda_G$. In the first case, μ is said to be of *Type* I; in the second case, μ is said to be of *Type* II.

Guivarc'h [**2**] shows that *if* G *is not amenable, then* μ *is necessarily of Type* I. The amenable case is discussed in Elie [**1**], who proves the following theorem. *Let* G *be amenable and almost connected. If* G *is unimodular, then there exists a transient, spread out measure of Type* II *on* G *if and only if* $G = \mathbb{R} \times K$ *for some compact group* K. Elie also completely analyses the (more complicated) nonunimodular case.

(6.48) Symmetry, the Wiener property, and polynomial growth.

Recall ([Ri], [BD]) that the *spectrum* $\mathrm{Sp}_A(a)$ (or $\mathrm{Sp}(a)$) of an element a in a Banach algebra A with identity 1 is the set of complex numbers α such that $(a - \alpha 1)$ is not invertible in A. If A does not have an identity—for example, if $A = L_1(G)$ with G nondiscrete—then $\mathrm{Sp}(a)$ is the spectrum of a in the algebra

$\tilde{A} = A + \mathbb{C}1$ obtained by adjoining an identity to A. The set $\mathrm{Sp}(a)$ is nonempty and compact in $\mathbb{C}$, and the spectral radius $\nu(a)$ of a is defined to be $\sup\{|\alpha|: \alpha \in \mathrm{Sp}(a)\}$, and equals $\lim_{n\to\infty} \|a^n\|^{1/n}$.

A Banach $*$-algebra A is said to be *symmetric* if $\mathrm{Sp}(x^*x) \subset [0, \infty)$ for all $x \in A$ [**Ri**, p. 233]. It is known that A is symmetric if and only if $\mathrm{Sp}(h) \subset \mathbb{R}$ whenever $h \in A$ is selfadjoint [**BD**].

Let G be a locally compact group. The group G is said to be *symmetric* if the Banach $*$-algebra $L_1(G)$ is symmetric. The class of symmetric groups is denoted by $[S]$. An interesting, and topical, question is that of determining which groups G are in $[S]$ (Bonic [**1**]). This question will also be referred to as "$[S]$." The question has amenability overtones: *every almost connected locally compact group in* $[S]$ *is amenable.* This result, due to Jenkins [**8**], uses the representation theory of semisimple Lie groups. Is there an easier proof available?

As Leptin and Poguntke [**1**] comment, "the history of $[S]$ is a line of destroyed hopes and wrong conjectures." For example, at one time it was hoped (Hulanicki [**1**]) that every amenable group was symmetric. However, Jenkins [**1**] produced an amenable group not in $[S]$. Indeed, Jenkins [**2**] showed that if a discrete group G contains FS_2 as a subsemigroup, then $G \notin [S]$ (Problem 6-24). Thus, for example, the (amenable) "$ax + b$" group S_2, as a discrete group, is not symmetric ((6.39)). (However, with its usual connected Lie group topology, $S_2 \in [S]$. Indeed, it is shown by Leptin and Poguntke that there exists exactly *one* simply connected, solvable Lie group of dimension less than or equal to 4 that is not symmetric.)

Does every group G of polynomial growth belong to $[S]$ (Gangoli [**1**])? Evidence in favour of this is provided by the following: $G \in [S]$ if one of (a), (b), (c), (d), or (e) holds:

(a) G is compact;

(b) G is discrete and nilpotent (Hulanicki [**5**]);

(c) G is a connected, nilpotent Lie group (Poguntke [**1**]);

(d) $G \in [FC]^-$ ((4.23) (Anusiak [**1**]);

(e) G is a motion group (Gangoli [**1**]).

However, it is shown in Fountain, Ramsay and Williamson [**1**] that there exists a discrete, nonsymmetric group of polynomial growth. This example is discussed in (6.51).

Perhaps the most remarkable result in recent years on $[S]$ is the following beautiful theorem of Ludwig [**1**]: $G \in [S]$ *if* G *is connected and of polynomial growth.* Ludwig's theorem is discussed in (6.50). He also establishes the symmetry of compact extensions of nilpotent groups. These results have recently been generalised by Losert [S7].

Jenkins [**8**] shows that every connected, reductive Lie group with noncompact, semisimple component is nonsymmetric.

Let G be exponentially bounded. Then there exists an important dense, $*$-subalgebra B of $L_1(G)$. References for B are Jenkins [**2**], Hulanicki [**7**], [**8**], [**9**], and Pytlik [**1**]. The elements of B are the "rapidly decreasing" functions on

G. The algebra B is "symmetric" in the sense that $\mathrm{Sp}_{L_1(G)}(f * f^{\sim}) \subset [0, \infty)$ for all $f \in B$, and is relevant to the problem of determining when $\mathrm{Sp}_{L_1(G)}(f) = \mathrm{Sp}(\pi_2(f))$ for $f = f^{\sim}$ in $L_1(G)$. (Here, π_2 is the left regular representation of G.) In this connection, see (6.49). The algebra B is also relevant to two other issues.

The first of these is the elegant "$C_c^{(r)}$-functional calculus theorem" (Kahane [**2**], Dixmier [**4**], Hulanicki [**8**])—see Problem 6-22.

The second of these concerns the study (Hulanicki [**8**]) of the algebra generated by the fundamental solution $\{p_t\}$ of the heat equation on a connected Lie group G. The functions p_t form a commuting, approximate identity for B. (This is generalised in Hulanicki and Pytlik [**1**].) Hulanicki obtains a Wiener-Tauberian type theorem when G has polynomial growth, using the above functional calculus theorem.

Another important property for a group, related to symmetry, is the *Wiener Property* (Leptin [**7**], [**8**]). A locally compact group G is said to have the *Wiener Property* (or, simply, is *Wiener*) if every closed, proper ideal of $L_1(G)$ is contained in the kernel of a nondegenerate, continuous, $*$-representation of $L_1(G)$ on a Hilbert space. (The classical Wiener theorem [**R**, Chapter 1, (4.4)] can be reformulated: $\mathbb{R}$ *is Wiener*.) A related condition is introduced in Leptin [**8**]: G is said to be *weakly Wiener* if every closed, proper ideal of $L_1(G)$ is contained in a primitive ideal of $L_1(G)$. It is proved in Leptin [**7**] that if G is symmetric, then G is Wiener if G is weakly Wiener.

From Leptin [**8**] and Ludwig [**1**], we have that *G is weakly Wiener if G has polynomial growth.* It follows from Ludwig's theorem on symmetry that *every connected group of polynomial growth is Wiener.* (See also Hulanicki, Jenkins et al. [**1**].) Ludwig shows that discrete exponentially bounded groups and discrete solvable groups are Wiener groups. Does polynomial growth imply Wiener (Hulanicki, Jenkins et al. [**1**])? Does there exist a discrete group which is not Wiener (Leptin [**8**])? Using a theorem of M. Dufflo, Leptin shows that nonamenable, connected Lie groups are never Wiener. However, not every amenable group is Wiener; Leptin and Poguntke [**1**] show that a certain 4-dimensional, solvable Lie group is not Wiener.

Other relevant papers are Hauenschild and Kaniuth [**1**], Ludwig [**2**], and Weit [**1**].

(6.49) A symmetric algebra associated with an exponentially bounded group (Hulanicki [**7**], [**8**]; Jenkins [**11**]). Let G be a locally compact group. A $*$-subalgebra A of $L_1(G)$ is said to be *symmetric* if $\mathrm{Sp}_{L_1(G)}(f * f^{\sim}) \subset [0, \infty)$ for all $f \in A$. A function $f \in L_1(G)$ is said to be *rapidly decreasing* if there exists $C \in \mathscr{C}_e(G)$ such that for all $r > 0$,

$$n^r \int_{G \sim C^n} |f(x)|\, d\lambda(x) \to 0 \quad \text{as } n \to \infty.$$

If G is exponentially bounded, it turns out that the set B of rapidly decreasing functions on G is a symmetric $*$-subalgebra of $L_1(G)$. We shall be content to prove that *if G is exponentially bounded then* $\{0\} \cup \mathrm{Sp}_{L_1(G)}(f) = \{0\} \cup \mathrm{Sp}(\pi_2(f))$ *for all hermitian* $f \in C_c(G)$ *(so that the subalgebra* $C_c(G)$ *of* B *is symmetric).* [Suppose that G is exponentially bounded. For $\alpha \in L_1(G)$ [$\mathbf{B}(L_2(G))$], let $\nu(\alpha)$ and $\mathrm{Sp}(\alpha)$ be the spectral radius and spectrum respectively of α in $L_1(G)$ [$\mathbf{B}(L_2(G))$]. Let $f \in C_c(G)$ be such that $f = f^{\sim}$. Let $C \in \mathscr{C}_e(G)$ contain the support of f, and let $f^{(n)}$ be the nth convolution power of f in $C_c(G)$. Then C^n contains the support of $f^{(n)}$, and using the Cauchy-Schwarz inequality, for $n > 1$,

$$\begin{aligned} \|f^{(n)}\|_1 &= \int_G |f^{(n)}(x)|\chi_{C^n}(x)\,d\lambda(x) \le \|\chi_{C^n}\|_2 \|f^{(n)}\|_2 \\ &\le \lambda(C^n)^{1/2}\|\pi_2(f^{(n-1)})\|\,\|f\|_2. \end{aligned} \tag{1}$$

Taking nth roots, letting $n \to \infty$, and using the exponential boundedness of G, we obtain

$$\begin{aligned} \nu(f) &= \lim_{n\to\infty} \|f^{(n)}\|_1^{1/n} \\ &\le \lim_{n\to\infty} \{[(\lambda(C^n))^{1/n}]^{1/2}\|\pi_2(f^{(n-1)})\|^{1/n}\|f\|_2^{1/n}\} \\ &= \nu(\pi_2(f)). \end{aligned}$$

The reverse inequality follows since $\|\pi_2\| \le 1$, and we have $\nu(f) = \nu(\pi_2(f))$.

Let A_f be the (commutative $*$-) subalgebra of $L_1(G)$ generated by f. Let X_f be the closure of A_f in $L_1(G)$, and Y_f the closure of $\pi_2(X_f)$ in $\mathbf{B}(L_2(G))$. Let F be a multiplicative linear functional on X_f, and define $F'\colon \pi_2(X_f) \to \mathbb{C}$ by $F'(\pi_2(g)) = F(g)$ $(g \in X_f)$. (The map F' is well defined since π_2 is a monomorphism.) If g, h are hermitian elements of A_f, then, since A_f is commutative,

$$\begin{aligned} |F'(\pi_2(g+ih))| &\le |F'(\pi_2(g))| + |F'(\pi_2(h))| = |F(g)| + |F(h)| \le \nu(g) + \nu(h) \\ &= \nu(\pi_2(g)) + \nu(\pi_2(h)) \le 2\nu(\pi_2(g+ih)), \end{aligned}$$

and so F' is continuous on $\pi_2(A_f)$. Extending F' by continuity to Y_f, it follows that $\mathrm{Sp}_{X_f}(f) \setminus \{0\} \subset \mathrm{Sp}_{Y_f}(\pi_2(f)) \setminus \{0\}$. The reverse conclusion is obvious, and since the preceding two spectra equal their own boundaries, we have [**Ri**, (1.6.12)]

$$\mathrm{Sp}_{X_f}(f) \cup \{0\} = \mathrm{Sp}(f) \cup \{0\}, \qquad \mathrm{Sp}_{Y_f}(f) \cup \{0\} = \mathrm{Sp}(\pi_2(f)) \cup \{0\}$$

so that $\{0\} \cup \mathrm{Sp}(f) = \{0\} \cup \mathrm{Sp}(\pi_2(f))$ as required.]

A much more general result than this is proved by Jenkins [**11**].

(6.50) Polynomial growth and the symmetry of $L_1(G)$. Our main objective in this section is to establish the symmetry result (v) of Ludwig. (Losert [S7] generalises (v).)

The following result (i) is valid for all connected groups of polynomial growth. However, we shall be content to deal only with the simply connected, Lie case.

(i) (Guivarc'h [**1**]). *Let G be a simply connected, Lie group of polynomial growth. Then there exists a series*

$$\{e\} = G_0 \subset G_1 \subset \cdots \subset G_n = G$$

of closed, normal subgroups G_i of G such that for each $k \geq 1$ and $x \in G_k/G_{k-1}$, there exists $C \in \mathscr{C}_e(G_k/G_{k-1})$ with $x \in C$ and $yCy^{-1} \subset C$ for all $y \in G/G_{k-1}$. [Let S be the radical of G, and $\mathfrak{s}$, $\mathfrak{g}$ the Lie algebras of S and G respectively. Then S is of Type R and using Lie's Theorem (B25) inductively, we can find real linear functions $\gamma_i : \mathfrak{s} \to i\mathbb{R}$ $(1 \leq i \leq r)$ and distinct subspaces W_k $(0 \leq k \leq r)$ of $\mathfrak{s}_{\mathbb{C}}$ such that

$$\{0\} = W_0 \subset W_1 \subset \cdots \subset W_r = \mathfrak{s}_{\mathbb{C}}$$

and, for each $k \geq 1$, $W_k/W_{k-1} = \{Z \in \mathfrak{s}_{\mathbb{C}}/W_{k-1} : (\operatorname{ad} X)_{\mathfrak{s}_{\mathbb{C}}/W_{k-1}}(Z) = \gamma_k(X)Z$ for all $X \in \mathfrak{s}\}$. We claim that W_k is an ideal in $\mathfrak{g}_{\mathbb{C}}$. The proof will be given for $k = 1$, the general case proceeding by induction and involving a very similar argument.

Let $Y \in \mathfrak{g}$, $Z \in W_1$, and $X \in \mathfrak{s}$. Noting that $\gamma_1(X)[Y, Z] = [Y, \gamma_1(X)Z] = [[Z, X], Y]$, we have, using Jacobi's identity,

$$(*) \qquad \begin{aligned} &\gamma_1([X, Y])Z - [X, [Y, Z]] + \gamma_1(X)[Y, Z] \\ &\quad = [[X, Y], Z] + [[Y, Z], X] + [[Z, X], Y] = 0 \end{aligned}$$

Now $[X, Y]$ belongs to the nil-radical of $\mathfrak{g}$ (B37), so that $T = \operatorname{ad}([X, Y]) \in L(\mathfrak{s}_{\mathbb{C}})$ is nilpotent. Since $\gamma_1([X, Y]) \in \operatorname{Sp}(T)$, it follows that $\gamma_1([X, Y]) = 0$, so that by $(*)$,

$$\operatorname{ad} X([Y, Z]) = \gamma_1(X)[Y, Z].$$

Thus $[Y, Z] \in W_1$ and W_1 is an ideal in $\mathfrak{g}_{\mathbb{C}}$ as required.

Since $\exp(\mathfrak{s})$ is dense in S (Appendix E), it follows that, for each $s \in S$, W_k is $(\operatorname{Ad} s)$-invariant, and $(\operatorname{Ad} s)_{W_k/W_{k-1}}$ is of the form zI, where $z \in \mathbb{T}$ and I is the identity of $L(W_k/W_{k-1})$.

Let V_k be the subspace of $\mathfrak{s}$ consisting of the "real parts" of elements of W_k and $Y_k = V_k/V_{k-1}$. Then V_k is an ideal in $\mathfrak{g}$ and

$$\{0\} = V_0 \subset V_1 \subset \cdots \subset V_r = \mathfrak{s}.$$

We now claim that each Lie algebra Y_k is abelian. For let X_j $(1 \leq j \leq 4)$ in V_k be such that $X_1 + iX_2$, $X_3 + iX_4 \in W_k$. Let $\gamma_k(X_1) = ia$, $\gamma_k(X_3) = ib$ $(a, b \in \mathbb{R})$. If $a = 0$, then $[X_1, X_3 + iX_4] \in W_{k-1}$ so that $[X_1, X_3] \in V_{k-1}$. Suppose that $a \neq 0$. Then $i[X_1, X_2] = [X_1, X_1 + iX_2] \in ia(X_1 + iX_2) + W_{k-1}$, whence, on taking real parts, $X_2 \in V_{k-1}$. Thus $[X_3, X_1] \in [X_3, X_1 + iX_2] + iV_{k-1} \subset ib(X_1 + iX_2) + W_{k-1} + iV_{k-1}$, whence $[X_3, X_1] \in V_{k-1}$. So in all cases, $[X_1, X_3] \in V_{k-1}$, and $Y_k = V_k/V_{k-1}$ is abelian.

Now as $\{(\operatorname{Ad} s)_{W_k/W_{k-1}} : s \in S\} \subset \mathbb{T}I$ and G/S is compact ((6.32), (6.39)), it follows that the group $\{(\operatorname{Ad} u)_{W_k/W_{k-1}} : u \in G\}$ is contained in a compact subgroup of $\operatorname{GL}(W_k/W_{k-1})$. It readily follows that H_k^-, where $H_k = \{(\operatorname{Ad} u)_{Y_k} : u \in G\}$, is compact. Let $\|\cdot\|_k$ be a norm on Y_k, relative to which the elements of H_k are isometries [**HR1**, (22.23)].

Let G_k be the connected, normal, Lie subgroup of G with V_k as Lie algebra. Then G_k is closed in G since G is simply connected (B17). Let $n = r + 1$ and $G_n = G$. The result of (i) is obviously true when $k = n$, since $G_n/G_{n-1} = G/S$ is compact. Suppose that $k < n$.

Then Y_k is the Lie algebra of G_k/G_{k-1}, and since Y_k is abelian, $\exp(Y_k) = G_k/G_{k-1}$. Let $x \in G_k/G_{k-1}$ and $X \in Y_k$ be such that $\exp X = x$. Let

$$U = \{Y \in Y_k \colon \|Y\|_k \leq \|X\|_k + 1\},$$

and $C = \exp(U)$. Then $C \in \mathscr{C}_e(G_k/G_{k-1})$, $x \in C$, and if $u \in G$ and $y = uG_{k-1}$, then (B12)

$$yCy^{-1} = \exp((\operatorname{Ad} u)_{Y_k}(U)) \subset \exp(U) = C.]$$

(ii) (Jenkins [**10**]). *Let $\varepsilon > 0$, let G be an exponentially bounded, locally compact group, $C \in \mathscr{C}(G)$, and let H be the closed subgroup of G generated by C. Then there exists $p \in L_1(G)$ such that $p(t) > 0$ for all $t \in H$ and $|p(ct) - p(t)| \leq \varepsilon p(t)$ for $t \in H$, $c \in C$.* [Find $U \in \mathscr{C}_e(G)$ such that $C \subset U = U^{-1}$. Then $H \subset \bigcup_{k=1}^{\infty} U^k$. Let $\eta > 0$. With $U^0 = \varnothing$, set

$$p = \sum_{k=0}^{\infty} (1+\eta)^{-k} \chi_{U^{k+1} \sim U^k}.$$

Then $p(t) > 0$ for all $t \in H$, and for $c \in C \cup C^{-1}$, $U^k \supset c^{-1}U^{k-1}$. Hence if $t \in U^{k+1} \sim U^k$, then $ct \notin U^{k-1}$, so that $ct \in U^{r+1} \sim U^r$ for some $r \geq (k-1)$. Hence, for this r,

$$p(ct) = (1+\eta)^{-r} \leq (1+\eta)^{-(k-1)} = (1+\eta)p(t). \tag{1}$$

Replacing t, c by ct, c^{-1} in (1), we have $p(t) \leq (1+\eta)p(ct)$, so that $|p(t) - p(ct)| \leq \eta(1+\eta)p(t)$. Now choose η so that $\eta(1+\eta) < \varepsilon$.

It remains to show that $p \in L_1(G)$. To this end,

$$\begin{aligned} \int_G |p(t)|\, d\lambda(t) &= \sum_{k=0}^{\infty} \int_{U^{k+1} \sim U^k} p(t)\, d\lambda(t) = \sum_{k=0}^{\infty} (1+\eta)^{-k} \lambda(U^{k+1} \sim U^k) \\ &\leq \sum_{k=0}^{\infty} (1+\eta)^{-k} \lambda(U^{k+1}) < \infty \end{aligned}$$

since $\lambda(U^{k+1})^{1/k} \to 1$ as $k \to \infty$.]

Jenkins obtains a converse to the above result.

We now introduce some notation. Recall [**Ri**, p. 42] that a left ideal I in an algebra A is called *modular* if for some $u \in A$, $A(1-u) \subset I$. Such an element u is said to be a *right modular unit* for I.

Let G be a locally compact group and $\mathscr{L}$ the set of proper, modular left ideals of $L_1(G)$.

A *bounded, sesquilinear, positive, hermitian functional* of $L_1(G)$ is a mapping $s \colon L_1(G) \times L_1(G) \to \mathbb{C}$ for which there exists $M > 0$ such that for all g, g',

$h \in L_1(G)$ and all $\alpha \in \mathbb{C}$, we have

$$s(g,g) \geq 0, \qquad s(g,h) = \overline{s(h,g)},$$
$$s(\alpha g + g', h) = \alpha s(g,h) + s(g',h), \quad \text{and} \quad |s(g,h)| \leq M\|g\|_1\|h\|_1.$$

The set of such functionals is denoted by $\mathscr{P}$. Of course, every $s \in \mathscr{P}$ satisfies the Cauchy-Schwarz inequality:

$$|s(f,g)| \leq s(f,f)^{1/2}s(g,g)^{1/2}. \tag{2}$$

Each positive, linear functional ϕ on $L_1(G)$ yields an element $s_\phi \in \mathscr{P}$, where $s_\phi(g,h) = \phi(h^\sim * g)$ and $h \to h^\sim$ is the involution on $L_1(G)$. Let P be the set of positive, linear functionals on $L_1(G)$.

For $s \in \mathscr{P}$, define $\|s\|$ by

$$\|s\| = \sup\{|s(g,h)|: g, h \in L_1(G), \|g\|_1 = \|h\|_1 = 1\}.$$

Clearly $\mathscr{P}$ is a G-set with action given by

$$(sx)(g,h) = s(x * g, x * h), \qquad (xs)(g,h) = s(g * x, h * x).$$

Clearly, $\|sx\| = \|s\| = \|xs\|$.

Now let $s \in \mathscr{P}$ and f, g, $h \in L_1(G)$. We claim that

$$\begin{aligned} s(f * g, h) &= \int f(x)s(x * g, h)\, d\lambda(x), \\ s(f * g, h) &= \int g(x)s(f * x, h)\, d\lambda(x). \end{aligned} \tag{3}$$

Indeed, if we define $\psi \in L_1(G)'$ by $\psi(k) = s(k,h)$ $(k \in L_1(G))$, then

$$\begin{aligned} s(f * g, h) &= \psi(f * g) = (g\psi)(f) = \int f(x)(g\psi)(x)\, d\lambda(x) \\ &= \int f(x)\psi(x * g)\, d\lambda(x) = \int f(x)s(x * g, h)\, d\lambda(x), \end{aligned}$$

giving the first equality of (3). The second equality follows similarly.

For $I \in \mathscr{L}$ and a subgroup H of G, define subspaces of $\mathscr{P}$:

$$\begin{aligned} \mathscr{P}_I &= \{s \in \mathscr{P}: s(f,g) = 0 \text{ for all } f \in I, g \in L_1(G)\}, \\ \mathscr{P}_I^H &= \{s \in \mathscr{P}_I: sx = s \text{ for all } x \in H\}. \end{aligned}$$

(iii) (Leptin [**7**], Ludwig [**1**]). (a) *Let* $I \in \mathscr{L}$. *If* $s_0 \in \mathscr{P}_I^G$, *then* $s_0(f_1 * f, g) = s_0(f, f_1^\sim * g)$ $(f_1, f, g \in L_1(G))$.

(b) $\mathscr{P}_I \neq \{0\}$ *for all* $I \in \mathscr{L}$.

(c) *The following three statements are equivalent*:

(α) $L_1(G)$ *is symmetric*;

(β) *if* $I \in \mathscr{L}$, *then there exists nonzero* $\phi \in P$ *such that* $\phi(I) = \{0\}$;

(γ) *if* $I \in \mathscr{L}$, *then* $\mathscr{P}_I^G \neq \{0\}$.

[(a) Let $s_0 \in \mathscr{P}_I^G$. If $x \in G$, then $s_0(x * f, g) = s_0x^{-1}(x * f, g) = s_0(f, x^{-1} * g)$.

Define ϕ_g, $\psi_f \in L_1(G)'$ by $\phi_g(f_1) = s_0(f_1, g)$, $\psi_f(g_1) = \overline{s_0(f, g_1)}$. Then $\phi_g(x * f) = s_0(x * f, g) = s_0(f, x^{-1} * g) = \overline{\psi_f(x^{-1} * g)}$, so that $s_0(f_1 * f, g) = \phi_g(f_1 * f) = (f\phi_g)(f_1)$, and using (1.1(3)),

$$\begin{aligned} s_0(f_1 * f, g) &= \int f_1(x)(f\phi_g)(x)\, d\lambda(x) = \int f_1(x)\phi_g(x * f)\, d\lambda(x) \\ &= \int f_1(x)\overline{\psi_f(x^{-1} * g)}\, d\lambda(x) = \overline{(g\psi_f)^*(f_1)} \\ &= \overline{\psi_f(f_1^{\sim} * g)} = s_0(f, f_1^{\sim} * g) \end{aligned}$$

as required.

(b) If $I \in \mathscr{L}$ and $\phi \in L_1(G)'$ is nonzero and such that $\phi(I) = \{0\}$, then $p_\phi \in \mathscr{P}_I \sim \{0\}$, where $p_\phi(f, g) = \phi(f)\overline{\phi}(g)$.

(c) Suppose that (α) holds, and let $I \in \mathscr{L}$. We prove that (β) holds. We will suppose that $L_1(G)$ does not have an identity, the argument when $L_1(G)$ does have an identity being similar. Let $\tilde{A}$ be the Banach algebra obtained by adjoining an identity 1 to $L_1(G)$. Let $J = I + \mathbb{C}(1-u)$, where u is a right, modular unit in $L_1(G)$ for I. Then $1 \notin J$, since otherwise $u \in I$. So J is a proper, left ideal of $\tilde{A}$. Using [**Ri**, (4.7.9), (4.7.11)], $\tilde{A}$ is also symmetric, and there exists a nonzero positive functional ψ on $\tilde{A}$ with $\psi(J) = \{0\}$. Since $\tilde{A} = L_1(G) + J$, it follows that $\phi = \psi|_{L_1(G)} \neq 0$. So (α) implies (β).

If $I \in \mathscr{L}$ and ϕ is as in (β) then $s_\phi \in \mathscr{P}_I^G$ since ϕ vanishes on the left ideal I, and $\phi((x * h)^{\sim}(x * g)) = \phi(h^{\sim} g)$. Also, $s_\phi \neq 0$ since $L_1(G)^2 = L_1(G)$ and $\phi \neq 0$. So (β) implies (γ). Conversely, suppose that (γ) holds. Let $I \in \mathscr{L}$ and $s \in \mathscr{P}_I^G \sim \{0\}$. Let u be a right modular unit for I. Then if f, $g \in L_1(G)$, we have, using (a), (2), and the fact that $L_1(G)(1-u) \subset I$,

$$\begin{aligned} (4) \qquad |s(f, g)| &= |s(f - f * u, g) + s(f * u, g)| = |s(u, f^{\sim} * g)| \\ &\leq s(u, u)^{1/2} s(f^{\sim} * g, f^{\sim} * g)^{1/2}. \end{aligned}$$

Since $s \neq 0$, it follows that

$$(5) \qquad s(u, u) > 0.$$

(Note that this inequality is true with s replaced by any $s' \in \mathscr{P}_I \sim \{0\}$.) Define $\phi \in L_1(G)'$ by $\phi(f) = s(f * u, u)$. Then by (a), $\phi(f^{\sim} * f) = s(f * u, f * u) \geq 0$. Thus $\phi \in P$. Now let $f \in I$. Then $\phi(f) = -s(f - f * u, u) + s(f, u) = 0$ so that $\phi(I) = \{0\}$. Finally, if $\{e_\delta\}$ is a bounded approximate identity for $L_1(G)$, then using (5),

$$\phi(e_\delta) = s(e_\delta * u, u) \to s(u, u) > 0$$

so that $\phi \neq 0$. So (γ) implies (β).

It remains to show that (β) implies (α). Let $z \in L_1(G)$ and $h = z^{\sim} * z$. Suppose that $-h$ is quasisingular, and let I be the proper, modular, left ideal $L_1(G)(1 + h)$. Let ϕ be as in (β). For all $f \in L_1(G)$, $f + f * h \in I$, so that $\phi(f) = -\phi(f * h)$. With $f = h$, we obtain $\phi(h) = -\phi(h^2)$, and since $\phi(h), \phi(h^2) \geq 0$, we have $\phi(h) = 0 = \phi(h^2)$. The Cauchy-Schwarz inequality yields that $\phi(f * h) = 0$ for all $f \in L_1(G)$ so that $\phi(f) = \phi(f + f * h - f * h) = 0$.

Since $\phi \neq 0$, a contradiction results. Hence $-h$ is quasiregular and [**Ri**, (4.7.5)] $L_1(G)$ is symmetric.]

(iv) (Ludwig [**1**]). *Let H and N be closed, normal subgroups of G and be such that $N \subset H$ and for each $z \in H/N$, there exists $C \in \mathscr{C}_e(H/N)$ with $z \in C$ and $yCy^{-1} \subset C$ for all $y \in G/N$. Let $I \in \mathscr{L}$. Then $\mathscr{P}_I^H \neq \varnothing$ if $\mathscr{P}_I^N \neq \varnothing$.* [We first establish the three inequalities (6), (7), (8) below for $s \in \mathscr{P}$. For f, $g \in L_1(G)$, we have, using (3) and (2), writing $|f(x)| = (|f(x)|^{1/2})^2$ and using the Cauchy-Schwarz inequality twice,

$$\begin{aligned} s(f*g, f*g) &= \int f(x)s(x*g, f*g)\,d\lambda(x) \\ &\leq \|f\|_1^{1/2}\left(\int |f(x)||s(x*g, f*g)|^2\,d\lambda(x)\right)^{1/2} \\ &\leq \|f\|_1^{1/2}\left(\int |f(x)|s(x*g, x*g)s(f*g, f*g)\,d\lambda(x)\right)^{1/2}. \end{aligned}$$

Dividing both sides of the preceding inequality by $(s(f*g, f*g))^{1/2}$ (when the latter is nonzero!) and squaring, it follows that

$$s(f*g, f*g) \leq \|f\|_1 \int |f(x)|(sx)(g,g)\,d\lambda(x). \tag{6}$$

Similarly, one can show that

$$s(f*g, f*g) \leq \|g\|_1 \int_G |g(x)|(xs)(f,f)\,d\lambda(x). \tag{7}$$

Further, if u is a right modular unit for I and $s \in \mathscr{P}_I$, then, using (6),

$$\begin{aligned} s(f,f) &= s(f*u, f*u) + s(f - f*u, f*u) + s(f, f - f*u) = s(f*u, f*u) \\ &\leq \|f\|_1 \int |f(x)|(sx)(u,u)\,d\lambda(x) \leq \|f\|_1^2 \sup_{x\in G}(sx)(u,u). \end{aligned} \tag{8}$$

Suppose that $\mathscr{P}_I^N \neq \varnothing$. Let $L = G/N$. By defining $s(xN)$ $(= s(Nx))$ to be sx $(s \in \mathscr{P}_I^N)$, we see that $\mathscr{P}_I^N$ is a right L-set. Let $K \in \mathscr{C}(H/N)$ and $\varepsilon > 0$. We show that there exists $\tilde{s} \in \mathscr{P}_I^N$ such that

(a) $\tilde{s}(u,u) \geq \frac{1}{2}$,

(b) $\|\tilde{s}\| \leq 1$,

(c) $|(\tilde{s}k - \tilde{s})(f,f)| \leq \varepsilon\|f\|_1^2$ $(k \in K,\ f \in L_1(G))$.

If $z \in H/N$, then, by hypothesis, there exists $C \in \mathscr{C}_e(H/N)$ such that $z \in C$ and $lCl^{-1} \subset C$ for all $l \in L$. Replacing C by C^2 if necessary, we can suppose that $z \in C^0$. A simple compactness argument now shows that there exists $U \in \mathscr{C}_e(H/N)$ with $K \subset U$ and $lUl^{-1} \subset U$ $(l \in L)$. Let V be the subgroup of H/N generated by U. Then V is an open and closed, normal subgroup of L. Since H/N is an $[FC]^-$ group, it is of polynomial growth (Problem 6-7) and so exponentially bounded. Hence by (ii), there exists $p \in L_1(V)$ such that $p(v) > 0$ for all $v \in V$ and

$$|p(u'v) - p(v)| \leq \varepsilon p(v) \qquad (u' \in U, v \in V). \tag{9}$$

Now let $s \in \mathscr{P}_I^N \sim \{0\}$ and, for $f, g \in L_1(G)$, define

$$s'(f,g) = \int_V p(v)(sv^{-1})(f,g)\, d\lambda_V(v). \tag{10}$$

It is easily checked that $s' \in \mathscr{P}_I^N$. Since the map $v \to sv^{-1}(u,u)$ is continuous, $p(v) > 0$ for all v, and $s(u,u) > 0$ (cf. (5)), we have $s'(u,u) > 0$, so that $s' \neq 0$. Let $b = \sup_{x \in G} s'x(u,u)$, and find $y \in G$ such that $s'y(u,u) \geq \frac{1}{2}b$. Finally, put $\tilde{s} = b^{-1}(s'y)$. We check that (a), (b) and (c) are satisfied. The inequality (a) is immediate. Further, using (8) with s replaced by $\tilde{s}$,

$$\begin{aligned} |\tilde{s}(f,g)| &\leq (\tilde{s}(f,f))^{1/2}(\tilde{s}(g,g))^{1/2} \leq \|f\|_1\|g\|_1 \sup_{x\in G}(\tilde{s}x)(u,u) \\ &= \|f\|_1\|g\|_1 \sup_{x \in G}(b^{-1}(s'yx)(u,u)) = \|f\|_1\|g\|_1. \end{aligned}$$

So (b) is true. It remains to prove (c).

Let $k \in K$, and let $\overline{y} = yN$. Using the substitution $v = \overline{y}k\overline{y}^{-1}w$, the fact that $\overline{y}U\overline{y}^{-1} \subset U$, and (9),

$$\begin{aligned} |(\tilde{s}k - \tilde{s})(f,f)| &= b^{-1}\left|\int_V p(v)(sv^{-1}\overline{y}k - sv^{-1}\overline{y})(f,f)\, d\lambda_V(v)\right| \\ &= b^{-1}\left|\int_V p(\overline{y}k\overline{y}^{-1}w)sw^{-1}\overline{y}(f,f)\, d\lambda_V(w)\right. \\ &\qquad \left. - \int_V p(v)sv^{-1}\overline{y}(f,f)\, d\lambda_V(v)\right| \\ &\leq b^{-1}\int_V |p(\overline{y}k\overline{y}^{-1}w) - p(w)|sw^{-1}\overline{y}(f,f)\, d\lambda_V(w) \\ &\leq \varepsilon b^{-1}\int_V p(w)sw^{-1}\overline{y}(f,f)\, d\lambda_V(w) = \varepsilon\tilde{s}(f,f) \leq \varepsilon\|f_1\|^2. \end{aligned}$$

Thus (c) is satisfied.

So $\tilde{s}$ belongs to the set $A_{K,\varepsilon}$, where

$$A_{K,\varepsilon} = \{r \in \mathscr{P}_I^N : \|r\| \leq 1, r(u,u) \geq \tfrac{1}{2}, |(rk - r)(f,f)| \leq \varepsilon\|f\|_1^2 \text{ for all } k \in K, f \in L_1(G)\}.$$

Further $A_{K,\varepsilon}$ is compact in the topology of pointwise convergence on $L_1(G) \times L_1(G)$. Allowing K to vary over $\mathscr{C}(H/N)$ and ε over $(0,\infty)$, the sets $A_{K,\varepsilon}$ have the finite intersection property. So we can find $r' \in \bigcap\{A_{K,\varepsilon} : K \in \mathscr{C}(H/N), \varepsilon > 0\}$. Clearly, $r'x = r'$ for all $x \in H$ and $r' \neq 0$. So $\mathscr{P}_I^H \neq \{0\}$.]

(v) (Ludwig [**1**]) *The algebra $L_1(G)$ is symmetric if either* (a), (b), *or* (c) *holds*:

(a) *G is a connected, Lie group of polynomial growth*;

(b) *G contains a closed, nilpotent, normal subgroup N with G/N compact*;

(c) *G is discrete, finitely-generated and has polynomial growth.* [Suppose that (a) holds. Suppose first that G is simply connected. Let $\{G_i\}$ be the sequence of (i). From (iv), we see that for $I \in \mathscr{L}$, $\mathscr{P}_I^{G_{i+1}} \neq \varnothing$ if $\mathscr{P}_I^{G_i} \neq \varnothing$. By (iii)(b), $\mathscr{P}_I^{G_0} = \mathscr{P}_I \neq \varnothing$. So $\mathscr{P}_I^G \neq \varnothing$ and $L_1(G)$ is symmetric by (iii)(c). Now drop the requirement that G be simply connected and let $\tilde{G}$ be the simply connected covering group of G. Then $\tilde{G}$ is of polynomial growth (since it is of Type R), and by the above, $L_1(\tilde{G})$ is symmetric. Then for some closed subgroup H of $\tilde{G}$, $G = \tilde{G}/H$. Now ((1.11)) the map Q_H is a continuous, *-homomorphism from $L_1(\tilde{G})$ onto $L_1(G)$. Since $\mathrm{Sp}(Q_H(f)) \subset \mathrm{Sp}(f) \cup \{0\}$ for all $f \in L_1(G)$, it follows that $L_1(G)$ is also symmetric.

Now suppose that (b) holds, and let Z be the centre of N. Then Z is a normal subgroup of G. For $x \in G$, let $\alpha_x \in \mathrm{Aut}\, Z$ be given by $\alpha_x(z) = xzx^{-1}$. Since α_x is the identity if $x \in N$, and as G/N is compact, it readily follows that if $z \in Z$, there exists $C \in \mathscr{C}_e(Z)$ with $z \in C$ and $xCx^{-1} \subset C$ for all $x \in G$. Using the subgroups $\{e\}$ and Z, we apply (iv) to deduce that for $I \in \mathscr{L}$, $\mathscr{P}_I^Z \neq \varnothing$. An easy induction argument involving the terms of the upper central series (Appendix A) for N and, finally, G, gives the symmetry of $L_1(G)$.

From Gromov's theorem ((6.40)) and [**M**, Theorem 4.16], we see that if (c) holds, then G has a normal, nilpotent subgroup of finite index. Now apply (b)].

We shall see below that the requirement in (c) that G be finitely-generated cannot be dropped.

(6.51) A discrete, nonsymmetric group of polynomial growth (Fountain et al. [**1**]). We shall produce a nonsymmetric, discrete group G that is locally finite (and so, a fortiori, is of polynomial growth). (Another example of a nonsymmetric group of polynomial growth is given by Hulanicki [**12**].) The group G is also an example of an amenable, nonsymmetric group (cf. Jenkins [**1**]). The author is grateful to John Williamson for a very helpful communication.

Let P be the group of permutations of $\mathbb{P}$ that leave all but a finite number of integers fixed. Let $x_n \in P$ be the finite product of transpositions:

$$x_n = (1, 2^n + 1)(2, 2^n + 2) \cdots (2^n, 2^n + 2^n).$$

Let G be the subgroup of P generated by $\{x_n : n \geq 1\}$. (The group G will turn out to be locally finite and nonsymmetric.) For each n, let G_n be the subgroup of G generated by $\{x_1, \ldots, x_n\}$.

A finite sequence $w = (x_{i_1}, \ldots, x_{i_k})$ is said to be a *word in* G. If $i_j \leq n$ for all j, then w is said to be a *word in* G_n. We also allow $w = \varnothing$ to be a word in G (and in G_n). The set of words in G $[G_n]$ is denoted by W $[W_n]$. If $w = (x_{i_1}, \ldots, x_{i_k})$, then the length $l(w)$ of w is defined to be k. (Of course, $l(\varnothing) = 0$.) If $w' = (x_{j_1}, \ldots, x_{j_r}) \in W$, then $ww' \in W$ is defined by $ww' = (x_{i_1}, \ldots, x_{i_k}, x_{j_1}, \ldots, x_{j_r})$. Associated with w is an element α_w of G, where $\alpha_w = x_{i_1} \cdots x_{i_k}$. (We take $\alpha_\varnothing = e$.) So $\alpha_{ww'} = \alpha_w \alpha_{w'}$. A nonempty word w is said to be *irreducible in* W $[W_n]$ if whenever $w' \in W$ $[W_n]$ is such that $\alpha_w = \alpha_{w'}$, then $l(w) \leq l(w')$. We now establish a number of useful properties of G.

(i)(a) *For all* n, $x_n^2 = e$ *and* $x_{n+1} \notin G_n$;
(b) $x_j x_k x_i x_k = x_k x_i x_k x_j$ *whenever* $i, j < k$;
(c) *each* G_n *is finite*;
(d) *every element of* G_n *is of one of the forms*:

(α) α_{w_1} *for some* $w_1 \in W_{n-1}$;

(β) $\alpha_{w_1} x_n \alpha_{w_2}$ *for some* $w_1, w_2 \in W_{n-1}$;

(γ) $\alpha_{w_1} x_n \alpha_{w_2} x_n$ *for some* $w_1, w_2 \in W_{n-1}$;

(e) *if* $w_1, w_2 \in W_{n-1}$ *are irreducible in* W_{n-1}, *then* $w_1 x_n w_2$ *is irreducible in* W_n;

(f) *if* $w \in W_n$ *is irreducible in* W_n, *then* w *is irreducible in* W;

(g) *if* $i_1, \ldots, i_k$ *are distinct, positive integers, then* $w = (x_{i_1}, \ldots, x_{i_k})$ *is irreducible in* W;

(h) *if* $w_1 = (x_{i_1}, \ldots, x_{i_n})$ *and* $w_2 = (x_{j_1}, \ldots, x_{j_m})$ *are such that* $i_1, \ldots, i_n$ *are all distinct,* $j_1, \ldots, j_m$ *are all distinct and* $w_1 \neq w_2$, *then* $\alpha_{w_1} \neq \alpha_{w_2}$.

[(a) Trivially, $x_n^2 = e$. The element $x_{n+1} \notin G_n$ since every element of G_n fixes $2^{n+1} + 1$.

(b) Let $\sigma, \tau \in P$ with $\tau^2 = e$. Then for each n,

$$\sigma\tau\sigma^{-1}(\sigma(n)) = \sigma(\tau(n)), \qquad \sigma\tau\sigma^{-1}(\sigma(\tau(n))) = \sigma(n)$$

so that $\sigma\tau\sigma^{-1}$ is the product of transpositions $(\sigma(n), \sigma(\tau(n)))$. Applying this result with $\sigma = x_j x_k$, $\tau = x_i$, we have, using (a) and the fact that $i, j < k$,

$$\begin{aligned} x_j x_k x_i x_k x_j &= (\sigma(1), \sigma(2^i + 1)) \cdots (\sigma(2^i), \sigma(2^{i+1})) \\ &= (x_k(1), x_k(2^i + 1)) \cdots (x_k(2^i), x_k(2^{i+1})) \\ &= x_k x_i x_k^{-1}. \end{aligned}$$

(c) The group G_n is finite since it can be regarded as a group of permutations on the finite set $[1; 2^{n+1}]$.

(d) Since $x_j = x_j^{-1}$ for all j ((a)), every element of G_{n-1} is of the form α_{w_1} for some $w_1 \in W_{n-1}$. Let $x \in G_n \sim G_{n-1}$. It remains to show that x is of the form (β) or (γ). Let $w_1 = (x_{i_1}, \ldots, x_{i_m})$, $w_2 = (x_{j_1}, \ldots, x_{j_k})$ belong to W_{n-1}. Using (b),

$$\begin{aligned} \alpha_{w_1} x_n \alpha_{w_2} &= x_{i_1} \cdots x_{i_{m-1}} (x_{i_m} x_n x_{j_1}) x_{j_2} \cdots x_{j_k} \\ &= x_{i_1} \cdots x_{i_{m-1}} (x_n x_{j_1} x_n x_{i_m} x_n) x_{j_2} \cdots x_{j_k} \\ &= x_{i_1} \cdots (x_{i_{m-1}} x_n x_{j_1} x_n) x_{i_m} x_n x_{j_2} \cdots x_{j_k} \\ &= x_{i_1} \cdots (x_{i_{m-2}} x_n x_{j_1} x_n) x_{i_{m-1}} x_{i_m} x_n x_{j_2} \cdots x_{j_k} \\ &= x_n x_{j_1} x_n (\alpha_{w_1} x_n x_{j_2} \cdots x_{j_k}). \end{aligned}$$

Repeating the above process,

$$\begin{aligned} \alpha_{w_1} x_n \alpha_{w_2} &= (x_n y_{j_1} x_n)(x_n x_{j_2} x_n) \cdots (x_n x_{j_k} x_n) \alpha_{w_1} x_n \\ &= x_n \alpha_{w_2} x_n \alpha_{w_1} x_n \end{aligned}$$

so that

$$\alpha_{w_1} x_n \alpha_{w_2} x_n = x_n \alpha_{w_2} x_n \alpha_{w_1}. \tag{1}$$

Since every element of G_n is of the form $\alpha_{v_1} x_n \alpha_{v_2} x_n \cdots \alpha_{v_r} x_n$, where $v_i \in W_{n-1}$, use of (1) shows that x is of the form (β) or (γ).

(e) Let $u = \alpha_{w_1} x_n \alpha_{w_2}$, and let w be irreducible in W_n such that $\alpha_w = u$. Let $l = l(w)$. Noting that the reduction to forms (β) and (γ) in (d) does not increase the number of elements involved, we can suppose that w is of the form $w_1' x_n w_2'$ or $w_1' x_n w_2' x_n$ ($w_1', w_2' \in W_{n-1}$). If w is the latter, then $x_n \alpha_{w_2} = (\alpha_{w_1})^{-1} \alpha_{w_1'} x_n \alpha_{w_2'} x_n = x_n \alpha_{w_2'} x_n (\alpha_{w_1})^{-1} \alpha_{w_1'}$ (by (1)) and we obtain $x_n \in G_{n-1}$. So w is of the form $w_1' x_n w_2'$. Suppose that $w_1 x_n w_2$ is not irreducible in W_n. Then $l(w_1 x_n w_2) > l(w_1' x_n w_2')$ so that either $l(w_1) > l(w_1')$ or $l(w_2) > l(w_2')$. Without loss of generality, we can suppose $l(w_2) > l(w_2')$. Since w_2' is irreducible in W_{n-1}, $\alpha_{w_2} \neq \alpha_{w_2'}$. From the equality $\alpha_{w_1} x_n \alpha_{w_2} = \alpha_{w_1'} x_n \alpha_{w_2'}$, we deduce that there exist $u_3, u_4 \in G_{n-1}$ with $u_3 x_n u_4 = x_n$ and $u_4 \neq e$. Since $u_4 \neq e$ and $u_4(r) = r$ if $r > 2^n$, there exists $N \in [1; 2^n]$ with $u_4(N) \neq N$. So $u_3 x_n u_4(N) = u_4(N) + 2^n$, while $x_n(N) = N + 2^n$. This is a contradiction.

(f) An argument along the same lines as the proof of (e) shows that if w is irreducible in W_n, then w is irreducible in W_{n+1}. Now use induction.

(g) Use (e) and (f) and induction on $\max\{i_1, \ldots, i_k\}$.

(h) The proof proceeds by induction on n. If $n = 1$ and $\alpha_{w_1} = \alpha_{w_2}$, then, since both w_1 and w_2 are irreducible ((g)), we have $m = 1$, so that $w_1 = w_2$. Hence the desired result is true when $n = 1$.

Now let $r > 1$ and suppose that the result is true for $n \leq (r-1)$. Suppose that $\alpha_{w_1} = \alpha_{w_2}$. Then $n = m$. Write $w_1 = z_1 x_{i_k} z_2$, $w_2 = z_3 x_{j_l} z_4$, where $i_k = \max\{i_p \colon 1 \leq p \leq n\}$, $j_l = \max\{j_p \colon 1 \leq p \leq m\}$. Let $u_i = \alpha_{z_i}$. If $i_k \neq j_l$, then a contradiction results from (a). So let $N = i_k = j_l$. Note that $u_i \in G_{N-1}$. Then $(u_3^{-1} u_1) x_N (u_2 u_4^{-1}) = x_N$, where $u_3^{-1} u_1, u_2 u_4^{-1} \in G_{N-1}$. Using (e), it follows that $u_3^{-1} u_1 = e = u_2 u_4^{-1}$, that is, $u_3 = u_1$, $u_4 = u_2$. By the induction hypothesis $z_1 = z_3$ and $z_2 = z_4$, so that $w_1 = w_2$. This is a contradiction.]

The nonsymmetry of $l_1(G)$ is proved using the notion of *capacity* in Banach algebras [**BD**, §45]. The spectral radius [spectrum] of an element a of a Banach algebra A is denoted by $\nu(a)$ [$\mathrm{Sp}(a)$]. The *capacity* $\mathrm{cap}(a)$ of a is defined to be $\mathrm{cap}(a) = \lim_{n\to\infty}\{\inf\{\|p(a)\|^{1/n} \colon p \text{ is a monic polynomial of degree } n\}\}$. (A monic polynomial of degree n is a polynomial $\sum_{r=0}^{n} \lambda_r x^r$ with $\lambda_r \in \mathbb{C}$, $\lambda_n = 1$.) The existence of the above limit follows from [**BD**, p. 251].

(ii) *Let A be a Banach $*$-algebra and a be a selfadjoint element of A such that $\nu(a) \leq 1$ and $\mathrm{cap}(a) > 1/2$. Then A is not symmetric.*

[Suppose that $\mathrm{Sp}(a)$ is real. Since $\nu(a) \leq 1$, $\mathrm{Sp}(a) \subset [-1, 1]$. Let T_n be the monic Chebyshev polynomial of degree n; so

$$T_n(x) = 2^{-n+1} \cos(n \cos^{-1}(x)) \qquad (-1 \leq x \leq 1).$$

Since $|T_n(x)| \leq 2^{-n+1}$, it follows that $\nu(T_n(a)) \leq 2^{-n+1}$. Now for fixed n, $(T_n(x))^m$ is monic of degree mn, so that

$$\mathrm{cap}(a) \leq \lim_{m\to\infty} \|(T_n(a))^m\|^{1/mn} = (\nu(T_n(a)))^{1/n} \leq 2^{-1+1/n}.$$

Hence $\mathrm{cap}(a) \leq 1/2$, giving a contradiction.]

(iii) *The Banach *-algebra $l_1(G)$ is not symmetric.*

[From (ii), it is sufficient to construct $f \in l_1(G)$ with $f^\sim = f$, $\|f\|_1 = 1$, and $\operatorname{cap}(f) > 1/2$. Let $a, b > 0$ and $r, N \in \mathbb{P}$ be such that $r < N$ and $Na + b = 1$. (The numbers a, b, r, N will be specified later.) Define

$$f = a\sum_{i=1}^{N} x_i + b\sum_{j=1}^{\infty} 2^{-j} x_{j+N}. \tag{2}$$

Using (i)(a), $f^\sim = f$ and $\|f\|_1 = Na + b = 1$.

Define recursively a sequence $\{Y_n\}$ of subsets of W by

$$\begin{aligned} Y_1 &= \{(x_{i_1}, \dots, x_{i_r})\colon 1 \le i_j \le N, i_j \ne i_k \text{ if } j \ne k\}, \\ Y_{n+1} &= \{w_1 x_{n+N} w_2\colon w_1, w_2 \in Y_n\}. \end{aligned}$$

Let $Z_n = \{\alpha_w\colon w \in Y_n\}$. We claim that $|Z_n| = |Y_n|$. By (i)(h), this is true for $n = 1$. Suppose that $m \ge 1$ and the result is true for $n = m$. Suppose that $u, v, u', v' \in Z_m$ are such that $u x_{m+N} v = u' x_{m+N} v'$. Then $(u^{-1}u')x_{m+N}(v'v^{-1}) = x_{m+N}$, and it follows from (i)(e) that $u^{-1}u' = e = v'v^{-1}$. Thus $u' = u$, $v' = v$, and by hypothesis the map $w \to \alpha_w$ is bijective from Y_{m+1} onto Z_{m+1}. So $|Y_n| = |Z_n|$ for all n. It also follows from (e) and (f) of (i) that the elements of every Y_n are irreducible in W.

Let l_n be the (common) length of the elements of Y_n. Then $l_1 = r$ and $l_{n+1} = 2l_n + 1$, and it follows that

$$l_n = 2^{n-1}(r+1) - 1. \tag{3}$$

Let $f^{(m)} = f * \cdots * f$ be the mth convolution power of f. Suppose that $x \in Z_{n+1}$ and that $x = z_1 z z_2$ with $f^{(l_n)}(z_i) \ne 0$, $f(z) \ne 0$. Now $x = \alpha_w$, where $w = w_1 x_{n+N} w_2$ is irreducible and $w_1, w_2 \in Y_n$. From (2), each $z_i \in \{x_1, x_2, \dots\}^{l_n}$, and $z = x_p$ for some p. As $w_1 x_{n+N} w_2$ is irreducible, we can find irreducible $v_i \in W$ with $\alpha_{v_i} = z_i$ and $l(v_i) = l_n$. Let k be the largest integer such that x_k occurs in the word $v_1 x_p v_2$. Now $v_1 x_p v_2$ is irreducible in W. So ((i)(d)) x_k occurs either once or twice in $v_1 x_p v_2$. Considering the equation $\alpha_{v_1} x_p \alpha_{v_2} = \alpha_{w_1} x_{n+N} \alpha_{w_2}$ and allowing both sides to act on a suitable integer (cf. the proof of (i)(e)), we see that $k = n + N$ and that x_k occurs only once. We see further that $z = x_{n+N}$ and $z_1 = \alpha_{w_1}$, $z_2 = \alpha_{w_2}$. It follows that if $m_n = f^{(l_n)}(Z_n)$ $(= \sum_{z \in Z_n} f^{(l_n)}(z))$, then, using (2)

$$\begin{aligned} m_{n+1} &= f^{(l_n)} * f * f^{(l_n)}(Z_{n+1}) \\ &= \sum\{f^{(l_n)}(z_1) f(z) f^{(l_n)}(z_2)\colon z_1, z, z_2 \in G, z_1 z z_2 \in Z_{n+1}\} \\ &= \sum\{f^{(l_n)}(z_1) f(x_{n+N}) f^{(l_n)}(z_2)\colon z_1, z_2 \in Z_n\} \\ &= f^{(l_n)}(Z_n) f(x_{n+N}) f^{(l_n)}(Z_n) = b 2^{-n} (m_n)^2. \end{aligned} \tag{4}$$

Now if $(x_{i_1}, \dots, x_{i_r}) \in Y_1$ and $x = x_{i_1} \cdots x_{i_r}$, then, using (2) and arguing as above, $f^{(r)}(x) = a^r$. So

$$m_1 = f^{(r)}(Z_1) = a^r |Z_1| = a^r |Y_1| = a^r N(N-1)\cdots(N-r+1). \tag{5}$$

From (4) and (5),

$$\begin{aligned} m_n &= b2^{-(n-1)}(m_{n-1})^2 = b2^{-(n-1)}b^2 2^{-2^1(n-2)}(m_{n-2})^4 \\ &= b^{(2^0+2^1+\cdots+2^{n-2})}2^{-[2^0(n-1)+2^1(n-2)+\cdots+2^{(n-2)}\cdot 1]}(m_1)^{2^{(n-1)}} \\ &= b^{(2^{n-1}-1)}2^{-[2^n-n-1]}[a^r N(N-1)\cdots(N-r+1)]^{2^{(n-1)}}. \end{aligned} \tag{6}$$

Since $f^{(m)}$ is supported by the set $\{x_n : n \geq 1\}^m$, we have $f^{(m)}(Y_n) = 0$ if $m < l_n$. Hence, if p is a monic polynomial of degree l_n, we have

$$\|p(f)\| > f^{(l_n)}(Z_n) = m_n. \tag{7}$$

Using (3), (7), and (6),

$$\operatorname{cap}(f) \geq \lim_{n\to\infty}(m_n)^{1/l_n} = [\tfrac{1}{4}ba^r N(N-1)\cdots(N-r+1)]^{1/(r+1)}. \tag{8}$$

We now have to choose a, b, r, N so that the right-hand side of (8) is $> 1/2$. Recall that the only constraints on a, b, r, N are

$$a, b > 0, \quad r, N \in \mathbb{P}, \quad r < N, \quad \text{and} \quad Na + b = 1. \tag{9}$$

For each $r \in \mathbb{P}$, $r > 1$ choose $N = r^2$, $a = 2/(3r^2)$, and $b = 1/3$. Then (9) is satisfied, and substituting these values in (8), we have

$$\begin{aligned} \operatorname{cap}(f) &\geq \lim_{r\to\infty}\{(1/12)\cdot r^2 a\cdot(r^2-1)a\cdots(r^2-r+1)a\}^{1/(r+1)} \\ &\geq \lim_{r\to\infty}\{(1/12)[(r^2-r+1)a]^r\}^{1/(r+1)} \\ &= \lim_{r\to\infty}\{(1/12)(2/3)^r[1-((r-1)/r^2)]^r\}^{1/(r+1)} = 2/3 > 1/2.] \end{aligned}$$

Other References

Barnes [1], Brooks [1], Hulanicki [13], Losert [2], Margulis [1], Tits [3].

Problems 6

Throughout, G is a locally compact group.

1. Determine $\lim_{n\to\infty}\lambda(C^n)^{1/n}$ when $G = F_2$ and $C = \{e, x, y, x^{-1}, y^{-1}\}$.

2. Let $A = \{\lim_{n\to\infty}\lambda(C^n)^{1/n} : C \in \mathscr{C}_e(G), \cup_{n=1}^{\infty}C^n = G\}$. Show that $k^r \in A$ if $k \in A$, $r \in \mathbb{P}$.

3. Let $H_{\mathbb{Z}}$ be the discrete group consisting of all matrices in the Heisenberg group with integer entries. (See Problem 4-29.) Show that $H_{\mathbb{Z}}$ is a finitely-generated, discrete nilpotent group. What is the degree of $H_{\mathbb{Z}}$?

4. Prove Lemma (6.16).

5. Give an example to show that a locally compact group G need not have polynomial growth if it contains a closed normal subgroup H with both H, G/H of polynomial growth.

6. Let θ_1, $\theta_2 \in \mathbb{R}$, $\theta_1 \neq 0$ and $\theta = \theta_1 + i\theta_2$. If $\theta_2 = 0$, let $\mathfrak{g}_\theta$ be the 2-dimensional, real Lie algebra with basis $\{X_1, X_2\}$ and multiplication determined by $[X_1, X_2] = \theta X_2$. If $\theta_2 \neq 0$, let $\mathfrak{g}_\theta = \operatorname{Span}\{X_1, X_2, X_3\}$ be the 3-dimensional, real Lie algebra with multiplication determined by $[X_1, X_2] = \theta_1 X_2 + \theta_2 X_3$, $[X_1, X_3] = \theta_1 X_3 - \theta_2 X_2$, $[X_2, X_3] = 0$. Let G_θ be the simply connected Lie group with $\mathfrak{g}_\theta$ as Lie algebra.

(i) Show that if $\theta_2 = 0$, then G_θ is the subgroup of $\mathrm{GL}(2, \mathbb{R})$ consisting of matrices of the form

$$\begin{bmatrix} e^{\theta t} & s \\ 0 & 1 \end{bmatrix} \qquad (s, t \in \mathbb{R}),$$

while if $\theta_2 \neq 0$, G_θ is the subgroup of $\mathrm{GL}(2, \mathbb{C})$ consisting of matrices of the form

$$\begin{bmatrix} e^{\theta t} & z \\ 0 & 1 \end{bmatrix} \qquad (t \in \mathbb{R}, z \in \mathbb{C}).$$

(ii) Let $t_0 \in \mathbb{R}$ be such that $|e^{\theta t_0}| < 1/3$ and

$$a = \begin{bmatrix} e^{\theta t_0} & 1 \\ 0 & 1 \end{bmatrix}, \qquad b = \begin{bmatrix} e^{\theta t_0} & -1 \\ 0 & 1 \end{bmatrix}.$$

Show that a, b generates a free, uniformly discrete semigroup S in two generators.

(iii) Let $\mathfrak{g}$ be a real Lie algebra, and suppose that for some $X \in \mathfrak{g}$, $\operatorname{Sp}(\operatorname{ad} X) \not\subset i\mathbb{R}$. Show that $\mathfrak{g}$ contains $\mathfrak{g}_\theta$ as a subalgebra for some θ.

(iv) Hence give another proof (not using (6.36)) to show that a connected Lie group G has polynomial growth if and only if it does not contain a free, uniformly discrete semigroup in two generators (cf. (6.39)).

7. Let $G \in [FC]^-$ ((4.23)). Prove that G has polynomial growth. (Hint: use the following result of Grosser and Moskowitz [**2**]: *if G is compactly generated, then the commutator subgroup of G has compact closure.*)

8. For $\alpha \in \mathbb{R}$, let $M(\alpha)$ be the semidirect product $\mathbb{C}^2 \times_\rho \mathbb{R}$, where

$$\rho(t)(z_1, z_2) = (e^{2\pi i t} z_1, e^{2\pi i \alpha t} z_2).$$

Show that $M(\alpha)$ has polynomial growth. Deduce that a simply connected, solvable Lie group of polynomial growth need not be of Type 1.

9. Let H be the locally compact group that, as a set, is $\mathbb{C} \times \mathbb{R}$, and that has multiplication $(z, r)(w, s) = (z + w, \operatorname{Im}(z\overline{w}) + r + s)$. The "diamond group" D is defined to be the semidirect product $H \times_\rho \mathbb{R}$, where $\rho(r)(w, s) = (e^{2\pi i r} w, s)$. Show that D is a solvable Lie group with polynomial growth.

10. Show that S_2, S_3^σ, and S_4 are not liminal (cf. (6.45)).

11. Show that S_2, S_3^σ, and S_4 are not unimodular, and deduce that a connected, solvable Lie group G has polynomial growth if and only if G/H is unimodular for every closed, normal subgroup H of G.

12. It is known that the sphere $\mathbb{S}^2$, the projective plane P_2, the torus $\mathbb{S}^1 \times \mathbb{S}^1$, and the Klein bottle K each admits a complete 2-dimensional Riemannian structure with nonnegative mean curvature. Check Milnor's theorem ((6.41)) directly for each of these four surfaces.

13. Prove (6.42(i)).

14. Let G be exponentially bounded and $\phi \geq 0$ be nonzero in $L_\infty(G)$. Let X_ϕ be the linear subspace of $L_\infty(G)$ spanned by $\{\phi x\colon x \in G\}$. Show that there exists a linear map $P\colon X_\phi \to \mathbb{C}$ such that $P(\phi) = 1$, and for $\psi \in X_\phi$, $x \in G$, $P(\psi) \geq 0$ if $\psi \geq 0$ and $P(\psi x) = P(\psi)$. [Hint: use Problem 13 above.]

15. Let G have polynomial growth and r be the degree of G.

(i) Show that if G is abelian or discrete nilpotent, then for each $C \in \mathscr{C}_e(G)$, there exists $m_C > 0$ such that

$$(1) \qquad m_C n^r \leq \lambda(C^n) \qquad (n \geq 1).$$

(ii) Suppose that (1) holds for all $C \in \mathscr{C}_e(G)$ and that G is compactly generated. Show that there exists a summing sequence $\{K_n\}$ for G such that $\{\lambda(K_n^{-1}K_n)/\lambda(K_n)\}$ is bounded (cf. (5.21)).

16. ((6.46.(i)) Let X be a compact Hausdorff space and $P\colon C(X) \to C(X)$ be a unit preserving, positive, linear projection from $C(X)$ onto a (closed) subspace B of $C(X)$. Let $\sim$ be the equivalence relation on X given by: $x \sim y \Leftrightarrow f(x) = f(y)$ for all $f \in B$. Let $H = \{\phi \in C(X)\colon \phi(x) = \phi(y)$ whenever $x \sim y$ in $X\}$. For $x \in X$, let $C_x = \{\mu \in PM(X)\colon \hat{\mu}(f) = f(x)$ for all $f \in B\}$.

(i) Show that if $\mu \in C_x$, then $\hat{\mu}(\phi) = \phi(x)$ for all $\phi \in H$, and the support $\mathbf{S}(\mu)$ of μ is contained in $\{y \in X\colon y \sim x\}$.

(ii) Show that for $x \in X$, we have $\hat{x} \circ P \in C_x$, and deduce that

$$P(\phi P\psi) = P((P\phi)(P\psi)) \quad (\phi, \psi \in C(X)).$$

(iii) Prove that B is a commutative, unital C^*-algebra with multiplication "$\times$", where $f \times g = P(fg)$ and $f^* = \overline{f}$ $(f, g \in B)$.

17. The "localisation conjecture" (Greenleaf [**2**, p. 69]) runs as follows: let G be a connected, separable, amenable locally compact group and $K = K^{-1} \in \mathscr{C}_e(G)$. Is it always true that given $\varepsilon > 0$ and $C \in \mathscr{C}_e(G)$, there exists $m_0 \in \mathbb{P}$ such that

$$\lambda(xK^m \triangle K^m)/\lambda(K^m) < \varepsilon \qquad (m \geq m_0, x \in C)?$$

(A weaker version of this is true if G is exponentially bounded ((6.8)).) By considering the "$ax + b$" group, show that the conjecture is false (c.f. (0.5)).

18. The locally compact group G is said to be *distal* if, whenever $x \in G \sim \{e\}$, then $e \notin (\mathrm{Cl}(x))^-$ ($\mathrm{Cl}(x)$ = conjugate class of x). Suppose that G is a connected Lie group. Show that G is distal if and only if G has polynomial growth.

19. Let G be a connected Lie group. Show that $\mathrm{Sp}(\mathrm{Ad}\, x) \subset \mathbb{T}$ if and only if G is of Type R.

20. Let G be a connected Lie group. Show that G is of Type R if and only if $\{|\mathrm{Tr}(\mathrm{Ad}\, x)|\colon x \in G\}$ is bounded.

21. Show that if G is abelian or compact, then $G \in [S]$ ((6.48)).

22. (i) Let A be a commutative Banach algebra with maximal ideal space X. Suppose that $f \in A$ is such that $\hat{f}$ is real-valued on X and that for some $k \in \mathbb{P}$, $\|e^{in\beta f} - 1\| = O(n^k)$ as $n \to \infty$ for every $\beta \in \mathbb{R}$. Let $C_c^r(\mathbb{R})$ be the space of functions $\phi \in C_c(\mathbb{R})$ such that, for all $n \leq r$, $D^n\phi$ exists and is continuous. Show that $C_c^{k+2}(\mathbb{R})$ "acts" on f in the sense that if $\phi \in C_c^{k+2}(\mathbb{R})$ and $\phi(0) = 0$, then there exists $g \in A$ such that $\hat{g} = \phi \circ \hat{f}$.

(ii) Let G be a locally compact group, and let $\omega\colon G \to [1, \infty)$ be measurable and such that $\omega(xy) \leq \omega(x)\omega(y)$ for all x, $y \in G$. Let

$$L_1(G, \omega) = \left\{ f \in L_1(G) \colon \|f\|_\omega = \int |f(x)|\omega(x)\, dx < \infty \right\}.$$

Show that with norm $\|\cdot\|_\omega$, involution $f \to f^\sim$, and convolution multiplication, $L_1(G, \omega)$ is a Banach $*$-algebra.

(iii) Let G be compactly generated and of polynomial growth. Let k be the degree of G and $p \in \mathbb{P}$ be such that $\frac{1}{2}k+1 \leq p < \frac{1}{2}k+2$. Let $f = f^\sim \in C_c(G)$, and let A_f be the closed subalgebra of $L_1(G)$ generated by f. Show that $C_c^{(p+2)}(\mathbb{R})$ "acts" on f in the sense of (i) (with $A = A_f$).

23. Let $G \in [S]$, and let H be a closed normal subgroup of G. Show that $G/H \in [S]$, and that $H \in [S]$ is H is open in G.

24. (i) Let A be a Banach $*$-algebra with involution $a \to a^\sim$ and identity element 1. Let $P(A)$ be the set of positive functionals F on A with $F(1) = 1$. Show that A is symmetric if and only if, for all $a \in A$, the spectrum $\operatorname{Sp}(a)$ of a is contained in $\{F(a)\colon F \in P(A)\}$. [Hint: use the result ([Ri], (4.7.11)): *if A is symmetric and B is a closed $*$-subalgebra of A containing* 1, *then the map $F \to F_{|B}$ maps $P(A)$ onto $P(B)$*.]

(ii) Let G be a discrete group containing FS_2 as a subsemigroup with generators a, b. For $f \in l_1(G)$ let $\nu(f)$ be the spectral radius of f. Let $f_0 = ba^3 + ba + i(ba^4 + b + 2ba^2)$.

(a) Show that $\nu(f_0) = \|f_0\|_1 = 6$.

(b) Show that if G is symmetric and if $f \in l_1(G)$ is such that $f^\sim * f = f * f^\sim$, then $\nu(t * f) \leq \nu(f)$ $(t \in G)$.

(c) Let $g_0 = a + a^{-1} + i(a + a^{-1})^2$ $(= (ba^2)^{-1} * f_0)$. Show that if G is symmetric, then $\nu(g_0) \leq 2\sqrt{5}$. Deduce that G is not symmetric.

25. Let $[\Psi]$ be the class of locally compact groups G of Problem 4-40.

(i) Show that $G \in [\Psi]$ if and only if, whenever π, ρ are nondegenerate $*$-representations of $L_1(G)$ such that $\ker \pi \subset \ker \rho$, then $\|\rho(f)\| \leq \|\pi(f)\|$ for all f of the form $h * h^\sim$ with $h \in C_c(G)$.

(ii) Show that $G \in [\Psi]$ if G has polynomial growth. [Hint: use Problem 22 above.]

26. Let G be a locally compact group and $\mu \in P(G)$. Prove that μ satisfies (B) of (6.46(iii)) if and only if the following Reiter-type condition holds: $\|x * \mu^n - \mu^n\|_1 \to 0$ *for all* $x \in G$.

CHAPTER 7

Sizes of Sets of Invariant Means

(7.0) Introduction. In this chapter, we are concerned with determining the sizes (cardinalities) of sets of invariant means on semigroups and locally compact groups. The general conclusion is remarkably simple: the size of such a set is "biggest possible" unless the semigroup or group possesses some strong property which clearly limits it. For example, *if a discrete group G is infinite and amenable, then* ((7.8)) $|\mathfrak{L}(G)| = 2^{2^{|G|}}$ $(= |l_\infty(G)'|)$: *if G is finite, then* $|\mathfrak{L}(G)| = 1$.

However, a word of caution is in order. Recall that, as a consequence of the results involved in the solution of the Banach-Ruziewicz Problem ((4.27)–(4.29).), there is an infinite, compact group G (e.g. $G = \mathrm{O}(5)$) with $|\mathfrak{L}(G)|$ smallest possible! However if we restrict attention to compact groups which are amenable as discrete, then the above general conclusion still applies.

There are four natural cardinality questions that arise in connection with left invariant means. Let G be a locally compact amenable group.

(1) *What is the cardinality $|\mathfrak{L}(G)|$ of the set of left invariant means on G?*

(2) *What is the cardinality $|\mathfrak{L}_t(G)|$ of topologically left invariant means on G?*

(3) *What is $|\mathfrak{L}(G) \sim \mathfrak{L}_t(G)|$, that is, how many (if any) left invariant means on G are* not *topologically left invariant?*

(4) *What is the cardinality $|\mathfrak{L}(S)|$ of the set of left invariant means on a left amenable semigroup S?* A related question is: *what is* $\dim \mathfrak{J}_l(S)$, *the dimension of the space of left invariant continuous linear functionals on $l_\infty(S)$?*

These are the four questions which we will study in this chapter.

We now briefly discuss these questions in turn.

The first question is unsolved in general and seems to be very difficult. For example, some (nonfinite) compact groups admit exactly one left invariant mean, while others admit many. The more tractable second problem, however, does provide a lower bound for $|\mathfrak{L}(G)|$ since $\mathfrak{L}(G) \supset \mathfrak{L}_t(G)$.

The second question has been completely settled recently. It has been shown that *if G is noncompact and $\mathfrak{m}$ is the smallest possible cardinality for a covering of G by compact subsets, then*

$$|\mathfrak{L}_t(G)| = 2^{2^{\mathfrak{m}}}.$$

(Of course, when G is compact, then $\mathfrak{L}_t(G) = \{\lambda\}$.) The main idea of the proof is to construct "many" left invariant, compact subsets of the maximal ideal space of $\mathrm{U}_r(G)$ and then to use a fixed-point theorem to construct left invariant means supported on these subsets. The result is proved in (7.6). A remarkable consequence of the result is the following ((7.8)): *if G is an infinite, amenable discrete group, then $|\mathfrak{L}(G)| = 2^{2^{|G|}}$.*

The third question, like the first, is unsolved in general. Progress has been made in the case where G is amenable as a discrete group. The reason for this is that we want to realise invariant means as probability measures on the maximal ideal space $\Phi(G)$ of $L_\infty(G)$, each supported on a compact invariant subset of $\Phi(G)$, and in order to ensure that such an invariant subset actually supports some $m \in \mathfrak{L}(G)$, we require the use of Day's Fixed-Point Theorem (cf. (2.25)). The latter result, of course, only works if G is amenable as discrete. We will show ((7.21)) that *if G is amenable as discrete, then $\mathfrak{L}(G) = \mathfrak{L}_t(G)$ if and only if G is discrete* and that *if, in addition, G is σ-compact and nondiscrete, then* ((7.20)) $|\mathfrak{L}(G) \sim \mathfrak{L}_t(G)| \geq 2^{\mathfrak{c}}$, *where $\mathfrak{c}$ is the cardinality of the continuum.*

The fourth problem is an example of a semigroup problem which is substantially harder than the group problem. Indeed, the exact cardinality of $\mathfrak{L}(S)$ was determined only recently. The determination uses the following cardinal $\mathfrak{m}(S)$, where

$$\mathfrak{m}(S) = \min\left\{\left|\bigcup_{i=1}^{n} s_i S_i\right| : n \geq 1,\ \{S_1, \ldots, S_n\} \text{ is a partition of } S,\ s_1, \ldots, s_n \in S\right\}.$$

If $\mathfrak{m}(S)$ is infinite, then $|\mathfrak{L}(S)| = 2^{2^{\mathfrak{m}(S)}} = \dim \mathfrak{I}_l(S)$. If $\mathfrak{m}(S)$ is finite, then $\mathfrak{I}_l(S)$ is finite-dimensional, and its dimension is the number of finite left ideal groups in S ((7.26), (7.27)). In many cases, $\mathfrak{m}(S)$ equals the more accessible cardinal $\min\{|sS| : s \in S\}$.

The central, rather technical, construction involved is effected in (7.25). The construction produces a "large" disjoint family of left thick subsets of S, and these in turn can be used to produce an even larger disjoint family of compact invariant subsets of βS, each "supporting" a left invariant mean. A number of other cardinalities associated with invariant means can be determined by modifying the construction. (See Problems 7.)

We start by proving a useful set-theoretic result. The result is closely related to [**HR1**, (16.8)].

(7.1) PROPOSITION. *Let A be an infinite set. For $B \subset A$, let $B^1 = B$ and $B^c = A \sim B$. Then there exists a family $\{N_\gamma : \gamma \in \Gamma\}$ of subsets of A such that*

(i) $|\Gamma| = 2^{|A|}$;

(ii) *if* $\gamma_1, \ldots, \gamma_m$ *are distinct elements of* Γ *and* $\varepsilon_i \in \{1, c\}$ *for* $1 \leq i \leq m$, *then*

$$\bigcap_{i=1}^{m} N_{\gamma_i}^{\varepsilon_i} \neq \varnothing.$$

PROOF. Let A' be a set disjoint from and equipotent with A, and let τ be a bijection from A onto A'. Let $\Gamma = \mathscr{P}(A)$ (so that $|\Gamma| = 2^{|A|}$), and for $\gamma \in \Gamma$, let $B(\gamma) = \tau(\gamma)^c \cup \gamma \in \mathscr{P}(A' \cup A)$, where $\tau(\gamma)^c = A' \sim \tau(\gamma)$. If $\gamma_1, \gamma_2 \in \Gamma$ and $\gamma_1 \neq \gamma_2$, then, since $B(\gamma_1) \sim B(\gamma_2) = (\gamma_1 \sim \gamma_2) \cup (\tau(\gamma_1)^c \sim \tau(\gamma_2)^c) = (\gamma_1 \sim \gamma_2) \cup \tau(\gamma_2 \sim \gamma_1)$, it follows that

$$B(\gamma_1) \sim B(\gamma_2) \neq \varnothing \qquad (\gamma_1 \neq \gamma_2). \tag{1}$$

Let $\mathscr{F}_1 = \mathscr{F}(A' \cup A)$, $\mathscr{F}_\gamma = \mathscr{F}(B(\gamma))$, and $M_\gamma = \mathscr{F}(\mathscr{F}_1 \sim \mathscr{F}_\gamma)$.

(Recall that $\mathscr{F}(X)$ is the family of finite subsets of a set X.) Since $\mathscr{F}(\mathscr{F}_1)$ is equipotent with A, we can find a bijection α from $\mathscr{F}(\mathscr{F}_1)$ onto A. We set $N_\gamma = \alpha(M_\gamma)$. To prove (ii) it is sufficient to show that if $\gamma_1, \ldots, \gamma_n$, $\delta_1, \ldots, \delta_m$ are distinct elements of Γ, then

$$M_{\gamma_1}^c \cap \cdots \cap M_{\gamma_n}^c \cap M_{\delta_1} \cap \cdots \cap M_{\delta_m} \neq \varnothing, \tag{2}$$

where $M_{\gamma_i}^c = \mathscr{F}(\mathscr{F}_1) \sim M_{\gamma_i}$.

For $1 \leq i \leq n$, $1 \leq j \leq m$, choose (using (1)) $x_{ij} \in B(\gamma_i) \sim B(\delta_j)$. Set $A_i = \{x_{ij} \colon 1 \leq j \leq m\}$ and consider $\Theta = \{A_1, \ldots, A_n\} \in \mathscr{F}(\mathscr{F}_1)$. For each i, $A_i \not\subset B(\delta_j)$ so that $\Theta \in M_{\delta_j}$. Further, for each i, $A_i \in \mathscr{F}_{\gamma_i}$ so that $\Theta \notin M_{\gamma_i}$. Thus $\Theta \in M_{\gamma_1}^c \cap \cdots \cap M_{\gamma_n}^c \cap M_{\delta_1} \cap \cdots \cap M_{\delta_m}$ as required. □

We now turn to our second problem. Let G be a locally compact amenable group. We will determine the cardinal $|\mathfrak{L}_t(G)|$. References for the results discussed here are Chou [**4**], [**9**], Granirer [**13**], Lau [**S8**], and Lau and Paterson [**S1**].

By (1.9), every $m \in \mathfrak{L}_t(G)$ can be regarded as a member of $\mathrm{U_r}(G)'$. Now $\mathrm{U_r}(G)$ is a commutative unital C^*-algebra and so can be identified with $C(X)$, where X is the maximal ideal space of $\mathrm{U_r}(G)$. The idea of the proof is to construct a "large disjoint" family of subsets of G each supporting a suitable function of $\mathrm{U_r}(G)$. Now G is a dense subset of X in the natural way, and using the above family of sets, we can produce a large disjoint family of compact invariant subsets of X. Each such set supports an invariant probability measure and this "belongs" to $\mathfrak{L}_t(G)$. (Here the left introversion of $\mathrm{U_r}(G)$ is important in order to apply the fixed-point theorem of (2.24). This would not work if we had used $L_\infty(G)$ or $C(G)$ in place of $\mathrm{U_r}(G)$.) This gives a lower bound for $|\mathfrak{L}_t(G)|$, and we then show that it is also an upper bound for $|\mathfrak{L}_t(G)|$.

Suppose that G is noncompact and let $\mathfrak{m}$ (or $\mathfrak{m}(G)$) be the smallest possible cardinality of a covering of G by compact subsets of G. Note that $\mathfrak{m}$ is infinite since G is noncompact. Let α be the smallest ordinal of cardinality $\mathfrak{m}$. Let $\{K_\beta \colon \beta \in \alpha\}$ be a family of compact subsets of G such that $\bigcup_{\beta \in \alpha} K_\beta = G$. We can suppose that the family is closed under finite unions.

(7.2) PROPOSITION. *Let $U \in \mathscr{C}_e(G)$. Then there exists a subset $\{x_{\beta\gamma}: 0 \leq \beta \leq \gamma < \alpha\}$ of G such that the family $\{UK_\gamma x_{\beta\gamma}: 0 \leq \beta \leq \gamma < \alpha\}$ is disjoint.*

PROOF. Let $\gamma_0 \in \alpha$ and suppose that elements $x_{\beta\gamma} (0 \leq \beta \leq \gamma < \gamma_0)$ have been constructed so that $UK_\gamma x_{\beta\gamma} \cap UK_{\gamma'} x_{\beta'\gamma'} = \varnothing$ whenever $(\beta, \gamma) \neq (\beta', \gamma')$. Let $\mathscr{B} = \{UK_\gamma x_{\beta\gamma}: 0 \leq \beta \leq \gamma < \gamma_0\}$ and $W = \bigcup \mathscr{B}$. We now construct by transfinite recursion the elements $x_{\beta\gamma_0}$. Suppose that the $x_{\beta\gamma_0}$ have been constructed for $\beta < \beta_0 \leq \gamma_0$, and let $Z = W \cup (\{UK_{\gamma_0} x_{\beta\gamma_0}: \beta < \beta_0\})$. Then $K_{\gamma_0}^{-1} U^{-1} Z$ admits a covering $\mathscr{B}'$ of compact sets, where $|\mathscr{B}'| < \mathfrak{m}$. So $K_{\gamma_0}^{-1} U^{-1} Z \neq G$, and we can pick $x_{\beta_0\gamma_0}$ in $G \sim K_{\gamma_0}^{-1} U^{-1} Z$. This completes the recursion. □

Recall that a subset E of G is called left thick ((1.20)) if whenever $F \in \mathscr{F}(G)$, there exists $x \in G$ such that $Fx \subset E$.

(7.3) PROPOSITION. *There exist a family $\{Z_\beta: \beta \in \alpha\}$ of left thick subsets of G and a set $\{\phi_\beta: \beta \in \alpha\}$ of functions in $\mathrm{U_r}(G)$ such that for each β, $\phi_\beta(Z_\beta) = \{1\}$ and $\phi_\beta(\bigcup_{\gamma \neq \beta} Z_\gamma) = \{0\}$. Further, if $K \subset \alpha$ is nonempty, then $\phi_K = \sum_{\beta \in K} \phi_\beta \in \mathrm{U_r}(G)$.*

PROOF. Let U and $x_{\beta\gamma}$ be as in (7.2). For each β, let $Z_\beta = \bigcup\{K_\gamma x_{\beta\gamma}: \beta \leq \gamma < \alpha\}$. Let $F \in \mathscr{F}(G)$. Since the family $\mathscr{A} = \{K_{\beta'}: \beta' \in \alpha\}$ covers G and is closed under finite unions, there exists γ such that $F \subset K_\gamma$. The family $\{K_\gamma \cup K_{\beta'}: \beta' \in \alpha\} \subset \mathscr{A}$ also covers G, and from the definition of $\mathfrak{m}$, has cardinality $= \mathfrak{m}$. So we can take $\gamma \geq \beta$. Then $Fx_{\beta\gamma} \subset Z_\beta$, that is, Z_β is left thick. It remains to define the functions ϕ_β.

To this end, let $V \in \mathscr{C}_e(G)$ be such that $V^3 \subset U$ and $V^{-1} = V$. By Urysohn's Lemma, we can find $f \in C_c(G)$ with $0 \leq f \leq 1$, $f(e) = 1$, $f(G \sim V) = \{0\}$. Let d be the pseudo-metric on G given by $d(x, y) = ||fx - fy||$, and, for each β, γ with $\beta \leq \gamma$, let $g_{\beta\gamma}(x) = 1 - d(x, K_\gamma x_{\beta\gamma})$ $(x \in G)$. Clearly, $g_{\beta\gamma} \in C(G)$, $0 \leq g_{\beta\gamma} \leq 1$, and $g_{\beta\gamma}(x) = 1$ for all $x \in K_\gamma x_{\beta\gamma}$. Now if $g_{\beta\gamma}(x) > 0$, then $d(x, y) < 1$ for some $y \in K_\gamma x_{\beta\gamma}$. This implies that $Vx \cap Vy \neq \varnothing$; for if $Vx \cap Vy = \varnothing$, then $yx^{-1} \notin V$, and $fx(x^{-1}) = 1$, $fy(x^{-1}) = 0$ giving $d(x, y) = 1$ and a contradiction. So if $g_{\beta\gamma}(x) > 0$, then $x \in V^{-1}VK_\gamma x_{\beta\gamma} = V^2 K_\gamma x_{\beta\gamma} \subset UK_\gamma x_{\beta\gamma}$. It follows, using (7.2), that ϕ_β, where $\phi_\beta = \sum_{\beta \leq \gamma} g_{\beta\gamma}$, is well defined and satisfies all the required conditions except possibly that of right uniform continuity. We now prove that $\phi_\beta \in \mathrm{U_r}(G)$. This will follow once we have shown

$$(1) \qquad ||\phi_\beta u - \phi_\beta|| \leq ||fu - f|| \qquad (u \in V)$$

since $f \in C_c(G) \subset \mathrm{U_r}(G)$.

Let $u \in V$ and suppose that $t \in G$ is such that $\phi_\beta(ut) - \phi_\beta(t) \neq 0$. If $\phi_\beta(ut) \neq 0$, then $ut \in V^2 K_\gamma x_{\beta\gamma}$ for some unique γ, giving $t \in u^{-1} V^2 K_\gamma x_{\beta\gamma} \subset UK_\gamma x_{\beta\gamma}$. So $ut, t \in UK_\gamma x_{\beta\gamma}$. Similarly, if $\phi_\beta(t) \neq 0$, then again, ut and t are in $UK_\gamma x_{\beta\gamma}$ for some unique γ. So in either case,

$$\begin{aligned} |\phi_\beta(ut) - \phi_\beta(t)| = |g_{\beta\gamma}(ut) - g_{\beta\gamma}(t)| &= |d(ut, K_\gamma x_{\beta\gamma}) - d(t, K_\gamma x_{\beta\gamma})| \\ &\leq d(ut, t) = ||fu - f||, \end{aligned}$$

establishing (1) and the first assertion of the proposition. For $K \neq \emptyset$, $K \subset \alpha$, let $\phi_K = \sum_{\beta \in K} \phi_\beta$. If $u \in V$, $t \in G$, $\beta \in K$, and either $\phi_\beta(ut) \neq 0$, $\phi_\beta(t) \neq 0$, then $\phi_{\beta'}(ut) = 0 = \phi_{\beta'}(t)$ for $\beta' \neq \beta$, and it follows that $||\phi_K u - \phi_K|| = \max_{\beta \in K} ||\phi_\beta u - \phi_\beta||$. From (1), $\phi_K \in \mathrm{U_r}(G)$. □

Now ((2.22)) the action $(x, F) \to xF$ of G on $\mathrm{U_r}(G)'$ is jointly continuous for the weak* topology. The maximal ideal space X of $\mathrm{U_r}(G)$ is, of course, a subset of $\mathrm{U_r}(G)'$ and is clearly (left) invariant for G. It is routine that G is a dense subset of X in the obvious way. Recall that $\mathfrak{m} = |\alpha|$.

(7.4) PROPOSITION. *There exists a disjoint family $\{C_p : p \in P\}$ of invariant, nonempty, compact subsets of X, where $|P| = 2^{2^\mathfrak{m}}$.*

PROOF. By (7.1), we can find a family $\{N_\gamma : \gamma \in \Gamma\}$ of subsets of α, where $|\Gamma| = 2^\mathfrak{m}$, such that if $\gamma_1, \ldots, \gamma_m$ are distinct elements of Γ and $\varepsilon_i \in \{1, c\}$, then $\bigcap_{i=1}^m N_{\gamma_i}^{\varepsilon_i} \neq \emptyset$. Let $P = \{1, c\}^\Gamma$ so that $|P| = 2^{2^\mathfrak{m}}$. Let Z_β, ϕ_β be as in (7.3), and for each $\gamma \in \Gamma$, let $E_\gamma = \bigcup\{Z_\beta : \beta \in N_\gamma\}$. For $A \subset G$ let $\hat{A}$ (or $A^\wedge$) be the closure of A in X and $A^c = (\bigcup_{\beta \in \alpha} Z_\beta) \sim A$. For $p \in P$, let

$$C_p = \bigcap\{(xE_\gamma^{p(\gamma)})^\wedge : x \in G,\ \gamma \in \Gamma\}.$$

Clearly C_p is compact. We show that C_p is invariant and nonempty. If $u \in C_p$, $x_0 \in G$, then for $y \in G$, $\gamma \in \Gamma$, we have $x_0 u \in x_0[(x_0^{-1} y E_\gamma^{p(\gamma)})^\wedge] = (yE_\gamma^{p(\gamma)})^\wedge$ since the map $v \to x_0 v$ is a homeomorphism on X. So $x_0 u \in C_p$ and C_p is invariant. Now let $x_1, \ldots, x_m \in G$, $\gamma_1, \ldots, \gamma_m \in \Gamma$, where the γ_i's are distinct. Let $\beta_0 \in \bigcap_{i=1}^n N_{\gamma_i}^{p(\gamma_i)}$. If $p(\gamma_i) = 1$, then $Z_{\beta_0} \subset E_{\gamma_i}^{p(\gamma_i)}$; if $p(\gamma_i) = c$, then, since the Z_β's are disjoint ((7.3)), we have $Z_{\beta_0} \subset (\bigcup_{\beta \in \alpha} Z_\beta) \sim E_{\gamma_i} = E_{\gamma_i}^{p(\gamma_i)}$. So $L = \bigcap_{j=1}^m x_j Z_{\beta_0} \subset x_i E_{\gamma_i}^{p(\gamma_i)}$ for each i. Since Z_{β_0} is left thick, there exists $x' \in G$ such that $x_i^{-1} x' \in Z_{\beta_0}$ $(1 \leq i \leq n)$ and so $x' \in L$. So $\bigcap_{i=1}^m (x_i E_{\gamma_i}^{p(\gamma_i)})^\wedge \supset L^\wedge \neq \emptyset$ and the sets $\{(xE_\gamma^{p(\gamma)})^\wedge : x \in G,\ \gamma \in \Gamma\}$ have the finite intersection property. Since X is compact, it follows that $C_p \neq \emptyset$.

It remains to show that if $p, q \in \{1, c\}^\Gamma$ with $p \neq q$, then $C_p \cap C_q = \emptyset$. To prove this, let $\gamma_0 \in \Gamma$ be such that $p(\gamma_0) \neq q(\gamma_0)$. Without loss of generality, we can suppose that $p(\gamma_0) = 1$, $q(\gamma_0) = c$. Then $C_p \subset E_{\gamma_0}^\wedge$, $C_q \subset (E_{\gamma_0}^c)^\wedge$. Let $\beta \in N_{\gamma_0}$. Since $E_{\gamma_0}^c = \bigcup_{\beta' \notin N_{\gamma_0}} Z_{\beta'}$, we see that $\phi_\beta(E_{\gamma_0}^c) = \{0\}$. Let $\phi = \phi_{N_{\gamma_0}}$ as in (7.3). Then $\phi(E_{\gamma_0}) = \{1\}$, $\phi(E_{\gamma_0}^c) = \{0\}$ so that $\hat{\phi}(E_{\gamma_0}^\wedge) = \{1\}$, $\hat{\phi}((E_{\gamma_0}^c)^\wedge) = \{0\}$, where $\hat{\phi}$ is the Gelfand transform of ϕ. So $\hat{\phi}$ is 1 on C_p and 0 on C_q giving $C_p \cap C_q = \emptyset$ as required. □

As in (2.25), for $m \in \mathfrak{M}(\mathrm{U_r}(G))$, let $\hat{m} \in M(X)$ be given by $\hat{m}(\hat{\phi}) = m(\phi)$ $(\phi \in \mathrm{U_r}(G))$. Then the map $m \to \hat{m}$ is an affine weak* continuous bijection from $\mathfrak{M}(\mathrm{U_r}(G))$ onto $\mathrm{PM}(X)$. It is easy to check that $m \in \mathfrak{L}(U_r(G))$ if and only if $\hat{m}$ is invariant, that is, if $\hat{m}(\hat{\phi}x) = \hat{m}(\hat{\phi})$ for all $\phi \in \mathrm{U_r}(G)$. Recall that $\mathbf{S}(\mu)$ is the support of a measure $\mu \in M(X)$.

(7.5) PROPOSITION. *There exists a subset E of $\mathfrak{L}_t(G)$ and a disjoint family $\mathscr{C}$ of compact, left invariant subsets of X such that*

(i) $|E| = 2^{2^{\mathfrak{m}}} = |\mathscr{C}|$;
(ii) $\mathbf{S}(\hat{m})$ *is contained in a member of* $\mathscr{C}$ *for each* $m \in E$.

PROOF. Let $\mathscr{C}$ be the set of C_p's of (7.4). For each p, let $K_p = \mathrm{PM}(C_p)$. As in the proof of (2.25), K_p can be regarded as a subset of $M(X) = C(X)'$ and is an affine left G-set in the standard way. From the comments preceding (7.4), the G-action is jointly continuous, and since G is amenable, there exists a G-fixed point $\mu_p \in K_p$. Let $m_p \in \mathfrak{M}(\mathrm{U_r}(G))$ be given by $\hat{m}_p = \mu_p$. Then $m_p \in \mathfrak{L}(\mathrm{U_r}(G)) = \mathfrak{L}_t(\mathrm{U_r}(G))$ by (1.8). Identifying $\mathfrak{L}_t(U_r(G))$ with $\mathfrak{L}_t(G)$ ((1.10)), we take $E = \{m_p : p \in P\}$. □

We now determine the cardinality of $\mathfrak{L}_t(G)$.

(7.6) THEOREM. $|\mathfrak{L}_t(G)| = 2^{2^{\mathfrak{m}}}$ (*G noncompact*).

PROOF. From (7.5), $|\mathfrak{L}_t(G)| \geq 2^{2^{\mathfrak{m}}}$. It remains to establish the reverse inequality.

To this end, let H be a noncompact, σ-compact open subgroup of G. Then there exists a compact normal subgroup K of H such that H/K is separable (cf. [**HR1**, (8.7)] or Problem 7-7). Let μ be the normalised Haar measure on K regarded as a probability measure on G. Let $\nu \in P(G)$ and $\theta = \nu * \mu$. Then $\theta \in P(G)$, and if $\phi \in \mathrm{U_r}(G)$, $x \in G$, $k \in K$, then

$$\begin{aligned}\phi * \theta(kx) &= (kx\phi)^\wedge(\nu * \mu) = (kx\phi * \nu)^\wedge(\mu) \\ &= (x\phi * \nu)^\wedge(\mu k) = (x\phi * \nu)^\wedge(\mu) = \phi * \theta(x).\end{aligned}$$

Thus $\phi * \theta$ can be regarded as a continuous function on the right coset space $G\backslash K$. Clearly, each $m \in \mathfrak{L}_t(G)$ is determined by its values on $\mathrm{U_r}(G) * \theta$, and so $\mathfrak{L}_t(G)$ can be regarded as a set of continuous linear functionals on some subspace A of $C(G\backslash K)$.

As K is compact, the smallest possible covering of $G\backslash K$ by compact sets has cardinality $\mathfrak{m}$. Further, since $H\backslash K$ is a separable, open subset of $G\backslash K$, every compact subset of $G\backslash K$ is separable. Hence, there exists a dense subset T of $G\backslash K$ of cardinality $\aleph_0\mathfrak{m} = \mathfrak{m}$. Since every function in $C(G\backslash K)$ is determined by its values on T, we have $|C(G\backslash K)| \leq \mathfrak{C}^{\mathfrak{m}} = 2^{\mathfrak{m}}$. So

$$|\mathfrak{L}_t(G)| \leq |A'| \leq |C(G\backslash K)'| \leq \mathfrak{C}^{2^{\mathfrak{m}}} = 2^{2^{\mathfrak{m}}}.$$

This establishes the theorem. □

(7.7) COROLLARY. $|\mathfrak{L}(G)| \geq 2^{2^{\mathfrak{m}}}$.

(7.8) COROLLARY. *If G is an infinite, amenable, discrete group, then*

$$|\mathfrak{L}(G)| = 2^{2^{|G|}}.$$

The preceding corollary extends to the case where G has a left action on an infinite set X and $\mathfrak{L}(G)$ is replaced by the set of G-invariant means on X. Indeed, Rosenblatt and Talagrand [**1**] show that *if G is amenable and $|G| \leq |X|$, then* $|\mathfrak{L}(X)| = 2^{2^{|X|}}$. An interesting open question is *what is the cardinal $|\mathfrak{L}(X)|$ when*

$|G| > |X|$? Recent progress on this issue has been made by Z. Yang [**S3**], who shows that under the assumption of the Continuum Hypothesis, there exists a locally finite (and so amenable) group G with $|G| = \mathfrak{c}$ and a denumerable left G-set X such that $|\mathfrak{L}(X)| = 1$.

We now turn to our third question: what is the cardinality of $\mathfrak{L}(G) \sim \mathfrak{L}_t(G)$? The next result provides a criterion for determining when a mean $m \in \mathfrak{L}(G)$ is not in $\mathfrak{L}_t(G)$. References for the result are Granirer [**15**], Rudin [**2**], and Rosenblatt [**5**]. We use the notations of (2.24). In particular, $\Phi(G)$ is the carrier space of $L_\infty(G)$.

Let G be a locally compact group. If means $m, n \in \mathfrak{M}(G)$ are such that the probability measures $\hat{m}$, $\hat{n}$ on $\Phi(G)$ are mutually singular, then we say that m and n are *mutually singular* or that m is *singular to* n ($\hat{m} \perp \hat{n}$). If $A \subset \mathfrak{M}(G)$ and m is singular to every element of A, then we will say that m is *singular to* A.

In our discussion of the cardinal $|\mathfrak{L}_t(G)|$, it was useful to use the closure $\hat{E}$ of a subset E of G in X, the maximal ideal space of $\mathrm{U_r}(G)$. When we consider the corresponding issue for $L_\infty(G)$, this does not work so well since if G is not discrete, $G \not\subset \Phi(G)$, the maximal ideal space of $L_\infty(G)$. The considerations below show that this difficulty can be readily overcome.

Let $\mathscr{M}(G)$ be the σ-algebra of λ-measurable subsets of G. If $E \in \mathscr{M}(G)$, then χ_E is an idempotent in $L_\infty(G)$ and so $\hat{\chi}_E$ is of the form $\chi_{\hat{E}}$, where $\hat{E}$ is an open and closed subset of $\Phi(G)$. (When G is discrete, then $\hat{E}$ is just the closure of E in βG $(= \Phi(G))$.) Some simple facts about $E \to \hat{E}$ are given in (7.11).

(7.9) PROPOSITION. *Suppose that D is a measurable subset of G for which $\lambda(G \sim D^{-1}) < 1$ and that $m \in \mathfrak{L}(G)$ is such that $m(\chi_D) = 0$. Then m is singular to $\mathfrak{L}_t(G)$.*

PROOF. The compact and noncompact cases are treated separately.

(i) *Suppose that G is compact.* Since $\mathfrak{L}_t(G) = \{\lambda\}$, we must show that $\hat{m} \perp \hat{\lambda}$.

Let $\Theta = \mathbf{S}(\hat{m}) \cap \mathbf{S}(\hat{\lambda})$ and $\nu = \hat{\lambda}|_\Theta$. Now for $E \in \mathscr{B}(\Phi(G))$, $x \in G$, we have, using the left invariance of Θ (2.25(ii)),

$$\hat{\lambda}(E \cap \Theta) = \hat{\lambda}(x(E \cap \Theta)) = \hat{\lambda}(xE \cap \Theta)$$

so that $\nu(xE) = \nu(E)$. If $\hat{\lambda}(\Theta) = 0$, then $\hat{m} \perp \hat{\lambda}$. Suppose, then, that $\hat{\lambda}(\Theta) \neq 0$. Then for some $c > 0$, $\nu = c\mu$, where $\mu \in P(\Phi(G))$. By (2.25(ii)), there exists $n \in \mathfrak{L}(G)$ with $\hat{n} = \mu$. Clearly $0 \leq n \leq c^{-1}\lambda$. Now n can be regarded as a finitely additive measure on $\mathscr{B}(G)$ by setting $n(A) = n(\chi_A)$ $(A \in \mathscr{B}(G))$. It follows as in the proof of (a) of (4.27(i)) that n is countably additive. The uniqueness of Haar measure then yields that $n = \lambda$ so that $\nu = c\hat{\lambda}$. So $\mathbf{S}(\hat{\lambda}) = \Theta$. Now since $\hat{m}(\hat{D}) = m(D) = 0$, the open set $\hat{D}$ cannot intersect $\mathbf{S}(\hat{m})$. Thus $\hat{D} \cap \Theta = \varnothing$, and since $\hat{\lambda}$ vanishes off Θ, we have $\lambda(D) = 0$. This is impossible since

$$\lambda(D) = \lambda(D^{-1}) = 1 - \lambda(G \sim D^{-1}) > 0.$$

(ii) *Now suppose that G is not compact.* Let $m_1 \in \mathfrak{L}_t(G)$. Noting that $\hat{m}$ is supported on $(G \sim D)^\wedge$, it suffices to prove that if $B = G \sim D$, then $m_1(B) = 0$.

To this end, let $\{C_n\}$ be a sequence in $\mathscr{C}(G)$ such that $0 < \lambda(C_n) \to \infty$, and define $\mu_n \in P(G)$ by setting

$$d\mu_n = (\chi_{C_n}/\lambda(C_n))\, d\lambda.$$

Then $\mu_n^{\sim} \in P(G)$, and for each $x \in G$, we have, using (1.1(3)),

$$\begin{aligned}\chi_B\mu_n^{\sim}(x) &= \int (x\chi_B)(t)\, d\mu_n^{\sim}(t) = \int (x\chi_B)(t^{-1})\, d\mu_n(t)\\ &= (\lambda(C_n))^{-1}\int_{C_n} \chi_B(t^{-1}x)\, d\lambda(t)\\ &= \lambda(xB^{-1}\cap C_n)/\lambda(C_n) \le \lambda(B^{-1})/\lambda(C_n).\end{aligned}$$

Since $\lambda(B^{-1}) = \lambda(G \sim D^{-1}) < 1$, we have

$$\|\chi_B\mu_n^{\sim}\|_\infty < (\lambda(C_n))^{-1}.$$

As $m_1(B) = m_1(\chi_B) = m_1(\chi_B\mu_n^{\sim})$, $\|m_1\| = 1$, and $(\lambda(C_n))^{-1} \to 0$, it follows that $m_1(B) = 0$ as required. □

The preceding result will be used in (7.17) with $D = (G \sim V)^{-1}$, where V is as in the next proposition.

(7.10) PROPOSITION. *Let G be a nondiscrete, σ-compact, locally compact group, and let $\varepsilon > 0$. Then there exists a dense, open subset V of G such that $\lambda(V) < \varepsilon$.*

PROOF. Let N, $\{U_n\}$ be as in Problem $7-7$, where $\lambda(U_n) < \varepsilon 2^{-n}$. Let $\{x_n\}$ be a sequence in G such that $\{x_nN \colon n \in \mathbb{P}\}$ is a dense subset of G/N. Then $\bigcup_{n=1}^{\infty}(x_nN)$ is a dense subset of G and we can take $V = \bigcup_{n=1}^{\infty}(x_nU_n)$. □

We now state some simple facts about the map $E \to \hat{E}$ (introduced in (7.8)).

(7.11) PROPOSITION. (i) *The map $E \to \hat{E}$ preserves finite intersections, finite unions, and complementation.*

(ii) *If $E \in \mathscr{M}(G)$ and $x \in G$, then $(x^{-1}E)^\wedge = x^{-1}(\hat{E})$.*

(iii) *If $\{E_\gamma \colon \gamma \in \Gamma\} \subset \mathscr{M}(G)$, then $\bigcap_{\gamma\in\Gamma}\hat{E}_\gamma$ is empty if and only if there exists a finite number of elements $\gamma_1, \ldots, \gamma_n$ of Γ for which*

$$\lambda(E_{\gamma_1} \cap \cdots \cap E_{\gamma_n}) = 0.$$

PROOF. (i) Let $E_1, E_2 \in \mathscr{M}(G)$. The equality $(E_1\cap E_2)^\wedge = \hat{E}_1 \cap \hat{E}_2$ follows since $\chi_{E_1}\chi_{E_2} = \chi_{E_1\cap E_2}$. To prove that $(E_1 \cup E_2)^\wedge = \hat{E}_1 \cup \hat{E}_2$, use the equality

$$\chi_{E_1\cup E_2} = \chi_{E_1} + \chi_{E_2} - \chi_{E_1}\chi_{E_2}.$$

From the equality $1 = \chi_{E_1} + \chi_{G\sim E_1}$, we deduce that $(G \sim E_1)^\wedge = \Phi(G) \sim \hat{E}_1$.

(ii) Use the equalities $\chi_E x = \chi_{x^{-1}E}$ and $(\chi_E x)^\wedge(p) = \chi_{\hat{E}}(xp)$ $(p \in \Phi(G))$.

(iii) Since $\Phi(G)$ is compact, $\bigcap_{\gamma\in\Gamma}\hat{E}_\gamma = \varnothing$ if and only if there exist $\gamma_1, \ldots, \gamma_n \in \Gamma$ such that $\hat{E}_{\gamma_1} \cap \cdots \cap \hat{E}_{\gamma_n} = \varnothing$. Now $\hat{E}_{\gamma_1} \cap \cdots \cap \hat{E}_{\gamma_n} = \varnothing$ if and only if $\chi_{E_{\gamma_1}\cap\cdots\cap E_{\gamma_n}} = 0$ in $L_\infty(G)$, that is, $\lambda(E_{\gamma_1} \cap \cdots \cap E_{\gamma_n}) = 0$. □

We now show that when G is amenable as a discrete group, then, in general, $\mathfrak{L}(G) \neq \mathfrak{L}_t(G)$ and $\mathfrak{L}(C(G)) \neq \mathfrak{L}_t(C(G))$. We require two preliminary propositions. The results (7.12)–(7.14) are due to Liu and van Rooij [**1**], although the assertion "$\mathfrak{L}(G) \neq \mathfrak{L}_t(G)$" in (7.14) was proved earlier in the work of Stafney [**S**], Granier [**15**] and Rudin [**2**].

(7.12) PROPOSITION. *Let G be a locally compact group that is amenable as a discrete group. Suppose that there exists $\phi \in C(G)$ satisfying the following conditions*:

(i) *given $x_1, \ldots, x_n \in G$, there exists a nonvoid open subset V of G such that*

$$\phi(x_i V) = \{1\} \qquad (1 \leq i \leq n); \tag{1}$$

(ii) *there exists $\nu \in P(G)$ such that*

$$\|\phi\nu\|_\infty \leq 1/2. \tag{2}$$

Then $\mathfrak{L}(B) \sim \mathfrak{L}_t(B)$ is not empty when $B = C(G)$ or when $B = L_\infty(G)$.

PROOF. Let E be the interior of the set $\{x \in G : \phi(x) = 1\}$. Let $x_1, \ldots, x_n \in G$ and choose V so that (1) is satisfied. Then

$$\lambda\left(\bigcap_{i=1}^n x_i^{-1} E\right) \geq \lambda(V) > 0.$$

So (cf. (7.4)) the set $\Psi = \bigcap_{x \in G} (xE)^\wedge$ is a closed, left invariant subset of $\Phi(G)$, and there exists $m \in \mathfrak{L}(G)$ such that $\mathbf{S}(\hat{m}) \subset \Psi$. The fact that $\phi\chi_E = \chi_E$ implies that $\hat{\phi}(\alpha) = 1$ for $\alpha \in \hat{E}$, and so, since $\Psi \subset \hat{E}$, $\hat{\phi}|_\Psi = 1$. Hence $m(\phi) = \hat{m}(\hat{\phi}) = 1$. But from (2), we have $|m(\phi\nu)| \leq 1/2$, so that $m \in \mathfrak{L}(G) \sim \mathfrak{L}_t(G)$. Finally, since $\phi \in C(G)$, it follows that $m|_{C(G)} \in \mathfrak{L}(C(G)) \sim \mathfrak{L}_t(C(G))$. □

We now construct a function ϕ satisfying (i) and (ii) of (7.12) when G is nondiscrete and noncompact. The cardinal $\mathfrak{m}$ of (7.1) will prove useful again.

(7.13) PROPOSITION. *Let G be a nondiscrete, noncompact, locally compact group. Then there exists an element ϕ of $C(G)$ satisfying conditions* (i) *and* (ii) *of* (7.12).

PROOF. The idea of the proof is to construct (by transfinite recursion) a certain disjoint family of relatively compact, open subsets of G (viz. the sets $U^{-1}US_\alpha a_\alpha$ below) and build up ϕ from simpler functions defined on these subsets.

Let $\mu \in P(G)$ be such that $d\mu/d\lambda = g$, where g is continuous with compact support. Define

$$\mathscr{A} = \{W : W \text{ is open in } G,\ W^- \text{ is compact, and } \lambda[(W^-)^{-1}] < \tfrac{1}{2}\|g\|_\infty^{-1}\}. \tag{1}$$

Since G is not discrete, the family $\mathscr{A}$ is an open cover for G. Note that:

(2) given $x_1, \ldots, x_n \in G$, there exists $W \in \mathscr{A}$ such that $x_i \in W$ $(1 \leq i \leq n)$.

We first cut down the size of $\mathscr{A}$, replacing it by a subset $\mathscr{D}$.

Choose a family $\mathscr{F}$ of compact subsets of G with $\bigcup \mathscr{F} = G$, where the cardinal number $\mathfrak{m}$ of $\mathscr{F}$ is smallest possible. Since G is not compact, the cardinal $\mathfrak{m}$ is infinite. Let $n \in \mathbb{P}$. If $Z_i \in \mathscr{F}$ $(1 \le i \le n)$, then, using (2), the compact set $Z_1 \times \cdots \times Z_n$ can be covered by a *finite* number of sets of the form $W^{(n)} = W \times \cdots \times W$ (n copies), where $W \in \mathscr{A}$. Since the sets $Z_1 \times \cdots \times Z_n$ cover G^n and as $\mathfrak{m}$ is infinite, we can find a family $\mathscr{A}_n$ of sets of the form $W^{(n)}$ $(W \in \mathscr{A})$ such that $\bigcup \mathscr{A}_n = G^n$ and $|\mathscr{A}_n| = \mathfrak{m}$.

Now define $\mathscr{D} = \{W \in \mathscr{A} : W^{(n)} \in \mathscr{A}_n \text{ for some } n\}$. Then (2) is satisfied with $\mathscr{A}$ replaced by $\mathscr{D}$, and $|\mathscr{D}| = \mathfrak{m}$. Well-ordering $\mathscr{D}$, we can write $\mathscr{D} = \{W_\beta : \beta \in \alpha\}$, where α is the first ordinal such that $|\alpha| = \mathfrak{m}$. Let $\beta \in \alpha$. Using (1), the regularity of λ, and Urysohn's Lemma, there exists $\psi_\beta \in C(G)$ with compact support such that

$$\begin{gathered}\psi_\beta \ge 0, \qquad \psi_\beta(x) = 1 \quad \text{if } x \in (W_\beta^-)^{-1}, \\ \int \psi_\beta(t)\, d\lambda(t) < \tfrac{1}{2}||g||_\infty^{-1}.\end{gathered} \tag{3}$$

Define $\phi_\beta(x) = \psi_\beta(x^{-1})$ $(x \in G)$. Then

$$\begin{aligned}|\phi_\beta \mu^\sim(x)| &= \left|\int \phi_\beta(t^{-1}x) g(t)\, d\lambda(t)\right| \\ &= \left|\int \psi_\beta(x^{-1}t) g(t)\, d\lambda(t)\right| = \left|\int \psi_\beta(t) g(xt)\, d\lambda(t)\right| \\ &\le ||g||_\infty \tfrac{1}{2} ||g||_\infty^{-1} = \tfrac{1}{2}.\end{aligned}$$

So

$$||\phi_\beta \mu^\sim||_\infty \le \tfrac{1}{2}. \tag{4}$$

We now translate the functions ϕ_β so that the resulting functions have disjoint support. To this end, let S_β be the support of ϕ_β, and let U be a relatively compact, open neighbourhood of e that contains the support of g. Using transfinite recursion, we now construct, for each β, an element $a_\beta \in G$ such that the sets $\{U^{-1}US_\beta a_\beta : \beta \in \alpha\}$ are disjoint.

Let $\beta_0 \in \alpha$ and suppose that a_γ has been defined for $\gamma < \beta_0$ such that the family $\{U^{-1}US_\gamma a_\gamma : \gamma < \beta_0\}$ is disjoint. If there is *no* $x \in G$ for which $U^{-1}US_{\beta_0}x \cap U^{-1}US_\gamma a_\gamma = \varnothing$ for all $\gamma < \beta_0$, then the family

$$\mathscr{G} = \{(S_{\beta_0}^{-1}U^{-1}UU^{-1}US_\gamma a_\gamma)^- : \gamma < \beta_0\}$$

is a covering of G by compact sets, and since $|\mathscr{G}| \le |\beta_0| < |\alpha| = \mathfrak{m}$, the definition of $\mathfrak{m}$ is contradicted. So the existence of the desired element a_{β_0} is assured, and by recursion, the required elements a_β $(\beta \in \alpha)$ exist.

The function ϕ that will satisfy (i) and (ii) of (7.12) is defined by

$$\phi = \sum_{\beta \in \alpha} a_\beta^{-1} \phi_\beta. \tag{5}$$

Since the support of each $a_\beta^{-1}\phi_\beta$ is contained in $U^{-1}US_\beta a_\beta$ and since

$$U^{-1}US_\beta a_\beta \cap U^{-1}US_\gamma a_\gamma = \varnothing \quad \text{if } \beta \ne \gamma,$$

it follows that ϕ is well defined, and is bounded on G. We now prove that ϕ is continuous on G. Let $x \in G$. If $U^{-1}Ux$ does not intersect any of the sets $S_\beta a_\beta$, then ϕ is 0 on $U^{-1}Ux$ and so is trivially continuous at x. If, however, $U^{-1}Ux \cap S_\beta a_\beta \neq \emptyset$ for some β, then $x \in U^{-1}US_\beta a_\beta$, and ϕ coincides with the continuous function $a_\beta^{-1}\phi_\beta$ on $U^{-1}US_\beta a_\beta$.

We now check (i) of (7.12). Let $x_1, \ldots, x_n \in G$. Recalling that $\mathscr{D}$ satisfies (2), we can find $\beta \in \alpha$ such that $x_i \in W_\beta$ $(1 \leq i \leq n)$. Let $V = [\bigcap_{i=1}^n x_i^{-1}W_\beta]a_\beta$. Then V is an open neighbourhood of a_β, and using (3), we have $W_\beta \subset S_\beta$, and

$$\phi(x_iV) \subset \phi(W_\beta a_\beta) = a_\beta^{-1}\phi_\beta(W_\beta a_\beta) = \phi_\beta(W_\beta) = \{1\}$$

as required.

To prove (ii) of (7.12), let $x \in G$. If $x \in US_\beta a_\beta$ for some β, then $U^{-1}x \subset U^{-1}US_\beta a_\beta$, and as g vanishes outside U,

$$|(\phi\mu^\sim)(x)| = \left|\int_U \phi(t^{-1}x)g(t)\,d\lambda(t)\right| = |(a_\beta^{-1}\phi_\beta\mu^\sim)(x)| \leq \tfrac{1}{2}$$

using (4). If, however, $x \notin US_\beta a_\beta$ for *all* β, then for all $t \in U$,

$$\phi(t^{-1}x) = \sum_\beta a_\beta^{-1}\phi_\beta(t^{-1}x) = 0.$$

So in this case, $\phi\mu^\sim(x) = 0$. It follows that $\|\phi\mu^\sim\|_\infty \leq 1/2$, and (ii) of (7.12) is established with $\nu = \mu^\sim$. □

(7.14) THEOREM. *Let G be a locally compact group that is amenable as a discrete group. Then $\mathfrak{L}(C(G)) = \mathfrak{L}_t(C(G))$ if and only if G is either compact or discrete. Further, if G is noncompact and nondiscrete, then $\mathfrak{L}(G) \neq \mathfrak{L}_t(G)$.*

PROOF. The first assertion follows using (7.12) and (7.13). (It is obvious that $\mathfrak{L}(C(G)) = \mathfrak{L}_t(C(G))$ if G is compact or discrete.) Now suppose that G is noncompact and nondiscrete, and find, by the first assertion of the theorem, an element $m_0 \in \mathfrak{L}(C(G)) \sim \mathfrak{L}_t(C(G))$. By Problem 1-13, we can find $n_0 \in \mathfrak{L}(G)$ with $n_0|_{C(G)} = m_0$. Clearly $n_0 \in \mathfrak{L}(G) \sim \mathfrak{L}_t(G)$. □

Note that (7.14) does not give any information about $\mathfrak{L}(G) \sim \mathfrak{L}_t(G)$ when G is compact. This will be remedied later ((7.21)). (Of course, when G is discrete, then $\mathfrak{L}(G) = \mathfrak{L}_t(G)$.)

Our next concern is to obtain an estimate for $|\mathfrak{L}(G) \sim \mathfrak{L}_t(G)|$. To this end, the following technical proposition gives, for a large class of groups, a family $\{E_\gamma : \gamma \in \Gamma\}$ of subsets of G that will produce elements of $\mathfrak{L}(G) \sim \mathfrak{L}_t(G)$. The proposition uses Mycielski's Theorem. References for the results (7.15)–(7.21) are Rosenblatt [**5**], [**7**], [**8**], [**9**] and Wells [**1**].

(7.15) PROPOSITION. *Let G be a nondiscrete, σ-compact, locally compact metric group. Then there exists a family $\{E_\gamma : \gamma \in \Gamma\}$ of measurable subsets of G, each of which has finite λ-measure, such that* (i) *and* (ii) *below hold:*

(i) $|\Gamma| = \mathfrak{c}$, *the cardinality of the continuum;*

(ii) *given* $n, m \in \mathbb{P}$ *and* x_{ij}, y_{ij} $(1 \leq i \leq n,\ 1 \leq j \leq m)$ *in* G *such that for each* i, *the points* $x_{i1}, \ldots, x_{im}, y_{i1}, \ldots, y_{im}$ *are distinct, then we have*

$$\lambda(\{\bigcap_{i,j}[x_{ij}E_{\gamma_i} \cap y_{ij}E^c_{\gamma_i}]\} \cap V) \neq 0 \tag{1}$$

for every nonvoid, open subset V *of* G *and for every* n*-tuple* $(\gamma_1, \ldots, \gamma_n) \in \Gamma^n$ *with* $\gamma_{i_1} \neq \gamma_{i_2}$ *for* $i_1 \neq i_2$.

PROOF. We derive the result by applying Mycielski's Theorem (MT) (Appendix D). The metric space X of the application is the set $\{E \in \mathscr{M}(G)\colon \lambda(E) < \infty\}$, where, as usual, E_1 and E_2 are identified if $\lambda(E_1 \,\triangle\, E_2) = 0$. The metric d on X is defined by

$$d(E_1, E_2) = \lambda(E_1 \,\triangle\, E_2).$$

Then [**DS**, III.7.1] (X, d) is a complete, separable metric space. Further, since G is not discrete and λ is regular, the space X contains elements of arbitrarily small measure, and as a result, (X, d) has no isolated points. So (X, d) satisfies the conditions of MT.

Since G is locally compact, σ-compact, and metric, it is separable, and so we can find a (countable) family $\mathscr{C} = \{C_r\colon r \in \mathbb{P}\}$ of compact subsets of G such that $\{C^0_r\colon r \in \mathbb{P}\}$ is a base for the topology of G and $C_r = (C^0_r)^-$. (So we need only consider in (1) the case when V is one of the C_r.)

Let $\mathscr{G}_p$ be the set of all $2p$-tuples $(K_1, \ldots, K_{2p})$, where $K_i \cap K_j = \varnothing$ if $i \neq j$ and each $K_i \in \mathscr{C}$. Let $\mathscr{S} = \bigcup_{p,q}(\mathscr{G}_p)^q$. A typical element S of $\mathscr{S}$ is of the form $(A_1, \ldots, A_q)$ with $A_i = (A_{i1}, \ldots, A_{ip}, B_{i1}, \ldots, B_{ip}) \in \mathscr{G}_p$. Note that $\mathscr{S}$ is countable. For such an S and $r \in \mathbb{P}$ let $U(r, s) \subset X^q$ be the set of all q-tuples $(E_1, \ldots, E_q)$ with the following property:

(2) there exist $2p$-tuples $(x_{i1}, x_{i2}, \ldots, x_{ip}, y_{i1}, \ldots, y_{ip}) \in A_{i1} \times \cdots \times A_{ip} \times B_{i1} \times \cdots \times B_{ip}$ $(1 \leq i \leq q)$ such that

$$\lambda(\{\bigcap_{i,j}[x_{ij}E_i \cap y_{ij}E^c_i]\} \cap C_r) = 0.$$

For the application of MT we will take the R_i to be the set $U(r, S)$, $(r, p \geq 1$, $S \in (\mathscr{G}_p)^q)$. The point about (2) is that we are picking out those q-tuples $(E_1, \ldots, E_q)$ that violate (1).

Suppose that we can show that each $U(r, S)$ is closed with empty interior in X^q. Then MT applies, and we have a subset A of X as in the statement of MT. Indexing A, we write $A = \{E_\gamma\colon \gamma \in \Gamma\}$. Let x_{ij}, y_{ij} $(1 \leq i \leq q,\ 1 \leq j \leq p)$ be elements of G with $x_{i1}, \ldots, x_{ip}, y_{i1}, \ldots, y_{ip}$ distinct for each i. Let $\gamma_1, \ldots, \gamma_q$ be distinct in Γ and $r \in \mathbb{P}$. We can find $S' = (A'_1, \ldots, A'_q) \in \mathscr{S}$ such that $x_{ij} \in A'_{ij}$, $y_{ij} \in B'_{ij}$, since $\{C^0_r\colon r \geq 1\}$ is a base for X. Since $(E_{\gamma_1}, \ldots, E_{\gamma_q}) \notin U(r, S')$ (by MT), (ii) of the present proposition is established.

It therefore only remains to show that every $U(r, S)$ is closed and has empty interior in X^q. As above, let $S = (A_1, \ldots, A_q)$.

For the purposes of this paragraph, a qp-tuple $x \in G^{qp}$ will be indexed:

$$x = (x_{11}, \ldots, x_{1p}, x_{21}, \ldots, x_{2p}, x_{31}, \ldots, x_{q1}, \ldots, x_{qp}).$$

Let $C = \bigcup_{ij}[A_{ij} \cup B_{ij}] \in \mathscr{C}(G)$, and for $E \in X$, let $K(E) = C^{-1}C_r \sim E$. Let $E_1, \ldots, E_q \in X$. Define maps Φ, Ψ from $G^{qp} \times G^{qp} \times X^q$ to $\mathbb{R}$ by

$$\Phi(x, y, E_1, \ldots, E_q) = \lambda[[\bigcap_{i,j}(x_{ij}E_i \cap y_{ij}E_i^c)] \cap C_r],$$

$$\Psi(x, y, E_1, \ldots, E_q) = \lambda\left[\left[\bigcap_{i,j}(x_{ij}E_i \cap y_{ij}K(E_i))\right] \cap C_r\right].$$

The map $E \to K(E)$ is continuous on X since $K(E) \triangle K(F) \subset E \triangle F$, and applying Problem 7-10 with $n = 2qp$, it follows that Ψ is continuous. Let

$$H = A_{11} \times \cdots \times A_{1p} \times A_{21} \times \cdots \times A_{qp} \times B_{11} \times \cdots \times B_{qp}$$

and $D = H \times X^q$. Now if $y_{ij} \in C$, then

$$y_{ij}E_i^c \cap C_r = y_{ij}K(E_i) \cap C_r,$$

and as $A_{ij} \cup B_{ij} \subset C$ for all i, j, Φ and Ψ coincide on D. It follows that $L = \Phi^{-1}(\{0\}) \cap D$ is closed in D. Since $D = H \times X^q$ with H compact, it follows that $P(L)$ is closed in X^q, where P is the projection map from D onto X^q. So $U(r, S) = P(L)$ is closed as required.

We now show that $U(r, S)$ has empty interior. Let $\varepsilon > 0$ and choose, by (7.10), a dense, open subset V' of G such that $\lambda(V') < \varepsilon$. Let $(E_1, \ldots, E_q) \in U(r, S)$. It is sufficient to construct elements $F_1, \ldots, F_q$ of X such that $(F_1, \ldots, F_q) \notin U(r, S)$ and $\lambda(E_i \triangle F_i) < \varepsilon$. We can suppose that $e \in C_r^0$.

By the definition of $\mathscr{S}$, for each i, the family $\{A_{i1}, \ldots, A_{ip}, B_{i1}, \ldots, B_{ip}\} \subset \mathscr{C}(G)$ is disjoint. So we can find an open neighbourhood W of e such that for each i, the family $\{A_{i1}W, \ldots, B_{ip}W\}$ is disjoint. Let

$$G_i = \bigcup_{k=1}^{p} [(A_{ik}W)^{-1} \cap V']$$

and

$$F_i = (E_i \cap (V')^c) \cup G_i.$$

Then $F_i \in X$, and since $E_i \triangle F_i \subset V'$, we have $\lambda(E_i \triangle F_i) < \varepsilon$.

It remains to check that $(F_1, \ldots, F_n) \notin U(r, S)$. For each pair (i, j), let $x_{ij} \in A_{ij}, y_{ij} \in B_{ij}$. Now

$$x_{ij}F_i \supset x_{ij}G_i \supset x_{ij}[(A_{ij}W)^{-1}] \cap x_{ij}V', \tag{3}$$

$$y_{ij}F_i^c \supset y_{ij}(G_i^c \cap V') \supset y_{ij}[(B_{ij}W)^{-1}] \cap y_{ij}V' \tag{4}$$

since $A_{ik}W \cap B_{ij}W = \varnothing$. Let

$$Z = \left(\bigcap_{i,j}[x_{ij}[(A_{ij}W)^{-1}] \cap y_{ij}[(B_{ij}W)^{-1}]]\right) \cap C_r^0.$$

Then Z is an open neighbourhood of e, and as every translate of V' is dense and open in G, we have $x_{ij}V' \cap Z$, $y_{ij}V' \cap Z$ dense and open in Z. From (3) and (4),

$$\left(\bigcap_{i,j}(x_{ij}F_i \cap y_{ij}F_i^c)\right) \cap C_r^0 \supset \bigcap_{i,j}[(Z \cap x_{ij}V') \cap (Z \cap y_{ij}V')]$$

and by the Baire Category Theorem for G, the finite intersection $(\bigcap_{i,j}(x_{ij}F_i \cap y_{ij}F_i^c)) \cap C_r$ contains a nonvoid, open set and so has positive λ-measure. Thus $(F_1, \ldots, F_n) \notin U(r, S)$ as required. □

(7.16) PROPOSITION. *Let G be a nondiscrete, σ-compact, locally compact metric group and let $\{E_\gamma : \gamma \in \Gamma\}$ be a family of measurable subsets of G of finite λ-measure satisfying properties* (i) *and* (ii) *of* (7.15). *Let B and R be the linear subspaces of $L_\infty(G)$ spanned by the sets $\{\chi_{xE_\gamma} : x \in G,\ \gamma \in \Gamma\}$ and $\{\chi_E : E \in \mathscr{M}(G), (E^-)^0$ is empty$\}$ respectively. Let $\phi \in C(G)$, $b \in B$, and $r \in R$. Write $b = \sum_{x,\gamma} b(x,\gamma)\chi_{xE_\gamma}$, where, for $x \in G$, $\gamma \in \Gamma$, $b(x,\gamma)$ belongs to $\mathbb{C}$ and is nonzero for only a finite number of pairs (x,γ). Then the following assertions are true:*

(i) *$\phi + b + r = 0$ in $L_\infty(G)$ if and only if $\phi = 0$, $b(x,\gamma) = 0$ for all pairs (x,γ) and $r = 0$ almost everywhere.*

(ii) *Let M be the sum of all the negative $b(x,\gamma)$. If $\phi + b + r \geq 0$ in $L_\infty(G)$, then ϕ is real-valued, and $\phi + M \geq 0$ on G.*

PROOF. (i) Suppose that $\phi + b + r = 0$ in $L_\infty(G)$.

If $E \in \mathscr{M}(G)$ and $(E^-)^0$ is empty, then E^c contains a dense open subset of G. Since the open, dense property is preserved under finite intersections, there exists a dense, open subset U of G such that $r(U) = \{0\}$.

Let V be any nonvoid, open subset of U. By (7.15), the set

$$E = \left[\bigcap\{xE_\gamma^c : b(x,\gamma) \neq 0\}\right] \cap V$$

is *not* λ-null. Since $b(E) = \{0\}$,

$$\phi|_E = (\phi + b + r)|_E = 0 \quad \text{almost everywhere.}$$

Since E is not null, we have $0 \in \phi(V)$. Since V was arbitrary and ϕ is continuous, ϕ vanishes on U and so on G. Thus $\phi = 0$.

It readily follows that $b = 0$ almost everywhere on U. Let $x_0 \in G$, $\gamma_0 \in \Gamma$. Then the set

$$E_1 = x_0E_{\gamma_0} \cap \left[\bigcap\{xE_\gamma^c : b(x,\gamma) \neq 0,\ (x_0,\gamma_0) \neq (x,\gamma)\}\right] \cap U$$

is *not* null, and $b(E_1) = \{b(x_0,\gamma_0)\}$. Since $b = 0$ almost everywhere on U, $b(x_0,\gamma_0) = 0$. So $b = 0$, and the nontrivial implication of (i) is proved.

(ii) Suppose that $\phi + b + r \geq 0$ in $L_\infty(G)$. Using arguments similar to those of (i), the functions ϕ and b are real-valued on G. By considering the set

$$\left[\bigcap\{xE_\gamma : b(x,\gamma) < 0\}\right] \cap \left[\bigcap\{xE_\gamma^c : b(x,\gamma) > 0\}\right] \cap V,$$

where V is as in the proof of (i), it follows that $\phi + M \geq 0$ on G. □

(7.17) THEOREM. *Let G be a nondiscrete, σ-compact, locally compact metric group that is amenable as a discrete group. Let $m \in \mathfrak{L}(C(G))$. Then there exists a set $\{m_\delta : \delta \in \Delta\} \subset \mathfrak{L}(G)$ with the following properties*:

(i) $|\Delta| = 2^{\mathfrak{c}}$;

(ii) $\mathbf{S}(\hat{m}_\delta) \cap \mathbf{S}(\hat{m}_\sigma) = \varnothing$ *when* $\delta \neq \sigma$ *in* Δ;

(iii) *if V is a dense open subset of G with $\lambda(V) < 1$, then $\mathbf{S}(\hat{m}_\delta) \subset (V^{-1})^\wedge$ for all δ*;

(iv) $m_\delta|_{C(G)} = m$ *for all* δ;

(v) m_δ *is singular to* $\mathfrak{L}_t(G)$ *for all* δ.

PROOF. Let $\{E_\gamma : \gamma \in \Gamma\}$, B, and R be as in (7.16). Then $\Delta = \{0,1\}^\Gamma$ satisfies (i). Let $A = C(G) + B + R$ and let $\delta \in \Delta$. Using (7.16), we can define a linear functional m'_δ on A, where

$$(1) \qquad m'_\delta(\phi + b + r) = m(\phi) + \sum_{x,\gamma} b(x,\gamma)\delta(\gamma)$$

for $\phi \in C(G)$, $b \in B$, and $r \in R$. If $\phi + b + r \geq 0$ in $L_\infty(G)$, then, by (7.16),

$$m'_\delta(\phi + b + r) \geq \inf\{\phi(x) + M : x \in G\} \geq 0.$$

Noting that $m'_\delta(1) = 1$, it follows that $m'_\delta \in \mathfrak{M}(A)$. It is easily checked that B and R are right invariant subspaces of $L_\infty(G)$ and that

$$\Big(\sum_{x,\gamma} b(x,\gamma)\chi_{xE_\gamma}\Big)x_0 = \Big(\sum_{x,\gamma} b(x_0x,\gamma)\chi_{xE_\gamma}\Big).$$

It follows using (1) that $m'_\delta \in \mathfrak{L}(A)$.

By Problem 1-13, there exists $m_\delta \in \mathfrak{L}(G)$ for which $m_\delta|_A = m'_\delta$. It remains to prove (ii), (iii), and (v).

Let V be an open, dense subset of G with $\lambda(V) < 1$. (The existence of such a V is assured by (7.10).) Let $D = (G \sim V)^{-1}$. Since $\chi_D \in R$, we have $m_\delta(\chi_D) = m'_\delta(\chi_D) = 0$, and applying (7.9), m_δ is singular to $\mathfrak{L}_t(G)$. This gives (v) and (iii), noting that $m_\delta(V^{-1}) = m_\delta(G \sim D) = 1$.

Finally, if $\sigma \in \Delta$ and $\delta \neq \sigma$, we can, without loss of generality, find $\gamma_0 \in \Gamma$ such that $\delta(\gamma_0) = 1$ and $\sigma(\gamma_0) = 0$. Let $E = E_{\gamma_0}$. From (1), it follows that $m_\delta(\chi_E) = 1$ while $m_\sigma(\chi_E) = 0$. So $\mathbf{S}(\hat{m}_\delta) \subset \hat{E}$ and $\mathbf{S}(\hat{m}_\sigma) \subset (\widehat{G \sim E}) = \Phi(G) \sim \hat{E}$, and (ii) is proved. □

(7.18) COROLLARY. *Let G be a nondiscrete, σ-compact, locally compact metric group that is amenable as a discrete group. Then the map $m \to m|_{C(G)}$ from $\mathfrak{L}(G)$ onto $\mathfrak{L}(C(G))$ is* not *one-to-one.*

The next result shows that much of (7.18) is still valid with the metric condition removed.

(7.19) THEOREM. *Let G be a nondiscrete, σ-compact, locally compact group that is amenable as a discrete group. Then there exists a subset P of $\mathfrak{L}(G)$ of* *cardinality* $2^{\mathfrak{c}}$, *such that* $\mathbf{S}(\hat{m}) \cap \mathbf{S}(\hat{n}) = \varnothing$ *whenever* $m \neq n$ *in* P. *Further, every element of* P *is singular to* $\mathfrak{L}_t(G)$.

PROOF. From Problem 7-7, there exists a compact normal subgroup H of G such that G/H is metrisable and nondiscrete. Further, G/H is amenable as a discrete group and is σ-compact.

Let $Q: G \to G/H$ be the quotient map. Now ((1.11)) the mapping

$$Q^*: L_\infty(G/H) \to L_\infty(G)$$

is an isometric $*$-homomorphism, where $(Q^*\psi)(x) = \psi(Q(x))$ for $\psi \in L_\infty(G/H)$ and $x \in G$. It is routine to verify that $Q_0 = Q^{**}|_{\Phi(G)}$ is a continuous map from $\Phi(G)$ into $\Phi(G/H)$ for which $Q_0(x\alpha) = Q(x)Q_0(\alpha)$ for all $x \in G$, $\alpha \in \Phi(G)$. Further $Q_0(\Phi(G)) = \Phi(G/H)$, since, with $L_\infty(G/H)$ identified by means of Q^* with a C^*-subalgebra of $L_\infty(G)$, every element of $\Phi(G/H)$ extends to an element of $\Phi(G)$.

Now G/H satisfies the hypotheses of (7.17) (with G replaced by G/H). So we can find a set $\{\hat{m}_\delta : \delta \in \Delta\} \subset \mathfrak{L}(G/H)$ such that the conclusions of (7.17) hold. For each δ, let $\Psi_\delta = Q_0^{-1}(\mathbf{S}(\hat{m}_\delta))$. Then $\{\Psi_\delta : \delta \in \Delta\}$ is a disjoint family of closed, left invariant subsets of $\Phi(G)$.

For each δ, find (2.25(iii)) an element $n_\delta \in \mathfrak{L}(G)$ such that $\mathbf{S}(\hat{n}_\delta) \subset \Psi_\delta$. Let $P = \{n_\delta : \delta \in \Delta\}$. It remains to prove that n_δ is singular to $\mathfrak{L}_t(G)$ for all δ.

To this end, let $\lambda_{G/H}$ be the canonical left Haar measure on G/H, determined, through Weil's formula, by λ and the normalised Haar measure on H. By (7.10), we can find a dense, open subset V of G/H with $\lambda_{G/H}(V) < 1$. Applying Weil's formula to the lower semicontinuous function $\chi_{Q^{-1}(V)}$, we obtain $\lambda(Q^{-1}(V)) = \lambda_{G/H}(V) < 1$. Since Q is open, $Q^{-1}(V)$ is dense in G. Trivially, $E = V^{-1} \in \mathscr{M}(G/H)$ and $F = Q^{-1}(E) \in \mathscr{M}(G)$. As $Q^*(\chi_F) = \chi_E$, it follows that $\hat{F} = Q_0^{-1}(\hat{E})$. Further, as $\mathbf{S}(\hat{m}_\delta) \subset (V^{-1})^\wedge = \hat{E}$, it follows that $\mathbf{S}(\hat{n}_\delta) \subset \Psi_\delta \subset \hat{F} = ((Q^{-1}(V))^{-1})^\wedge$. Applying (7.9) with $D = G \sim (Q^{-1}(V))^{-1}$ we obtain that n_δ is singular to $\mathfrak{L}_t(G)$. □

(7.20) COROLLARY. *Let* G *be a nondiscrete,* σ*-compact, locally compact group that is amenable as a discrete group. Then*

$$|\mathfrak{L}(G) \sim \mathfrak{L}_t(G)| \geq 2^{\mathfrak{c}}.$$

We can now give the promised improvement to (7.14).

(7.21) THEOREM. *Let* G *be a locally compact group that is amenable as a discrete group. Then* $\mathfrak{L}(G) = \mathfrak{L}_t(G)$ *if and only if* G *is discrete.*

PROOF. Use (7.20) and (7.14). □

We finally turn to our fourth problem. Let S be a left amenable semigroup. What is the cardinal $|\mathfrak{L}(S)|$? We know, of course, that if S is an (infinite) group, then $|\mathfrak{L}(S)| = 2^{2^{|S|}}$. As semigroups are so much more complicated than groups, we are not going to get such a simple result for them. We introduce the cardinal $\mathfrak{m}$ (or $\mathfrak{m}(S)$) which will replace $|S|$ in the group case. In many cases, $\mathfrak{m}$ equals the more natural cardinal $\mathfrak{p}(S)$, where $\mathfrak{p}(S) = \min\{|sS| \colon s \in S\}$. References for the work discussed here are Day [**4**], Granirer [**1**]–[**6**], Chou [**2**], [**9**], Klawe [**1**], [**3**], Paterson [**9**], and Yang [**S1**].

(7.22) Definitions. Let S be a semigroup. The cardinal $\mathfrak{m}(S)$ (or simply $\mathfrak{m}$) is defined by:

$$\mathfrak{m}(S) = \min\left\{ \left| \bigcup_{i=1}^{n} s_i S_i \right| : n \geq 1,\ \{S_1, \dots, S_n\} \text{ is a partition of } S, \right.$$

$$\left. s_1, \dots, s_n \in S \right\}.$$

The significance of this cardinal for us is that if S is left amenable and $\mathfrak{m}$ ($= \mathfrak{m}(S)$) is infinite, then

$$|\mathfrak{L}(S)| = 2^{2^{\mathfrak{m}}}.$$

The semigroup S is called *almost left cancellative* (a.l.c.) if $\mathfrak{m} = |S|$. For justification of this nomenclature, observe that if S is a.l.c. and $s \in S$, then, applying the above definition with $n = 1$, $s_1 = s$, we have $|sS| = |S|$, so that, in a rough sense, S is "close" to being left cancellative inasmuch as multiplication by s does not "collapse" S too much. If S is left cancellative and infinite, and $\{S_1, \dots, S_n\}$ is a partition of S, then $|S| = |S_i|$ for some i, and it easily follows that S is a.l.c. We now show that if $\mathfrak{m}$ is infinite, there exists an a.l.c. subsemigroup T of S with $|T| = \mathfrak{m}(S)$ and $\mathfrak{L}(T)$ closely related to $\mathfrak{L}(S)$. Note that in the statement of the following result, we can take T to be the subsemigroup of S generated by A (since $|A|$ is infinite).

(7.23) PROPOSITION. *Let $\mathfrak{m}$ be infinite, $\{S_1, \dots, S_n\}$ be a partition of S and $s_1, \dots, s_n \in S$ be such that $A = \bigcup_{i=1}^{n} s_i S_i$ has cardinality $\mathfrak{m}$. Let T be a subsemigroup of S such that $A \subset T$ and $|A| = |T|$ ($= \mathfrak{m}$). Then*

(i) *T is a.l.c. with $\mathfrak{m}(T) = \mathfrak{m}$;*

(ii) *$m(A) > 0$ for all $m \in \mathfrak{L}(S)$, and the map $m \to (m|_T)/m(T)$ is one-to-one from $\mathfrak{L}(S)$ into $\mathfrak{L}(T)$;*

(iii) *$|\mathfrak{L}(S)| \leq |\mathfrak{L}(T)|$.*

PROOF. (i) Let $\{T_1, \dots, T_m\}$ be a partition of T and $t_1, \dots, t_m$ belong to T. For $1 \leq i \leq n$, $1 \leq j \leq m$, let

$$A_{ij} = S_i \cap s_i^{-1} T_j.$$

Suppose that $A_{ij} \cap A_{kl} \neq \varnothing$. Since $\{S_1, \dots, S_n\}$ is disjoint, we have $i = k$, and since $s_i^{-1} T_j \cap s_i^{-1} T_l = s_i^{-1}(T_j \cap T_l) \neq \varnothing$ and $\{T_1, \dots, T_m\}$ is disjoint, we have $j = l$. Further, if $s \in S$, then $s \in S_{i'}$ for some i', and $s_{i'} s \in T_{j'}$ for some j' (since $s_{i'} S_{i'} \subset A \subset T$), so that $s \in A_{i'j'}$. It follows that $\{A_{ij} \colon 1 \leq i \leq n,\ 1 \leq j \leq m\}$ is a partition of S, and so

$$\mathfrak{m} \leq \left| \bigcup_{i,j} t_j s_i A_{ij} \right| \leq \left| \bigcup_{j=1}^{m} t_j T_j \right| \leq |T| = \mathfrak{m}.$$

It follows that $\mathfrak{m}(T) = \mathfrak{m} = |T|$ and T is a.l.c.

(ii) Suppose that $m \in \mathfrak{L}(S)$. Since $\bigcup_{i=1}^{n} s_i^{-1}A = S$ and $m(S) > 0$, we have $m(A) > 0$ and so $m(T) > 0$. By Problem 1-21, $m_T = (m|_T)/m(T)$ belongs to $\mathfrak{L}(T)$. We now show that the map $m \to m_T$ is one-to-one from $\mathfrak{L}(S)$ into $\mathfrak{L}(T)$.

Suppose that $m, n \in \mathfrak{L}(S)$ is such that $m_T = n_T$. Let $p \in \mathfrak{J}_l(S)$ be given by $p = n(T)m - m(T)n$. Then $p|_T = 0$. By (2.31), $|p|$, evaluated in the L-space $l_\infty(S)'$"="$M(\beta S)$, belongs to $\mathfrak{J}_l(S)$. We claim that $|p|\,|_T = 0$.

To see this, observe that $p(l_\infty(T)) = \{0\}$. One can argue, using L-space theory [**Sch**, p. 72], that for $\phi \geq 0$ in $l_\infty(T)$,

$$|p|(\phi) = \sup\{|p(\psi)| \colon \psi \in l_\infty(S),\ 0 \leq |\psi| \leq \phi\} = 0,$$

so that, as required, $|p|\,|_T = 0$. (Alternatively, one can argue directly in terms of measures on βS.)

If $|p| \neq 0$, then $|p|$ is a positive multiple of some $m' \in \mathfrak{L}(S)$, and we know that $m'(T) > 0$. So $|p| = 0$. Since $|p| = p^+ + p^-$, we have $p^+ = 0 = p^-$, and hence $p = p^+ - p^- = 0$. Thus $n(T)m = m(T)n$, and evaluating at S gives $m = n$.

So (ii) is proved, and (iii) is an immediate consequence of (ii). □

Our next result asserts that if S is left amenable with $\mathfrak{m}$ infinite, then T of the above proposition can be chosen to be left thick ((1.20)).

(7.24) PROPOSITION. *Let S be left amenable with $\mathfrak{m}$ infinite, and let A be as in* (7.23). *Then there exists a left thick subsemigroup T of S with $A \subset T$ and* $|T| = \mathfrak{m}$.

PROOF. Let $m \in \mathfrak{L}(S)$. Let

$$k = \sup\{m(RA) \colon R \text{ is a countable subset of } S\}.$$

For each $n \in \mathbb{P}$, we can find a countable subset R_n of S such that $m(R_nA) \geq k - n^{-1}$. Let $R = \bigcup\{R_n \colon n \geq 1\}$. Then R is countable, and for each n,

$$k - n^{-1} \leq m(R_nA) \leq m(RA) \leq k.$$

So $m(RA) = k$. Since $\mathfrak{m}$ is infinite, $|RA| \leq \mathfrak{m}$. Now let T be the semigroup generated by $Y = (A \cup RA)$. Then $|T| = \mathfrak{m}$ and $A \subset T$. The proposition will be proved once we have shown that $m(Y) = 1$. (For then $m(T) = 1$ and T is left thick in S by (1.21).)

Suppose, on the contrary, that $m(Y) < 1$. Let $Z = S \sim Y$. Then $m(Z) > 0$. So $n \in l_\infty(S)'$, where

$$n(E) = m(E \cap Z),$$

is nonzero and ≥ 0. It is sufficient to show that $n \in \mathfrak{J}_l(S)$. For then $n/m(Z) \in \mathfrak{L}(S)$ with $0 = n(Y) = n(A)$, and (7.23(ii)) is contradicted.

Now for $E \subset S$, $s \in S$,

$$\begin{aligned} n(s^{-1}E) &= m(s^{-1}E \cap Z) = m(s^{-1}E) - m(s^{-1}E \cap Y) \\ &= m(E) - m(s^{-1}E \cap Y) = n(E) + [m(E \cap Y) - m(s^{-1}E \cap Y)]. \end{aligned}$$

We therefore have to show

$$(1) \qquad m(E \cap Y) = m(s^{-1}E \cap Y).$$

Now

$$s^{-1}E \cap s^{-1}Y \subset (s^{-1}E \cap Y) \cup (s^{-1}Y \sim Y),$$

and

$$s^{-1}E \cap Y \subset (s^{-1}E \cap s^{-1}Y) \cup (Y \sim s^{-1}Y),$$

and, applying m to both of the preceding inclusions and noting that

$$m(s^{-1}E \cap s^{-1}Y) = m(s^{-1}(E \cap Y)) = m(E \cap Y),$$

we see that (1) will follow once we have shown

$$(2) \qquad m(s^{-1}Y \triangle Y) = 0.$$

To prove (2), observe first that $m(Y) = k = m(sY)$, since

$$k = m(RA) \leq m(Y) \leq m(sY) = m((\{s\} \cup sR)A) \leq k.$$

Similarly, $m(sY \cup Y) = m(Y)$. Thus

$$k = m(sY) + m(Y \sim sY) = m(Y) + m(sY \sim Y),$$

and as $k = m(sY) = m(Y)$, we have $m(Y \triangle sY) = 0$. Now

$$0 = m(Y \triangle sY) = m(s^{-1}(Y \triangle sY)) = m(s^{-1}Y \triangle s^{-1}(sY)),$$

while $m(s^{-1}(sY) \triangle Y) = m(s^{-1}(sY)) - m(Y) = k - k = 0$. The equality (2) now follows. □

We now state and prove the following key result that will enable us to determine $|\mathfrak{L}(S)|$ when S is left amenable with $\mathfrak{m}(S)$ infinite.

(7.25) PROPOSITION. *Let S be an a.l.c. semigroup with $\mathfrak{m}$ $(= |S|)$ infinite. Let α be the smallest ordinal of cardinality $\mathfrak{m}$. Then there exists a disjoint family $\{\Theta_\varepsilon : \varepsilon \in \alpha\}$ of left thick subsets of S.*

PROOF. The set $\mathscr{F}(S)$ of finite subsets of S has cardinality $\mathfrak{m}$ since $\mathfrak{m}$ is infinite. We can thus well-order $\mathscr{F}(S)$: $\mathscr{F}(S) = \{F_\beta : \beta \in \alpha\}$. Using transfinite recursion, we will construct a family $\{\Delta_\varepsilon^\beta : \varepsilon, \beta \in \alpha,\ \varepsilon \leq \beta\}$ of subsets of S having the following properties:

(i) $\Delta_\varepsilon^\beta \cap \Delta_\eta^\beta = \varnothing$ if $\varepsilon \neq \eta$;
(ii) $|\Delta_\varepsilon^\beta| \leq |\beta|$ if β is infinite and Δ_ε^β is finite if β is finite $(\varepsilon \leq \beta)$;
(iii) $\Delta_\varepsilon^\beta \subset \Delta_\varepsilon^{\beta_1}$ whenever $\varepsilon \leq \beta < \beta_1$;
(iv) $\bigcap\{s^{-1}(\Delta_\varepsilon^\beta) : s \in F_\beta\} \neq \varnothing$ whenever $\varepsilon \leq \beta$.

Suppose that $\gamma \in \alpha$ and that sets Δ_ε^β have been constructed so that (i)–(iv) are satisfied for $\beta, \beta_1 < \gamma$. Let

$$K = \bigcup\{\Delta_\varepsilon^\beta : \varepsilon \leq \beta < \gamma\}.$$

If γ is finite, then K is finite by (ii), so that $|K| < \mathfrak{m}$. If γ is infinite, then, again using (ii), $|K| \leq |\gamma|^3 = |\gamma| < \mathfrak{m}$. So in both cases, $|K| < \mathfrak{m}$. Let $F_\gamma = \{s_1, \dots, s_n\}$. We have to construct the $\Delta_\varepsilon^\gamma$.

To this end, we first construct, by transfinite recursion, a family $\{\Gamma_\varepsilon^\gamma : \varepsilon \le \gamma\}$ of subsets of $S \sim K$ such that

(a) $\Gamma_\varepsilon^\gamma = \{s_{\varepsilon,1}^\gamma, \ldots, s_{\varepsilon,n}^\gamma\}$ for some $s_{\varepsilon,i}^\gamma$;

(b) $\Gamma_\varepsilon^\gamma \cap \Gamma_\eta^\gamma = \varnothing$ if $\varepsilon \neq \eta$;

(c) for each ε, there exists $s_\varepsilon^\gamma \in S$ with $s_i s_\varepsilon^\gamma = s_{\varepsilon,i}^\gamma$ $(1 \le i \le n)$.

Suppose that $\delta \le \gamma$ and that sets $\{\Gamma_\varepsilon^\gamma : \varepsilon < \delta\}$ have been constructed so that (a), (b), and (c) are valid for $\varepsilon, \eta < \delta$. Let $E = \bigcup\{\Gamma_\varepsilon^\gamma : \varepsilon < \delta\}$ and $L = K \cup E$. Then $|L| < \mathfrak{m}$ since both $|E| < \mathfrak{m}$, $|K| < \mathfrak{m}$, and $\mathfrak{m}$ is infinite.

Now define recursively a disjoint family $\{S_1, \ldots, S_n\}$ of subsets of S as follows: $S_1 = s_1^{-1}(L)$ and $S_i = [s_i^{-1}(L)] \sim (\bigcup_{r=1}^{i-1} S_r)$ for $i > 1$. Note that $\bigcup_{i=1}^n s_i^{-1}(L) = \bigcup_{i=1}^n S_i$. Suppose that $\bigcup_{i=1}^n S_i = S$. Then $|\bigcup_{i=1}^n s_i S_i| \le |L| < \mathfrak{m}$, and we contradict the fact that S is a.l.c. So $\bigcup_{i=1}^n S_i \neq S$. Let $s_\delta^\gamma \in S \sim (\bigcup_{i=1}^n S_i)$, and $s_{\delta,i}^\gamma = s_i s_\delta^\gamma$ $(1 \le i \le n)$. Clearly, $s_{\delta,i}^\gamma \notin L$, and it is obvious that (a), (b), and (c) are true for $\varepsilon, \eta \le \delta$. This completes the construction of the sets $\Gamma_\varepsilon^\gamma$.

Now set

$$\Delta_\varepsilon^\gamma = \left(\bigcup\{\Delta_\varepsilon^\beta : \varepsilon \le \beta < \gamma\}\right) \cup \Gamma_\varepsilon^\gamma \qquad (\varepsilon < \gamma),$$
$$\Delta_\gamma^\gamma = \Gamma_\gamma^\gamma.$$

We now check that the conditions (i)–(iv) are satisfied for $\beta, \beta_1 \le \gamma$. Clearly, we need only check (i), (ii) and (iv) when $\beta = \gamma$. Suppose that $\varepsilon, \eta \le \gamma$ with $\varepsilon \neq \eta$. We can assume that $\varepsilon < \eta$. If $\varepsilon \le \beta_0 < \gamma$, $\eta \le \beta_1 < \gamma$, and $\beta' = \max\{\beta_0, \beta_1\}$, then by (iii) and (i), $\Delta_\varepsilon^{\beta_0} \cap \Delta_\eta^{\beta_1} \subset \Delta_\varepsilon^{\beta'} \cap \Delta_\eta^{\beta'} = \varnothing$. Noting that $\Gamma_\varepsilon^\gamma \cup \Gamma_\eta^\gamma \subset X \sim K$, we have, using (b),

$$\begin{aligned}\Delta_\varepsilon^\gamma \cap \Delta_\eta^\gamma = &\left[\left(\bigcup\{\Delta_\varepsilon^{\beta_0} : \varepsilon \le \beta_0 < \gamma\}\right) \cup \Gamma_\varepsilon^\gamma\right] \\ &\cap \left[\left(\bigcup\{\Delta_\eta^{\beta_1} : \eta \le \beta_1 < \gamma\}\right) \cup \Gamma_\eta^\gamma\right] = \varnothing.\end{aligned}$$

So (i) follows for $\beta = \gamma$. When we note that $\Gamma_\varepsilon^\gamma$ is finite and that (c) holds, it readily follows that (ii) and (iv) hold for $\beta = \gamma$. This completes the construction of the sets Δ_ε^β.

For each $\varepsilon \in \alpha$, let

$$\Theta_\varepsilon = \bigcup\{\Delta_\varepsilon^\beta : \varepsilon \le \beta,\ \beta \in \alpha\}.$$

Since $\Delta_\varepsilon^\beta \cap \Delta_\eta^{\beta_1} = \varnothing$ if $\varepsilon \neq \eta$, it follows that $\{\Theta_\varepsilon : \varepsilon \in \alpha\}$ is a disjoint family. It remains to show that each Θ_ε is left thick.

Let $F \in \mathscr{F}(S)$ and $A = \{\beta \in \alpha : F \subset F_\beta\}$. Then $|A| = \mathfrak{m} = |S|$. Since $|\varepsilon| < |S|$, there exists $\beta \in A$ with $\varepsilon \le \beta$. Then by (iv),

$$\bigcap\{s^{-1}\Theta_\varepsilon : s \in F_\beta\} \supset \bigcap\{s^{-1}(\Delta_\varepsilon^\beta) : s \in F_\beta\} \neq \varnothing.$$

Let $x \in \bigcap\{s^{-1}\Theta_\varepsilon : s \in F_\beta\}$. Then $Fx \subset \Theta_\varepsilon$, and so Θ_ε is left thick in S. □

We now obtain the precise cardinality of $\mathfrak{L}(S)$ when S is left amenable with $\mathfrak{m}$ infinite.

(7.26) THEOREM. *Let S be a left amenable semigroup with $\mathfrak{m}$ infinite. Then* $|\mathfrak{L}(S)| = 2^{2^{\mathfrak{m}}}$.

PROOF. Let A and T be as in (7.24). Then T is a.l.c. with $\mathfrak{m}(T) = \mathfrak{m}$ and so by (7.25), there exists a disjoint family $\{\Theta_\varepsilon : \varepsilon \in A\}$ of left thick subsets of T. If $F \in \mathscr{F}(S)$, then, as T is left thick in S, there exists $s \in S$ with $Fs \subset T$, and since $Fs \in \mathscr{F}(T)$ and Θ_ε is left thick in T, we can find $t \in T$ with $Fst \subset \Theta_\varepsilon$. Thus Θ_ε is also left thick in S.

Now apply an argument similar to that in the proof of (7.4). We take $A = \alpha$ and form N_γ's, $P = \{1, c\}^\Gamma$, and $E_\gamma = \bigcup\{\Theta_\varepsilon : \varepsilon \in N_\gamma\}$. We do have to be careful with the C_p's, since multiplication on S is not necessarily left cancellative. We define

$$C_p = \bigcap\{[x^{-1}(E_\gamma^{p(\gamma)})]^\wedge : x \in S\}.$$

(Here, $A^\wedge$ is the closure of $A \subset S$ in βS.) One easily checks that the C_p's form a disjoint family of nonempty, compact, invariant subsets of βS. Now apply Day's Fixed-Point Theorem to give that $|\mathfrak{L}(S)| \geq 2^{2^{\mathfrak{m}}}$. The reverse inequality follows from (7.23(ii)): $|\mathfrak{L}(S)| \leq |\mathfrak{L}(T)| \leq |l_\infty(T)'| = 2^{2^{|T|}} = 2^{2^{\mathfrak{m}}}$ (cf. Problem 7-1). □

It readily follows that in the situation of the above theorem, $\dim \mathfrak{J}_l(S) = 2^{2^{\mathfrak{m}}}$.

It remains to determine $|\mathfrak{L}(S)|$ and $\dim \mathfrak{J}_l(S)$ when S is left amenable and $\mathfrak{m}$ is finite. In this case, $\mathfrak{J}_l(S)$ is finite-dimensional so that $\dim \mathfrak{J}_l(S)$ is a more appropriate cardinal than $|\mathfrak{L}(S)|$.

(7.27) THEOREM. *Let S be a left amenable semigroup with $\mathfrak{m}$ finite. Then S contains a finite ideal, and if $\Lambda(S)$ is the set of finite, left ideal groups in S, then*

$$\dim \mathfrak{J}_l(S) = |\Lambda(S)| < \infty.$$

PROOF. Let $\{S_1, \ldots, S_n\}$ be a partition of S and $s_1, \ldots, s_n \in S$ be such that $A = \bigcup_{i=1}^n s_i S_i$ is finite. Let $m \in \mathfrak{L}(S)$. Then $m(\{a\}) > 0$ for some $a \in A$, and since $m(\{sa\}) \geq m(\{a\})$ for all $s \in S$ and $m(S) < \infty$, we have that Sa is a finite left ideal of S. By (1.20), S contains a finite left ideal group L. Further m_L belongs to $\mathfrak{L}(S)$, where

$$m_L(E) = |E \cap L|/|L| \qquad (E \subset S).$$

Since $m_L(A) > 0$, $L \cap A \neq \varnothing$. Since A is finite, the set $\Lambda(S)$ of finite, left ideal groups in S is finite. Let $L_1, \ldots, L_m$ be the elements of $\Lambda(S)$. Clearly, L_1 is a minimal left ideal of S, and every minimal left ideal of S is of the form $L_1 x$ ($x \in S$). Now each $L_1 x$ is also a group, where, if e_1 is the identity of L_1, then the identity of $L_1 x$ is $(xe_1)^{-1}x$ (with $(xe_1)^{-1}$ the inverse of xe_1 in L_1). It follows that the kernel K of S ((1.16)) is the finite ideal $\bigcup_{i=1}^m L_i$. Now every $n \in \mathfrak{L}(S)$ can be regarded as a mean on K, since $n(K) = 1$. Noting that $L_i \cap L_j = \varnothing$ if $i \neq j$, we readily see that

$$n = \sum_{i=1}^m n(L_i) m_{L_i}.$$

Since $\mathfrak{L}(S)$ spans $\mathfrak{J}_l(S)$ (cf. (2.2)) and $\{m_{L_i}: 1 \leq i \leq m\}$ is a linearly independent subset of $\mathfrak{J}_l(S)$, we see that $\dim \mathfrak{J}_l(S) = m$. □

References

Banach [**1**], Chou [**2**], [**4**], [**9**], Day [**4**], Granirer [**1**], [**2**], [**3**], [**4**], [**6**], [**13**], [**15**], Greenleaf [**2**], Johnson [**4**], Klawe [**1**], [**3**], Liu and van Rooij [**1**], Luthar [**1**], [**2**], Rosenblatt [**5**], [**7**], [**8**], [**9**], Rudin [**2**], [**4**], Wells [**1**], and Yang [**S1**].

Further Results

(7.28) Exposed points and left invariant means. Let X be a Banach space and K a convex subset of X'. A point α_0 of K is said to be *exposed* (cf. Klee [**2**]) if there exists $\xi \in X$ such that

$$\operatorname{Re} \alpha(\xi) < \operatorname{Re} \alpha_0(\xi)$$

for all $\alpha \in K \sim \{\alpha_0\}$. The set of exposed points of K is denoted by $\operatorname{Exp} K$. It is easy to prove that $\operatorname{Exp} K \subset \operatorname{Ext} K$ and that if K is compact and generates a finite-dimensional subspace of X', then $\operatorname{Exp} K \neq \varnothing$. As Granirer [**13**] points out, "a set without exposed points might be considered 'big'." The results (i), (ii), and (iii) below are due to Granirer [**13**], who derives them using very general theorems. For prototypes see Fairchild [**1, 2**], Chou [**5**], and Granirer [**6**]. As usual, Σ is either a locally compact group or a discrete semigroup.

(i) *Let Σ be noncompact, σ-compact, left amenable, and with no finite, left ideals. Let $\{\phi_n\}$ be a sequence in $L_\infty(\Sigma)$ and H a compact, G_δ-subset of $\Phi(\Sigma)$. Let*

$$\mathfrak{K} = \{m \in \mathfrak{L}_t(\Sigma) \colon \mathbf{S}(\hat{m}) \subset H,\ m(\phi_n) = 0 \text{ for all } n \geq 1\}.$$

Then $\operatorname{Exp} \mathfrak{K} = \varnothing$, *and if $\mathfrak{K} \neq \varnothing$, then $\mathfrak{K}$ is not norm separable.* [Suppose that $\operatorname{Exp} \mathfrak{K} \neq \varnothing$ and let $m \in \operatorname{Exp} \mathfrak{K}$. Find $\phi_0 \in L_\infty(\Sigma)$ with

$$\operatorname{Re} n(\phi_0) < \operatorname{Re} m(\phi_0) \qquad (n \in \mathfrak{K} \sim \{m\}). \tag{1}$$

Since H is a G_δ-set, we can find a sequence $\{U_n\}$ of open subsets of $\Phi(\Sigma)$ such that $U_{n+1}^- \subset U_n$ and $H = \bigcap_{n=1}^\infty U_n$. Now as $\Phi(\Sigma)$ is the maximal ideal space of an L_∞-space, it is zero-dimensional. So we can suppose that each U_n is open and closed. Let $\psi_n \in L_\infty(\Sigma)$ be such that $\hat{\psi}_n = \chi_{U_n}$.

Let $\{K_n\}$ be an increasing sequence in $\mathscr{C}(\Sigma)$ with $\bigcup_{n=1}^\infty K_n^0 = \Sigma$. From the proof of (4.4) in the group case, we have:

(∗) *given $\varepsilon > 0$ and $n \in \mathbb{P}$, there exists a net $\{g_\delta\}$ in $P(\Sigma)$ with $\hat{g}_\delta \to m$ weak* and, for all δ, $\|x * g_\delta - g_\delta\|_1 < \varepsilon$ $(x \in K_n)$.*

Using (∗) and the fact that $m(\psi_n) = 1$, we can find a sequence of functions $\{f_n\}$ in $P(\Sigma)$ such that

(a) $\|x * f_n - f_n\|_1 < n^{-1}$ $(x \in K_n)$;

(b) $|(\hat{f}_n - m)(\phi_r)| < n^{-1}$ $(1 \le r \le n)$;
(c) $|(\hat{f}_n - m)(\phi_0)| < n^{-1}$;
(d) $|\hat{f}_n(\psi_r) - 1| < n^{-1}$ $(1 \le r \le n)$.

Suppose that $\{\hat{f}_{\alpha(\delta)}\}$ is a subnet of $\{\hat{f}_n\}$ that is weak* convergent to some $p \in \mathfrak{M}(\Sigma)$. From (a) and (4.3), $p \in \mathfrak{L}_t(\Sigma)$. From (b), $p(\phi_r) = m(\phi_r) = 0$ for all $r \ge 1$. Further, from (d), $\hat{p}(U_n) = p(\psi_n) = 1$, and since $U_n = U_n^-$, it follows that $\mathbf{S}(\hat{p}) \subset \bigcap_{n=1}^{\infty} U_n = H$. So $p \in \mathfrak{K}$. From (c), $p(\phi_0) = m(\phi_0)$, and using (1), $p = m$. Hence $\hat{f}_n \to m$ weak*. It follows that $\{f_n\}$ is a weak Cauchy sequence in $L_1(\Sigma)$, and since the latter space is weakly sequentially complete [**DS**, IV.8.6], $m = \hat{f}$ for some $f \in P(\Sigma)$. If Σ is a discrete semigroup, then, by (1.20), there exists a finite, left ideal in Σ, and a contradiction results. If, on the other hand, Σ is a locally compact group, then m can be regarded as a left, Haar measure of total mass 1 on Σ, and this implies that G is compact, so that a contradiction results. Thus $\operatorname{Exp}\mathfrak{K} = \emptyset$ as required.

Now suppose that $\mathfrak{K} \ne \emptyset$ is norm separable, and let $\{m_n : n \in \mathbb{P}\}$ be a norm-dense subset of $\mathfrak{K}$. Let $m \in \mathfrak{K}$. Then for each n, we can find a sequence $\{\phi_{n,k}\}$ in $L_\infty(\Sigma)$ with $||\phi_{n,k}||_\infty = 1$ and

$$(2) \qquad \lim_{k\to\infty} |(m_n - m)(\phi_{n,k})| = ||m_n - m||.$$

Now find a sequence $\{f_n\}$ in $P(\Sigma)$ satisfying (a), (b), (d), and the condition

(e) $|(\hat{f}_n - m)(\phi_{r,s})| < n^{-1}$ $(1 \le r, s \le n)$.

Let $\{\hat{f}_{\alpha(\delta)}\}$ be a subnet of $\{\hat{f}_n\}$ weak* convergent to some p. As earlier, $p \in \mathfrak{K}$. From (e), $m(\phi_{r,s}) = p(\phi_{r,s})$ for all r, s. Using (2), for each n,

$$||m_n - p|| \ge \lim_{s\to\infty} |(m_n - p)(\phi_{n,s})| = ||m_n - m||.$$

So any sequence in the set $R = \{m_n : n \in \mathbb{P}\}$ that converges in norm to p also converges in norm to m. The norm density of R in $\mathfrak{K}$ now implies that $m = p$. The rest of the argument proceeds as in the above proof that $\operatorname{Exp}\mathfrak{K} = \emptyset$.]

We now look at the case where Σ is a discrete semigroup and the requirement that H be a G_δ-set is removed. The conditions on the semigroup are strengthened.

(ii) *Let S be a countable, left amenable semigroup that is not n-ELA for any n ((2.30)). Let $\{\phi_n\}$ be a sequence in $l_\infty(S)$, H a compact subset of βS, and*

$$\mathfrak{Q} = \{m \in \mathfrak{L}(S) : \mathbf{S}(\hat{m}) \subset H,\ m(\phi_n) = 0 \text{ for all } n \ge 1\}.$$

Then $\operatorname{Exp}\mathfrak{Q} = \emptyset$, *and if* $\mathfrak{Q} \ne \emptyset$, *then* $\mathfrak{Q}$ *is not norm separable.* [Suppose that $m \in \operatorname{Exp}\mathfrak{Q}$ and let $\phi_0 \in l_\infty(S)$ be such that

$$\operatorname{Re} n(\phi_0) < \operatorname{Re} m(\phi_0) \qquad (n \in \mathfrak{G} \sim \{m\}).$$

Replacing H by $\mathbf{S}(\hat{m})$, we can suppose that $SH \subset H$ ((2.25(ii))). Let $\{K_n\}$ be as in the proof of (i) and $r \in \mathbb{P}$. Find $g_r \in P(S)$ such that $||x * g_r - g_r||_1 < r^{-1}$ $(x \in K_r)$. Let $\{\alpha_\delta\}$ be a net in $\operatorname{co} H$ $(\subset l_\infty(S)' = M(\beta S))$ with $\alpha_\delta \to m$ weak*. Now each g_r is a convex combination of points $s \in S$. Using (2.25(i)) and the fact that $m \in \mathfrak{L}(S)$, we have $\lim_\delta (g_r\alpha_\delta) = m$ and $\|x(g_r\alpha_\delta) - g_r\alpha_\delta\| < r^{-1}$ $(x \in K_r)$. By choosing suitable elements $g_r\alpha_\delta$, we can find a sequence $\{\beta_n\}$ in $\operatorname{co} H$ such that

(a′) $\|x\beta_n - \beta_n\| < n^{-1}$ $(x \in K_n)$,

(b′) $|(\beta_n - m)(\phi_r)| < n^{-1}$ $(0 \le r \le n)$.

A version of the argument in (i) now gives that $\beta_n \to m$ weak* in $l_\infty(S)'$. (Note that since $\mathbf{S}(\widehat{\beta_n}) \subset H$, it follows that if m is a weak* cluster point of $\{\beta_n\}$ in $l_\infty(S)'$, then $\mathbf{S}(\hat{m}) \subset H$.) A theorem of Grothendieck [**Sch**, p. 131] entails that $\beta_n \to m$ weakly in $l_\infty(S)'$. Now if $\alpha_1, \dots, \alpha_r$ are distinct in H and $\lambda_1, \dots, \lambda_r \in \mathbb{C}$, then, identifying $l_\infty(S)$ with $C(\beta S)$ and using Tietze's Extension Theorem, we obtain

$$\left\| \sum_{i=1}^{r} \lambda_i \alpha_i \right\| = \sum_{i=1}^{r} |\lambda_i|$$

so that $l_1(H)$ is isometrically embedded in $l_\infty(S)'$. Since $\beta_n \in l_1(H)$ and $l_1(H)$ is weakly sequentially complete, we have $m \in l_1(H)$. So $\hat{m}$ is not continuous, and by (2.37), S is n-ELA for some n. This is a contradiction.

To prove that $\mathfrak{Q}$ is not norm separable (if it is nonempty), the reader is invited to combine the above construction with the corresponding proof given in (i).]

A natural question (raised in Granirer [**13**]) involves a locally compact group analogue of (ii). Talagrand [**6**] proves the following result. *Let G be an infinite, locally compact group that is amenable as a discrete group, and let F be a nonvoid, compact, left invariant subset of $\Phi(G)$. Then the set*

$$\{m \in \mathfrak{L}(G) \colon \mathbf{S}(\hat{m}) \subset F\}$$

has cardinality not less than $2^{\mathfrak{c}}$ and is not norm separable.

A substantial amount of information is known about the extreme points of sets of invariant means. References are Chou [**1**], Nillsen [**1**], [**2**], and Talagrand [**1**], [**4**], [**7**]. See Yang [**S1**] for recent work on exposed points.

(7.29) Nonmeasurability of some $L_\infty(G)$-convolutions. (Wells [**1**], Rudin [**4**]). Let G be a locally compact group and $\psi \in L_\infty(G)$, $p \in \Phi(G)$. Let $\phi = p\psi$. Then $\phi \in l_\infty(G)$ and $\|\phi\| = \sup_{x\in G} |\phi(x)| \le \|\psi\|_\infty$. In the striking words of Rudin: "*What else can one say about ϕ?...* the answer is: *Nothing!*" The paper by Wells deals with the case of the circle group. Rudin proves that if G is infinite and metric, then there exists $\psi \in L_\infty(G)$ with $\|\psi\|_\infty = 1$ and such that *every* $\phi \in l_\infty(G)$ with $|\phi(x)| \le 1$ for all $x \in G$ is of the form $p\psi$ for some $p \in \Phi(G)$. In particular, $L_\infty(G)$, like $C(G)$ ((2.33)), is not, in general, left introverted.

We now prove Rudin's result when G is metric, nondiscrete, and σ-compact. The proof below is due to J. M. Rosenblatt.

Let G be a nondiscrete, σ-compact, locally compact metric group. Then there exists $\psi \in L_\infty(G)$ with $\|\psi\|_\infty = 1$ such that

$$\{p\psi \colon p \in \Phi(G)\} = \{\phi \in l_\infty(G) \colon \|\phi\|_\infty \le 1\}.$$

[Let $\{E_\gamma : \gamma \in \Gamma\}$ be the family of subsets of G given by (7.15). Let $\{\gamma_n\}$ be a sequence in Γ with $\gamma_n \neq \gamma_m$ if $n \neq m$. Let $A_n = E_{\gamma_{2n}}$, $B_n = E_{\gamma_{2n+1}}$. Define $\psi_1, \psi_2 \in L_\infty(G)$ by setting

$$\psi_1 = \sum_{n=1}^{\infty} 2^{-n}\chi_{A_n}, \qquad \psi_2 = \sum_{n=1}^{\infty} 2^{-n}\chi_{B_n}.$$

Let $\psi = \psi_1 \exp(2\pi i \psi_2)$. Then $\psi \in L_\infty(G)$. For each N, $\lambda(\bigcap_{n=1}^{N} A_n) > 0$ (by (1) of (7.15)), so that if $x \in A_n$ $(1 \leq n \leq N)$, then

$$1 \geq ||\psi||_\infty \geq \psi_1(x) \geq \sum_{n=1}^{N} 2^{-n}.$$

It follows that $||\psi||_\infty = 1$.

Let $\phi \in l_\infty(G)$ with $||\phi||_\infty \leq 1$. Let $\phi_1 = |\phi|$, and find $\phi_2 : G \to [0,1]$ such that $\phi = \phi_1 \exp(2\pi i \phi_2)$. If $p \in \Phi(G)$ then p is a multiplicative linear functional on $L_\infty(G)$, and as $\psi x = (\psi_1 x)\exp(2\pi i(\psi_2 x))$ $(x \in G)$ and $p\psi(x) = p(\psi x)$, we obtain $p\psi = (p\psi_1)\exp(2\pi i(p\psi_2))$. It therefore suffices to show that we can find p so that $p\psi_i = \phi_i$ $(i = 1, 2)$.

To achieve this we first express each ϕ_i in the form $\sum_{n=1}^{\infty} 2^{-n}\chi_{C_n}$. In fact, a simple recursion argument shows that we can construct a sequence $E_0, E_1, \ldots$ of subsets of G such that

$$0 \leq \phi_1 - \sum_{n=0}^{N} 2^{-n}\chi_{E_n} \leq 2^{-N} \qquad (N \in \mathbb{P}).$$

(Indeed, we can take $E_N = \{x \in G : (\phi_1 - \sum_{n=0}^{N-1} 2^{-n}\chi_{E_n})(x) \geq 2^{-N}\}$ with $E_0 = \varnothing$). Clearly

$$\phi_1 = \sum_{n=1}^{\infty} 2^{-n}\chi_{E_n}.$$

Similarly, we can find a sequence $\{F_n\}$ of subsets of G such that

$$\phi_2 = \sum_{n=1}^{\infty} 2^{-n}\chi_{F_n}.$$

Using (1) of (7.15), we take p to be a point in the set

$$\bigcap_{n\geq 1}\left\{\left[\bigcap_{x\in E_n}(x^{-1}A_n)^\wedge\right] \cap \left[\bigcap_{x\in E_n^c}(x^{-1}A_n^c)^\wedge\right] \cap \left[\bigcap_{x\in F_n}(x^{-1}B_n)^\wedge\right] \cap \left[\bigcap_{x\in F_n^c}(x^{-1}B_n^c)^\wedge\right]\right\}.$$

If $x \in E_n$, then, as $p \in (x^{-1}A_n)^\wedge$, we have $p\chi_{A_n}(x) = (\chi_{x^{-1}A_n})^\wedge(p) = 1$. If $x \in E_n^c$, then we have $p\chi_{A_n}(x) = (\chi_{x^{-1}A_n})^\wedge(p) = (\chi_G - \chi_{x^{-1}A_n^c})^\wedge(p) = 1 - 1 = 0$. So $p\chi_{A_n} = \chi_{E_n}$, and

$$p\psi_1 = \sum_{n=1}^{\infty} 2^{-n}(p\chi_{A_n}) = \sum_{n=1}^{\infty} 2^{-n}\chi_{E_n} = \phi_1.$$

Similarly, $p\psi_2 = \phi_2$ as required.]

(7.30) Locally compact groups that are amenable as discrete. In view of theorems such as (7.14) and (7.19), locally compact groups that are amenable as discrete groups are of special interest. The following result is based on Rosenblatt [**7**] and Chou [**11**]. The condition (iii) is a fixed-point theorem for the groups G under consideration, and the fact that (iii) implies (i) is of note in view of the essential use of "(i) implies (iii)" in the chapter.

Let G be a nondiscrete, σ-compact, locally compact group. Then the following three statements are equivalent:

(i) *G is amenable as a discrete group:*

(ii) *G is amenable and for each $\phi \in L_\infty(G, \mathbb{R})$, we have*

$$\sup\{m(\phi) \colon m \in \mathfrak{L}(G)\} = \inf_{A \in \mathscr{F}(G)} \left[\operatorname{ess\,sup}(|A|^{-1} \sum_{y \in A} \phi y) \right]; \tag{1}$$

(iii) *if C is a left invariant, nonempty, compact subset of $\Phi(G)$, then there exists $m \in \mathfrak{L}(G)$ with $\mathbf{S}(\hat{m}) \subset C$.*

[We first prove that (i) implies (ii). Suppose that (i) holds. Let $\phi \in L_\infty(G, \mathbb{R})$, and let a, b be respectively the left- and right-hand sides of (1). Obviously, $a \le b$.

We now prove that $b \le a$. Let $\{A_\delta\}$ be a summing net for G regarded as a discrete group ((4.15)). For $A \in \mathscr{F}(G)$, let

$$\mu_A = (|A|)^{-1} (\sum_{y \in A} y) \in l_1(G) \subset M(G).$$

Let $\mu_\delta = \mu_{A_\delta}$. Then

$$\operatorname{ess\,sup}[(\phi\mu_A)\mu_\delta] \le |A_\delta|^{-1} \Big(\sum_{x \in A_\delta} \operatorname{ess\,sup}[(\phi\mu_A)x] \Big) = \operatorname{ess\,sup}(\phi\mu_A),$$

and

$$||(\phi\mu_A\mu_\delta - \phi\mu_\delta)||_\infty \le ||\phi||_\infty ||\mu_A\mu_\delta - \mu_\delta||_1 \to 0$$

using (3) of (4.15). It follows that given $\varepsilon > 0$, there exists δ_0 such that for all $\delta \ge \delta_0$,

$$b \le \operatorname{ess\,sup} \phi\mu_\delta \le (\operatorname{ess\,sup}(\phi\mu_A\mu_\delta)) + \varepsilon \le \operatorname{ess\,sup} \phi\mu_\delta + 2\varepsilon$$

and from the definition of b,

$$\lim_\delta [\operatorname{ess\,sup}(\phi\mu_\delta)] = b.$$

Note that by replacing ϕ by $-\phi$ we deduce the existence of $\lim_\delta[\operatorname{ess\,inf}(\phi\mu_\delta)]$.

Let B be the linear span in $L_\infty(G, \mathbb{R})$ of 1 and the functions ϕx ($x \in G$). Let $\psi \in B$. Then for some $\beta, \beta_i \in \mathbb{R}$, $x_i \in G$,

$$\psi = \beta 1 + \sum_{i=1}^{n} \beta_i (\phi x_i).$$

Since

$$\left\| \psi\mu_\delta - \left(\beta 1 + \left(\sum_{i=1}^{n} \beta_i\right) \phi\mu_\delta\right) \right\|_\infty \leq \sum_{i=1}^{n} |\beta_i| \cdot \|\phi\| \, \|x_i\mu_\delta - \mu_\delta\|_1 \to 0,$$

it follows that

$$(2) \quad \begin{aligned} \lim_\delta [\operatorname{ess\,sup}(\psi\mu_\delta)] &= \beta + \lim_\delta \left[\operatorname{ess\,sup}\left(\left(\sum_{i=1}^{n} \beta_i\right) \phi\mu_\delta\right)\right], \\ \lim_\delta [\operatorname{ess\,inf}(\psi\mu_\delta)] &= \beta + \lim_\delta \left[\operatorname{ess\,inf}\left(\left(\sum_{i=1}^{n} \beta_i\right) \phi\mu_\delta\right)\right]. \end{aligned}$$

If $\psi = \gamma 1 + \sum_{i=1}^{m} \gamma_i(\phi y_i)$ ($\gamma, \gamma_i \in \mathbb{R}$, $y_i \in G$), then $0 = (\beta - \gamma)1 + \sum_{i=1}^{n} \beta_i(\phi x_i) - \sum_{i=1}^{m} \gamma_i(\phi y_i)$, and by taking either "ess sup" or "ess inf" and using (2), we obtain

$$\beta - \gamma + \left[\sum_{i=1}^{n} \beta_i - \sum_{i=1}^{m} \gamma_i\right] b = 0.$$

It follows that we can define a linear functional n on B by setting

$$n(\psi) = \beta + \left(\sum_{i=1}^{n} \beta_i\right) b.$$

Clearly $n(\phi) = b$, $n(\psi x) = n(\psi)$, and $\|n\| = 1 = n(1)$. By Day's Theorem, we can find $m_0 \in \mathfrak{L}(G)$ with $m_0|_B = n$. Then

$$b = m_0(\phi) \leq a$$

as required, and (i) implies (ii).

We now prove that (ii) implies (iii). Suppose that (ii) holds. Let C be a left invariant, nonempty, compact subset of $\Phi(G)$. Let

$$\mathscr{E} = \{E \colon E \text{ is a Borel subset of } G \text{ and } C \subset \hat{E}\}.$$

Since G is zero-dimensional and every open and closed subset of $\Phi(G)$ is of the form $\hat{E}$ for some Borel set $E \subset G$, it follows that every open subset of $\Phi(G)$ containing C also contains some E with $E \in \mathscr{E}$. Let $E \in \mathscr{E}$, $A \in \mathscr{F}(G)$, and $W = \bigcap_{x \in A} x^{-1}E$. Since

$$C \subset \bigcap_{x \in A} (x^{-1}E)^\wedge = \hat{W},$$

it follows that $\lambda(W) > 0$, and if $w \in W$, then $[|A|^{-1} \sum_{x \in A} \chi_E x](w) = 1$, and so $\operatorname{ess\,sup}(|A|^{-1} \sum_{x \in A} \chi_E x) = 1$. Using (ii), we can find $m_E \in \mathfrak{L}(G)$ such that $m_E(E) = 1$. Now since $E_1 \cap E_2 \in \mathscr{E}$ if $E_1, E_2 \in \mathscr{E}$, it follows that $\mathscr{E}$ is a net (under intersection). Thus $\{m_E\}$ ($E \in \mathscr{E}$) becomes a net, and if m is a weak cluster point of this net in $\mathfrak{L}(G)$, then, using regularity, $\hat{m}(C) = 1$. So $\mathbf{S}(\hat{m}) \subset C$ and (ii) implies (iii).

It remains to show that (iii) implies (i). Suppose that (iii) holds. Assume for the moment that G is metric. Let $\phi \in l_\infty(G)$ be real-valued, and find ((7.29)) $\psi \in L_\infty(G, \mathbb{R})$ and $p \in \Phi(G)$ such that

$$p\psi = \phi.$$

Let C be the left invariant, compact subset $(Gp)^-$ of $\Phi(G)$. By (iii), we can find $m \in \mathfrak{L}(G)$ with $\mathbf{S}(\hat{m}) \subset C$. Let $\mu = \hat{m} \in \mathrm{PM}(C)$. Then we can find a net $\{\nu_\delta\}$, where each ν_δ is a convex combination of point masses δ_ξ $(\xi \in C)$, such that $\nu_\delta \to \mu$ weak* in $\mathrm{PM}(C)$. Now if $\{x_\sigma\}$ is a net in G with $x_\sigma p \to \xi$, then $\delta_{x_\sigma} p \to \delta_\xi$ weak* in $\mathrm{PM}(C)$. It follows that we can find a net $\{a_\delta\}$, with each a_δ a convex combination of elements of G, such that $a_\delta p \to \mu$ weak* in $\mathrm{PM}(C)$. Then for $x \in G$, in an obvious notation,

$$a_\delta \phi(x) = a_\delta(p\psi)(x) = (a_\delta p)(\psi x) \to \hat{\mu}(\hat{\psi x}) = \hat{\mu}(\hat{\psi}).$$

The "real" version of (2.13), with B, $L_\infty(G)$ both replaced by the space $l_\infty(G,\mathbb{R})$ of real-valued, bounded functions on G, then yields an element of $l_\infty(G,\mathbb{R})'$ that extends in the usual way to a left invariant mean on $l_\infty(G)$. So G is amenable as a discrete group.

Now remove the requirement that G be metric. To show that G is amenable as a discrete group, it is sufficient to show that every countable subgroup H of G is amenable as discrete. Let $H = \{e\} \cup \{h_n : n \in \mathbb{P}\}$ be a (countable) subgroup of G with $h_n \neq e$ for all n. By choosing U_n in Problem 7-7 so that $h_n \notin U_n$, the resulting compact subgroup N is such that $N \cap H = \{e\}$. Let $L = G/N$. Then G/N is nondiscrete, σ-compact, and metric, and contains a copy of H as a subgroup. The amenability of H will follow once we have shown that L satisfies (iii); for then, by the above, L will be amenable as discrete, and so also will be its subgroup H.

Let D be a left invariant, nonempty, compact subset of $\Phi(L)$. Let $Q_0 : \Phi(G) \to \Phi(L)$ be as in the proof of (7.19). Then $C = Q_0^{-1}(D)$ is left invariant in $\Phi(G)$, and if $\mu \in \mathrm{PM}(C)$ is left invariant for G, then $\mu \circ Q_0^{-1} \in \mathrm{PM}(D)$ and is left invariant for L. Thus L satisfies (iii) as required.

So (iii) implies (i).]

(7.31) Amenability and the radicals of certain second dual Banach algebras. (Day [**4**], Civin and Yood [**1**], Granirer [**2**], [**3**], [**4**], [**6**], [**13**], [**14**], Granirer and Rajagopalan [**1**], Klawe [**1**], Duncan and Hosseiniun [**1**]). Let Σ be either a discrete semigroup or a locally compact group. The second dual $L_1(\Sigma)''$ is a Banach algebra under the Arens product. Let the radical of a Banach algebra A be denoted by $\mathrm{rad}(A)$.

As we shall see, left invariant means can be used to produce elements in $\mathrm{rad}(L_1(\Sigma)'')$. Recall that $\mathfrak{J}_{lt}(\Sigma)$ is the space of topologically left invariant, linear functionals on $L_1(\Sigma)' = L_\infty(\Sigma)$. Define

$$J(\Sigma) = \{f \in \mathfrak{J}_{lt}(\Sigma) : f(1) = 0\}.$$

(i) $J(\Sigma)$ *is a closed, linear subspace of* $\mathrm{rad}(L_1(\Sigma)'')$.

[Obviously, $J(\Sigma)$ is a closed subspace of $L_1(\Sigma)''$. To show that $J(\Sigma) \subset \mathrm{rad}(L_1(\Sigma)'')$, it is sufficient to prove that $J(\Sigma)$ is a left ideal in $L_1(\Sigma)''$ and $J(\Sigma)^2 = \{0\}$. Let $p \in L_\infty(\Sigma)'$ and $f \in \mathfrak{J}_{lt}(\Sigma)$. Then $pf = p(1)f$. Clearly $pf(1) = 0$ if $f(1) = 0$, so that $J(\Sigma)$ is a left ideal in $L_1(\Sigma)''$. Since $p(1) = 0$ if $p \in J(\Sigma)$, we also have $J(\Sigma)^2 = \{0\}$ as required.]

(ii) *Let S be a left amenable semigroup.*

(a) *If $\mathfrak{m}(S)$ is infinite, then*

$$\dim(\operatorname{rad}(l_1(S)'')) \geq 2^{2^{\mathfrak{m}(S)}}.$$

(b) *If $\mathfrak{m}(S)$ is finite and $\Lambda(S)$ is the set of finite, left ideal groups in S, then*

$$1 + \dim(\operatorname{rad}(l_1(S)'')) \geq |\Lambda(S)|.$$

[Fix $m_0 \in \mathfrak{L}(S)$. Since $(m - m_0) \in J(S)$ for each $m \in \mathfrak{L}(S)$, we have

$$\dim J(S) \geq \dim[\operatorname{Span}\{(m - m_0) \colon m \in \mathfrak{L}(S)\}].$$

(a) now follows from (7.26) and (b) from (7.27).]

(iii) *Let G be an amenable, noncompact, locally compact group, and let $\mathfrak{m}$ be as in* (7.0). *Then*

$$\dim(\operatorname{rad} L_1(G)'') \geq 2^{2^{\mathfrak{m}}}.$$

[$J(G)$ is of codimension 1 in $\mathfrak{J}_{lt}(G)$, and by (7.6) and (2.2), $\dim \mathfrak{J}_{lt}(G) = 2^{2^{\mathfrak{m}}}$.]

It readily follows that if G is as in (iii) then $\operatorname{rad} L_1(G)''$ is not norm separable. Similar results (Granirer [**6**], [**13**]) hold for the radical of $\mathrm{U_r}(G)'$. (The latter space is a Banach algebra from (2.8).)

(iv) Granirer [**14**], developing ideas of S. Gulick [**1**], shows that if G is locally compact and nondiscrete, then $\operatorname{rad}(L_1(G)'')$ is not norm-separable. The case in which G is discrete and nonamenable still seems to be open.

Problems 7

1. Let X be an infinite set. Prove that

$$|\beta X| = 2^{2^{|X|}} = l_\infty(X)'.$$

2. Construct explicitly sets Z_β and functions ϕ_β satisfying (7.3) when $G = \mathbb{R}$.

3. Let G be a noncompact, locally compact group and X the maximal ideal space of $\mathrm{U_r}(G)$. Show that there exists a family $\mathscr{G}$ of *minimal*, compact, left invariant subsets of X with $|\mathscr{G}| = 2^{2^{\mathfrak{m}}}$.

4. Let G be as in Problem 3 above. Suppose, further, that G is amenable. Show that $\mathfrak{L}_t(G)$ contains at least $2^{2^{\mathfrak{m}}}$ extreme points.

5. Let G be a σ-compact, noncompact, locally compact amenable group. This problem gives another proof that $|\mathfrak{L}_t(G)| \geq 2^{\mathfrak{c}}$ ((7.6)).

(i) Show that there exists a summing sequence $\{U_n\}$ for G in $\mathscr{C}_e(G)$ with $\lambda(U_{n+1}) \geq (n+1)\lambda(U_n)$;

(ii) Define $\tau \colon L_\infty(G) \to l_\infty$ by

$$\tau(\phi)(n) = (\lambda(U_{n+1} \sim U_n))^{-1} \int_{U_{n+1} \sim U_n} \phi(x)\, d\lambda(x).$$

Show that τ is linear, surjective, and of norm 1.

(iii) Let $\mathfrak{M}_c = \{m \in \mathfrak{M}(\mathbb{P}) \colon m(\phi) = \lim_{n\to\infty} \phi(n)$ for every convergent sequence $\phi \in l_\infty\}$. Show that $\tau^*(\mathfrak{M}_c) \subset \mathfrak{L}_t(G)$.

(iv) Deduce that $|\mathfrak{L}_t(G)| \geq 2^{\mathfrak{c}}$ (a special case of (7.6)).

6. Let Σ, $\mathfrak{K}$ be as in (7.28(i)). Assume the Continuum Hypothesis. Show that $|\mathfrak{K}| \geq 2^{\mathfrak{c}}$. (Use the Čech-Pospísil Theorem [**HR2**, (28.58)]. It runs as follows: *if X is an infinite compact Hausdorff space, if $\mathfrak{u}(x)$ is the smallest possible cardinal that a family of open subsets of X with intersection $\{x\}$ can have and if $\mathfrak{m}$ is a cardinal such that $\mathfrak{u}(x) \geq \mathfrak{m}$ for all $x \in X$, then $|X| \geq 2^{\mathfrak{m}}$.*)

7 (cf. [**HR1**, (8.7)]). Let G be nondiscrete and σ-compact, and let $\{U_n\}$ be a sequence of neighbourhoods of e in G. Show that there exists a compact normal subgroup N of G such that $N \subset \bigcap_{n=1}^\infty U_n$, and G/N is a nondiscrete, separable metric group.

8. Construct a dense open subset V of $\mathbb{R}$ such that $\lambda(V) < \varepsilon$ (cf. (7.10)).

9. Let G be a nondiscrete, σ-compact locally compact metric group. Show that there exists a family $\mathcal{I}$ of closed, right invariant ideals of $L_\infty(G)$ with $|\mathcal{I}| = 2^{\mathfrak{c}}$.

10. Let (X, d) be as in the proof of (7.15). Let $E \in X$, $n \in \mathbb{P}$, and define $T_n \colon G^n \times X^n \to \mathbb{R}$ by

$$T_n(x_1, \ldots, x_n, E_1, \ldots, E_n) = \lambda(x_1 E_1 \cap \cdots \cap x_n E_n \cap E).$$

Show that T_n is continuous.

11. Give examples of a.l.c. semigroups.

12. Let S be a left amenable semigroup and

$$\mathfrak{p}\ (= \mathfrak{p}(S)) = \min\{|sS| \colon s \in S\}.$$

(So $\mathfrak{m} \leq \mathfrak{p} \leq |S|$ where $\mathfrak{m}$ is as in (7.22).) Prove that if S satisfies (v) or if $\mathfrak{p}$ is infinite and S satisfies either of the conditions (i), (ii) then $\mathfrak{m} = \mathfrak{p}$. Prove also that if S is infinite and S satisfies either of the conditions (iii), (iv), then S is a.l.c. (and so $\mathfrak{p} = \mathfrak{m} = |S|$).

(i) S is right reversible;

(ii) S is amenable;

(iii) S is left cancellative;

(iv) S is right cancellative and left amenable;

(v) S is ELA ((2.27)).

Does $\mathfrak{p} = \mathfrak{m}$ in general?

13. Show that if S is an infinite, left amenable semigroup that is either left or right cancellative, then

$$|\mathfrak{L}(S)| = 2^{2^{|S|}}.$$

The next three problems can all be solved by adapting the proof of (7.25).

14. Let S be an amenable semigroup and

$$\mathfrak{n} = \min\{|sSs|: s \in S\}.$$

Recall that $\mathfrak{I}(S)$ is the set of invariant means on S. Show that if $\mathfrak{n}$ is infinite, then $|\mathfrak{I}(S)| = 2^{2^{\mathfrak{n}}}$, and that if $\mathfrak{n}$ is finite then S contains a finite ideal and $|\mathfrak{I}(S)| = 1$.

15. Let G be an amenable discrete group and $\mathfrak{I}^*(G)$ the set of inversion invariant means on G (Pr. 2–7). Show that if G is infinite, then

$$|\mathfrak{I}^*(G)| = 2^{2^{|G|}}.$$

16. Let G be an infinite amenable discrete group. For $F \in \mathscr{F}(G)$, let

$$C_F = \{x \in G: FxF \cap (FxF)^{-1} = \varnothing\}.$$

(i) Show that if, for each $F \in \mathscr{F}(G)$, there exists $x \in G$ such that $FxF \cap (FxF)^{-1} = \varnothing$, then for all F,

$$|C_F| = |G|.$$

(ii) Show that (a), (b), and (c) are equivalent:
(a) $|\mathfrak{I}(G) \sim \mathfrak{I}^*(G)| = 2^{2^{|G|}}$;
(b) $\mathfrak{I}(G) \neq \mathfrak{I}^*(G)$;
(c) for each $F \in \mathscr{F}(G)$, there exists $x \in G$ such that $FxF \cap (FxF)^{-1} = \varnothing$.

17. Let G be an infinite abelian group. Show that $\mathfrak{I}(G) \neq \mathfrak{I}^*(G)$ if and only if the set $B = \{x^2: x \in G\}$ is infinite.

18. Suppose that $\mathfrak{m} = \mathfrak{p}$, where $\mathfrak{m}$, $\mathfrak{p}$ are as in Problem 12 above. Assume GCH, the Generalised Continuum Hypothesis. Prove that if

$$\mathfrak{r} = \min\{|A|: A \subset S,\ m(A) = 1 \text{ for all } m \in \mathfrak{L}(S)\}$$

then $\mathfrak{p} = \mathfrak{r}$.

19. Let S be a semigroup and $\mathscr{P}_r(S)$ be the family of right thick subsets of S ((1.20)). Show that $\mathscr{P}_r(S)$ is a filter if and only if S contains a finite, left ideal group. Show that $\mathscr{P}_r(S)$ is an ultrafilter in S if and only if S has a zero.

20. Let G be an infinite, amenable group containing a proper, normal subgroup H of finite index. Let $J(G)$ be as in (7.31). Show that

$$\dim(\mathrm{rad}(l_1(G)'')/J(G)) \geq 2^{2^{|G|}}.$$

(So $J(G)$ is not *all* of the radical of $l_1(G)''$.)

21. Let S be a countable, left amenable, left cancellative semigroup. Prove that $\mathfrak{L}(S)$ has an exposed point if and only if $S = S_k \times G$ for some $k \in \{2, \ldots, \infty\}$, where S_k is the semigroup with elements e_i $(1 \leq i < k)$ such that $e_i e_j = e_j$ for all i, j, and G is a finite group.

APPENDIX A

Nilpotent, Solvable, and Semidirect Product Groups

Let G be a group. A *normal series* of G is a finite sequence

$$G = G_1 \triangleright G_2 \triangleright \cdots \triangleright G_r = \{e\} \tag{1}$$

of subgroups G_i of G, each G_{i+1} being a normal subgroup of G_i. A normal series (1) for G is said to be a *central series* of G if, for each i, G_i is a normal subgroup of G and, for $1 \leq i < r$, $G_i/G_{i+1} \subset Z(G/G_{i+1})$, or, equivalently, $[G, G_i] \subset G_{i+1}$ (in the notation below).

The group G is said to be *nilpotent* if it has a central series. The group G is said to be *solvable* if it has a normal series

$$G_1 \triangleright G_2 \triangleright \cdots \triangleright G_r = \{e\}$$

such that every G_i/G_{i+1} is abelian.

Obviously, every nilpotent group is solvable. *Commutators* are often useful, particularly when discussing nilpotent and solvable groups. For $x, y \in G$, the commutator $[x, y]$ is defined $[x, y] = x^{-1}y^{-1}xy$. If $A, B \subset G$, then $[A, B]$ is the subgroup of G generated by $\{[a, b] \colon a \in A,\ b \in B\}$. Let G be a nilpotent group. Then G has two central series of particular importance, the *upper central series* and the *lower central series.*

The upper central series for G is the series

$$\{e\} = Z_0G \triangleleft Z_1G \triangleleft \cdots \triangleleft Z_sG = G,$$

where the Z_iG are defined recursively: $Z_0G = \{e\}$ and $Z_{i+1}G \supset Z_iG$ is specified by $Z_{i+1}G/Z_iG = Z(G/Z_iG)$.

The lower central series is the series

$$G = C^1G \triangleright C^2G \triangleright \cdots \triangleright C^{r+1}G = \{e\},$$

where the subgroups C^iG are defined recursively: $C^1G = G$ and $C^{i+1}G = [C^iG, G]$. The subgroup C^2G is the *derived group* of G and is the subgroup of G generated by the set of commutators $\{[x, y] \colon x, y \in G\}$. The *class* of G is the smallest integer r for which $C^{r+1}(G) = \{e\}$.

A useful fact about the lower central series for G is that for all i, j, $[C^iG, C^jG] \subset C^{i+j}G$ (M. Hall [**1**, Corollary 10.3.5]).

The following result is given in [**M**, Theorem 9.16].

Let G be a finitely generated, nilpotent group, and H a subgroup of G. Then H is also finitely generated.

For solvable groups, the *derived series* plays a fundamental role. The derived series $\{D_iG\}$ of a group G is defined recursively: $D_1G = G$ and $D_{i+1}G = [D_iG, D_iG]$. It is readily checked that each D_iG is a normal subgroup of G and that $D_iG/D_{i+1}G$ is abelian.

The group G is solvable if and only if $D_rG = \{e\}$ for some r. Solvable groups behave well under extensions [**M**, Theorem 10.07]. Indeed

(*) *if G is a group and N is a normal subgroup of G, then G is solvable if and only if both N and G/N are solvable.*

Let G, H and K be groups. The group G is said to be an *extension of K by H* if H is a normal subgroup of G and $G/H \cong K$. The semidirect product construction discussed below is a particularly important source of group extensions for our purposes.

Let H, K be groups and $\rho\colon K \to \operatorname{Aut} H$ a homomorphism, were $\operatorname{Aut} H$ is the automorphism group of H. We define a product on the set $H \times K$ as follows:

$$(h, k)(h', k') = (h\rho(k)(h'), kk').$$

With this product, the set $H \times K$ becomes a group, which is called the *semidirect product $H \times_\rho K$* of H and K [**HR1**, (2.6)]. It is easily checked that $G = H \times_\rho K$ is an extension of K by H.

The identity of $H \times_\rho K$ is, with an obvious abuse of notation, the element (e, e), and the inverse of an element $(h, k) \in H \times_\rho K$ is $(\rho(k^{-1})(h^{-1}), k^{-1})$.

Semidirect products also occur in the topological setting. Let H and K be locally compact groups and $\operatorname{Aut} H$ the group of topological group automorphisms of H. Then $\operatorname{Aut} H$ is itself a topological group in the natural way ((3.3)). Let $\rho\colon K \to \operatorname{Aut} H$ be a homomorphism. The map ρ is continuous if and only if the map $(k, h) \to \rho(k)(h)$ is continuous from $K \times H$ into H [**Ho**, Chapter 3, §3]. Suppose that ρ is continuous. The semidirect product $H \times_\rho K$ is then a locally compact group under the product topology.

Let $G = H \times_\rho K$. Then H and K are canonically identified with closed subgroups, also denoted H, K, of G, where H is normal, $H \cap K = \{e\}$, and $G = HK$. Conversely, if a locally compact group G contains closed subgroups H, K with these three properties, then G is canonically isomorphic to a semidirect product $H \times_\rho K$: here, ρ is given by

$$\rho(k)(h) = khk^{-1}.$$

A number of naturally occurring locally compact (Lie) groups are semidirect products of locally compact groups. An important example for us is the "$ax+b$" group S_2, the affine group of R. Here, S_2 is the group of transformations of $\mathbb{R}$ of the form $x \to ax + b$, where $a, b \in \mathbb{R}$ with $a > 0$, the product being that of composition. Associating the transformation $x \to ax + b$ with the pair (b, a), we identify S_2 with the locally compact group $\mathbb{R} \times (0, \infty)$ with group product given by

$$(b, a)(b', a') = (b + ab', aa').$$

(Note that $(0, \infty)$ is a locally compact abelian group with multiplication as product.) It is clear that S_2 is the semidirect product $\mathbb{R} \times_\rho (0, \infty)$, where $\rho(a)(b) = ab$. The map $t \to e^t$ is an isomorphism from $\mathbb{R}$ onto $(0, \infty)$, so that S_2 can be regarded as the semidirect product $\mathbb{R} \times_\sigma \mathbb{R}$, where $\sigma(t)(s) = e^t s$.

APPENDIX B

Lie Groups

The purpose of this appendix is to describe briefly the concepts and results from Lie theory needed in the text. Proofs are almost entirely omitted. Our main sources for the theory here are [**He**] and [**B**], and results for which no references are given will normally be found in one or other of these texts. On a number of occasions, we will also require results from [**Ho**], [**J**], [**HS**], and [**SW**]. The book [**SW**] gives an excellent introduction to the subject.

B1. Charts and manifolds. Let M be a topological space and $m \in M$. An *n-dimensional chart* is a pair (U, ϕ), where U is an open subset of M and ϕ is a homeomorphism from U onto an open subset of $\mathbb{R}^n$. Sometimes, a chart (U, ϕ) will simply be referred to as U, reference to ϕ being left implicit. If (U, ϕ) is an n-dimensional chart and, for $p \in U$, $\phi(p) = (x_1(p), \ldots, x_n(p))$, then the functions $x_1, \ldots, x_n$ are called *coordinate functions* (for the chart) and the numbers $x_i(p)$ are called the *local coordinates* of p.

A *C^∞-structure* on M of dimension n is a family $\mathscr{A} = \{(U_\alpha, \phi_\alpha) \colon \alpha \in A\}$ of n-dimensional charts in M such that

(i) $\bigcup_{\alpha \in A} U_\alpha = M$;

(ii) if $\alpha, \beta \in A$, then the map $\phi_\alpha \circ \phi_\beta^{-1}$ is a C^∞-map from $\phi_\beta(U_\alpha \cap U_\beta)$ onto $\phi_\alpha(U_\alpha \cap U_\beta)$;

(iii) the family $\mathscr{A}$ is a maximal family of n-dimensional charts in M satisfying (i) and (ii).

We note that if $\mathscr{A}$ is a family of n-dimensional charts in M such that (i) and (ii) are satisfied, then, by including all those n-dimensional charts in M that intersect the members of $\mathscr{A}$ in a "C^∞-way," we obtain a C^∞-structure on M (containing $\mathscr{A}$). A *C^∞-manifold* (or simply a *manifold*) is a separable, Hausdorff space M equipped with a C^∞-structure $\mathscr{A}$. In such a case, we normally refer to M as a manifold, the structure $\mathscr{A}$ involved being left implicit. A *local chart* on a manifold M is just a member of the C^∞-structure $\mathscr{A}$.

A topological space is said to be *0-dimensional* if it is discrete. An *analytic structure* on M of dimension n is defined as in (i), (ii), and (iii) above, the maps $\phi_\alpha \circ \phi_\beta^{-1}$ being required to be analytic (in the sense that their coordinate functions are given, locally, by power series expansions in real variables). An *analytic manifold* is defined in the obvious way. Every analytic manifold is also a C^∞-manifold in the natural way.

Every manifold M is obviously a locally compact Hausdorff space. It is routine to check that M is connected if and only if M is pathwise connected.

If M, N are [analytic] manifolds, then $M \times N$ is also an [analytic] manifold in the obvious way. Trivially, every open subset of an n-dimensional [analytic] manifold is itself an n-dimensional [analytic] manifold.

Let M and N be manifolds and $\Phi\colon M \to N$. The map Φ is said to be a *C^∞-map* if, whenever $m \in M$, there exist in M, N local charts (U, ϕ), (V, ψ) at m and $\Phi(m)$ respectively such that $\Phi(U) \subset V$ and $\psi \circ \Phi \circ \phi^{-1}\colon \phi(U) \to \psi(V)$ is a C^∞-function. *Analytic* maps are defined in a similar way. A bijective map Ψ between [analytic] manifolds is called an [*analytic*] *diffeomorphism* if both Ψ and Ψ^{-1} are [analytic] C^∞-maps.

The set of real-valued, differentiable functions on M is denoted by $C^\infty(M)$. The set $C^\infty(M)$ is an algebra over $\mathbb{R}$ under the usual pointwise operations.

Since M is a separable, locally compact Hausdorff space, it is *paracompact*, that is, if $\mathscr{B}$ is an open cover of M, then there exists an open cover $\{U_\alpha\}$ with the following properties:

(i) each U_α is contained in some member of $\mathscr{B}$;

(ii) $\{U_\alpha\}$ is *locally finite* in the sense that each $m \in M$ has a neighbourhood that intersects only finitely many of the sets U_α.

Further M admits *partitions of unity* in the sense that if $\{U_\alpha\colon \alpha \in A\}$ is a locally finite, open cover of M with U_α^- compact for all α, then there exists a set $\{\phi_\alpha\colon \alpha \in A\}$ in $C^\infty(M)$ such that

(i) ϕ_α has compact support contained in U_α $(\alpha \in A)$;

(ii) $\phi_\alpha \geq 0$ $(\alpha \in A)$;

(iii) $\sum_{\alpha \in A} \phi_\alpha = 1$.

B2. Tangent spaces and vector fields. Let M be an n-dimensional manifold, and $m \in M$. The *tangent space* TM_m at m is the real vector space of linear functionals $F\colon C^\infty(M) \to \mathbb{R}$ such that

$$F(\phi\psi) = F(\phi)\psi(m) + \phi(m)F(\psi) \tag{1}$$

for all $\phi, \psi \in C^\infty(M)$.

If $a < 0 < b$ in $\mathbb{R}$ and $u\colon (a, b) \to M$ is a C^∞-path with $u(0) = m$, then we can define an element $F_u \in TM_m$ by setting

$$F_u(\phi) = \left[\frac{d}{dt}(\phi(u(t)))\right]_{t=0}.$$

It can be shown that *every* element of TM_m is of the form F_u for some C^∞-path u.

Going into local coordinates at m, so that an open neighbourhood of m is identified with an open subset of $\mathbb{R}^n$, we can define $(\partial/\partial x_i)_m \in TM_m$ by $(\partial/\partial x_i)_m(\phi) = \partial\phi(m)/\partial x_i$. It turns out that the set $\{(\partial/\partial x_i)_m\colon 1 \leq i \leq n\}$ is a basis for the vector space TM_m, so that the latter is n-dimensional.

Let M and N be manifolds and $\Phi\colon M \to N$ a C^∞-map. Then Φ can be "differentiated" at a point $m \in M$ to give a linear map $\Phi_*(m)\colon TM_m \to TM_{\Phi(m)}$, where, for $F \in TM_m$,

$$\Phi_*(m)F(\phi) = F(\phi \circ \Phi) \qquad (\phi \in C^\infty(N)).$$

The map $\Phi_*(m)$ is called the *differential* of Φ at m.

If $u\colon (a,b) \to M$ is as above, then $F_u = u_*(0)((d/dt)_0)$. If $t_0 \in (a,b)$, then we sometimes identify $u_*(t_0)$ with $u_*(t_0)((d/dt)_{t_0})$. If P is also a manifold and $\Phi\colon M \to N$ and $\Psi\colon N \to P$ are C^∞-maps, then the following "chain rule" is easily established:

$$(\Psi \circ \Phi)_*(m) = \Psi_*(\phi(m)) \circ \Phi_*(m).$$

A *vector field* on M is a *derivation* X on $C^\infty(M)$, that is, a map $X\colon C^\infty(M) \to C^\infty(M)$ that is linear, and satisfies the equality

$$X(\phi\psi) = (X(\phi))\psi + \phi(X(\psi)) \qquad (\phi, \psi \in C^\infty(M)).$$

Obviously, the set $\Delta(M)$ of vector fields on M is a linear subspace of $L(C^\infty(M))$, the space of linear transformations on $C^\infty(M)$. Further, $\Delta(M)$ is a *Lie algebra* in the sense that if $[X,Y]$ is defined to be $XY - YX$, then $[X,Y] \in \Delta(M)$ whenever $X, Y \in \Delta(M)$, and the following properties hold:

(i) $[X,X] = 0$ $(X \in \Delta(M))$;

(ii) (Jacobi's identity)

$$[X,[Y,Z]] + [Y,[Z,X]] + [Z,[X,Y]] = 0$$

for all $X, Y, Z \in \Delta(M)$.

The space $\Delta(M)$ is a "$C^\infty(M)$-module" under the map $(\phi, X) \to \phi X$, where, for $\phi \in C^\infty(M)$, $X \in \Delta(M)$, we define $\phi X \in \Delta(M)$ by

$$(\phi X)(\psi) = \phi(X(\psi)) \qquad (\psi \in C^\infty(M)).$$

[Thus, $(\phi X)(\psi)$ is the pointwise product of the functions ϕ and $X(\psi)$.]

For $m \in M$, $X \in \Delta(M)$, define $X_m\colon C^\infty(M) \to \mathbb{R}$ by

$$X_m(\phi) = (X(\phi))(m).$$

One readily checks that the map $X \to X_m$ is a linear map from $\Delta(M)$ into M_m. The range of this map is actually the whole of M_m.

In terms of coordinate functions $x_1, \ldots, x_n$ for a local chart U, each map $\partial/\partial x_i$, where $(\partial/\partial x_i)(\phi) = \partial\phi/\partial x_i$, belongs to $\Delta(U)$, and this $(\partial/\partial x_i)_m$ is the same as the $(\partial/\partial x_i)_m$ introduced earlier.

It is sometimes convenient to identify $X \in \Delta(M)$ with the function $m \to X_m$ from M into $\bigcup_{p \in M} M_p$. In terms of local coordinates for a local chart (U, ϕ), there exist functions $\phi_i \in C^\infty(U)$ such that

$$X_m = \sum_{i=1}^{n} \phi_i(m) \left(\frac{\partial}{\partial x_i}\right)_m \qquad (m \in U).$$

B3. Covering spaces and the fundamental group (Massey [**1**], Spanier [**1**], [**SW**]). We will discuss the theory of covering spaces for manifolds only, although it applies more generally under suitable local connectedness hypotheses.

Let M be a connected manifold. A *covering space* of M is a pair (P, p), where P is a connected manifold and $p: P \to M$ is continuous, surjective, and satisfies the following: if $m \in M$, there exists a neighbourhood U of m such that $p^{-1}(U)$ is the union of a disjoint family of open subsets, each of which is mapped homeomorphically onto U by p. The map p is called a *covering projection.* Sometimes, P itself is referred to as a covering space for M, reference to p being left implicit.

Clearly, if (P, p) is a covering space of M, then p is open and surjective. Equally obvious is the fact that p is a *local homeomorphism* in the sense that if $x \in P$, then there exists an open neighbourhood of x mapped homeomorphically onto an open neighbourhood of $p(x)$. In the natural way, we can define a manifold structure on P so that p is a C^∞-mapping. Of course, P is an analytic manifold and p an analytic map if M is analytic.

The study of covering spaces essentially involves the fundamental group, which we now briefly discuss.

Let m_0, $m_1 \in M$ and α, β be paths joining m_0 to m_1. The paths α, β are said to be *homotopic* (written $\alpha \sim \beta$) if there exists a continuous map $h: [0,1] \times [0,1] \to M$ such that $h(0, s) = m_0$, $h(1, s) = m_1$, $h(t, 0) = \alpha(t)$, and $h(t, 1) = \beta(t)$ for all $s, t \in [0,1]$. Homotopy is an equivalence relation on the set of paths joining m_0 to m_1. The equivalence class containing α is denoted by $[\alpha]$.

Let α be a path in M joining m_0 to m_1 and γ be a path in M joining m_1 to $m_2 \in M$. The product $\alpha\gamma$ is the path in M defined by:

$$\alpha\gamma(t) = \begin{cases} \alpha(2t) & (0 \leq t \leq \frac{1}{2}), \\ \gamma(2t-1) & (\frac{1}{2} \leq t \leq 1). \end{cases}$$

If α', γ' are paths in M with $\alpha \sim \alpha'$ and $\gamma \sim \gamma'$, then $\alpha\gamma \sim \alpha'\gamma'$. So we can define a homotopy equivalence class $[\alpha][\gamma]$ by $[\alpha][\gamma] = [\alpha\gamma]$.

Now define a path α^* by setting $\alpha^*(t) = \alpha(1-t)$ $(0 \leq t \leq 1)$. If $\alpha \sim \alpha'$, then $\alpha^* \sim (\alpha')^*$, so that it makes sense to define $[\alpha]^* = [\alpha^*]$.

The set $\pi_1(M, m_0)$ of homotopy equivalence classes of paths starting and finishing at m_0 is a group, with product and inversion given by $([\alpha], [\beta]) \to [\alpha\beta]$ and $[\alpha] \to [\alpha]^*$.

Recall that every connected manifold is pathwise connected. Hence, if $m_1 \in M$, we can join up m_0 to m_1 by a path in M, and using this path to relate $\pi_1(M, m_0)$ to $\pi_1(M, m_1)$, we see that these two groups are naturally isomorphic. Thus $\pi_1(M, m_0)$ is essentially independent of the "base point" m_0, and so is usually written $\pi_1(M)$, reference to the base point m_0 being left implicit. The group $\pi_1(M)$ is called the *fundamental group* of M. The manifold M is called *simply connected* if $\pi_1(M)$ is trivial.

Let (P, p) be a covering space of M. Then (P, p) has *unique path lifting* in the sense that, *whenever $x_0 \in P$ and α is a path in M with $\alpha(0) = p(x_0)$, then there* *exists a unique path α' in P with $\alpha'(0) = x_0$ and $p \circ \alpha' = \alpha$.* A related result is the following: *if α', β' are paths in P with $\alpha'(0) = \beta'(0)$ and $p \circ \alpha' \sim p \circ \beta'$, then $\alpha' \sim \beta'$ (so that, in particular, $\alpha'(1) = \beta'(1)$).* The proofs involve a fairly routine compactness argument.

The *simply connected covering space* $(\tilde{M}, \pi)$ is constructed as follows. Two paths α, β in M starting at the base point m_0 are said to be *equivalent* if $\alpha(1) = \beta(1)$ and α and β are homotopic. This gives an equivalence relation on the set of paths in M starting at m_0. We take $\tilde{M}$ to be the resulting set of equivalence classes, and define $\pi([\alpha]) = \alpha(1)$, where $[\alpha]$ is the equivalence class containing a path α. If U is an open neighbourhood of $\alpha(1)$ in M, let $\tilde{U}_\alpha$ be the set of equivalence classes $[\alpha\beta]$, where β is a path in U starting at $\alpha(1)$. It is easy to check that the family of all such sets $\tilde{U}_\alpha$ is a base for a topology on $\tilde{M}$ and that $(\tilde{M}, \pi)$ is a covering space of M that is simply connected.

Let (P, p) be a covering space for M, and fix $x_0 \in P$ such that $p(x_0) = m_0$. Let $\tilde{m} \in \tilde{M}$ and $\alpha \in \tilde{m}$. So α is a path in M. We can lift α to a unique path α' in P starting at x_0. By a result above, $\alpha'(1)$ is independent of the choice of α. The map $\tilde{m} \to \alpha'(1)$ is a continuous map $F\colon \tilde{M} \to P$ such that the following diagram commutes:

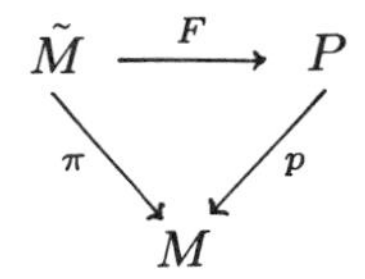

If P is itself simply connected, then the procedure leading to the definition of F is reversible, and F is a homeomorphism. Thus, in the obvious sense, $\tilde{M}$ is the *unique* simply connected covering space of M.

A *covering transformation* of $\tilde{M}$ is a homeomorphism T of $\tilde{M}$ such that $\pi \circ T = \pi$. The set $G(\tilde{M})$ of covering transformations of $\tilde{M}$ obviously forms a group of transformations of $\tilde{M}$. Using the unique path-lifting property of $(\tilde{M}, \pi)$, we see that if $T_1, T_2 \in G(\tilde{M})$ and $a \in \tilde{M}$ is such that $T_1(a) = T_2(a)$, then $T_1 = T_2$.

The group $G(\tilde{M})$ is isomorphic to $\pi_1(M)$. Indeed, let f be a path in M starting and ending at m_0. Then the map $T_{[f]}$, where $T_{[f]}([\alpha]) = [f\alpha]$ is a covering transformation of $\tilde{M}$, and the map $[f] \to T_{[f]}$ is an isomorphism from $\pi_1(M)$ onto $G(\tilde{M})$.

B4. Lie algebras. A vector space $\mathfrak{g}$ over a field $\mathbb{F}$, where $\mathbb{F}$ is $\mathbb{R}$ or $\mathbb{C}$, is called a *Lie algebra* if there is given a bilinear map $(X, Y) \to [X, Y]$ on $\mathfrak{g}$ such that

(i) $[X, X] = 0$ for all $X \in \mathfrak{g}$;

(ii) (Jacobi's identity) $[X, [Y, Z]] + [Y, [Z, X]] + [Z, [X, Y]] = 0$ for all $X, Y, Z \in \mathfrak{g}$.

Let $\mathfrak{g}$ be a Lie algebra. If $X, Y \in \mathfrak{g}$, then $[X, Y]$ is the *(Lie) product* of X and Y. It easily follows from (i) that $[X, Y] = -[Y, X]$ for all $X, Y \in \mathfrak{g}$.

If $\mathfrak{a}$, $\mathfrak{b}$ are subspaces of $\mathfrak{g}$, then $[\mathfrak{a}, \mathfrak{b}]$ is the subspace of $\mathfrak{g}$ spanned by the elements $[X, Y]$ ($X \in \mathfrak{a}$, $Y \in \mathfrak{b}$). A *subalgebra* [*ideal*] of $\mathfrak{g}$ is a subspace $\mathfrak{s}$ of $\mathfrak{g}$ such that $[\mathfrak{s}, \mathfrak{s}] \subset \mathfrak{s}$ $[[\mathfrak{g}, \mathfrak{s}] \subset \mathfrak{s}]$. The Lie algebra $\mathfrak{g}$ is said to be *abelian* if $[\mathfrak{g}, \mathfrak{g}] = \{0\}$.

Let $\mathfrak{h}$ also be a Lie algebra over $\mathbb{F}$. A linear map $\Phi\colon \mathfrak{g} \to \mathfrak{h}$ is called a (*Lie*) *homomorphism* if $[\Phi(X), \Phi(Y)] = \Phi([X, Y])$ for all $X, Y \in \mathfrak{g}$. The map Φ is called a (*Lie*) *isomorphism* if it is a bijective homomorphism. An isomorphism from $\mathfrak{g}$ onto $\mathfrak{g}$ is called an *automorphism* of $\mathfrak{g}$.

If $\mathfrak{s}$ is an ideal in $\mathfrak{g}$, then $\mathfrak{g}/\mathfrak{s}$ is also a Lie algebra in the natural way, and the quotient map $X \to X + \mathfrak{s}$ is a homomorphism from $\mathfrak{g}$ onto $\mathfrak{g}/\mathfrak{s}$.

The direct sum $\mathfrak{g} \oplus \mathfrak{h}$ of $\mathfrak{g}$ and $\mathfrak{h}$ is a Lie algebra in the obvious way. If $\mathbb{F} = \mathbb{R}$, then the complexification $\mathfrak{g}_{\mathbb{C}}$ of $\mathfrak{g}$ is also a Lie algebra over $\mathbb{C}$ in the obvious way.

Now let V be a vector space over $\mathbb{F}$. Then the space $L(V)$ of $\mathbb{F}$-linear transformations on V is a Lie algebra, the Lie product being given by

$$[A, B] = AB - BA.$$

An important theorem of Ado asserts that *if $\mathfrak{g}$ is a finite-dimensional Lie algebra over $\mathbb{F}$, then there exists a finite-dimensional vector space W over $\mathbb{F}$ such that $\mathfrak{g}$ is isomorphic to a Lie subalgebra of $L(W)$.*

Proofs of Ado's Theorem are given in [**B**, Chapter 1, §7.3] and [**J**, Chapter 6].

A *representation* of the Lie algebra $\mathfrak{g}$ on an $\mathbb{F}$-vector space V is a Lie homomorphism $Q\colon \mathfrak{g} \to L(V)$. The space V is called a $\mathfrak{g}$-*module*, and we often write Xv for $Q(X)(v)$ ($X \in \mathfrak{g}$, $v \in V$). Direct sums of $\mathfrak{g}$-modules are themselves $\mathfrak{g}$-modules in the obvious way.

An important representation of $\mathfrak{g}$ on $\mathfrak{g}$ is the *adjoint representation* $\mathrm{ad}_{\mathfrak{g}}$ (or simply ad), where

$$\mathrm{ad}_{\mathfrak{g}} X(Y) = [X, Y] \qquad (Y \in \mathfrak{g}).$$

(That ad is a representation of $\mathfrak{g}$ follows easily using Jacobi's identity.) We note that ker ad is the *center* $Z(\mathfrak{g})$ of $\mathfrak{g}$, where

$$Z(\mathfrak{g}) = \{Z \in \mathfrak{g}\colon [X, Z] = 0 \text{ for all } X \in \mathfrak{g}\}.$$

It is sometimes convenient to regard ad as a representation on $\mathfrak{g}_{\mathbb{C}}$: simply set $\mathrm{ad}\, X(Y + iZ) = \mathrm{ad}\, X(Y) + i\, \mathrm{ad}\, X(Z)$. The context will always make clear if ad is being regarded as a representation on $\mathfrak{g}$ or $\mathfrak{g}_{\mathbb{C}}$.

A nonzero $\mathfrak{g}$-module V is called *irreducible* if the only $\mathfrak{g}$-invariant subspaces of V are $\{0\}$ and V.

B5. Derivations, automorphisms, and semidirect products. A *derivation* of a Lie algebra $\mathfrak{g}$ is a map $D \in L(\mathfrak{g})$ such that $D([X, Y]) = [DX, Y] + [X, DY]$ for all $X, Y \in \mathfrak{g}$. The set of derivations on $\mathfrak{g}$ is denoted by $\mathrm{Der}(\mathfrak{g})$. It is easy to check that $\mathrm{Der}\, \mathfrak{g}$ is a subalgebra of the Lie algebra $L(\mathfrak{g})$. Jacobi's identity shows that $\mathrm{ad}\, X \in \mathrm{Der}\, \mathfrak{g}$ for all $X \in \mathfrak{g}$, so that ad is a homomorphism from $\mathfrak{g}$ into $\mathrm{Der}\, \mathfrak{g}$. An element D of $\mathrm{Der}\, \mathfrak{g}$ is called an *inner derivation* if $D \in \mathrm{ad}\, \mathfrak{g}$. If $X \in \mathfrak{g}$, $D \in \mathrm{Der}\, \mathfrak{g}$, then $[D, \mathrm{ad}\, X] = \mathrm{ad}(DX)$, so that $\mathrm{ad}\, \mathfrak{g}$ is an ideal of $\mathrm{Der}\, \mathfrak{g}$.

The set of automorphisms of $\mathfrak{g}$ is denoted by Aut $\mathfrak{g}$. Clearly, $\mathrm{Aut}\, \mathfrak{g}$ is a subgroup of $\mathrm{GL}(\mathfrak{g})$, the group of invertible elements of $L(\mathfrak{g})$. If $\mathfrak{g}$ has finite dimension n, then $\mathrm{GL}(\mathfrak{g})$ is a locally compact group under the relative topology that it inherits as a subset of $L(\mathfrak{g})$ ("=" $M_n(\mathbb{F})$), and $\operatorname{Aut}\mathfrak{g}$ is a closed subgroup of $\mathrm{GL}(\mathfrak{g})$.

Let $D \in \operatorname{Der}\mathfrak{g}$. By induction, Leibniz's rule holds for D:

$$D^n([X,Y]) = \sum_{r=0}^{n} \binom{n}{r} [D^r(X), D^{n-r}(Y)]. \tag{1}$$

It follows that $e^D = \sum_{n=0}^{\infty} D^n/n! \in \operatorname{Aut}\mathfrak{g}$. If $T \in L(\mathfrak{g})$ is such that $e^{tT} \in \operatorname{Aut}\mathfrak{g}$ for all $t \in \mathbb{R}$, then $T = \lim_{t\to 0}(e^{tT} - 1)/t \in \operatorname{Der}\mathfrak{g}$. Thus *if* $T \in L(\mathfrak{g})$, *then* $T \in \operatorname{Der}\mathfrak{g}$ *if and only if* $e^{tT} \in \operatorname{Aut}\mathfrak{g}$ *for all* $t \in \mathbb{R}$.

Now let $\mathfrak{h}$ be a Lie algebra over $\mathbb{F}$, and $\Phi\colon \mathfrak{h} \to \operatorname{Der}\mathfrak{g}$ a homomorphism. The *semidirect product* $\mathfrak{g} \times_\Phi \mathfrak{h}$ is the Lie algebra that, as a vector space, is $\mathfrak{g} \times \mathfrak{h}$, and whose product is given by

$$[(X,Y),(X',Y')] = ([X,X'] + \Phi(Y)(X') - \Phi(Y')(X), [Y,Y'])$$

$(X, X' \in \mathfrak{g},\ Y, Y' \in \mathfrak{h})$. Canonically, $\mathfrak{g}[\mathfrak{h}]$ is obviously an ideal [subalgebra] of $\mathfrak{g} \times_\Phi \mathfrak{h}$.

Conversely, if $\mathfrak{g}_1$ is a Lie algebra containing $\mathfrak{g}$ as an ideal and $\mathfrak{h}$ as a subalgebra such that $\mathfrak{g}_1$ is the vector space direct sum $\mathfrak{g}\oplus\mathfrak{h}$, then $\mathfrak{g}_1$ is canonically isomorphic to the semidirect product $\mathfrak{g} \times_\Phi \mathfrak{h}$, where $\Phi(Y)(X) = [Y,X]$ for $Y \in \mathfrak{h}$, $X \in \mathfrak{g}$.

B6. Lie groups. A *Lie group* is a group G that is an analytic manifold such that the map $(x,y) \to xy^{-1}$ from $G \times G$ into G is analytic. Thus the product map $(x,y) \to xy$ and the inversion map $x \to x^{-1}$ are analytic. Every Lie group G is, of course, a locally compact group. Since G is a manifold, the identity e of G has a connected neighbourhood, so that G_e is an open (and closed) subgroup of G.

We now state the "approximation" theorem, which effectively reduces the study of connected, locally compact groups to that of connected Lie groups.

(Montgomery and Zippin [**1**, (4.6)]). *Let G be a connected, locally compact group and U a neighbourhood of e. Then there exists a compact, normal subgroup K of G such that $K \subset U$ and G/K is a* (*connected*) *Lie group.*

Let G be a connected Lie group, and let $(\tilde{G}, \pi)$ be the simply connected covering space of G (B3).

The manifold $\tilde{G}$ can be made into a Lie group as follows. Let α, β be paths in G starting at e and define paths $\alpha \cdot \beta$ and α^{-1} in G by

$$\alpha \cdot \beta(t) = \alpha(t)\beta(t), \quad \alpha^{-1}(t) = (\alpha(t))^{-1} \qquad (t \in [0,1]).$$

Then $\tilde{G}$, regarded as the set of equivalence classes of paths starting at e in G, becomes a group, with product and inversion given by

$$[\alpha]\cdot[\beta] = [\alpha\cdot\beta], \qquad [\alpha]^{-1} = [\alpha^{-1}].$$

From the definition of π, it is clear that π is a homomorphism. By considering how the analytic structure on $\tilde{G}$ is induced by that on G, one readily checks that

$\tilde{G}$ is a Lie group. The group $\tilde{G}$ is called the *simply connected covering group* of G [**SW**, Chapter 8].

A useful observation is that *the kernel* $\ker \pi$ *of* π *is a discrete subgroup of* $Z(\tilde{G})$. [Since π is a local homeomorphism, $\ker \pi$ is discrete. The fact that $\ker \pi \subset Z(\tilde{G})$ is a consequence of the fact that for each $n \in \ker \pi$, the set $\{xnx^{-1}: x \in \tilde{G}\}$ is a connected subset of $\ker \pi$ and so has to be the singleton $\{n\}$.]

Let G, H be Lie groups and U a neighbourhood of e in G. A map $\phi: U \to H$ is called a *local homomorphism* if there exists a neighbourhood V of e in G such that $V^2 \subset U$ and $\phi(xy) = \phi(x)\phi(y)$ for all $x, y \in V$.

For the following result, see, for example, [**HS**, Chapter 1, §2].

B7. *Let G and H be Lie groups with G simply connected. Let U be a neighbourhood of e in G and $\phi: U \to H$ a continuous local homomorphism. Then there exists a continuous homomorphism $\phi': G \to H$ such that ϕ and ϕ' coincide on a neighbourhood of e.*

B8. The Lie algebra of a Lie group (Kirillov [**2**, (6.3)]; [**SW**, Chapter 5]). Let G be a Lie group of dimension n and $\mathfrak{g}$ the tangent space at e for the manifold G. For each $x \in G$, define a linear map $R_x: C^\infty(G) \to C^\infty(G)$ by

$$R_x\phi(y) = \phi(xy) \qquad (y \in G).$$

Recall (B2) that $\Delta(G)$ is the Lie algebra of vector fields on G. Let

$$\mathfrak{g}_L = \{X \in \Delta(G): R_xX = XR_x \text{ for all } x \in G\}.$$

The elements of $\mathfrak{g}_L$ are known as the *left invariant vector fields of* G.

It is elementary to check that $\mathfrak{g}_L$ is a Lie subalgebra of $\Delta(G)$.

For $A \in \mathfrak{g}$, define $\Phi(A) \in L(C^\infty(G))$ as follows:

$$\Phi(A)\phi(x) = A(R_x\phi) \qquad (\phi \in C^\infty(G), x \in G).$$

(The fact that $\Phi(A)\phi \in C^\infty(G)$ follows from the fact that, in terms of coordinate functions $y_1, \ldots, y_n$ about e, the map $(x, y) \to R_x\phi(y)$ is C^∞ and A is of the form $\sum_{i=1}^n \alpha_i(\partial/\partial y_i)_e$ with $\alpha_i \in \mathbb{R}$.) It is straightforward to check that Φ is a linear isomorphism from $\mathfrak{g}$ onto $\mathfrak{g}_L$, the inverse Φ^{-1} being the map that associates $X \in \mathfrak{g}_L$ with the element X_e of $\mathfrak{g}$. Defining $[A, B] = \Phi^{-1}([\Phi(A), \Phi(B)])$ for $A, B \in \mathfrak{g}$, the space $\mathfrak{g}$ becomes a Lie algebra.

Kirillov gives three other ways of defining the Lie product on $\mathfrak{g}$, each useful in the appropriate context. However, the above way is adequate for our purposes.

It is straightforward to check that the Lie algebra of the simply connected covering space $\tilde{G}$ of G is canonically isomorphic to $\mathfrak{g}$.

B9. The exponential map for Lie groups. A *one-parameter subgroup* of the Lie group G is an analytic homomorphism $\alpha: \mathbb{R} \to G$. Thus such a map α satisfies the equalities: $\alpha(0) = e$, $\alpha(t + s) = \alpha(t)\alpha(s)$ $(t, s \in \mathbb{R})$. A fundamental result is that *for every $B \in \mathfrak{g}$, there exists a unique one-parameter subgroup α_B of G such that $(\alpha_B)_*(0) = B$.*

We can now define the exponential map $\exp_G : \mathfrak{g} \to G$ by

$$\exp_G(B) = \alpha_B(1).$$

We sometimes write exp or $\exp_{\mathfrak{g}}$ for $\exp_G$ when the group G is understood. Here are some basic properties of exp.

(1) $\exp(tB) = \alpha_B(t)$, so that $t \to \exp tB$ is a one-parameter subgroup of G;

(2) $\exp : \mathfrak{g} \to G$ is analytic;

(3) $\exp_*(0) = I$, the identity map on $\mathfrak{g}$, the tangent space at 0 to $\mathfrak{g}$ being canonically identified with $\mathfrak{g}$;

(4) there exist neighbourhoods U_0' and U_0 of 0 and e in $\mathfrak{g}$ and G respectively such that $\exp|_{U_0'}$ is a bijection onto U_0 and the inverse log of $\exp|_{U_0'}$ is analytic;

(5) if $A, B \in \mathfrak{g}$, then there exists $\eta > 0$ such that for all $t \in (-\eta, \eta)$, we have

(a) $\exp tA \exp tB = \exp(t(A+B) + \frac{1}{2}t^2[A,B] + a_3(t))$,

(b) $\exp tA \exp tB \exp -tA = \exp(tB + t^2[A,B] + b_3(t))$,

(c) $\exp -tA \exp -tB \exp tA \exp tB = \exp(t^2[A,B] + c_3(t))$

where $a_3(t)$, $b_3(t)$, $c_3(t) \in \mathfrak{g}$, and $\|a_3(t)\|$, $\|b_3(t)\|$, $\|c_3(t)\|$ are $O(t^3)$ as $t \to 0$. (Here $\|\cdot\|$ is some norm on the finite-dimensional vector space $\mathfrak{g}$.)

B10. *Suppose that $\mathfrak{g}$ is the vector space direct sum $\bigoplus\{\mathfrak{g}_i : 1 \le i \le k\}$ of subspaces $\mathfrak{g}_i$ and define $\Phi' : \mathfrak{g} \to G$ by*

$$\Phi'\left(\sum_{i=1}^{k} X_i\right) = (\exp X_1)(\exp X_2)\cdots(\exp X_k)$$

($X_i \in \mathfrak{g}$). *Then Φ' is analytic, and $\Phi'_*(0) = I$, the identity map on $\mathfrak{g}$. (So Φ' is an analytic diffeomorphism from an open neighbourhood of* 0 *in $\mathfrak{g}$ onto an open neighbourhood of e in G.)*

B11. *Let G, H be Lie groups with Lie algebras $\mathfrak{g}$, $\mathfrak{h}$ respectively.*

(i) *If $T : G \to H$ is an analytic homomorphism, then $T_*(e) : \mathfrak{g} \to \mathfrak{h}$ is a Lie algebra homomorphism, and the following diagram commutes:*

$$\text{(1)} \qquad \begin{array}{ccc} \mathfrak{g} & \xrightarrow{T_*(e)} & \mathfrak{h} \\ {\scriptstyle \exp}\downarrow & & \downarrow{\scriptstyle \exp} \\ G & \xrightarrow[T]{} & H \end{array}$$

(ii) *If $\Phi : \mathfrak{g} \to \mathfrak{h}$ is a homomorphism, then there exists a neighbourhood U_0 of e in G as in (4) of B9 and an analytic local homomorphism $S : U_0 \to H$ given by the formula:*

$$S = \exp_{\mathfrak{h}} \circ \Phi \circ (\exp_{\mathfrak{g}})^{-1}.$$

Further, $S_*(e) = \Phi$.

(iii) *Every continuous homomorphism from G into H is analytic.*

B12. The adjoint representation. Let G be a Lie group with Lie algebra $\mathfrak{g}$. For $x \in G$, let α_x be the inner automorphism of G given by $\alpha_x(y) = xyx^{-1}$. Clearly, α_x is analytic. Let $\operatorname{Ad} x = (\alpha_x)_*(e) \in \operatorname{GL}(\mathfrak{g})$. We sometimes write

Ad_G in place of Ad. By considering matrix entries, it is obvious that $\mathrm{GL}(\mathfrak{g})$ is a Lie group in a natural way. Since $\alpha_a \circ \alpha_b = \alpha_{ab}$ $(a, b \in G)$, the chain rule applies to give $\mathrm{Ad}(ab) = (\mathrm{Ad}\, a)(\mathrm{Ad}\, b)$. It is readily checked that Ad is an analytic homomorphism. Since each α_x is a homomorphism, it follows that $\mathrm{Ad}\, G \subset \mathrm{Aut}\, \mathfrak{g}$.

We now give three useful properties of Ad.

(i) $a(\exp X)a^{-1} = \exp(\mathrm{Ad}\, a(X))$ $(a \in G, X \in \mathfrak{g})$;

(ii) $\mathrm{Ad}(\exp Y) = e^{\mathrm{ad}\, Y}$ $(Y \in \mathfrak{g})$;

(iii) *if G is connected, then* $\ker \mathrm{Ad} = Z(G)$.

B13. Haar measure of G. Let G be a connected Lie group. We require the following expression for the modular function Δ on G [**B**, Chapter 3, §3.16]:

$$\Delta_G(a) = |\det \mathrm{Ad}\, a|^{-1} \qquad (a \in G).$$

B14. Lie subgroups. Let M, N be analytic manifolds with $M \subset N$, and $I\colon M \to N$ the identity map: $I(p) = p$ $(p \in M)$. Then M is said to be a *submanifold* of N if I is analytic, and, for each $p \in M$, the map $I_*(p)\colon TM_p \to TN_p$ is one-to-one. One readily shows, using the Inverse Function Theorem, that if M, N have dimensions m, n respectively and M is a submanifold of N, then, given $p \in M$, there exists a local chart V in N with coordinate functions $x_1, \ldots, x_n$ for which $p \in V$ and $x_i(p) = 0$ $(1 \le i \le n)$, and such that

$$U = \{q \in V : x_j(q) = 0 \text{ for } m + 1 \le j \le n\}$$

is a chart in M with local coordinate functions $x_1|_U, \ldots, x_m|_U$.

Note that the topology of a submanifold M of N need not be the relative topology induced by N. An example of this is given by the case where $N = \mathbb{T} \times \mathbb{T}$ and

$$M = \{(e^{2\pi i a t}, e^{2\pi i b t})\colon t \in \mathbb{R}\},$$

where a, $b \in \mathbb{R} \sim \{0\}$, a/b is irrational, and the topology of M is that of $\mathbb{R}$, a point $t \in \mathbb{R}$ being associated with the point $(e^{2\pi i a t}, e^{2\pi i b t}) \in M$.

Now let G be a Lie group and H a Lie group that algebraically is a subgroup of G. The Lie group H is called a *Lie subgroup* of G if H is a submanifold of G.

In the above example, M is a Lie subgroup of the Lie group $N = \mathbb{T} \times \mathbb{T}$. Thus the topology of a Lie subgroup H of a Lie group G need not be the relative topology and H need not be closed in G.

The next result relates Lie subgroups of G to the subalgebras of the Lie algebra $\mathfrak{g}$ of G.

B15. *Let G be a Lie group with Lie algebra $\mathfrak{g}$. If H is a Lie subgroup of G with Lie algebra $\mathfrak{h}$, then the subset*

$$\mathfrak{h}' = \{X \in \mathfrak{g}\colon \textit{the map } t \to \exp_{\mathfrak{g}}(tX) \textit{ is continuous from } \mathbb{R} \textit{ into } H\}$$

is a subalgebra of $\mathfrak{g}$ that is canonically isomorphic to $\mathfrak{h}$, and, with $\mathfrak{h}$, $\mathfrak{h}'$ identified, we have $\exp_{\mathfrak{h}} = (\exp_{\mathfrak{g}})|_{\mathfrak{h}}$. Further, the map $H \to \mathfrak{h}$ is a bijection from the set of connected, Lie subgroups of G onto the set of subalgebras of $\mathfrak{g}$.

B16. *Let H be a closed subgroup of a Lie group G. Then there exists a unique analytic manifold structure on H such that H is a Lie subgroup of G; further, the manifold topology on H is the relative topology induced by G.*

We note that if G_1, G_2 are closed subgroups of G with Lie algebras $\mathfrak{g}_1$, $\mathfrak{g}_2$ respectively, then $G_1 \cap G_2$ is also a closed subgroup of G, and, using B15, the Lie algebra of $G_1 \cap G_2$ is $\mathfrak{g}_1 \cap \mathfrak{g}_2$.

B17. More on Lie subgroups. *Let G be a Lie group and H a Lie subgroup of G. Let $\mathfrak{g}$, $\mathfrak{h}$ be the Lie algebras of G and H respectively. Then*

(i) $\mathfrak{h} = \{X \in \mathfrak{g}\colon \exp tX \in H$ *for all* $t \in \mathbb{R}\}$;

(ii) *if G and H are connected, then H is a normal subgroup of G if and only if $\mathfrak{h}$ is an ideal in $\mathfrak{g}$;*

(iii) *if G is simply connected and H is connected and is a normal subgroup of G, then H is closed in G.*

B18. Quotient Lie groups. Let G be a Lie group and H a closed, normal subgroup of G. Let $\mathfrak{g}, \mathfrak{h}$ be the Lie algebras of G and H respectively. Since H_e is a Lie subgroup of G_e, it follows from B17(ii) that $\mathfrak{h}$ is an ideal of $\mathfrak{g}$.

Let $\pi\colon G \to G/H$ be the quotient map and $\mathfrak{m}$ a complementary subspace for $\mathfrak{h}$ in $\mathfrak{g}$. Observing that the Lie group topology of H is the relative topology and using B10, we can find an open neighbourhood U of 0 in $\mathfrak{m}$ such that $Q_U = \pi \circ (\exp|_U)$ is a homeomorphism onto an open neighbourhood of the identity in G/H. Then (U, Q_U^{-1}) is a chart, and by "translating" this chart by members of G/H, we obtain an analytic structure on G/H for which G/H is a Lie group. Clearly $T = \exp \circ Q_U^{-1}$ is a *local cross section* for $V = Q_U(U)$ in the sense that $T\colon V \to G$ is analytic and $\pi \circ T(v) = v$ $(v \in V)$. Further the homomorphism π is analytic, and, using B11(i) and B17(i), one readily shows that $\ker \pi_*(e) = \mathfrak{h}$ and that the Lie algebra of G/H is, canonically, $\mathfrak{g}/\mathfrak{h}$.

Now let L be a Lie group and $\Phi\colon G \to L$ a continuous (and therefore analytic) homomorphism. As in the case of $\Phi = \pi$ above, the Lie algebra of $\ker \Phi$ is $\ker \Phi_*(e)$. By considering the connected Lie subgroup of L whose Lie algebra is $\Phi_*(e)(\mathfrak{g})$, we see that when G is connected, $\Phi(G)$ is a Lie subgroup of L with $\Phi_*(e)(\mathfrak{g})$ as Lie algebra, and the canonical group isomorphism from $G/\ker \Phi$ onto the Lie subgroup $\Phi(G)$ is actually analytic.

B19. Matrix groups. Let $\mathbb{F}$ be either $\mathbb{R}$ or $\mathbb{C}$ and $n \in \mathbb{P}$. Let $M_n(\mathbb{F})$ be the algebra of $n \times n$ matrices with entries in $\mathbb{F}$ and $\mathrm{GL}(n, \mathbb{F})$ the group of invertible elements of $M_n(\mathbb{F})$. The determinant, trace, and transpose of a matrix $A \in M_n(\mathbb{F})$ are denoted by $\det A, \operatorname{tr} A$, and A' respectively. If $\mathbb{F} = \mathbb{C}$, we define $A^* = \overline{A}'$.

The group $\mathrm{GL}(n, \mathbb{R})$ is an open subset of $M_n(\mathbb{R})$ $(= \mathbb{R}^{n^2})$ and is clearly a Lie group in the relative topology. It is standard (and easy to prove) that the one-parameter subgroups of $\mathrm{GL}(n, \mathbb{R})$ are all of the form $t \to e^{tA}$ $(A \in M_n(\mathbb{R}))$ so that the Lie algebra of $\mathrm{GL}(n, \mathbb{R})$ as a set is just $M_n(\mathbb{R})$ with $\exp A = e^A$ $(A \in M_n(\mathbb{R}))$. Using the power series expansion for exp in $M_n(\mathbb{R})$, together with

(5) of B9, we see that the Lie algebra vector space structure on $M_n(\mathbb{R})$ is the usual one, and that the Lie product on $M_n(\mathbb{R})$ is given by

$$[A, B] = AB - BA.$$

Similarly, $\mathrm{GL}(n, \mathbb{C})$ is a Lie group with $M_n(\mathbb{C})$ as Lie algebra. Every closed subgroup of $\mathrm{GL}(n, \mathbb{R})$ (or $\mathrm{GL}(n, \mathbb{C})$) is also a Lie group (B16) and using (i) of B17, the Lie algebras of the (closed) subgroups discussed below can be determined. For the connectivity properties of these subgroups see [**He**, Chapter 10, §2].

The group $\mathrm{SL}(n, \mathbb{R})$ is defined

$$\mathrm{SL}(n, \mathbb{R}) = \{A \in \mathrm{GL}(n, \mathbb{R})\colon \det A = 1\}.$$

The group is connected, and its Lie algebra is $\mathrm{sl}(n, \mathbb{R})$, where

$$\mathrm{sl}(n, \mathbb{R}) = \{A \in M_n(\mathbb{R})\colon \operatorname{tr} A = 0\}.$$

The quotient group $\mathrm{SL}(2, \mathbb{R})/\{-I, I\}$ is denoted by $\mathrm{PSL}(2, \mathbb{R})$. Since $\{-I, I\}$ is discrete, the Lie algebra of $\mathrm{PSL}(2, \mathbb{R})$ is the same as that of $\mathrm{SL}(2, \mathbb{R})$, viz., $\mathrm{sl}(2, \mathbb{R})$. The group $\mathrm{PSL}(2, \mathbb{R})$ is discussed in more detail in Chapter 3.

The group $\mathrm{SL}(n, \mathbb{C})$ is defined in the obvious way.

The (compact) *orthogonal group* $\mathrm{O}(n, \mathbb{R})$ (or $\mathrm{O}(n)$) is defined

$$\mathrm{O}(n, \mathbb{R}) = \{A \in \mathrm{GL}(n, \mathbb{R})\colon A' = A^{-1}\}.$$

The identity component of $\mathrm{O}(n, \mathbb{R})$ is $\mathrm{SO}(n, \mathbb{R})$, where

$$\mathrm{SO}(n, \mathbb{R}) = \{A \in \mathrm{O}(n, \mathbb{R})\colon \det A = 1\} = \mathrm{O}(n, \mathbb{R}) \cap \mathrm{SL}(n, \mathbb{R}).$$

The Lie algebra of $\mathrm{SO}(n, \mathbb{R})$ is $\mathrm{so}(n, \mathbb{R})$, where

$$\mathrm{so}(n, \mathbb{R}) = \{A \in M_n(\mathbb{R})\colon A' = -A\}.$$

The *unitary group* $\mathrm{U}(n)$ is defined

$$\mathrm{U}(n) = \{A \in \mathrm{GL}(n, \mathbb{C})\colon AA^* = I = A^*A\}.$$

The group $\mathrm{U}(n)$ is connected. The subgroup $\mathrm{SU}(n)$ of $\mathrm{U}(n)$ is defined

$$\mathrm{SU}(n) = \{A \in \mathrm{U}(n)\colon \det A = 1\} = \mathrm{U}(n) \cap \mathrm{SL}(n, \mathbb{C}).$$

The Lie algebra of $\mathrm{SU}(n)$ is $\mathrm{su}(n)$, where

$$\mathrm{su}(n) = \{X \in M_n(\mathbb{C})\colon X^* = -X,\ \operatorname{tr} X = 0\}.$$

Of particular importance for us is $\mathrm{SU}(2)$. It is readily checked that the map

$$(x_1, x_2, x_3, x_4) \to \begin{bmatrix} x_1 + ix_2 & -x_3 + ix_4 \\ x_3 + ix_4 & x_1 - ix_2 \end{bmatrix}$$

is a homeomorphism from $\mathbb{S}^3$ onto $\mathrm{SU}(2)$, so that $\mathrm{SU}(2)$ is simply connected. Further, $\mathrm{so}(3, \mathbb{R})$ is isomorphic to $\mathrm{su}(2)$, so that $\mathrm{SU}(2)$ can be identified with the simply connected covering group of $\mathrm{SO}(3, \mathbb{R})$. (The canonical homomorphism from $\mathrm{SU}(2)$ onto $\mathrm{SO}(3, \mathbb{R})$ is given explicitly in [**HR2**, (29.35), (29.36)].)

B20. The Lie group $\operatorname{Aut} G$**.** Let G be a connected Lie group. Suppose, first, that G is also simply connected. If $\Phi \in \operatorname{Aut} G$, then Φ is an analytic diffeomorphism on G and $\Phi_*(e) \in \operatorname{Aut} \mathfrak{g}$. Conversely, if $T \in \operatorname{Aut} \mathfrak{g}$, then, using B7, (ii) of B11, and the invertibility of T, we can find a unique $\Phi \in \operatorname{Aut} G$ with $\Phi_*(e) = T$. Using the chain rule, the map $\Phi \to \Phi_*(e)$ is an isomorphism from $\operatorname{Aut} G$ onto $\operatorname{Aut} \mathfrak{g}$, where $\operatorname{Aut} G$ is given its canonical topology and $\operatorname{Aut} \mathfrak{g}$ is given the relative topology it inherits as a subset of $\mathrm{GL}(\mathfrak{g})$. Since $\operatorname{Aut} \mathfrak{g}$ is a closed subgroup of $\mathrm{GL}(\mathfrak{g})$, it is a Lie group, and so the above isomorphism enables us to impose a Lie group structure on the topological group $\operatorname{Aut} G$.

Now suppose that G is not necessarily simply connected, and let $(\tilde{G}, \pi)$ be the simply connected covering group of G. Each $\Phi \in \operatorname{Aut} G$ induces a continuous local homomorphism between neighbourhoods of e in $\tilde{G}$, and since $\tilde{G}$ is simply connected, this local homomorphism extends to a (unique) element $\tilde{\Phi}$ of $\operatorname{Aut} \tilde{G}$. The map $\Phi \to \tilde{\Phi}$ is an isomorphism of $\operatorname{Aut} G$ onto the closed subgroup

$$K = \{\Psi \in \operatorname{Aut} \tilde{G} \colon \Psi(\ker \pi) = \ker \pi\},$$

of the Lie group $\operatorname{Aut} \tilde{G}$, and $\operatorname{Aut} G$ inherits the Lie group structure of K.

Since the Lie algebra of $\operatorname{Aut} \mathfrak{g}$ is $\operatorname{Der} \mathfrak{g}$ (B5) and $\operatorname{Aut} G$ is isomorphic to a Lie subgroup of $\operatorname{Aut} \tilde{G} \cong \operatorname{Aut} \mathfrak{g}$, it follows that *the Lie algebra of* $\operatorname{Aut} G$ *can be identified with a subalgebra of* $\operatorname{Der} \mathfrak{g}$*, and can be equated with* $\operatorname{Der} \mathfrak{g}$ *if* G *is simply connected.*

B21. Connected abelian Lie groups. Let G be a connected, abelian Lie group. Then G *is of the form* $\mathbb{R}^r \times \mathbb{T}^s$. [The Lie algebra of G is abelian (B9) and so is the Lie algebra of the abelian Lie group $\mathbb{R}^n$ for some n. Thus $\mathbb{R}^n$ is the simply connected covering group of G, so that G is a quotient $\mathbb{R}^n/D$ with D a discrete subgroup of $\mathbb{R}^n$. The group D is isomorphic to some $\mathbb{Z}^s$, giving the desired result.]

If $G = \mathbb{T}^n$, then the covering projection $p \colon \mathbb{R}^n \to \mathbb{T}^n$ can be taken to be the obvious one:

$$p(x_1, \ldots, x_n) = (e^{2\pi i x_1}, \ldots, e^{2\pi i x_n}).$$

It follows from B20 that $\operatorname{Aut} \mathbb{T}^n$ can be identified with the group

$$\{T \in \mathrm{GL}(n, \mathbb{R}) \colon T(\mathbb{Z}^n) = \mathbb{Z}^n\},$$

which is readily shown to be the group of $n \times n$, integer-valued matrices of determinant ± 1 [**HR1**, (26.18)]. In particular, $\operatorname{Aut} \mathbb{T}^n$ *is discrete.*

B22. Lie semidirect products. Let $H \times_\rho K$ be a semidirect product of connected Lie groups H, K. Since $\operatorname{Aut} H$ is a Lie group and $\rho \colon K \to \operatorname{Aut} H$ is continuous, the map ρ is actually analytic, and it follows that the map $(x, y) \to \rho(y)x$ from $H \times K$ into H is analytic. Further, $G = H \times_\rho K$ is a Lie group when given the product manifold stucture.

Let $\mathfrak{h}$, $\mathfrak{k}$, $\mathfrak{g}$ be the Lie algebras of H, K, $H \times_\rho K$ respectively. Then $\mathfrak{h}$ is an ideal and $\mathfrak{k}$ a subalgebra of $\mathfrak{g}$, and using B17, $\mathfrak{h} \cap \mathfrak{k} = \{0\}$. From dimensional considerations, $\mathfrak{g} = \mathfrak{h} \oplus \mathfrak{k}$. Thus (B5) $\mathfrak{g}$ is a semidirect product $\mathfrak{h} \times_\Phi \mathfrak{k}$.

It remains to specify $\Phi\colon \mathfrak{k} \to \operatorname{Der}\mathfrak{h}$ in terms of ρ. Recalling that $\rho(y)(x) = yxy^{-1} = \alpha_y(x)$, we have, for $Y \in \mathfrak{k}$, using B12,

$$(\rho(\exp tY))_*(e) = (\alpha_{\exp tY})_*(e)|_{\mathfrak{h}} = \operatorname{Ad}(\exp tY)|_{\mathfrak{h}} = e^{t\,\operatorname{ad} Y}|_{\mathfrak{h}}$$

so that $\rho_*(e)(Y) = (\operatorname{ad} Y)|_{\mathfrak{h}} = \Phi(Y)$.

The following gives a partial converse to the above result. A related result is given in B44.

Let G be a simply connected Lie group with Lie algebra $\mathfrak{g}$. Suppose that $\mathfrak{g}$ is a semidirect product $\mathfrak{h} \times_\Phi \mathfrak{k}$, and let H, K be the connected Lie subgroups of G with Lie algebras $\mathfrak{h}$, $\mathfrak{k}$ respectively. Then H, K are simply connected, closed subgroups of G, and G is a semidirect product $H \times_\rho K$.

B23. Nilpotent and solvable Lie groups and Lie algebras. A Lie group G is nilpotent [solvable] if it is nilpotent [solvable] in the sense of Appendix A.

Let G be a nilpotent Lie group with lower central series $\{C^iG\}$. Clearly $[G, (C^iG)^-] \subset (C^{i+1}G)^-$ so that $\{(C^iG)^-\}$ is a central series of *closed* subgroups of G. Similarly, if H is a solvable Lie group and $\{D_iH\}$ is the derived series for H, then $\{(D_iH)^-\}$ is a normal series with $(D_iH)^-$ normal in H and $(D_iH)^-/(D_{i+1}H)^-$ abelian.

Now let $\mathfrak{g}$ be a Lie algebra. The *lower central series* $\{\mathscr{C}^i\mathfrak{g}\}$ is the decreasing sequence of ideals in $\mathfrak{g}$ defined inductively as follows: $\mathscr{C}^1\mathfrak{g} = \mathfrak{g}$, $\mathscr{C}^{i+1}\mathfrak{g} = [\mathscr{C}^i\mathfrak{g}, \mathfrak{g}]$. The *derived series* $\{\mathscr{D}_i\mathfrak{g}\}$ of $\mathfrak{g}$ is the decreasing sequence of ideals in $\mathfrak{g}$ defined inductively as follows: $\mathscr{D}_1\mathfrak{g} = \mathfrak{g}$, $\mathscr{D}_{i+1}\mathfrak{g} = [\mathscr{D}_i\mathfrak{g}, \mathscr{D}_i\mathfrak{g}]$. Each $\mathscr{C}^i\mathfrak{g}$, $\mathscr{D}_i\mathfrak{g}$ is *characteristic* in the sense of being invariant for every automorphism of $\mathfrak{g}$.

The Lie algebra $\mathfrak{g}$ is said to be *nilpotent* [*solvable*] if, for some p, $\mathscr{C}^p\mathfrak{g} = \{0\}$ [$\mathscr{D}_p\mathfrak{g} = \{0\}$]. Since $\mathscr{D}_i\mathfrak{g} \subset \mathscr{C}^i\mathfrak{g}$, every nilpotent Lie algebra is solvable.

The *class* of a nilpotent Lie algebra $\mathfrak{g}$ is the smallest integer $p \geq 0$ such that $\mathscr{C}^{p+1}\mathfrak{g} = \{0\}$.

An important source for nilpotent and solvable Lie groups and Lie algebras is the space $\Delta(n)$ of upper triangular matrices in $M_n(\mathbb{R})$. For example, let

$$\begin{aligned} G_1 &= \{A \in \Delta(n)\colon \text{for some } k \neq 0, A_{ii} = k \ (1 \leq i \leq n)\}, \\ G_2 &= \{A \in \Delta(n)\colon A_{ii} \neq 0 \ (1 \leq i \leq n)\}, \\ \Delta'(n) &= \{A \in \Delta(n)\colon \text{for some } k, A_{ii} = k \text{ for all } i\}. \end{aligned}$$

Then $\Delta(n)$ [$\Delta'(n)$] is a solvable [nilpotent] Lie subalgebra of $M_n(\mathbb{R})$, while G_2 [G_1] is a solvable [nilpotent] (closed) Lie subgroup of $\mathrm{GL}(n, \mathbb{R})$.

Note that if A, $B \in \Delta(n)$, then $[A, B]_{ii} = 0$ for all i. Thus $[\Delta(n), \Delta(n)] \subset \Delta^0(n)$, where $\Delta^0(n)$ is the nilpotent Lie subalgebra of $M_n(\mathbb{R})$ given by

$$\Delta^0(n) = \{A \in \Delta(n)\colon A_{ii} = 0 \ (1 \leq i \leq n)\}.$$

This latter result is true in greater generality (B35).

B24. *Let G be a connected Lie group with Lie algebra $\mathfrak{g}$. Then*
(i) *for each i, C^iG [D_iG] is a Lie subgroup of G with $\mathscr{C}^i\mathfrak{g}$ [$\mathscr{D}_i\mathfrak{g}$] as Lie algebra;*
(ii) *G is nilpotent [solvable] if and only if $\mathfrak{g}$ is nilpotent [solvable].*

We now state some of the properties of solvable and nilpotent Lie algebras.

For the rest of this Appendix, every Lie algebra is assumed to be finite dimensional.

The next result shows that, under reasonable circumstances, solvable Lie algebras can be realised as algebras of upper triangular matrices.

B25 (Lie's Theorem). *Let V be a nontrivial, finite-dimensional vector space over $\mathbb{C}$, and let $\mathfrak{g}$ be a solvable Lie subalgebra of $L(V)$. Then*

(i) *there exists $v \in V$ that is an eigenvector for all $T \in \mathfrak{g}$;*

(ii) *there exists a basis for V with respect to which every $T \in \mathfrak{g}$ is an upper triangular matrix.*

Using the exponential map, the Lie group version of Lie's Theorem, stated below, easily follows.

B26. *Let G be a connected, solvable Lie subgroup of $\mathrm{GL}(V)$ where V is as in B25. Then there exists a basis for V with respect to which every element of G is an upper triangular matrix.*

The next result gives a version of B25 for nilpotent Lie algebras. Recall that $T \in L(V)$ is *nilpotent* if $T^k = 0$ for some $k \in \mathbb{P}$.

B27 (Engel's Theorem). *Let V be a nontrivial, finite-dimensional vector space over $\mathbb{F}$ $(= \mathbb{R}$ or $\mathbb{C})$ and $\mathfrak{g}$ a Lie subalgebra of $L(V)$ whose elements are all nilpotent. Then*

(i) *there exists $v \neq 0$ in V such that $Tv = 0$ for all $T \in \mathfrak{g}$;*

(ii) *there exists a basis for V with respect to which every $T \in \mathfrak{g}$ is upper triangular with all diagonal elements equal to 0;*

(iii) *$\mathfrak{g}$ is a nilpotent Lie algebra and generates a nilpotent associative subalgebra of $L(V)$.*

We omit the obvious Lie group version of Engel's Theorem.

The next two results are easy corollaries to Engel's Theorem.

B28. *A Lie algebra $\mathfrak{g}$ over $\mathbb{F}$ is nilpotent if and only if $\operatorname{ad} X$ is nilpotent for all $X \in \mathfrak{g}$.*

B29. *Let $\mathfrak{g}$ be a Lie algebra over $\mathbb{F}$. Suppose that $\mathfrak{h}$ is an ideal of $\mathfrak{g}$ such that $\mathfrak{g}/\mathfrak{h}$ is nilpotent and, for each $X \in \mathfrak{g}$, $\operatorname{ad} X|_{\mathfrak{h}}$ is nilpotent. Then $\mathfrak{g}$ is nilpotent.*

The next result is simple to prove.

B30. *Let $\mathfrak{g}$ be a Lie algebra and $\mathfrak{h}$ be an ideal of $\mathfrak{g}$. Then $\mathfrak{g}$ is solvable if and only if $\mathfrak{h}$ and $\mathfrak{g}/\mathfrak{h}$ are solvable.*

B31. Weight spaces and the Jordan canonical form. Let W be a finite-dimensional vector space over $\mathbb{C}$, and let $\mathfrak{g}$ be a Lie subalgebra of $L(W)$.

A *weight* of $\mathfrak{g}$ is a (linear) functional $\psi\colon \mathfrak{g} \to \mathbb{C}$ such that the *weight space* W_ψ, where

$$W_\psi = \{v \in W\colon \text{for each } T \in \mathfrak{g}, (T - \psi(T)I)^n v = 0 \text{ for some } n \in \mathbb{P}\},$$

is not $\{0\}$.

When $T \in L(W)$ and $\mathfrak{g} = \mathbb{C}T$, then the weights of $\mathfrak{g}$ can be identified, in the obvious way, with the eigenvalues $\lambda_1, \ldots, \lambda_k$ of T. The theory of the Jordan canonical form (e.g., Herstein [**1**]) shows that if W_i is the weight space associated with λ_i, then W_i is T-invariant, W is the direct sum of the weight spaces W_i, and, for each i, a basis can be chosen for W_i such that, with respect to this basis, $T|_{W_i}$ is an upper triangular matrix of the form:

$$\text{(1)} \qquad \begin{bmatrix} \lambda_i & & & \\ 0 & \lambda_i & * & \\ \vdots & & & \\ 0 & \ldots & 0 & \lambda_i \end{bmatrix}.$$

The *semisimple part of* T is the element $S \in L(W)$ given by $S\xi_i = \lambda_i \xi_i$ ($\xi_i \in W_i$, $1 \le i \le k$). The *nilpotent part of* T is the transformation $N = T - S$. From (1), we see that N is nilpotent; further

$$\text{(2)} \qquad T = N + S, \ N \text{ is nilpotent, and } NS = SN.$$

There exists [**SW**, p. 237] a polynomial p in one variable such that $p(0) = 0$ and $p(T) = S$.

We note that the semisimple [nilpotent] part of $\operatorname{ad} T$ ($\in L(L(W))$) is $\operatorname{ad} S$ [$\operatorname{ad} N$].

We will have occasion to use the weight spaces associated with D, where D is a derivation on a Lie algebra $\mathfrak{g}$ over $\mathbb{C}$. Let $\alpha_1, \ldots, \alpha_r$ be the eigenvalues of D, and let $\mathfrak{g}_i$ be the weight space associated with α_i. Then we have the following result.

For each pair i, j, we have $[\mathfrak{g}_i, \mathfrak{g}_j] \subset \mathfrak{g}_k$ if $\alpha_i + \alpha_j = \alpha_k$, and $[\mathfrak{g}_i, \mathfrak{g}_j] = \{0\}$ if $\alpha_i + \alpha_j \notin \{\alpha_1, \ldots, \alpha_r\}$.

The results above on weight spaces leading up to the matrix representation (1) extend to the context of nilpotent Lie subalgebras of $L(W)$.

B32. *Let $\mathfrak{g}$ be a nilpotent Lie subalgebra of $L(W)$, and let $\psi_1, \ldots, \psi_n$ be the weights of $\mathfrak{g}$ with corresponding weight spaces $W_1, \ldots, W_n$. Then W is the direct sum of the subspaces W_i, and each W_i is $\mathfrak{g}$-invariant and has a basis with respect to which the matrix of $T|_{W_i}$ ($T \in \mathfrak{g}$) has the upper triangular form:*

$$\begin{bmatrix} \psi_i(T) & & & \\ 0 & \psi_i(T) & * & \\ \vdots & & & \\ 0 & \ldots & 0 & \psi_i(T) \end{bmatrix}$$

The above theorem is still valid if W is a finite-dimensional vector space over $\mathbb{R}$ and the eigenvalues of the every $\operatorname{ad} X$ ($X \in \mathfrak{g}$) are assumed to be real. Similarly, the "real" version of the following result holds.

B33. *Let $\mathfrak{h}$ be a nilpotent Lie algebra of derivations on a Lie algebra $\mathfrak{g}$ over $\mathbb{C}$. Let Ψ be the set of weights of $\mathfrak{h}$ ($\subset L(\mathfrak{g})$), and, for $\psi \in \Psi$, let $\mathfrak{g}_\psi$ be the weight space associated with ψ. Then each $\mathfrak{g}_\psi$ is $\mathfrak{h}$-invariant, $\mathfrak{g} = \bigoplus\{\mathfrak{g}_\psi : \psi \in \Psi\}$, and there exists a basis for $\mathfrak{g}_\psi$ such that the matrix of $T|_{\mathfrak{g}_\psi}$ ($T \in \mathfrak{h}$) is upper triangular with diagonal entries equal to $\psi(T)$. Finally, if ϕ, $\psi \in \Psi$, then $[\mathfrak{g}_\phi, \mathfrak{g}_\psi] \subset \mathfrak{g}_{\phi+\psi}$ if $\phi + \psi \in \Psi$ and is $\{0\}$ otherwise.*

The following interesting result on the centre of nilpotent Lie groups appears in [**Ho**, p. 188].

B34. *Let G be a connected, nilpotent Lie group. Then the centre $Z(G)$ of G is connected.*

B35. Radicals of Lie algebras and Lie groups. Let $\mathfrak{g}$ be a Lie algebra over $\mathbb{F}$. The following result is of fundamental importance.

There exists a largest solvable [nilpotent] ideal $\mathfrak{r}$ $[\mathfrak{n}]$ of $\mathfrak{g}$.

The ideal $\mathfrak{r}$ $[\mathfrak{n}]$ is called the *radical* [*nil-radical*] of $\mathfrak{g}$.

Let G be a Lie group with $\mathfrak{g}$ as Lie algebra. Then the *radical* [*nil-radical*] of G is the connected Lie subgroup R $[N]$ of G with $\mathfrak{r}$ $[\mathfrak{n}]$ as Lie algebra. Note that R $[N]$ is a solvable [nilpotent] subgroup of G (B24). By the maximality property of $\mathfrak{r}$ $[\mathfrak{n}]$, we see that $R^- = R$ $[N^- = N]$. Thus R $[N]$ is the maximal (closed) connected, solvable [nilpotent] subgroup of G, and, of course, is normal in G.

More generally, $\mathfrak{r}$, $\mathfrak{n}$ $[R, N]$ are *characteristic* in the sense of being invariant for every element of $\operatorname{Aut}\mathfrak{g}$ (or $\operatorname{Der}\mathfrak{g}$) $[\operatorname{Aut} G]$.

B36. *Let $\mathfrak{g}$ be a finite-dimensional Lie algebra over $\mathbb{F}$, and let $D \in \operatorname{Der}\mathfrak{g}$. Then $D\mathfrak{g} \cap \mathfrak{r} \subset \mathfrak{n}$. In particular, $D\mathfrak{r} \subset \mathfrak{n}$.*

B37. In the above notations, $[\mathfrak{g}, \mathfrak{r}] \subset \mathfrak{n}$.

B38. Semisimplicity and the Killing form. Let $\mathfrak{g}$ be a (finite-dimensional) Lie algebra over $\mathbb{F}$. The Lie algebra $\mathfrak{g}$ is called *semisimple* if its radical is $\{0\}$. It is easy to check that $\mathfrak{g}$ is semisimple if and only if it contains no nontrivial abelian ideals. In particular, if $\mathfrak{g}$ is semisimple, then $Z(\mathfrak{g}) = \{0\}$, and so $\operatorname{ad}_\mathfrak{g}$ is one-to-one. Also, if $\mathfrak{h}$ is a Lie algebra with radical $\mathfrak{r}$, then $\mathfrak{h}/\mathfrak{r}$ is semisimple (B30).

A Lie group G is called *semisimple* if its radical is the trivial group $\{e\}$. Clearly, G is semisimple if and only if its Lie algebra is semisimple (B35). Further, $Z(G)_e = \{e\}$, so that $Z(G)$ is a discrete subgroup of G if G is semisimple.

Recall that a *bilinear form* on a vector space V over a field $\mathbb{F}$ is a bilinear map $\Phi : V \times V \to \mathbb{F}$. A bilinear form Φ on V is called *symmetric* if $\Phi(X, Y) = \Phi(Y, X)$ ($X, Y \in V$) and is called *nondegenerate* if it satisfies

$$\{0\} = \{X \in V : \Phi(X, Y) = 0 \text{ for all } Y \in V\}.$$

The *Killing form* of the Lie algebra $\mathfrak{g}$ is the bilinear form $B : \mathfrak{g} \times \mathfrak{g} \to \mathbb{F}$ defined as follows:

$$B(X, Y) = \operatorname{tr}(\operatorname{ad} X \operatorname{ad} Y).$$

If $\mathfrak{a} \subset \mathfrak{g}$, define

$$\mathfrak{a}^\perp = \{X \in \mathfrak{g} \colon B(X, A) = 0 \text{ for all } A \in \mathfrak{a}\}.$$

Clearly, $\mathfrak{a}^\perp$ is a subspace of $\mathfrak{g}$.

Obviously, B is symmetric. We now state some simple results involving B. The radical of $\mathfrak{g}$ is denoted by $\mathfrak{r}$.

B39. (i) *If* $T \in \operatorname{Aut} \mathfrak{g}$ *and* $D \in \operatorname{Der} \mathfrak{g}$, *then*

$$B(TX, TY) = B(X, Y), \quad B(DX, Y) = -B(X, DY) \qquad (X, Y \in \mathfrak{g});$$

(ii) $B(X, Y) = B(Y, X)$ *and* $B(X, [Y, Z]) = B(Y, [Z, X]) = B(Z, [X, Y])$ *for all* X, Y, $Z \in \mathfrak{g}$;

(iii) *if* $\mathfrak{a}$ *is an ideal of* $\mathfrak{g}$, *then* $\mathfrak{a}^\perp$ *is an ideal of* $\mathfrak{g}$;

(iv) $[\mathfrak{g}, \mathfrak{g}]^\perp = \mathfrak{r}$;

(v) $\mathfrak{g}$ *is semisimple if and only if* B *is nondegenerate*;

(vi) *if* $\mathfrak{g}$ *is semisimple and* $\mathfrak{a}$ *is an ideal of* $\mathfrak{g}$, *then*

$$\mathfrak{g} = \mathfrak{a} \oplus \mathfrak{a}^\perp, \qquad (\mathfrak{a}^\perp)^\perp = \mathfrak{a}.$$

A Lie algebra $\mathfrak{h}$ is called *simple* if $\mathfrak{h}$ is nonabelian and the only ideals of $\mathfrak{h}$ are $\{0\}$ and $\mathfrak{h}$. The next result is a consequence of (vi) above.

B40. *Let* $\mathfrak{g}$ *be semisimple. Then* $\mathfrak{g}$ *is the direct sum* $\mathfrak{g}_1 \oplus \cdots \oplus \mathfrak{g}_n$, *where* $\mathfrak{g}_1, \ldots, \mathfrak{g}_n$ *are the simple ideals in* $\mathfrak{g}$. *Further, every ideal of* $\mathfrak{g}$ *is a direct sum of* $\mathfrak{g}_i$*'s.*

The above result can be expressed as asserting the "complete reducibility" of $\mathfrak{g}$ as a $\mathfrak{g}$-module under the adjoint representation. The following much more general result holds [**B**, Chapter 1, §6], [**SW**, Chapter 12, §4].

B41. *Let* $\mathfrak{g}$ *be semisimple, and let* V *be a finite-dimensional* $\mathfrak{g}$*-module. Then* V *is a direct sum* $\bigoplus_{i=1}^r V_i$ *of irreducible* $\mathfrak{g}$*-submodules* V_i.

Let $\mathfrak{g}$ be a Lie algebra. Since $\operatorname{ad} \mathfrak{g}$ is a Lie subalgebra of $L(\mathfrak{g})$, there exists a unique connected Lie subgroup of $\mathrm{GL}(\mathfrak{g})$ with $\operatorname{ad} \mathfrak{g}$ as Lie algebra. This Lie subgroup is denoted by $\operatorname{Int} \mathfrak{g}$. Note that if G is a connected Lie group with $\mathfrak{g}$ as Lie algebra, then (B12) we have $\operatorname{Int} \mathfrak{g} = \operatorname{Ad} G$ $(= G/\ker \operatorname{Ad})$.

B42. *Let* $\mathfrak{g}$ *be semisimple. Then*

(i) $[\mathfrak{g}, \mathfrak{g}] = \mathfrak{g}$;

(ii) *every ideal and every quotient algebra of* $\mathfrak{g}$ *is semisimple*;

(iii) $\operatorname{Der} \mathfrak{g} = \operatorname{ad} \mathfrak{g}$, *and if* I *is the identity map on* $\mathfrak{g}$, *then* $\operatorname{Int} \mathfrak{g} = (\operatorname{Aut} \mathfrak{g})_I$;

(iv) *if* G *is a connected, semisimple Lie group with* $\mathfrak{g}$ *as Lie algebra, then* $(\operatorname{Aut} G)_i = I(G)$ ((3.3)).

The next result is known as the *Levi-Malcev Theorem*. Proofs of this important theorem are given in [**J**, Chapter 3, §9] and [**B**, Chapter 1, §6.8]. The subalgebra $\mathfrak{s}$ below can be taken to be *any* maximal, semisimple subalgebra of $\mathfrak{g}$. The decomposition $\mathfrak{s} \oplus w$ of the theorem is called the *Levi-Malcev Decomposition*.

B43. *Let* $\mathfrak{r}$ *be the radical of the Lie algebra* $\mathfrak{g}$. *Then there exists a (semisimple) subalgebra* $\mathfrak{s}$ *of* $\mathfrak{g}$ *such that* $\mathfrak{g}$ *is the vector space direct sum* $\mathfrak{s} \oplus \mathfrak{r}$. *(Thus* $\mathfrak{g}$ *is a semidirect product* $\mathfrak{r} \times_\Phi \mathfrak{s}$ *in the canonical way* (B5).)

It follows from the above result and B22 that *if* G *is a simply connected Lie group with radical* R, *then* G *is a semidirect product* $R \times_\rho S$.

B44. *Let* G *be a simply connected Lie group,* H *a closed, connected, normal subgroup of* G, *and* $Q: G \to G/H$ *the quotient map. Then there exists an analytic map* $\rho: G/H \to G$ *such that* $Q \circ \rho$ *is the identity map on* G/H. *For any such map* ρ, *the map* $(h, m) \to h\rho(m)$ *is an analytic diffeomorphism from* $H \times G/H$ *onto* G. *Finally, both* H *and* G/H *are simply connected.*

The next result is a corollary of the preceding theorem.

B45. *Let* G *be a connected Lie group containing a closed, simply connected normal subgroup* H *with* G/H *simply connected. Then* G *is simply connected.*

We now turn to two other results involving semidirect products. The first of these essentially appears in [**Ho**, Chapter 12, Theorem 2.2].

B46. *Let* S *be a simply connected, solvable Lie group with* $\mathfrak{s}$ *as Lie algebra. Then there exists a basis* $\{X_1, \ldots, X_n\}$ *for* $\mathfrak{s}$ *such that if* $\mathfrak{s}_i = \operatorname{Span}\{X_1, \ldots, X_i\}$, *then* $\mathfrak{s}_i$ *is an ideal in* $\mathfrak{s}_{i+1}$, *and if* H_i *is the connected Lie subgroup of* G *with* $\mathfrak{s}_i$ *as Lie algebra, then* H_i *is closed in* G *and*

(i) *for each* i, *the map* $\sum_{j=1}^{i} \alpha_j X_j \to (\exp \alpha_1 X_1) \cdots \exp(\alpha_i X_i)$ *(where* $\alpha_j \in \mathbb{R}$) *is an analytic diffeomorphism from* $\mathfrak{s}_i$ *onto* H_i;

(ii) H_{i+1} *is a semidirect product* $H_i \times_{\rho_i} L_{i+1}$, *where* $L_{i+1} = \exp(\mathbb{R}X_{i+1})$ $(0 \leq i \leq (n-1))$.

The next result is proved in [**Ho**, Chapter 12, Theorem 3.2].

B47. *Let* G *be a locally compact group, and suppose that* S *is a simply connected, solvable, closed, normal subgroup of* G *with* G/S *compact. Then there exists a compact subgroup* K *of* G *such that* G *is a semidirect product* $S \times_\rho K$.

B48. Compact Lie algebras. Let $\mathfrak{g}$ be a Lie algebra over $\mathbb{R}$, and recall (B41) that $\operatorname{Int}\mathfrak{g}$ is the connected Lie subgroup of $\operatorname{GL}(\mathfrak{g})$ with $\operatorname{ad}\mathfrak{g}$ as Lie algebra. The algebra $\mathfrak{g}$ is said to be *compact* if $\operatorname{Int}\mathfrak{g}$ is compact (in its Lie group topology). More generally, a subalgebra $\mathfrak{k}$ of $\mathfrak{g}$ is said to be *compactly embedded* in $\mathfrak{g}$ if the connected Lie subgroup K of $\operatorname{GL}(\mathfrak{g})$ with $\operatorname{ad}_{\mathfrak{g}}\mathfrak{k}$ as Lie algebra is compact. It is easily checked that $\mathfrak{k}$ is compactly embedded in $\mathfrak{g}$ if and only if K is compact in the relative topology induced by $\operatorname{GL}(\mathfrak{g})$, and that if $\mathfrak{k}$ is compactly embedded in $\mathfrak{g}$, then $\mathfrak{k}$ is a compact Lie algebra.

We shall see that the Lie algebra $\mathfrak{g}$ is compact if and only if it is the Lie algebra of a compact Lie group.

B49. *Let* $\mathfrak{g}$ *be over the field* $\mathbb{R}$ *with Killing form* B (B38). *Then*

(i) *if* $\mathfrak{g}$ *is compact, then* $\mathfrak{g}$ *is the direct sum* $Z(\mathfrak{g}) \oplus [\mathfrak{g}, \mathfrak{g}]$, *and* $[\mathfrak{g}, \mathfrak{g}]$ *is compact and semisimple*;

(ii) *if* $\mathfrak{g}$ *is semisimple, then* $\mathfrak{g}$ *is compact if and only if* $B(X, X) < 0$ *for all nonzero* $X \in \mathfrak{g}$;

(iii) *if* $\mathfrak{g}$ *is semisimple and* $\mathfrak{k}$ *is a compactly embedded subalgebra of* $\mathfrak{g}$, *then* $B(X, X) < 0$ *for all nonzero* $X \in \mathfrak{k}$.

Here are two easy consequences of the preceding result.

B50. *Let* $\mathfrak{g}$ *be a Lie algebra over* $\mathbb{R}$. *Then* $\mathfrak{g}$ *is compact if and only if it is the Lie algebra of a compact (connected) Lie group.*

A more general version of B51 is proved in [**HR2**, (29.44)]: the Lie group requirement is unnecessary.

B51. *Let* G *be a compact, connected, solvable Lie group. Then* G *is abelian.*

The next result is proved in [**He**, Chapter 2, Theorem 6.9]. See also [**HS**, pp. 221–224].

B52. *The simply connected covering group* $\tilde{G}$ *of a compact, connected, semisimple Lie group* G *is itself compact.*

It immediately follows using B50 that if *G is a connected, semisimple Lie group with compact Lie algebra, then G is compact.*

For the remainder of this Appendix, we will be discussing the theory of semisimple Lie algebras. Our account is primarily based on the treatment of this topic given by Helgason [**He**, Chapter 3 and Chapter 6], though [**Ho**] and [**SW**] have also influenced our treatment.

We shall discuss briefly the theory of Cartan subalgebras, roots and root spaces, complexifications and compact forms, the Cartan decomposition, the Cartan involution, and the Iwasawa decomposition associated with a semisimple Lie algebra $\mathfrak{g}_0$ over $\mathbb{R}$.

For the remainder of this Appendix, $\mathfrak{g}$ *will be a (finite-dimensional) semisimple Lie algebra over* $\mathbb{C}$.

B53. Cartan subalgebras and roots. A subalgebra $\mathfrak{h}$ of $\mathfrak{g}$ is said to be a *Cartan subalgebra* if

(i) $\mathfrak{h}$ is nilpotent;

(ii) if $Y \in \mathfrak{g}$ and $[Y, \mathfrak{h}] \subset \mathfrak{h}$, then $Y \in \mathfrak{h}$.

A fundamental result [**He**, Chapter 3, §3] asserts that there exists a Cartan subalgebra $\mathfrak{h}$ for $\mathfrak{g}$. Using property (i), $\mathrm{ad}_{\mathfrak{g}}\, \mathfrak{h}$ is a nilpotent subalgebra of $L(\mathfrak{g})$. The weights (B31) of $\mathrm{ad}_{\mathfrak{g}}\, \mathfrak{h}$ are called the *roots* of $\mathfrak{g}$, and the corresponding weight spaces are called the *root spaces* of $\mathfrak{g}$. (It can be shown that the roots and root spaces are essentially independent of the choice of the Cartan subalgebra $\mathfrak{h}$.) The root space associated with a root α is denoted by $\mathfrak{g}^{\alpha}$. Using (i) and (ii), it turns out that 0 is a root and $\mathfrak{g}^0 = \mathfrak{h}$ (and $\mathfrak{h}$ is abelian).

The set of roots of $\mathfrak{g}$ is denoted by R, and we set $\Delta = R \sim \{0\}$. If $\alpha \in R$, $H \in \mathfrak{h}$, we write $\alpha(H)$ in place of $\alpha(\mathrm{ad}_{\mathfrak{g}} H)$.

We now give some elementary properties of the roots and root spaces of $\mathfrak{g}$ (cf. B33). The Killing form of $\mathfrak{g}$ is, as usual, denoted by B.

B54. (i) $\mathfrak{g}$ *is the vector space direct sum*

$$(\bigoplus_{\alpha\in\Delta} \mathfrak{g}^\alpha) \oplus \mathfrak{h};$$

(ii) $[\mathfrak{g}^\alpha, \mathfrak{g}^\beta] \subset \mathfrak{g}^{\alpha+\beta}$ *if* $\alpha+\beta \in R$ *and is* $\{0\}$ *otherwise*;
(iii) *if* $\alpha, \beta \in R$ *and* $\alpha+\beta \neq 0$, *then* $B(X,Y) = 0$ *for* $X \in \mathfrak{g}^\alpha$, $Y \in \mathfrak{g}^\beta$;
(iv) $B|_{\mathfrak{h}\times\mathfrak{h}}$ *is nondegenerate*;
(v) $-\alpha$ *is a root if* α *is a root.*

Now each $\alpha \in \Delta$ is a nonzero, linear functional on $\mathfrak{h}$, and since $B|_{\mathfrak{h}\times\mathfrak{h}}$ is nondegenerate, there exists a unique element $H_\alpha \in \mathfrak{h}$ such that

$$\alpha(H) = B(H_\alpha, H) \qquad (H \in \mathfrak{h}).$$

Clearly, $H_\alpha = -H_{-\alpha}$, and $H_{\alpha+\beta} = H_\alpha + H_\beta$ if $\alpha, \beta, \alpha+\beta \in \Delta$.

B55. (i) *Let* $\alpha \in \Delta$. *Then*
(a) *if* $X \in \mathfrak{g}^\alpha$, $Y \in \mathfrak{g}^{-\alpha}$, *then* $[X,Y] = B(X,Y)H_\alpha$;
(b) Δ *spans* $\mathfrak{h}'$, *the dual space of* $\mathfrak{h}$;
(c) $\mathfrak{g}^\alpha$ *is one-dimensional*;
(d) $\alpha(H_\alpha) > 0$ *and* $\beta(H_\alpha) \in \mathbb{R}$ *for all* $\beta \in \Delta$.

(ii) *There exists a subset* $\{X_\alpha \colon \alpha \in \Delta\}$ *of* $\mathfrak{g}$ *and a subset* $\{N_{\alpha,\beta} \colon \alpha, \beta \in \Delta\}$ *of* $\mathbb{R}$ *such that*
(a) $\mathfrak{g}^\alpha = \mathbb{C}X_\alpha$;
(b) $[X_\alpha, X_{-\alpha}] = H_\alpha$, $B(X_\alpha, X_{-\alpha}) = 1$, *and* $[H, X_\alpha] = \alpha(H)X_\alpha$ $(\alpha \in \Delta$, $H \in \mathfrak{h})$;

(c) $$[X_\alpha, X_\beta] = \begin{cases} 0 & \text{if } \alpha+\beta \notin R, \\ N_{\alpha,\beta}X_{\alpha+\beta} & \text{if } \alpha+\beta \in \Delta; \end{cases}$$

(d) $N_{\alpha,\beta} = -N_{-\alpha,-\beta}$.

B56. Complexifications and real forms. Let $\mathfrak{g}_0$ be a Lie algebra over $\mathbb{R}$ and $\mathfrak{g} = (\mathfrak{g}_0)_{\mathbb{C}}$. Every element of $\mathfrak{g}$ is uniquely of the form $A + iB$ $(A, B \in \mathfrak{g}_0)$, and the map $\sigma \colon \mathfrak{g} \to \mathfrak{g}$, where

$$\sigma(A+iB) = A - iB \qquad (A, B \in \mathfrak{g}_0)$$

has the following properties:

$$\begin{gathered} \sigma^2 = I, \qquad \sigma(X+Y) = \sigma(X) + \sigma(Y), \\ \sigma(\alpha X) = \overline{\alpha}\sigma(X), \qquad \sigma([X,Y]) = [\sigma(X), \sigma(Y)], \end{gathered} \tag{1}$$

for all $\alpha \in \mathbb{C}$, $X, Y \in \mathfrak{g}$. Clearly

$$\mathfrak{g}_0 = \{X \in \mathfrak{g} \colon \sigma(X) = X\}.$$

Conversely, suppose that $\mathfrak{g}$ is a Lie algebra over $\mathbb{C}$ and that $\tau\colon \mathfrak{g} \to \mathfrak{g}$ is a *conjugation*, that is, a map satisfying the conditions of (1). If we set $\mathfrak{g}_0 = \{X \in \mathfrak{g}\colon \tau(X) = X\}$, then $\mathfrak{g} \cong (\mathfrak{g}_0)_{\mathbb{C}}$ in the obvious way, and $\tau(A+iB) = A - iB$ $(A, B \in \mathfrak{g}_0)$.

It is easy to prove that if τ is a conjugation of $\mathfrak{g}$, then $B(\tau X, \tau Y) = \overline{B(X,Y)}$ $(X, Y \in \mathfrak{g})$.

A *real form* of $\mathfrak{g}$ is a Lie algebra over $\mathbb{R}$ whose complexification is isomorphic to $\mathfrak{g}$. From the above, the real forms of $\mathfrak{g}$ are in a natural, one-to-one correspondence with the conjugations of $\mathfrak{g}$.

A *compact* real form of a Lie algebra $\mathfrak{g}$ over $\mathbb{C}$ is, of course, a real form of $\mathfrak{g}$ that is a compact Lie algebra (B48). The next result enables us to associate with a semisimple Lie algebra $\mathfrak{g}_0$ (over $\mathbb{R}$) a *compact* semisimple Lie algebra $\mathfrak{a}$ over $\mathbb{R}$ through the mediation of $(\mathfrak{g}_0)_{\mathbb{C}}$.

B57. *Let $\mathfrak{g}$ be a semisimple Lie algebra over $\mathbb{C}$. Then $\mathfrak{g}$ has a compact real form.*

Indeed, using B54 and B55, $\mathfrak{a}$ is a compact real form of $\mathfrak{g}$, where $\mathfrak{a}$ is the set of elements of the form

$$\sum_{\alpha\in\Delta} \lambda_\alpha(iH_\alpha) + \sum_{\alpha\in\Delta} \mu_\alpha(X_\alpha - X_{-\alpha}) + \sum_{\alpha\in\Delta} \nu_\alpha(i(X_\alpha + X_{-\alpha})),$$

where for each $\alpha \in \Delta$, $\lambda_\alpha, \mu_\alpha, \nu_\alpha \in \mathbb{R}$.

If $\mathfrak{g}_0$ is a semisimple Lie algebra over $\mathbb{R}$, then a direct sum decomposition $\mathfrak{k}_0 \oplus \mathfrak{p}_0$ of $\mathfrak{g}_0$, with the properties given in (ii) of the next result, is called a *Cartan decomposition* of $\mathfrak{g}_0$.

To motivate the Cartan decomposition, suppose that G_0 is a closed subgroup of $\mathrm{GL}(n,\mathbb{R})$ and that the transpose $U' \in G_0$ whenever $U \in G_0$. Then the map $U \to (U')^{-1}$ is an involutory automorphism Θ of G_0. Let $\mathfrak{g}_0$ be the Lie algebra of G_0. Then $\theta = d\Theta(I)$ is an involutary automorphism of $\mathfrak{g}_0 \subset L(\mathbb{R}^n)$, with $\theta(X) = -X'$. We obtain a Cartan decomposition of $\mathfrak{g}_0$ by setting $\mathfrak{k}_0 = \{X \in \mathfrak{g}_0\colon \theta(X) = X\}$, $\mathfrak{p}_0 = \{X \in \mathfrak{g}_0\colon \theta(X) = -X\}$. Note that $\mathfrak{k}_0$ is the Lie algebra of the maximal compact subgroup $K_0 = \mathrm{O}(n,\mathbb{R}) \cap G_0$ of G_0. (See Knapp [**S2**, p. 3ff].) The result below shows that a similar result holds for general $\mathfrak{g}_0$.

B58. *Let $\mathfrak{g}_0$ be a semisimple Lie algebra over $\mathbb{R}$, $\mathfrak{g}$ its complexification, and σ the conjugation of $\mathfrak{g}$ associated with $\mathfrak{g}_0$. Let B be the Killing form of $\mathfrak{g}$. Then*

(i) *if $\mathfrak{a}$ is a compact real form of $\mathfrak{g}$, then there exists an automorphism ϕ of $\mathfrak{g}$ such that $\sigma(\phi(\mathfrak{a})) = \phi(\mathfrak{a})$;*

(ii) *there exist sets $\mathfrak{k}_0$, $\mathfrak{p}_0$ and $\mathfrak{g}_k$, where $\mathfrak{k}_0$ is a subalgebra of $\mathfrak{g}_0$, $\mathfrak{p}_0$ is a vector subspace of $\mathfrak{g}_0$, and $\mathfrak{g}_k$ is a compact real form of $\mathfrak{g}$ such that:*

(a) $\mathfrak{k}_0 = \mathfrak{g}_0 \cap \mathfrak{g}_k$, $\mathfrak{p}_0 = \mathfrak{g}_0 \cap (i\mathfrak{g}_k)$;

(b) $\sigma(\mathfrak{g}_k) = \mathfrak{g}_k$;

(c) $\mathfrak{g}_0 = \mathfrak{k}_0 \oplus \mathfrak{p}_0$;

(d) $[\mathfrak{k}_0, \mathfrak{p}_0] \subset \mathfrak{p}_0$, $[\mathfrak{p}_0, \mathfrak{p}_0] \subset \mathfrak{k}_0$;

(e) *if $X \in \mathfrak{g}_k \sim \{0\}$, $Y \in \mathfrak{p}_0 \sim \{0\}$, then $B(X,X) < 0$, $B(Y,Y) > 0$;*

(f) $\mathfrak{k}_0$ *is a maximal, compactly embedded subalgebra of* $\mathfrak{g}_0$;

(iii) *any two compact real forms of* $\mathfrak{g}$ *are isomorphic.*

A simple example of the Iwasawa decomposition below is provided by $\mathfrak{g}_0 = \mathrm{sl}(m, \mathbb{R})$. Here, $\mathfrak{k}_0 = \mathrm{so}(m, \mathbb{R})$, $\mathfrak{a}_0$ is the algebra of diagonal matrices in $\mathfrak{g}_0$, and $\mathfrak{n}_0$ is the algebra of upper triangular matrices with zeros on the diagonal.

B59. The Iwasawa Decomposition Theorem. *There exist an abelian subalgebra* $\mathfrak{a}_0$ *of* $\mathfrak{g}_0$ *and a nilpotent subalgebra* $\mathfrak{n}_0$ *of* $\mathfrak{g}_0$ *such that* $\mathfrak{a}_0 + \mathfrak{n}_0$ *is a solvable subalgebra of* $\mathfrak{g}_0$, *and* $\mathfrak{g}_0$ *is the vector space direct sum*

$$\mathfrak{k}_0 \oplus \mathfrak{a}_0 \oplus \mathfrak{n}_0.$$

B60. More on the Iwasawa decomposition. (i) *If* $\mathfrak{g}_0$ *is not compact, then* $\mathfrak{a}_0 \neq \{0\}$ *and* $\mathfrak{n}_0 \neq \{0\}$.

(ii) *Let* $X \in \mathfrak{n}_0$. *Then* $\mathrm{ad}_{\mathfrak{g}_0} X$ *is nilpotent.*

We now state the Lie group version of the Iwasawa decomposition [**He**, Chapter 6, §5].

B61. *Let* G *be a connected, semisimple Lie group with* $\mathfrak{g}_0$ *as Lie algebra. Let* K_0, A_0, *and* N_0 *be the connected Lie subgroups of* G *with* $\mathfrak{k}_0$, $\mathfrak{a}_0$, *and* $\mathfrak{n}_0$ *as Lie algebras respectively. Then* A_0, N_0 *are simply connected, and the map* $\Phi\colon K_0 \times A_0 \times N_0 \to G$, *where*

$$\Phi(k, a, n) = kan$$

is an analytic diffeomorphism from $K_0 \times A_0 \times N_0$ *onto* G.

Of particular importance (especially for representation theory) is the case where $Z(G)$ is finite. Indeed, $\mathrm{Ad}_G K_0$, as a group, equals the Lie subgroup of $\mathrm{GL}(\mathfrak{g}_0)$ with $\mathrm{ad}_{\mathfrak{g}_0} \mathfrak{k}_0$ as Lie algebra (since both groups are generated by $\mathrm{Ad}_G(\exp \mathfrak{k}_0)$) and so is compact (B58(ii)(f)). The canonical bijection from $K_0/(K_0 \cap Z(G))$ onto $\mathrm{Ad}_G K_0$ is bicontinuous, noting that both groups have isomorphic Lie algebras. Hence $K_0/(K_0 \cap Z(G))$ is compact, and since $Z(G)$ is finite, K_0 is also compact. The maximality property of $\mathfrak{k}_0$ yields that $K_0 Z(G)$ is a maximal compact subgroup of G. (In fact, $K_0 Z(G) = K_0$ [**He**, Chapter 6, §1].)

The following more general result is proved in [**Ho**, Chapter 15].

B62. *Every connected Lie group contains a maximal, compact subgroup.*

The final result of this section appears in [**J**, Chapter 3, Theorem 17] and [**He**, Chapter 9, §7]. Note that if $\mathfrak{g}$ in the result is noncompact, then (B60) ensures that there exists a nonzero element $X \in \mathfrak{g}$ with $\mathrm{ad}\, X$ nilpotent.

B63. *Let* $\mathfrak{g}$ *be a semisimple Lie algebra over* $\mathbb{R}$, *and let* $X \in \mathfrak{g}$ *be nonzero, and such that* $\mathrm{ad}\, X$ *is nilpotent in* $L(\mathfrak{g})$. *Then there exist elements* H', E', $F' \in \mathfrak{g}$ *with* $E' = X$ *such that*

$$[H', E'] = 2E', \qquad [H', F'] = -2F', \qquad [E', F'] = H'$$

(*so that the subalgebra* $(\mathbb{R}H' + \mathbb{R}E' + \mathbb{R}F')$ *of* $\mathfrak{g}$ *is canonically isomorphic to* $\mathrm{sl}(2, \mathbb{R})$—*cf.*(3.2)).

APPENDIX C

Existence of Borel Cross-Sections

The following result is due to Federer and Morse [**1**]. See also Mackey [**1**] and Feldman and Greenleaf [**1**].

THEOREM. *Let X, Y be Hausdorff spaces with X separable. Let $g\colon X \to Y$ be continuous and $C \in \mathscr{C}(X)$. Then there exists a Borel set $B \subset C$ such that $g(B) = g(C)$ and $g|_B$ is one-to-one.*

PROOF. The argument of (e), p. 166 of Kelley [**2**] shows that there is a continuous map h from the Cantor set K onto C. For each $n \in \mathbb{P}$, let

$$K_n = \{t \in K\colon g(h(t)) \notin g(h([0, t - n^{-1}] \cap K))\}. \tag{1}$$

(Here, of course, K is identified with the Cantor discontinuum.) Clearly, K_n is an open subset of K. Let $B_n = h(K_n)$. Since K_n (and so B_n) is σ-compact, it follows that B_n is a Borel subset of X, and so $B = \bigcap_{n=1}^{\infty} B_n$ is a Borel subset of X. We now check that B satisfies the conclusions of the theorem.

Let $b_1, b_2 \in B$ with $b_1 \neq b_2$. Since h is continuous, there exists $n \in \mathbb{P}$ such that $n^{-1} < |t_1 - t_2|$ whenever $t_1, t_2 \in K$ are such that $h(t_i) = b_i$. Choosing t_1, t_2 with $h(t_i) = b_i$ and $t_i \in K_n$, we see that $g(b_1) \neq g(b_2)$. So $g|_B$ is one-to-one. Now let $y_0 \in g(C)$ and define

$$c_0 = \inf\{t \in K\colon g(h(t)) = y_0\}.$$

It is easy to check that $c_0 \in K_n$ for all n. Hence $h(c_0) \in B$ and $g(h(c_0)) = y_0$. This concludes the proof.

COROLLARY. *If X is also σ-compact, then there exists a Borel set $B' \subset X$ such that $g(B') = g(X)$ and $g|_{B'}$ is one-to-one.*

APPENDIX D

Mycielski's Theorem

We state below a very useful category theorem due to Mycielski. The theorem is fundamental for (7.15); it can also be used to construct "large" free groups in $\mathrm{SO}(n, \mathbb{R})$ (Wagon [**S2**]). References for the theorem are Mycielski [**3**], [**S2**] and Kuratowski [**1**]. The reader is referred to (6.5) of Wagon [**S2**] for a short, ingenious proof. The theorem is true without the separability requirement, but then requires the Axiom of Choice.

THEOREM (MYCIELSKI). *Let (X, d) be a complete, separable, metric space with no isolated points. Let $\{m_i\}$ be a sequence of positive integers, and for each i, let R_i be a closed subset of X^{m_i} with empty interior. Then there exists a subset A of X such that $|A| = \mathfrak{c}$ and if $i \geq 1$ and $a_1, \ldots, a_{m_i}$ are distinct points of A, then $(a_1, \ldots, a_{m_i}) \notin R_i$.*

APPENDIX E

On the Density of the Exponential Map

The following result seems to have been proved first by Dixmier [**2**]. The proof we give is based on Hofman and Mukherjea [**1**].

THEOREM. *Let S be a solvable, connected Lie group with Lie algebra $\mathfrak{s}$. Then* $\exp \mathfrak{s}$ *is dense in* S.

PROOF. The proof proceeds by induction on $n = \dim S$. The result is trivially true if $n = 0$. Let $n \geq 1$. Suppose that the result is true for all solvable, connected Lie groups of dimension $< n$.

By Lie's Theorem (B25) we can find $Z \in \mathfrak{s}_{\mathbb{C}} \sim \{0\}$ and a function $\alpha: \mathfrak{s} \to \mathbb{C}$ such that

$$\operatorname{ad} A(Z) = \alpha(A)Z \qquad (A \in \mathfrak{s}).$$

Write $Z = X + iY$, where $X, Y \in \mathfrak{s}$, and let $\mathfrak{w} = \mathbb{R}X + \mathbb{R}Y$. Then $\mathfrak{w}$ is either one- or two-dimensional, and is an ideal in $\mathfrak{s}$.

Let $(\widetilde{S}, p)$ be the simply connected covering group of S. Since $p \circ \exp_{\widetilde{S}} = \exp_S$, we can (and shall) suppose that $S = \widetilde{S}$.

Let W be the connected Lie subgroup of S with $\mathfrak{w}$ as Lie algebra. By B17(iii) and B44, W is a simply connected, closed, normal subgroup of S.

Let U be a nonempty, open subset of S. It is sufficient to show that

$$U \cap \exp_S(\mathfrak{s}) \neq \varnothing. \tag{1}$$

Since $\dim(S/W) < n$ and S/W is solvable, we have, by hypothesis, that $\exp_{S/W}(\mathfrak{s}/\mathfrak{w})$ is dense in S/W. Let $Q: S \to S/W$, $\Phi: \mathfrak{s} \to \mathfrak{s}/\mathfrak{w}$ be the canonical maps. Since Q is open, we can find $K \in \mathfrak{s}/\mathfrak{w}$ such that $\exp_{S/W}(K) \in Q(U)$. Since the following diagram commutes (B11(i)):

$$\begin{array}{ccc} S & \xleftarrow{\exp_S} & \mathfrak{s} \\ {\scriptstyle Q}\downarrow & & \downarrow{\scriptstyle \Phi} \\ S/W & \xleftarrow[\exp_{S/W}]{} & \mathfrak{s}/\mathfrak{w}, \end{array}$$

we can find $A \in \mathfrak{s}$ such that $\exp_S A \in UW$. Let $H = (\exp_S \mathbb{R}A)W$. Then H is a (solvable) connected Lie subgroup of G with $\mathfrak{h} = \mathbb{R}A + \mathfrak{w}$ as Lie algebra.

Now $U \cap H$ is open in H (with its Lie subgroup topology), and $U \cap H \supset U \cap (\exp_S A)W^{-1} \neq \varnothing$. So (1) will follow if we can show that

$$\text{(2)} \qquad \exp_H \mathfrak{h} \ (= \exp_S \mathfrak{h}) \text{ is dense in } H.$$

As earlier, we can take H to be simply connected. Since $\mathfrak{h}$ is either $\mathfrak{w}$ (if $A \in \mathfrak{w}$) or a semidirect product $\mathfrak{w} \times_\Psi \mathbb{R}$, we see (B22) that H is either W or a semidirect product $W \times_\rho \mathbb{R}$.

We can rule out the cases $H = W = \mathbb{R}$, $H = W = \mathbb{R}^2$, since in those cases, H is *exponential*, that is, exp is surjective. Suppose that W is two-dimensional (solvable) and not $\mathbb{R}^2$. Then we can find a basis $\{X, Y\}$ for $\mathfrak{w}$ such that $[X, Y] = Y$. So $\mathfrak{w}$ is $\mathfrak{s}_2$ ((6.35)) and $W = S_2$, the "$ax + b$" group. Thus ((6.36)) the multiplication of W is given by

$$(s, t)(s', t') = (s + e^t s', t + t'),$$

where $s, s', t, t' \in \mathbb{R}$. We thus have to consider the following (semidirect product) possibilities for H, where M is $\mathbb{R}$ or $\mathbb{R}^2$:

$$M \times_\rho \mathbb{R}, \qquad S_2 \times_\rho \mathbb{R}.$$

We consider these cases in turn.

(i) *Suppose that* $H = M \times_\rho \mathbb{R}$. Then $\rho\colon \mathbb{R} \to \operatorname{Aut} M$ is a one-parameter subgroup in $\operatorname{Aut} M = \mathrm{GL}(M)$, and so we can find $A \in L(M)$ such that $\rho(t) = e^{tA}$ ($t \in \mathbb{R}$). Thus the product in H is given by

$$\text{(3)} \qquad (m, t)(m', t') = (m + e^{tA} m', \ t + t').$$

The Lie algebra $\mathfrak{h}$ of H is (cf. B22) $M \times_\Phi \mathbb{R}$, where $\Phi(s) = sA$.

We now deal with four subcases.

(a) *Suppose that* $A = 0$. Then, trivially, H is $\mathbb{R}^2$ or $\mathbb{R}^3$ and exp is surjective.

(b) *Suppose that* $M = \mathbb{R}^2$, $A \neq 0$, *and A is nilpotent.* Using the Jordan canonical form, we can take $A = \left[\begin{smallmatrix} 0 & 1 \\ 0 & 0 \end{smallmatrix}\right]$. Then $e^{tA} = \left[\begin{smallmatrix} 1 & t \\ 0 & 1 \end{smallmatrix}\right]$, and (3) becomes

$$((m_1, m_2), t)((m_1', m_2'), t') = ((m_1 + m_1' + t m_2', m_2 + m_2'), t + t'),$$

where $m_1, m_2, m_1', m_2' \in \mathbb{R}$. Identifying $((m_1, m_2), t)$ with the 3×3 matrix

$$B = \begin{bmatrix} 1 & t & m_1 \\ 0 & 1 & m_2 \\ 0 & 0 & 1 \end{bmatrix},$$

we see that H is the "Heisenberg group" (Problem 4-29). Since

$$B = \exp((B - I) - \frac{1}{2}(B - I)^2)$$

(using the power series expansion for $\log(1 + x)$), we see that H is exponential. (More generally, every connected, nilpotent Lie group is exponential [**Ho**, Chapter 12, §2.0].)

(c) *Suppose that $M = \mathbb{R}^2$, A is not nilpotent and not invertible.* Then the eigenvalues of A are $0, k$, where $k \in \mathbb{R} \sim \{0\}$. So we can take $A = \left[\begin{smallmatrix} k & 0 \\ 0 & 0 \end{smallmatrix}\right]$. An elementary calculation shows that

$$e^{tA} = \begin{bmatrix} e^{kt} & 0 \\ 0 & 1 \end{bmatrix},$$

so that (3) becomes

$$((m_1, m_2), t)((m_1', m_2'), t') = ((m_1 + e^{kt} m_1', m_2 + m_2'), t + t').$$

Writing $((m_1, t), m_2)$ in place of $((m_1, m_2), t)$, we see that M is isomorphic to $S_2 \times \mathbb{R}$. Since $\mathbb{R}$ is exponential, we can take $H = S_2$. This case is covered by (d) below.

(d) *Suppose that A is invertible.* Suppose that $M = \mathbb{R}^2$. Let $s \to (m(s), t(s))$ be a one-parameter subgroup of H. The homomorphism property satisfied by the preceding map gives (using (3)):

$$m(r + s) = m(r) + e^{t(r)A} m(s), \tag{4}$$

$$t(r + s) = t(r) + t(s) \tag{5}$$

for $r, s \in \mathbb{R}$. From (5), there exists $k \in \mathbb{R}$ such that $t(r) = rk$. If $k = 0$, then, using (4), we obtain $c \in M$ such that

$$(m(r), t(r)) = (rc, 0) \qquad (r \in \mathbb{R}).$$

Suppose that $k \neq 0$. Then from (4) and noting that $m(0) = 0$,

$$m'(r) = \lim_{s \to 0} (m(r + s) - m(r))/s = \lim_{s \to 0} e^{krA}(m(s) - m(0))/s = e^{krA} m'(0).$$

It follows that if $a = (kA)^{-1}(m'(0))$, then $m(r) = e^{krA} a - a$. The one-parameter subgroups of H are thus of two kinds: those of the form $r \to (rc, 0)$ $(c \in M)$ and those of the form $r \to (e^{krA} a - a, kr)$ $(k \in \mathbb{R} \sim \{0\},\ a \in M)$. Observing that $(e^{krA} - I)$ is invertible whenever $2n\pi i$ is not an eigenvalue of krA for any $n \in \mathbb{Z}$, we see that (2) holds.

The case $M = \mathbb{R}$ $(H = S_2)$ is an easier version of the above.

(ii) *Now suppose that H is of the form $S_2 \times_\rho \mathbb{R}$.* Then the Lie algebra $\mathfrak{h}$ of H is of the form $\mathfrak{s}_2 \times_\Phi \mathbb{R}$. Let $Z = (0, 1) \in \mathfrak{h}$ and $D = \Phi(Z) \in \operatorname{Der} \mathfrak{s}_2$. Thus (B22) $D(A) = [Z, A]$ $(A \in \mathfrak{s}_2)$. Let $\{X, Y\}$ be a basis for $\mathfrak{s}_2$ such that $[X, Y] = Y$. Since $[\mathfrak{s}_2, \mathfrak{s}_2] = \mathbb{R}Y$, we have $DY = bY$ for some $b \in \mathbb{R}$. Thus $bY = DY = D([X, Y]) = [DX, Y] + bY$, so that $[DX, Y] = 0$ and $DX = aY$ for some $a \in \mathbb{R}$. Let $T = Z + aY - bX$. Then $[T, X] = 0 = [T, Y]$, and since $\{X, Y, T\}$ is a basis for $\mathfrak{h}$, we see that H is of the form $S_2 \times \mathbb{R}$. This case obviously reduces to the S_2 case of (d) above. $\square$

Some Abbreviations

The following standard textbooks are referred to throughout the present work, and the reference to each is abbreviated as follows:

[BD] Bonsall, F. F. and Duncan, J., *Complete normed algebras*, Springer-Verlag, Berlin–Heidelberg–New York, 1973.

[B] Bourbaki, N., *Elements of Mathematics. Lie groups and Lie algebras.* Part I: Chapters 1–3, Hermann, Paris, 1975. (English translation of: *Éléments de mathématique, Groupes et algèbres de Lie*, Hermann, Paris, 1971, 1973.)

[CP1] Clifford, A. H. and Preston, G. B., *The algebraic theory of semigroups*, Vol. 1, Amer. Math. Soc., Providence, R. I., 1961.

[CP2] ——, *The algebraic theory of semigroups*, Vol. 2, Amer. Math. Soc., Providence, R. I., 1967.

[D1] Dixmier, J., *Von Neumann algebras*, North-Holland Publishing Company, Amsterdam–New York–Oxford, 1981. (English translation of: *Les algèbres d'opérateurs dans l'espace hilbertien (algèbres de von Neumann)*, Gauthier-Villars, Paris, 1969.)

[D2] ——, *C^*-algebras*, North-Holland Publishing Company, Amsterdam–New York–Oxford, 1977. (English translation of: *Les C^*-algèbres et leurs représentations*, Gauthier-Villars, Paris, 1969.)

[DS] Dunford, N. and Schwartz, J. T., *Linear operators*, Part 1, Interscience Publishers, New York, 1958.

[HS] Hausner, M. and Schwartz, J. T., *Lie groups • Lie algebras*, Gordon and Breach, New York, 1968.

[He] Helgason, S., *Differential geometry, Lie groups and symmetric spaces*, Academic Press, New York–San Francisco–London, 1978.

[HR1] Hewitt, E. and Ross, K. A., *Abstract harmonic analysis.* I, Springer-Verlag, Berlin–Heidelberg–New York, 1963.

[HR2] ——, *Abstract harmonic analysis.* II, Springer-Verlag, Berlin–Heidelberg–New York, 1970.

[**Ho**] Hochschild, G., *The structure of Lie groups*, Holden-Day, San Francisco, 1965.

[**J**] Jacobson, N., *Lie algebras*, Interscience Publishers, New York, 1962.

[**M**] MacDonald, I. D., *The theory of groups*, Oxford Univ. Press, Oxford, 1968.

[**R**] Reiter, H., *Classical harmonic analysis and locally compact groups*, Oxford Univ. Press, Oxford, 1968.

[**Ri**] Rickart, C. E., *General theory of Banach algebras*, Van Nostrand, New York, 1960.

[**SW**] Sagle A. and Walde, R., *Introduction to Lie groups and Lie algebras*, Academic Press, New York–San Francisco–London, 1973.

[**Sch**] Schaefer, H. H., *Banach lattices and positive operators*, Springer-Verlag, Berlin–Heidelberg–New York, 1974.

Bibliography

Adelson-Welsky, G. and Sreider, Y. U. A.

1. *The Banach mean on groups*, Uspekhi Mat. Nauk. **12** (1957), 131–136.

Adian, S. I.

1. *The Burnside problem*, Springer-Verlag, Berlin–Heidelberg–New York, 1979.

Adler, A. and Hamilton, J.

1. *Invariant means via the ultrapower*, Math. Ann. **202** (1973), 71–76.

Agnew, R. P.

1. *Linear functionals satisfying prescribed conditions*, Duke Math. J. **4** (1938), 55–77.

Agnew, R. P. and Morse, A. P.

1. *Extensions of linear functionals, with applications to limits, integrals, measures and densities*, Ann. of Math. (2) **39** (1938), 20–30.

Akemann, C. A.

1. *Projections onto separable C^*-subalgebras of a W^*-algebra*, Bull. Amer. Math. Soc. **73** (1967), 925.
2. *Operator algebras associated with Fuchsian groups*, Houston J. Math. **7** (1981), 295–301.

Akemann, C. A. and Ostrand, P. A.

1. *Computing norms in group C^*-algebras*, Amer. J. Math. **98** (1976), 1015–1048.

Akemann, C. A. and Walter, M. E.

1. *Non-abelian Pontryagin duality*, Duke Math. J. **39** (1972), 451–463.
2. *The Riemann-Lebesgue property for arbitrary locally compact groups*, Symposia Math. **22** (1976), 283–289.
3. *Unbounded negative definite functions*, Canad. J. Math. **33** (1981), 862–871.

Akemann, C. A. and Wright, S.

1. *Compact and weakly compact derivations of C^*-algebras*, Pacific J. Math. **85** (1979), 253–259.

Alaoglu, L. and Birkhoff, G.

1. *General ergodic theorems*, Ann. of Math. (2) **41** (1940), 293–309.

Anantharaman-Delaroche, C.

1. *Action moyennable d'un groupe localement compact sur une algèbre de von Neumann*, Math. Scand. **45** (1979), 289–304.
2. *Sur la moyennabilité des actions libres d'un groupe localement compact dans une algèbre de von Neumann*, C. R. Acad. Sci. Paris Sér. A **289** (1979), 605–607.
3. *Action moyennable d'un groupe localement compact sur une algèbre de von Neumann*. II, Math. Scand. **50** (1982), 251–268.

Anker, J.-Ph.

1. *Sur la propriété P_**, Monatsh. Math. **90** (1980), 87–90.

Anusiak, Z.

1. *Symmetry of L_1-group algebra of locally compact groups with relatively compact classes of conjugated elements*, Bull. Acad. Polon Sci. Ser. Sci. Math. Astronom. Phys. **18** (1970), 329–332.

Appel, K. I. and Djorup, F. M.

1. *On the group generated by the free semigroup*, Proc. Amer. Math. Soc. **15** (1964), 838–841.

Argabright, L. N.

1. *Invariant means on topological semigroups*, Pacific J. Math. **16** (1966), 193–203.
2. *Invariant means and fixed points—a sequel to Mitchell's paper*, Trans. Amer. Math. Soc. **130** (1968), 127–130.

Argabright, L. N. and Wilde, C. O.

1. *Semigroups satisfying a strong Følner condition*, Proc. Amer. Math. Soc. **18** (1967), 587–591.

Aribaud, F.

1. *Un théorème ergodique pour les espaces L^1*, J. Funct. Anal. **5** (1970), 395–411.
2. *Sur la moyenne temporelle d'un système dynamique*, Illinois J. Math. **17** (1973), 90–110.

Arveson, W. B.

1. *Subalgebras of C^*-algebras*, Acta Math. **123** (1969), 141–224.

Auslander, L.

1. *Differential geometry*, Harper and Row, London, 1967.

Auslander, L. and Brezin, J.

1. *Uniform distribution in solvmanifolds*, Adv. in Math. **7** (1971), 111–144.

Auslander, L. and Konstant, B.

1. *Polarization and unitary representations of solvable groups*, Invent. Math. **14** (1971), 255–354.

Auslander, L. and Moore, C. C.

1. *Unitary representations of solvable Lie groups*, Mem. Amer. Math. Soc. **62** (1966).

Avez, A.

1. *Limite de quotients pour des marches aléatoires sur les groupes*, C. R. Acad. Sci. Paris Sér. A **276** (1973), 317–320.

Azencott, R.

1. *Espaces de Poisson des groupes localement compacts*, Lecture Notes in Math., vol. 148, Springer-Verlag, Berlin–Heidelberg–New York, 1970.

Bachelis, G. F. and Gilbert, J. E.

1. *Banach spaces of compact multipliers and their dual spaces*, Math. Z. **125** (1972), 285–297.

Bachelis, G. F., Parker, W. A., and Ross, K. A.

1. *Local units in $L^1(G)$*, Proc. Amer. Math. Soc. **31** (1972), 312–313.

Baer, R.

1. *Supersolvable groups*, Proc. Amer. Math. Soc. **6** (1955), 16–32.

Baggett, L.

1. *On the continuity of Mackey's extension process*, J. Funct. Anal. **56** (1984), 233–250.

Baggett, L. W. and Sund, T.

1. *The Hausdorff dual problem for connected groups*, J. Funct. Anal. **43** (1981), 60–68.

Baggett, L. and Taylor, K.

1. *Groups with completely reducible regular representation*, Proc. Amer. Math. Soc. **72** (1978), 593–600.

Baker, A. C. and Baker, J. W.

1. *Algebras of measures on a locally compact semigroup*, J. London Math. Soc. **1** (1969), 249–259.
2. *Algebras of measures on a locally compact semigroup*. II, J. London Math. Soc. **2** (1970), 651–659.
3. *Algebras of measures on a locally compact semigroup*. III, J. London Math. Soc. **4** (1972), 685–695.

Baker, J. W. and Butcher, R. J.

1. *The Stone-Čech compactification of a topological semigroup*, Math. Proc. Cambridge Philos. Soc. **80** (1976), 103–107.

Baker, J. W. and Milnes, P.

1. *The ideal structure of the Stone-Čech compactification of a group*, Math. Proc. Cambridge Philos. Soc. **82** (1977), 401–409.

Banach, S.

1. *Sur le problème de mésure*, Fund. Math. **4** (1923), 7–33.
2. *Théorie des opérations lineaires*, Warsaw, 1932.

Banach, S. and Tarski, A.

1. *Sur la décomposition des ensembles de points en parts respectivement congruents*, Fund. Math. **6** (1924), 244–277.

Banascewski, B.

1. *Extension of invariant linear functionals: Hahn-Banach in the topos of M-sets*, J. Pure Appl. Algebra **17** (1980), 227–248.

Barnes, B. A.

1. *Ideal and representation theory of the L^1-algebra of a group with polynomial growth*, Colloq. Math. **45** (1981), 301–315.

Bass, H.

1. *The degree of polynomial growth of finitely generated nilpotent groups*, Proc. London Math. Soc. **25** (1972), 603–614.

Belluce, L. P. and Kirk, W. A.

1. *Non-expansive mappings and fixed-points in Banach spaces*, Illinois J. Math. **11** (1967), 474–479.

Berberian, S. K.

1. *Measure and integration*, Chelsea, New York, 1970.

Berg, C. and Christensen, J. P. R.

1. *Sur la norme des opérateurs de convolution*, Invent. Math. **23** (1974), 173–178.
2. *On the relation between amenability of locally compact groups and the norm of convolution operators*, Math. Ann. **208** (1974), 149–153.

Berglund, J. F. and Hofmann, K. H.

1. *Compact semitopological semigroups and weakly almost periodic functions*, Lecture Notes in Math., vol. 42, Springer-Verlag, Berlin–Heidelberg–New York, 1967.

Berglund, J. F., Junghenn, H. D., and Milnes, P.

1. *Compact right topological semigroups and generalizations of almost periodicity*, Lecture Notes in Math., vol. 663, Springer-Verlag, Berlin–Heidelberg–New York, 1978.

2. *Universal mapping properties of semigroup compactifications*, Semigroup Forum **15** (1978), 375–386.

Bewley, T.

1. *Extension of the Birkhoff and von Neumann ergodic theorems to general semigroups*, Ann. Inst. H. Poincaré Sect. B **7** (1971), 283–291.

Birgé, L. and Raugi, A.

1. *Fonctions harmoniques sur les groupes moyennables*, C. R. Acad. Sci. Paris Sér. A **278** (1974), 1287–1289.

Bishop, R. L. and Crittenden, R. J.

1. *Geometry of manifolds*, Academic Press, New York and London, 1964.

Blum, J. R. and Cogburn, R.

1. *On ergodic sequences of measures*, Proc. Amer. Math. Soc. **51** (1975), 359–365.

Blum, J. and Eisenberg, B.

1. *Generalised summing sequences and the mean ergodic theorem*, Proc. Amer. Math. Soc. **42** (1974), 423–429.

Blum, J. R., Eisenberg, B., and Hahn, L.-S.

1. *Ergodic theory and the measure of sets in the Bohr group*, Acta Sci. Math. (Szeged) **34** (1973), 17–24.

Blum, J. R. and Reich, J. I.

1. *A mean ergodic theorem for families of contractions in Hilbert space*, Proc. Amer. Math. Soc. **61** (1976), 183–185.

Boidol, J., Leptin, H., Schürman, J., and Vahle, D.

1. *Raüme primitiver Ideale von Gruppenalgebren*, Math. Ann. **236** (1978), 1–13.

Boll, C.

1. *Comparison of experiments in the infinite case*, Thesis, Stanford University, 1955.

Bombal, F. and Vera, G.

1. *Medias en espacios localmente convexos y semireflexividad*, Collectanea Math. **24** (1973), 3–31.
2. *Funciones vectoriales casi convergentes*, Seminario Matematico de Barcelona, 1975.

Bon, J.-L.

1. *Les groupoïdes amenables*, Thesis, Univ. d'Orléans, Tours, 1980.

Bondar, J. V.

1. *A conditional confidence principle*, Ann. Statist. **5** (1977), 881–891.

Bondar, J. V. and Milnes, P.

1. *Amenability: a survey of Hunt-Stein and related conditions on groups*, Z. Wahrsch. Verw. Gebiete **57** (1981), 103–128.

Bonic, R. A.

1. *Symmetry in group algebras of discrete groups*, Pacific J. Math. **11** (1961), 73–94.

Borel, A.

1. *Linear algebraic groups*, Benjamin, New York and Amsterdam, 1969.

Borel, A. and Tits, J.

1. *Groupes reductifs*, Inst. Hautes Études Sci. Publ. Math. **27** (1965), 55–150.

Bourbaki, N.

1. *Integration*, Actualites Sci. Indust., No. 1306, Hermann, Paris, 1963, Chapters 7 and 8.

Bożejko, M.

1. *Some aspects of harmonic analysis on free groups*, Colloq. Math. **41** (1979), 265–271.
2. *Uniformly amenable discrete groups*, Math. Ann. **251** (1980), 1–6.

Bożejko, M. and Pełczyński, A.

1. *An analogue in commutative harmonic analysis of the uniform bounded approximation property of Banach space*, Seminaire D'Analyse Fonctionnel, Exposé No. IX, 1978–79.

Brenner, J. L.

1. *Quelques groupes libres de matrices*, C. R. Acad. Sci. Paris **241** (1955), 1689–1691.

Brickell, F. and Clark, R. S.

1. *Differentiable manifolds*, Van Nostrand Reinhold, London, 1970.

Brooks, R.

1. *Amenability and the spectrum of the Laplacian*, Bull. Amer. Math. Soc. (N.S.) **6** (1982), 87–89.

Brown, I. D. and Guivarc'h, Y.

1. *Espaces de Poisson des groupes de Lie*, Ann. Sci. École Norm. Sup. **7** (1974), 175–180.

Bruck, R. E.

1. *A simple proof of the mean ergodic theorem for non-linear contractions in Banach spaces*, Israel J. Math. **32** (1979), 107–116.

Bruhat, F. and Tits, J.

1. *Groupes algébriques sur un corps local: cohomologie galoisienne, décompositions d'Iwasawa et de Cartan*, C. R. Acad. Sci. Paris **263** (1966), 867–869.

Brunel, A., Crepel, P., Guivarc'h, Y., and Keane, M.

1. *Marches aléatoires récurrentes sur les groupes localement compacts*, C. R. Acad. Sci. Paris Sér. A **275** (1972), 1359–1361.

Brunel, A. and Revuz, D.

1. *Sur la théorie du renouvellement pour les groupes non abeliens*, Israel. J. Math. **20** (1975), 46–56.

Bunce, J. W.

1. *Representations of strongly amenable C^*-algebras*, Proc. Amer. Math. Soc. **32** (1972), 241–246.
2. *Characterizations of amenable and strongly amenable C^*-algebras*, Pacific J. Math. **43** (1972), 563–572.
3. *Finite operators and amenable C^*-algebras*, Proc. Amer. Math. Soc. **56** (1976), 145–151.
4. *The similarity problem for representations of C^*-algebras*, Proc. Amer. Math. Soc. **81** (1981), 409–414.

Bunce, J. W. and Paschke, W. L.

1. *Derivations on a C^*-algebra and its double dual*, J. Funct. Anal. **37** (1980), 235–247.

Butcher, R. J.

1. *The Stone-Čech compactification of a topological semigroup*, Thesis, University of Sheffield, Sheffield, 1975.

Byers, W.

1. *On a theorem of Preissman*, Proc. Amer. Math. Soc. **24** (1970), 50–51.

Calderon, A. P.

1. *A general ergodic theorem*, Ann. of Math. (2) **58** (1953), 182–191.
2. *Sur les mesures invariantes*, C. R. Acad. Sci. Paris **240** (1955), 1960–1962.

Cartier, P.

1. *Espaces de Poisson des groupes localement compacts*, Séminaire Bourbaki, No. 370, 1969–70.

Cecchini, C.

1. *Operators on $VN(G)$ commuting with $A(G)$*, Colloq. Math. **43** (1980), 137–142.

Cecchini, C. and Zappa, A.

1. *Some results on the center of an algebra of operators on $VN(G)$ for the Heisenberg group*, Canad. J. Math. **33** (1981), 1469–1486.

Chatard, J.

1. *Applications des propriétés de moyenne d'un groupe localement compact à la théorie ergodique*, Ann. Inst. H. Poincaré. Sect. B **6** (1970), 307–326.

Chen, S.

1. *A remark on a question of Margulis*, Duke Math. J. **43** (1976), 805–808.
2. *On the fundamental group of a compact negatively curved manifold*, Proc. Amer. Math. Soc. **71** (1978), 119–122.
3. *The fundamental group of a compact manifold of non-positive curvature*, Preprint.
4. *On non-amenable groups*, Internat. J. Math. Sci. **1** (1978), 529–532.

Chew, J.

1. *Left invariant measures in topological semigroups*, J. Austral. Math. Soc. **20** (1975), 142–145.

Choi, M. D.

1. *A simple C^*-algebra generated by two, finite-order unitaries*, Canad. J. Math. **31** (1979), 867–880.

Choi, M. D. and Effros, E. G.

1. *Separable nuclear C^*-algebras and injectivity*, Duke Math. J. **43** (1976), 309–322.
2. *Injectivity and operator spaces*, J. Funct. Anal. **24** (1977), 156–209.
3. *Nuclear C^*-algebras and the approximation property*, Amer. J. Math. **100** (1978), 61–79.

Chou, C.

1. *Minimal sets and ergodic measures for $\beta N \sim N$*, Illinois J. Math. **13** (1969), 778–788.
2. *On the size of the set of left invariant means on a semigroup*, Proc. Amer. Math. Soc. **23** (1969), 199–205.
3. *On a conjecture of E. Granirer concerning the range of an invariant mean*, Proc. Amer. Math. Soc. **26** (1970), 105–107.
4. *On topological invariant means on a locally compact group*, Trans. Amer. Math. Soc. **151** (1970), 443–456.
5. *On a geometric property of the set of invariant means of a group*, Proc. Amer. Math. Soc. **30** (1971), 296–302.
6. *The multipliers of the space of almost convergent sequences*, Illinois J. Math. **16** (1972), 687–694.
7. *Weakly almost periodic functions with zero means*, Bull. Amer. Math. Soc. **80** (1974), 297–299.
8. *Weakly almost periodic functions and almost convergent functions on a group*, Trans. Amer. Math. Soc. **206** (1975), 175–200.
9. *The exact cardinality of the set of invariant means on a group*, Proc. Amer. Math. Soc. **55** (1976), 103–106.

10. *Minimal sets, recurrent points and discrete orbits in $\beta N \backslash N$*, Illinois J. Math. **22** (1978), 54–63.
11. *Locally compact groups which are amenable as discrete groups*, Proc. Amer. Math. Soc. **76** (1979), 46–50.
12. *Uniform closures of Fourier-Stieltjes algebras*, Proc. Amer. Math. Soc. **77** (1979), 99–102.
13. *Minimally weakly almost periodic groups*, J. Funct. Anal. **36** (1980), 1–17.
14. *Elementary amenable groups*, Illinois J. Math. **24** (1980), 396–407.
15. *Topological invariant means on the von Neumann algebra $VN(G)$*, Trans. Amer. Math. Soc. **273** (1982), 207–229.

Chou, C. and Duran, J. P.

1. *Multipliers for the space of almost convergent functions on a semigroup*, Proc. Amer. Math. Soc. **39** (1973), 125–128.

Christensen, E.

1. *On non self-adjoint representations of C^*-algebras*, Amer. J. Math. **103** (1981), 817–833.

Christensen, J. P. R. and Pachl, J. K.

1. *Measurable functionals on function spaces*, Ann. Inst. Fourier (Grenoble) **31** (1981), 137–152.

Chuaqui, R. B.

1. *Measures invariant under a group of transformations*, Pacific J. Math. **68** (1977), 313–329.

Civin, P. and Yood, B.

1. *The second conjugate space of a Banach algebra as an algebra*, Pacific J. Math. **11** (1961), 847–870.

Clausing, A.

1. *Representations of semigroups and extension of invariant linear functionals*, Semigroup Forum **15** (1978), 327–341.

Cohen, J. M.

1. *Operator norms on free groups*, Boll. Un. Mat. Ital. B (6) **1** (1982), 1055–1065.
2. *Cogrowth and amenability of discrete groups*, J. Funct. Anal. **48** (1982), 301–309.

Coifman, R. R. and Weiss, G.

1. *Analyse harmonique non-commutative sur certains espaces homogènes*, Lecture Notes in Math., vol. 242, Springer-Verlag, Berlin–Heidelberg–New York, 1971.
2. *Operators associated with representation of amenable groups*, Studia Math. **47** (1973), 285–303.
3. *Some examples of transference methods in harmonic analysis*, Sympos. Math. **22** (1977), 33–45.

Comfort, W. W. and Hill, P.

1. *On extending non-vanishing semicharacters*, Proc. Amer. Math. Soc. **17** (1966), 936–941.

Comfort, W. W. and Itzkowitz, G. L.

1. *Density character in topological groups*, Math. Ann. **226** (1977), 223–227.

Connes, A.

1. *Une classification des facteurs de type* III, Ann. Sci. École Norm Sup. **6** (1973), 133–252.
2. *On hyperfinite factors of type* III_0 *and Krieger's factors*, J. Funct. Anal. **18** (1975), 318–327.
3. *Classification of injective factors*, Ann. of Math. (2) **104** (1976), 73–115.
4. *On the cohomology of operator algebras*, J. Funct. Anal. **28** (1978), 248–253.

Connes, A. and Woods, E. J.

1. *A construction of approximately finite-dimenisonal non-ITPFI factors*, Canad. Math. Bull. **23** (1980), 227–230.

Converse, G., Namioka, I., and Phelps, R. R.

1. *Extreme invariant positive operators*, Trans. Amer. Math. Soc. **137** (1969), 375–385.

Conway, J. B., Duncan, J., and Paterson, A. L. T.

1. *Monogenic inverse semigroups and their* C^**-algebras*, Proc. Roy. Soc. Edinburgh Sect. A **98** (1984), 13–24.

Conze, J. P. and Dang-Ngoc, N.

1. *Non-commutative ergodic theorems*, Bull. Amer. Math. Soc. **83** (1977), 1297–1299.
2. *Ergodic theorems for non-commutative dynamical systems*, Invent. Math. **46** (1978), 1–16.

Cowling, M. G.

1. *The Kunze-Stein phenomenon*, Ann. of Math. (2) **107** (1978), 209–234.
2. *Some applications of Grothendieck's theory of topological tensor products in harmonic analysis*, Math. Ann. **232** (1978), 273–285.
3. *An application of Littlewood-Paley theory in harmonic analysis*, Math. Ann. **241** (1979), 83–96.

Cowling, M. G. and Fournier, J. J. F.

1. *Inclusions and noninclusion of convolution operators*, Trans. Amer. Math. Soc. **221** (1976), 59–95.

Crombez, G.

1. *An elementary proof about the order of elements in a discrete group*, Proc. Amer. Math. Soc. **85** (1982), 59–60.

Crombez, G. and Govaerts, W.

1. *A characterization of certain weak*-closed subalgebras of* $L_\infty(G)$, J. Math. Anal. Appl. **73** (1979), 430–434.

Cuntz, J.

1. *Simple C^*-algebras generated by isometries*, Comm. Math. Phys. **57** (1977), 173–185.
2. *K-theoretic amenability for discrete groups*, J. Reine Angew. Math. **344** (1983), 180–195.

Dacunha-Castelle, D., Heyer, H., and Roynette, B.

1. *École d'eté de probabilités de Saint-Four* VII—1977, Springer-Verlag, Berlin–Heidelberg–New York, 1978.

Darsow, W. F.

1. *Positive definite functions and states*, Ann. of Math. (2) **60** (1954), 447–453.

Das, G.

1. *Banach and other limits*, J. London Math. Soc. **7** (1973), 501–507.

Davis, H.

1. *On the mean value of Haar measurable almost periodic functions*, Duke Math. J. **34** (1967), 201–214.

Day, M. M.

1. *Ergodic theorems for abelian semigroups*, Trans. Amer. Math. Soc. **51** (1942), 399–412.
2. *Means for the bounded functions and ergodicity of the bounded representations of semigroups*, Trans. Amer. Math. Soc. **69** (1950), 276–291.
3. *Amenable groups*, Bull. Amer. Math. Soc. **56** (1950), 46.
4. *Amenable semigroups*, Illinois J. Math. **1** (1957), 509–544.
5. *Fixed point theorems for compact convex sets*, Illinois J. Math. **5** (1961), 585–596.
6. *Convolutions, means and spectra*, Illinois J. Math. **8** (1964), 100–111.
7. *Correction to my paper "Fixed point theorems for compact convex sets"*, Illinois J. Math. **8** (1964), 713.
8. *Amenability and equicontinuity*, Studia Math. **31** (1968), 481–494.
9. *Semigroups and amenability*, Semigroups (K. W. Folley, ed.), Academic Press, New York–San Francisco–London, 1969, pp. 5–53.
10. *Normed linear spaces*, 3rd ed., Springer-Verlag, Berlin–Heidelberg–New York, 1973.
11. *Lumpy subsets in left amenable locally compact abelian groups*, Pacific J. Math. **62** (1976), 87–93.
12. *Invariant renorming*, Fixed-Point Theory and Its Applications (S. Swaminathan, ed.), Academic Press, New York–San Francisco–London, 1976, pp. 51–62.

13. *Left thick to left lumpy—a guided tour*, Pacific J. Math. **101** (1982), 71–92.

De Leeuw, K. and Glicksberg, I.

1. *Applications of almost periodic compactifications*, Acta Math. **105** (1961), 63–97.

De Marr, R.

1. *Common fixed-points for commuting contraction mappings*, Pacific J. Math. **13** (1963), 1139–1141.

Dean, D. and Raimi, R. A.

1. *Permutations with comparable sets of invariant means*, Duke Math. J. **27** (1960), 467–479.

Deeds, J. B.

1. *The Stone-Čech operator and its associated functionals*, Studia Math. **29** (1967), 5–17.

Dekker, T. J.

1. *On reflections in Euclidean spaces generating free products*, Indag. Math. **7** (1959), 57–60.

Dekker, T. J. and de Groot, J.

1. *Decompositions of a sphere*, Fund. Math. **43** (1956), 185–194.

Delaroche, C. and Kirillov, A.

1. *Sur les relations entre l'espace dual d'un groupe et la structure de ses sous-groupes fermés*, Séminaire Bourbaki (1967/68), No. 343.

Derighetti, M. A.

1. *Sur la propriété P_1*, C. R. Acad. Sci. Paris Sér A **283** (1976), 317–319.
2. *Relations entre les convoluteurs d'un groupe localement compact et ceux d'un sous-groupe fermé*, Bull. Sci. Math. **106** (1982), 69–84.
3. *A propos des convoluteurs d'un groupe quotient*, Bull. Sci. Math. **107** (1983), 3–23.

Derriennic, Y. and Guivarc'h, Y.

1. *Théorème de renouvellement pour les groupes non moyennables*, C. R. Acad. Sci. Paris Sér. A-B **277** (1973), 613–615.

Derriennic, Y. and Lin, M.

1. *On invariant measures and ergodic theorems for positive operators*, J. Funct. Anal. **13** (1973), 252–267.

Devi, S. L.

1. *Banach limits and infinite matrices*, J. London Math. Soc. **12** (1976), 397–401.

Dieudonné, J.

1. *Sur une propriété des groupes libres*, J. Reine Angew. Math. **204** (1960), 30–34.

2. *Sur le produit de composition.* II, J. Math. Pures Appl. **39** (1960), 275–292.
3. *Foundations of modern analysis*, Academic Press, New York and London, 1960.

Dijk, G. van
1. *The property P_1 for semisimple p-adic groups*, Indag. Math. **34** (1972), 82–85.

Dixmier, J.
1. *Les moyennes invariantes dans les semi-groupes et leurs applications*, Acta Sci. Math. (Szeged) **12** (1950), 213–227.
2. *L'application exponentielle dans les groupes de Lie résolubles*, Bull. Soc. Math. France **85** (1957), 113–121.
3. *Sur les représentations unitaires des groupes de Lie nilpotents.* V, Bull. Soc. Math. France **87** (1959), 65–79.
4. *Opérateurs de rang fini dans les représentations unitaires*, Inst. Hautes Études Sci. Publ. Math. **6** (1960), 305–317.

Doplicher, S., Kastler, D., and Størmer, E.
1. *Invariant states and asymptotic abelianness*, J. Funct. Anal. **3** (1969), 419–434.

Douglas, R. G.
1. *Extremal measures and subspace density*, Michigan Math. J. **11** (1964), 243–246.
2. *On the inversion invariance of invariant means*, Proc. Amer. Math. Soc. **16** (1965), 642–644.
3. *On the measure-theoretic character of an invariant mean*, Proc. Amer. Math. Soc. **16** (1965), 30–36.
4. *Generalised group algebras*, Illinois J. Math. **10** (1966), 309–321.

Douglass, S. A.
1. *Ergodic theorems for amenable semigroups*, Thesis, University of California at Los Angeles.
2. *On a concept of summability in amenable semigroups*, Math. Scand. **23** (1968), 96–102.

Downing, D. J. and Ray, O. W.
1. *Uniformly Lipschitzian semigroups in Hilbert space*, Canad. Math. Bull. **25** (1982), 210–214.

Dubins, L. E.
1. *Finitely additive conditional probabilities, conglomerality and disintegrations*, Ann. Probab. **3** (1975), 89–99.

Duncan, J. and Hosseiniun, S. A. R.
1. *The second dual of a Banach algebra*, Proc. Roy. Soc. Edinburgh Sect. A **84** (1979), 309–325.

Duncan, J. and Namioka, I.

1. *Amenability of inverse semigroups and their semigroup algebras*, Proc. Roy. Soc. Edinburgh Sect. A **80** (1978), 309–321.

Duncan, J. and Paterson, A. L. T.

1. *C^*-algebras of inverse semigroups*, Proc. Edinburgh Math. Soc. **28** (1985), 41–58.

Duncan, J. and Williamson, J. H.

1. *Spectra of elements in the measure algebra of a free group*, Proc. Roy. Irish Acad. **82A** (1982), 109–120.

Dunford, N.

1. *An individual ergodic theorem for non-commutative transformations*, Acta Sci. Math. (Szeged) **14** (1951), 1–4.

Dunkl, C. F. and Ramirez, D. E.

1. *Weakly almost periodic functions*, Sympos. Math. **22** (1971), 11–20.
2. *Fourier-Stieltjes transforms and weakly almost periodic functionals for compact groups*, Pacific J. Math. **39** (1971), 637–639.
3. *Helson sets in compact and locally compact groups*, Michigan Math. J. **19** (1972), 65–69.
4. *Existence and non-uniqueness of invariant means in $\mathfrak{L}^\infty(\hat{G})$*, Proc. Amer. Math. Soc. **32** (1972), 525–530.
5. *Bounded projections on Fourier-Stieltjes transforms*, Proc. Amer. Math. Soc. **31** (1972), 122–126.
6. *Weakly almost periodic functionals carried by hypercosets*, Trans. Amer. Math. Soc. **164** (1972), 427–434.
7. *C^*-algebras generated by Fourier-Stieltjes transforms*, Trans. Amer. Math. Soc. **164** (1972), 435–441.
8. *Subalgebras of the dual of the Fourier algebra of a compact group*, Math. Proc. Cambridge Philos. Soc. **71** (1972), 329–333.
9. *Weakly almost periodic functionals on the Fourier algebra*, Trans. Amer. Math. Soc. **185** (1973), 501–514.

Duran, J. P.

1. *Almost convergence, summability and ergodicity*, Canadian J. Math. **26** (1974), 372–387.

Dye, H. A.

1. *On the ergodic mixing theorem*, Trans. Amer. Math. Soc. **118** (1965), 123–130.

Dzinotyiweyi, H. A. M.

1. *Algebras of functions and invariant means on semigroups*, Preprint.
2. *Non-separability of quotient spaces of function algebras on topological semigroups*, Trans. Amer. Math. Soc. **272** (1982), 223–235.

Eberlein, P.

1. *Some properties of the fundamental group of a fuchsian manifold*, Invent. Math. **19** (1973), 5–13.

Eberlein, W.

1. *Abstract ergodic theorems and weak almost periodic functions*, Trans. Amer. Math. Soc. **67** (1949), 217–240.

Edwards, R. E.

1. *Functional analysis: Theory and applications*, Holt, Rinehart and Winston, New York, 1965.

Effros, E. G.

1. *Property* Γ *and inner amenability*, Proc. Amer. Math. Soc. **47** (1975), 483–486.

Effros, E. G. and Hahn, F.

1. *Locally compact transformation groups and* C^**-algebras*, Mem. Amer. Math. Soc. **75** (1967).

Effros, E. G. and Lance, E. C.

1. *Tensor products of operator algebras*, Adv. in Math. **25** (1977), 1–34.

Effros, E. G. and Størmer, E.

1. *Positive projections and Jordan structure in operator algebra*, Math. Scand. **45** (1979), 127–138.

Elie, L.

1. *Renouvellement sur les groupes moyennables*, C. R. Acad. Sci. Paris Sér. A **284** (1977), 555–558.

Elliott, G. A.

1. *On approximately finite-dimensional von Neumann algebras*, Math. Scand. **39** (1976), 91–101.
2. *On approximately finite-dimensional von Neumann algebras.* II, Canad. Math. Bull. **21** (1978), 415–418.

Elliott, G. A. and Woods, E. J.

1. *The equivalence of various definitions for a properly infinite von Neumann algebra to be approximately finite-dimensional*, Proc. Amer. Math. Soc. **60** (1976), 175–178.

Ellis, R.

1. *Locally compact transformation groups*, Duke Math. J. **24** (1957), 119–125.
2. *A note on the continuity of the inverse*, Proc. Amer. Math. Soc. **8** (1957), 372–373.

Emerson, W. R.

1. *Sequences of sets with ratio properties in locally compact groups and asymptotic properties of a class of integral operators*, Thesis, University of California, Berkeley, 1967.
2. *Ratio properties in locally compact abelian groups*, Trans. Amer. Math. Soc. **133** (1968), 179–204.
3. *Large symmetric sets in amenable groups and the individual ergodic theorem*, Amer. J. Math. **96** (1974), 242–247.
4. *The pointwise ergodic theorem for amenable groups*, Amer. J. Math. **96** (1974), 472–487.
5. *Averaging strongly subadditive set functions in unimodular amenable groups*. I, Pacific J. Math. **61** (1975), 391–400.
6. *Averaging strongly subadditive set functions in unimodular amenable groups*. II, Pacific J. Math. **64** (1976), 353–368.
7. *Characterizations of amenable groups*, Trans. Amer. Math. Soc. **241** (1978), 183–194.
8. *The Hausdorff paradox for general group actions*, J. Funct. Anal. **32** (1979), 213–227.

Emerson, W. R. and Greenleaf, F. P.

1. *Covering properties and Følner conditions for locally compact groups*, Math. Z. **102** (1967), 370–384.
2. *Asymptotic behaviour of products $C^p = C + \cdots + C$ in locally compact abelian groups*, Trans. Amer. Math. Soc. **145** (1969), 171–204.

Eymard, P.

1. *Algèbres A_p et convoluteurs de L^p*, Séminaire Bourbaki, No. 367 (1969–70).
2. *Moyennes invariantes et représentations unitaires*, Lecture Notes in Math., vol. 300, Springer-Verlag, Berlin–Heidelberg–New York, 1972.
3. *Invariant means and unitary representations*, Harmonic Analysis on Homogeneous Spaces, Proc. Sympos. Pure Math., Vol. **26** Amer. Math. Soc., Providence, R. I., 1973, pp. 373–376.

Eymard, P., Faraut, J., Schiffman, G., and Takahashi, R. (editors)

1. *Analyse harmonique sur les groupes de Lie*, Séminaire Nancy-Strasbourg 1973–1975, Lecture Notes in Math., vol. 497, Springer-Verlag, Berlin–Heidelberg–New York, 1977.

Fairchild, L. R.

1. *Extreme invariant means and minimal sets in the Stone-Čech compactification of a semigroup*, Thesis, University of Illinois, 1970.
2. *Extreme invariant means without minimal support*, Trans. Amer. Math. Soc. **172** (1972), 83–93.

Fan, K.

1. *Orbits of semigroups of contractions and groups of isometries*, Abh. Math. Sem. Univ. Hamburgh **45** (1976), 245–250.
2. *Extension of invariant linear functionals*, Proc. Amer. Math. Soc. **66** (1977), 23–29.

Federer, H. and Morse, A. P.

1. *Some properties of measurable functions*, Bull. Amer. Math. Soc. **49** (1943), 270–277.

Feichtinger, H. G.

1. *Results on Banach ideals and spaces of multipliers*, Math. Scand. **41** (1977), 315–324.

Feldman, J. and Greenleaf, F. P.

1. *Existence of Borel transversals in groups*, Pacific J. Math. **25** (1968), 455–462.

Feldman, J., Hahn, P., and Moore, C. C.

1. *Orbit structure and countable sections for actions of continuous groups*, Adv. in Math. **28** (1978), 186–230.

Feldman, J. and Moore, C. C.

1. *Ergodic equivalence relations, cohomology and von Neumann algebras*. II, Trans. Amer. Math. Soc. **234** (1977), 325–359.

Felix, R., Henrichs, R. W., and Skudlarek, H. L.

1. *Topological Frobenius reciprocity for projective limits of Lie groups*, Math. Z. **165** (1979), 19–28.

Fell, J. M. G.

1. *The dual spaces of C^*-algebras*, Trans. Amer. Math. Soc. **94** (1960), 365–403.
2. *A new proof that nilpotent groups are CCR*, Proc. Amer. Math. Soc. **13** (1962), 93–99.
3. *Weak containment and induced representations of groups*, Canad. J. Math. **14** (1962), 237–268.
4. *Weak containment and induced representations of groups*. II, Trans. Amer. Math. Soc. **110** (1964), 424–447.

Fickett, J. and Mycielski, J.

1. *A problem of invariance for Lebesgue measure*, Colloq. Math. **42** (1979), 123–125.

Figà-Talàmanca, A.

1. *Translation invariant operators in L^p*, Duke Math. J. **32** (1965), 495–501.

Finetti, B. de

1. *La prevision: ses lois logiques, ses sources subjectives*, Ann. Inst. H. Poincaré **7** (1937), 1–68.

Fisher, M. J.

1. *Recognition and limit theorems for L_p multipliers*, Studia Math. **50** (1974), 31–41.
2. *Multipliers on p-Fourier algebras*, Studia Math. **54** (1975), 109–116.
3. *On the algebra of multipliers of a p-Fourier algebra*, Amer. J. Math. **98** (1976), 171–181.

Flory, V.

1. *Estimating norms in C^*-algebras of discrete groups*, Math. Ann. **224** (1976), 41–52.

Foguel, S. R.

1. *The ergodic theory of Markov processes*, Van Nostrand, New York, 1969.

Følner, E.

1. *Generalization of a theorem of Bogolioúboff to topological abelian groups*, Math. Scand. **2** (1954), 5–18.
2. *On groups with full Banach mean value*, Math. Scand. **3** (1955), 243–254.
3. *Notes on groups with and without full Banach mean value*, Math. Scand. **5** (1957), 5–11.

Fong, H.

1. *On invariant functions for positive operators*, Colloq. Math. **22** (1970), 75–84.

Fong, H. and Sucheston, L.

1. *On the ratio ergodic theorem for semigroups*, Pacific J. Math. **39** (1971), 659–667.

Fountain, J. B., Ramsay, R. W. and Williamson, J. H.

1. *Functions of measures on compact groups*, Proc. Roy. Irish Acad. Sect. A **76** (1976), 235–251.

Freedman, D. A. and Purves, R. A.

1. *Bayes' method for bookies*, Ann. Math. Statist. **40** (1969), 1177–1186.

Frémond, C. and Suer-Pontier, M.

1. *Caractérisation des groupes localement compacts de type (T) ayant la propriété de point fixe*, Ann. Inst. H. Poincaré Sect. B **7** (1971), 293–298.

Frey, A. H.

1. *Studies in amenable semigroups*, Thesis, University of Washington, Seattle, Washington, 1960.

Friedman, N. A.

1. *Introduction to ergodic theory*, Van Nostrand, New York, 1970.

Frolik, Z.

1. *Baire spaces and some generalizations of complete metric spaces*, Czech. Math. J. **11** (1961), 237–238.

Furstenberg, H.

1. *A Poisson formula for semisimple Lie groups*, Ann. of Math. (2) **77** (1963), 335–386.
2. *Non-commuting random products*, Trans. Amer. Math. Soc. **108** (1963), 377–428.
3. *Boundary theory and stochastic processes on homogeneous spaces*, Harmonic Analysis on Homogeneous Spaces, Proc. Sympos. Pure Math., Vol. **26** Amer. Math. Soc., Providence, R. I., 1973, pp. 192–229.
4. *Random walks on Lie groups*, Harmonic Analysis and Representations of Semisimple Lie Groups (J. A. Wolf, M. Cahen and M. De Wilde, eds.), Reidel, Dordrecht, Holland, 1980, pp. 467–489.

Gait, J.

1. *Transformation groups with no equicontinuous minimal set*, Compositio Math. **25** (1972), 87–92.

Gangoli, R.

1. *On the symmetry of L_1 algebras of locally compact motion groups and the Wiener Tauberian theorem*, J. Funct. Anal. **25** (1977), 244–252.

Gaudry, G. I.

1. *Quasimeasures and operators commuting with convolution*, Pacific J. Math. **18** (1966), 461–476.
2. *Multipliers of type (p, q)*, Pacific J. Math. **18** (1966), 477–488.

Gerl, P.

1. *Uber die Anzahl der Darstellungen von Worten*, Monatsh. Math. **75** (1971), 205–214.
2. *Probability measures on semigroups*, Proc. Amer. Math. Soc. **40** (1973), 527–532.
3. *Diskrete mittelbare Gruppen*, Monatsh. Math. **77** (1973), 307–318.
4. *Gleichverteilung auf lokal kompakten Gruppen*, Math. Nachr. **71** (1976), 249–260.
5. *Wahrscheinlichkeitsmasse auf diskreten Gruppen*, Arch. Math. **31** (1978), 611–619.

Ghatage, P.

1. *C^*-algebras generated by weighted shifts*, Indiana Univ. Math. J. **28** (1979), 1007–1012.

Gilbert, J. E.

1. *Convolution operators on $L^p(G)$ and properties of locally compact groups*, Pacific J. Math. **24** (1968), 257–268.
2. *L^p convolution operators and tensor products of Banach spaces*, Bull. Amer. Math. Soc. **80** (1974), 1127–1132.
3. *Harmonic analysis and the Grothendieck fundamental theorem*, Sympos. Math. **22** (1977), 393–420.

Gillman, L. and Jerison M.

1. *Rings of continuous functions*, Van Nostrand, New York, 1960.

Glicksberg, I.

1. *Weak compactness and separate continuity*, Pacific J. Math. **11** (1961), 205–216.
2. *On convex hulls of translates*, Pacific J. Math. **13** (1963), 153–164.

Godement, R.

1. *Les fonctions de type positif et la theorie des groupes*, Trans. Amer. Math. Soc. **63** (1948), 1–84.

Goldberg, K. and Newman, M.

1. *Pairs of matrices of order two which generate free groups*, Illinois J. Math. **1** (1957), 446–448.

Golod, E. S.

1. *On nil algebras and finitely approximable groups*, Izv. Akad. Nauk SSSR Ser. Mat. **28** (1964), 273–276.

Golod, E. S. and Shafarevitch, I. R.

1. *On towers of class fields*, Izv. Akad. Nauk SSSR Ser. Mat. **28** (1964), 261–272.

Gootman, E. C.

1. *Weak containment and weak Frobenius reciprocity*, Proc. Amer. Math. Soc. **54** (1976), 417–422.

Gootman, E. C. and Rosenberg, J.

1. *The structure of crossed product C^*-algebras: a proof of the generalized Effros-Hahn conjecture*, Invent. Math. **52** (1979), 283–298.

Goto, M.

1. *Faithful representations of Lie groups*. I, Math. Japonicae **1** (1948), 107–118.

Gottschalk, W. H. and Hedlund, G. A.

1. *Topological dynamics*, Amer. Math. Soc., Providence, R. I., 1955.

Granirer, E. E.

1. *On left amenable semigroups which admit countable left-invariant means*, Bull. Amer. Math. Soc. **69** (1963), 101–105.
2. *On amenable semigroups with a finite-dimensional set of invariant means*. I, Illinois J. Math. **7** (1963), 32–48.
3. *On amenable semigroups with a finite-dimensional set of invariant means*. II, Illinois J. Math. **7** (1963), 49–58.
4. *A theorem on amenable semigroups*, Trans. Amer. Math. Soc. **111** (1964), 367–369.
5. *Extremely amenable semigroups*, Math. Scand. **17** (1965), 177–197.

6. *On the invariant mean on topological semigroups and on topological groups*, Pacific J. Math. **15** (1965), 107–140.
7. *Extremely amenable semigroups*, Bull. Amer. Math. Soc. **72** (1966), 1028–1032.
8. *On the range of an invariant mean*, Trans. Amer. Math. Soc. **125** (1966), 384–394.
9. *On Baire measures on D-topological spaces*, Fund. Math. **60** (1967), 1–22.
10. *Extremely amenable semigroups.* II, Math. Scand. **20** (1967), 93–113.
11. *Functional analytic properties of extremely amenable semigroups*, Trans. Amer. Math. Soc. **137** (1969), 53–75.
12. *On finite equivalent invariant measures for semigroups of transformations*, Duke Math. J. **38** (1971), 395–408.
13. *Exposed points of convex sets and weak sequential convergence*, Mem. Amer. Math. Soc. **123** (1972).
14. *The radical of $L^\infty(G)^*$*, Proc. Amer. Math. Soc. **41** (1973), 321–324.
15. *Criteria for compactness and for discreteness of locally compact amenable groups*, Proc. Amer. Math. Soc. **40** (1973), 615–624.
16. *Properties of the set of topologically invariant means on P. Eymard's W^*-algebra $VN(G)$*, Indag. Math. **36** (1974), 116–121.
17. *Weakly almost periodic and uniformly continuous functionals on the Fourier algebra of any locally compact group*, Trans. Amer. Math. Soc. **189** (1974), 371–382.
18. *A characterization of discreteness for locally compact groups in terms of the Banach algebras $A_p(G)$*, Proc. Amer. Math. Soc. **54** (1976), 189–192.
19. *Density theorems for some linear subspaces and some C^*-subalgebras of $VN(G)$*, Sympos. Math. **22** (1977), 61–70.
20. *On group representations whose C^*-algebra is an ideal in its von Neumann algebra*, Ann. Inst. Fourier (Grenoble) **29** (1979), 37–52.
21. *An application of the Radon-Nikodym property in harmonic analysis*, Boll. Un. Mat. Ital. B (5) **18** (1981), 663–671.
22. *On group representations whose C^*-algebra is an ideal in its von Neumann algebra.* II, J. London Math. Soc. **26** (1982), 308–316.
23. *Geometric and topological properties of certain w^* compact convex subsets of double duals of Banach spaces, which arise from the study of invariant means*, Preprint.

Granirer, E. E. and Lau, A. T.

1. *Invariant means on locally compact groups*, Illinois J. Math. **15** (1971), 249–257.

Granirer, E. E. and Leinert, M.

1. *On some topologies which coincide on the unit sphere of the Fourier-Stieltjes algebra $B(G)$ and of the measure algebra $M(G)$*, Rocky Mountain J. Math. **11** (1981), 459–472.

Granirer, E. E. and Rajagopalan, M.

1. *A note on the radical of the second conjugate algebra of a semigroup algebra*, Math. Scand. **15** (1964), 163–166.

Green, P.

1. *C^*-algebras of transformation groups with smooth orbit space*, Pacific J. Math. **72** (1977), 71–97.

Greenleaf, F. P.

1. *Følner's condition for locally compact groups*, unpublished manuscript, 1967.
2. *Invariant means on topological groups*, Van Nostrand, New York, 1969.
3. *Amenable actions of locally compact groups*, J. Funct. Anal. **4** (1969), 295–315.
4. *Ergodic theorems and the construction of summing sequences in amenable locally compact groups*, Comm. Pure Appl. Math. **26** (1973), 29–46.
5. *Concrete methods for summing almost periodic functions and their relation to uniform distribution of semigroup actions*, Colloq. Math. **41** (1979), 105–116.

Greenleaf, F. P. and Emerson, W. R.

1. *Group structure and the pointwise ergodic theorem for connected amenable groups*, Adv. in Math. **14** (1974), 153–172.

Greenleaf, F. P., Moskowitz, M., and Rothschild, L. P.

1. *Unbounded conjugacy classes in Lie groups and location of central measures*, Acta. Math. **132** (1974), 225–244.
2. *A unipotent group associated with certain linear groups*, Colloq. Math. **43** (1980), 41–45.

Grigorchuk, R. I.

1. *Burnside's problem on periodic groups*, Functional Anal. Appl. **14** (1980), 41–43.
2. *Symmetric random walks on discrete groups*, Multi-Component Random Systems (Dobrushin and Sinai, eds.), M. Dekker, New York–Basel, 1980, pp. 285–325.

Gromoll, D. and Wolf, J. A.

1. *Some relations between the metric structure and the algebraic structure of the fundamental group in manifolds of nonpositive curvature*, Bull. Amer. Math. Soc. **77** (1971), 545–552.

Gromov, M.

1. *Groups of polynomial growth and expanding maps*, Inst. Hautes Études Sci. Publ. Math. **53** (1981), 53–73.

Grosser, S. and Moskowitz, M.

1. *On central topological groups*, Trans. Amer. Math. Soc. **127** (1967), 317–340.

2. *Compactness conditions in topological groups*, J. Reine Angew. Math. **246** (1971), 1–40.

Grossman, M. W.

1. *A categorical approach to invariant means and fixed-point properties*, Semigroup Forum **5** (1972), 14–44.
2. *Uniqueness of invariant means on certain introverted spaces*, Bull. Austral. Math. Soc. **9** (1973), 109–120.
3. *Invariant means and fixed-point properties on completely regular spaces*, Bull. Austral. Math. Soc. **16** (1977), 203–212.

Grothendieck, A.

1. *Critères de compacité dans les espaces fonctionelles generaux*, Amer. J. Math. **74** (1952), 168–186.
2. *Produits tensoriels topologiques et espaces nucléaires*, Mem. Amer. Math. Soc. **16** (1955).
3. *Topological vector spaces*, Gordon and Breach, London, 1973.

Guivarc'h, Y.

1. *Croissance polynomiale et périodes des fonctions harmonique*, Bull. Soc. Math. France **101** (1973), 333–379.
2. *Fonctions propres des opérateurs de convolution et loi des grandes nombres*, Preprint.

Gulick, S. L.

1. *Commutativity and ideals in the biduals of topological algebras*, Pacific J. Math. **18** (1966), 121–137.

Haagerup, U.

1. *An example of a non nuclear C^*-algebra which has the metric approximation property*, Invent. Math. **50** (1979), 279–293.
2. *The Grothendieck inequality for bilinear forms on C^*-algebras*, Preprint, 1981.
3. *All nuclear C^*-algebras are amenable*, Invent. Math. **74** (1983), 305–319.
4. *The reduced C^*-algebra of the free group on two generators*, 18th Scandinavian Congr. Math., Birkhäuser, Boston, 1981, pp. 321–325.
5. *Injectivity and decomposition of completely bounded maps*, Preprint.
6. *A new proof of the equivalence of injectivity and hyperfiniteness for factors on a separable Hilbert space*, J. Funct. Anal. **62** (1985), 160–201.

Hahn, P.

1. *The σ-representations of amenable groupoids*, Rocky Mountain J. Math. **9** (1979), 631–639.

Hajian, A. and Ito, Y.

1. *Weakly wandering sets and invariant measures for a group of transformations*, J. Math. Mech. **18** (1969), 1203–1216.

Hakeda, J.

1. *On property P of von Neumann algebras*, Tôhoku Math. J. **19** (1967), 238–242.

Hakeda, J. and Tomiyama, J.

1. *On some extension properties of von Neumann algebras*, Tôhoku Math. J. **19** (1967), 315–323.

Hall, M.

1. *The theory of groups*, Macmillan, New York, 1959.

Hall, P.

1. *Nilpotent groups*, Queen Mary College Mathematics Notes, 1969.

Halmos, P. R.

1. *Measure theory*, Van Nostrand, Princeton, N. J., 1950.

Harpe, P. de la

1. *Moyennabilité de quelques groupes topologiques de dimension infinie*, C. R. Acad. Sci. Paris Sér. A **277** (1973), 1037–1040.
2. *Moyennabilité du groupe unitaire et propriété P de Schwartz des algèbres de von Neumann*, Lecture Notes in Math., vol. 725 (P. de la Harpe, ed.), Springer-Verlag, Berlin–Heidelberg–New York, 1979, pp. 220–227.

Hart, G.

1. *Absolutely continuous measures on semigroups*, Thesis, University of Kansas, 1970.

Hauenschild, W. and Kaniuth, E.

1. *The generalized Wiener theorem for groups with finite-dimensional irreducible representations*, J. Funct. Anal. **31** (1979), 13–23.

Hausdorff, F.

1. *Bemerkung über den Inhalt von Punktmengen*, Math. Ann. **75** (1914), 428–433.
2. *Grundzüge der Mengenlehre*. Veit, Leipzig, 1914.

Heath, D. and Sudderth, W.

1. *On finitely additive priors, coherence, and extended admissability*, Ann. Statist. **6** (1978), 333–345.

Henrichs, R. W.

1. *Die Frobeniuseigenschaft FP für diskrete Gruppen*, Math. Z. **147** (1976), 191–199.
2. *Weak Frobenius reciprocity and compactness conditions in topological groups*, Pacific J. Math. **82** (1979), 387–406.

Herstein, I. N.

1. *Topics in algebra*, Waltham, Mass.–Toronto–London, 1964.

2. *Non-commutative rings*, The Carus Mathematical Monographs, Math. Assoc. Amer., Washington, D. C., 1968.

Herz, C.

1. *The spectral theory of bounded functions*, Trans. Amer. Math. Soc. **94** (1960), 181–232.
2. *Remarques sur la note précédente de M. Varopoulos*, C. R. Acad. Sci. Paris Sér. A **260** (1965), 6001–6004.
3. *Le rapport entre l'algèbre A_p d'un groupe et d'un sous-groupe*, C. R. Acad. Sci. Paris Sér. A–B **271** (1970), 244-246.
4. *The theory of p-spaces with an application to convolution operators*, Trans. Amer. Math. Soc. **154** (1971), 69–82.
5. *Harmonic synthesis for subgroups*, Ann. Inst. Fourier (Grenoble) **23** (1973), 91–123.
6. *Une généralisation de la notion de transformée de Fourier-Stieltjes*, Ann. Inst. Fourier (Grenoble) **23** (1974), 145–157.
7. *On the asymmetry of norms of convolution operators.* I, J. Funct. Anal. **23** (1976), 11–22.

Hewitt, E.

1. *A new proof of Plancherel's theorem for locally compact groups*, Acta Sci. Math. (Szeged) **24** (1963), 219–227.
2. *Invariant means of compact semigroups*, unpublished manuscript, 1964.

Hewitt, E. and Stromberg, K.

1. *Real and abstract analysis*, Springer-Verlag, New York–Heidelberg–Berlin, 1965.

Hey, H. J. and Ludwig, J.

1. *Der Satz von Helson-Reiter für spezielle nilpotente Lie-Gruppen*, Math. Ann. **239** (1979), 207–218.

Heyneman, R. G.

1. *Dualitiy in ergodic theory*, Pacific J. Math. **12** (1962), 1329–1341.

Hiai, F. and Sato, R.

1. *Mean ergodic theorems for semigroups of positive linear operators*, J. Math. Soc. Japan **29** (1977), 123–134.

Hicks, N. J.

1. *Notes on differential geometry*, Van Nostrand Math. Studies, No. 3, Van Nostrand, Princeton, N. J., 1965.

Hille, E. and Phillips, R. S.

1. *Functional analysis and semigroups*, Amer. Math. Soc., Providence, R. I., 1957.

Hindman, N. and Pym, J.

1. *Free groups and semigroups on βN*, Semigroup Forum **30** (1984), 177–193.

Hirsch, M. W. and Thurston, W. P.

1. *Foliated bundles, invariant measures and flat manifolds*, Ann. of Math. (2) **101** (1975), 369–390.

Hochschild, G.

1. *On the cohomology groups of an associative algebra*, Ann. of Math. (2) **46** (1945), 58–67.
2. *On the cohomology theory for associative algebras*, Ann. of Math. (2) **47** (1946), 568–579.
3. *Cohomology and representations of associative algebras*, Duke Math. J. **14** (1947), 921–948.

Hochster, M.

1. *Subsemigroups of amenable groups*, Proc. Amer. Math. Soc. **21** (1969), 363–364.

Hofmann, K. H. and Mukherjea, A.

1. *On the density of the image of the exponential function*, Math. Ann. **234** (1978), 263–273.

Holmes, R. D. and Lau, A. T.

1. *Non-expansive actions of topological semigroups and fixed points*, J. London Math. Soc. **5** (1972), 330–336.

Howie, J. M.

1. *An introduction to semigroup theory*, Academic Press, London–New York–San Francisco, 1976.

Huff, R. E.

1. *Invariant functionals and fixed-point theorems*, Thesis, University of North Carolina, 1969.
2. *Some applications of a general lemma on invariant means*, Illinois J. Math. **14** (1970), 216–221.
3. *Existence and uniqueness of fixed points for semigroups of affine maps*, Trans. Amer. Math. Soc. **152** (1970), 99–106.

Hulanicki, A.

1. *On symmetry in group algebras*, Bull. Acad. Polon. Sci. Sér. Sci. Math. Astronom. Phys. **11** (1963), 1–2.
2. *Groups whose regular representation weakly contains all unitary representations*, Studia Math. **24** (1964), 37–59.
3. *On the spectral radius of hermitian elements in group algebras*, Pacific J. Math. **18** (1966), 277–287.
4. *Means and Følner condition on locally compact groups*, Studia Math. **27** (1966), 87–104.
5. *On symmetry of group algebras of discrete nilpotent groups*, Studia Math. **35** (1970), 207–219.

6. *On positive functionals on a group algebra multiplicative on a subalgebra*, Studia Math. **37** (1971), 163–171.
7. *On the spectrum of convolution operators on groups with polynomial growth*, Invent. Math. **17** (1972), 135–142.
8. *Subalgebra of $L_1(G)$ associated with Laplacian on a Lie group*, Colloq. Math. **31** (1974), 259–287.
9. *On L_p-spectra of the Laplacian on a Lie group with polynomial growth*, Proc. Amer. Math. Soc. **44** (1974), 482–484.
10. *On the spectrum of the Laplacian on the affine group of the real line*, Studia Math. **54** (1976), 199–204.
11. *Growth of the L^1 norm on the convolution powers of functions on nilpotent groups of class* 2, Sympos. Math. **22** (1977), 439–447.
12. *A solvable group with polynomial growth and non-symmetric group algebra*, Preprint.
13. *Invariant subsets of non-synthesis Leptin algebras and non-symmetry*, Colloq. Math. **43** (1980), 127–136.

Hulanicki, A., Jenkins, J. W., Leptin, H., and Pytlik, T.

1. *Remarks on Wiener's Tauberian theorems for groups with polynomial growth*, Colloq. Math. **35** (1976), 293–304.

Hulanicki, A. and Pytlik, T.

1. *On commutative approximate identities and cyclic vectors of induced representations*, Studia Math. **48** (1973), 188–199.

Hulanicki, A. and Ryll-Nardzewski, C.

1. *Invariant extensions of the Haar measure*, Colloq. Math. **42** (1979), 223–227.

Humphreys, J. E.

1. *Linear algebraic groups*, Springer-Verlag, Berlin–Heidelberg–New York, 1975.
2. *Arithmetic groups*, Lecture Notes in Math., vol. 789, Springer-Verlag, Berlin–Heidelberg–New York, 1980.

Hunt, G. A. and Stein, C.

1. *Most stringent tests of statistical hypotheses*, unpublished manuscript.

Husain, T.

1. *Amenability of locally compact groups and vector-valued function spaces*, Sympos. Math. **16** (1975), 417–431.

Husain, T. and Wong, J. C. S.

1. *Invariant means on vector-valued functions.* I, Ann. Scuola Norm. Sup. Pisa **27** (1973), 717–729.
2. *Invariant means on vector-valued functions.* II, Ann. Scuola Norm. Sup. Pisa **27** (1973), 729–742.

Itzkowitz, G. L.

1. *Continuous measures, Baire category and uniform continuity in topological groups*, Pacific J. Math. **54** (1974), 115–125.

Iverson, P.

1. *Mean value of compact convex sets, Day's fixed-point theorem and invariant subspaces*, J. Math. Anal. Appl. **57** (1977), 1–11.

Iwasawa, K.

1. *Some types of topological groups*, Ann. of Math. (2) **50** (1949), 507–558.

Jech, T.

1. *Set theory*, Academic Press, New York–San Francisco–London, 1978.

Jenkins, J. W.

1. *An amenable group with a non-symmetric group algebra*, Bull. Amer. Math. Soc. **75** (1969), 357–360.
2. *Symmetry and non-symmetry in the group algebras of discrete groups*, Pacific J. Math. **32** (1970), 131–145.
3. *Amenable subsemigroups of locally compact groups*, Proc. Amer. Math. Soc. **25** (1970), 766–770.
4. *On the spectral radius of elements in a group algebra*, Illinois J. Math. **15** (1971), 551–554.
5. *Free semigroups and unitary group representations*, Studia Math. **43** (1972), 27–39.
6. *Growth of connected locally compact groups*, J. Funct. Anal. **12** (1973), 113–127.
7. *A characterization of growth in locally compact groups*, Bull. Amer. Math. Soc. **79** (1973), 103–106.
8. *Non-symmetric group algebras*, Studia Math. **45** (1973), 295–307.
9. *Følner's condition for exponentially bounded groups*, Math. Scand. **35** (1974), 165–174.
10. *A fixed-point theorem for exponentially bounded groups*, J. Funct. Anal. **22** (1976), 346–353.
11. *Representations of exponentially bounded groups*, Amer. J. Math. **98** (1976), 29–38.
12. *On group actions with non-zero fixed points*, Pacific J. Math. **91** (1980), 363–371.

Jerison, M.

1. *A property of extreme points of compact convex sets*, Proc. Amer. Math. Soc. **5** (1954), 782–783.
2. *On the set of all generalised limits of bounded sequences*, Canad. J. Math. **9** (1957), 79–89.

Johnson, B. E.

1. *Separate continuity and measurability*, Proc. Amer. Math. Soc. **20** (1969), 420–422.

2. *Cohomology in Banach algebras*, Mem. Amer. Math. Soc. **127** (1972).
3. *Approximate diagonals and cohomology of certain annihilator Banach algebras*, Amer. J. Math. **94** (1972), 685–698.
4. *Some examples in harmonic analysis*, Studia Math. **48** (1973), 181–188.
5. *Perturbations of Banach algebras*, Proc. London Math. Soc. **34** (1977), 439–458.
6. *A counterexample in the perturbation theory of C^*-algebras*, Canad. Math. Bull. **25** (1982), 311–316.

Johnson, B. E., Kadison, R. V., and Ringrose, J. R.

1. *Cohomology of operator algebras.* III: *Reduction to normal cohomology*, Bull. Soc. Math. France **100** (1972), 73–96.

Jones, V. F. R.

1. *An invariant for group actions*, Algèbres d'Opérateurs (Sém., Les Plans-sur-Bex, 1978), Lecture Notes in Math., vol. 725, Springer-Verlag, Berlin, 1979, pp. 237–253.

Junco, A. D. and Rosenblatt, J.

1. *Counterexamples in ergodic theory and number theory*, Math. Ann. **245** (1979), 185–197.

Junghenn, H. D.

1. *Some general results on fixed points and invariant means*, Semigroup Forum **11** (1975), 153–164.
2. *Amenability of function spaces on thick subsemigroups*, Proc. Amer. Math. Soc. **75** (1979), 37–41.
3. *Topological left amenability of semidirect products*, Canad. Math. Bull. **24** (1981), 79–85.
4. *Amenability induced by amenable homomorphic images*, Semigroup Forum **24** (1982), 11–23.

Kadison, R. V. and Ringrose, J. R.

1. *Cohomology of operator algebras.* I: *Type* I *von Neumann algebras*, Acta Math. **126** (1971), 227–243.
2. *Cohomology of operator algebras.* II: *Extended cobounding and the hyperfinite case*, Ark. Mat. **9** (1971), 55–63.

Kahane, J.-P.

1. *Sur les fonctions presque-périodiques généralisées dont le spectre est vide*, Studia Math. **21** (1962), 231–236.
2. *Séries de Fourier absolument convergentes*, Springer-Verlag, Berlin–Heidelberg–New York, 1970.

Kamowitz, H.

1. *Cohomology groups of commutative Banach algebras*, Trans. Amer. Math. Soc. **102** (1962), 352–372.

Kaniuth, E.

1. *Die Struktur der regulären Darstellung lokalkompakter Gruppen mit invarianter Umgebungsbasis der Eins*, Math. Ann. **194** (1971), 225–248.

Kaufman, R.

1. *Remark on invariant means*, Proc. Amer. Math. Soc. **18** (1967), 120–122.

Kazhdan, D. A.

1. *Connection of the dual space of a group with the structure of its closed subgroups*, Functional Anal. Appl. **1** (1967), 63–65.

Kelley, J.

1. *Averaging operators on $C_\infty(X)$*, Illinois J. Math. **2** (1958), 214–223.
2. *General topology*, Van Nostrand, New York, 1969.

Kerstan, J. and Matthes, K.

1. *Gleichverteilungseigenschaften von Faltung von Verteilungsgesetzen auf lokalkompakten abelschen Gruppen.* I, Math. Nachr. **37** (1968), 267–312.

Kesten, H.

1. *Full Banach mean-values on countable groups*, Math. Scand. **7** (1959), 146–156.
2. *Symmetric random walks on groups*, Trans. Amer. Math. Soc. **92** (1959), 336–354.

Kharaghani, H.

1. *The evolution of bounded linear functionals with application to invariant means*, Pacific J. Math. **78** (1978), 369–374.
2. *Left thick subsets of a topological semigroup*, Illinois J. Math. **22** (1978), 41–48.

Khelemskii, A. Y. and Sheinberg, M. V.

1. *Amenable Banach algebras*, Functional Anal. Appl. **13** (1979), 32–37.

Kieffer, J. C.

1. *Invariance, minimax sequential estimation and continuous time processes*, Ann. Math. Statist. **28** (1957), 573–601.

Kinzl, F.

1. *Absolut stetige Masse auf lokalkompakten Halbgruppen*, Monatsh. Math. **87** (1979), 109–121.
2. *Gleichverteilung auf diskreten Halbgruppen*, Semigroup Forum **18** (1979), 105–118.

Kirillov, A. A.

1. *Unitary representations of nilpotent Lie groups*, Uspekhi Mat. Nauk. **1061** (1962), 57–110.
2. *Elements of the theory of representations*, Springer-Verlag, Berlin–Heidelberg–New York, 1976.

Kister, J. M.

1. *Uniform continuity and compactness in topological groups*, Proc. Amer. Math. Soc. **13** (1962), 37–40.

Klawe, M.

1. *On the dimension of left invariant means and left thick subsets*, Trans. Amer. Math. Soc. **231** (1977), 507–518.
2. *Semidirect product of semigroups in relation to amenability, cancellation properties and strong Følner conditions*, Pacific J. Math. **73** (1977), 91–106.
3. *Dimensions of the sets of invariant means of semigroups*, Illinois J. Math. **24** (1980), 233–243.

Klee, V.

1. *Invariant extensions of linear functionals*, Pacific J. Math. **4** (1954), 37–46.
2. *Extremal structure of convex sets*. II, Math. Z. **69** (1958), 90–104.

Komlósi, S.

1. *Mean ergodicity in G-semifinite von Neumann algebras*, Acta Sci. Math. (Szeged) **41** (1979), 327–334.

König, D.

1. *Sur les correspondances multivoques des ensembles*, Fund. Math. **8** (1926), 114–134.

Köthe, G.

1. *Topological vector spaces*. I, Springer-Verlag, Berlin–Heidelberg–New York, 1969.

Kotzmann, E. and Rindler, H.

1. *Central approximate units in a certain ideal of $L^1(G)$*, Proc. Amer. Math. Soc. **57** (1976), 155–158.

Kovacs, I. and Szücs, J.

1. *Ergodic type theorems in von Neumann algebras*, Acta Sci. Math. (Szeged) **27** (1966), 233–246.

Kuipers, L. and Niederreiter, H.

1. *Uniform distribution of sequences*, Interscience, New York, 1974.

Kunze, R. A. and Stein, E. M.

1. *Uniformly bounded representations and harmonic analysis of the 2×2 unimodular group*, Amer. J. Math. **82** (1960), 1–62.

Kuratowski, K.

1. *Applications of the Baire category method to the problem of independent sets*, Fund. Math. **81** (1973–1974), 65–72.

Lai, H. C.

1. *Restrictions of Fourier transforms on A^p*, Tôhoku Math. J. **26** (1974), 453–460.

Laison, D. and Laison, G.

1. *Weak almost periodicity on C^*-algebras*, Math. Ann. **227** (1977), 135–143.

Lance, C.

1. *On nuclear C^*-algebras*, J. Funct. Anal. **12** (1973), 157–176.
2. *Tensor products of non-unital C^*-algebras*, J. London Math. Soc. **12** (1976), 160–168.

Landsberg, M. and Schirotzek, W.

1. *Extremal and invariant extensions of linear functionals*, Math. Nachr. **71** (1976), 191–202.

Lane, D. A. and Sudderth, W. D.

1. *Diffuse models for sampling and predictive inference*, Ann. Statist. **6** (1978), 1318–1336.

Lau, A. T.

1. *Functional analytic properties of topological semigroups and N-extreme amenability*, Trans. Amer. Math. Soc. **152** (1970), 431–439.
2. *Topological semigroups with invariant means in the convex hull of multiplicative means*, Trans. Amer. Math. Soc. **148** (1970), 69–84.
3. *Invariant means on dense subsemigroups of topological groups*, Canad. J. Math. **23** (1971), 797–801.
4. *Action of topological semigroups, invariant means and fixed points*, Studia Math. **43** (1972), 139–156.
5. *Invariant means on almost periodic functions and fixed point properties*, Rocky Mountain J. Math. **3** (1973), 69–76.
6. *Invariant means on subsemigroups of locally compact groups*, Rocky Mountain J. Math. **3** (1973), 77–81.
7. *Invariant means on almost periodic functions and equicontinuous actions*, Proc. Amer. Math. Soc. **49** (1975), 379–382.
8. *Semigroups of operators on dual Banach spaces*, Proc. Amer. Math. Soc. **54** (1976), 393–396.
9. *W^*-algebras and invariant functionals*, Studia Math. **56** (1976), 253–261.
10. *Extension of invariant linear functionals: a sequel to Fan's paper*, Proc. Amer. Math. Soc. **63** (1977), 259–262.
11. *Characterizations of amenable Banach algebras*, Proc. Amer. Math. Soc. **70** (1978), 156–160.
12. *Actions of semitopological semigroups and extension properties*, Preprint.
13. *Uniformly continuous functionals on the Fourier algebra of any locally compact group*, Trans. Amer. Math. Soc. **251** (1979), 39–59.

14. *Invariantly complemented subspaces of $L_\infty(G)$ and amenable locally compact groups*, Illinois J. Math. **26** (1982), 226–235.
15. *Some fixed point theorems and their applications to W^*-algebras*, Fixed-Point Theory and Its Applications, Academic Press, New York–San Francisco–London, 1976, pp. 121–129.

Lau, A. T. and Wong, C. S.
1. *Common fixed points for semigroups of mappings*, Proc. Amer. Math. Soc. **41** (1973), 223–228.

Lawson, J. D.
1. *Additional notes on continuity in semitopological semigroups*, Semigroup Forum **12** (1976), 265–280.

Lawson, H. B. and Yau, S. T.
1. *Compact manifolds of nonpositive curvature*, J. Differential Geom. **7** (1972), 211–228.

Lebesgue, H.
1. *Leçons sur l'intégration et la recherche des fonctions primitives*, Gauthier-Villars, Paris, 1904.

Le Cam, L.
1. *Sufficiency and approximate sufficiency*, Ann. Math. Statist. **35** (1964), 1419–1455.
2. *On the information contained in additional observations*, Ann. Statist. **2** (1974), 630–649.

Lee, T.
1. *Embedding theorems in group C^*-algebras*, Canad. Math. Bull. **26** (1983), 157–166.

Lehman, E.
1. *Testing statistical hypotheses*, Wiley, New York, 1959.

Leinert, M.
1. *Faltungoperatoren auf gewissen diskreten Gruppen*, Studia Math. **52** (1974), 149–158.
2. *Abschätzung von Normen gewinner Matrizen und eine Anwendung*, Math. Ann. **24** (1979), 13–19.

Leptin, H.
1. *Faltungen von Borelschen Massen mit L^p-Funktionen auf lokalkompakten Gruppen*, Math. Ann. **163** (1966), 111–117.
2. *On a certain invariant of a locally compact group*, Bull. Amer. Math. Soc. **72** (1966), 870–874.
3. *Zur harmonischen Analyse klassenkompakter Gruppen*, Invent. Math. **5** (1968), 249–254.

4. *Sur l'algèbre de Fourier d'un groupe localement compact*, C. R. Acad. Sci. Paris Sér. A-B **266** (1968), 1180–1182.
5. *On locally compact groups with invariant means*, Proc. Amer. Math. Soc. **19** (1968), 489–494.
6. *Harmonische Analyse auf gewissen nilpotenten Lieschen Gruppen*, Studia Math. **48** (1973), 201–205.
7. *On group algebras of nilpotent Lie groups*, Studia Math. **47** (1973), 37–49.
8. *Ideal theory in group algebras of locally compact groups*, Invent. Math. **31** (1976), 259–278.
9. *Symmetrie in Banachschen Algebren*, Arch. Math. (Basel) **27** (1976), 394–400.
10. *Lokal kompakte Gruppen mit symmetrischen Algebren*, Sympos. Math. **22** (1977), 267–281.

Leptin, H. and Poguntke, D.
1. *Symmetry and nonsymmetry for locally compact groups*, J. Funct. Anal. **33** (1979), 119–134.

Liapounoff, A. A.
1. *On completely additive vector functions*, Izv. Akad. Nauk SSSR Ser. Mat. **4** (1940), 465–478.

Lin, M.
1. *Semigroups of Markov operators*, Boll. Un. Mat. Ital. **6** (1972), 20–44.

Lindenstrauss, J.
1. *A short proof of Liapounoff's convexity theorem*, J. Math. Mech. **15** (1966), 971–972.

Liu, T. S. and Rooij, A. van
1. *Invariant means on a locally compact group*, Monatsh. Math. **78** (1974), 356–359.

Liukkonen, J.
1. *Characters and centroids of [SIN] amenable groups*, Mem. Amer. Math. Soc. **148** (1974), 145–151.

Liukkonen, J. and Mosak, R.
1. *Harmonic analysis and centers of group algebras*, Trans. Amer. Math. Soc. **195** (1974), 147–163.
2. *The primitive dual space of $[FC]^-$ groups*, J. Funct. Anal. **15** (1974), 279–296.

Ljapin, E. S.
1. *Semigroups*, Amer. Math. Soc. Transl., vol. 3, Amer. Math. Soc., Providence, R. I., 1963.

Lloyd, S. P.

1. *On extreme averaging operators*, Proc. Amer. Math. Soc. **14** (1963), 305–310.
2. *A mixing condition for extreme left invariant means*, Trans. Amer. Math. Soc. **125** (1966), 461–481.
3. *Subalgebras in a subspace of $C(X)$*, Illinois J. Math. **14** (1970), 259–267.
4. *On the mean ergodic theorem of Sine*, Proc. Amer. Math. Soc. **56** (1976), 121–126.

Loebl, R. I.

1. *Injective von Neumann algebras*, Proc. Amer. Math. Soc. **44** (1974), 46–48.

Lohoué, N.

1. *Estimations L^p des coefficients de représentations et opérateurs de convolution*, Adv. in Math. **38** (1980), 178–221.

Loomis, L. H.

1. *An introduction to abstract harmonic analysis*, Van Nostrand, New York, 1953.

Lorch, E. R.

1. *A calculus of operators in reflexive vector spaces*, Trans. Amer. Math. Soc. **45** (1939), 217–234.
2. *Bicontinuous linear transformations in certain vector spaces*, Bull. Amer. Math. Soc. **45** (1939), 564–569.
3. *The integral representation of weakly almost periodic transformations in reflexive Banach spaces*, Trans. Amer. Math. Soc. **49** (1941), 18–40.

Lorentz, G. C.

1. *A contribution to the theory of divergent sequences*, Acta Math. **80** (1948), 167–190.

Losert, V.

1. *Some properties of groups without property P_1*, Comment. Math. Helv. **54** (1979), 133–139.
2. *A characterization of groups with the one-sided Wiener property*, J. Reine Angew. Math. **331** (1982), 47–52.

Losert, V. and Rindler, H.

1. *Uniform distribution and the mean ergodic theorem*, Invent. Math. **50** (1978), 65–74.
2. *Almost invariant sets*, Bull. London Math. Soc. **13** (1981), 145–148.

Ludwig, J.

1. *A class of symmetric and a class of Wiener algebras*, J. Funct. Anal. **31** (1979), 187–194.
2. *Polynomial growth and ideals in group algebras*, Manuscripta Math. **30** (1980), 215–221.

Luschgy, H.

1. *Invariant extensions of positive operators and extreme points*, Math. Z. **171** (1980), 75–81.

Luthar, I. S.

1. *Uniqueness of the invariant mean on an abelian semigroup*, Illinois J. Math. **3** (1959), 28–44.
2. *Uniqueness of the invariant mean on abelian topological semigroups*, Trans. Amer. Math. Soc. **104** (1962), 403–411.

Lyndon, R. C. and Schupp, P. E.

1. *Combinatorial group theory*, Springer-Verlag, Berlin–Heidelberg–New York, 1977.

Mackey, G. W.

1. *Induced representations of locally compact groups.* I, Ann. of Math. (2) **55** (1952), 101–139.
2. *Induced representations of locally compact groups.* II: *The Frobenius reciprocity theorem*, Ann. of Math. (2) **58** (1953), 193–221.
3. *Borel structure in groups and their duals*, Trans. Amer. Math. Soc. **85** (1957), 134–165.

Maddox, I. J.

1. *Nörlund means and almost convergence*, J. London Math. Soc. **17** (1978), 317–330.
2. *On strong almost convergence*, Math. Proc. Cambridge Philos. Soc. **85** (1979), 345–350.

Magnus, W., Karrass, A., and Solitar, D.

1. *Combinatorial group theory*, Interscience, New York–London–Sydney, 1966.

Mah, P. F.

1. *Summability in amenable semigroups*, Trans. Amer. Math. Soc. **156** (1971), 391–403.

Marczewski, E.

1. *Problem* 169, *The Scottish Book*, 1937–38. (Boston, Birkhäuser, 1981).

Margulis, G. A.

1. *Quotient groups of discrete subgroups and measure theory*, Functional Anal. Appl. **12** (1978), 295–305.
2. *Some remarks on invariant means*, Monatsh. Math. **90** (1980), 233–235.
3. *Finitely additive measures on Euclidean spaces*, Ergodic Theory Dynamical Systems **2** (1982), 383–396.

Massey, W. S.

1. *Algebraic topology: An introduction*, Springer-Verlag, New York–Heidelberg–Berlin, 1967.

Mauceri, G.

1. *Square integrable representations and the Fourier algebra of a unimodular group*, Pacific J. Math. **73** (1977), 143–154.

Mauceri, G. and Picardello, M. A.

1. *Non compact unimodular groups with purely atomic Plancherel measures*, Proc. Amer. Math. Soc. **78** (1980), 77–84.

Michelle, L. de and Soardi, P. M.

1. *Existence of Sidon sets in discrete FC groups*, Proc. Amer. Math. Soc. **55** (1976), 457–460.

Millet, A. and Sucheston, L.

1. *On the existence of σ-finite invariant measures for operators*, Israel J. Math. **33** (1979), 349–367.

Milnes, P.

1. *Compactifications of semitopological semigroups*, J. Austral. Math. Soc. **15** (1973), 488–503.
2. *Extension of continuous functions on topological semigroups*, Pacific J. Math. **58** (1975), 553–562.
3. *Left mean ergodicity, fixed points and invariant means*, J. Math. Anal. Appl. **65** (1978), 32–43.
4. *Amenable groups for which every topologically left invariant mean is right invariant*, Rocky Mountain J. Math. **11** (1981), 261–266.

Milnes, P. and Bondar, J. V.

1. *A simple proof of a covering property of locally compact groups*, Proc. Amer. Math. Soc. **73** (1979), 117–118.

Milnes, P. and Pym, J. S.

1. *Counter-example in the theory of continuous functions on semigroups*, Pacific J. Math. **66** (1976), 205–209.

Milnor, J.

1. *A note on curvature and fundamental group*, J. Differential Geom. **2** (1968), 1–7.
2. *Problem* 5603, Amer. Math. Monthly **75** (1968), 685–686.
3. *Growth of finitely generated solvable groups*, J. Differential Geom. **2** (1968), 447–449.
4. *On fundamental groups of complete affinely flat manifolds*, Adv. in Math. **25** (1977), 178–187.

Mitchell, T.

1. *Invariant means on semigroups and the constant functions*, Thesis, Illinois Institute of Technology, 1964.
2. *Constant functions and left invariant means on semigroups*, Trans. Amer. Math. Soc. **119** (1965), 244–261.

3. *Fixed points and multiplicative left invariant means*, Trans. Amer. Math. Soc. **122** (1966), 195–202.
4. *Function algebras, means and fixed points*, Trans. Amer. Math. Soc. **130** (1968), 117–126.
5. *Fixed points of reversible semigroups of non-expansive mappings*, Kodai Math. Sem. Rep. **22** (1970), 322–323.
6. *Topological semigroups and fixed points*, Illinois J. Math. **14** (1970), 630–641.
7. *Common fixed-points for equi-continuous semigroups of mappings*, Proc. Amer. Math. Soc. **33** (1972), 146–150.
8. *Invariant means and analytic actions*, Pacific J. Math. **85** (1979), 145–153.

Monk, J. D.

1. *Introduction to set theory*, McGraw-Hill, New York, 1969.

Montgomery, D. and Zippin, L.

1. *Topological transformation groups*, Interscience, New York, 1955.

Moore, C. C.

1. *Distal affine transformation groups*, Amer. J. Math. **90** (1968), 733–751.
2. *Amenable subgroups of semisimple groups and proximal flows*, Israel J. Math. **34** (1979), 121–138.

Moore, C. C. and Rosenberg, J.

1. *Comments on a paper of Brown and Guivarc'h*, Ann. Sci. École Norm. Sup. **5** (1975), 379–382.
2. *Groups with T_1 primitive ideal spaces*, J. Funct. Anal. **22** (1976), 204–224.

Moore, R. T.

1. *Measurable, continuous and smooth vectors for semigroups and group representations*, Mem. Amer. Math. Soc. **78** (1968).

Moran, W.

1. *Separate continuity and support of measures*, J. London Math. Soc. **44** (1969), 320–324.
2. *Invariant means on $C(G)$*, J. London Math. Soc. **2** (1970), 133–138.

Morse, A. P.

1. *Squares are normal*, Fund. Math. **36** (1949), 35–39.

Morse, A. P. and Agnew, R.

1. *Extensions of linear functionals, with applications to limits, integrals, measures and densities*, Ann. of Math. (2) **39** (1938), 20–30.

Mosak, R. D.

1. *The L^1- and C^*-algebras of $[FIA]_B^-$ groups and their representations*, Trans. Amer. Math. Soc. **163** (1972), 277–310.

Moulin-Ollagnier, J. and Pinchon, D.

1. *Une nouvelle démonstration du théorème de E. Følner*, C. R. Acad. Sci. Paris Sér. A **287** (1978), 557–560.

Mycielski, J.

1. *On the decompositions of Euclidean spaces*, Bull. Acad. Polon. Sci. Sér. Sci. Math. **4** (1956), 417–418.
2. *Commentary on a paper of Banach*, in S. Banach, Oevres avec des Commentaires, Vol. 1, Editions scientifiques de Pologne, Warsaw, 1967.
3. *Almost every function is independent*, Fund. Math. **81** (1973–1974), 43–48.
4. *Remarks on invariant measures in metric spaces*, Colloq. Math. **32** (1974), 105–112.
5. *Two problems on geometric bodies*, Amer. Math. Monthly **84** (1977), 116–118.
6. *Finitely additive invariant measures.* I, Colloq. Math. **42** (1979), 309–318.

Mycielski, J. and Świerczkowski, S.

1. *On free groups of motions and decompositions of the Euclidean space*, Fund. Math. **45** (1958), 283–291.

Myers, S.

1. *Riemannian manifolds with positive mean curvature*, Duke Math. J. **8** (1941), 401–404.

Nagel, R. J.

1. *Mittelergodische Halbgruppen linearer Operatoren*, Ann. Inst. Fourier (Grenoble) **23** (1973), 78–87.

Nagy, B. Sz.

1. *On uniformly bounded linear transformations in Hilbert space*, Acta Sci. Math. (Szeged) **11** (1947), 152–157.

Namioka, I.

1. *Følner's conditions for amenable semigroups*, Math. Scand. **15** (1964), 18–28.
2. *On a recent theorem by H. Reiter*, Proc. Amer. Math. Soc. **17** (1966), 1101–1102.
3. *On certain actions of semigroups on L-spaces*, Studia Math. **29** (1967), 63–77.
4. *Right topological groups, distal flows and a fixed-point theorem*, Math. Systems Theory **6** (1972), 193–209.
5. *Separate continuity and joint continuity*, Pacific J. Math. **51** (1974), 515–531.

Natarajan, S.

1. *Contributions to ergodic theory*, Thesis, The Indian Statistical Institute, Calcutta, 1968.
2. *Invariant measures for families of transformations*, Preprint.

Nelson, E. and Stinespring, W. F.

1. *Representations of elliptic operators in an enveloping algebra*, Amer. J. Math. **81** (1959), 547–560.

Neumann, J. von

1. *Zur allgemeinen Theorie des Masses*, Fund. Math. **13** (1929), 73–116.
2. *Almost periodic functions in a group*. I, Trans. Amer. Math. Soc. **36** (1934), 445–492.
3. *Invariant measures*, Institute for Advanced Studies, Princeton, N. J., 1940–1941.

Nillsen, R.

1. *Discrete orbits in $\beta N \sim N$*, Colloq. Math. **33** (1975), 71–81.
2. *Nets of extreme Banach limits*, Proc. Amer. Math. Soc. **55** (1976), 347–352.

Oberlin, D. M.

1. *Translation-invariant operators of weak type*, Pacific J. Math. **85** (1979), 155–164.

Ocneanu, A.

1. *Action des groupes moyennables sur les algèbres de von Neumann*, C. R. Acad. Sci. Paris Sér. A **291** (1980), 399–401.
2. *A Rohlin type theorem for groups acting on von Neumann algebras*, Proc. Conf. on Operator Theory (Timişoara/Herculane, 1979), 1980, pp. 49–65, and Topics on Modern Operator Theory (Timişoara/Herculane, 1980), Adv. Appl., vol. 2, Birkhäuser, Bäsel-Boston, Mass., 1981, pp. 247–258.

Ol′shanskii, A. Yu

1. *Infinite groups with cyclic subgroups*, Soviet Math. Dokl. **20** (1979), 343–346.
2. *An infinite simple Noetherian group without torsion*, Math. USSR-Izv. **15** (1980), 531–588.
3. *An infinite group with subgroups of prime orders*, Izv. Akad. Nauk SSSR Ser. Mat. **44** (1980), 309–321.
4. *On the problem of the existence of an invariant mean on a group*, Russian Math. Surveys **35** (1980), 180–181.

Ornstein, D. S. and Weiss, B.

1. *Ergodic theory of amenable group actions*, I: *the Rohlin Lemma*, Bull. Amer. Math. Soc. (N.S.) **2** (1980), 161–164.

Paalman de Miranda, A. B.

1. *Topological semigroups*, Mathematisch Centrum, Amsterdam, 1970.

Palmer, T. W.

1. *Classes of non abelian, non compact locally compact groups*, Rocky Mountain J. Math. **8** (1978), 683–741.

Paschke, W. L.

1. *The crossed product of a C^*-algebra by an endomorphism*, Proc. Amer. Math. Soc. **80** (1980), 113–118.
2. *Inner amenability and conjugation operators*, Proc. Amer. Math. Soc. **71** (1978), 117–118.

Paschke, W. L. and Salinas, N.

1. *C^*-algebras associated with free products of groups*, Pacific J. Math. **82** (1979), 211–221.

Paterson, A. L. T.

1. *Amenability and locally compact semigroups*, Math. Scand. **42** (1978), 271–288.
2. *Weak containment and Clifford semigroups*, Proc. Roy. Soc. Edinburgh Sect. A **81** (1978), 23–30.
3. *Amenable groups for which every topological left invariant mean is invariant*, Pacific J. Math. **84** (1979), 391–397.
4. *On invariant means which are not inversion invariant*, J. London Math. Soc. **19** (1979), 312–318.
5. *The size of the set of left invariant mean on an ELA semigroup*, Proc. Amer. Math. Soc. **72** (1978), 62–64.
6. *A non-probabilistic approach to Poisson spaces*, Proc. Roy. Soc. Edinburgh **93A** (1983), 181–188.
7. *Amenability and translation experiments*, Canad. J. Math. **35** (1983), 49–58.
8. *Unitary groups and harmonic analysis*, J. Funct. Anal. **53** (1983), 203–223.
9. *The cardinality of the set of left invariant means on a left amenable semigroup*, Illinois J. Math. **29** (1985), 567–583.
10. *Nonamenability and Borel paradoxical decompositions for locally compact groups*, Proc. Amer. Math. Soc. **96** (1986), 89–90.

Pathak, P. K. and Shapiro, H. S.

1. *A characterization of certain weak*-closed subalgebras of L^∞*, J. Math. Anal. Appl. **58** (1977), 174–177.

Peisakoff, M. P.

1. *Transformation parameters*, Thesis, Princeton University, 1950.

Peck, J. E. L.

1. *An ergodic theorem for a noncommutative semigroup of linear operators*, Proc. Amer. Math. Soc. **2** (1951), 414–421.

Peters, J.

1. *Representing positive definite B-invariant functions on* $[FC]_B^-$ *groups*, Monatsh. Math. **80** (1975), 319–324.

Phelps, R. R.

1. *Extreme positive operators and homomorphisms*, Trans. Amer. Math. Soc. **108** (1963), 265–274.
2. *Lectures on Choquet's theorem*, Van Nostrand, New York, 1966.

Pickel, B. S.

1. *Informational futures of amenable groups*, Soviet Math. Dokl. **16** (1975), 1037–1041.

Plante, J.

1. *A generalization of the Poincaré-Bendixson theorem for foliations of codimension* 1, Topology **12** (1973), 177–181.
2. "Foliations with measure preserving holonomy", Ann. Math. **102** (1975), 327–361.
3. "A generalization of the Poincaré-Bendixson theorem for foliations of codimension one", Topology **12** (1973), 177–181.

Poguntke, D.

1. *Nilpotente Liesche Gruppen haben symmetrische Gruppenalgebren*, Math. Ann. **227** (1977), 51–59.

Porada, E.

1. *Croissance de la mesure de Haar dans les groupes nilpotents de Lie*, Preprint.

Preissmann, A.

1. *Quelques propriétés globales des espaces de Riemann*, Comment. Math. Helv. **15** (1942), 175–216.

Pukanszky, L.

1. *Unitary representations of solvable Lie groups*, Ann. Sci. École Norm. Sup. **4** (1971), 457–608.
2. *The primitive ideal space of solvable Lie groups*, Invent. Math. **22** (1973), 75–118.
3. *Characters of connected Lie groups*, Acta Math. **133** (1974), 81–137.
4. *Lie groups with completely continuous representations*, Bull. Amer. Math. Soc. **81** (1975), 1061–1063.

Pytlik, T.

1. *On the spectral radius of elements in group algebras*, Bull. Acad. Polon. Sci. Sér. Sci. Math. **21** (1973), 899–902.

Racher, G.

1. *On amenable and compact groups*, Monatsh. Math. **92** (1981), 305–311.

Raimi, R. A.

1. *Mean values and Banach limits*, Proc. Amer. Math. Soc. **8** (1957), 1029–1036.
2. *On a theorem of E. Følner*, Math. Scand. **6** (1958), 47–49.
3. *On Banach's generalised limits*, Duke Math. J. **26** (1959), 17–28.
4. *Convergence, density and τ-density of bounded sequences*, Proc. Amer. Math. Soc. **14** (1963), 708–712.
5. *Invariant means and invariant matrix methods of summability*, Duke Math. J. **30** (1963), 81–94.
6. *Minimal sets and ergodic measures in $\beta N\backslash N$*, Bull. Amer. Math. Soc. **70** (1964), 711–712.
7. *Homeomorphisms and invariant measures for $\beta N\backslash N$*, Duke Math. J. **33** (1966), 1–12.
8. *Translation properties of finite partitions of the positive integers*, Fund. Math. **61** (1968), 253–258.

Rajagopalan, M.

1. *L^p conjecture for locally compact groups*, Thesis, Yale University, 1963.
2. *L^p-spaces of abelian semi-groups*, Duke Math. J. **32** (1965), 263–266.
3. *L^p-conjecture for locally compact groups*. I, Trans. Amer. Math. Soc. **125** (1966), 216–222.
4. *A note on the l^p-space of a semigroup*, J. London Math. Soc. **41** (1966), 697–700.
5. *L^p-conjecture for locally compact groups*. II, Math. Ann. **169** (1967), 331–339.

Rajagopalan, M. and Ramakrishnan, P. V.

1. *On a conjecture of Granirer and strong Følner condition*, J. Indian Math. Soc. **37** (1973), 85–92.
2. *Uses of βS in invariant means and extremely left amenable semigroups*, Memphis State University Report 77–79.

Rajagopalan, M. and Witz, K. G.

1. *On invariant means which are not inverse invariant*, Canad. J. Math. **20** (1968), 222–224.

Rajagopalan, M. and Zelazko, W.

1. *L_p conjecture for solvable locally compact groups*, J. Indian Math. Soc. **29** (1965), 87–93.

Ramamohana Rao, C.

1. *Invariant means on spaces of continuous or measurable functions*, Trans. Amer. Math. Soc. **114** (1965), 187–196.

Reiter, H. J.

1. *Investigations in harmonic analysis*, Trans. Amer. Math. Soc. **73** (1952), 401–427.

2. *Über L^1-Räume auf Gruppen.* I, Monatsh. Math. **58** (1954), 73–76.
3. *Über L^1-Räume auf Gruppen.* II, Monatsh. Math. **58** (1954), 172–180.
4. *Contributions to harmonic analysis.* I, Acta Math. **96** (1956), 253–263.
5. *Beiträge zur harmonischen Analyse.* II, Math. Ann. **133** (1957), 298–302.
6. *Contributions to harmonic analysis.* III, J. London Math. Soc. **32** (1957), 477–483.
7. *Contributions to harmonic analysis.* IV, Math. Ann. **135** (1958), 467–476.
8. *The convex hull of translates of a function in L^1*, J. London Math. Soc. **35** (1960), 5–16.
9. *Beiträge zur harmonischen Analyse.* V, Math. Ann. **140** (1960), 422–441.
10. *Une propriété analytique d'une certaine classe de groupes localement compacts*, C. R. Acad. Sci. Paris **254** (1962), 3627–3629.
11. *Sur les groupes de Lie semisimples connexes*, C. R. Acad. Sci. Paris **255** (1962), 2883–2884.
12. *Contributions to harmonic analysis.* VI, Ann. of Math. (2) **77** (1963), 552–562.
13. *Sur la propriété (P_1) et les functions de type positif*, C. R. Acad. Sci. Paris **258** (1964), 5134–5135.
14. *On some properties of locally compact groups*, Indag. Math. **27** (1965), 697–701.
15. *Subalgebras of $L^1(G)$*, Indag. Math. **27** (1965), 691–696.
16. *Zwei Anwendungen der Bruhatschen Funktion*, Math. Ann. **163** (1966), 118–121.
17. *Sur certains idéaux dans $L^1(G)$*, C. R. Acad. Sci. Paris Sér. A **267** (1968), 882–885.
18. *L^1-algebras and Segal algebras*, Lecture Notes in Math., vol. 231, Springer-Verlag, Berlin–Heidelberg–New York, 1971.
19. *Über den Satz von Wiener und lokalkompakte Gruppen*, Comment. Math. Helv. **49** (1974), 333–364.

Renaud, P. F.

1. *General ergodic theorems for locally compact groups*, Amer. J. Math. **93** (1971), 52–64.
2. *Centralizers of the Fourier algebra of an amenable group*, Proc. Amer. Math. Soc. **32** (1972), 539–542.
3. *Equivalent types of invariant means on locally compact groups*, Proc. Amer. Math. Soc. **31** (1972), 495–498.
4. *Invariant means on a class of von Neumann algebras*, Trans. Amer. Math. Soc. **170** (1972), 285–291.

Revuz, D.

1. *Markov chains*, North-Holland, Amsterdam and Oxford, 1975.

Rickert, N. W.

1. *Some properties of locally compact groups*, J. Austral. Math. Soc. **7** (1967), 433–454.
2. *Amenable groups and groups with the fixed point property*, Trans. Amer. Math. Soc. **127** (1967), 221–232.
3. *Convolution of L_2-functions*, Colloq. Math. **19** (1968), 301–303.

Riemersma, M.

1. *On a theorem of Glicksberg and fixed point properties of semigroups*, Indag. Math. **33** (1971), 340–345.

Rindler, H.

1. *Gleichverteilte Folgen von Operatoren*, Compositio Math. **29** (1974), 201–212.
2. *Uniform distribution on locally compact groups*, Proc. Amer. Math. Soc. **57** (1976), 130–132.
3. *Gleichverteilte Folgen in lokalkompakten Gruppen*, Monatsh. Math. **82** (1976), 207–235.

Ringrose, J. R. and Kadison, R. V.

1. *Cohomology of operator algebras.* II: *Extending cobounding and the hyperfinite case*, Ark. Mat. **9** (1971), 55–63.

Robertson, L. C.

1. *A note on the structure of Moore groups*, Bull. Amer. Math. Soc. **75** (1969), 594–599.

Robinson, G. B.

1. *Invariant integrals over a class of Banach spaces*, Pacific J. Math. **4** (1954), 123–150.

Roelcke, W., Asam, L., Dierolf, S., and Dierolf, P.

1. *Discontinuous translation-invariant linear forms on $\mathscr{K}(G)$*, Math. Ann. **239** (1979), 219–222.

Rooij, A. C. M. van

1. *Invariant means in a non-Archimedean valued field*, Indag. Math. **29** (1967), 220–228.
2. *Non-Archimedean functional analysis*, Marcel Dekker, New York and Basel, 1978.

Rosen, W. G.

1. *On invariant means over topological semigroups*, Thesis, University of Illinois, Urbana, Illinois, 1954.
2. *On invariant means over compact semigroups*, Proc. Amer. Math. Soc. **7** (1956), 1076–1082.

Rosenberg, J.

1. *The C^*-algebras of some real and p-adic solvable groups*, Pacific J. Math. **65** (1976), 175–192.
2. *Amenability of crossed products of C^*-algebras*, Comm. Math. Phys. **57** (1977), 187–191.

Rosenblatt, J. M.

1. *Invariant linear functionals and counting conditions*, Thesis, University of Washington, 1972.
2. *A generalization of Følner's condition*, Math. Scand. **33** (1973), 153–170.
3. *Equivalent invariant measures*, Israel J. Math. **17** (1974), 261–270.
4. *Invariant measures and growth conditions*, Trans. Amer. Math. Soc. **193** (1974), 33–53.
5. *Invariant means and invariant ideals in $L_\infty(G)$ for a locally compact G*, J. Funct. Anal. **21** (1976), 31–51.
6. *Uniform distribution in compact groups*, Mathematika **23** (1976), 198–207.
7. *Invariant means for the bounded measurable functions on a nondiscrete locally compact group*, Math. Ann. **220** (1976), 219–228.
8. *The number of extensions of an invariant mean*, Compositio Math. **33** (1976), 147–159.
9. *Invariant means on the continuous bounded functions*, Trans. Amer. Math. Soc. **236** (1978), 315–324.
10. *A distal property of groups and the growth of connected locally compact groups*, Mathematika **26** (1979), 94–98.
11. *Finitely additive invariant measures.* II, Colloq. Math. **42** (1979), 361–363.
12. *Strongly equivalent invariant measures*, Math. Proc. Cambridge Philos. Soc. **88** (1980), 33–43.
13. *The Haar integral is the unique invariant mean for automorphisms of the torus*, Preprint.
14. *Uniqueness of invariant means for measure preserving transformations*, Trans. Amer. Math. Soc. **265** (1981), 623–636.

Rosenblatt, J. M. and Talagrand, M.

1. *Different types of invariant means*, J. London Math. Soc. **24** (1981), 525–532.

Rosenblatt, M.

1. *Markov processes. Structure and asymptotic behavior*, Springer-Verlag, Berlin–Heidelberg–New York, 1971.

Rosenthal, H. P.

1. *Projections onto translation invariant subspaces of $L^p(G)$*, Mem. Amer. Math. Soc. **63** (1966).

Ross, K. A.

1. *A note on extending semi-characters on semigroups*, Proc. Amer. Math. Soc. **10** (1959), 579–583.
2. *Extending characters on semigroups*, Proc. Amer. Math. Soc. **12** (1961), 988–990.

Roynette, B.

1. *Marches aleatoires sur les groupes de Lie*, École d'Eté de Probabilités de Saint-Flour VII-1977, Lecture Notes in Math., vol. 678, Springer-Verlag, Berlin–Heidelberg–New York, 1978, pp. 238–379.

Rubel, L. A. and Shields, A. L.

1. *Invariant subspaces of L^∞ and H^∞*, J. Reine Angew. Math. **272** (1975), 32–44.

Rudin, W.

1. *Averages of continuous functions on compact spaces*, Duke Math. J. **25** (1958), 195–204.
2. *Invariant means on L^∞*, Studia Math. **44** (1972), 219–227.
3. *Real and complex analysis*, Tata McGraw-Hill, New Delhi, 1974.
4. *Homomorphisms and translations in $L^\infty(G)^*$*, Adv. in Math. **15** (1975), 79–90.
5. *Principles of mathematical analysis*, McGraw-Hill, New York, 1976.

Ruppert, W.

1. *Rechtstopologische Halbgruppen*, J. Reine Angew. Math. **261** (1973), 123–133.
2. *Rechtstopologische Intervallhalbgruppen und Kreishalbgruppen*, Manuscripta Math. **14** (1974), 183–193.
3. *Über kompakte rechtstopologische Gruppen mit gleichgradig stetigen Links translationen*, Sitzungsber. Österreich Akad. Wiss. Math.-Natur. Kl. **184** (1975), 159–169.
4. *Notes on compact semigroups with identity*, Semigroup Forum **14** (1977), 199–234.

Ryll-Nardzewski, C.

1. *Remark on Raimi's theorem on translations*, Fund. Math. **61** (1968), 257–258.

Sachdeva, V.

1. *On finite invariant measure for semigroups of operators*, Canad. Math. Bull. **14** (1971), 197–206.

Sakai, K.

1. *Amenable transformation groups*, Sci. Rep. Kagoshima Univ. **22** (1973), 1–7.
2. *Extremely amenable transformation semigroups*, Proc. Japan Acad. **49** (1973), 424–427.

3. *Amenable transformation groups.* II, Proc. Japan Acad. **49** (1973), 428–431.
4. *Extremely amenable transformation semigroups.* II, Proc. Japan Acad. **50** (1974), 374–377.
5. *Følner's conditions for amenable transformation semigroups*, Sci. Rep. Kagoshima Univ. **23** (1974), 7–13.
6. *On amenable transformation semigroups.* I, J. Math. Kyoto Univ. **16** (1976), 555–596.
7. *On amenable transformation semigroups.* III, Sci. Rep. Kagoshima Univ. **25** (1976), 31–51.
8. *On amenable transformation semigroups.* II, J. Math. Kyoto Univ. **16** (1976), 597–626.

Sakai, S.

1. *C^*-algebras and W^*-algebras*, Springer-Verlag, Berlin–Heidelberg–New York, 1971.

Sato, R.

1. *Ergodic properties of bounded L_1-operators*, Proc. Amer. Math. Soc. **39** (1973), 540–546.
2. *Invariant measures for semigroups*, Studia Math. **53** (1975), 129–134.
3. *On the existence of positive invariant functions for semigroups of operators*, Tôhoku Math. J. **27** (1975), 187–196.
4. *Invariant measures for bounded amenable semigroups of operators*, Proc. Japan Acad. **52** (1976), 215–218.
5. *Invariant measures for ergodic semigroups*, Pacific J. Math. **71** (1977), 173–192.
6. *On abstract mean ergodic theorems*, Tôhoku Math. J. **30** (1978), 575–581.

Schaefer, P.

1. *Matrix transformations of almost convergent sequences*, Math. Z. **112** (1969), 321–325.

Schikhof, W. H.

1. *Non-Archimedean invariant means*, Compositio Math. **30** (1975), 169–180.

Schmidt, K.

1. *Asymptotically invariant sequences and an action of* SL(2, **Z**) *on the 2-sphere*, Israel J. Math. **37** (1980), 193–208.
2. *Amenability, Kazhdan's property T, strong ergodicity, and invariant means for ergodic actions*, Ergodic Theory Dynamical Systems **1** (1981), 223–236.

Schwartz, J.

1. *Two finite, non-hyperfinite, non-isomorphic factors*, Comm. Pure Appl. Math. **16** (1963), 19–26.

Scott, W. R.

1. *Group theory*, Prentice-Hall, Englewood Cliffs, N. J., 1964.

Seever, G. L.

1. *Nonnegative projections on $C_0(X)$*, Pacific J. Math. **17** (1966), 159–166.

Series, C.

1. *The Rohlin tower theorem and hyperfiniteness for actions of continuous groups*, Israel J. Math. **30** (1978), 99–122.

Serre, J.-P.

1. *Trees*, Springer-Verlag, Berlin–Heidelberg–New York, 1980.

Sherman, J.

1. *Paradoxical sets and amenability in groups*, Thesis, University of California, Los Angeles, 1975.
2. *A new characterization of amenable groups*, Trans. Amer. Math. Soc. **254** (1979), 365–389.

Sierpiński, W.

1. *Sur l'equivalence des ensembles par décomposition en deux parties*, Fund. Math. **35** (1948), 151–158.
2. *On the congruence of sets and their equivalence by finite decomposition*, Congruence of Sets and Other Monographs, Chelsea, New York, 1954.

Silverman, R. J.

1. *Invariant extensions of linear operators*, Thesis, University of Illinois, Urbana, 1952.
2. *Means on semigroups and the Hahn-Banach extension property*, Trans. Amer. Math. Soc. **83** (1956), 222–237.
3. *Invariant linear functions*, Trans. Amer. Math. Soc. **81** (1956), 411–424.
4. *Invariant means and cones with vector interiors*, Trans. Amer. Math. Soc. **88** (1958), 75–79.

Simons, S.

1. *Banach limits, infinite matrices and sub-linear functionals*, J. Math. Anal. Appl. **26** (1969), 640–655.

Sine, R.

1. *A mean ergodic theorem*, Proc. Amer. Math. Soc. **24** (1970), 438–439.
2. *Sequential convergence to invariance in $BC(G)$*, Proc. Amer. Math. Soc. **55** (1976), 313–318.

Skudlarek, H. L.

1. *On a two-sided version of Reiter's condition and weak containment*, Arch. Math. (Basel) **31** (1979), 605–619.

Sleijpen, G. L. G.

1. *Convolution measure algebras on semigroups*, Thesis, Catholic University, Toernooiveld, Nijmegen, 1976.

2. *Emaciated sets and measures with continuous translations*, Proc. London Math. Soc. **37** (1978), 98–119.
3. *Locally compact semigroups and continuous translations of measures*, Proc. London Math. Soc. **37** (1978), 75–97.

Smith, R. R.

1. *Completely bounded maps between C^*-algebras*, J. London Math. Soc. **27** (1983), 157–166.

Sorensen, J. R.

1. *Left amenable semigroups and cancellation*, Notices Amer. Math. Soc. **11** (1964), 1763.
2. *Existence of measures that are invariant under a semigroup of transformations*, Thesis, Purdue University, 1966.

Spanier, E. H.

1. *Algebraic topology*, McGraw-Hill, New York, 1966.

Specht, W.

1. *Zur Theorie der messbaren Gruppen*, Math. Z. **74** (1960), 325–366.

Stegeman, J. D.

1. *On a property concerning locally compact groups*, Indag. Math. **27** (1965), 702–703.

Stegmeir, U.

1. *Center of group algebras*, Math. Ann. **243** (1979), 11–16.

Stinespring, W. F.

1. *Positive functions on C^*-algebras*, Proc. Amer. Math. Soc. **6** (1955), 211–216.

Stone, C.

1. *Ratio limit theorems for random walks on groups*, Trans. Amer. Math. Soc. **125** (1966), 86–100.

Stone, M.

1. *Necessary and sufficient conditions for convergence in probability to invariant posterior distributions*, Ann. Math. Statist. **41** (1970), 1349–1353.

Størmer, E.

1. *Invariant states of von Neumann algebras*, Math. Scand. **30** (1972), 253–256.

Stromberg, K.

1. *The Banach-Tarski paradox*, Amer. Math. Monthly **86** (1979), 151–161.

Sucheston, L.

1. *An ergodic application of almost convergent sequences*, Duke Math. J. **30** (1963), 417–422.
2. *On existence of finite invariant measures*, Math. Z. **86** (1964), 327–336.

3. *On the ergodic theorem for positive operators.* I, Z. Wahrsch. Verw. Gebiete **8** (1966), 1–11.

Sullivan, D.

1. *For $n > 3$, there is only one finitely additive rotationally invariant measure on the n-sphere defined on all Lebesgue measurable subsets*, Bull. Amer. Math. Soc. (N.S.) **4** (1981), 121–123.

Swaminathan, S. (editor)

1. *Fixed-point theory and its applications*, Academic Press, New York–San Francisco–London, 1976.

Takahashi, W.

1. *Fixed point theorem for amenable semigroup of non-expansive mappings*, Kodai Math. Sem. Rep. **21** (1969), 383–386.
2. *Invariant ideals for amenable semigroups of Markov operators*, Kodai Math. Sem. Rep. **23** (1971), 121–126.
3. *Invariant functions for amenable semigroups of positive contractions on* L^1, Kodai Math. Sem. Rep. **23** (1971), 131–143.
4. *Ergodic theorems for amenable semigroups of positive contractions on* L^1, Sci. Rep. Yokohama Nat. Univ. Sect. I **19** (1972), 5–11.
5. *A nonlinear ergodic theorem for an amenable semigroup of non-expansive mappings in a Hilbert space*, Proc. Amer. Math. Soc. **81** (1981), 253–256.

Takenouchi, O.

1. *Sur une classe de fonctions continues de type positif sur un groupe localement compact*, Math. J. Okayama Univ. **4** (1955), 143–173.

Takesaki, M.

1. *On the singularity of a positive linear functional on operator algebra*, Proc. Japan Acad. **35** (1959), 365–366.
2. *On the cross norm of the direct product of C^*-algebras*, Tôhoku Math. J. **161** (1964), 111–122.
3. *Conditional expectations in von Neumann algebras*, J. Funct. Anal. **9** (1972), 306–321.
4. *Duality for crossed products and the structure of von Neumann algebras of type* III, Acta Math. **131** (1973), 249–310.

Takesaki, M. and Tatsuuma, N.

1. *Duality and subgroups*, Ann. of Math. (2) **93** (1971), 344–364.

Talagrand, M. M.

1. *Moyennes de Banach extrémales*, C. R. Acad. Sci. Paris Sér. A **282** (1976), 1359–1362.
2. *Capacités invariantes*, C. R. Acad. Sci. Paris Sér. A **284** (1977), 481–484.
3. *Capacités invariantes: le cas alterné*, C. R. Acad. Sci. Paris Sér. A **284** (1977), 543–546.

4. *Geometrie des simplexes de moyennes invariantes*, J. Funct. Anal. **34** (1979), 304–337.
5. *Some functions with a unique invariant mean*, Proc. Amer. Math. Soc. **82** (1981), 253–256.
6. *Moyennes invariantes s'annulant sur les ideaux*, Compositio Math. **42** (1981), 213–216.
7. *Ergodic measures, almost periodic points and discrete orbits*, Illinois J. Math. **25** (1981), 504–507.

Tarski, A.

1. *Problème* 38, Fund. Math. **7** (1925), 381.
2. *Algebraische Fassung des Massproblems*, Fund. Math. **31** (1938), 47–66.
3. *Über das absolute Mass linearer Punktmengen*, Fund. Math. **30** (1938), 218–234.
4. *Cardinal algebras*, Oxford University Press, Oxford, 1969.

Templeman, A. A.

1. *Ergodic theorems for general dynamical systems*, Soviet Math. Dokl. **8** (1967), 1213–1216.

Thoma, E.

1. *Ein Charakterisierung diskreter Gruppen vom Type* I, Invent. Math. **6** (1968), 190–196.

Tits, J.

1. *Automorphismes à déplacement borné des groupes de Lie*, Topology **3** (1964), 97–107.
2. *Free subgroups in linear groups*, J. Algebra **20** (1972), 250–272.
3. *Travaux de Margulis sur les sous-groupes discrets de groupes de Lie*, Séminaire Bourbaki, No. 482 (1975/76), 174–190.
4. *Groupes à croissance polynomiale*, Séminaire Bourbaki **33** (1980/81), no. 572.

Tomiyama, J.

1. *On the projection of norm one in W^*-algebras*, Proc. Japan Acad. **33** (1957), 608–612.
2. *On the projection of norm one in W^*-algebras*. III, Tôhoku Math. J. **11** (1959), 125–129.
3. *On the tensor products of von Neumann algebras*, Pacific J. Math. **30** (1969), 263–270.
4. *Tensor products and projections of norm one in von Neumann algebras*, Lecture notes, University of Copenhagen, 1970.

Torgersen, E. N.

1. *Comparison of translation experiments*, Ann. Math. Statist. **43** (1972), 1383–1399.

Truitt, C. C. B.

1. *An extension to Orlicz spaces of theorems of M. M. Day on "Convolutions, means and spectra"*, Thesis, University of Illinois, Urbana, Illinois, 1967.

Urbanik, K.

1. *On a theorem on Zelazko on L_p-algebras*, Colloq. Math. **8** (1961), 121–123.

Vera, G.

1. *On a general notion of summability*, Rev. Real Acad. Ciencias Exactas Fisicas y Naturales de Madrid **70** (1976), 297–335.

Voiculescu, D.

1. *Amenability and Katz algebras*, Algèbres d'Opérateurs et Leurs Applications en Physique Mathématique (Proc. Colloq. Marseille, 1977), Colloq. Internat. CNRS, vol. 274, CNRS, Paris, 1979, pp. 451–457.

Wagon, S.

1. *Circle-squaring in the twentieth century*, Math. Intelligencer **3** (1981), 176–181.
2. *Invariance properties of finitely additive measures in* $\mathbf{R}^n$, Illinois J. Math. **25** (1981), 74–86.
3. *The use of shears to construct paradoxes in* $\mathbf{R}^2$, Proc. Amer. Math. Soc. **85** (1982), 353–359.

Walker, R. C.

1. *The Stone-Čech compactification*, Springer-Verlag, Berlin–Heidelberg–New York, 1974.

Walter, M. E.

1. *W^*-algebras and non-abelian harmonic analysis*, J. Funct. Anal. **11** (1972), 17–38.
2. *A duality between locally compact groups and certain Banach algebras*, J. Funct. Anal. **17** (1974), 131–160.
3. *Convolution on the reduced dual of a locally compact group*, Math. Scand. **37** (1975), 145–166.
4. *On the structure of the Fourier-Stieltjes algebra*, Pacific J. Math. **58** (1975), 267–281.

Warner, G.

1. *Harmonic analysis on semisimple Lie groups*, Vol. I and II, Springer-Verlag, Berlin–Heidelberg–New York, 1972.

Wasserman, S.

1. *Injective W^*-algebras*, Math. Proc. Cambridge Philos. Soc. **82** (1977), 39–47.

Watatani, Y.

1. *The character groups of amenable group C^*-algebras*, Math. Japonica **1** (1979), 141–144.
2. *Haagerup's multiplications on group C^*-algebras*, Math. Japonica **2** (1979), 223–224.

Weit, Y.

1. *On the one-sided Wiener's Theorem for the motion group*, Ann. of Math. (2) **111** (1980), 415–422.

Wells, B.

1. *Homomorphisms and translates of bounded functions*, Duke Math. J. **41** (1974), 35–39.

Wesler, O.

1. *Invariance theory and a modified minimax principle*, Ann. Math. Statist. **30** (1959), 1–20.

Wilde, C. O.

1. *On amenable semigroups and applications of the Stone-Čech compactification*, Thesis, University of Illinois, Urbana, Illinois, 1964.
2. *Characterisation of finite amenable transformation semigroups*, J. Austral. Math. Soc. **15** (1973), 86–93.

Wilde, C. O. and Argabright, L.

1. *Invariant means and factor semigroups*, Proc. Amer. Math. Soc. **18** (1967), 226–228.

Wilde, C. O. and Jayachandran, T.

1. *Amenable transformation semigroups*, J. Austral. Math. Soc. **12** (1971), 502–510.

Wilde, C. O. and Witz, K. G.

1. *Invariant means and the Stone-Čech compactification*, Pacific J. Math. **21** (1967), 577–586.

Williamson, J. H.

1. *Harmonic analysis on semigroups*, J. London Math. Soc. **42** (1967), 1–41.

Willis, G. A.

1. *Derivations from group algebras, factorization in cofinite ideals and topologies on $B(X)$*, Thesis, University of Newcastle upon Tyne, 1980.
2. *Approximate units in finite codimensional ideals of group algebras*, J. London Math. Soc. **26** (1982), 143–154.

Witz, K. G.

1. *Applications of a compactification for bounded operator semigroups*, Illinois J. Math. **8** (1964), 685–696.

Wolf, J. A.

1. *Growth of finitely generated solvable groups and curvature of Riemannian manifolds*, J. Differential Geom. **2** (1968), 421–446.

Wolff, M.

1. *Vectorvertige invariante Masse von rechtsamenablen Halbgruppen positiver operatoren*, Math. Z. **120** (1971), 265–271.

Wong, J. C. S.

1. *Topologically stationary locally compact groups and amenability*, Trans. Amer. Math. Soc. **144** (1969), 351–363.
2. *Topological left invariant means on locally compact groups and fixed points*, Proc. Amer. Math. Soc. **27** (1971), 572–578.
3. *Invariant means on locally compact semigroups*, Proc. Amer. Math. Soc. **31** (1972), 39–45.
4. *An ergodic property of locally compact amenable semigroups*, Pacific J. Math. **48** (1973), 615–619.
5. *Topological semigroups and representations*. I, Trans. Amer. Math. Soc. **200** (1974), 89–109.
6. *Absolutely continuous measures on locally compact semigroups*, Canad. Math. Bull. **18** (1975), 127–131.
7. *A characterization of topological left thick subsets in locally compact left amenable semigroups*, Pacific J. Math. **62** (1976), 295–303.
8. *A characterisation of locally compact amenable semigroups*, Preprint.
9. *Amenability and substantial semigroups*, Canad. Math. Bull. **19** (1976), 231–234.
10. *Abstract harmonic analysis of generalised functions on locally compact semigroups with applications to invariant means*, J. Austral. Math. Soc. **23** (1977), 84–94.
11. *Convolution and separate continuity*, Pacific J. Math. **75** (1978), 601–611.
12. *On topological analogues of left-thick subsets in semigroups*, Preprint.
13. *Characterisations of extremely amenable semigroups*, Math. Scand. **48** (1981), 101–108.
14. *On the relation between left thickness and topological left thickness in semigroups*, Proc. Amer. Math. Soc. **86** (1982), 471–476.
15. *On left thickness of subsets in semigroups*, Proc. Amer. Math. Soc. **84** (1982), 403–407.

Wong, J. C. S. and Riazi, A.

1. *Characterisations of amenable locally compact semigroups*, Pacific J. Math. **108** (1983), 479–496.

Wong, R. Y. T.

1. *On affine maps and fixed points*, J. Math. Anal. Appl. **29** (1970), 158–162.

Wood, J. C.

1. *A note on the fundamental group of a manifold of negative curvature*, Math. Proc. Cambridge Philos. Soc. **83** (1978), 415–417.

Woodward, G.

1. *Invariant means and ergodic sets in Fourier analysis*, Pacific J. Math. **54** (1974), 281–299.

Wordingham, J. R.

1. *The left regular *-representation of an inverse semigroup*, Proc. Amer. Math. Soc. **86** (1982), 55–58.

Wulbert, D. E.

1. *Averaging projections*, Illinois J. Math. **13** (1969), 689–693.

Yau, S. T.

1. *On the fundamental group of compact manifolds of non-positive curvature*, Ann. of Math. (2) **93** (1971), 579–585.

Yeadon, F. J.

1. *Fixed points and amenability—a counterexample*, J. Math. Anal. Appl. **45** (1974), 718–720.

Yood, B.

1. *Fixed points for semigroups of linear operators*, Proc. Amer. Math. Soc. **2** (1951), 225–233.

Yoshizawa, H.

1. *Some remarks on unitary representations of the free group*, Osaka J. Math. **3** (1951), 55–63.

Yosida, K.

1. *A theorem concerning the semisimple Lie groups*, Tôhoku Math. J. **44** (1938), 81–84.

Zelazko, W.

1. *On algebras L_p of locally compact groups*, Colloq. Math. **8** (1961), 112–120.
2. *A note on L_p-algebras*, Colloq. Math. **10** (1963), 53–56.
3. *On the Burnside problem for locally compact groups*, Sympos. Math. **16** (1975), 409–416.

Želobenko, D. P.

1. *Compact Lie groups and their representations*, Transl. Math. Mono., vol. 40, Amer. Math. Soc., Providence, R. I., 1973.

Zimmer, R. J.

1. *Hyperfinite factors and amenable ergodic actions*, Invent. Math. **41** (1977), 23–32.

2. *Amenable ergodic actions, hyperfinite factors and Poincaré flows*, Bull. Amer. Math. Soc. **83** (1977), 1078–1080.
3. *On the von Neumann algebra of an ergodic group action*, Proc. Amer. Math. Soc. **66** (1977), 289–293.
4. *Uniform subgroups and ergodic actions of exponential Lie groups*, Pacific J. Math. **78** (1978), 267–272.
5. *Amenable pairs of groups and ergodic actions and the associated von Neumann algebras*, Trans. Amer. Math. Soc. **243** (1978), 271–286.
6. *Amenable ergodic actions and an application to Poisson boundaries of random walks*, J. Funct. Anal. **27** (1978), 350–372.
7. *Ergodic theory, group representations and rigidity*, Bull. Amer. Math. Soc. (N.S.) **6** (1982), 383–416.

Supplementary Bibliography ([S])

Alspach, D. E., Matheson, A. and Rosenblatt, J.M., *Projections onto translation-invariant subspaces of* $L_1(G)$, J. Funct. Anal. **59** (1984), 254–292.

———, *Erratum to "Projections onto translation-invariant subspaces of* $L_1(G)$", J. Funct. Anal. **69** (1986), 141.

Anker, J.-P., *Applications de la p-induction en analyse harmonique*, Comment. Math. Helv. **58** (1983), 622–645.

Armstrong, T. and Prikry, K., *K-finiteness and K-additivity of measures on sets and left invariant measures on discrete groups*, Proc. Amer. Math. Soc. **80** (1980), 105–112.

Arsac, G., *Sur l'espace de Banach engendré par les coefficients d'une représentation unitaire*, Publ. Dép. Math. (Lyon) **13** (1976), 1–101.

Ayyaswamy, S. K. and Ramakrishnan, P. V., *Solutions to some problems in amenable semigroups*, Proc. Roy. Soc. Edinburgh Sect. A **104** (1986), 343–348.

Bade, W. G., Curtis Jr., P. C. and Dales, H. G., *Amenability and weak amenability for Beurling and Lipschitz algebras*, Proc. London Math. Soc. **55** (1987), 359–378

———, *The continuity of derivations of Banach algebras*, J. Funct. Anal. **16** (1974), 372–387.

Baggett, L., *Measures invariant under a linear group*, Proc. Amer. Math. Soc. **94** (1985), 179–186.

Balcerzyk, S. and Mycielski, J., *On the existence of free subgroups in topological groups*, Fund. Math. **44** (1957), 303–308.

Balibrea, F. and Vera, G., *On the sublinear functional associated to a family of invariant means*, Manuscripta Math. **55** (1986), 101–109.

Banach, S. and Kuratowski, C., *Sur une généralisation du problème de la mesure*, Fund. Math. **14** (1929), 127–131.

Barnes, B. A., *When is a representation of a Banach *-algebra Naimark-related to a *-representation*? Pacific J. Math. **72** (1977), 5–25.

Baronti, M., *Banach algebras A_p of locally compact groups*, Riv. Mat. Univ. Parma **11** (1985), 399–407.

Batty, C. J. K., *Semiperfect C^*-algebras and the Stone-Weierstrass problem*, J. London Math. Soc. **34** (1986), 97–110.

Bédos, E. and Harpe, P. de la, *Moyennabilité interieure des groupes: definitions et exemples*, Enseign. Math. **32** (1986), 139–157.

Bekka, M. E. B., A characterization of locally compact amenable groups by means of tensor products. Preprint.

Bezuglyi, S. I. and Golodets, V. Ya., *Outer conjugation of actions of countable amenable groups on a measure space*, Izv. Akad. Nauk. SSSR Ser. Mat. **50** (1986), 643–660.

Bondy, J. A. and Murty, U. S. R., *Graph theory with applications*, The MacMillan Press Ltd., London, 1976.

Bondar, J. V. and Milnes, P., *A converse to the Hunt-Stein Theorem*, Preprint.

Bourgain, J., *Translation-invariant forms on $L^p(G)$ $(1 < p < \infty)$*, Ann. Inst. Fourier (Grenoble) **36** (1986), 97–104.

Božejko, M., *Remark on Herz-Schur multipliers on free groups*, Math. Ann. **258** (1981), 11–15.

——, *Positive definite bounded matrices and a characterization of amenable groups*, Proc. Amer. Math. Soc. **95** (1985), 357–360.

Bunce, J. W. and Paschke, W. L., *Quasi-expectations and amenable von Neumann algebras*, preprint.

Burckel, R. B., *Weakly almost periodic functions on semigroups*, Gordon and Breach, New York, 1970.

Canniere, J. de and Haagerup, U., *Multipliers of the Fourier algebras of some simple Lie groups and their discrete subgroups*, Amer. J. Math. 107 (1985), 455–500.

Carling, L. N., *On the restriction map of the Fourier-Stieltjes algebra $B(G)$ and $B_p(G)$*, J. Funct. Anal. **25** (1977), 236–243.

Chen, S., *On non-amenable groups*, Internat. J. Math. Math. Sci. **1** (1978), 529–532.

Choda, M., *The factors of inner amenable groups*, Math. Japonica **24** (1979), 145–152.

——, *Effect of inner amenability on strong ergodicity*, Math. Japonica **28** (1983), 109–115.

Choda, H. and Choda, M., *Fullness, simplicity and inner amenability*, Math. Japonica **24** (1979), 235–246.

Choda, H. and Echigo, M., *A new algebraical property of certain von Neumann algebras*, Proc. Japan Acad. **39** (1963), 651–655.
——, *A remark on a construction of finite factors.* I, II, Proc. Japan Acad. **40** (1964), 474–481.

Choi, M. D. and Effros, E. G., *Nuclear C^*-algebras and injectivity: The general case*, Indiana Univ. Math. J. **26** (1977), 443–446.

Chojnacki, W., *Sur un théorème de Day, un théorème de Mazur-Orlicz et une généralisation de quelques théorèmes de Silverman*, Colloq. Math. **50** (1986), 257–262.

Chou, C., *Ergodic group actions with nonunique invariant means*, Proc. Amer. Math. Soc. **100** (1987), 647–650

Chou, C., Lau, A. T. and Rosenblatt, J., *Approximation of compact operators by sums of translations*, Illinois J. Math. **29** (1985), 340–350.

Christensen, J. P. R., *Joint continuity of separately continuous functions*, Proc. Amer. Math. Soc. **82** (1981), 455–461.

Cole, A. J. and Świerczkowski, S., *On a class of non-measurable groups*, Proc. Cambridge Philos. Soc. **57** (1961), 227–299.

Connes, A., *On the equivalence between injectivity and semidiscreteness for operator algebras*, Algèbres d'Opérateurs et Leurs Applications en Physique Mathématique, No. 274, CNRS, 1979, pp. 107–112.
——, *Classifications des facteurs*, Proc. Sympos. Pure Math., vol. 38, Part 2, Amer. Math. Soc., Providence, R.I., 1982, pp. 43–109.

Connes, A., Feldman, J., and Weiss, B., *An amenable equivalence relation is generated by a single transformation*, J. Ergodic Theory Dynamical Systems **1** (1981), 431–450.

Connes, A. and Weiss, B., *Property T and asymptotically invariant sequences*, Israel J. Math. **37** (1980), 209–210.

Cowling, M., *La synthèse des convoluteurs de L^p de certains groupes pas moyennables*, Boll. Un. Mat. Ital. **14A** (1977), 551–555; **6** (1982), 317.

Cuntz, J., *Simple C^*-algebras generated by isometries*, Comm. Math. Phys. **57** (1977), 173–185.
——, *K-theoretic amenability for discrete groups*, J. Reine Angew. Math. **344** (1983), 180–195.

Curtis Jr., P. C. and Loy, R. J., *Homological properties of amenable Banach algebras*, Preprint.

Dani, S. G., *On invariant finitely additive measures for automorphism groups of tori*, Trans. Amer. Math. Soc. **287** (1985), 189–199.

———, *A note on invariant finitely additive measures*, Proc. Amer. Math. Soc. **93** (1985), 67–72.

Dekker, T. J., *Decompositions of sets and spaces*. I, II, III, Indag. Math. **18** (1956), 581–589, 590–595; **19** (1957), 104–107.

———, *Paradoxical decompositions of sets and spaces*, Academisch Proefschrift, Amsterdam : van Soest, 1958.

Deligne, P. and Sullivan, D., *Division algebras and the Hausdorff-Banach-Tarski Paradox*, End. Math. **29** (1983), 145–150.

Derighetti, A., *Some results on the Fourier-Stieltjes algebra of a locally compact group*, Comment. Math. Helv. **45** (1970), 219–228.

———, *On the Property P_1 of locally compact groups*, Comment. Math. Helv. **46** (1971), 226–239.

———, *Sur certaines propriétés des représentations unitaires des groupes localement compacts*, Comment. Math. Helv. **48** (1973), 328–339.

———, *Sulla nozione di contenimento debole e la proprietà di Reiter*, Rend. Sem. Mat. Fis. Milano **44** (1974), 47–54.

———, *Sur les représentations unitaires des sous-groupes fermés d'un groupe localement compact*, Symposia Math., vol. 22, Academic Press, London, 1977, pp. 133–143.

———, *Some remarks on $L_1(G)$*, Math. Z. **164** (1978), 189–194.

———, *A propos des convoluteurs d'un groupe quotient*, Bull. Sci. Math. France **107** (1983), 3–23.

———, *Quelques observations concernant les ensembles de Ditkin d'un groupe localement compact*, Preprint.

Diestel, J. and Uhl Jr., J. J., *Vector measures*, Math. Surveys, no. 15, Amer. Math. Soc., Providence, R.I., 1977.

Drinfeld, V. G., *Solution of the Banach-Ruziewicz problem on* $\mathbf{S}^2$ *and* $\mathbf{S}^3$, Functional Anal. Appl. **18** (1984), 77–78.

Duncan, J. and Paterson, A. L. T., "C^*-algebras of Clifford semigroups", to appear, Proc. Roy. Soc. Edinburgh.

Dzinotyiweyi, H. A. M., *Almost convergence and weakly almost periodic functions on a semigroup*, Trans. Amer. Math. Soc. **277** (1983), 125–132.

———, *The analogue of the group algebra for topological semigroups*, Res. Notes in Math., vol. 98, Pitman, Boston and London, 1984.

———, *The cardinality of the set of invariant means on a locally compact topological semigroup*, Compositio Math. **54** (1985), 41–49.

———, *Sizes of quotient spaces of certain function algebras on topological semigroups*, Compositio Math. **55** (1985), 303–311.

Enock, M. and Schwartz, J.-M., *Moyennabilité des groups localement compacts et algèbres de Kac*, C. R. Acad. Sci. Paris Sér. I Math. **300** (1985), 625–626.

——, *Algèbres de Kac moyennables*, Pacific J. Math. **25** (1986), 363–379.

Eymard, P., *L'Algèbre de Fourier d'un groupe localement compact*, Bull. Soc. Math. France **92** (1964), 181–236.

——, *Sur les moyennes invariantes et les représentations unitaires*, C. R. Acad. Sci. Paris Sér. A **272** (1971), 1649–1652.

——, *Initiation à la théorie des groupes moyennables*, Analyse Harmonique sur les Groupes de Lie, Lectures Notes in Math., vol. 497, Springer-Verlag, Berlin, 1975, pp. 89–107.

Fan, K., *Invariant subspaces for a semigroup of linear operators*, Indag. Math. **27** (1985), 447–551.

Figà-Talamanca, A. and Picardello, M., *Multiplicateurs de $A(G)$ qui ne sont pas dans $B(G)$*, C. R. Acad. Sci. Paris Sér. A **277** (1973), 117–119.

Flory, V., *On the Fourier algebra of a locally compact amenable group*, Proc. Amer. Math. Soc. **29** (1971), 603–606.

Forrest, B., *Invariant means on function spaces on semitopological semigroups*, Thesis, University of Alberta, 1983.

——, *Amenability and ideals in the Fourier algebra of a locally compact group*, Thesis, University of Alberta, 1987.

——, *Amenability and derivations of the Fourier algebra*, Preprint.

——, *Amenability and bounded approximate identities in ideals of $A(G)$*, Preprint.

——, *Complemented ideals in the Fourier algebra of a locally compact group*, Preprint.

Fox, J., *Frobenius reciprocity and extensions of nilpotent Lie groups*, Trans. Amer. Math. Soc. **298** (1986), 123–144.

Ganesan, S., *A necessary and sufficient condition for a connected amenable group to have polynomial growth*, Proc. Amer. Math. Soc. **93** (1985), 176–178.

——, *P-amenable locally compact groups*, Monatsh. Math. **101** (1986), 301–308.

Ganesan, S. and Jenkins, J. W., *An Archimedean property for groups with polynomial growth*, Proc. Amer. Math. Soc. **88** (1983), 550–554.

Ghys, E. and Carrière, Y., *Relations d'equivalence moyennables sur les groupes de Lie*, C. R. Acad. Sci. Paris Sér. I Math. **300** (1985), 677–680.

Glasner, S., *Proximal flows*, Lectures Notes in Math., vol. 517, Springer-Verlag, Berlin and New York, 1976.

Glicksberg, I. and de Leeuw, K., *Applications of almost periodic compactifications*, Acta Math. **105** (1961), 63–97.

Granirer, E. E., *On some properties of the Banach algebras $A_p(G)$ for locally compact groups*, Proc. Amer. Math. Soc. **95** (1985), 375–381.

____, *A strong containment property for discrete amenable groups of automorphisms on W^*-algebras*, Trans. Amer. Math. Soc. **297** (1986), 753–761.

____, *Some results on $A_p(G)$ submodules of $PM_p(G)$*, Colloq. Math. **51** (1987), 155–163.

____, *On some spaces of linear functionals on the algebras $A_p(G)$ for locally compact groups*, Colloq. Math. **52** (1987), 119–132.

____, *On amenable groups of autormorphisms on von Neumann algebras*, Preprint.

Green, P., *The local structure of twisted covariance algebras*, Acta Math. **140** (1978), 191–250.

Grigorchuk, R. I., *On Milnor's problem of group growth*, Soviet Math. Dokl. **28** (1983), 23–26.

____, *Degrees of growth of finitely generated groups and the theory of invariant means*, Izv. Akad. Nauk. SSSR Ser. Mat. **48** (1984), 939–985.

____, *Superamenability and the occurrence problem of free semigroups*, Funct. Anal. Appl. **21** (1987), 74–75.

Groenbaek, N., *A characterization of weakly amenable Banach algebras*, Preprint.

Grosser, M., Losert, V., and Rindler, H., *"Double multipliers" und asymptotisch invariante approximerende Einheiten*, Anz. Österreich Akad. Wiss. Math.-Natur. Kl. **117** (1980), 7–11.

Grosvenor, J. R., *A relation between invariant means on Lie groups and invariant means on their discrete subgroups*, Trans. Amer. Math. Soc. **288** (1985), 813–825.

Guichardet, A., *Sur l'homologie et la cohomologie des algèbres de Banach*, C. R. Acad. Sci. Paris Sér A **262** (1966), 38–41.

____, *Sur l'homologie et la cohomologie des groupes localement compacts*, C. R. Acad. Sci. Paris Sér A **262** (1966), 118–120.

____, *Tensor products of C^*-algebras*, Aarhus Univ. Lecture Notes Series, No. 12, 1969.

Haagerup, U., *The reduced C^*-algebra of the free group on two generators*, Prog. Math., vol. 11, Birkhäuser Verlag, 1980.

____, *Solution of the similarity problem for cyclic representations of C^*-algebras*, Ann. of Math. (2) **118** (1983), 215–240.

____, *Connes' bicentralizer problem and uniqueness of the injective factor of type* III_1, Acta Math. **158** (1987), 95–148.

Hahn, P., *The σ-representation of amenable groupoids*, Rocky Mountain J. Math. **9** (1979), 631–639.

Harpe, P. de la and Skandalis, G., *Un résultat de Tarski sur les actions moyennables de groupes et les partitions paradoxales*, Enseign. Math. **32** (1986), 121–138.

Henle, J. and Wagon, S., *Problem* 6353, Amer. Math. Monthly **90** (1983), 64.

Herz, C., *Remarques sur la note précédente de M. Varopoulos*, C. R. Acad. Sci. Paris Sér A **260** (1965), 6001–6004.

——, *Synthèse spectrale pour les sous-groupes par rapport aux algèbres* A_p, C. R. Acad. Sci. Paris Sér. A **271** (1970), 316–318.

Jarosz, K., *Perturbations of Banach algebras*, Lectures Notes in Math., vol. 1120, Springer-Verlag, Berlin and New York, 1985.

Johnson, B. E., *An introduction to the theory of centralizers*, Proc. London Math. Soc. **14** (1964), 299–320.

——, *The Wedderburn decomposition of Banach algebras with finite-dimensional radical*, Amer. J. Math. **90** (1968), 866–876.

——, *Introduction to cohomology in Banach algebras*, Algebras in Analysis, Academic Press, New York, 1975, pp. 84–100.

——, *Low dimensional cohomology of Banach algebras*, Proc. Sympos. Pure Math., vol. 38, Part 2, Amer. Math. Soc., Providence, R.I., 1982, pp. 253–259.

——, *Derivations from* $L^1(G)$ *into* $L^1(G)$ *and* $L^\infty(G)$ (Proc. Internat. Conf. on Harmonic Analysis, Luzembourg, 1987), Lecture Notes in Math., Springer-Verlag, Berlin, Heidelberg and New York (to appear).

Jolissaint, P., *Croissance d'un groupe de génération finie et fonctions lisses sur son dual*, C. R. Acad. Sci. Paris Sér. I Math. **300** (1985), 601–604.

Julg, P, and Valette, A., *K-moyennabilité pour les groupes opérant sur les arbres*, C. R. Acad. Sci. Paris Sér. I Math. **296** (1983), 977–980.

——, *K-theoretic Amenability for* $\mathrm{SL}_2(\mathbf{Q}_p)$ *and the action on the associated tree*, J. Funct. Anal. **58** (1984), 194–215.

——, *Twisted coboundary operator on a tree and the Selberg principle*, J. Operator Theory **16** (1986), 285–304.

Kaimanovich, V. A. and Vershik, A. M., *Random walks on discrete groups: Boundary and entropy*, Ann. Probab. **11** (1983), 457–490.

Kaniuth, E., *Weak containment and tensor products of group representations.* II, Math. Ann. **270** (1985), 1–15.

Kasparov, G. G., *The operator K-functor and extensions of* C^**-algebras*, Izv. Akad. Nauk. SSSR Ser. Mat. **44** (1980), 571–636.

Khelemskii, A. Y., *On the homological dimension of normal modules over Banach algebras*, Math. USSR-Sb. **10** (1970), 399–412.

——, *On a method for calculating and estimating the global homological dimension of Banach algebras*, Math. USSR-Sb. **16** (1972), 125–138.

——, *Flat Banach modules and amenable algebras*, Trudy Moskov. Mat. Obshch. **47** (1984), 179–218.

Knapp, A. W., *A decomposition theorem for bounded uniformly continuous functions on a group*, Amer. J. Math. **88** (1966), 902–914.

——, *Representation theory of semisimple groups*, Princeton Univ. Press, Princeton, N.J., 1986.

Krasa, S., *Non-uniqueness of invariant means for amenable group actions*, Monatsh. Math. **100** (1985), 121–125.

Kugler, W., *Über die Einfachheit gewisser verallgemeinerter L^1-Algebren*, Arch. Math. (Basel) **26** (1975), 82–88.

Lai, H. C., *On some properties of $A^p(G)$-algebras*, Proc. Japan Acad. **45** (1969), 572–576.

Lai, H. C. and Chen, I., *Harmonic analysis on the Fourier algebras $A_{1,p}(G)$*, J. Austral. Math. Soc. Ser. A **30** (1981), 438–452.

Lance, E. C., *Tensor products and nuclear C^*-algebras*, Proc. Sympos. Pure Math., vol. 38, Part 1, Amer. Math. Soc., Providence, R.I., 1982, pp. 379–399.

Larsen, R., *An introduction to the theory of multipliers*, Springer-Verlag, Berlin and New York, 1971.

Lau, A. T., *Amenability and invariant subspaces*, J. Austral. Math. Soc. Ser. A **18** (1974), 200–204.

——, *Closed convex invariant subsets of $L_p(G)$*, Trans. Amer. Math. Soc. **232** (1977), 131–142.

——, *Operators which commute with convolutions in subspaces of $L_\infty(G)$*, Colloq. Math. **39** (1978), 351–359.

——, *The second conjugate algebra of the Fourier algebra of a locally compact group*, Trans. Amer. Math. Soc. **267** (1981), 53–63.

——, *Analysis on a class of Banach algebras with applications to harmonic analysis on locally compact groups and semigroups*, Fund. Math. **118** (1983), 161–175.

——, *Finite-dimensional invariant subspaces for a semigroup of linear operators*, J. Math. Anal. Appl. **97** (1983), 374–379.

——, *Invariantly complemented subspaces and groups with fixed point property*, Nonlinear Functional Analysis and Its Applications, Reidel, Dordrecht and Boston, Mass., 1986, pp. 305–311.

——, *Continuity of Arens multiplication on the dual space of bounded uniformly continuous functions on locally compact groups and topological semigroups*, Math. Proc. Cambridge Philos. Soc. **99** (1986), 273–283.

——, *Uniformly continuous functionals on Banach algebras*, Colloq. Math. **51** (1987), 195–205.

Lau, A. T. and Losert, V., *Weak* closed complemented invariant subspaces of $L_\infty(G)$ and amenable locally compact groups*, Pacific J. Math. **123** (1986), 149–159.

Lau, A. T. and Paterson, A. L. T., *The exact cardinality of the set of topological left invariant means on an amenable locally compact group*, Proc. Amer. Math. Soc. **98** (1986), 75–80.

———, *Operator theoretic characterizations of* [IN]*-groups and inner amenability*, Proc. Amer. Math. Soc. (to appear).

———, *The class of locally compact groups G for which $C^*(G)$ is amenable*, Preprint.

Lau, A. T., Paterson, A. L. T. and Wong, J. C. S., *Invariant subspace theorms for amenable groups*, Proc. Edinburgh Math. Soc., to appear.

Lau, A. T. and Wong, J. C. S., *Finite-dimensional invariant subspaces for measurable semigroups of linear operators*, J. Math. Anal. Appl. **127** (1987), 548–558.

———, *Invariant subspaces for algebras of linear operators and amenable locally compact groups*, Preprint.

Leeuw, K. de and Glicksberg, I., *Applications of almost periodic compactifications*, Acta Math. **105** (1961), 63–97.

Leinert, M., *Convoluteurs de groupes discrets*, C. R. Acad. Sci. Paris Sér. A **271** (1971), 630–631.

Leptin, H., *Verallgemeinerte L^1-Algebren*, Math. Ann. **159** (1965), 51–76.

Lipsman, R., *Representation theory of almost connected groups*, Pacific J. Math. **42** (1972), 453–467.

Liu, T.-S., van Rooij, A. and Wang, J.-K., *Projections and approximate identities for ideals in group algebras*, Trans. Amer. Math. Soc. **175** (1973), 469–482.

Lodkin, A. A., *Approximation for actions of amenable groups and transversal automorphisms*, Operator Algebras and their Connections with Topology and Ergodic Theory, Lecture Notes in Math., vol. 1132, Springer-Verlag , Berlin and New York, 1985.

Losert, V., *Uniformly distributed sequences on compact, separable, nonmetrissable groups*, Acta Sci. Math. (Szeged) **40** (1978), 107–110.

———, *On the existence of uniformly distributed sequences in compact topological spaces.* I, Trans. Amer. Math. Soc. **246** (1978), 463–471.

———, *On the existence of uniformly distributed sequences in compact topological spaces.* II, Preprint.

———, *Some properties of groups without Property P_1*, Comment. Math. Helv. **54** (1979), 133–139.

———, *On tensor products of Fourier algebras*, Arch. Math. **43** (1984), 370–372.

———, *Properties of the Fourier algebra that are equivalent to amenability*, Proc. Amer. Math. Soc. **92** (1984), 347–354.

——, *On the structure of groups with polynomial growth.* Math. Z. **195** (1987), 109–117.

Losert, V. and Rindler, H., *Uniform distribution and the mean ergodic theorem*, Invent. Math. **50** (1978/9), 65–74.

——, *Asymptotically central functions and invariant extensions of Dirac measure*, Probability Measures on Groups VII (Oberwolfach, 1983), Lecture Notes in Math., vol. 1064, Springer-Verlag, Berlin and New York, 1984, pp. 368–378.

——, *Conjugation-invariant means*, Colloq. Math. **51** (1987), 221–225.

Mackey, G. W., *Ergodic theory and virtual groups*, Math. Ann. **166** (1966), 187–207.

Maharam, D., *On measure in abstract sets*, Trans. Amer. Math. Soc. **51** (1942), 413–433.

Maxones, W. and Rindler, H., *Asymptotisch gleichverteilte Massfolgen in Gruppen von Heisenberg-Typ*, Sitzungsber. Wien, Math.- Natur. Kl. **2** (1976), 485–504.

——, *Bemerkungen zu einer Arbeit von P. Gerl "Gleichverteilung auf lokalkompakten Gruppen"*, Math. Nachr. **79** (1977), 193–199.

McKennon, K., *Multipliers, positive functionals, positive-definite functions and Fourier-Stieltjes transforms*, Mem. Amer. Math. Soc. **111** (1971).

Meisters, G. H. and Schmidt, W. M., *Translation-invariant linear forms on $L^2(G)$ for compact abelian groups*, J. Funct. Anal. **11** (1972), 407–424.

Michelle, L. de and Soardi, P. M., *Symbolic calculus in $A_p(G)$*, Atti Accad. Naz. Lincei Rend. Cl. Sci. Fis. Natur. **57** (1974), 24–30, 31–35.

Miličic, D., *Representations of almost connected groups*, Proc. Amer. Math. Soc. **47** (1975), 517–518.

Milnes, P., *Counterexample to a conjecture of Greenleaf*, Canad. Math. Bull. **13** (1970), 497–499.

——, *Minimal and distal functions on some non-abelian groups*, Math. Scand. **49** (1981), 86–94.

——, *Almost periodic functions on semitopological semigroups*, Rocky Mountain J. Math. **12** (1982), 265–269.

——, *The Bohr almost periodic functions do not form a linear space*, Math. Z. **188** (1984), 1–2.

——, *On Bohr almost periodicity*, Math. Proc. Cambridge Philos. Soc. **99** (1986), 489–493.

——, *Minimal and distal functions on semidirect products of groups*. II, Math. Scand. **58** (1986), 215–226.

Milnes, P. and Paterson, A. L. T., *Ergodic sequences and a subspace of $B(G)$*, Rocky Mountain J. Math. (to appear).

Moore, C. C., *Ergodic theory and von Neumann algebras*, Proc. Sympos. Pure Math., vol. 38, Part 2, Amer. Math. Soc., Providence, R.I., 1982, pp. 179–226.

Müller-Römer, P., *Kontrahierende Erweiterungen und kontrahierbare Gruppen*, J. Reine Angew. Math. **283/284** (1976), 238–264.

Mycielski, J., *On the paradox of the sphere*, Fund. Math. **42** (1955), 348–355.

——, *Independent sets in topological algebras*, Fund. Math. **55** (1964), 139–147.

Namioka, I., *Affine flows and distal points*, Math. Z. **184** (1983), 259–269.

Nebbia, C., *Multipliers and asymptotic behaviour of the Fourier algebra of non-amenable groups*, Proc. Amer. Math. Soc. **84** (1982), 549–554.

Niederreiter, H., *On a paper by Blum, Eisenberg and Hahn concerning ergodic theory and the distribution of sequences in the Bohr group*, Acta Sci. Math. (Szeged) **37** (1975), 103–108.

Nielsen, O. A., *The failure of the topological Frobenius property for nilpotent Lie groups*, Math. Scand. **45** (1979), 305–310.

——, *Unitary representations and coadjoint orbits of low-dimensional nilpotent Lie groups*, Queen's University Paper No. 63 (1983).

Nillsen, R., *Invariant subspaces of some function spaces on a locally compact group*, J. Funct. Anal. **64** (1985), 338–357.

Ocneanu, A., *Actions of discrete amenable groups on von Neumann algebras*, Lecture Notes in Math., vol. 1138, Springer-Verlag, Berlin and New York, 1985.

Paterson, A. L. T., *The class of locally compact groups G for which $C^*(G)$ is amenable* (Proc. Internat. Conf. on Harmonic Analysis, Luxembourg, 1987), Lecture Notes in Math., Springer-Verlag, Berlin, Heidelberg and New York (to appear).

——, *An explicit formula for the Plancherel measure on nilpotent Lie groups*, Math. Ann. (to appear).

Pedersen, C. K., *C^*-algebras and their automorphism groups*, Academic Press, London, 1979.

Peressini, A. L., *Ordered topological vector spaces*, Harper and Row, New York, 1967.

Pier, J.-P., *Quasi-invariance intérieure sur les groupes localement compacts*, G.E.M.L. Actualités Mathématiques, Actes du 6e Congrès du Groupement des Mathématiciens d'Expression Latine, Luxembourg, 1982, pp. 431–436.

——, *Le phénomène de la moyennabilité*, Rend. Sem. Mat. Fis. Milano **53** (1983), 319–332.

——, *Amenable locally compact groups*, Wiley, New York, 1984.

——, *Invariance et moyennabilité en analyse harmonique*, Nieuw. Arch. Wisk. **3** (1985), 267–274.

____, *Invariance intérieure sur les groupes localement compacts*, Rev. Roumaine Math. Pures Appl. **32** (1987), 375–396.

____, *Amenable Banach algebras.* Pitman research notes in mathematics series, Longman Scientific and Technical, Harlow, England, 1988.

Poguntke, D., *Auflösbare Liesche Gruppen mit symmetrischen L^1-Algebren* , J. Reine Angew Math. **358** (1985), 20–42.

Popa, S., "*A short proof of injectivity implies hyperfiniteness for finite von Neumann algebras*", J. Operator Theory 16 (1986), 261–272.

Powers, R. T., *Simplicity of the C^*-algebra associated with the free group of two generators*, Duke Math. J. **42** (1975), 151–156.

Promislow, D., *Non-existence of invariant measures*, Proc. Amer. Math. Soc. **88** (1983), 89–92.

Pukanszky, L., *Unitary representations of Lie groups and generalized symplectic geometry*, Proc. Sympos. Pure Math., vol. 38, Part 1, Amer. Math. Soc., Providence, R.I., 1982, pp. 435–466.

Racher, G., *The nuclear multipliers from $L^1(G)$ into $L^\infty(G)$ for amenable G*, J. London Math. Soc. **33** (1986), 339–346.

Ramakrishnan, P. V. and Ayyaswamy, S. K., *Amenable semigroups and densities*, Math. Today **4** (1986), 1–10.

Rindler, H., *Ein Gleichverteilungsbegriff für mittelbare Gruppen*, Österreich. Akad. Wiss. Math.-Natur Kl. Sitzungber. II **182** (1973), 107–119.

____, *Zur Eigenschaft P_1 lokalkompakter Gruppen*, Indag. Math. **35** (1973), 142–147.

____, *Ein Problem aus der Theorie der Gleichverteilung.* II, Math. Z. **135** (1973), 73–92.

____, *Über ein Problem von Reiter und ein Problem von Derighetti zur Eigenschaft P_1 lokalkompakter Gruppen*, Comment. Math. Helv. **48** (1973), 492–497.

____, *Ein Problem aus der Theorie der Gleichverteilung.* I Teil, Monatsh. Math. **78** (1974), 51–67.

____, *Zur $L^1(G)$-Gleichverteilung auf abelschen und kompakten Gruppen*, Arch. Math. (Basel) **26** (1975), 209–213.

____, *Uniformly distributed sequences in quotient groups*, Acta Sci. Math. (Szeged) **38** (1976), 153–156.

____, *Uniform distribution on locally compact groups*, Proc. Amer. Math. Soc. **57** (1976), 130–132.

____, *Approximierende Einheigen in Idealem von Gruppen algebren*, Anz. Österreich. Akad. Wiss. Math.-Natur. Kl. **5** (1976), 37–39.

____, *Eine Charakterisierung Gleichverteilter Folgen*, Arch. Math. (Basel) **32** (1979), 185–188.

____, *Groups of measure preserving transformations*, Math. Ann. **279** (1987) (to appear).

Rindler, H. and Schoissengeier, J., *Gleichverteilte Folgen und differenzierbare Funktionen*, Monatsh. Math. **84** (1977), 125–131.

Robinson, R. M., *On the decomposition of spheres*, Fund. Math. **34** (1947), 246–260.

Rogers, C. A. et al., *Analytic sets*, Academic Press, London and New York, 1980.

Rosenblatt, J. M., *Ergodic and mixing random walks on locally compact groups*, Math. Ann. **257** (1981), 31–42.

——, *Translation-invariant linear forms on $L_p(G)$*, Proc. Amer. Math. Soc. **94** (1985), 226–228.

——, *Ergodic group actions*, Arch. Math. (Basel) **47** (1986), 263–269.

Rosenblatt, J. M. and Yang, Z., *Functions with a unique mean value*. Preprint.

Ruppert, W., *On semigroup compactifications of topological groups*, Proc. Royal Irish Academy Sect. A **79** (1979), 179–200.

——, *Compact semitopological semigroups: an intrinsic theory*, Lecture Notes in Math., vol. 1079, Springer-Verlag, Berlin and New York, 1984.

Saeki, S., *Discontinuous translation invariant functionals*, Trans. Amer. Math. Soc. **282** (1984), 403–414.

Sakai, K., *Amenability of semigroups with amenable homomorphic images*, Sci. Rep. Kagoshima Univ. **34** (1985), 1–10.

Schmetterer, L., *Introduction to mathematical statistics*, Springer-Verlag, Berlin, Heidelberg and New York, 1974.

Schwartz, J. T., *W^*-algebras*, Gordon and Breach, New York, 1967.

Skarantharajah, J., *Amenable actions of locally compact groups on coset spaces*, M.Sc. Thesis, University of Alberta, 1986.

Snell, R. C., *The range of invariant means on locally compact groups and semigroups*, Proc. Amer. Math. Soc. **37** (1973), 441–447.

——, *The range of invariant means on locally compact abelian groups*, Canad. Math. Bull. **17** (1974), 567–573.

Stafney, J. D., *Arens multiplication and convolution*, Pacific J. Math. **14** (1964), 1423–1447.

Sutherland, C. E., *On a construction of unitary cocycles and the representation theory of amenable groups*, Ergodic Theory Dynamical Systems **3** (1983), 129–133.

——, *Cartan subalgebras, transverse measures and non-type* 1 *Plancherel formulae*, J. Funct. Anal. **60** (1985), 281–308.

——, *A Borel parametrization of Polish groups*, Preprint.

Sutherland, C. E. and Takesaki, M., *Actions of discrete amenable groups and groupoids on von Neumann algebras*, Publ. Res. Inst. Math. Sci. **21** (1985), 1087–1120.

Szücs, J., *On G-finite W^*-algebras*, Acta Sci. Math. (Szeged) **48** (1985), 477–481.

Takesaki, M. and Tatsuuma, N., *Duality and subgroups*. II, J. Funct. Anal. **11** (1972), 184–190.

Talagrand, M., *Géométrie des simplexes de moyennes invariantes*, J. Funct. Anal. **34** (1979), 304–337.

____, *Closed convex hull of set of measurable functions, Riemann measurable functions and measurability of translations*, Ann. Inst. Fourier (Grenoble) **32** (1982), 39–69.

____, *Invariant means on an ideal*, Trans. Amer. Math. Soc. **288** (1985), 257–272.

____, Pettis integral and measure theory. *Mem. Amer. Math. Soc.* 59 (1984), no. 307.

Tarski, A., *On the congruence of sets and their equivalence by finite decomposition.*

Thomsen, K. E., *Invariant states for positive operator semigroups*, Studia Math. **81** (1985), 285–291.

Varopoulos, N. Th., *Tensor algebras and harmonic analysis*, Acta. Math. **119** (1986), 15–21.

Wagon, S., *Partitioning intervals, spheres and balls into congruent pieces*, Canad. Math. Bull. **26** (1983), 337–340.

____, *The Banach-Tarski paradox*, Cambridge Univ. Press, Cambridge, 1985.

Weit, Y., *On the one-sided Wiener's theorem for the motion group on* $\mathbf{R}^N$, Israel J. Math. **55** (1986), 111–120.

Willis, G., *Translation invariant functionals on $L^p(G)$ when G is not amenable*, J. Austral. Math. Soc. A **41** (1986), 237–250.

Wong, J. C. S., *Uniform semigroups and fixed point properties*, Preprint.

____, *On the semigroup of probability measures of a locally compact semigroup*, Canad. Math. Bull. **30** (1987), 142–146.

____, *A note on topological left thickness in locally compact semigroups*, Preprint.

____, *On Følner conditions and Følner numbers for semigroups*, Preprint.

Woodward, G. S., *Translation-invariant linear forms on $C_0(G)$, $C(G)$, $L^p(G)$ for non-compact groups*, J. Funct. Anal. **16** (1974), 205–220.

Wu, T. S., *Left almost periodicity does not imply right almost periodicity*, Bull. Amer. Math. Soc. **72** (1966), 314–316.

Yang, Z., *Exposed points of left invariant means*, Pacific J. Math. **125** (1986), 487–494.

——, *On Følner numbers and Følner-type conditions for amenable semigroups*, Illinois J. Math. **31** (1987), 496–517.

——, *Action of amenable groups and uniqueness of invariant means*, Preprint.

——, *On the set of invariant means*, Preprint.

Yuan, C. K., *The existence of inner invariant means on* $L^\infty(G)$, J. Math. Anal. Appl. **130** (1988), 514–524.

Sketched Solutions to Problems

Problems 0

0-1. To show that the map is onto, given μ, define m on simple functions $\phi = \sum \alpha_i \chi_{E_i}$ ($\{E_i\}$ disjoint) by $m(\phi) = \sum \alpha_i \mu(E_i)$. The space of such functions is norm dense in $l_\infty(X)$.

0-2. This is a variation on Problem 0-1; cf. [**DS**, IV.8.16].

0-4. $\mathfrak{I}_l(G)^\perp \cap L_\infty(G)^\wedge = A^\wedge$, where A is the closure of Span$\{\phi - \phi x\colon \phi \in L_\infty(G),\, x \in G\}$ in $L_\infty(G)$ ((2.1)).

0-6. $2\lambda(xA_n \cap A_n) = 2\lambda(A_n) - \lambda(xA_n \,\triangle\, A_n)$.

0-7. Let m be symmetrically invariant. For $n \geq 1$, we have, for $t \in G$, $t^{2n}m + t^{-2n}m = 2m$, and combining with $t^{2n-1}m + t^{-2n+1}m = 2m$ gives $t^{2n+1}m = t^{2n-1}m + 2(tm - m)$ whence $t^{2n+1}m = 2n(tm - m)$. Now use the boundedness of m to infer $tm = m$. (Such means arise in connection with positive linear operators on C^*-algebras (G. Robertson).)

0-8 (Milnes [**S1**]). $\lambda(A_n) = \int_{n^{-1}}^{n} y^{-2}\, dy \int_{-ny}^{ny} dx = 4n \log n$. Let $u = (b, a) \in S_2$. Deal with different cases of u depending on how uA_n intersects A_n. For example, if $0 < a \leq 1$ and $0 \leq b \leq 2a$, then for large enough n, $uA_n \cap A_n$ is the trapezium, with vertices $(b-1, n^{-1})$, $(1, n^{-1})$, (an^2, an), and $(b - an^2, an)$, so that $\lambda(uA_n \cap A_n) = [2n \log y + b/y]_{n^{-1}}^{an}$, $\lambda(uA_n \cap A_n)/\lambda(A_n) \to 1$, and Problem 0-6 applies.

0-9 (Mitchell [**2**]). Choose g_δ so that $\hat{g}_\delta \to m \in \mathfrak{L}(G)$. (See (2.13) for the converse.)

0-10. If $m(C_0(G)) \neq \{0\}$, show that $m(G) = \infty \neq 1$!

0-11. $\hat{f}_\delta(\chi)(\chi(x) - 1) \to 0$ for each $\chi \in \Gamma$. If $\chi \neq 1$, then $\chi(x) \neq 1$ for some $x \in G$.

0-12. Use the uniqueness of normalised Haar measure on G.

0-13 (Lorentz [**1**]). Let $\phi r(n) = \phi(r+n)$. If $\phi \in \mathrm{AC}$, then $\|p^{-1}\sum_{r=1}^{p}\phi r - l1\|_\infty \to 0$. Now apply $m \in \mathfrak{L}(G)$. Examples of elements of AC are z^n ($z \in \mathbb{C}$, $|z| \le 1$), all almost periodic functions. If $B = \{k^2 \colon k \in \mathbb{P}\}$, then $A = \{\{a_n\} \colon a_n = 0 \text{ if } n \notin B,\ a_n \in \{0,1\} \text{ if } n \in B\}$ is a nonseparable subset of AC.

0-15. Construct $\{x_n\}$ in G with $x_nE \cap x_mE = \varnothing$ if $n \ne m$.

0-16. If $D \in \mathscr{C}_e(G)$, then there exist N_1, N_2 such that $C \subset D^{N_1}$, $D \subset C^{N_2}$ [**HR1**, (7.9)].

0-18. Let G be abelian and $F = \{a_1, \ldots, a_k\} \subset G$. A typical element of F^r is of the form $a_1^{r_1} \cdots a_k^{r_k}$, $r_i \ge 0$, $\sum_{i=1}^{k} r_i = n$. There are $p(n) = \binom{n+k-1}{n}$ ways of partitioning n into a sum of k nonnegative integers—see (6.16), Problem 6-4. For the free abelian group on two generators, $|E^s| = (s+1)^2$ (Milnor [**1**]).

0-19. Let $C = [-2,2] \times [\frac{1}{2}, 2] \in \mathscr{C}_e(S_2)$. Show $C^n \supset [-2,2] \times [2^{-n}, 2]$ so that $\lambda(C^n) \ge 4[2^n - \frac{1}{2}]$. Use Problem 0-16.

References for many of the results below are Day [**4**] , [**9**].

0-21. G is an upwards directed union of finite (and so amenable) groups.

0-22. The group is locally finite.

0-23. An extension of an amenable group by an amenable group is amenable. For the solvable case, use the derived series (Appendix A).

0-24 (Dunkl-Ramirez [**7**]). Let H be a finite group containing every G_i as a subgroup. Show that every finite subset of G generates a subgroup of some H^m ($m \in \mathbb{P}$). Now apply Problem 0-21 (cf. Problem 3-1).

0-25. G_2 is a semidirect product $\mathbb{R}^2 \times_\rho \mathrm{O}(2,\mathbb{R})$, $D_iG_2 \subset \mathbb{R}^2 \times_\rho D_i(\mathrm{O}(2,\mathbb{R}))$, and $[\mathrm{O}(2,\mathbb{R}), \mathrm{O}(2,\mathbb{R})] \subset \mathrm{SO}(2,\mathbb{R})$ is abelian. So G_2 is solvable, and G_1 is a subgroup of G_2.

0-26. Use Zorn's Lemma and (0.16).

0-27. Consider G/N, where N is as in Problem 0-26.

0-29. No—S_2 is amenable and yet contains FS_2 (cf. (6.38)) (Hochster [**1**], Appel and Djorup [**1**], Milnes [**S1**]).

0-31. Try a right zero semigroup ($ef = f$ for all e, f).

0-33. $x^{-1}I = S$ if $x \in I$.

0-34 (Sz. Nagy [**1**], Dixmier [**1**]). This is the simplest version of the "similarity problem", and the proof is close to the corresponding proof for compact groups [**HR1**, (22-23)]. Let $m \in \mathfrak{N}(G)$, and define $\langle \xi, \eta \rangle = m(x \to (\pi(x)\xi, \pi(x)\eta))$. Then $\langle\ ,\ \rangle$ is "equivalent" to $(\ ,\ )$ in the sense that for some invertible, $(\ ,\ )$-positive $A \in B(\mathfrak{H})$, we have $\langle \xi, \eta \rangle = (A\xi, A\eta)$. Take $\pi'(x) = A\pi(x)A^{-1}$. See

Bunce [4], Christensen [1] and Problem 1-40 for the C^*-version of the similarity problem.

Problems 1

1-1. E.g. $\phi\mu(\nu) = \phi(\mu * \nu) = \iint \phi(xy)\,d\mu(x)\,d\nu(y) = \int d\nu(y) \int \phi(xy)\,d\mu(x) = \int \phi\mu(y)\,d\nu(y)$.

1-2. If $\phi \in C_c(G)$, then there exist $C, D \in \mathscr{C}(G)$, $C \subset D^0$, ϕ vanishing outside C in G. If $x_\delta \to e$ in G, then $x_\delta^{-1}C \subset D$ eventually and compactness gives eventually $\|\phi x_\delta - \phi\| = \sup\{|(\phi x_\delta - \phi)(y)| \colon y \in D\} \to 0$. So $C_0(G) = C_c(G)^- \subset \mathrm{U}_r(G)$.

1-3 (Kister [1], Itzkowitz [1]). Suppose that G is neither discrete nor compact. Let $C \in \mathscr{C}_e(G)$, and let $\{x_n\}$ be a sequence in G with $Cx_n \cap Cx_m = \varnothing$ if $n \neq m$. Find $\{V_n\}$ in $\mathscr{C}_e(G)$ such that $V_1 = V_1^{-1} \in \mathscr{C}_e(G)$, $V_1^3 \subset C$, $V_n \subset V_1$ $(n \geq 1)$, $\lambda(V_n) \to 0$. Let $\phi_n \in C(G)$ be such that $0 \leq \phi_n \leq 1$, $\phi_n(x_n) = 1$, $\phi_n(G \sim V_n x_n) = \{0\}$. Then $\phi = \sum_{n=1}^{\infty} \phi_n \in C(G) \sim \mathrm{U}_r(G)$.

1-4. Any $m \in \mathfrak{L}_t(G)$ is determined by its values on $C(G)$, and if G is compact, then $\mathscr{L}(C(G)) = \{\lambda\}$ by uniqueness of Haar measure.

1-5. Use (1.13).

1-7. Adapt (0.16(4)) and use (1.12).

1-8 (Chou [4]). If R is a transversal for the *left* H-cosets in G and $_R\phi(rh) = \phi(h)$ $(r \in R, h \in H)$, then $\beta(p)(_R\phi) = p(\phi)$, and β is one-to-one. The map γ is defined in the same way as β. Then $|\mathfrak{L}(C(H))| \leq |\mathfrak{L}(C(G))|$, and when the uniform structures are equivalent, $|\mathfrak{L}(U(H))| \leq |\mathfrak{L}(U(G))|$.

1-10. Suppose that G is not compact and that μ is a countably additive left invariant mean (c.a.l.i.m.) on $\mathscr{B}(G)$. If $C \in \mathscr{C}(G)$, then $\mu(C) = 0$ (cf. Problem 0-15). Since $\mu(G) = 1$, G cannot be σ-compact. Construct a sequence $\{G_n\}$ of open, compactly generated subgroups of G with $G_n \subsetneqq G_{n+1}$; let $H = \bigcup_{n=1}^{\infty} G_n$. If H admits a c.a.l.i.m., then since it is σ-compact and noncompact, we have a contradiction. We can obtain a c.a.l.i.m. ν on H by setting $\nu(E) = \int \phi_E\,d\mu$, where $E \in \mathscr{B}(H)$, $\phi_E(x) = \int_E \beta(x^{-1}h)\,d\lambda_H(h)$, and β is a Bruhat function for H ((1.11)) c.f. the proof of (1.12). This gives a contradiction.

1-11. Apply Day's Fixed-Point Theorem to the natural action of S on $\mathrm{PM}(X)$, the (weak* compact, convex) set of probability measures on X.

1-12. For (i) consider the natural left action of G on $\mathfrak{M}(G)$. For (ii), $\mathfrak{L}(C(G)) \neq \varnothing$ since $C(G) \subset l_\infty(G)$.

1-13. For $m \in \mathfrak{L}(B)$, apply Day's Fixed-Point Theorem to $\{m' \in \mathfrak{M}(G) \colon m'|_B = m\}$.

1-14. Apply the R.H. version of Day's Theorem to the set of left fixed-points in K.

1-15. Silverman [**1**]–[**4**]; cf. Problem 1-13.

1-16. The "Rees product" theory of the kernel (M. Rosenblatt [**1**, Chapter 5]) suggests the following class of compact semigroups. Let Y be a compact Hausdorff space, G be a compact group, and $\phi\colon Y \to G$ be continuous. Let $T = G \times Y$ with multiplication $(g, y)(g', y') = (g\phi(y)g', y')$. Now apply (1.17).

1-17 (Mitchell [**2**]). $A \subset \mathbb{N}$ is (left) thick $\Leftrightarrow A = \bigcup_{n=1}^{\infty} [a_n; b_n]$, where $\sup_n (b_n - a_n) = \infty$.

1-18. Use (1.21) and Problem 0-15.

1-19. Suppose that L is a left amenable, left ideal of S. Using Day's Fixed-Point Theorem, find $m \in \mathfrak{M}(S)$ with $lm = m$ $(l \in L)$. Then $m \in \mathfrak{L}(S)$.

1-20 (Klawe [**2**]). T is a homomorphic image of S and so T is left amenable. For the case of U, let $K = \{M_0 \in \mathfrak{M}(U)\colon m_0(\phi) = m_0((\phi u_0)\rho(t_0))$ for all $\phi \in l_\infty(U)$, $u_0 \in U$, $t_0 \in T\}$. If $m \in \mathfrak{L}(S)$, then the functional $\phi \to m((u,t) \to \phi(u))$ is in K. A fixed-point for the left action $t \to \rho(t)^*$ of T on K gives an element of $\mathfrak{L}(U)$.

1-21. Day [**4**].

1-22. Dixmier [**1**].

1-23 (Frey [**1**]). Using (1.27) and (1.28), a subsemigroup T of S is left amenable $\Leftrightarrow T$ is left reversible. If T is not left amenable, find $s, t \in T$ with $sT \cap tT = \varnothing$. Then s, t generate FS_2 in T. Conversely use Problem 0-28.

1-24 (Day [**9**]). We can suppose that S generates G. Show that $G = SS^{-1}$, and deduce that S is left thick in G and so is left amenable (Problem (1-21)).

1-25. (a) and (b), (i), (ii) are standard; e.g., see Howie [**1**]. Use (ii) to prove (iii) and (iv) of (b). (c), due to Duncan and Namioka [**1**], follows from (b) and (1.25).

1-26 (Day [**4**]). The functional $mn \in \mathfrak{I}(G)$ if $m \in \mathfrak{L}(G)$, $n \in \mathfrak{R}(G)$.

1-27 (Day [**9**]). $\Rightarrow$ Modify the last paragraph of (1.21). If $C \in \mathscr{C}(G)$, C not null, $\mu_C = \chi_C/\lambda(C)$, and $\lambda(C \sim Et^{-1}) \geq \varepsilon$ for all $t \in G$, then $\|\chi_E \mu_C\|_\infty \leq 1 - (\varepsilon/\lambda(C))$. For $\Leftarrow$, modify the first part of the proof of (1.21) to produce $m_0 \in \mathfrak{M}(\mathrm{U_r}(G))$ in the set $K = \{n \in \mathfrak{M}(\mathrm{U_r}(G))\colon n(\chi_E \mu) = 1$ for all $\mu \in P(G)\}$. (Note that if $C \subset D \in \mathscr{C}(G)$, then $\chi_E \mu_C \geq (1 - \varepsilon)$ if $\chi_E \mu_D(x) \geq [1 - \varepsilon(\lambda(C)/\lambda(D))]$ for $x \in G$.) Apply the Fixed Point Theorem (2.24) to find $n_0 \in \mathfrak{L}(\mathrm{U_r}(G)) \cap K$. Finally use (1.7) to extend n_0 to the desired $m \in \mathfrak{L}_t(G)$.

1-28. Use 1.27.

1-29 (Klawe [2]). Let $U = \mathbb{Z} \times_\rho \mathbb{Z}$, where $\rho(n)(\mathbb{Z}) = \{0\}$ for all n. Show that U is not left reversible.

1-30. Use the natural maps $j: C(G) \to M(G)'$, $k: M(G)' \to L_1(G)' = L_\infty(G)$, where $j(\phi)(\mu) = \int \phi \, d\mu$ ($\mu \in M(G)$) and k restricts elements of $M(G)'$ to $L_1(G)'$. Then $j^*(\mathfrak{L}(M(G)')) \subset \mathfrak{L}(C(G))$, $k^*(\mathfrak{L}(G)) \subset \mathfrak{L}(M(G)')$.

[*Note*: The preceding suggests how amenability can be defined for a separately continuous, locally compact, Hausdorff semigroup S. We define S to be left amenable if there exists a left S-invariant state on $M(S)'$ (cf. Namioka [**3**], Jenkins [**3**], Lau [**6**]). Another approach to amenability for S, based on an analogue of $L_1(G)$ studied by Baker and Baker [**1**], [**2**], [**3**] and Sleijpen [**1**], [**2**], [**3**] is developed by Paterson [**1**]. See also Wong [**10**] and Kinzl [**1**]. Versions of topological left thickness for Borel subsets of S are developed in Day [**9**], [**11**], [**13**], Junghenn [**3**], Kharaghani [**2**], and Wong [**2**], [**4**], [**6**], [**7**], [**9**], and [**12**]. This depends on $M(S)$ being a Banach algebra under convolution, a substantial fact following from the work of Glicksberg [**1**] and Johnson [**1**]. (See also Moran [**1**] and Wong [**11**].)]

See Johnson [**2**] for Problems 31–34 below.

1-31. Every Banach B-module X is a Banach A-module in the natural way, and if $D: B \to X'$ is a derivation, then $D \circ \Phi$ is a derivation on A.

1-32. Suppose that A is amenable. Then A/J is amenable by Problem 1-31. To prove that J is amenable, we need only consider a neo-unital Banach J-module X and a derivation $D: J \to X'$ ((1.30(ii))). Then X is a Banach $\Delta(J)$-module, D extends to a $\Delta(J)$-derivation, and we can use the canonical homomorphism from A into $\Delta(J)$ together with the amenability of A to deduce that D is inner.

1-33. $A \hat{\otimes} B$ has a bounded approximate identity (using (1.30(i))). To show that a derivation $D: A \hat{\otimes} B \to X'$ (X neo-unital) is inner, extend D to derivation, also denoted D, on $\Delta(A \hat{\otimes} B) \supset (A \otimes 1) \cup (1 \otimes B)$. By subtracting an inner derivation from D, we can suppose that $D(1 \otimes B) = \{0\}$. Using the fact that $(a \otimes 1)(1 \otimes b) = a \otimes b = (1 \otimes b)(1 \otimes a)$, we have $D(A \otimes 1) \subset Y^\perp$, where $Y = \{(1 \otimes b)\xi - \xi(1 \otimes b) : \xi \in X, b \in B\}$. Now subtract (using the amenability of A) an inner derivation from D so that we can take $D((A \otimes 1) \cup (1 \otimes B)) = \{0\}$. Then $D = 0$.

What about the converse?

1-35 (Bunce [**2**], Lau [**11**]). Let $h' \in X'$ be any extension of h. The derivation $a \to (ah' - h'a)$ is $Y^\perp$-valued and so is of the form $a \to (ah'' - h''a)$ ($a \in A$) for $h'' \in Y^\perp$. Take $k = h' - h''$. [*Note*: The converse of the result is also true.]

1-36 (Johnson [**3**]). (i) Suppose that N is a virtual diagonal for A. Then there exists a bounded net $\{\alpha_\sigma\}$ in $A \otimes A$, with each α_σ of the form $\sum_{i=1}^n a_i \otimes b_i$, such that $\hat{\alpha}_\sigma \to N$ weak*. Since π^{**} (so $((a \otimes b)\hat{}\,)) = ab$ and π^{**} is weak* continuous, the second virtual diagonal condition gives $(f_\sigma a-a)^\wedge \to 0$, $(af_\sigma -a)^\wedge \to 0$ weak* in A'', where $f_\sigma = \pi^{**}(\alpha_\sigma)$. A routine argument (c.f. (0.8)) gives a bounded approximate identity $\{e_\delta\}$ for A. (ii) Let $N, \alpha_\sigma, f_\sigma, e_\delta$ be as above. Each e_δ is a convex combination of f_σ's. Let β_δ be the corresponding combination of α_σ's. We can suppose that $\beta_\delta \to M$ weak* in $(A\hat{\otimes}A)''$. Now check that M is a virtual diagonal for A. Using the neo-unital condition, $D(e_\delta)(\xi) \to 0$ for all $\xi \in X$. Since $\beta_\delta(a\otimes b \to D(ab)(\xi)) = D(e_\delta)(\xi)$ and $\beta_\delta \to M$ weak*, we have $M(a\otimes b \to D(ab)(\xi)) = 0$. (iii) $\Rightarrow$ Let $\{e_\delta\}$ be a bounded approximate identity for A. We can suppose that $(e_\delta \otimes e_\delta)^\wedge \to n$ weak* in $(A\hat{\otimes}A)''$. The map $a \to (an - na)$ is a derivation from A into the dual module $\ker \pi^{**}$, and since A is amenable, there exists $m \in \ker\pi^{**}$ such that $an-na = am-ma$ $(a \in A)$. Then $(n-m)$ is a virtual diagonal for A. $\Leftarrow$ Let X be a Banach A-module and $D\colon A \to X'$ be a derivation. By (i), we can suppose that X is neo-unital. Let M be a virtual diagonal for A as in (ii). The virtual diagonal conditions translate to: (a) for all $g \in A', a_0 \in A$, we have $M(a\otimes b \to g(aba_0)) = g(a_0) = M(a\otimes b \to g(a_0ab))$, and (b) for all $h \in (A\hat{\otimes}A)'', a_0 \in A$, we have $M(a\otimes b \to h(a\otimes ba_0)) = M(a\otimes b \to h(a_0a\otimes b)$. Let $a_0 \in A$, $\xi \in X$. Applying (b) with $h(a\otimes b) = (Da)b(\xi)$ and then (a) with $g(a) = Da(\xi)$ gives $M(a\otimes b \to (Da)ba_0(\xi)) = M(a\otimes b \to (D(a_0ab) - a_0aDb)(\xi)) = Da_0(\xi) - M(a\otimes b \to (D(ab) - (Da)b(\xi a_0))$. Now use (ii) (with ξa_0 in place of ξ) to obtain $Da_0 = a_0\alpha - \alpha a_0$ where $\alpha(\xi) = M(a\otimes b \to -(Da)b(\xi))$. [Note: The net $\{\beta_\delta\}$ of (i) can also be arranged so that $\|a\beta_\delta - \beta_\delta a\| \to 0$ for all $a \in A$. Such a net is called by Johnson an *approximate diagonal* and is an analogue of the "Reiter type" net of (4.1). Virtual diagonals are analogous to invariant means, while virtual diagonals arising as weak* cluster points of approximate diagonals are analogous to topologically invariant means.]

1-37 (Lau and Paterson [**S3**]). The case $H = \{e\}$ is (1.30(iv)). For general H, modify that proof. On needs: G/H amenable $\Leftrightarrow$ there exists a left invariant mean on the space $U_r(G/H)$ of functions in $C(G/H)$ which are right uniformly continuous under the natural G-action. To this end, modify (1.7). (See Eymard [**2**]).

1-38 (Johnson [**2**], Bunce [**3**], cf. (2.35)). Suppose that $A = C_l^*(G)$ is amenable (G discrete). Let $Y = A$, $X = \mathbf{B}(l_2(G))$, and $h \in A'$ be the tracial state: $h(T) = (T\delta_e, \delta_e)$. By Problem 1-35, there exists an extension $k \in \mathbf{B}(l_2(G))'$ of h with $ak = ka$ for all $a \in A$. Then $\phi \to k(L_\phi)$, where L_ϕ is the multiplication operator associated with $\phi \in l_\infty(G)$, is a nonzero, right invariant functional, and G is amenable.

1-39 (Bunce [**2**], Rosenberg [**2**]). Let $S(A)$ be the state space of A and $g \in S(A)$. Applying strong amenability to the derivation $D\colon A \to A'$ given by $D(a) = ag - ga$, we can find h in the weak* closure of $\text{co}\{-D(u)u^*\colon u \in \mathbf{U}(A)\}$ such that $D(a) = ah - ha$ $(a \in A)$, and so $a\eta = \eta a$, where $\eta = (g - h)$. Now if $u \in \mathbf{U}(A)$, then $g - (-(Du)u^*) = ugu^* \in S(A)$, so that $\eta \in S(A)$. So η is a tracial state.

Note: Rosenberg proves that $\mathcal{O}_n$ ($n \geq 2$) is not strongly amenable by showing that it does not admit a tracial state.

1-40 (Bunce [**1**], [**2**]). $\mathbf{B}(\mathfrak{K})$ is a dual Banach A-module with operations $aT = \pi(a)T$, $Ta = T\pi(a^*)^*$ (cf. (2.35)). If $D\colon A \to \mathbf{B}(\mathfrak{K})$ is the derivation given by $D(a) = (\pi(a) - \pi(a^*)^*)$, then we can find T_0 in the weak* closure of $\operatorname{co}\{-D(u)u^*\colon u \in \mathbf{U}(A)\}$ with $D(a) = \pi(a)T_0 - T_0\pi(a^*)^*$ $(a \in A)$. Note that $D(u)u^* = \pi(u)\pi(u)^* - I$, so that $R = (I - T_0)$ is in the weak* closure of $\operatorname{co}\{\pi(u)\pi(u)^*\colon u \in \mathbf{U}(A)\}$. Now show that for all $u \in \mathbf{U}(A)$, $\pi(u)\pi(u)^*$, and hence R, is $\geq \|\pi\|^{-2}I$, so that R is positive and invertible. Since $\pi(a) - \pi(a^*)^* = \pi(a)T_0 - T_0\pi(a^*)^*$, we have $\pi(a)R = R\pi(a^*)^*$. Then $R^{1/2}$ implements the similarity between π and a *-representation of A.

1-41 (Green [**S**], Lau and Paterson [**S3**], Paterson [**S1**]). Let X be a neo-unital Banach $C^*(G)$-module and $D\colon C^*(G) \to X'$ a derivation. Then D extends to $\Delta(C^*(G))$. There is a canonical *-homomorphism from $C^*(H)$ onto a C^*-algebra $A \subset \Delta(C^*(G))$. If $H \in \mathscr{A}$, then $D|_A$ is inner. So regarding D as a G-derivation, $D|_H$ is inner. So D is inner by Problem 1-37.

1-42 (Lau and Paterson [**S3**], Paterson [**S1**]). Suppose that $H \in \mathscr{A}$. Let B be a C^*-algebra. We have to show that $A_{\max} = C^*(H) \otimes_{\max} B$ equals $A_{\min} = C^*(H) \otimes_{\min} B$. Realise $A_{\max}$ on the Hilbert space $\mathfrak{K}$ of its universal representation, and $A_{\min}$ on a Hilbert space tensor product $\mathfrak{K}_1 \otimes \mathfrak{K}_2$. Associated with these realisations are unitary representations π, π' of H on $\mathfrak{K}$ and $\mathfrak{K}_1 \otimes \mathfrak{K}_2$ respectively, and each gives a faithful representation of $C^*(H)$. The representations π, π' induce representations Φ, Φ' of G, and $C = \Phi(C^*(G)) \cong \Phi'(C^*(G))$ (Fell [**3**]). Note that Φ is realised on $L_2(G\backslash H, \mathfrak{K})$ and is given by $(\Phi(x)f)(u) = \pi(s(u)xs(ux)^{-1})f(ux)$ $(x \in G,\ u \in G\backslash H)$, where $s\colon G\backslash H \to G$ is a Borel cross-section for the quotient map (Kirillov [**2**]). Now C is nuclear (since $G \in \mathscr{A}$), and one uses this to show that the two C^*-algebras generated by $\Phi(C^*(G))(1 \otimes B)$, $\Phi'(C^*(G))(1 \otimes B)$ are isometrically $*$-isomorphic. The isomorphism extends to their multiplier algebras, and using the formulae for Φ and Φ' and the normality of H, we see that $A_{\max}, A_{\min}$ are canonically embedded in the multiplier algebras and are isomorphic.

Problems 2

2-1. Use (2.1), (2.2), and the fact that $D(G) = \{\phi \in L_\infty(G)\colon p(\phi) = 0$ for all $p \in \mathfrak{L}(G)\}$.

2-2 (Emerson [**7**], J. C. S. Wong [**1**]). Suppose that for all $\phi \in$ the set $B_{\mathbf{R}}$ of real-valued functions in B, we have $\inf_{x \in G}(\phi\mu - \phi\nu) \leq 0$ $(\mu, \nu \in P(G))$. For $\phi \in B_{\mathbf{R}}$, let B_ϕ be the subspace of $B_{\mathbf{R}}$ spanned by functions $\phi\mu$ $(\mu \in P(G))$ and 1. Show that if $\phi_i \in B_\phi$ $(1 \leq i \leq n)$ and $\mu_i, \nu_i \in P(G)$, then $\sum_{i=1}^n(\phi_i\mu_i - \phi_i\nu_i)$ is of the form $\gamma(\phi\mu - \phi\nu)$ for some $\mu, \nu \in P(G)$, $\gamma > 0$.

Deduce that $\sup_{x\in G}\sum_{i=1}^{n}(\phi_i\mu_i - \phi_i\nu_i)(x) \geq 0$ and apply versions of (2.3) and (2.13) to obtain $\mathfrak{L}_t(B) \neq \varnothing$.

2-4 (Emerson [**8**]). Suppose that the "ess inf" condition holds. Show that if $\psi = \sum_{i=1}^{n}(\chi_{x_i^{-1}E_i} - \chi_{y_i^{-1}E_i})$, then $\operatorname{ess\,inf}_{x\in G}\psi(x) \leq 0$. Show that the same holds if ψ is replaced by $\sum_{i=1}^{n} r_i(\chi_{x_i^{-1}E_i} - \chi_{y_i^{-1}E_i})$ ($r_i \in \mathbb{Q}$) and apply (2.3) to deduce that G is amenable.

2-5 (Emerson [**7**]). Adapt the proof of Problem 2-4, and use Problem 2-2 above.

2-6. For the second part, the two interpretations of Ff coincide on S and hence on $l_1(S)$. For the third part, take S to be the unit ball of A and $B = \{f|_S : f \in A'\}$.

2-7. Take $m = \frac{1}{2}(n + n^*)$ where n is an invariant mean on G (Problem 1-26).

2-9. Show first that the norm and pointwise topologies coincide on the norm closure of ϕS ($\phi \in \mathrm{AP}(S)$).

2-10 (Jenkins [**3**], Lau [**6**]). For the first part, follow the proof of (1.27) using $U(G)$, $U(S)$ in place of $l_\infty(G)$, $l_\infty(S)$ to obtain $m_0 \in \mathfrak{L}(U(G))$. If $\mu \in P(G)$, $\mathbf{S}(\mu) \subset S$, then the map $\phi \to m_0(\mu\phi\mu)$ is in $\mathfrak{L}_t(G)$ ((1.7)), and is 1 on S. For the second part, use the first part and adapt the proof of (1.28), using (2.23) in place of Day's Fixed-Point Theorem.

2-11. Use (2.15) to obtain an S-fixed-point in the weak* closure of $\operatorname{co} S\hat{y}_0$. This weak* closure is $(c_{y_0})\hat{}$.

2-14 (Furstenberg [**3**]). $\Leftarrow$ Let $\mathscr{A}$ be the set of separable, unital right invariant C^*-subalgebras of $\mathrm{U_r}(G)$. Since $\{\phi x : x \in G\}$ is norm separable for $\phi \in \mathrm{U_r}(G)$, $\mathrm{U_r}(G)$ is the (upwards directed) union of $\mathscr{A}$. Every $A \in \mathscr{A}$ is of the form $C(X)$, where X is a compact metric left G-space. So $\mathfrak{L}(A) \neq \varnothing$. This gives G amenable (cf. (0.16(4))).

2-15 (Milnes [**1**], Kharaghani [**1**]). Let $\phi \in B$ and, for $a > 0$, $S_a = \{f \in B' : \|f\| \leq a\}$ with the weak*-topology. The map $(t, f) \to f\phi(t)$ is separately continuous on the locally compact space $S \times S_a$, and by using a well-known result of Glicksberg [**1**], if $\mu \in \mathrm{PM}(S)$, then the linear functional F_μ, where

$$F_\mu(f) = \int f\phi(t)\,d\mu(t), \quad (f \in B') \tag{1}$$

is continuous in the bounded B-topology and so weak* continuous [**DS**, V.5.6]. So $F_\mu = \widehat{\psi}$ for some $\psi \in B$. Now show that $\psi = \phi\mu$, so that B is right invariant for $\mathrm{PM}(S)$. Taking $f \in \mathfrak{L}(B)$ and $\phi\mu$ for ψ in (1), we obtain $f(\phi\mu) = f(\phi)$.

2-16 (Lau [**5**]). (i) Show that the map $(p, m) \to pm$ is jointly continuous on $K_e = \mathfrak{M}(\mathrm{AP}(S))$ (cf. (2.36(ii))). It follows that the set of maps $m \to pm$ ($p \in K_e$) is equicontinuous.

(ii) $\Leftarrow$ Let $K \in \mathscr{G}_e$, $\phi \in A_f(K)$. We claim ϕS is equicontinuous. Indeed, let $\mathscr{U}$ be the uniformity on K, $k \in K$ and $\varepsilon > 0$. Let

$$W = \{(k_1, k_2) \in K \times K \colon |\phi(k_1) - \phi(k_2)| < \varepsilon\}.$$

By equicontinuity, there exists an open neighbourhood U of k such that $(sk, su) \in W$ whenever $s \in S$, $u \in U$. So $\sup_{s \in S} \sup_{u \in U} |\phi(su) - \phi(sk)| \leq \varepsilon$. So ϕS is equicontinuous in $C(K)$. Now use the Arzelà-Ascoli Theorem [**DS**, (IV.6.7)] to give ϕS conditionally compact in $C(K)$. So (with k_0 as in (2.22)), $\hat{k}_0 \phi \in \mathrm{AP}(S)$.

2-17. This fixed-point theorem will be used in proving the Hunt-Stein Theorem in (4.25). Proof of (b): let $F \in A_f(K)$, $p \in \mathfrak{M}(A_f(K))$. It is sufficient to show that $pF \in L_\infty(G)$. For some $\phi \in K$, we have $pF(x) = F(x\phi)$. Prove that $pF \in L_\infty(G)$ first when F is of the form $F_f + \alpha 1$, where $\alpha \in \mathbb{C}$, $f \in L_1(Z)$ and $F_f = \hat{f}|_K$. For the general case, the set of such functions $F_f + \alpha 1$ is norm dense in $A_f(K)$ (Phelps [**2**, p. 31]).

2-18 (Takahashi [**1**], De Marr [**1**]). (i) Find $x_0, x_1 \in M$ with $\|x_0 - x_1\| = \sigma = \operatorname{diam} M$. Let $M_0 \supset \{x_0, x_1\}$ ($M_0 \subset M$) be maximal for the property: $\|x - y\| = 0$ or σ if $x, y \in M_0$. Show $M_0 = \{x_0, x_1, \ldots, x_m\}$ for some $m \in \mathbb{P}$ (x_i distinct) and take $u = (m+1)^{-1} \sum_{i=0}^{m} x_i$.

(ii) Let X_1 be a minimal, nonempty, S-invariant, compact convex subset of K and M a minimal, nonempty, S-invariant compact subset of X. Show that $sM = M$ for all $s \in S$. If $|M| > 1$, then there exists u in $\operatorname{co} M$ such that $\sup_{m \in M} \|u - m\| = \rho < \operatorname{diam} M$. Then $X_0 = \{y \in X_1 \colon \sup_{m \in M} \|y - m\| \leq \rho\}$ contradicts the minimality of X_1. So $|M| = 1$.

2-19 (Lau [**5**]). Let Z be the a.p. compactification of S so that $AP(S)^\wedge = C(Z)$. Considering the canonical homomorphism from S into Z, the left reversibility of S implies the same for Z. Now apply (1.17).

2-20. (i) Take P to be the projection of $B \oplus C$ onto the first component. P is continuous by the Closed Graph Theorem or Banach's Isomorphism Theorem.

(ii) (c) The subgroup N is closed (using the "$\mu * x$" version of (1.2)). Let $A = \{\phi \in L_\infty(G) \colon x\phi = \phi \text{ for all } x \in N\} \supset B$. Then A is a right invariant von Neumann subalgebra of $L_\infty(G)$. Suppose that $B \neq A$. Then we can find $\mu \in L_1(G)$ of compact support K such that $\|\hat{\mu}_{|B}\| < \frac{1}{2}$, $\|\hat{\mu}_{|A}\| > \frac{1}{2}$. Let $B_0 = \{\phi_{|K} \colon \phi \in C(G) \cap B\}$, $A_0 = \{\phi_{|K} \colon \phi \in C(G) \cap A\}$. Using (a), $\|\hat{\mu}_{|B}\| = \|\hat{\mu}_{|B_0}\|$, $\|\hat{\mu}_{|A}\| = \|\hat{\mu}_{|A_0}\|$. Derive a contradiction by showing that B_0, A_0 have the same norm closures in $C(K)$. To this end, if $k_1, k_2 \in K$ and $\phi(k_1) = \phi(k_2)$ for all $\phi \in B_0$ then, since B_0 is right invariant, $k_1\phi(x) = k_2\phi(x)$ ($x \in G, \phi \in B_0$); then $k_2^{-1}k_1 \in N$ and $k_1\phi = k_2\phi$ ($\phi \in A_0$). The Stone-Weierstrass Theorem then gives that the norm closures of B_0, A_0 are the same.

(iii) $\Rightarrow$ We can suppose $B \neq \{0\}$. Let N be as in (ii). Let $X = L_1(G)$, $S = N$ and apply (2.15) to obtain a projection $P \colon L_\infty(G) \to \mathfrak{J}_l(X) = B$. $\Leftarrow$Consider $B = \mathbb{C}1$. [References: Lau [**14**], Crombez and Govaerts [**1**], Pathak and Shapiro [**1**], Rosenthal [**1**], Takesaki and Tatsuuma [**1**].]

2-21 (Milnes [**1**], Ruppert [**S1**]). Suppose that $p \in \beta\mathbb{Z} \sim \mathbb{Z}$ and the map $q \to pq$ is continuous on $\beta\mathbb{Z}$. We can suppose that there is a net $n_\delta \to \infty$ in $\mathbb{Z}$ with $n_\delta \to p$ in $\beta\mathbb{Z}$, and that $-n_\delta \to p_0$ for some $p_0 \in \beta\mathbb{Z}$. Then $p(-n_\delta)(\chi_{(-\infty,-1]}) = 0$ for all δ, while $pp_0(\chi_{(-\infty,-1]}) = \lim_\delta p_0(\chi_{(-\infty,-1]}n_\delta) = 1$, giving a contradiction. (Ruppert has a much more general theorem. See also Lau [**S8**].)

2-22 (Granirer [**13**]). $\Rightarrow$ The unit ball of $\mathfrak{J}_l(S)$ is weakly compact. Apply the Dunford-Pettis Property [**DS**, VI.8.13] to the identity operator on the L-space $\mathfrak{J}_l(G)$ to obtain $\mathfrak{J}_l(G)$ finite-dimensional.

2-23 (von Neumann [**1**], cf. [**HR1**, (17.22)]). Adapt the proof of (2.32).

2-24 (Ellis [**1**], [**2**]). By (2.33(ii)), the multiplication in H is jointly continuous. For inversion, show first that if $x_\delta \to x$ in H and $\{x_\delta^{-1}\}$ is contained in a compact set, then $x_\delta^{-1} \to x^{-1}$. So C^{-1} is closed if $C \in \mathscr{C}(H)$. If H is separable and $V \in \mathscr{C}_e(H)$, show by the Baire Category Theorem that $(V^{-1})^0 \neq \varnothing$. Show then that $C^{-1} \in \mathscr{C}(H)$ if $C \in \mathscr{C}(H)$. Deduce that inversion is continuous in this case. The general case can be reduced to this case.

2-25 (van Rooij [**1**]). (a) Use Zorn's Lemma and (2.34(iii)).

(b) Let $n \in \mathfrak{L}(B_m)$, and let C be the set of functions $\phi \in l_\infty(\mathbb{P}, K)$ such that $\phi - T\phi \in B_m$, $n(\phi - T\phi) = 0$. Show (using $\|I - T\| \leq 1$ and (2.34(iii))) that $C = B_m$. For $\phi \in B_m$, let $\phi_n'(k) = (\sum_{r=1}^k \phi(r)) - kn(\phi)$ $(k \in \mathbb{P})$. Then $\phi_n' - T\phi_n' = -T\phi + n(\phi)1$, and $\phi_n' \in B_m$. If $p \in \mathfrak{L}(B_m)$, $\phi \in B_m$, show $\phi_n' - \phi_p' = (p(\phi) - n(\phi))j \in B_m$, implying $n = p$ (using (2.34(iii))).

2-26. (ii) Let $T \in \mathbf{B}(\mathfrak{H})$ and θ be as in (2.35(C)). Show that $U \to T\theta(U)$ is in $\mathrm{U_r}(H)$. (We can take $\theta(R) = (R\xi, \eta)$ $(R \in \mathbf{B}(\mathfrak{H}))$ for some $\xi, \eta \in \mathfrak{H}$.)

2-27 (de la Harpe [**2**]). Let $M = \bigcup_{n=1}^\infty M_n$. Since $\mathfrak{L}(\mathbf{U}(M_n) \neq \varnothing$, we have $\mathfrak{L}(\mathrm{U_r}(\mathbf{U}(M))) \neq \varnothing$ (cf. (0.16(4))). Now show that $\mathbf{U}(M)$ is strongly dense in $\mathbf{U}(A) = H$ so that $\mathfrak{L}(\mathrm{U_r}(H)) \neq \varnothing$. So $\mathfrak{L}(X(H)) \neq \varnothing$ (preceding problem) and A has Property P. (Assuming the result that Property P$=$ AFD and using (2.35(C)(ii)), we have a remarkable result due to de la Harpe: *A has Property P* $\Leftrightarrow \mathfrak{L}(\mathrm{U_r}(H)) \neq \varnothing$.)

If $m \in \mathfrak{L}(C(H))$ define $\Phi\colon \mathbf{B}(\mathfrak{H}) \to \mathbf{B}(\mathfrak{H})$ by

$$(\Phi(T)\xi, \eta) = m(U \to (U^*TU\xi, \eta)).$$

(Note that $U \to (U^*TU\xi, \eta)$ is of the form $U \to T\theta(U^*)$ and $\in C(H)$.) Then Φ is continuous, takes I into I, and vanishes on commutators. But I is a sum of commutators.

2-28 (Tomiyama [**2**]). If A is abelian, then H is amenable as a discrete group and (2.35(C)) applies. Now suppose that A is of Type 1. Realise A so that A^c is abelian. So A^c has Property P, implying that A also has Property P (cf. (2.35(F)(d))).

2-29. Let G be discrete and of Type 1. By a result of Schwartz ((2.35 (G), (H))) and Problem 2-28, G is amenable. (Thoma [**1**] gives much more: *G is an abelian extension of a finite group.*)

2-30 (Zimmer [**6**]). For each $n \in \mathbb{P}$ there exists a finite subset F_n of E_1' such that every point of E_1' is of distance $< 1/n$ from some point of F_n for a metric d on E_1' giving the relative weak* topology. For each $\beta \in F_n$ let $W_\beta = X \times \{\alpha \in E_1' : d(\alpha, \beta) < 1/n\}$. Let $p\colon W_\beta \cap \tilde{A} \to X$ be the projection map. Then $Y_\beta = p(W_\beta \cap \tilde{A})$ is analytic, and by the von Neumann Selection Theorem (Auslander and Moore [**1**, Proposition 2.15]; Rogers et al. [**S**]), there is a Borel subset X_β of Y_β with $\mu(Y_\beta \sim X_\beta) = 0$ and a Borel cross-section r_β^n for p on X_β. Now piece together the r_β^n's to find the desired Borel map b_n with $(x, b_n(x))$ in some $W_\beta \cap \tilde{A}$ a.e. $(\beta \in F_n)$.

2-31 (cf. Lance [**S**]). Using the nuclearity of A, we obtain a representation π of $A \otimes_{\min} A^c$ on $\mathfrak{H}$ given by $\pi(a \otimes b) = ab$. There exists a representation π' of $A \otimes_{\min} \mathbf{B}(\mathfrak{H})$ on a Hilbert space $\mathrm{K} \supset \mathfrak{H}$ such that $\pi'(a \otimes b)|_{\mathfrak{H}} = \pi(a \otimes b)$ for all $a \in A$, $b \in A^c$. (To show this, we can assume that π is cyclic and so comes via the GNS construction from a state on $A \otimes_{\min} A^c$. Extend this state to one on $A \otimes_{\min} \mathbf{B}(\mathfrak{H})$ and consider the resulting GNS construction.) If P is the orthogonal projection from K onto $\mathfrak{H}$, then the map $T \to (P\pi'(1 \otimes T))|_{\mathfrak{H}}$ is a norm one projection from $\mathbf{B}(\mathfrak{H})$ onto A^c. For the last part, realise A through its universal representation and use A^c injective $\Leftrightarrow$ A^{cc} $(= A'')$ is injective.

2-32. (i) (Effros [**1**]). Adapt (0.6).

(ii) (Paschke [**2**]) $\Rightarrow$ Let m be a nontrivial inner invariant mean on G. Find $\{f_\delta\}$ in $P(G)$ such that $\hat{f}_\delta \to m$, $\|x * f_\delta * x^{-1} - f_\delta\|_1 \to 0$ for all $x \in G$. By subtracting $f_\delta(e)\delta_e$ from f and scaling, we can suppose that $f_\delta(e) = 0$ for all δ. By considering $\|(P_e - T)\xi\|_2$ for $T \in C_\pi(G)$ and $\xi = f_\delta^{1/2}, e$, show that $d(P_e, C_\pi(G)) \geq 1/2$. $\Leftarrow$ For all $T \in C_\pi(G)$, we have $P_eT = TP_e = (Te, e)P_e$. The functional ϕ on $C_\pi(G) + \mathbb{C}P_e$ given by $\alpha(T + \gamma P_e) = (Te, e)$ $(\gamma \in \mathbb{C})$ is a state. Extend α to a state β on $\mathbf{B}(l_2(G))$. Since $\beta(\pi(x)) = 1 = \|\pi(x)\|$, we have $\beta(\pi(x)W\pi(x)^*) = \beta(W)$ for all $W \in \mathbf{B}(l_2(G))$. (To see this, use the GNS construction for β.) Noting that $\pi(x)L_\phi\pi(x)^{-1} = L_{x\phi x^{-1}}$ $(\phi \in l_\infty(G))$ and $L_{\chi_{\{e\}}} = P_e$, the map $\phi \to \beta(L_\phi)$ is a nontrivial inner invariant mean on $l_\infty(G)$.

2-33 (Effros [**1**]). The operators in VN(G) can be identified with $l_2(G)$ functions by the map $T \to Te$, and the $l_2(G)$-norm corresponds to the $\|\cdot\|_2$-norm in VN(G). Let $x_1, \ldots, x_n \in G$. Apply the Property Γ definition with $T_j = \pi_2(x_j)$. One checks (using $\pi_r(G) \subset VN(G)^c$) that if $g = U^*e$, then $g(e) = 0$ and $\|\pi_2(x_j)g - \pi_r(x_j^{-1})g\|_2 < \varepsilon$. Show that $h = |g|^2 \in P(G)$ and use the h's to produce an inner invariant mean $\neq \delta_e$ in the obvious way. [*Note*: The converse to this result is important, and seems to be open. The main use of Property Γ and its refinements is to distinguish between II_1 factors ([**D1**], [**Pt.3**, Ch. 7, 7].]

2-34 (Lau and Paterson [**S2**]). If m is an inner invariant mean on $L_\infty(G)$, then the operator $\phi \to m(\phi)1$ belongs to A^c_∞. Conversely, suppose that T is a nonzero, compact operator in A^c_∞. Then [**Sch**, Chapter IV, §1] T has a (compact) modulus $|T|$, where for $\phi \geq 0$, $|T|(\phi) = \sup\{|T\psi|: \psi \in L_\infty(G), |\psi| \leq \phi\}$. Then $|T| \in A^c_\infty$ and $f = |T|(1)$ is positive, $\neq 0$, and inner invariant. Let $\varepsilon > 0$ be such that $X = f^{-1}([\varepsilon, \infty))$ is not locally null, and let $P: L_\infty(G) \to L_\infty(X)$ be the restriction map. Show that $L_\infty(X)$ is a right Banach G-module under conjugation action, and for all $\Phi \in L_\infty(G)$, $\tilde{\phi} = P(|T|(\phi)) \in A$, the space of functions $\psi \in C(X)$ which are almost periodic for this action. Then $\phi \to n(\tilde{\phi})/n(\tilde{1})$ is inner invariant on G, where n is a G-invariant mean on A. To show that such an n exists, let τ be the (locally convex) topology on A' determined by seminorms p_ϕ $(\phi \in A)$ where $p_\phi(\alpha) = \sup_{x\in G} |\alpha(\phi x)|$. Then $\mathfrak{M}(A)$ is weakly compact for τ (by the Mackey-Arens Theorem (Köthe [**1**], §21, 4)) and the Ryll-Nardzewski Theorem ((2.36)) yields a G-fixed element $n \in \mathfrak{M}(A)$.

2-35. (i) cf. (1.20).

(ii) (Lau [**2**]) $\Rightarrow$ Let $G = \{\gamma_0, \ldots, \gamma_{n-1}\}$ be a finite left ideal group in βS with γ_0 as identity. Let $A_i = \{s \in S: s\gamma_0 = \gamma_i\}$. Then $S = \bigcup_{i=1}^n A_i$ (disjoint), $A_iA_j \subset A_k$ $(\gamma_i\gamma_j = \gamma_k)$ and A_0 is ELA. Let $s_i \in A_i$ and $W = \{s_0, \ldots, s_{n-1}\}$. Show that if $F \in \mathscr{F}(S)$, then there exists $t_0 \in A_0$ such that $F(Wt_0) = Wt_0$.

(iii) Sorenson [**2**], Granirer [**13**], Lau [**2**].

(iv) Try $S = G \times T$, G a finite group, T ELA.

2-36 (Mitchell [**3**], Granirer [**11**], Ljapin [**1**]).

2-37 (Sorenson [**2**], Granirer [**10**], Lau [**2**]). Show that $\mathrm{PM}(K) = \mathfrak{L}(S)^\wedge$, where $K = \{p \in \beta S: Sp = \{p\}\}$. Clearly, $\mathrm{PM}(K) \subset \mathfrak{L}(S)^\wedge$. Conversely, if $s \in S$, then $R_s = \{t \in S: st = t\}$ is a right ideal of S and so if $m \in \mathfrak{L}(S)$, $\hat{m}(\hat{R}_s) = 1$, and $\mathbf{S}(\hat{m}) \subset \bigcap_{s\in S} \hat{R}_s = K$.

2-38 (Wilde and Witz [**1**]).

2-39 (Ky Fan [**S**]). Let $\mathscr{Y}_n$ be the set of n-dimensional linear subspaces of X contained in Y. For each $Z \in \mathscr{Y}_n$, show that $(\xi + H) \cap Z$ is a singleton $\{P_Z(\xi)\}$. Then P_Z is a projection in $\mathbf{B}(X)$ with range Z and kernel H. Let $\mathbf{L}(X)$ be the locally convex space of linear transformations on X with the pointwise convergence topology. Show that $C = \{P_Z: Z \in \mathscr{Y}_n\}$ is a norm compact, convex subset of $\mathbf{L}(X)$ that is invariant under the group action $(xT)(\xi) = x(T(x^{-1}\xi))$ $(x \in G, T \in \mathbf{L}(X))$. Apply Day's Fixed-Point Theorem to obtain an invariant P_L. [The above result has semigroup versions. Ky Fan gives an example to which his semigroup result applies with X a Hilbert space. Natural, geometric examples are readily constructed with X a Euclidean space $\mathbb{R}^k$. Lau, Paterson and Wong [**S**] have removed the finite-dimensional requirements from the result, and this paper, together with Lau [**10**], [**S6**], Lau and Wong [**S**] , characterizes amenability in terms of the "fixed-subspace" property.]

Problems 3

3-1. For each $z \in F_2 \sim \{e\}$, there exists a homomorphism $Q_z\colon F_2 \to S_z$, where S_z is a finite permutation group and $Q_z(z) \neq e$. Take the S_z's to be the G_α's. Then $\prod_{\alpha\in A} G_\alpha$ contains F_2 and so is not amenable.

3-2 (Rosenblatt [**6**]). Let S_n be the group of permutations of n elements and $G = \prod_{n\in\mathbb{P}} S_n$. Then the direct limit H of the system of finite groups $S_1 \times \cdots \times S_n$ is amenable as discrete and is dense in the compact group G. As in the preceding solution, the group G is not amenable as discrete.

3-3 (cf. Dixmier [**1**]). Suppose that $|A| > 1$. Let $\Phi\colon G = *_{\alpha\in A}\mathbb{Z}_{n_\alpha} \to H = \bigoplus_{\alpha\in A} Z_{n_\alpha}$ be the canonical epimorphism. Then by the Kurosh Subgroup Theorem, $K = \ker\Phi$ is of the form $F * (*_{i\in I} L_i)$, where L_i is conjugate in G to a subgroup of one of the free factors $\mathbb{Z}_{n_\alpha}$ and F is a free group. Using the fact that H is abelian, show that $K = F$. Every commutator of G is in F, and we can show in the appropriate circumstances that F is nonabelian by exhibiting two noncommuting commutators in G.

3-4 (Cole and Świerczkowski [**S**]). Use Problem 3-3. If $I = \{1,2\} = J$ and $x_1^2 = x_2^2 = e$, show that $H = \{(x_1x_2)^n\colon n \in \mathbb{Z}\}$ is a normal subgroup of G with abelian quotient, so that G is amenable.

3-5. SU(2) is the simply connected covering group of SO$(3,\mathbb{R})$, and we can apply (3.1). A nice example of a pair of elements ϕ, ψ in SO$(3,\mathbb{R})$ generating F_2 was developed by Świerczkowski and is presented in Wagon [**S2**, Theorem 2.1]: ϕ, ψ are the anticlockwise rotations through $\cos^{-1}(1/3)$ about the z- and x-axes respectively.

3-6. Aut $\mathbb{T}^n$ is the group of $n \times n$ integer-valued matrices of determinant ± 1 [**HR1**, (26.18)] and so is discrete. Thus $(\operatorname{Aut} G)_{\mathfrak{i}} = I_{G_e}(G) =$ trivial group.

3-7 (Granirer [**22**]). $\Leftarrow$ $G/\operatorname{rad} G$ has a compact, semisimple Lie algebra and so is compact (Appendix B52).

3-8. If G is nilpotent, then it is of Type R ($\operatorname{Sp}(\operatorname{ad} X) = \{0\}$). Every compact G is of Type R ($\{\|\exp t \operatorname{ad} X\|\colon t \in \mathbb{R}\}$ is bounded). For more examples see Problems 6-7–6-9. For the "$ax + b$" group case, see (6.36). Suppose that G is of Type R and not amenable. Then using the Levi-Malcev Theorem and (3.3(ii)), $\mathfrak{g} \supset \mathrm{sl}(2,\mathbb{R})$. Finally SL$(2,\mathbb{R})$ is not of Type R ((6.29)), giving a contradiction.

3-9. Use (3.9), (0.13).

3-10. $\Leftarrow$ Let $A = A^{-1} \in \mathscr{F}(G)$, and let L be the subgroup of G generated by A. Let $Q\colon G \to G/H$ be the quotient map. Let $x_1, \ldots, x_r$ be a left transversal of L with respect to $L \cap H$. Write $ax_j = x_{i(a)}h_{ij}(a)$, $a \in A$, $h_{ij}(a) \in H$. Let M be the (finite) subgroup of H generated by the $h_{ij}(a)$'s, and show that $L \subset \{x_1, \ldots, x_r\}Mx_1^{-1}$, which is finite.

3-11. We get an easy negative answer by defining

$$\mu(E)=\begin{cases}\lambda_1(E) & \text{if } E\in\mathscr{M}_b(\mathbb{R}),\\ \infty & \text{otherwise.}\end{cases}$$

3-12. Let F_2 be free on x,y. For $u\in\{x,y,x^{-1},y^{-1}\}$ let E_u be the set of elements in F_2 whose reduced form begins with u. Let $C=\{y^{-1},y^{-2},\dots\}$ and obtain, in the notation of (3.12), the p.d.:

$$A_1=E_x,\quad A_2=E_{x^{-1}},\quad B_1=E_y,\quad B_2=E_{y^{-1}}\sim C,\quad B_3=C\cup\{e\},$$
$$x_1=x^{-1},\quad x_2=e,\quad y_1=y^{-1},\quad y_2=e,\quad y_3=y^{-1}.$$

3-13. *Proof of* (3.13(ii)) (cf. Jech [**1**]'s proof of the Cantor-Bernstein Theorem). There exist subsets E of B and D of A such that $A\cong E$, $D\cong B$. Find partitions A_i,E_i $(1\le i\le m)$, D_j,B_j $(1\le j\le n)$ of A,E,D,B and elements s_i,t_j such that $s_iA_i=E_i$, $t_jD_j=B_j$. Define 1-1 functions $f\colon A\to E$, $g\colon B\to D$ by $f(a)=s_ia$ $(a\in A_i)$, $g(b)=t_j^{-1}b$ $(b\in B_j)$. Let $F=g\circ f\colon A\to A$, and define recursively sets A_n',B_n' by $A_0=A$, $A_{n+1}=F(A_n)$, $B_0'=D$, $B_{n+1}'=F(B_n')$. Then $g'\colon A\to D$ is a bijection, where $g'(x)=F(x)$ if $x\in A_n\sim B_n'$ for some n and is x otherwise. Composing with g^{-1} gives a partition $\{A',A''\}$ of A and a bijection $h\colon A\to B$, where $h(a)=f(a)$ if $a\in A'$ and $h(a)=g^{-1}(a)$ if $a\in A''$. Use the sets $A'\cap A_i$, $A''\cap D_j$, and elements s_i,t_j to implement $A\cong B$.

3-14 (Wagon [**S2**, p. 113]). In the notation of (3.15), let $\alpha=\tau(A)$, $\beta=\tau(B)$. Show, using (3.15(v)) and (3.13(ii)), that $n\alpha=n\beta$ for some $n\in\mathbb{P}$, and then apply the Division Theorem. For the last part, note that if $B\subset\mathbb{S}^n$, $B^0\ne\varnothing$, then $\mathbb{S}^n$ is contained in the union of a finite number of translates of B.

3-15 (Paterson [**10**]). (i) If A_i,B_i,s_i,t_i give a Borel p.d. for H, then A_iT, B_iT, s_i, t_i give a Borel p.d. for G, where T is a right transversal for H in G.

(ii) $\Rightarrow$ Let H be a nonamenable, σ-compact, open subgroup of G. Then G has a Borel p.d. if H has. Let K be a compact normal subgroup of H with H/K separable. It is sufficient to show that $M=H/K$ has a Borel p.d. Let L be a compact open subgroup of M/M_e, and let M' be the inverse image of L in M. If M' is not amenable, use (3.8) and the fact that F_2 has a p.d. to produce a p.d. for M. If M' is amenable, there does not exist an M-invariant mean on M/M' and Tarski's Theorem applies.

3-16. Use Problem 3-15.

3-18. The existence of a p.d. for $\mathbb{S}^3$ follows from the $n=2$ case by identifying $\mathrm{SO}(3,\mathbb{R})$ as a subgroup of $\mathrm{SO}(4,\mathbb{R})$ by allowing action on the first three coordinates only. The general case follows by induction.

Problems 4

4-2 (Banach [1], Emerson [7]). (i) (d) If $\mu \in P(G)$, $\mu_N = (N+1)^{-1} \sum_{n=0}^{N} x^n \mu$, then $\|(\phi x - \phi)\mu_N\|_\infty \to 0$. (ii) $\Rightarrow$ Let $\phi, \psi \in D(G)$. Let $\{\mu_\delta\}$ be a net in $P(G)$ such that $\|\xi * \mu_\delta - \mu_\delta\|_1 \to 0$ for all $\xi \in P(G)$. Show $\|\phi\mu_\delta\|_\infty, \|\psi\mu_\delta\|_\infty \to 0$ so that $\|(\phi + \psi)\mu_\delta\|_\infty \to 0$. For $\Leftarrow$, cf. Problem 2-1 and use (b).

4-3 (Müller-Römer [S]). Let $F \in \mathscr{F}(G)$, $\mu \in P(G)$, $\varepsilon > 0$. Find $C \in \mathscr{C}_e(G)$ such that $\|x * \mu - \mu\| < \varepsilon$ $(x \in C)$. If $\alpha \in \operatorname{Aut} G$ is such that $\alpha(F) \subset C$, then, if $\mu_\alpha = \mu \circ \alpha$, $\phi \in L_\infty(G)$, and $x \in F$, we have

$$|(x * \mu_\alpha - \mu_\alpha)^\wedge(\phi)| = |(\alpha(x) * \mu - \mu)^\wedge(\phi \circ \alpha^{-1})| < \varepsilon\|\phi\| \qquad (x \in F).$$

Thus G satisfies a Reiter condition. $\mathbb{R}$ and $\mathbb{T}$ are contractible; any nontrivial discrete amenable group is not contractible.

4-4 (Day [6], Willis [S]). (i) Suppose not and let $x_0, x_1, \ldots, x_n \in G$, $x_0 = e$, $\mu = (n+1)^{-1} \sum_{i=0}^{n} \delta_{x_i}$. From $\|\pi_p(\mu)\| = 1$ deduce that there exists a sequence $\{f_k\}$ in $P_p(G)$ such that

$$\lim_k \|(n+1)^{-1} \sum_{i=0}^{n} x_i * f_k\|_p = 1.$$

Now show using uniform convexity that $\|x_i * f_k - f_k\|_p \to 0$ so that G is amenable.

(ii) Let $\alpha\colon L_p(G) \to \mathbb{C}$ be linear and left invariant. Let x_i $(1 \le i \le n)$ be such that $\|\pi_p(\mu)\| < 1$, where $\mu = n^{-1} \sum_{i=1}^{n} \delta_{x_i}$. Then $(I - \pi_p(\mu))$ is invertible so that if $f \in L_p(G)$, there exists $g \in L_p(G)$ such that $f = n^{-1}(\sum_{i=1}^{n}(g - x_i * g))$. Now apply α to obtain $\alpha = 0$. [There are always discontinuous invariant linear functionals on $L_1(G)$ (G infinite). See Willis [S] for references and more information.]

4-5. Let $f(x) = \lambda(Cx \cap D)$. Then using $\int g(z^{-1})\, d\lambda(z) = \int \Delta(z^{-1}) g(z)\, d\lambda(z)$ $(g \in L_1(G))$,

$$\begin{aligned} |f(x) - f(y)| &= \Big|\int (\chi_{Cx \cap D} - \chi_{Cy \cap D})(z)\, d\lambda(z)\Big| \\ &\le \Big(\sup_{z \in D^{-1}} \Delta(z^{-1})\Big)\, \lambda(x^{-1}C^{-1} \,\triangle\, y^{-1}C^{-1}). \end{aligned}$$

Also $\lambda(x^{-1}C^{-1} \,\triangle\, y^{-1}C^{-1}) = \|yx^{-1} * \chi_{C^{-1}} - \chi_{C^{-1}}\|_1 \to 0$ as $x \to y$.

4-6 (Skudlarek [1]). cf. (4.5).

4-7. Try K_n of the form $[-a_n, a_n] + C_n$, where $a_n \to \infty$ and C_n is a Cantor type set in an interval large enough and far enough away from $[-a_n, a_n]$.

4-8 (Emerson [2]). If $\phi_n = \chi_{K_n}/\lambda(K_n)^{1/2}$, then $\lambda(xK_n \cap K_n)/\lambda(K_n) = \phi_n * \phi_n^\dagger(x)$. Further $\{\phi_n * \phi_n^\dagger\}$ is a sequence of positive-definite functions on G converging to 1 pointwise. The convergence is uniform on compacta [**HR2**, (32.42)].

4-9 (Emerson [**2**]). Show, using dominated convergence, that

$$\lambda(V_n)^{-1} \int_U \lambda(uV_n \cap V_n)\, d\lambda(u) \to \lambda(U)$$

and that, by Fubini, the preceding integral is $\leq \lambda(V_n^{-1})$.

4-10 (Chou [**4**], Emerson [**3**]). $\Leftarrow$ (G not unimodular). Let $\varepsilon > 0$, $U \in \mathscr{C}(G)$, $L \subset U$, and $\lambda(xU \,\Delta\, U)/\lambda(U) < \frac{1}{2}\varepsilon$ $(x \in C)$. For $y \in G$ let $U_y = (Uy) \cup (Uy)^{-1}$. Show that

$$\lambda(xU_y \,\Delta\, U_y)/\lambda(U_y) < \tfrac{1}{2}\varepsilon + 2(\lambda(U^{-1})/\lambda(U))\Delta(y)^{-1},$$

and choose y so that $K = U_y \cup L \cup L^{-1}$ "works."

For the unimodular case, the symmetric version of (4.10) holds (using (4.8)). Now apply Problem 4-9 (last part).

4-11 (Leptin [**1**], [**2**], [**5**]). Take the "inf" over K of each side of the equality:

$$\lambda(C^2K)/\lambda(K) = (\lambda(C(CK))/\lambda(CK))(\lambda(CK)/\lambda(K)).$$

4-12 (Emerson [**7**]). The converse is open.

4-13 (Bondar [**1**]). $\Rightarrow$ $\lambda(\bigcap_{c\in C} c(C^{-1}K))/\lambda(C^{-1}K)$ is close to 1 if $\lambda(C^{-1}K)/\lambda(K)$ is.

4-14 (Lorentz [**1**], Chou [**4**]). (ii) $\Rightarrow$ Use (4.17). (iii) If $m \in \mathfrak{L}(\mathrm{AC}_l(G))$, show that $m = n|_{\mathrm{AC}_l(G)}$ for some $n \in \mathfrak{L}_t(\mathrm{U_r}(G))$, and then use (ii).

[*Notes*: In his paper [**8**], C. Chou determines the "multiplier" algebra $\{\phi \in \mathrm{U}(G) \colon \phi\,\mathrm{AC}_l(G) \subset \mathrm{AC}_l(G)\}$: *the latter equals* $\mathbb{C}1 + \mathrm{AC}_{l,0}(G)$, where $\mathrm{AC}_{l,0}(G) = \{\phi \in \mathrm{U}(G) \colon m(|\phi|) = 0 \text{ for all } m \in \mathfrak{L}(G)\}$ $(\subset \mathrm{AC}_l(G))$. The space $\mathrm{AC}_{\mathrm{r},0}(G)$ is defined in the obvious way, and Chou shows that if $\mathrm{AC}_0(G) = \mathrm{AC}_{l,0}(G) \cap \mathrm{AC}_{\mathrm{r},0}(G)$, then $\mathrm{AC}_0(G)/(\mathrm{AC}_0(G) \cap \mathrm{WP}(G))$ *contains a linear, isometric copy of* l_∞ *for a very large class of* σ*-compact groups* G. So $\mathrm{AC}_l(G)$ is, in general, very much larger than $\mathrm{WP}(G)$. Also, $\mathrm{U}(G)$ is, in general, much larger than $\mathrm{AC}_l(G)$ (Granirer [**13**], Chou [**8**]). Almost convergent functions also arise naturally in the study of the Ergodic Mixing Theorem for dynamical systems (Dye [**1**]). See also Problem 5-8.]

4-16 (Wu [**S**], Knapp [**S1**], Milnes [**S2**]–[**S5**]). Let $\varepsilon > 0$. Suppose that ϕ is in $\mathrm{U}_l(G) \cap \mathrm{BAP}(G)$. Find $V \in \mathscr{C}_e(G)$ such that $\|v\phi - \phi\| < \varepsilon$ $(v \in V)$. Let $C \in \mathscr{C}(G)$ be such that (1) holds, and $c_1, \ldots, c_n \in G$ be such that $\bigcup_{i=1}^n c_iV \supset C$. Then $G\phi \subset$ the union of the balls, centre $c_i\phi$, radius 2ε, and $G\phi$ is totally bounded. So $\phi \in \mathrm{AP}(G)$. Trivially, $\mathrm{U}(G) \cap \mathrm{BAP}(G) \supset \mathrm{AP}(G)$.

Now let $\psi \in \mathrm{BAP}(G)$. Choose an increasing sequence $\{C_n\}$ in $\mathscr{C}(G)$ such that for each $x \in G$, $\exists c_n^x \in C_n$ such that $\|x\psi - c_n^x\psi\| < n^{-1}$. Find $\{\nu_n\}$ in $P(G)$ such that $\|\nu_n * x - \nu_n\|_1 < n^{-1}$ $(x \in C_n)$. Then for $x \in G$, $|\psi\nu_n(x) - \hat{\nu}_n(\psi)| \leq n^{-1}(1 + \|\psi\|)$. We can suppose $\hat{\nu}_n(\psi) \to k$. Then $\|\psi\nu_n - k1\|_\infty \to 0$, and $\psi \in \mathrm{AC}_l(G)$ (cf. Problem 4-14). [*Note*: Examples of functions in $\mathrm{BAP}(G) \sim \mathrm{AP}(G)$ have been given by Wu and Milnes.]

4-17. Let $g \in L_p(G)$ $g \geq 0$. Define a measure μ_g on $\mathscr{B}(G)$ by $\mu_g(E) = \int_E g\,d\lambda$. By (4.19), $\mu_g(G) < \infty$. This gives $L_p(G) \subset L_1(G)$. Show that this forces G to be compact. (References: Leptin [**1**], Lohoué [**1**], Rajagopalan [**1**]–[**5**], Rajagopalan and Zelazko [**1**], Rickert [**3**], Urbanik [**1**], Zelazko [**1**]–[**3**], [**HR2**, pp. 469–472]).

4-18 (Heath and Sudderth [**1**], Bondar). $\Rightarrow$ Suppose that $m \in \mathfrak{L}_t(G)$. Let μ_δ, x_δ, and $\mathfrak{L}(\{x_\delta\})$ be as in (4.16)–(4.17). We can suppose that $m \in \mathfrak{L}(\{x_\delta\})$. Let $\nu \in P(G)$, ν with compact support K. Show that

$$\begin{aligned}
&|(\mu_\delta * x_\delta)^\wedge(\nu \circ_2 \phi - \nu^\sim \circ_1 \phi)| \\
&\quad = (\lambda(K_\delta x_\delta))^{-1} \left| \int_K d\nu(x) \int_G [\chi_{K_\delta x_\delta}(s)\phi(s, xs) \right. \\
&\qquad\qquad\qquad\qquad \left. - \chi_{x^{-1}K_\delta x_\delta}(s)\phi(s, xs)]\,d\lambda(s) \right| \\
&\quad \leq \sup_{x \in K} (\lambda(xK_\delta \,\triangle\, K_\delta)/\lambda(K_\delta))\|\phi\|_\infty.
\end{aligned}$$

4-19 (Heath and Sudderth [**1**], Bondar). Show that $m_1 = m$, $\widehat{\mathbf{P}(s)}({}_s\phi) = \mu \circ_2 \phi(s)$, $\widehat{\mathbf{Q}(x)}(\phi_x) = \mu^\sim \circ_1 \phi(x)$, and use Problem 4-18. (m_1 is called the *marginal* of $(m, \mathbf{P})$.)

4-20 ([**R**, p. 175], Derighetti [**1**], Glicksberg [**2**]). Adapt the proof of (4.18). (The converse of the result is also true, and the normality condition can be removed.) For the last part, note that the map $g \to g^\sim$ is isometric from C_f onto $\operatorname{co}\{x * f^\sim : x \in G\}$.

4-21 (cf. Greenleaf [**2**, p. 79]). Note that $f * \mu_\delta \in L_1(G)$ and $\mu_\delta = \chi_{K_\delta}/\lambda_H(K_\delta) \in L_1(H) \subset M(G)$. The LHS of (1) is just $\|Q_H f\|_1$, and the RHS is $\lim \|f * \mu_\delta\|_1$. By Problem 4-20, (1) will follow once we have shown $d(0, C_f(H)) = \lim \|f * \mu_\delta\|_1$. Suppose first that $f \in C_c(G)$, with support C. If $\nu \in \operatorname{co} H$, then

$$\|f * \nu * \mu_\delta - f * \mu_\delta\|_1 \leq \|f\|_1 \|\nu * \mu_\delta - \mu_\delta\|_1 \to 0,$$

so that $\limsup \|f * \mu_\delta\|_1 \leq d(0, C_f(H))$. For the reverse inequality, it is sufficient to show that $\|f * \mu\|_1 \geq d(0, C_f(H))$ for $\mu \in P(H)$ with compact support K. Show that for $\varepsilon > 0$ there exists a measurable partition $E_1, \ldots, E_n$ of K and elements $h_i \in E_i$ such that if $F(x) = \sum_{i=1}^n f * h_i(x)\mu(E_i)$, then $|f * \mu(x) - F(x)| < \varepsilon$ $(x \in CK)$. Now approximate $\|f * \mu\|_1 = \int_{CK} |f * \mu(x)|\,d\lambda(x)$ by $\int_{CK} |F(x)|\,d\lambda(x)$, giving $\|f * \mu\|_1 \geq d(0, C_f(H))$ as required. The equality (1) for general $f \in L_1(G)$ follows by approximation.

For the last part, take $G = \mathbb{R}$, $H = \mathbb{Z}$, and $K_N = [-N; N]$.

4-22 (Rindler [**1**]–[**3**], [**S1**]ff.). (i) Use (4.18).

(ii) $\Rightarrow$ Let $L^0 = \{f \in L^1(G) \colon \int f\,d\lambda = 0\}$, and let $\{f_n : n \in \mathbb{P}\}$ be a norm dense subset of L^0. There exist elements $z_r^m \in G$ $(1 \leq r \leq M_m)$ such that $\|M_m^{-1} \sum_{r=1}^{M_m} f_k * z_r^m\|_1 < m^{-1}$ for $1 \leq k \leq m$. Construct sequences N_i, K_i such that $N_1 = 1$, $N_{i+1} - N_i = K_i M_i$, $M_i/N_i \to 0$. For $n \in \mathbb{P}$ write $n = N_m + jM_m + i$

$(j < K_m, i < M_m)$, and set $a_n = z_i^m$. Then $\{a_n\}$ is a u.d. sequence for G. First show this for $f \in L^0$, and deduce the general case as in the second paragraph of the proof of (4.18). $\Leftarrow$ Show that if $\{x_n\}$ is a u.d. sequence in G, then G is the closure of the subgroup of G generated by $\{x_n : n \geq 1\}$.

4-23. $n^{-1} \sum_{r=1}^{n} \phi * x_r(y) \to \int \phi \, d\lambda$ for all $y \in G$. Now use the Dominated Convergence Theorem to obtain that $\{x_r\}$ is u.d. Property (1) is the classical definition of a u.d. sequence in a compact group—see Kuipers and Niederreiter [**1**]. For the second part, $\Rightarrow$ follows by taking ϕ to be a suitable exponential. For $\Leftarrow$, (1) is trivial if ϕ is a trigonometric polynomial, and the general case follows by approximating ϕ by such a polynomial. Using Weyl's criterion, one can easily show that if θ is irrational, then $\{n\theta\}$ (mod 1) is u.d. in the sense of (1).

4-25 (References as for (4.20)). Apply the Riesz-Thorin Convexity Theorem [**DS**, VI.10.11] to the function $t \to \log \|\pi_{1/t}(\mu)\|$ $(0 \leq t \leq 1)$.

4-27. *G abelian*: The map $\pi_2(f) \to \hat{f}$ from $\pi_2(C_l^*(G))$ into $C_0(\hat{G}) = C^*(G)$ is isometric by Plancherel's Theorem. *G compact*: Use the Peter-Weyl Theorem.

4-29. The support of μ is, by [**D2**, (8.6.8)], the same as the set $A = \{\rho \in \hat{G} : \rho$ is weakly contained in $\pi_2\}$. Now $\pi \in \hat{G}$ is in $A \Leftrightarrow$ every state associated with π is a weak* limit of states associated with π_2 [**D2**, (3.4.10)]. Now show that $A = \hat{G} \Leftrightarrow C^*(G) = C_l^*(G)$. When G is abelian, μ is a Haar measure on $\hat{G}$ and so has dense support in $\hat{G}$. Now let G be the Heisenberg group. For information about this group, see, for example, Warner [**1**], Kirillov [**1**], and Nielsen [**S2**]. As a set, $\hat{G}$ can be identified with $(\mathbb{R} \sim \{0\}) \cup \mathbb{R}^2$. Both $\mathbb{R} \sim \{0\}, \mathbb{R}^2$ have their normal topologies, but the closure of $\mathbb{R} \sim \{0\}$ in $\hat{G}$ is the whole of $\hat{G}$. The Plancherel measure can be identified with $|s| \, ds$ on $\mathbb{R} \sim \{0\}$. (Nielsen gives Plancherel measures for the low-dimensional, simply connected, nilpotent Lie groups. See also Paterson [**S2**].)

4-30 (Paterson [**2**]). Show that $\chi_H \in S(G) \sim \operatorname{co}\{S_l(G) \cup \{1\}\}$ ([**HR2**, (32.43)]). The argument proceeds by contradiction using $\chi_H \notin S_l(H)$. This requires showing that $\phi_{|H} \in S_l(H)$ if $\phi \in S_l(G)$. To prove this, approximate ϕ by functions $\psi \in C_c(G) \cap P(G)$ and then apply Godement's Theorem ((4.20)) to $\psi_{|H}$.

4-31. $\Rightarrow$ Since G is amenable, π_0 is weakly contained in π_2, and by Property (T) there exists $\xi \in L_2(G)$, $\|\xi\| = 1$, such that $1 = (\pi_2(x)\xi, \xi)$ $(x \in G)$. But then $1 \in C_0(G)$, forcing G to be compact.

4-32. (Argabright and Wilde [**1**]). Suppose that S satisfies (SFC). If $C \in \mathscr{F}(S)$, $\varepsilon > 0$, and non-void $K \in \mathscr{F}(S)$ is such that $|K \sim xK|/|K| < \frac{1}{2}\varepsilon$ $(x \in C)$, show that the Reiter condition $\|x * f - f\|_1 < \varepsilon$ $(x \in C)$ holds, where $f = \chi_K/|K|$. For the second part, if S is left amenable, then $|K \sim xK| = |xK \sim K|$ in the left cancellative case and (4.9) applies, while (1.19) applies in the finite case.

4-33. (Granirer [**3**]). Let $\{x_1, x_2, \dots\}$ be an enumeration of T_0. Using the Reiter condition, construct inductively a sequence $\{f_n\}$ in $P(S)$, where each f_n has finite support A_n, such that $\|x * f_n - f_n\|_1 < n^{-1}$ for x belonging to $\{x_1, \dots, x_n\} \cup A_1 \cup \dots \cup A_{n-1}$. Then $T_0 \cup (\bigcup_{n=1}^{\infty} A_n)$ generates a left amenable countable subsemigroup.

4-34. (Talagrand [**5**]). Let ϕ be Riemann measurable on G. Construct, for $\varepsilon > 0$, functions $\psi, \psi' \in C(G)$ such that

$$\psi \le \phi \le \psi', \qquad \int (\psi' - \psi)\, d\lambda < \varepsilon.$$

Then (Problem 0-12) $|m(\phi) - \int \phi\, d\lambda| < \varepsilon$ for all $m \in \mathfrak{L}(G)$, giving $m(\phi) = \int \phi\, d\lambda$.

4-36. G has to be unimodular.

4-37. The map $f \to \int f\, d\lambda$ extends to a multiplicative linear functional on $C_l^*(G)$. The kernel of the functional is a proper, closed, two-sided ideal in $C_l^*(G)$. [*Note*: $C_l^*(F_2)$ is simple (Powers [**S**]).]

4-38. (Reiter [**17**], Johnson [**2**], Rindler [**S2**], [**S9**]). Suppose that G is amenable. We can suppose that $\{f_\delta\}$, $\{\mu_\delta\}$ are nets in $P(G)$ as in (4.1), (1.5). By expressing $f \in L^0(G)$ as a member of $\operatorname{Span} P(G)$, we have $\|f * f_\delta\|_1 \to 0$. Then $\{\mu_\delta - f_\delta\}$ is a bounded right approximate identity for $L^0(G)$. Conversely, if $\{u_\delta\}$ is a bounded right approximate identity for $L^0(G)$ and we fix $\nu \in P(G)$, then for all $\mu \in P(G)$, $(\mu\nu - \nu) \in L^0(G)$, and $\|\mu\xi_\delta - \xi_\delta\|_1 \to 0$, where $\xi_\delta = \nu u_\delta - \nu$. Since $\hat{\xi}(1) = -1$, any weak* cluster point of $\{\hat{\xi}_\delta\}$ is a nonzero element of $\mathfrak{J}_l(G)$, and (2.2) applies. [*Note*: Johnson [**4**] showed that $L^0(F_2)$ does not even possess an (unbounded) right approximate identity. Liu, van Rooij, and Wang [**S**] have shown that if G is amenable, then every cofinite, closed ideal in $L_1(G)$ has a bounded approximate identity. Willis [**1**], [**2**] has shown that if $L_1(G)$ contains a cofinite, closed, left ideal that has a bounded right approximate identity, then G is amenable.]

4-39. If $A(G) = B(G)$, then $1 \in A(G) \subset C_0(G)$, giving G compact. Conversely, if G is compact, then $1 = 1 * 1^\dagger \in A(G)$, giving $B(G) = B(G)1 \subset A(G)$.

4-40. (Boidal et al [**1**]). (i) Follow the C^*-algebra version in [**D2**, (3.1)]—the obvious version of [**D2**, (2.11.4)] holds. So a set $\mathscr{S} \subset \operatorname{Prim}_* A$ is closed if and only if $\mathscr{S} = \{I \in \operatorname{Prim}_* A \colon I \supset \bigcap \mathscr{S}\}$.

(ii) The bijectiveness of Ψ_G follows from the natural correspondence of $L_1(G)^\wedge$ with $C^*(G)^\wedge$ [**D2**, (2.7.4)].

(iii) Suppose that $G \in [\Psi]$, $\mathscr{S} = \{I \in \operatorname{Prim} C^*(G) \colon I \supset \ker \pi_2\}$. Then $\mathscr{S}$ is closed in $\operatorname{Prim} C^*(G)$. So $\Psi_G(\mathscr{S})$ is closed in $\operatorname{Prim}_* L_1(G)$. Now $\bigcap \Psi_G(\mathscr{S}) = \bigcap_{I \in \mathscr{S}} (I \cap L_1(G)) = \{0\}$ since π_2 is faithful on $L_1(G)$, and so $\Psi_G(\mathscr{S}) = \{J \in \operatorname{Prim}_* L_1(G) \colon J \supset \{0\}\} = \operatorname{Prim}_* L_1(G)$. Hence $\bigcap \mathscr{S} = \{0\}$ and $C^*(G) = C_l^*(G)$, giving G amenable.

4-41 (Greenleaf [**3**]). Suppose that G satisfies the weak Frobenius property. With $N = \{e\}$, $\pi \in \hat{G}$, we have U^σ unitarily equivalent to π_2. Thus π_2 weakly contains π for all $\pi \in \hat{G}$ and so $C^*(G) = C_l^*(G)$ and G is amenable. Conversely, suppose that G is amenable, and let N be a closed normal subgroup of G, $\pi \in \hat{G}$ and $\sigma = \pi|_N$. Let $\mathfrak{H}$ be the Hilbert space of π, and let $s\colon G\backslash N \to G$ be a Borel function such that $s(Ng) \in Ng$ for all $g \in G$. Then U^σ is realised on $Y = L_2(G\backslash N, \mathfrak{H}) = L_2(G\backslash N) \otimes \mathfrak{H}$, and for all $f \in Y$, $g \in G$, we have

$$U^\sigma(g)f(x) = \sigma(s(x)gs(xg)^{-1})f(xg) \qquad (x \in G\backslash N).$$

(Kirillov [**2**, §13]). Now the map α is an isometry of Y, where $\alpha(f)(x) = \pi(s(x))(f(x))$, and implements a unitary equivalence between U^σ and $U^J \otimes \pi$, where J is the trivial one-dimensional representation of N. (This is a special case of a result of Mackey [**1**].) If I is the trivial one-dimensional representation of G, then $I \otimes \pi = \pi$, and it follows that we can take $\pi = I$, $\sigma = J$. Now U^J is the "quasi-regular representation" of G on $L_2(G\backslash N)$; to prove that U^J weakly contains I, use the fact that G/N has the weak containment property. [*Note*: Greenleaf actually proves a more general "weak Frobenius" characterisation of amenability, in which the "separability" and "normal" requirements above are removed. Weak Frobenius properties were first discussed by Fell [**4**]. Gootman [**1**] examines the Frobenius Reciprocity Theorem for simply connected Lie groups. See also Henrichs [**1**], [**2**] and Nielsen [**S1**].]

4-42 (Renaud [**4**]). The function $t \to (tf)g \otimes \overline{(tf)g}$ is continuous and of compact support from G into $L_2(G)\hat{\otimes}L_2(G)$. Hence its Bochner integral h exists and belongs to $L_2(G)\hat{\otimes}L_2(G)$, and a direct calculation with an application of Fubini's Theorem gives $\Phi(h)(x) = (f * f^\dagger)(x)g * g^\dagger(x)$. It follows that

$$\begin{aligned}
&\|(f * f^\dagger)(g * g^\dagger) - g * g^\dagger\| \\
&\le \left\| \int (tf - f(t))g \otimes \overline{(tf)g}\, d\lambda(t) + \int f(t)g \otimes \overline{[tf - f(t)]g}\, d\lambda(t) \right\| \\
&\le \left(\int \|(tf - f(t))g\|_2^2\, d\lambda(t) \right)^{1/2} \left(\int \|(tf)g\|_2^2\, d\lambda(t) \right)^{1/2} \\
&\quad + \left(\int \|g\|_2^2 |f(t)|^2\, d\lambda(t) \right)^{1/2} \left(\int \|(tf - f(t))g\|_2^2\, d\lambda(t) \right)^{1/2}.
\end{aligned}$$

Now argue

$$\begin{aligned}
\int \|(tf - f(t))g\|_2^2\, d\lambda(t) &= \int d\lambda(x) \int |f(xt) - f(t)|^2 |g(x)|^2\, d\lambda(t) \\
&\le \sup_{x \in C} \|x^{-1} * f - f\|_2^2.
\end{aligned}$$

An easy Fubini-type argument gives $\int \|(tf)g\|_2^2\, d\lambda(t) = 1$ and (i) follows. [*Note*: This argument is a special case of the use of certain maps P, M that play a useful role in the $A_p(G)$ theory—these maps have their origin in the work of Varopoulos.]

Using (i) we can take each ϕ_δ to be of the form $g * g^\dagger$ ("$C \to e$") to obtain $\|\phi\phi_\delta - \phi_\delta\| \to 0$ for each $\phi \in C_c(G) \cap P(A(G))$. The latter fact is true for all $\phi \in P(A(G))$: consider $\phi\phi_0$, where ϕ_0 is a fixed element of $C_c(G) \cap P(A(G))$. A topologically invariant state m on $\mathrm{VN}(G)$ can be obtained by taking a weak* cluster point of $\{\hat{\phi}_\delta\}$ in $\mathrm{VN}(G)'$. [*Note*: Renaud shows that G is discrete if and only if there exists exactly one topologically invariant state on $\mathrm{VN}(G)$. Granirer [**16**] shows that the set of topologically invariant states on $\mathrm{VN}(G)$, where G is nondiscrete, is not norm separable; cf. (7.28).]

4-43 (Lance [**1**], Effros and Lance [**1**]). By replacing $u_{C,\varepsilon}$ of the proof of (4.34(ii)) by $(\chi_K * \chi_K^*)/\lambda(K)$, we see that there exists a bounded approximate identity of positive-definite functions $\{\phi_\delta\}$ for $A(G)$, with $\phi_\delta(e) = 1$, $\phi_\delta \in C_c(G)$ for all δ. Define $\Phi_\delta \in \mathbf{B}(\mathrm{VN}(G))$ by $\Phi_\delta(T) = \phi_\delta T$. Then $\{\Phi_\delta\}$ is a net of weak* continuous, unit preserving maps, and $\Phi_\delta(T) \to T$ weak* for all $T \in \mathrm{VN}(G)$. Since $\Phi_\delta(\pi_2(x)) = \phi_\delta(x)\pi_2(x)$ and ϕ_δ has finite support, the range of Φ_δ is the finite-dimensional space $\mathrm{Span}\{\pi_2(x)\colon \phi_\delta(x) \neq 0\}$. To show that Φ_δ is completely positive, realise $M_n \otimes \mathrm{VN}(G) = M_n(\mathrm{VN}(G))$ on $L_2(G) \oplus \cdots \oplus L_2(G)$. Let $f_{ij} \in \mathrm{Span}\,\pi_2(G)$, $A = [f_{ij}] \in M_n(\mathrm{VN}(G))$, and $h = (h_1, \ldots, h_n) \in C_c(G)^n$. It is sufficient to show that $(\Phi_\delta(A^*A)h, h) \geq 0$. In fact

$$(\Phi_\delta(A^*A)h, h) = \sum_{l,u} \sum_{(i,m),(j,n)} \left(\overline{f_{lim} h_i(x_{lim}^{-1}u)}\right) (f_{ljn} h_j(x_{ljn}^{-1}u))\phi_\delta(x_{lim}^{-1}x_{ljn}) \geq 0$$

by the positive-definiteness of ϕ_δ, where for some $f_{ijk} \in \mathbb{C}$, $x_{ijk} \in G$, we have $f_{ij} = \sum_k f_{ijk}\pi_2(x_{ijk})$. [*Note*: What happens if G is locally compact?]

4-44 (Coifman and Weiss [**1**], [**2**], [**3**]). Let $\varepsilon > 0$ and find ((4.13)) nonnull $V \in \mathscr{C}(G)$ such that $\lambda(CV)/\lambda(V) < 1 + \varepsilon$. For $v \in V$, show, using $R_{v^{-1}}R_v = I$, that $\|R_k f\|_p \leq M\|R_v R_k f\|_p$, and deduce, using Fubini's Theorem, that

$$\|R_k f\|_p^p \leq (M^p/\lambda(V)) \int_X d\mu(x) \int_V |R_v R_k f(x)|^p \, d\lambda(v).$$

Next show that for a.e. x,

$$\int_V |R_v R_k f(x)|^p \, d\lambda(v) = \int_V |T_k g_x(v)|^p \, d\lambda(v),$$

where $g_x(u) = R_u f(x)\chi_{CV}(u)$, and deduce, using Fubini's Theorem, that

$$\begin{aligned} \|R_k f\|_p^p &\leq (M^p/\lambda(V)) \int_X d_{\mu(x)} \|T_k\|^p \int_{CV} |R_u f(x)|^p \, d\lambda(u) \\ &\leq \left[\frac{M^{2p}\|T_k\|^p \lambda(CV)}{\lambda(V)}\right] \|f\|_p^p. \end{aligned}$$

Problems 5

5-2 Let m be the unique invariant mean on $\mathrm{WP}(G)$. Then $X_f = \mathbb{C}1$, $X_0 = \{\phi \in \mathrm{WP}(G) \colon m(\phi) = 0\}$, and $P\phi = m(\phi)1$.

5-3 (Greenleaf [**2**], [**4**], Greenleaf and Emerson [**1**]). In the notation of (5.19), (5.20), $G = G' = N \times_\rho H$, where $N = H = \mathbb{R}$. Take $C_m = [-m, m] \times [-m, m]$. Then $(C_m)_H = [-m, m]$. Choose R_m to satisfy $m/R_m \to 0$. Then

$$(C_m)^*_N = [-m\exp(m + R_m), m\exp(m + R_m)].$$

Put $L'_m = [-m, m]$. The subsequence $\{L'_{\alpha(m)}\}$ has to satisfy $(m\exp(m + R_m))/\alpha(m) \to 0$. Then

$$\begin{aligned} K_m &= j(L'_{\alpha(m)} \times [-R_m, R_m]) \\ &= \{(e^x y, x) \colon -R_m \le x \le R_m,\ -\alpha(m) \le y \le \alpha(m)\}. \end{aligned}$$

5-4 (Emerson [**4**]). (iv) Let $J_n[0,1) \cap ([a_n, a_{n+1}] + n\mathbb{Z})$ and show that each $x \in [0, 1)$ belongs to J_n for infinitely many n. Then show that if $x \in J_m$, $r \in D_m$, then $r.\ x \le 3m^{-1}$, and obtain $A_m f(x) \ge m^{1/4}/(6\sqrt{3})$ for large enough m.

5-5 (Blum and Eisenberg [**1**], Blum and Cogburn [**1**], Blum, Eisenberg, and Hahn [**1**], Milnes and Paterson [**S**]). (ii) If G is abelian, then

$$B_I(G) = (\mathrm{Span}\{\gamma \colon \gamma \in \hat{G}\})^- = \mathrm{AP}(G).$$

(iii) $\underline{(c) \Rightarrow (a)}$. Let $\pi \in \hat{G}$, $\xi, \eta \in \mathfrak{H}_\pi$. The function $x \to (\pi(x)\xi, \eta)$ is in $B_I(G)$ and so

$$(\pi(\mu_n)\xi, \eta) = \hat{\mu}_n(x \to (\pi(x)\xi, \eta)) \to m(x \to (\pi(x)\xi, \eta)).$$

So $\pi(\mu_n) \to$ some $T \in B(\mathfrak{H}_\pi)$ in the weak operator topology. The invariance of m yields that $T \in \pi(G)^c$ and irreducibility gives $T = P^\pi \in \{0, I\}$. Now use disintegration theory to obtain $\pi'(\mu_n) \to P^{\pi'}$ for every continuous representation π' of G. (Note that the separability requirement can be dropped if G is either abelian or compact.)

(iv) Use (iii)(c).

(v) The finite-dimensional representations in $\hat{H}$ are the characters $U_{a,b}$ ($a, b \in \mathbb{R}$), where $U_{a,b}(x) = \exp(2\pi i(ax_2 + bx_3))$, and the infinite-dimensional ones are of the form U_a ($a \in \mathbb{R} \sim \{0\}$) acting on $L_2(\mathbb{R})$, where

$$(U_a(x)f)(t) = \exp(2\pi i(x_1 - x_2 t)a)f(t - x_3).$$

In an obvious way, $\mathrm{AP}(H) = \mathbb{C} \otimes \mathrm{AP}(\mathbb{R}^2)$. By considering functions of the form $x \to (U_a(x)f, g)$ we obtain the $\mathrm{AP}_0(\mathbb{R})\check{\otimes}C_0(\mathbb{R}^2)$ part.

For related results, see the Milnes-Paterson paper. Blum and Cogburn [**1**] and Niederreiter [**S**] show that if $G = \mathbb{Z}$, then there are many nonsumming sequences $\{A_n\}$ of finite subsets of G such that $\{\mu_n\}$ is ergodic, where $\mu_n = \chi_{A_n}/|A_n|$. For example, we could take $A_n = \{x_1, \ldots, x_n\}$ where, for some fixed $\alpha \in (1, \infty) \sim \mathbb{Z}$, $x_r = [r^\alpha]$.

5-6 (Takahashi [**2**]). $\|T\| \le 1$ for all $T \in S$. Let $\phi \in C_{\mathbf{R}}(X)$, the set of real-valued functions in $C(X)$, and $\varepsilon > 0$. There exists $\psi \in C_{\mathbf{R}}(X)$, $T\psi = \psi$ $(T \in S)$, and a convex combination $\sum_{i=1}^{n} \alpha_i T_i$ in $l_1(S)$ such that $\|\psi - \sum_{i=1}^{n} \alpha_i T_i \phi\| < \varepsilon$. Let $h_x(T) = T\phi(x)$. Then $h_x \in l_\infty(S)$. Also $\|\psi - \sum_{i=1}^{n} \alpha_i T T_i \phi\| < \varepsilon$ $(T \in S)$ so that $\|\psi(x)1 - \sum_{i=1}^{n} \alpha_i T_i h_x\| < \varepsilon$. Applying $m \in \mathfrak{I}(S)$ gives $|\psi(x) - m(h_x)| < \varepsilon$. So $m(h_x) = \psi(x)$. Set $P\phi(x) = m(h_x)$. [More references: Derriennic and Lin [**1**], Foguel [**1**],Fong [**1**], Fong and Sucheston [**1**], Hiai and Sato [**1**], Lin [**1**], Nagel [**1**], Sachdeva [**1**], Sato [**1**]–[**5**], Sucheston [**2**], [**3**], Takahashi [**3**], [**4**], Wolff [**1**].]

5-7 (Sine [**1**]). $\Leftarrow$ Let $f \in X'$, $A_n(T) = (n+1)^{-1} \sum_{r=0}^{n} T^r$. Every weak* cluster point of $\{A_n(T)^* f\}$ is T^*-invariant, and there is only one such by the separation condition. So for some T-invariant $Q(f)$, $A_n(T)^* f \to Q(f)$ weak*. Let B^* be the unit ball of X'. Show that the topology on $Q(B^*)$ generated by the T-invariant elements of X coincides with the weak* topology. Show then that for $x \in X$, the functional $f \to Qf(x)$ is weak* continuous and so is of the form $(Px)^\wedge$. Then $A_n(T) \to P$ in the weak (and indeed strong) operator topology.

Not surprisingly, this delightful result has been extended to amenable semigroups of operators—Lloyd [**4**], Nagel [**1**], and Sato [**3**].

5-8 (Namioka [**3**], Lloyd [**2**], Sucheston [**1**]). (i) For the last part, use the facts that for $m, n \in \mathfrak{L}(S)$, we have $nm = m$, $mn = n$ (so that $Q_m Q_n = Q_n$, $Q_n Q_m = Q_m$) and $F = \ker(I - Q_m)$.

(ii) $K = F \cap \mathrm{PM}(Y)$ is a weak* compact, convex subset of $M(Y)$. We can suppose that $\xi \in K$. If $\xi \in \operatorname{Ext} K$, then $Q_m(\eta) \in F$, $0 \le Q_m(\eta) \le \xi$, and the extremeness of ξ gives $Q_m(\eta) = \eta(1)\xi = Q_n(\eta)$. Now suppose that $\xi \in L = \operatorname{co}(\operatorname{Ext} K)$, and write ξ as a convex combination $\sum_{i=1}^{n} \alpha_i \xi_i$, $\xi_i \in \operatorname{Ext} K$. By regarding the ξ_i, η as functions in $L_1(\xi)$, one easily shows that there exist $\eta_i \in M(Y)$ such that $0 \le \eta_i \le \alpha_i \xi_i$, $\eta = \sum_{i=1}^{n} \eta_i$. Since $Q_m(\eta_i) = Q_n(\eta_i)$, we have $Q_m(\eta) = Q_n(\eta)$. For general $\xi \in K$, we can suppose, by the norm density of $C(Y)$ in $L_1(\xi)$, that $d\nu/d\xi = f \in C(Y)$. By the Krein-Milman Theorem, there exists a net $\{\xi_\delta\}$ in L with $\xi_\delta \to \xi$ weak*. If ν_δ is such that $d\nu_\delta/d\xi_\delta = f$, then $\nu_\delta \to \nu$ weak* and $Q_m(\nu_\delta) = Q_n(\nu_\delta)$. By weak* continuity, $Q_m(\nu) = Q_n(\nu)$ as required.

(iii) We can suppose that $0 \le \phi \le 1$. So $0 \le \nu \le \mu$, where $d\nu/d\mu = \phi$. Then $m(\nu\psi) = n(\nu\psi)$ for all $m, n \in \mathfrak{L}(S)$. [*Note*: Another remarkable ergodic mixing theorem is proved by Dye [**1**].]

Problems 6

6-1. The limit is 3.

6-2. If $k = \lim_{n\to\infty} \lambda(C^n)^{1/n}$, then $k^r = \lim_{n\to\infty} \lambda((C^r)^n)^{1/n}$. What else can be said about A? What if G is the "$ax+b$" group?

6-3 (Milnor [**1**]). The class of $H_{\mathbb{Z}}$ is 2. With $G = H_{\mathbb{Z}}$, $G_1/G_2 = \mathbb{Z} \times \mathbb{Z}$, $G_2/G_3 = \mathbb{Z}$ so that ((6.17)) the degree $d = 1 \cdot 2 + 2 \cdot 1 = 4$.

6-4. Use induction.

6-5. Take $G = S_2$.

6-6 (Jenkins [**6**]). (ii) The freeness of S follows by showing that the $(1,2)$ component w of the matrix $u \in S$ satisfies $|w-1| < \frac{1}{2}$ if $u \in aS$ and $|w+1| < \frac{1}{2}$ if $u \in bS$. (iii) If for some $X \in \mathfrak{g}$, $\theta \in \mathrm{Sp}(\mathrm{ad}\, X) \cap (\mathbb{R} \sim \{0\})$ and Y is a θ-eigenvector of $\mathrm{ad}\, X$, then $\mathrm{Span}\{X, Y\} = \mathfrak{g}_\theta$. Otherwise let $X_1, X_2, X_3 \in \mathfrak{g}$, $\theta_1, \theta_2 \in \mathbb{R}$, be such that $[X_1, X_2 + iX_3] = (\theta_1 - i\theta_2)(X_2 + iX_3)$, $X_i \neq 0$, $\theta_i \neq 0$. Now use Jacobi to obtain $\mathfrak{g}_\theta = \mathrm{Span}\{X_1, X_2, X_3\}$.

6-7. Use (6.20(i)).

6-8 (Auslander and Moore [**1**]). If α is irrational, then $M(\alpha)$ is a Mautner group and so is not of Type 1.

6-9. Leptin [**8**].

6-11 (Guivarc'h [**1**]). $\Leftarrow$ Suppose not. Then ((6.37)) G has a homomorphic image of the form $H/Z(H)$, where $H \in \{S_2, S_3^\sigma, S_4\}$. Show that $Z(H)$ is discrete, and (using B13) that $H/Z(H)$ is not unimodular. [Note: the solvability requirement can be removed.]

6-12 (Milnor [**1**]). The groups $\pi_1(\mathbb{S}^2) = \{e\}$, $\pi_1(P_2) = \mathbb{Z}_2$, and $\pi_1(\mathbb{S}^1 \times \mathbb{S}^1) = \mathbb{Z} \times \mathbb{Z}$ clearly have polynomial growth. Now $G = \pi_1(K)$ is the group generated by two elements a, b with the one relation $a^2 b^2 = e$. The subgroup H of G generated by $\{a^2, ab\}$ is abelian and normal, and G/H is finite. So G has polynomial growth by (6.22(i)). [Question: *For which Riemannian manifolds M is $\pi_1(M)$ amenable*?]

6-13 (Jenkins [**9**], [**10**], [**11**]). Let $C = C^{-1} \in \mathscr{C}_e(G)$ be such that $|\mu|(G \sim C) = 0$, $\phi\chi_C \neq 0$ in $L_\infty(G)$. Let $a_n = \int_{C^n} \phi\, d\lambda$. Then $a_n^{1/n} \to 1$. Let $\{a_{n_k}\}$ be a subsequence such that $a_{n_k+2}/a_{n_k} \to 1$. Let $A_k(s) = (\int_{C^{n_k+1}} \phi(sx)\, d\lambda(x))/a_{n_k}$. Show that $A_k(s) \to 1$ uniformly on C, and using Fubini, that

$$\mu(G) = \lim_{k\to\infty} \int_C A_k(s)\, d\mu(s) = \lim_{k\to\infty} \left(\left(\int_{C^{n_k+1}} \phi\mu(x)\, d\lambda(x) \right) \Big/ a_{n_k} \geq 0 \right. .$$

6-14 (Jenkins [**11**]). Define $P(\phi x) = 1$. Use Problem 6-13 with $\mu = \sum_{i=1}^n \alpha_i \delta_{x_i}$ to show that P extends linearly to X_ϕ.

6-15. For (i), use (3) of (6.43) and the proof of (6.17). For (ii), if $C = C^{-1} \in \mathscr{C}_e(G)$ generates G, then we can take $K_n = C^{\alpha(n)}$ for some subsequence $\{\alpha(n)\}$ using (6.8) and (i).

6-16 (Wulbert [**1**], Lloyd [**2**], Seever [**1**]). (i) We can identify H with $C(X/\sim)$, and, by the Stone-Weierstrass Theorem, H is the closed subalgebra of $C(X)$ generated by B. The first assertion is true if $\hat{\mu}_{|_H} \in \operatorname{Ext}\{\hat{\nu}_{|_H} : \nu \in C_x\}$ ("=" $\{\hat{u} : u \in X/\sim\}$), and true for general $\mu \in C_x$ by the Krein-Milman Theorem. For the second assertion, if $y_0 \in \mathbf{S}(\mu) \sim \{y \in X : y \sim x\}$, obtain a contradiction by considering $f \geq 0$ in H, $f(y_0) = 1$, $f(y) = 0$ if $y \sim x$. (ii) Note that $P\psi$, $(P\phi)(P\psi) \in H$; show, using the second assertion of (i), that

$$\hat{x} \circ P(\phi P\psi) = P\phi(x)P\psi(x) = \hat{x} \circ P((P\phi)(P\psi)).$$

[Note: a C^*-algebra version of this result is proved in Choi and Effros [**2**].]

6-17 (Milnes [**S1**]). Take K to be the polygon connecting, in order, the points $(n, 1/n), (n, 1), (n^2, n), (-n^2, n), (-n, 1), (-n, 1/n)$, and back to $(n, 1/n)$ $(n \geq 2)$. Show that there exists $\varepsilon > 0$ such that

$$\lambda((0,2)K^m \cap K^m)/\lambda(K^m) \leq 1 - \varepsilon \quad \text{for all } m$$

and hence contradict the localisation conjecture.

6-18 (Rosenblatt [**10**]). $\Rightarrow$ If G is not of Type R, we can find $y \in \exp \mathfrak{g}$ ($\mathfrak{g}$ the Lie algebra of G), $Y_1, Y_2 \in \mathfrak{g}$, $Y_1 + iY_2 \neq 0$, and $\lambda \in \mathbb{C}, |\lambda| < 1$, such that $\operatorname{Ad} y(Y_1 + iY_2) = \lambda(Y_1 + iY_2)$. Then $(\operatorname{Ad} y)^n(X) \to 0$ for some $X \neq 0$ in $\{Y_1, Y_2\}$. If $\exp X = x$, show that $y^n x y^{-n} \to e$, contradicting the distality of G.

$\Leftarrow$ Suppose that G is not distal. Using (6.32), show that the radical S of G is also not distal. There exists a basis for $\mathfrak{s}_{\mathbb{C}}$ ($\mathfrak{s}$ the Lie algebra of S) such that $\operatorname{Ad} S \subset L$, the Lie group of upper triangular, complex matrices with diagonal entries of modulus 1. Then L is not distal, and neither is the subgroup N of L with diagonal entries all 1. Now N is nilpotent, and an induction argument based on the central series for L gives a contradiction. [See Moore [**1**] for more on the distal property.]

6-19 (Auslander and Moore [**1**], Jenkins [**6**]).

6-20 (Greenleaf, Moskowitz, and Rothschild [**2**]). $\Leftarrow$ Suppose that for some $M > 0$, $|\operatorname{Tr}(\operatorname{Ad} x)| \leq M$ for all $x \in G$. Suppose that $x \in G$ and $\operatorname{Sp}(\operatorname{Ad} x) = \{\lambda_1, \ldots, \lambda_n\} \not\subset \mathbb{T}$. Let $r = \max_{1 \leq i \leq n} |\lambda_i|$. We can take $\lambda_1, \ldots, \lambda_k$ to be those eigenvalues of modulus r. We can suppose $r > 1$. Let $\lambda_j = re^{i\theta_j}$ $(1 \leq j \leq k)$. Show that for $N \in \mathbb{P}$, $|\sum_{j=1}^k e^{iN\theta_j}| \to 0$ as $N \to \infty$. Now use the Kronecker Approximation Theorem [**HR1**, (26.15)] to infer that for arbitrarily large N, $|\sum_{j=1}^k e^{iN\theta_j}| \geq k - 1/2$. This gives a contradiction.

6-22 (Dixmier [**4**], Hulanicki [**8**], Kahane [**2**]). (i) Let $a > 0$ be such that $[-a, a]$ contains $\operatorname{Sp} f$ and the support of ϕ. Define $\hat{\phi} : \mathbb{Z} \to \mathbb{C}$ by

$$\hat{\phi}(n) = (2a)^{-1} \int_{-a}^{a} \phi(s) \exp(-i\pi n s/a)\, ds.$$

Show, using integration by parts, that $\hat{\phi}(n) = O(|n|^{-(k+2)})$ $(n \to \infty)$. Then take

$$g = \sum_{-\infty}^{\infty} \widehat{\phi}(n)(e^{in\pi f/a} - 1) \in A.$$

(ii) Use the fact that if $f, g \in L_1(G, \omega)$, then

$$|f(t)g(t^{-1}x)|\omega(x) \leq |(f\omega)(t)|\,|(g\omega)(t^{-1}x)|.$$

(iii) By (6.49), it makes sense to talk of $\phi \circ \hat{f}$. It is sufficient from (i) to show that $\|e^{in\beta f} - 1\|_1 = O(n^p)$ as $n \to \infty$ for each $\beta \in \mathbb{R}$. We can suppose that $\beta = 1$. Define $\alpha, \omega \colon G \to [1, \infty)$ by $\alpha(x) = \min\{n \colon x \in C^n\}$, $\omega = \exp\alpha$, where $C \in \mathscr{C}_e(G)$ is such that f vanishes outside C and $\bigcup_{n=1}^{\infty} C^n = G$. Let $M > 0$ be such that $\lambda(C^m) \leq Mm^k$ $(m \geq 1)$. Then show that for $n, m \geq 1$,

$$\|e^{inf} - 1\|_1 \leq M^{1/2} n^{k/2} \|T\|\,\|f\|_2 + e^{-m}\exp(n\|f\|_\omega) \leq C_1 n^p$$

for some $C_1 > 0$, where

$$T = in\sum_{k=0}^{\infty} (in\pi_2(f))^k/(k+1)! \in \mathbf{B}(L_2(G)).$$

Now show $\|T\| \leq n$ and choose m so that $|m - n\|f\|_\infty| < 1$ to obtain the desired result.

6-23 (cf. Palmer [**1**]). $L_1(G/H)$ is a homomorphic image of $L_1(G)$ ((1.11)) and so $G \in [S] \Rightarrow G/H \in [S]$. If H is an open subgroup of G, then $L_1(H)$ can be regarded as a closed $*$-subalgebra of $L_1(G)$, and $(G \in [S] \Rightarrow H \in [S])$ follows from [**Ri**, (4.7.7)].

6-24 (Jenkins [**4**], [**5**], [**8**]). (i) It is sufficient to show that if $a \in A$ and a does not have a left inverse in A, then $F_0(a) = 0$ for some $F_0 \in P(A)$. For such an a, $0 \in \mathrm{Sp}(a^\sim a)$, and since $|F(a)|^2 \leq F(a^\sim a)$, it is sufficient to obtain $F_0 \in P(A)$ with $F_0(a^\sim a) = 0$. Let M be a maximal, closed $*$-subalgebra of A containing $a^\sim a$. Then $\phi(a^\sim a) = 0$ for some ϕ in the carrier space of M, and $\phi \in P(M)$ by symmetry of M and so extends to the desired $F_0 \in P(A)$. (ii) (a) Straight calculation shows that $\|f_0\|_1^n = 6^n$ so that $\nu(f_0) = \lim_{n\to\infty} \|f_0^n\|_1^{1/n} = 6$. (b) Let $A = l_1(G)$. For $F \in P(A)$, $|F(t * f)|^2 \leq F((t * f)^\sim * (t * f)) = F(f^\sim * f) \leq \nu(f^\sim * f) \leq \nu(f)^2$. By (i), $\nu(t * f) \leq \nu(f)$. (c) For $F \in P(A)$, $|F(g_0)| \leq [\nu(a + a^{-1})^2 + \nu(a + a^{-1})^4]^{1/2} \leq 2\sqrt{5}$. By (i), $\nu(g_0) \leq 2\sqrt{5}$ if G is symmetric. This is impossible by (b) and (a), since $g_0^\sim * g_0 = g_0 * g_0^\sim$ and $6 = \nu(f_0) = \nu(ba^2 * g_0) \leq \nu(g_0)$.

6-25 (Boidal et al. [**1**]). (i) $\Leftarrow$ If π, ρ are as in (i) and π_*, ρ_* are the extensions of π, ρ to $C^*(G)$, then $\|\rho(f)\| \leq \|\pi(f)\|$ for all $f \in L_1(G)$, and it follows that $\ker\pi_* \subset \ker\rho_*$. Let $\mathscr{S}$ be a closed subset of $\mathrm{Prim}\, C^*(G)$, and let $\pi' = \bigoplus_{I\in\mathscr{S}} \pi_I$, where $\pi_I \in C^*(G)^\wedge$ has kernel I. If $\rho \in \hat{G}$ has kernel $\supset (\cap\mathscr{S}) \cap L_1(G)$, then $\|\rho(f)\| \leq \|\pi'(f)\|$. It follows that $\Psi_G(\mathscr{S}) = \{J \in \mathrm{Prim}_*\, L_1(G) \colon J \supset (\cap\mathscr{S}) \cap L_1(G)\}$ is closed in $\mathrm{Prim}_*\, L_1(G)$.

(ii) Let π, ρ be nondegenerate, $*$-representations of $L_1(G)$ with $\ker \pi \subset \ker \rho$, and let $h \in C_c(G), f = h * h^{\sim}$. Suppose that $\|\rho(f)\| > \|\pi(f)\|$. We can suppose that G is compactly generated. Let $\phi \in C_c^\infty(\mathbb{R})$ with $\phi([0, \|\pi(f)\|]) = \{0\}$, $\phi(\|\rho(f)\|) = 1$. By Problem 6-22, there exists $g \in A_f$ such that $\hat{g} = \phi \circ \hat{f}$. Now show that $g \in \ker \pi \sim \ker \rho$, and derive a contradiction. It follows from (i) that $G \in [\Psi]$.

6-26 (Rosenblatt [**S1**]). $\Rightarrow \|(x * \mu - \mu)\mu^n\|_1 \to 0$. $\Leftarrow$ If $f \in L^0(G)$ is of the form $(g * x - g)$ for some $g \in L_1(G)$, then $\|f * \mu^n\|_1 \to 0$. Let X be the span of the set of such functions f. By examining $X^\perp \subset L_\infty(G)$, show that X is dense in $L^0(G)$.

Problems 7

7-1. With $X = A$ and N_γ as in (7.1), the set $T = \{\alpha \in \beta X\colon$ for some $\varepsilon \in \{1, c\}^\Gamma, \alpha \in \bigcap_\gamma (N_\gamma^{\varepsilon(\gamma)})^-\}$ has cardinality $\geq 2^{2^{|X|}}$. The inequality $|l_\infty(X)'| \leq 2^{2^{|X|}}$ follows as in (7.6). [For another proof that $|\beta X| = 2^{2^{|X|}}$, see Gilman and Jerison [**1**].]

7-2. Chop up $[0, \infty)$ into intervals $I_{11}, I_{21}, I_{22}, I_{31}, I_{32}, I_{33}, I_{41}, \ldots$, where $I_{11} = [0, 3]$, I_{21}, I_{22} are $[3, 7]$, $[7, 11]$, $I_{31} = [11, 16], \ldots$. On each $I_{ij} = [a_{ij}, b_{ij}]$, let ϕ_{ij} be the tent function

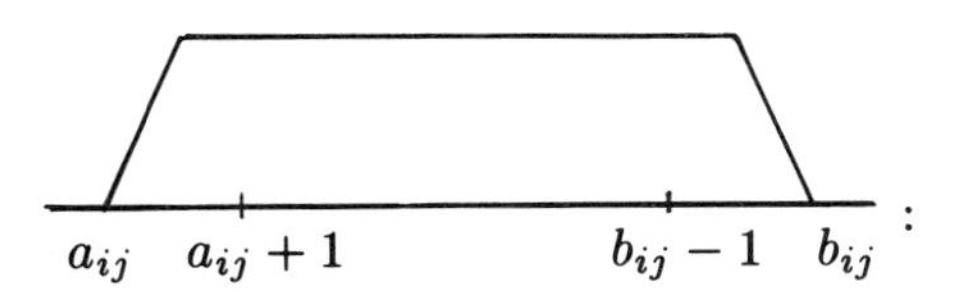

:

$\max_{x \in \mathbf{R}} \phi_{ij}(x) = 1$ and ϕ_{ij} vanishes outside I_{ij}. Take $\phi_j = \sum_{i=j}^\infty \phi_{ij}, Z_j = \bigcup_{i=j}^\infty [a_{ij} + 1, b_{ij} - 1]$.

7-3,4. Let $\mathscr{C}, E$ be as in (7.5). Each member of $\mathscr{C}$ contains a minimal, compact, left invariant subset. If $C \in \mathscr{C}$, then $\operatorname{Ext}\{m \in \mathfrak{L}_t(G)\colon \hat{m}(C) = 1\} \subset \operatorname{Ext} \mathfrak{L}_t(G)$. [Note: the amenability of G is not used in the construction of $\mathscr{C} = \{C_p\colon p \in P\}$ in (7.4).]

7-5 (Chou [**4**]). (iii) Let $m \in \mathfrak{M}_c$, $f = \chi_U/\lambda(U)$ ($U \in \mathscr{C}(G)$, U not null). Show that

$$\begin{aligned} |\tau(\phi f - \phi)(n)| &\leq \lambda(U_{n+1} \sim U_n)^{-1} \|\phi\|_\infty \sup_{t \in U} \lambda(t(U_{n+1} \sim U_n)) \mathbin{\triangle} (U_{n+1} \sim U_n)) \\ &\leq \|\phi\|_\infty [((n+1)/n) \sup_{t \in U} [\lambda(U_{n+1} \mathbin{\triangle} tU_{n+1})/\lambda(U_{n+1})] + 2/n] \\ &\to 0 \text{ as } n \to \infty. \end{aligned}$$

So for $m \in \mathfrak{M}_c$, $\tau^*(m)(\phi f - \phi) = m(\tau(\phi f - \phi)) = 0$.

(iv) By Gilman and Jerison [**1**], $\beta\mathbf{P}$ is homeomorphic to a compact subset E of $\beta\mathbf{P} \sim \mathbf{P}$. Since $E \subset \mathfrak{M}_c$, $|\mathfrak{L}_t(G)| \geq |\mathfrak{M}_c| \geq |E| = 2^{\mathfrak{c}}$.

7-6 (Granirer [13]). Take $X = \mathfrak{K}$. It is sufficient to show that $\mathfrak{u}(m) > \aleph_0$ for each $m \in \mathfrak{K}$. Suppose that the contrary holds for some m, and construct a sequence $\{f_n\}$ in $P(\Sigma)$ such that $\hat{f}_n \to m$ weak*. Derive a contradiction (cf. (7.28(i))).

7-7. Let $\{C_n\}$ be a sequence in $\mathscr{C}(G)$ such that $\bigcup_{n=1}^{\infty} C_n = G$, $C_n \subset C_{n+1}$. We can suppose $\lambda(U_n) \to 0$. Let $\{V_n\}$ be a sequence in $\mathscr{C}_e(G)$ such that $V_n = V_n^{-1}$, $V_n^2 \subset V_{n-1} \cap U_n$, and $xV_nx^{-1} \subset V_{n-1}$ $(x \in C_n)$ for $n > 1$. Take $N = \bigcap_{n=1}^{\infty} V_n$.

7-8. Let $\{r_n : n \in \mathbb{P}\}$ be an enumeration of $\mathbb{Q}$ and

$$V = \bigcup_{n=1}^{\infty} (r_n - \varepsilon 2^{-n-1}, r_n + \varepsilon 2^{-n-1}).$$

7-9 (Rudin [2]). By forming intersections of translates of the sets $\hat{E}_\gamma, \hat{E}_\gamma^c$ of (7.15), we obtain a disjoint family $\mathscr{A}$ of compact, nonempty, left invariant subsets of $\Phi(G)$ with $|\mathscr{A}| = 2^{\mathfrak{c}}$. For $C \in \mathscr{A}$, let $I_C = \{\phi \in L_\infty(G) : \hat{\phi}(C) = \{0\}\}$. Take $\mathscr{I} = \{I_C : C \in \mathscr{A}\}$.

7-10. We can suppose $E = G$. It is sufficient to show that S_n, where $S_n(x_1, \ldots, x_n, E_1, \ldots, E_n) = \chi_{x_1E_1 \cap \cdots \cap x_nE_n}$, is continuous from $G^n \times X^n \to L_1(G)$. If $n = 1$,

$$\|S_1(x_1, E_1) - S_1(y_1, F_1)\|_1 \le \|x_1 * \chi_{E_1} - y_1 * \chi_{E_1}\|_1 + \|\chi_{E_1} - \chi_{F_1}\|_1,$$

and S_1 is continuous. Now use induction.

7-11. Let $\mathscr{A}$ be the class of a.l.c. semigroups. If S is a left cancellative, infinite semigroup and F is a semigroup with $|F| < |S|$, then $S \times F \in \mathscr{A}$. If $\{S_n\}$ is a sequence of finite semigroups with S_n a subsemigroup of S_{n+1} and $|K_n| \to \infty$, where K_n is the kernel of S_n, then $\bigcup_{n=1}^{\infty} S_n$ is a.l.c. (Use (1.19).) If V, W are semigroups with $V \in \mathscr{A}$, V infinite and such that there exists an epimorphism $Q: V \to W$ and a cardinal $\mathfrak{m}$ such that $|Q^{-1}(\{w\})| \le \mathfrak{m} < |V|$ for all $w \in W$, then W is also a.l.c.

7-12 (Klawe [3], Paterson [9]). Suppose that (i) holds and $\mathfrak{p}$ is infinite. Let $\{S_1, \ldots, S_n\}$ be a partition of S, $s_1, \ldots, s_n \in S$. Find $u \in \bigcap_{i=1}^{n} Ss_i$. Then $|uS| \le |\bigcup_{i=1}^{n} s_iS_i| = \max_{1 \le i \le n} |s_iS_i|$. So $\mathfrak{p} \le \mathfrak{m}$. Trivially, $\mathfrak{m} \le \mathfrak{p}$. (ii) follows from (i). Suppose, now, that S is infinite. Trivially, S is a.l.c. if (iii) holds. Now suppose that (iv) holds, and S is infinite. It is sufficient to show $\mathfrak{m} = |S|$. Let $m \in \mathfrak{L}(S)$. With S_i, s_i as above, show $m(B) > 0$, where $B = \bigcup_{i=1}^{n} s_iS_i$. If $|B| < |S|$, then $m(B) = 0$ and a contradiction follows. To show the former, construct a sequence $\{y_n\}$ in S such that $\{y_nB\}$ is a disjoint family—the right cancellative property is used here. Now suppose that (v) holds. With S_i, s_i as above, let $Z = \{s \in S : s_is = s \ (1 \le i \le n)\}$. Then ((2.29)) Z is a right ideal of S. Then $s_iS \supset S_i \cap Z$ and $\mathfrak{p} \le |Z| \le |\bigcup_{i=1}^{n} s_iS_i|$. So. $\mathfrak{p} = \mathfrak{m}$. For the final part, if G is a finite group, then $\mathfrak{p} = |G|$, $\mathfrak{m} = 1$ so that $\mathfrak{p} \neq \mathfrak{m}$ in general. What about infinite semigroups?

7-13. Use (7.26) and Problem 7-12 above.

7-14 (Klawe [3]). There exists $s_0 \in S$ such that if $T_0 = s_0 S s_0$, then $|T_0| = \mathfrak{n}$. Every $m \in \mathfrak{I}(S)$ is supported on T_0. Suppose that $\mathfrak{n}$ is infinite. Let $\mathscr{F}(T_0) = \{F_\beta \colon \beta \in \alpha\}$, where α is the smallest ordinal of cardinality $\mathfrak{n}$. Construct subsets Δ_ε^β of T_0 such that (i), (ii), and (iii) of (7.25) are satisfied, and (iv) of (7.25) is replaced by $\bigcap\{(x^{-1}\Delta_\varepsilon^\beta y^{-1}) \colon x, y \in F_\beta\} \neq \varnothing$ for each pair ε, β with $\varepsilon \leq \beta$. In place of the $\Gamma_\varepsilon^\gamma$ of (7.25), construct pairwise disjoint sets $\overline{\Gamma}_\varepsilon^\gamma = \{x_{\varepsilon,i,j}^\gamma \colon 1 \leq i, j \leq n\} \subset T_0 \sim K$ such that for each ε, there exists $z_\varepsilon^\gamma \in T_0$ such that $x_i z_\varepsilon^\gamma x_j = x_{\varepsilon,i,j}^\gamma$ $(1 \leq i, j \leq n)$. Modify the rest of (7.25) to produce a disjoint family $\{D_p \colon p \in P\}$ of compact (two-sided) invariant subsets of βS, each supporting some $\hat{m}_p$, $m_p \in \mathfrak{I}(S)$, with $|P| = 2^{2^{\mathfrak{n}}}$. If $\mathfrak{n}$ is finite, show that T_0 is the unique left (right) ideal group in S.

7-15 (Chou [9]). Follow the proof of Problem 7-14 with $S = G = T_0$, $\mathfrak{n} = |G|$. We require that the sets Δ_ε^β satisfy the additional condition: (v) $(\Delta_\varepsilon^\beta)^{-1} = \Delta_\varepsilon^\beta$.

7-16–17 (Douglas [2], Kaufman [1], Rajagopalan and Witz [1], Paterson [4], Rosenblatt and Talagrand [1]). (i) Show that if $F \in \mathscr{F}(G)$, then C_F is left thick in G and so has cardinality $|G|$ (Problem 1-18).

(ii) (b)$\Rightarrow$(c). If (c) does not hold, then there exists $F_0 \in \mathscr{F}(G)$ such that $(F_0 x F_0) \cap (F_0 x F_0)^{-1} \neq \varnothing$ for all $x \in G$. For $a, b, c, d \in F_0$, let $A_{a,b,c,d} = \{x \in G \colon axb = (cxd)^{-1}\}$. If $E \subset A_{a,b,c,d}$, $m(E) = m(E^{-1})$ for $m \in \mathfrak{I}(G)$ and $\mathfrak{I}(G) = \mathfrak{I}^*(G)$.

(c)$\Rightarrow$(a). Proceed as in Problem 7-15, replacing (v) of its solution by $\Delta_\varepsilon^\beta \cap (\Delta_\eta^{\beta_1})^{-1} = \varnothing$ whenever $\varepsilon \leq \beta < \alpha$, $\eta \leq \beta_1 < \alpha$. This produces a subset Ψ of $\mathfrak{I}(G) \sim \mathfrak{I}^*(G)$ with $|\Psi| = 2^{2^{|G|}}$.

For Problem 7-17, use Problem 7-16 and the observation $(FxF) \cap (FxF)^{-1} \neq \varnothing$ for all $x \in G \Leftrightarrow B \subset F^{-4} \Leftrightarrow B$ is finite.

7-18. Let $A \subset S$ be such that $A \subset S, m(A) = 1$ for all $m \in \mathfrak{L}(S)$, and $|A| = \mathfrak{r}$. Then $2^{2^{\mathfrak{p}}} = |\mathfrak{L}(S)| \leq |l_\infty(A)'| \leq 2^{2^{|A|}}$, giving $\mathfrak{p} \leq \mathfrak{r}$. Obviously, $\mathfrak{r} \leq \mathfrak{p}$. [*Note*: Z. Yang [S4] has shown that $\mathfrak{m} = \mathfrak{r}$ in general.]

7-19 (Rajagopalan and Ramakrishnan [2], Klawe [3]). $\Leftarrow$ If G is a finite left ideal group in S, then every right thick subset of $S \supset$ the right thick subset GS. $\Rightarrow$ Let $a \in S$ be such that $T = Sa$ has smallest possible cardinality. Then show that if T_1, T_2 are right thick subsets of T, then $T_1 \cap T_2 \neq \varnothing$. Next show that T is finite, by assuming the contrary and constructing (using transfinite induction) two disjoint right thick subsets of T. This gives the first assertion of the Problem.

7-20 (Civin and Yood [**1**]). Let $J_H(G) = \{F \in l_\infty(G)'' \colon hF = F$ for all $h \in H, F(\chi_{xH}) = 0$ for all $x \in G\}$. Show that $J_H(G)$ is a left ideal with square $\{0\}$ and containing $J(G)$, and that $J_H(G)/J(G)$ contains a copy of $J(H)$. This follows using the map $F \to F^H + \mathrm{J}(G)$, where $F \in \mathrm{J}(H)$ and $F^H \in \mathrm{J}(G)$ is given by: $F^H(\phi) = F(\phi_{|_H})$ $(\phi \in l_\infty(G))$.

7-21 (Granirer [**3**], [**13**]). $\Rightarrow$ Using (7.28(i)) and (1.20), S contains a finite left ideal group. Now use the left cancellative property and the theory of the kernel ((1.16)).

Index of Terms

References for terms in this index are section numbers, except for references to the problems (found at the end of each chapter), which begin with the abbreviation Pr, followed by the chapter number and the problem number. Capital letters A, B, C, D, E refer to the appendices.

Index of Symbols

ABCDEFGHIJ-898